RECENT DEVELOPMENTS IN FLUVIAL SEDIMENTOLOGY

Contributions from the Third International Fluvial Sedimentology Conference

Edited by

FRANK G. ETHRIDGE(1) ROMEO M. FLORES(2), AND
MICHAEL D. HARVEY(1)
(1) Department of Earth Resources
Colorado State University
Fort Collins, Colorado 80523
and (2) U.S. Geological Survey
Denver, Colorado 80225

Index editor

JEAN N. WEAVER
U.S. Geological Survey
Denver, Colorado 80225

SOCIETY OF ECONOMIC PALEONTOLOGISTS AND MINERALOGISTS

Barbara H. Lidz, Editor of Special Publications

Special Publication No. 39

Tulsa Oklahoma, U.S.A. ISBN: 0-918985-67-6 *APRIL 1987*

A Publication of
The Society of Economic Paleontologists and Mineralogists
a division of
The American Association of Petroleum Geologists

PREFACE

This volume brings together 36 of the manuscripts that were presented as oral or poster papers at the Third International Fluvial Sedimentology Conference hosted by Colorado State University in August, 1985. The conference was co-sponsored by the Society of Economic Paleontologists and Mineralogists and the U.S. Geological Survey with additional support and cooperation from Conoco, Inc., Rocky Mountain Energy, Mobil Corporation, the Rocky Mountain Section-SEPM, and the College of Forestry and Natural Resources and the Department of Earth Resources of Colorado State University. The 59 authors and co-authors are from seven countries including Canada, China, Italy, Norway, Republic of South Africa, United Kingdom, and the United States. Fluvial systems and sedimentary rock sequences discussed range in age from Holocene to Precambrian and include such diverse areas as North America (20), Europe (7), Asia (4), Africa (2), New Zealand (1) and Iceland (1).

The principal objective of the volume was to document recent developments in the application of facies analysis to the reconstruction of the architecture of fluvial systems. Reconstruction of architecture is an integral part of the overall process of basin analysis. The volume begins with an overview of the development of fluvial facies models since 1977 (Miall). In this review Miall emphasizes the development of a number of variants on existing meandering models, development of anastomosed channel and floodplain facies models, recognition of mid-channel bar complexes, and documentation of alluvial architecture from two- and three-dimensional outcrop data in conjunction with chronostratigraphic markers.

Development of facies models and their use in the reconstruction of fluvial architecture begins with a knowledge of hydraulics, sediment transport, and deposition in modern fluvial systems including: framework or channel subsystems, non-framework or floodplain and backswamp subsystems, and fluvial-lacustrine subsystems. The first section of this volume contains four papers that deal with various aspects of sediment transport and hydraulics in modern rivers including: microforms and ingress of matrix fines in gravel-bed rivers (Reid and Frostick), measurement of magnetic particle transport in a gravel-bed river (Custer, Ergenzinger, Anderson and Bugosh), sediment transport and deposition at channel confluences (Best), and grain size distributions and sediment sorting within bedforms in a sand-bed river (Love, Gutjahr and Robinson-Cook). The second section contains eight papers that consider sediment deposition in modern channel floodplain and fluvial-lacustrine environments. Of special interest to sedimentologists working in the rock record are two papers describing sedimentation in large rivers, the Kosi (Wells and Dorr) and the Brahmaputra (Bristow). One paper describes sediments in coarse-grained counterpoint bars (S.A. Smith). Two papers deal with inclined heterolithic stratification (D.G. Smith) and aeolian deposition (Shepherd) in fine-grained point bars. Two papers document floodplain deposition in coarse-grained (Desloges and Church) and fine-grained (Farrell) fluvial systems. The final paper in this section examines the effect of variations in sediment and water inflow on deltaic sedimentation (Bogen).

Reconstruction of the hydrologic conditions at the time of deposition involves the application of geomorphic principles (Maizels; Goodwin and Diffendal) and the application of vertebrate taphonomy, an emerging subdiscipline (Koster; Behrensmeyer). As well as providing insights into hydrologic conditions, these techniques assist in reconstruction of facies and architectural elements of ancient fluvial sequences.

The broader aspects of the application of facies models to architectural reconstruction are reviewed in six papers that use data obtained from vertical sections, and from two- and three-dimensional outcrop exposures. These papers consider coarse-grained meanderbelt deposits of post-Pleistocene age (Campbell and Hendry), Pleistocene glacial outwash gravels (Dawson and Bryant), Plio-Pleistocene fluvial deposits (Billi, Magi, and Sargi), non-glacial, Pliocene, confined upland valley deposits (Morrison and Hein), Cenozoic nonmarine sediments associated with active volcanic basins (G.A. Smith), and fine-grained overbank deposits of upper Paleozoic age (Ghosh).

The importance of tectonic uplift and subsidence as major controls on architecture is reviewed in six papers that deal with compressional and extensional terrains, including intermontane and foreland basins. Papers include a theoretical consideration and discussion of Pleistocene and Holocene fluvial and alluvial fan deposits (Alexander and Leeder), a comparison of the architecture of fluvial systems of Eocene and Plio-Pleistocene age in two intermontane basins (Kraus and Middleton), an analysis of extra- and intrabasin controls on the alluvial architecture of Eocene deposits (Groll and Steidtmann), a study of the architecture of deposits of a single, radial distributary system of Lower Miocene age (Nichols), an analysis of subsidence rates on architecture of Cretaceous deposits (Shuster and Steidtmann), and a tectonic model for the inferred reversal in dispersal directions in a Devonian axial drainage system that evolved during oblique continental collision (Lawrence and Williams).

The ultimate goal of any architectural analysis is to enhance the exploration for, and exploitation of economic deposits. These aspects are covered in seven papers that deal with coal, hydrocarbons and gold in fluvial deposits. The origin and distribution of coal seams in fluvial deposits are covered in two papers that investigate the evolution of Eocene fluvial systems (Warwick and Flores) and allocyclic and autocyclic mechanisms that controlled the accumulation of thick coals in Cretaceous and Tertiary alluvial sequences (Flores and Pillmore). Hydrocarbon-bearing alluvial sequences are discussed in three papers that develop methods for predicting the lateral discontinuity of reservoir sandstones (Fielding and Crane; Qui, Xue and Xiao), and a detailed reservoir model for Late Mississippian hydrocarbon-bearing fluvial deposits (Melvin). The final two papers in the volume present two numerical models (Nami and James) and a facies model (Kingsley) for predicting the distribution of gold in Precambrian deposits.

Frank G. Ethridge
Romeo M. Flores
Michael D. Harvey

ACKNOWLEDGMENTS

The editors of this volume acknowledge with thanks technical reviews of the manuscripts by C. D. Atkinson, A. K. Behrensmeyer, J. L. Best, W. L. Bilodeau, R. C. Blakey, R. H. Blodgett, J. Bogen, R. J. Bottjer, T. M. Bown, C. S. Bristow, J. P. Buchanan, D. J. Cant, V. V. Cavaroc, Jr., H. G. Churnet, S. Crews, T. A. Cross, J. R. K. Daniel, R. A. Davis, M. R. Dawson, B. G. Els, C. R. Fielding, T. D. Fouch, P. F. Friend, L. Frostick, D. Germanoski, S. K. Ghosh, R. G. Goodwin, M. W. Green, F. J. Hein, R. Hereford, J. M. Hook, J. F. Hubert, S. Y. Johnson, R. C. Kochel, E. H. Koster, M. J. Kraus, R. A. Kuhnle, M. Leeder, J. Lewin, J. C. Lorenz, A. D. Miall, L. T. Middleton, H. H. Mills, W. E. L. Minter, J. Mossa, M. Nami, K. Nibbelink, G. J. Nichols, L. Palmer, C. L. Pillmore, J. Pitlick, K. L. Prestegaard, Y. Qui, R. A. Rahmani, J. Schmitt, R. G. Shepherd, M. W. Shuster, R. Slingerland, G. A. Smith, B. R. Turner, C. Turner-Peterson, N. Tyler, P. D. Warwick, N. Wells, W. A. Wescott, P. Wilcock, B. P. J. Williams, and G. P. Williams.

Financial support for editing this volume was provided by a Grant from the Resources for Scholarly Programs administered by the College of Forestry and Natural Resources, Colorado State University. Brian T. Dennison and Andrea Podgorny assisted with numerous day-to-day editorial tasks. Barbara H. Lidz, SEPM Special Publications Editor, provided much needed editorial guidance and advice. Laura N. R. Roberts, Michael Brownfield and Gary D. Stricker of the U.S. Geological Survey assisted in the indexing. Stricker provided computer expertise for the indexing. Sarah Andrews designed and drew the cover illustration.

RECENT DEVELOPMENTS IN FLUVIAL SEDIMENTOLOGY
TABLE OF CONTENTS

PART V. TECTONIC INFLUENCES

PART VI. ECONOMIC APPLICATIONS

RECENT DEVELOPMENTS IN THE STUDY OF FLUVIAL FACIES MODELS

ANDREW D. MIALL
Department of Geology, University of Toronto, Toronto, Ontario, M5S 1A1 CANADA

ABSTRACT: Several hundred papers on fluvial sedimentology have appeared since the first fluvial conference 9 years ago (Calgary, 1977). Many variants of existing models have been published, including several sub-models of high-sinuosity rivers based on sediment load grain size, and the anastomosed model has become well documented and widely known. The main advance in the study of low-sinuosity rivers is the recognition of mid-channel macroform bar complexes comparable in their medium- to long-term geomorphic significance to point bars.

Increasing attention is gradually being paid to the overbank environment, with the development of floodplain facies models, and an improved understanding of the conditions necessary for coal formation.

Few new syntheses of facies models have been attempted. The current research trend is toward documentation of three-dimensional architecture, both on the local scale (channels and complex bars) and on the regional scale, using two- and three-dimensional lateral outcrop profiles and employing chronostratigraphic markers to tie sections together.

INTRODUCTION

Since the first International Symposium on Fluvial Sedimentology was held in Calgary, October 1977, there have been at least six other major symposia that have been devoted partly or entirely to fluvial deposits. These events are listed below with references to the resulting publications:

1. Recent and ancient nonmarine depositional environments. Rocky Mountain Section, SEPM, Casper, Wyoming, June 1979 (Ethridge and Flores, 1981).
2. Modern and ancient fluvial systems: Sedimentology and processes (Second International Symposium). Sponsored by IAS, Keele, U.K., September 1981 (Collinson and Lewin, 1983).
3. Fluvial sedimentology in western North America. Cordilleran Section, GSA, Anaheim, California, April 1982 (Nilsen, 1984).
4. Rudites formed by unidirectional flow. Symposium, Eleventh International Congress on Sedimentology, Hamilton, Canada, August 1982 (Koster and Steel, 1984).
5. Modern and ancient fluvial sedimentation. General session, Eleventh International Congress on Sedimentology, Hamilton, Canada, August 1982 (no proceedings publication).
6. Third International Fluvial Sedimentology Conference. Fort Collins, Colorado, August 1985 (this volume).

Together with the proceedings volume from the first conference (Miall, 1978a), these represent a substantial body of literature (133 full length papers, plus those in this volume). Not included in this listing, of course, are the many individual papers that have appeared in journals during the same 9-year period. The fluvial afficionado must also contend with review articles (Collinson, 1978a; Rust, 1979; Walker and Cant, 1979, 1984; Miall 1980, 1985; Cant, 1982; Rust and Koster, 1984; Bridge, 1985), books (Turner, 1980; Galloway and Hobday, 1983; Nilsen, 1985;) and short courses (Miall, 1981; Flores, 1983; Flores and others, 1985).

In the first proceedings volume I contributed an historical article which attempted to trace the history of fluvial sedimentology under six main headings: descriptive fluvial geomorphology, quantitative fluvial geomorphology, sediment transport and textural studies, bedforms and paleocurrents, fluvial facies models, and paleohydraulics (Miall, 1978b). In this article I focus on within-basin (autocyclic) fluvial sedimentary controls and facies models in an attempt (made at the editors' suggestion) to bring this part of the subject up to date.

Has all this activity gotten us anywhere? Are we any further ahead than we were in 1977?

OVERVIEW

A quick answer to the questions above is, no, we have not advanced very far. A noticeable characteristic of the Calgary meeting was the presentation of several detailed attempts to build suites of facies models (Friend, 1978; Miall, 1978c; Rust, 1978b; Cant, 1978; Boothroyd and Nummedal, 1978; Heward, 1978) and several other papers providing critiques of facies model concepts (Jackson, 1978; Collinson, 1978b). None of the subsequent symposia have yielded the same focussed examination of the facies model idea. The paper by Friend (1983) at the Keele meeting, providing a classification of channel architecture, is the only post-1977 symposium paper to attempt a generalized, theoretical examination of fluvial facies models. Several non-symposia papers providing critiques of the ideas in this area include Miall (1980, 1985) and Bridge (1985). Many symposium papers and individual journal articles have contained proposals for new facies models or variants on existing ones. Many of these are discussed below. No new syntheses (as opposed to critiques) have appeared since 1978, however. Dott and Bourgeois (1983) stated recently that facies models for braided rivers have "multiplied like rabbits" but, in fact, the rabbits have been celibate for 9 years!

Numerous case studies published during the last 9 years have demonstrated a great variability in fluvial facies characteristics. Many departures from the norms, as they existed in 1977, have become apparent and several entirely new models have been proposed for specific geomorphic settings. These developments necessitate an entirely new approach (Miall, 1985). As Jackson (1978), Bridge (1985) and others have noted, many of our criteria for interpreting channel morphology from facies characteristics are based on over-simplified or erroneous ideas. One of the main mistakes is the assumption that channel morphology and processes can be reconstructed from a few basic observations, such as the lithofacies assemblage and sequence in a vertical profile, or the presence of epsilon crossbedding. Miall (1985) and Bridge (1985) cited examples of fluvial processes that are common to most or all rivers and pointed out the gradations that exist between various end-member conditions, making simple model classifications meaningless.

Miall (1985) suggested that we abandon the concept of the generalized facies model, such as the "meandering stream" model (except as a teaching aid). Facies models may be built to explain specific river systems or their deposits and are useful as summaries of specific ancient environments (for specific parts of specific stratigraphic units), but each such model will be virtually unique because of the variability of fluvial processes. At a more detailed level, however, it is now possible to recognize a total of at least eight basic "architectural elements" that are common to *all* rivers. Some of these elements contain much more three-dimensional variability than is generally recognized. We should focus now on building facies models for specific, commonly occurring components of fluvial environments, especially such macroform elements as channel-fills, lateral accretion deposits and within-channel bar complexes (see details in Miall, 1985). Better field description of such elements based, if possible, on detailed two- or three-dimensional *lateral* profiles, should permit more accurate reconstructions of local contemporary fluvial geomorphology than existing methods.

The remainder of this paper consists of an examination of some of the important developments that have been published since the first symposium in 1977.

RECENT DEVELOPMENTS

Two of the most important developments are our realization of the complex controls on channel morphology and an understanding of the wide range of scales of the depositional components of fluvial systems. These points are discussed first, followed by a consideration of some of the new data and ideas dealing with specific types of river. The last section, on alluvial architecture, is an attempt to pull some of these threads together.

Channel patterns.—

A simple fourfold channel classification (braided, meandering, anastomosing, straight) was proposed by Rust (1978a), but has never been entirely satisfactory as a basis for facies modelling. The terms braided and meandering refer to different parameters, and it is possible for a river to be defined by both terms; anastomosed rivers vary widely in morphology, particularly in sinuosity. Several workers have illustrated a spectrum of channel types showing a great variability in sinuosity, channel multiplicity, channel stability, and frequency of bars and islands (Mollard, 1973; Galay and others, 1973; Brice and others, 1978; Schumm, 1981, 1985). These morphological parameters are to some extent independent and may combine to form a variety of channel patterns intermediate between Rust's (1978a) four basic types. Channel morphology changes downstream, in response to changes in valley slope, sediment load, bank materials, climate or tectonic regime, and the same controls may cause changes through time in the morphology of a particular river reach (Schumm, 1969; Burnett and Schumm, 1983; Peterson, 1984; Carson, 1984a,b,c). It is therefore unwise to assume that fluvial style will remain constant throughout a given stratigraphic unit.

Many facies criteria such as channel body width/depth ratio, geometry of epsilon crossbedding, vertical profile, and proportion of overbank fines are not very sensitive indicators of fluvial style. Their interpretation commonly is ambiguous because similiar deposits may result from different combinations of the geomorphic variables listed above. This is why the proposed architectural-element approach is thought to hold promise, because it focusses on a much more rigorous field documentation than has been typical of most research during the past nine years.

Bounding surfaces and the element hierarchy.—

It has been known for some time that a hierarchy of bedforms of different scales and migrating at different speeds, even in different directions, can occur together on the same bed at any one time (e.g., Allen, 1968, 1973). Brookfield (1977, 1984) applied these concepts very successfully to eolian environments. He showed that complex eolian bedforms contain a hierarchy of internal bounding surfaces that define dune-forming processes at three different physical scales. First-order bounding surfaces delimit major, giant forms, called draas. Second- and third-order surfaces define successively smaller and more ephemeral dune-forming events. Allen (1983) demonstrated how the same ideas can be applied to fluvial deposits. For reasons not stated, he reversed the numbering of the hierarchy. This may cause some confusion although, as noted below, the result is an open-ended nomenclature that permits the definition of larger scale levels in the hierarchy as required.

According to Allen (1983, p. 249–250), "First-order contacts bound such entities as individual trough cross-bedding sets, or bundles of plane-bedded laminae genetically associated with cross strata. . . . Second-order contacts bound clusters of sedimentation units of the kinds delineated by first-order contacts." In the terminology of McKee and Weir (1953), a first-order bounding surface is a set boundary and a second-order surface may be a coset boundary, although Allen (1983) pointed out that different structure types could occur between two second-order surfaces. He showed that second-order surfaces commonly define individual "complexes" (Allen's, 1983, term) or architectural elements, and Miall and Middleton (in prep.) demonstrated that they also occur within elements as a type of large-scale reactivation surface.

Third-order surfaces are sub-horizontal erosion surfaces which typically define major channel migration or avulsion events. Miall (1985) and Miall and Middleton (in prep.) noted that groups of channels could define fourth-order surfaces.

Documentation of these bounding surfaces is an essential prerequisite to any detailed facies analysis. Correct hierarchical classification requires laterally extensive exposure, particularly for third- and fourth-order surfaces and those second-order surfaces that define individual elements. Such classification is difficult when dealing with cores, because the small-scale features of the various types of surface may be identical.

Anastomosed rivers.—

Until the late seventies, the "meandering" and "braided" fluvial styles were the only two that had been studied in detail by sedimentologists. Owing mainly to the work of D. G. Smith, a new model, that of the anastomosed river, became popular at the time of the first fluvial conference (Smith and Smith, 1980; Smith and Putnam, 1980). The term anastomosed is used "for an interconnected network of low-gradient, relatively deep and narrow, straight to sinuous channels with stable banks composed of fine-grained sediment (silt/clay) and vegetation. . . . Separating the channels are floodplains consisting of vegetated islands, natural levees, and wetlands. Occasionally, crevasse channels and splay deposits occur in wetland areas" (Smith and Smith, 1980, p. 157–158). Channels have low width/depth ratios and, for the most part, aggrade vertically. They may correspond to the ribbon channel bodies of Friend and others (1979).

Because the channels do not sweep across their floodplains, as do braided and meandering systems, wetland areas were thought by Smith and Smith (1980) to be prime sites for long-term peat accumulation, and several coal deposits have now been interpreted as the product of an anastomosed fluvial environment (Smith and Putnam, 1980; Leblanc Smith and Eriksson, 1979; Cairncross, 1980; Flores and Hanley, 1984). The model has also been applied to interpretations of subsurface petroleum pools (Putnam and Oliver, 1980; Putnam, 1982a,b). The difficulty of defining and correlating narrow sand bodies in the subsurface, however, has caused some to be skeptical of these interpretations (Wightman and others, 1981). The frequency of crevasse-splay development may be an inhibiting factor in the accumulation of thick coals, and many coal sedimentologists now cite other environments and floodplain processes as more favorable for coal development (e.g., Flores, 1981, 1984; McCabe, 1984; see also discussion below).

One of the prime conditions for the generation of the anastomosed model seems to be a low gradient. This may reflect a stable or rising base level control, which causes a "backwater effect" upstream, and encourages aggradation. This condition can be met in enclosed basins where the river flows out of the basin over a tectonic barrier. It may also occur in some coastal plain basins open to marine base level, such as foreland basins or divergent margin coastlines (e.g., King and Martini, 1984).

Smith (1983) focused on anastomosed rivers in humid climates, but Rust (1981), Rust and Legun (1983) and Nanson and Rust (1985) showed that a similar fluvial style could occur in arid regions.

High sinuosity rivers.—

Jackson's (1978) critique of facies models was a major contribution to fluvial studies. Preliminary field work by him led to a tentative classification of high-sinuosity rivers (those with a sinuosity >1.5) into five classes, based mainly on the grain size of the sediment load. Several of these categories were poorly known at the time of writing, but various studies of modern rivers and ancient deposits during the last 9 years have confirmed, with some modification, Jackson's original work. These developments have been summarized elsewhere (Miall, 1985) and only the main points are noted below.

It must be emphasized that high-sinuosity reaches can occur in almost any fluvial setting. Such reaches tend to migrate laterally, developing point bars with the well known epsilon crossbedded internal structure. The occurrence of this structure is therefore not the simple diagnostic criterion for fluvial style that it is often assumed to be but must be evaluated in context with the other associated facies.

Modern high-sinuosity rivers with a fine-grained sediment load have been described by Nanson (1980) and Jackson (1981) and ancient examples by Miall (1979) and Stewart (1983). Point bars consist of simple epsilon sets with banks or benches formed as a result of flow separation eddies on tight bends (Nanson, 1980; Nanson and Page, 1983). Subordinate bedforms on the epsilon foresets are small or absent. Point bar dip is steep (as much as 25°).

Gravel-bed, high-sinuosity rivers are not common, but our knowledge of them has improved considerably during the last 9 years. Modern examples have been studied by Gustavson (1978) and Forbes (1983). Point bar structures are difficult to define in the deposits because of the complexities of small- to medium-scale bedforms and minor channels. A well exposed ancient example was described by Ori (1979), however.

High-sinuosity rivers with medium to coarse sand or pebbly sand beds have been studied by several workers. Such rivers are similar to the classic examples described by Bernard and Major (1963), McGowen and Garner (1970) and Puigdefabregas (1973), and more recent work has not introduced any major modifications to these earlier models.

Low-sinuosity rivers.—

Few studies have been carried out on modern, multiple-channel, low-sinuosity ("braided") rivers during the last 9 years. Blodgett and Stanley (1980) and Crowley (1983) reexamined the Platte River, long regarded as the classic sandy braided river, following the work of Smith (1970, 1971, 1972). Both the two recent papers focussed on the linguoid "bars" which characterize much of the lower Platte River. Crowley (1983) demonstrated that in at least some cases these bedforms form part of much larger structures 200–400 m (650–1300 ft) in length and 0.7–1.5 m (2–5 ft) high, which he interpreted as macroforms.

Other depositional processes and facies sequences were described by Bluck (1979, 1980). He demonstrated that coarse bar deposits commonly develop preferentially in the upstream part of bends and may migrate downstream. This produces coarsening-upward sequences, the reverse of that usually expected in fluvial deposits.

Smith and Smith (1984) described the Williams River, a stream showing steady discharge that becomes characteristically braided, with linguoid bedforms and sand flats similar to those of the Platte, where abundant sand bedload is added by eolian processes. This nicely confirms Carson's

(1984a,b,c) ideas regarding the control of sediment load on fluvial style.

Ephemeral rivers.—

Important new details of facies associations and bed geometries have been presented by Tunbridge (1981, 1984) and Wells (1983) based on well exposed ancient examples. Tunbridge (1984) and Stear (1983) showed that point bar deposits, with epsilon crossbedding, could develop in ephemeral streams. Stear (1983) illustrated channel sandstones showing both the "ribbon" and "sheet" geometries, as defined by Friend and others, (1979).

Deposits of some recent floods have been described by Lucchitta and Suneson (1981), Sneh (1983) and Stear (1985), and include useful data on sediment characteristics. Parkash and others (1983) described the sedimentology of a modern low-gradient "terminal" alluvial fan, on which flow decreases to zero at the distal edges due to evaporation and infiltration. Some ancient deposits that were possibly formed in this type of environment were described by Friend (1978).

Large rivers.—

Few thorough sedimentological studies of any modern giant rivers have been undertaken since the seminal contribution by Coleman (1969) on the Brahmaputra. Bristow (this volume) re-examined this river and emphasized the importance of lateral accretion as one of the main depositional mechanisms in this river. He showed that the giant "sand waves" described by Coleman (1969) have downstream accretionary faces dipping at only 4°. They are therefore not simple foreset bedforms, as many sedimentologists have assumed, but are probably complex bar-like deposits bounded by second-order surfaces comparable to the macroforms described below.

Recognition of the deposits of giant rivers (including delta distributaries) requires identification of bar complexes or channel scour surfaces with appropriately large relief. A few recent studies have achieved this. Jones and McCabe (1980) described complex crossbed sets 40 m (130 ft) thick which they attributed to deposition on alternate bars in a deep deltaic distributary channel. Several studies of the Athabasca Oil Sands, Alberta (Mossop and Flach, 1983; Flach and Mossop, 1985) documented point bars 25 m (82 ft) thick, also interpreted as the deposits of distributary channels. Deep fluvial channels in the Morrison Formation of New Mexico, earlier described by Campbell (1976), have been re-examined by Miall and Middleton (in prep.) and interpreted using the concept of the bounding surface hierarchy and architectural-element analysis. Large macroform bar complexes bounded by second-order surfaces and channel-bounding surfaces of third and fourth order are present. This work has led to the realization that many ancient deposits were formed by much deeper rivers than have hitherto been realized.

Other fluvial models.—

As noted elsewhere (e.g., Miall, 1980), fluvial sedimentologists have simply not studied enough rivers to be aware of depositional styles in the full range of tectonic and climatic conditions. For example, in the Amazon Basin trunk and tributary rivers have different discharge and sediment load characteristics reflecting contrasting geology and climate of the source areas (Baker, 1978). The sedimentology of these rivers is strongly influenced by the presence of dense forest but has not received detailed study. Rivers on the forearc coastal plain of Guatemala are subject to catastrophic flash floods during volcanic eruptions, but facies assemblages and bed geometries are not well known (Kuenzi and others, 1979; Vessell and Davies, 1981). Interestingly, a common characteristic of both these areas is the damming of the mouths of tributaries by the abundant sediment carried in the trunk streams. The tributaries back up to form lakes, with fluvial input from the tributaries forming Gilbertian deltas.

A fluvial style that is receiving increasing attention is that of the tidally-influenced river. Point bar sets consist of sand-mud couplets, commonly with abundant bioturbation. This distinctive facies occurs in the giant point bars of the Athabasca Oil Sands (Mossop and Flach, 1983; Flach and Mossop, 1985) and has now been identified in several other modern and ancient units (Koster, 1984; Smith, 1985b). The term "inclined heterolithic strata" (IHS) has been coined for this facies, which is the product of stage fluctuation during the tidal cycle (Koster, 1984, 1985; Smith, 1985a).

The fan-delta environment, which includes a substantial fluvial component, has been much studied in recent years and is discussed at length elsewhere (Koster and Steel, 1984).

Floodplain environments.—

The predominantly fine-grained deposits of the floodplain have received much less detailed study than the coarser "framework" gravel-sand units of channels. Several commentators noted this at the first fluvial conference, and few sedimentologists have focussed their attention on this environment in subsequent years. The major exception to this trend comprises those geologists concerned with the stratigraphy and sedimentology of coal (e.g., Horne and others, 1978; Flores, 1981, 1984; Fielding, 1984; Rahmani and Flores, 1984).

One of the most distinctive characteristics of the anastomosed fluvial environment is the presence of large, stable floodplain areas, typically (at least in the majority of descriptions) wetlands with peat marshes, ponds and crevasse-splays. Recent references to this model are discussed above, although even these studies are concerned mainly with the composition and geometry of the channel and crevasse-splay units rather than the fine-grained deposits of the floodplain. Similarly Bridge (1984), in a paper on facies sequences in overbank environments, deals exclusively with the types of fining-upward and coarsening-upward sand-dominated sequences found in levee and crevasse-splay deposits.

Several studies of coal-bearing sequences subdivide floodplain deposits into a series of distinct facies assemblages. For example, Ethridge and others (1981) recognized crevasse-splay, levee, abandoned channel, well-drained swamp, poorly-drained swamp and lacustrine assemblages. A similar range of facies was described by Gersib and McCabe (1981). Both these studies focus on vertical profiles, and on the building of generalized facies models, but do not deal in detail with floodplain architecture. Wing (1984) subdivided plant-bearing floodplain fines into two main facies assemblages and was able to construct a separate depositional model for each, based on lithofacies, plant assemblages and unit geometry. Tabular units were interpreted as the deposits of gradually aggrading swamps, whereas lenticular bodies represent the fill of abandoned channels. This is one of the few studies to have exploited the floodplain deposits themselves for their potential to yield architectural information. As noted below, floodplain sequences may hold the key for unlocking a vast amount of architectural detail, but few workers have made use of this potential.

Coal is commonly interpreted as an integral part of the floodplain succession in fluvial and delta plain environments. McCabe (1984) pointed out, however, that in an active clastic environment characterised by crevassing and avulsion, floodplains receive so much detrital input that high quality, low-ash coals are unlikely to form. Raised peat swamps may develop where high rainfall permits an elevated water table to be maintained as rapid peat accumulation outpaces channelized clastic aggradation. The relief of these swamps prevents detrital influx and constrains channel migration and avulsion (Flores, 1984; McCabe, 1984).

Alluvial architecture.—

In order to reconstruct the three-dimensional geometry or architecture of a fluvial unit, it is necessary to study lengthy lateral exposures or to correlate isolated outcrops accurately. Correlation is a problem in fluvial deposits because of rapid lateral facies changes and numerous internal erosion surfaces. Four types of medium- to long-distance correlation can be carried out under certain conditions, however, by focussing on floodplain deposits. First, paleomagnetic reversal stratigraphy is proving to be extremely valuable for providing local chronostratigraphic datum planes, particularly in the more laterally extensive and possibly less episodically preserved floodplain sequences. Behrensmeyer and Tauxe (1982) provide an excellent example of an architectural interpretation based on this type of control. They were able to distinguish and relate to each other interbedded trunk river and tributary deposits and to calculate local sedimentation rates.

Second, the same kind of refined lateral control may be possible using tuff beds. These are rarely preserved in channel deposits because of the high transport energy but may form a widespread marker horizons on the floodplain. Allen and Williams (1982) provided an example of a unit less than 30 m (98 ft) thick that can be subdivided into seven "chronosomes" (Schultz's, 1982, term), permitting a very detailed reconstruction of architectural evolution.

Paleosols are the third type of floodplain marker horizon. These can be distinguished from each other on the basis of petrological and geochemical signatures and have the added advantage of yielding a considerable amount of evidence for local environmental conditions, sedimentation rates and local flooding and avulsion frequencies (Bown and Kraus, 1981; Kraus, 1985). Coal seams have a similar utility for local lithostratigraphic correlation.

Several workers have attempted classifications of alluvial architecture based on such criteria as channel body geometry and internal fill style, and the ratio of channel to overbank net thicknesses. Friend and others (1979) recognized that channels have a variety of shapes, that their fill may be simple or complex, and that both vertical and lateral accretion may occur. He suggested a classification into ribbon and sheet geometries, containing simple or complex (multistory) fills. Later Friend (1983) subdivided the "sheet" category into sheet flood and mobile channel belt types, recognizing that mobility may be the product of several processes, such as those commonly associated with the meandering and braiding styles. He proposed an architectural classification which combined these details with a subdivision into bedload (coarse) and mixed to suspended load (coarse to fine) streams, yielding five broad categories of architectural style.

A slightly different approach was adopted by Galloway (1981). He used Schumm's (1977) classification of rivers into bedload, mixed load and suspended load types and, for each of the three types, illustrated characteristic channel-fill geometries, lithofacies assemblages and sequences. He erected schematic architectural cross sections for four contrasting fluvial systems in the Cenozoic record of the Gulf Coast, using subsurface data to reconstruct such details as channel-fill geometry and interconnection.

Very few studies of carefully documented regional architecture have been published. Those by Behrensmeyer and Tauxe (1982) and by Allen and Williams (1982) are referred to above. Blakey and Gubitosa (1984) classified sandstone bodies in the Chinle Formation of the Colorado Plateau using the approach of Friend and others (1979) and discussed architectural interpretations in terms of climatic, tectonic, and other sedimentary controls. They did not, however, provide any detailed regional architectural cross sections comparable to those in the two previous papers. Behrensmeyer (this volume) extended her earlier work on the Siwalik sediments of Pakistan.

Recently, several important studies have documented in considerable detail the internal architecture of ancient complex channel-fill and bar deposits. Okolo (1983) and Hopkins (1985) described some well exposed channels filled by vertical aggradation. Allen and Matter (1982) and Okolo (1983) described other, more complex channel fills. Other writers, including Allen (1983), Haszeldine (1983a,b), Kirk (1983), Ramos and Sopeña (1983) and Miall and Middleton (in prep.) have provided detailed two- and three-dimensional cross sections of large compound bars, revealing a range of complex structures bounded by and containing second-order bounding surfaces. These studies provide the first data on the internal geometry and facies architecture of compound bars comparable to the modern side bars of Collinson (1970) and the sand flats of Cant and Walker (1978). These are macroforms (Jackson, 1975), comparable to and gradational with point bars in terms of their scale, complexity and depositional longevity. The discovery of the truly three-dimensional nature of these deposits is the most significant recent contribution to fluvial facies studies. Conventional vertical profile analysis would tend to treat the component crossbed sets within these bars as the product of the migration of unrelated trains of bedforms. The discovery of the interbedded second-order bounding surfaces shows the correct scale of the feature, demonstrating that the individual bedforms are components of much larger structures, usually several meters high. This implies much deeper channels than the conventional interpretation would have indicated. Studies to date reveal a considerable lateral complexity to these bars and an absence of any characteristic vertical profile, facts which have clear implications for future directions in fluvial research.

Miall (1985) attempted to synthesize these new data into an improved method of field facies analysis and classification, termed architectural-element analysis. He showed how the eight basic elements could be combined to form a variety of fluvial styles but emphasized the limitation of the fixed-model approach. To some extent, every river and every ancient fluvial unit is unique. Modelling is therefore simpler at the level of the component elements, which do seem to show some constant characteristics throughout rivers of all kinds.

CONCLUSIONS

It is becoming clear throughout the entire field of clastic sedimentology that vertical profile analysis, the basis of most facies models for more than 20 years, has reached its limit of usefulness. The results of such analysis are commonly of little help in sorting out the various types of autocyclic and allocyclic processes that may act simultaneously in a fluvial basin (Miall, 1980). It is time to recognize that there are severe limits to the level of sophistication in environmental interpretation that is possible with subsurface data, which consists largely of vertical profiles. A new and interesting field of research is opening up for the outcrop geologist in the study of alluvial architecture, but it remains to be seen how far the results can be applied to the practical business of subsurface fuel and mineral exploration.

ACKNOWLEDGMENTS

Comments by E. H. Koster, R. M. Flores and an anonymous reviewer have enabled me to make several useful improvements.

Research in fluvial sedimentology is supported by Operating Grants from the Natural Sciences and Engineering Research Council of Canada.

REFERENCES

ALLEN, J. R. L., 1968, Current ripples, their relation to patterns of water and sediment motion: North Holland Publishing Co., Amsterdam, 433 p.

———, 1973, Features of cross-stratified units due to random and other changes in bed-forms: Sedimentology, v. 20, p. 189–202.

———, 1983, Studies in fluviatile sedimentation: bars, bar-complexes and sandstone sheets (low-sinuosity braided streams) in the Brownstones (L. Devonian), Welsh Borders: Sedimentary Geology, v. 33, p. 237–293.

———, AND WILLIAMS, B. P. J., 1982, The architecture of an alluvial suite: Rocks between the Townsend Tuff and Pickard Bay Tuff Beds (Early Devonian), Southwest Wales: Philosophical Transactions of the Royal Society, London, Series B, v. 297, p. 51–89.

ALLEN, P., AND MATTER, A., 1982, Oligocene meandering stream sedimentation in the eastern Ebro Basin, Spain: Ecologae Geologica Helvetica, v. 75, p. 33–50.

BAKER, V. R., 1978, Adjustment of fluvial systems to climate and source terrain in tropical and subtropical environments, *in* Miall, A. D., ed., Fluvial Sedimentology: Canadian Society of Petroleum Geologists Memoir 5, p. 211–230.

BEHRENSMEYER, A. K., AND TAUXE, L., 1982, Isochronous fluvial systems in Miocene deposits of northern Pakistan: Sedimentology, v. 29, p. 331–352.

BERNARD, H. A., AND MAJOR, C. J., 1963, Recent meander belt deposits of the Brazos River, an alluvial "sand" model (abs.): American Association of Petroleum Geologists Bulletin, v. 37, p. 350.

BLAKEY, R. C., AND GUBITOSA, R., 1984, Controls of sandstone body geometry and architecture in the Chinle Formation (Upper Triassic) Colorado Plateau: Sedimentary Geology, v. 38, p. 51–86.

BLODGETT, R. H., AND STANLEY, K. O., 1980, Stratification, bed-forms, and discharge relations of the Platte braided river system, Nebraska: Journal of Sedimentary Petrology, v. 50, p. 139–148.

BLUCK, B. J., 1979, Structure of coarse-grained braided stream alluvium: Transactions of the Royal Society, Edinburgh, Earth Sciences, v. 70, p. 181–221.

———, 1980, Structure, generation and preservation of upward fining braided stream cycles in the Old Red Sandstone of Scotland: Transactions of the Royal Society, Edinburgh, Earth Sciences, v. 71, p. 29–46.

BOOTHROYD, J. C., AND NUMMEDAL, D., 1978, Proglacial braided outwash: a model for humid alluvial fan deposits, *in* Miall, A. D., ed., Fluvial Sedimentology: Canadian Society of Petroleum Geologists Memoir 5, p. 641–668.

BOWN, T. M., AND KRAUS, M. J., 1981, Lower Eocene alluvial paleosols (Willwood Formation, northwest Wyoming, U.S.A.) and their significance for paleoecology, paleoclimatology, and basin analysis: Palaeogeography, Palaeoclimatology, Palaeoecology, v. 34, p. 1–30.

BRICE, J. C., AND BLODGETT, J. C. and others, 1978, Countermeasures for hydraulic problems at bridges: Federal Highway Administration Report, FHWA-RD-78-163, v. 2, Washington, D.C., 558 p.

BRIDGE, J. S., 1984, Large scale facies sequences in alluvial overbank environment: Journal of Sedimentary Petrology, v. 54, p. 583–588.

———, 1985, Paleochannel patterns inferred from alluvial deposits: A critical evaluation: Journal of Sedimentary Petrology, v. 55, p. 579–589.

BROOKFIELD, M. E., 1977, The origin of bounding surfaces in ancient aeolian sandstones: Sedimentology, v. 24, p. 303–332.

———, 1984, Eolian facies, *in* Walker, R. G., ed., Facies Models: Geoscience Canada Reprint Series No. 1, second edition, p. 91–103.

BURNETT, A. W., AND SCHUMM, S. A., 1983, Alluvial-river response to neotectonic deformation in Louisiana and Mississippi: Science, v. 222, p. 49–50.

CAIRNCROSS, B., 1980, Anastomosing river deposits: paleoenvironmental control on coal quality and distribution, northern Karoo Basin: Transactions of the Geological Society of South Africa, v. 83, p. 327–332.

CAMPBELL, C. V., 1976, Reservoir geometry of a fluvial sheet sandstone: American Association of Petroleum Geologists Bulletin, v. 60, p. 1009–1020.

CANT, D. J., 1978, Development of a facies model for sandy braided river sedimentation: Comparison of the South Saskatchewan River and the Battery Point Formation, *in* Miall, A. D., ed., Fluvial Sedimentology: Canadian Society of Petroleum Geologists Memoir 5, p. 627–639.

———, 1982, Fluvial facies models, *in* Scholle, P. A., and Spearing, D., eds., Sandstone Depositional Environments: American Association of Petroleum Geologists Memoir 31, p. 115–137.

———, AND WALKER, R. G., 1978, Fluvial processes and facies sequences in the sandy braided South Saskatchewan River, Canada: Sedimentology, v. 25, p. 625–648.

CARSON, M. A., 1984a, The meandering-braided river threshold: A reappraisal: Journal of Hydrology, v. 73, p. 315–334.

———, 1984b, Observations on the meandering-braided river transition, the Canterbury Plains, New Zealand, Part One: New Zealand Geographer, v. 40, p. 12–17.

———, 1984c, Observations on the meandering-braided river transition, the Canterbury Plains, New Zealand, Part Two: New Zealand Geographer, v. 40, p. 89–99.

COLEMAN, J. M., 1969, Brahmaputra River: Channel processes and sedimentation: Sedimentary Geology, v. 3, p. 129–239.

COLLINSON, J. D., 1970, Bedforms of the Tana River: Norway, Geo-

grafiska Annaler, v. 52A, p. 31–55.
———, 1978a, Alluvial sediments, *in* Reading, H. G., ed., Sedimentary Environments and Facies: Blackwell, Oxford, p. 15–60.
———, 1978b, Vertical sequence and sandbody shape in alluvial sequences, *in* Miall, A. D., ed., Fluvial Sedimentology: Canadian Society of Petroleum Geologists Memoir 5, p. 577–586.
——— and Lewin, J., eds., 1983, Modern and ancient fluvial systems: International Association of Sedimentologists Special Publication 6, 575 p.
Crowley, K. D., 1983, Large-scale bed configurations (macroforms), Platte River Basin, Colorado and Nebraska: Primary structures and formative processes: Geological Society of America Bulletin, v. 94, p. 117–133.
Dott, R. H., Jr., and Bourgeois, J., 1983, Hummocky stratification: Significance of its variable bedding sequences, reply to discussion by R. G. Walker and others: Geological Society of America Bulletin, v. 94, p. 1245–1251.
Ethridge, F. G., and Flores, R. M., eds., 1981, Recent and ancient nonmarine depositional environments: Models for exploration: Society of Economic Paleontologists and Mineralogists Special Publication 31, 349 p.
———, Jackson, T. J., and Youngberg, A. D., 1981, Floodbasin sequence of a fine-grained meander belt subsystem: The coal-bearing Lower Wasatch and Upper Fort Union Formations, southern Powder River Basin, Wyoming, *in* Ethridge, F. G., and Flores, R. M., eds., Recent and Ancient Nonmarine Depositional Environments, models for exploration: Society of Economic Paleontologists and Mineralogists Special Publication 31, p. 191–209.
Fielding, C. R., 1984, A coal depositional model for the Durham Coal Measures of NE England: Journal of the Geological Society of London, v. 141, p. 919–932.
Flach, P. D., and Mossop, G. D., 1985, Depositional environments of Lower Cretaceous McMurray Formation, Athabasca Oil Sands, Alberta: American Association of Petroleum Geologists Bulletin, v. 69, p. 1195–1207.
Flores, R. M., 1981, Coal deposition in fluvial paleoenvironments of the Paleocene Tongue River Member of the Fort Union Formation, Powder River area, Powder River Basin, Wyoming and Montana, *in* Ethridge, F. G., and Flores, R. M., eds., Recent and Ancient Nonmarine Depositional Environments: Models for Exploration: Society of Economic Paleontologists and Mineralogists Special Publication 31, p. 169–190.
———, ed., 1983, Fluvial systems: Their economic and field applications: American Association of Petroleum Geologists Field Seminar, 383 p.
———, 1984, Comparative analysis of coal accumulation in Cretaceous alluvial deposits, southern United States Rocky Mountain basins, *in* Stott, D. F., and Glass, D. J., eds., The Mesozoic of Middle North America: Canadian Society of Petroleum Geologists Memoir 9, p. 373–385.
———, Ethridge, F. G., Miall, A. D., Galloway, W. E., and Fouch, T. D., 1985, Recognition of fluvial depositional systems and their resource potential: Society of Economic Paleontologists and Mineralogists Short Course 19, 290 p.
———, and Hanley, J. H., 1984, Anastomosed and associated coal-bearing fluvial deposits: Upper Tongue River Member, Palaeocene Fort Union Formation, northern Powder River Basin, Wyoming, U.S.A., *in* Rahmani, R. A., and Flores, R. M., eds., Sedimentology of Coal and Coal-Bearing Sequences: International Association of Sedimentologists Special Publication 7, p. 85–103.
Forbes, D. L., 1983, Morphology and sedimentology of a sinuous gravel-bed channel system, lower Babbage River, Yukon coastal plain, Canada, *in* Collinson, J. D., and Lewin, J., eds., Modern and Ancient Fluvial Systems: International Association of Sedimentologists Special Publication 6, p. 195–206.
Friend, P. F., 1978, Distinctive features of some ancient river systems, *in* Miall, A. D., ed., Fluvial Sedimentology: Canadian Society of Petroleum Geologists Memoir 5, p. 531–542.
———, 1983, Toward the field classification of alluvial architecture or sequence, *in* Collinson, J. D., and Lewin, J., eds., Modern and Ancient Fluvial Systems: International Association of Sedimentologists Special Publication 6, p. 345–354.
———, Slater, M. J., and Williams, R. C., 1979, Vertical and lateral building of river sandstone bodies, Ebro Basin, Spain: Journal of the Geological Society of London, v. 136, p. 39–46.
Galay, V. J., Kellerhals, R., and Bray, D. I., 1973, Diversity of river types in Canada, *in* Fluvial Processes and Sedimentation: Proceedings Hydrology Symposium, Edmonton, National Research Council, p. 217–250.
Galloway, W. E., 1981, Depositional architecture of Cenozoic Gulf Coastal Plain fluvial systems, *in* Ethridge, F. G., and Flores, R. M., eds., Recent and Ancient Nonmarine Depositional Environments: Society of Economic Paleontologists and Mineralogists Special Publication 31, p. 127–156.
———, and Hobday, D. K., 1983, Terrigenous clastic depositional systems: Springer-Verlag, New York, 423 p.
Gersib, G. A., and McCabe, P. J., 1981, Continental coal-bearing sediments of the Port Hood Formation (Carboniferous), Cape Linzee, Nova Scotia, Canada, *in* Ethridge, F. G., and Flores, R. M., eds., Recent and Ancient Nonmarine Depositional Environments: Models for Exploration: Society of Economic Paleontologists and Mineralogists Special Publication 31, p. 95–108.
Gustavson, T. C., 1978, Bed forms and stratification types of modern gravel meander lobes, Nueces River, Texas: Sedimentology, v. 25, p. 401–426.
Haszeldine, R. S., 1983a, Fluvial bars reconstructed from a deep, straight channel, Upper Carboniferous coalfield of northeast England: Journal of Sedimentary Petrology, v. 53, p. 1233–1248.
———, 1983b, Descending tabular cross-bed sets and bounding surfaces from a fluvial channel in the Upper Carboniferous coalfield of northeast England, *in* Collinson, J. D., and Lewin, J., eds., Modern and Ancient Fluvial Systems: International Association of Sedimentologists Special Publication 6, p. 449–456.
Heward, A. P., 1978, Alluvial fan sequence and megasequence models, with examples from Westphalian D-Stephanian B Coalfields, northern Spain, *in* Miall, A. D., ed., Fluvial Sedimentology: Canadian Society of Petroleum Geologists Memoir 5, p. 669–702.
Hopkins, J. C., 1985, Channel-fill deposits formed by aggradation in deeply scoured, superimposed distributaries of the Lower Kootenai Formation: Journal of Sedimentary Petrology, v. 55, p. 42–52.
Horne, J. C., Ferm, J. C., Caruccio, F. T., and Baganz, B. P., 1978, Depositional models in coal exploration and mine planning in Appalachian region: American Association of Petroleum Geologists Bulletin, v. 62, p. 2379–2411.
Jackson, R. G., II, 1975, Hierarchial attributes and a unifying model of bedforms composed of cohesionless material and produced by shearing flow: Geological Society of America Bulletin, v. 86, p. 1523–1533.
———, 1978, Preliminary evaluation of lithofacies models for meandering alluvial streams, *in* Miall, A. D., ed., Fluvial Sedimentology: Canadian Society of Petroleum Geologists Memoir 5, p. 543–576.
———, 1981, Sedimentology of muddy fine-grained channel deposits in meandering streams of the American middle west: Journal of Sedimentary Petrology, v. 51, p. 1169–1192.
Jones, C. M., and McCabe, P. J., 1980, Erosion surfaces within giant fluvial cross-beds of the Carboniferous in northern England: Journal of Sedimentary Petrology, v. 50, p. 613–620.
King, W. A., and Martini, I. P., 1984: Morphology and recent sediments of the lower anastomosing reaches of the Attawapiskat River, James Bay, Ontario: Sedimentary Geology, v. 37, p. 295–320.
Kirk, M., 1983, Bar developments in a fluvial sandstone (Westphalian 'A'), Scotland: Sedimentology, v. 30, p. 727–742.
Koster, E. H., 1984, Sedimentology of a foreland coastal plain, Upper Cretaceous Judith River Formation at Dinosaur Provincial Park: Canadian Society of Petroleum Geologists Field Trip Guidebook, 115 p.
———, 1985, Channel processes, vertebrate taphonomy and architecture of an Upper Cretaceous coastal plain sequence, southeast Alberta: Third International Fluvial Sedimentology Conference, Fort Collins, Abstracts, p. 24.
———, and Steel, R. J., eds., 1984, Sedimentology of gravels and conglomerates: Canadian Society of Petroleum Geologists Memoir 10, 441 p.
Kraus, M. J., 1985, Changes in sediment accumulation rates determined from Lower Eocene alluvial paleosols: Third International Fluvial Sedimentology Conference, Fort Collins, Abstracts, p. 25.
Kuenzi, W. D., Horst, O. H., and McGehee, R. V., 1979, Effect of volcanic activity on fluvial-deltaic sedimentation in a modern arc-trench gap, southwestern Guatemala: Geological Society of America Bulletin, v. 90, pt. 1, p. 827–838.

LEBLANC SMITH, G., AND ERIKSSON, K. A., 1979, A fluvioglacial and glaciolacustrine deltaic depositional model for Permo-Carboniferous coals of the northeastern Karoo Basin, South Africa: Palaeogeography, Palaeoclimatology, Palaeoecology, v. 27, p. 67–84.

LUCCHITTA, I., AND SUNESON, N., 1981, Flash flood in Arizona—observations and their application to the identification of flash-flood deposits in the geologic record: Geology, v. 9, p. 414–418.

MCCABE, P. J., 1984, Depositional environments of coal and coal-bearing strata, *in* Rahmani, R. A., and Flores, R. M., eds., Sedimentology of Coal and Coal-Bearing Sequences: International Association of Sedimentologists Special Publication 7, p. 13–42.

MCGOWEN, J. H., AND GARNER, L. E., 1970, Physiographic features and stratification types of coarse-grained point bars: Modern and ancient examples: Sedimentology, v. 14, p. 77–112.

MCKEE, E. D., AND WEIR, G. W., 1953, Terminology for stratification and cross-stratification in sedimentary rocks: Geological Society of America Bulletin, v. 64, p. 381–390.

MIALL, A. D., ed., 1978a, Fluvial Sedimentology: Canadian Society of Petroleum Geologists Memoir 5, 859 p.

———, 1978b, Fluvial Sedimentology: An historical review, *in* Miall, A. D., ed., Fluvial Sedimentology: Canadian Society of Petroleum Geologists Memoir 5, p. 1-47.

———, 1978c, Lithofacies types and vertical profile models in braided river deposits: A summary, *in* Miall, A. D., ed., Fluvial Sedimentology: Canadian Society of Petroleum Geologists Memoir 5, p. 597–604.

———, 1979, Tertiary fluvial sediments in the Lake Hazen intermontane basin, Ellesmere Island, Arctic Canada: Geological Survey of Canada Paper 79–9, 25 p.

———, 1980, Cyclicity and the facies model concept in fluvial deposits: Bulletin of Canadian Petroleum Geology, v. 28, p. 59–80.

———, 1981, Analysis of fluvial depositional systems: American Association of Petroleum Geologists Education Course Note Series 20, 75 p.

———, 1985, Architectural-element analysis: A new method of facies analysis applied to fluvial deposits: Earth Science Reviews, p. 261–308.

MOLLARD, J. D., 1973, Airphoto interpretation of fluvial features, in Fluvial Processes and Sedimentation: Proceedings of Hydrology Symposium, Edmonton, National Research Council, p. 241–380.

MOSSOP, G. D., AND FLACH, P. D., 1983, Deep channel sedimentation in the Lower Cretaceous McMurray Formation, Athabasca Oil Sands, Alberta: Sedimentology v. 30, p. 493–509.

NANSON, G. C., 1980, Point bar and floodplain formation of the meandering Beatton River, northeastern British Columbia, Canada: Sedimentology v. 27, p. 3–30.

——— AND PAGE, K., 1983, Lateral accretion of fine-grained concave benches on meandering rivers, *in* Collinson, J. D., and Lewin, J., eds., Modern and Ancient Fluvial Systems: International Association of Sedimentologists Special Publication 6, p. 133–144.

——— AND RUST, B. R., 1985, Channel patterns and mud deposition in an arid zone river: Cooper Creek, central Australia: Third International Sedimentology Conference, Fort Collins, Abstracts, p. 31.

NILSEN, T. H., ed., 1984, Fluvial sedimentation and related tectonic framework, western North America: Sedimentary Geology, v. 38, 523 p. (special issue).

———, ed., 1985, Modern and ancient alluvial fan deposits: Hutchinson Ross Benchmark v. 87, Van Nostrand Reinhold Co., New York, 372 p.

OKOLO, S. A., 1983, Fluvial distributary channels in the Fletcher Bank Grit (Namurian R2b), at Ransbottom, Lancashire, England, *in* Collinson, J. D., and Lewin, J., eds., Modern and Ancient Fluvial Systems: International Association of Sedimentologists Special Publication 6, p. 421–433.

ORI, G. G., 1979, Barre de meandro nelle alluvioni ghiaiose del fiume reno (Bologna): Bollettino Societa Geologica Italiana, v. 98, p. 35–54.

PARKASH, B., AWASTHI, A. K., AND GOHAIN, K., 1983, Lithofacies of the Markanda terminal fan, Kurukshetra district, Haryana, India, *in* Collinson, J. D., and Lewin, J., eds., Modern and Ancient Fluvial Systems: International Association of Sedimentologists Special Publication 6, p. 337–344.

PETERSON, F., 1984, Fluvial sedimentation on a quivering craton: Influence of slight crustal movements on fluvial processes, Upper Jurassic Morrison Formation, western Colorado Plateau: Sedimentary Geology, v. 38, p. 21–50.

PUIGDEFABREGAS, C., 1973, Miocene point-bar deposits in the Ebro Basin, northern Spain: Sedimentology, v. 20, p. 133–144.

PUTNAM, P. E., 1982a, Fluvial channel sandstones within Upper Mannville (Albian) of Lloydminster Area, Canada—geometry, petrography, and paleogeographic implications: American Association of Petroleum Geologists, v. 66, p. 436–459.

———, 1982b, Aspects of the petroleum geology of the Lloydminster heavy oil fields, Alberta and Saskatchewan: Bulletin of Canadian Petroleum Geology, v. 30, p. 81–111.

———, AND OLIVER, T. A., 1980, Stratigraphic traps in channel sandstones in the Upper Mannville (Albian) of east-central Alberta: Bulletin of Canadian Petroleum Geology, v. 28, p. 489–508.

RAHMANI, R. A., AND FLORES, R. M., eds. 1984, Sedimentology of coal and coal-bearing sequences: International Association of Sedimentologists Special Publication 7, 412 p.

RAMOS, A., AND SOPEÑA, A., 1983, Gravel bars in low-sinuosity streams (Permian and Triassic, central Spain), *in* Collinson, J. D., and Lewin, J., eds., Modern and Ancient Fluvial Systems: International Association of Sedimentologists Special Publication 6, p. 301–312.

RUST, B. R., 1978a, A classification of alluvial channel systems, *in* Miall, A. D., ed., Fluvial Sedimentology: Canadian Society of Petroleum Geologists Memoir 5, p. 187–198.

———, 1978b, Depositional models for braided alluvium, *in* Miall, A. D., ed., Fluvial Sedimentology: Canadian Society of Petroleum Geologists Memoir 5, p. 605–625.

———, 1979, Facies models 2. Coarse alluvial deposits, *in* Walker, R. G., ed., Facies Models: Geoscience Canada Reprint Series No. 1, p. 9–21.

———, 1981, Sedimentation in an arid-zone anastomosing fluvial system: Cooper's Creek, central Australia: Journal of Sedimentary Petrology, v. 51, p. 745–755.

———, AND KOSTER, E. H., 1984, Coarse alluvial deposits, *in* Walker, R. G., ed., Facies Models, second edition: Geoscience Canada Reprint Series No. 1, p. 53–69.

———, AND LEGUN, A. S., 1983, Modern anastomosing-fluvial deposits in arid central Australia, and a Carboniferous analogue in New Brunswick, Canada; *in* Collinson, J. D., and Lewin, J., eds., Modern and Ancient Fluvial Systems: International Association of Sedimentologists Special Publication 6, p. 385–392.

SCHULTZ, E. H., 1982, The chronosome and supersome: Terms proposed for low-rank chronostratigraphic units: Bulletin of Canadian Petroleum Geology, v. 30, p. 29–33.

SCHUMM, S. A., 1969, River Metamorphosis: Journal of the Hydraulics Division, American Society of Civil Engineers, v. 95, p. 255–273.

———, 1977, The fluvial system: John Wiley and Sons, 338 p.

———, 1981, Evolution and response of the fluvial system, sedimentological implications, *in* Ethridge, F. G., and Flores, R. M., eds., Recent and Ancient Nonmarine Depositional Environments: Models for Exploration: Society of Economic Paleontologists and Mineralogists Special Publication 31, p. 19–30.

———, 1985, Patterns of alluvial rivers: Annual Review of Earth and Planetary Sciences, v. 13, p. 5–27.

SMITH, D. G., 1983, Anastomosed fluvial deposits: Modern examples from Western Canada, *in* Collinson, J. D., and Lewin, J., eds., Modern and Ancient Fluvial Systems: International Association of Sedimentologists Special Publication 6, p. 155–168.

———, 1985a, Modern mesotidal-influenced meandering river point bar deposits: Examples similar to the Athabasca Oil Sands in the McMurray Formation, northeast Alberta, Canada: Symposium on Modern and Ancient Clastic Tidal Deposits, Utrecht, Abstracts, p. 129.

———, 1985b, Epsilon cross-stratification (ECS) in modern and ancient lateral accretion channel (point bar) deposits: Alluvial, deltaic and tidal systems compared: Third International Fluvial Sedimentology Conference, Fort Collins, Abstracts, p. 35.

———, AND PUTNAM, P. E., 1980, Anastomosed river deposits: Modern and ancient examples in Alberta, Canada: Canadian Journal of Earth Sciences, v. 17, p. 1396–1406.

———, AND SMITH, N. D., 1980, Sedimentation in anastomosed river systems: Examples from alluvial valleys near Banff, Alberta: Journal of Sedimentary Petrology, v. 50, p. 157–164.

SMITH, N. D., 1970, The braided stream depositional environment: Comparison of the Platte River with some Silurian clastic rocks, north central Appalachians: Geological Society of America Bulletin, v. 81, p. 2993-3014.

———, 1971, Transverse bars and braiding in the lower Platte River, Nebraska: Geological Society of America Bulletin, v. 82, p. 3407–3420.

———, 1972, Some sedimentological aspects of planar cross-stratification in a sandy braided river: Journal of Sedimentary Petrology, v. 42, p. 624–634.

———, AND SMITH, D. G., 1984, Williams River: An outstanding example of channel widening and braiding caused by bed-load addition: Geology, v. 12, p. 78–82.

SNEH, A., 1983, Desert stream sequences in the Sinai Peninsula: Journal of Sedimentary Petrology, v. 53, p. 1271–1280.

STEAR, W. M., 1983, Morphological characteristics of ephemeral stream channel and overbank splay sandstone bodies in the Permian Lower Beaufort Group, Karoo Basin, South Africa, *in* Collinson, J. D., and Lewin, J., eds., Modern and Ancient Fluvial Systems: International Association of Sedimentologists Special Publication 6, p. 405–420.

———, 1985, Comparison of the bedform distribution and dynamics of modern and ancient sandy ephemeral flood deposits in the southwestern Karoo region, South Africa: Sedimentary Geology, v. 45, p. 209–230.

STEWART, D. J., 1983, Possible suspended-load channel deposits from the Wealden Group (Lower Cretaceous) of southern England, *in* Collinson, J. D., and Lewin, J., eds., Modern and Ancient Fluvial Systems: International Association of Sedimentologists Special Publication 6, p. 369–384.

TUNBRIDGE, I. P., 1981, Sandy high-energy flood sedimentation—Some criteria for recognition, with an example from the Devonian of S.W. England: Sedimentary Geology, v. 28, p. 79–96.

———, 1984, Facies model for a sandy ephemeral stream and clay playa complex: The Middle Devonian Trentishoe Formation of North Devon, U.K.: Sedimentology, v. 31, p. 697–716.

TURNER, P., 1980, Continental red beds: Developments in Sedimentology 29, Elsevier, 562 p.

VESSELL, R. K., AND DAVIES, D. K., 1981, Non-marine sedimentation in an active forearc basin, *in* Ethridge, F. G., and Flores, R. M., eds., Recent and Ancient Nonmarine Depositional Environments: Models for Exploration: Society of Economic Paleontologists and Mineralogists Special Publication 31, p. 31–45.

WALKER, R. G., AND CANT, D. J., 1979, Sandy fluvial systems, *in* Walker, R. G., ed., Facies Models: Geoscience Canada Reprint Series No. 1, p. 23–31.

——— AND CANT, D. J., 1984, Sandy fluvial systems, *in* Walker, R. G., ed., Facies Models, second edition: Geoscience Canada Reprint Series No. 1, p. 71–89.

WELLS, N. A., 1983, Transient streams in sand-poor red beds: Early-Middle Eocene Kuldana Formation of northern Pakistan, *in* Collinson, J. D., and Lewin, J., eds., Modern and Ancient Fluvial Systems: International Association of Sedimentologists Special Publication 6, p. 393–403.

WIGHTMAN, D. M., TILLEY, B. J., AND LAST, W. M., 1981, Stratigraphic traps in channel sandstones in the Upper Mannville (Albian) of east-central Alberta: Discussion: Bulletin of Canadian Petroleum Geology, v. 29, p. 622–625.

WING, S. L., 1984, Relation of paleovegetation to geometry and cyclicity of some fluvial carbonaceous deposits: Journal of Sedimentary Petrology, v. 54, p. 52–66.

PART I
SEDIMENT TRANSPORT AND HYDRAULICS

TOWARD A BETTER UNDERSTANDING OF BEDLOAD TRANSPORT

IAN REID
Department of Geography, Birkbeck College, Malet Street, London WC1E 7HX, UK.
AND
LYNNE E. FROSTICK
Department of Geology, Royal Holloway & Bedford New College, Egham, Surrey TW20 0EX, UK.

ABSTRACT: The development of three separate devices at Birkbeck College, London, and their deployment in a gravel-bed stream allows us to pinpoint entrainment thresholds, to measure bedload continuously and to quantify the ingress of matrix fines. This provides a comprehensive picture of processes and gives clear indication of the reasons why gravel beds appear to react unpredictably to flood flows and why a universal bedload equation has yet to be developed. We show that ubiquitous bed microforms (e.g., pebble clusters) delay initial motion so that stream power is 5 times greater than it is when bedload transport ceases. We also show that low-flow ingress of matrix fines increases the difference to 16 times by increasing shear strength of the stream bed. Our continuous monitoring devices indicate a regular pulsation of bedload transport (mean wave period 1.7 hrs) that *cannot* be attributed to deficiencies in the samplers because it is corroborated by distinctly different and independent instruments; the pulses are not explained by changes in hydraulic parameters. We suggest that they are kinematic waves of mobile bed particles, the first dynamic manifestation of this phenomenon in a natural stream. Our data give a clear indication of those facets of water-sediment interaction that must be investigated in order to arrive at a *general* bedload transport equation for gravel-bed streams, and the data highlight the naive assumptions currently made in paleohydraulics when using clast size as an indicator of paleo-flow conditions.

INTRODUCTION

Predictive modelling of bedload transport in alluvial channels has made faltering progress over the last 50 years, yet still remains unsatisfactory. A host of equations have been developed. The best known are those of Schoklitsch (1934), Meyer-Peter and Muller (1948), Einstein (1950), Bagnold (1956) and Yalin (1963), yet none have proved sufficiently universal for general application to streams outside the one that was used to derive the empirical constants appropriate to each case. This has been shown very clearly by Parker and others (1982), by Bathurst (pers. commun.), and by several others.

The reasons why a general bedload equation is elusive are many. There is the obvious fact that each river and, indeed, each reach of a single river has its own hydraulic and sedimentological character. Yet, perhaps surprising in hindsight, the bed of a stream has almost invariably been envisaged as a neat arrangement of particles, the primary purpose being that any modelling of the mechanisms of entrainment can be simplified (e.g., Meland and Norrman 1966). When complicated bed arrangements have been acknowledged (e.g., Einstein, 1950; Proffitt and Sutherland, 1983; Ikeda, 1984), there has been a tendency to seek an *average* response of particles to an applied stress. In these circumstances, we should not expect to be able to equate actual and predicted bedload except in general, *average* terms. There is, however, an increasing acknowledgment of the significance of both bed microforms and the changing integrity of the armor-layer in explaining why bedload discharge is often out of phase with changing hydraulic conditions (e.g., Reid and Frostick, 1984; Brayshaw, 1985; Naden, pers. commun.; Ashworth & Ferguson, pers. commun.).

Most damaging to the development of a bedload equation of general value has been the explicit belief that the bed of a stream will respond to applied stress in the same way, time after time, and that bedload begins and ends at the same threshold value of applied force. Both of these assumptions have been shown to be inappropriate, at least for one river (Reid and others, 1985). There has also been a presumption, yet to be tested, that bedload transport rates rise, and continue to rise, as some function of increasing stream energy, whether expressed as stream power or bed shear stress. Again, the availability of an increasing amount of reliable field data suggests that this is too simple an expectation of such a complex process as bedload transport (Reid and Frostick, 1986).

The problem of finding a universal bedload transport equation, or indeed a set of like equations, lies in improving our understanding of the factors that control particle movement in natural channels. There is also a need to improve available instrumentation. Only by obtaining reliable field observations can we hope to be able to identify the limitations of present approaches.

INSTRUMENTATION

At Birkbeck College, London, we have developed three separate instruments to look at different aspects of bedload transport and bed sediments. All three are permanently installed in a stream bed and two are fully automated recording devices. The first is a pit sampler that continuously weighs bedload. Technical specifications of the device are given in Reid and others (1980) and examples of its applications appear in Reid and others (1985). Three of the samplers can be seen in the process of installation in Figure 1C. The second instrument also records bedload movement but this time at particle scale. It is an electromagnetic sensor that acts like a metal detector and is similar to that of Ergenzinger and Custer (1983) which was developed simultaneously. Components of the system can be seen in Figure 1A and an installation in the bed of Turkey Brook in Figure 1B. The device registers the movement of particles tagged with ferrite and is capable of pinpointing the entrainment of individual clasts that have been seeded in the stream bed. A full technical description is given by Reid and others (1984). The third device is non-recording and is a set of boxes that sits in the stream bed below an artificially fixed armor layer (Fig. 1D). Each box is filled with

FIG. 1.—A) Components of a Birkbeck electromagnetic bedload particle sensor. Shown is the search head (black and elongate), signal demodulator and amplifier, recorder, and concrete channel into which the search head fits when installed. B) Birkbeck electromagnetic sensor installed in the bed of Turkey Brook. C) Birkbeck automatic recording pit samplers being installed in the bed of Turkey Brook. D) Matrix traps showing fixed armor layer lids temporarily removed and compartmentalized boxes containing simulated sub-armor layers of different particle caliber. The bridge is used to gain access without disturbing the bed and is removed between visits.

a different grade of gravel which is intended to represent a sub-armor framework with a specific size distribution. The boxes act as passive traps for fine matrices that pass through the armor layer and become lodged in the interstices of the sub-armor framework. These matrix traps are described in Frostick and others (1984).

FIELD SITE

We recorded the patterns of bedload transport at Turkey Brook, a gravel-bed stream just north of London, England, over a period of 4 years. Our installations were placed in straight reaches of the brook to avoid the complications that are associated with secondary flows at meander bends (e.g., Thorne and others, 1985). The channel bed is incised so that bankfull stage ranges between 0.5 and 1.4 m (1.6 and 4.6 ft). In these experimental reaches, bedforms consist of low amplitude (generally <10 cm) alternating bars. The bed material of this stream is ovate flint gravel derived from Pleistocene outwash deposits which clothe part of the catchment. The armor layer of the bed has a D_{50} of 22 mm. The sub-armor layer has a D_{50} of 16 mm, a value very close to the D_{50} of the bedload which is 11 mm (Reid and Frostick, 1986).

The catchment is underlain by impervious Eocene London Clay. Because of this, flood hydrographs are flashy and bedload-generating events are frequent. Our records suggest that more than 12 such events can be expected in a year, almost all occurring in winter months (Reid and Frostick, 1984).

RESULTS AND DISCUSSION

The relationships between bedload transport and hydraulic parameters for Turkey Brook show considerable scatter about a general trend (Fig. 2, top left). Similar patterns have been observed in other streams, for example, the Snake and Clearwater Rivers (Fig. 2, data from Emmett, 1976); the East Fork River (Fig. 2, data from Leopold and Emmett, 1976, 1977); the Torlesse stream (Hayward, 1979); and Virginia Creek (Tacconi and Billi, pers. commun.).

We set out to find explanations for some of the residuals from the general trend line as well as to evaluate the trend line itself as a general statement about the relationship between hydraulics and sediment motion.

Our first observation was that the incidence of bedload in relation to the flood wave is highly variable (Fig. 3A). Bedload sometimes starts moving on the rising limb of a flood hydrograph; however, sometimes no movement occurs until after the flood peak. On these occasions bedload transport is confined to the recession limb of the flood. Often, bedload transport rates fall off as stream stage rises, the exact opposite of what might be expected from a simple consideration of sediment response to increasing stream

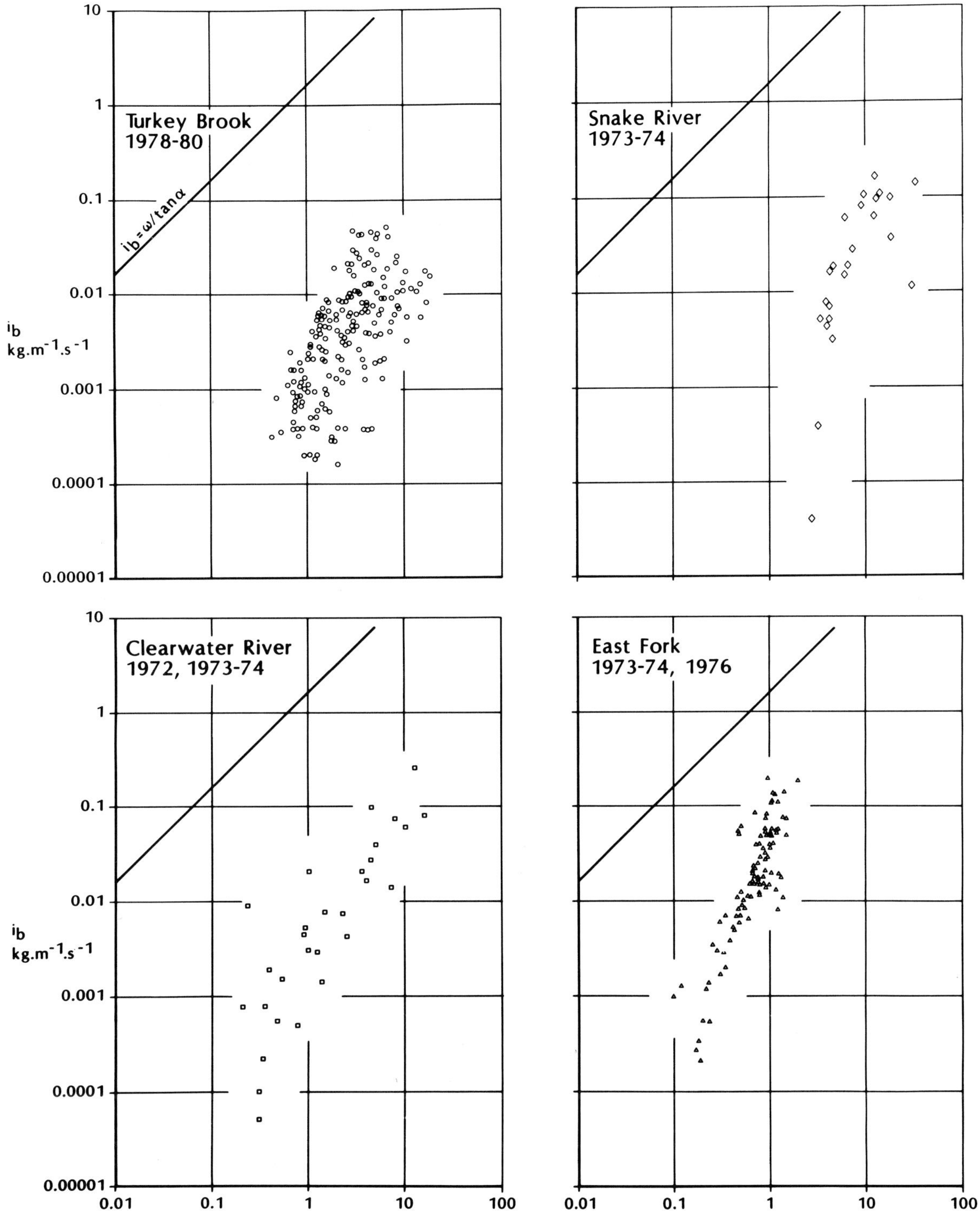

FIG. 2.—Instantaneous values of bedload transport rates per unit width, i_b, plotted agaisnt unit stream power, ω, for three gravel-bed rivers and one sand-bed river (East Fork). ib = ω/tan α is Bagnold's (1973) relationship for 100 percent efficiency, in which α is the internal angle of friction for loose granular material. Data for the Snake and Clearwater Rivers come from Emmett (1976); that for the East Fork River comes from Leopold and Emmett (1976, 1977). Diagram after Reid and Frostick (1986).

power. What transport rates do *not* do is follow a simple pattern that mirrors the flood hydrograph. Our work at Turkey Brook suggests that the most significant causes of this mismatch between bedload and water discharges include the effects of pebble clusters and other bed microforms on entrainment, the fact that there are significant differences between traction thresholds at the beginning and end of sediment motion, and the fact that bedload transport is

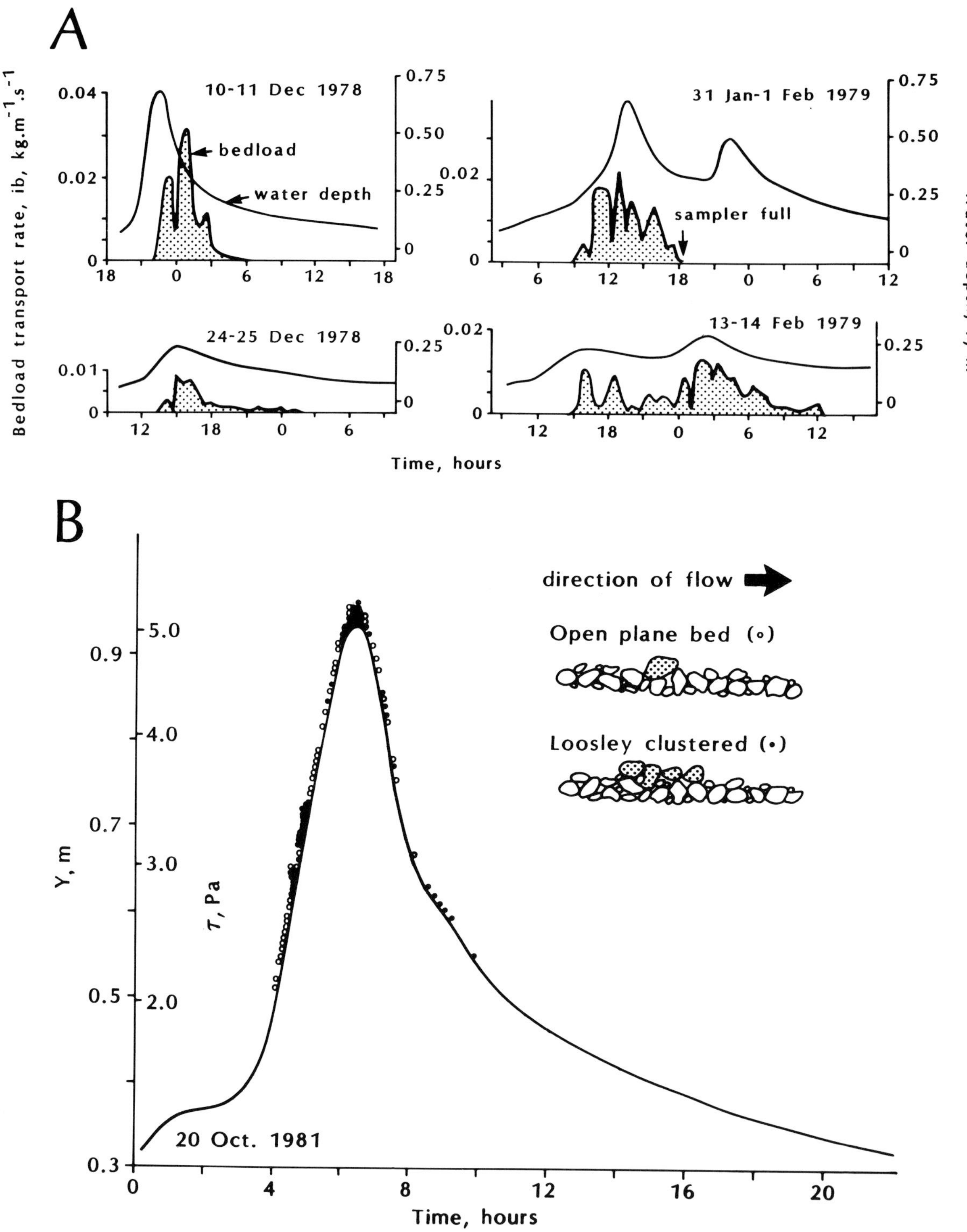

FIG. 3.—A) Bedload discharge per unit width and water depth during four selected floods on Turkey Brook. B) Entrainment of particles seeded in two characteristic positions in the bed microtopography of Turkey Brook during one selected flood (open and closed circles). Particle motion was detected with two Birkbeck electromagnetic sensors. Y is water depth and τ is average bed shear stress. Both relate to the continuous line.

characterized by pulses which are governed by grain interaction rather than by gross hydraulic parameters.

Effect of pebble clusters.—

The arrangement of particles on the stream bed influences traction thresholds. Flume experiments carried out by Fenton and Abbott (1977), among others, have shown that the relative exposure of individual grains has a significant effect on entrainment. Our field experiments carried out in Turkey Brook provide evidence that the distinctive bed microtopography associated with congregating particles also plays an important role in controlling bedload transport in natural channels. A detailed analysis of several gravel-bed rivers has shown that pebble clusters can occupy up to 50 percent of the stream bed, with average figures around 10–20 percent (Brayshaw, 1985). They therefore constitute a major element of the stream bed. Typical clusters center on a single, large obstacle clast around which smaller particles accumulate. They can also be closely spaced clasts of like size, however, as shown in Figure 3B. We have monitored the movement of tagged particles seeded in different characteristic parts of various types of cluster bedform with our electromagnetic sensor (Reid and others, 1984) and have compared them with the movement of particles occupying the open plane-bed position envisaged in the classical treatment of entrainment mechanics of White (1940) and others.

Figure 3B illustrates the incidence of initial motion from two such bed positions during the course of one flood event. It can be seen that the open plane-bed particles are the first to be entrained and that shear stress has to rise by about 30 percent before the first of the loosely clustered particles moves. Whereas open plane-bed particles generally move before those in clusters, because they do not benefit from the mutual protection that comes from both the generation of flow separation (Brayshaw and others, 1983) and the mechanical interlock of one particle with another, there is considerable overlap in the pattern of entrainment between the two groups. Greater distinction emerges by combining the data of more than one flood. This has been done in Figure 4.

The lessons to be learned by the sedimentologist considering the general question of bedload transport are various. First, there are several thresholds for initial sediment motion that are not necessarily related to the size of the bed material but are a function of clast position within bed microforms. These cause the release of material from the bed at varying times during flood flow. Because this is the case, there will not be a simple relationship between increasing bed shear stress and sediment transport rate. Herein lies the possible reason for differences between sediment transport in sand-bed and gravel-bed streams, at least where plane-bed conditions prevail in both. There are no significantly differentiated storage areas at the microscale in a plane sand-bed (i.e., no obvious particle clusters). The release of sediment from its rest position is a more singular event and,

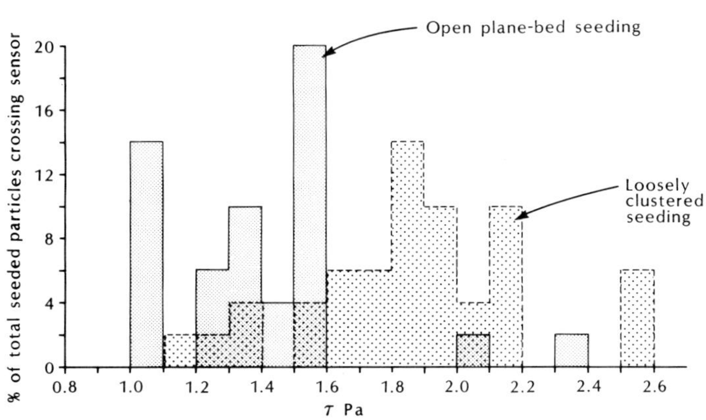

FIG. 4.—Histograms of average bed shear stress, τ, at the moment of particle entrainment from two characteristic bed positions in Turkey Brook during three floods.

TABLE 1.—AVERAGE BED SHEAR STRESS, τ_0 AND τ_0', AND UNIT STREAM POWER, ω_0 AND ω_0', AT THE BEGINNING AND END OF BEDLOAD TRANSPORT DURING FLOODS IN TURKEY BROOK.

	Beginning		End	
Flood Date	τ_0 Pa	ω_0 W·m^{-2}	τ_0' Pa	ω_0' W·m^{-2}
10 Dec 1978	6.65	18.31	1.07	1.12
24 Dec	0.82	0.65	0.68	0.43
28 Dec	1.04	0.71		
25 Jan 1979	2.25	2.05		
31 Jan	2.31	2.13		
13 Feb	1.56	1.40	0.74	0.67
13 Mar	2.61	2.86		
26 May	1.56	1.29		
9 Dec	3.31	5.36	0.89	0.85
27 Dec	1.98	2.26		
12 Mar 1980	3.18	4.12	0.78	0.72
Mean	2.20	3.74	0.79	0.76

hence, it is more predictable. In this context, it is interesting to note that scatter in the stream power-bedload transport plot for the East Fork River is generally less than it is for the three gravel-bed streams shown in Figure 2. The second lesson to be learned is that the microforms of gravel beds will delay entrainment (Reid and Frostick, 1984). The degree of influence inevitably depends upon the fraction of the bed occupied by pebble clusters. It must be borne in mind, however, that the sphere of influence of such bodies, whether comparatively streamlined or bluff, will extend well beyond their spatial extent (Brayshaw and others, 1983; Best and Brayshaw, 1985). It is this fact that leads us to suspect that their control on bedload is considerable.

Traction thresholds.—

Continuous monitoring of sediment movement at Turkey Brook has revealed distinctly different traction thresholds for the beginning and end of bedload transport during a flood wave. Expressed in terms of stream power, the average traction threshold at the beginning of general bed motion is some 5 times higher than that when general bed motion stops (Table 1). This might be anticipated from a general consideration both of the effects of shielding that are inherent in pebble clusters and of the differential resistance to motion that is implicit in the change from static to dynamic friction as particles move from an initial point of rest. It does mean, however, that a single traction threshold (a characteristic of all published bedload equations) may be entirely inappropriate to the problem of finding a general solution to bedload prediction.

In addition to the different traction thresholds within a flood, there are also differences between floods which compound the problems of bedload prediction. The recent history of flow has to be taken into account when considering initial entrainment thresholds. We have shown that if no floods occur for extended periods of time (say, a month or more), fine particles carried by low flows infiltrate the interstices of the bed and strengthen the structure very much like mortar in a stone wall (Frostick and others, 1984; Reid and Frostick, 1985). Once this happens, the stream power needed to initiate sediment motion may be as high as 5 times the average value for close-spaced floods. This is as high as 25 times the stream power prevailing on average at the finish of bedload transport. Such an effect of matrix accumulation can be seen by noting the traction thresholds of the floods dated 10 December 1978, 9 December 1979, and 12 March 1980 in Table 1. All three follow extended periods of low flow in a situation where the rate of matrix accumulation has been shown to exceed 0.5 $kg.m^{-2}.day^{-1}$ (Frostick and others, 1984). The overall effect can be seen in Figure 2A where bedload transport for the flood 10–11 December 1978 is confined to the recession limb because the energy of the rising limb of this first-of-the-season event is used to loosen the matrix-filled bed.

Bedload pulses.—

The continuous records of bedload transport obtained from the Birkbeck bedload samplers and magnetic sensors installed in Turkey Brook provide irrefutable proof of the existence of regular pulses in bedload discharge that are not a response to similarly phased changes in hydraulic conditions. Pulses have been observed in other rivers, for example, the Inn (Muhlhoffer, 1933), the Rhine (Einstein, 1937), and Slate Creek (Emmett, 1975), but until now there has always been some suspicion that they might be a product of the sampling devices used. The Birkbeck samplers have no such instrumental deficiencies. Indeed, we have two different kinds of machine amounting to eight independent pieces of equipment that indicate the regularity of bedload pulses.

The pulses in bedload discharge that we record are not a result of random inputs of transportable material such as occurs with bank collapse, etc. This has been suggested as the cause of phase differences in sediment and water discharges for the Roaring River in Colorado after a dam-burst flood (Pitlick and Thorne, pers. commun.). Examples of the pulse phenomenon operating at different scales in Turkey Brook are given in Figure 3. First, the entrainment of individual particles from all seeding positions within the bed microtopography can be seen to be spasmodic (Fig. 3B). There is an interesting indication in the diagram that the motion of clustered particles is more spasmodic than those occupying open plane-bed positions. This suggests that the gravel is released in waves as pebble clusters break, reform and rebreak. On a broader scale, there is the pulsing nature of bulk sediment transport that is evident in Figure 3A. This is perhaps an averaged expression of the activity going on at particle scale. In Turkey Brook the mean period of bedload pulses is 1.7 hrs. This must be compared with values as low as 0.1 hr (River Inn, Muhlhoffer, 1933) and as high as 20 hrs (River Rhine, Einstein, 1937). In the absence of obvious migrating bedforms, we have suggested that these bedload pulses are an expression of the kinematic waves of sediment that were postulated by Langbein and Leopold (1968).

CONCLUSIONS

Taking into account all of the factors discussed above, it is not surprising that plots of bedload discharge as a function of some hydraulic parameter such as stream power have a wide scatter. It is also not surprising that a predictive equation developed with one set of data cannot be transferred easily to predict bedload transport successfully in other river channels. At present, our data make it difficult to envisage a method by which instantaneous measures of bedload discharge might be predicted. Only by integrating values throughout the period of sediment mobility can we give a good prediction of average bedload discharge during the passing of a flood. This not only allows for differences in traction thresholds at the start and finish of sediment motion but also 'irons out' bedload pulses. The results obtained for Turkey Brook using such integrated values are encouraging. There is a fair degree of agreement between predicted and actual average bedload discharge (Fig. 5).

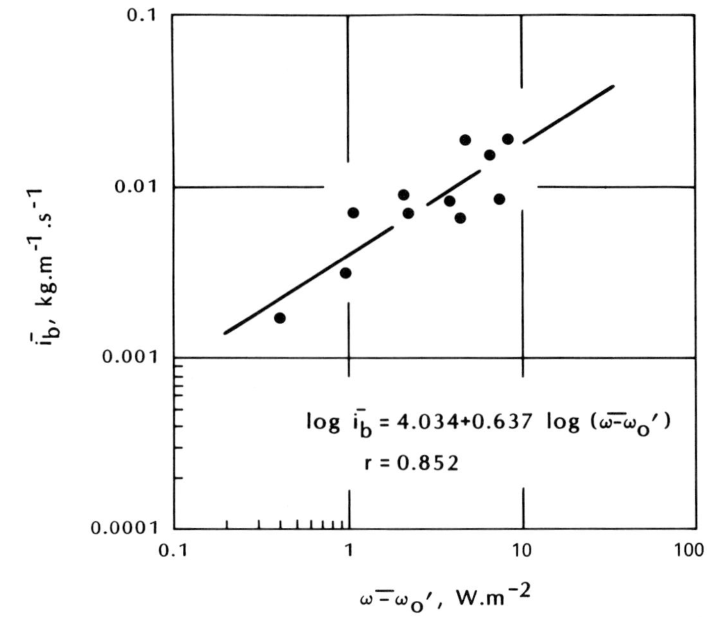

FIG. 5.—Average bedload discharge per unit width, $\bar{i}_b$, plotted against average unit stream power in excess of the mean value of the threshold associated with the end of bedload transport and for the flood period that follows the attainment of the mean value of the threshold of initial sediment motion, $\overline{\omega - \omega_0'}$, for 11 floods on Turkey Brook (after Reid and Frostick, 1986).

We have some way to go before we can generalize from our findings in Turkey Brook. We need many more continuous bedload discharge data sets from different rivers in order to investigate the causes of differences in behavior between channels. Only by getting such data can we hope to provide convenient nomograms that will give easy solutions to bedload problems in a wide range of alluvial channels.

ACKNOWLEDGMENTS

Financial support was provided by the University of London's Central Research Fund. John Layman and Andrew Brayshaw were intimately involved with development of the field sensors and with the experimental program and were both funded by NERC Research Training Awards.

REFERENCES

BAGNOLD, R. A., 1956, The flow of cohesionless grains in fluids: Philosophical Transactions of the Royal Society of London, v. A249, p. 235–297.

———, 1973, The nature of saltation and of bedload transport in water: Proceedings, Royal Society of London, v. A332, p. 473–504.

Best, J. L., and Brayshaw, A. C., 1985, Flow separation—a physical process for the concentration of heavy minerals within alluvial channels: Journal of the Geological Society of London, v. 142, p. 747–755.

Brayshaw, A. C., 1985, Bed microtopography and entrainment thresholds in gravel-bed rivers: Geological Society of America Bulletin, v. 96, p. 218–223.

———, Frostick, L. E., and Reid, I., 1983, The hydrodynamics of particle clusters and sediment entrainment in coarse alluvial channels: Sedimentology, v. 30, p. 137–143.

Einstein, H. A., 1937, Die eichung des im Rhein verwendeten Geschiebefangers: Schweizer Bauzeitung, v. 110, p. 29–32.

———, 1950, The bedload function for sediment transportation in open-channel flows: U.S. Department of Agriculture Technical Bulletin 1026, 71 p.

Emmett, W. W., 1975, The channels and waters of the upper Salmon River area, Idaho: U.S. Geological Survey Professional Paper 870A, 116 p.

———, 1976, Bedload transport in two large gravel-bed rivers, Idaho and Washington: Proceedings, Third Federal Inter-Agency Sedimentation Conference, Denver, Colorado, March 1976, p. 4-101 to 4-114.

Ergenzinger, P. J., and Custer, S. G., 1983, Determination of bedload transport using naturally magnetic tracers: First experiences at Squaw Creek, Gallatin County, Montana: Water Resources Research, v. 19, p. 187–193.

Fenton, J. D., and Abbott, J. E., 1977, Initial movement of grains on a stream bed: The effect of relative protrusion: Proceedings of the Royal Society of London, v. A352, p. 523–537.

Frostick, L. E., Lucas, P. M., and Reid, I., 1984, The infiltration of fine matrices into coarse-grained alluvial sediments and its implications for stratigraphical interpretation: Journal of the Geological Society of London, v. 141, p. 955–965.

Hayward, J. A., 1979, Mountain stream sediments, *in* Murray, D. L., and Ackroyd, P., eds., Physical Hydrology: New Zealand Hydrological Society, Wellington, p. 193–212.

Ikeda, H., 1984, Flume experiments on the causes of superior mobility of sediment mixtures: Annual Report of the Institute of Geoscience, University of Tsukuba 10, p. 53–56.

Langbein, W. B., and Leopold, L. B., 1968, River channel bars and dunes—theory of kinematic waves: U.S. Geological Survey Professional Paper 422L, 20 p.

Leopold, L. B. and Emmett, W. W., 1976, Bedload measurements, East Fork River, Wyoming: Proceedings, National Academy of Science, United States of America, v. 73, p. 1000–1004.

———, and ———, 1977, 1976, 1977 bedload measurements, East Fork River, Wyoming: Proceedings, National Academy of Science, United States of America, v. 74, p. 2644–2648.

Meland, N., and Norrman, J. O., 1966, Transport velocities of single particles in bed-load motion: Geografiska Annaler 48A, p. 165–182.

Meyer-Peter, E., and Müller, R., 1948, Formulas for bedload transport: International Association of Hydraulic Structures Research Report, Second Meeting, Stockholm, p. 39–64.

Muhlhoffer, L., 1933, Unterschungen uber die Schwebstoff und geschiebefuhrung des Inn nacht Kirchbichl (Tirol): Die Wasserwirtschaft, v. 26, p. 1–4.

Parker, G., Klingeman, P. C., and McLean, D. G., 1982, Bedload and size distribution in paved gravel-bed streams: Journal of Hydraulics Division, American Society of Civil Engineers, v. 108, p. 544–571.

Proffitt, G. T., and Sutherland, A. J., 1983, Transport of non-uniform sediments: Journal of Hydraulic Research, v. 21, p. 33–43.

Reid, I., Brayshaw, A. C., and Frostick, L. E., 1984, An electromagnetic device for automatic detection of bedload motion and its field applications: Sedimentology, v. 31, p. 269–276.

———, and Frostick, L. E., 1984, Particle interaction and its effect on the thresholds of initial and final motion in coarse alluvial channels, *in* Koster, E. H., and Steel, R. J., eds., Sedimentology of Gravels and Conglomerates: Canadian Society of Petroleum Geologists Memoir, v. 10, p. 61–68.

———, and ———, 1985, Role of settling, entrainment and dispersive equivalence and of interstice trapping in placer formation: Journal of the Geological Society of London, v. 142, p. 739–746.

———, and ———, 1986, Dynamics of bedload transport in Turkey Brook, a coarse-grained alluvial channel: Earth Surface Processes and Landforms, v. 11, p. 1–11.

———, ———, and Layman, J. T., 1985, The incidence and nature of bedload transport during flood flows in coarse-grained alluvial channels: Earth Surface Processes and Landforms, v. 10, p. 33–44.

———, Layman, J. T., and Frostick, L. E., 1980, The continuous measurement of bedload discharge: Journal of Hydraulic Research, v. 18, p. 243–249.

Schoklitsch, A., 1934, Der Geschiebetrieb und Geschiebefracht: Wasserkraft und Wasserwirtschaft, v. 4, p. 37.

Thorne, C. R., Zevenbergen, L. W., Pitlick, J. C., Rais, S., Bradley, J. B., and Julien, P. Y., 1985, Direct measurements of secondary currents in a meandering sand-bed river: Nature, v. 316, p. 746–747.

White, C. M., 1940, The equilibrium of grains on the bed of a stream: Proceedings, Royal Society of London, v. A174, p. 322–338.

Yalin, M. S., 1963, An expression for bed-load transportation: Journal of Hydraulics Division, American Society of Civil Engineers, v. 89, p. 221–251.

ELECTROMAGNETIC DETECTION OF PEBBLE TRANSPORT IN STREAMS: A METHOD FOR MEASUREMENT OF SEDIMENT-TRANSPORT WAVES

STEPHEN G. CUSTER AND NICHOLAS BUGOSH
Department of Earth Science, Montana State University, Bozeman 59717
PETER E. ERGENZINGER
Institut fur Physiche Geographie, Freie Universitat, Berlin, West Germany
AND
BRET C. ANDERSON
Department of Geology, Colorado School of Mines, Golden 80401

ABSTRACT: An electromagnetic coil placed in the bed of a stream can provide instantaneous real-time electronic data on the motion of coarse, naturally-magnetic, bed material. Data collected in May 1982 show that sediment transport of material larger than 32 mm occurred on the bed during falling stage, began somewhat gradually, and ended nearly instantaneously. Because transport data were not averaged, transport waves of material larger than coarse pebbles can be seen on the record. During early parts of the transport event, the period of the waves was less than 1 min. By the end of the event, waves were occurring at approximately 6–8 min intervals. In 1985, the device detected the motion of coarse pebbles and cobbles in one part of a stream, while other parts of the stream showed no motion. If background noise problems can be overcome, the device has the potential to provide insight into the nature and location of pebble and cobble transport on the bed of gravelly streams and the conditions which lead to that transport.

INTRODUCTION

The process of bedload transport in gravel-bed streams is of interest to scientists the world over from both a practical and theoretical standpoint (e.g., Hey and others, 1982). An important focal point for research has been the nature of transport in gravel-bed streams. Answers to a variety of questions are currently being sought. What are the hydraulic conditions necessary to initiate transport? What is the nature of transport as an event begins and ends? During transport, do the gravel particles move in kinematic waves such as those described by Langbein and Leopold (1968), or do they move continuously past a point? Where does transport take place on a stream bottom?

A variety of devices and techniques have been developed to seek answers to these questions. Painted, tagged, and radioactively or magnetically labeled clasts have been used as tracers (Hubbell and Sayne, 1964; Langbein and Leopold, 1968; Buttler, 1977; Oldfield and others, 1981; Reid, and others, 1984). Mechanical samplers, such as the Helley-Smith sampler, have also been used to measure coarse bedload transport (Hubbell, 1964; Emmett, 1980). Large-scale traps and vortex samplers have been installed in stream bottoms (Leopold and Emmett, 1976; Klingeman and Emmett, 1982). Flumes have also been used to examine gravel transport (White and Day, 1982; Kuhnle and Southard, 1985). Pressure sensitive devices have been used to monitor bedload transport in streams during floods (Reid and others, 1985). Instruments have been devised to monitor the motion of individual particles on the stream bottom during transport as well. In 1973, Tywonivk and Warnock used acoustical techniques to relate sound, motion of bed material, water velocity, and location of bed-material motion. Artificial magnetic tracers have also been installed in stream beds to examine bedload transport quantities and transport waves (Ergenzinger and Custer, 1982; Reid and others, 1984).

This list of techniques is not exhaustive, but does serve to point out the diversity of approaches used to investigate bedload transport in gravel-bed streams. The proliferation of techniques is related in part to the difficulties associated with monitoring transport (Reid and others, 1985; Raemy and Jaeggi, 1981). Floodwaters are dangerous, turbid, and unpredictable. Visual observation of transport is difficult due to turbidity. Adequate sampling is often dangerous to the researcher because it must be done during peak stage. Natural events which must be sampled occur at times which may be unexpected and may occur at night. These problems make measurement of transport phenomena such as kinematic waves and thresholds difficult (Reid and Frostick, this volume).

Most approaches in use today average sampling measurements over a discrete amount of time because of mechanical or other practical constraints. Such averaging can obscure the transport patterns. For example, if 30-sec sediment transport averages are used, transport features with durations of less than 30 sec or even a minute are lost. A device was introduced by Ergenzinger and Custer (1983) which reduces such problems and facilitates the investigation of such phenomena in streams. The instrument consists of 9000 windings of wire around an iron core cast in concrete and embedded in a stream bottom which contains naturally magnetic pebbles and cobbles. As large, naturally magnetic clasts pass over the detector, they induce an electrical field which can be detected. The technique does not require introduced labeled particles and is capable of counting the passage of individual naturally magnetic pebbles and cobbles over the detector. The device produces a real-time record of transport with minimal count averaging. These data, when combined with hydraulic data, make it possible to study the nature of transport as coarse particle motion begins and ends and to examine sediment transport waves.

This paper presents new data from experiments run at Squaw Creek, Montana (Fig. 1) in 1982 and 1985. During these flood events, a strip chart record of pebbles and cobbles passing a point was produced. The real-time records of transport provide insight into the nature of transport. The initiation and termination of transport can be seen, as can

FIG. 1.—A. Location of Squaw Creek, Gallatin County, Montana, USA (T. 4 S., R. 4 E., sec. 34). B. View upstream of the detector. C. View of 1-m-long detector installed in the stream.

sediment transport waves. The data on sediment transport waves can be compared to data presented by Reid and others (1985) and Kuhnle and Southard (1985).

SITE AND EVENT CHARACTERISTICS

The experiment was run where Squaw Creek joins the Gallatin River in Montana (Fig. 1). A 1-m long detector was installed downstream of a log sill below a bridge. Approximately 30 percent of the bed material is sufficiently magnetic to produce a signal. Laboratory experiments indicate that signals are generated by particles larger than 32 mm. Motion of particles smaller than this are not detected. Particle size in gravel bars is quite variable. Particles as large as 64 mm are known to have moved. Median particle size from gravel bars at the site ranges from 128 mm to 11 mm (Table 1). A representative median particle size is approximately 12 mm. Other site characteristics have been described in Ergenzinger and Custer (1983). Since the runoff events to be described are new, some of the stream characteristics for the event need to be established.

Discharge for the recorded events peaked at 3.4 m^3/sec in 1982 and 3.7 m^3/sec in 1985. The hydrograph for the first bedload transport event of the year in 1982 is shown in Figure 2. The figure also shows the time at which the coarse bedload began moving and stopped moving. Note that coarse pebbles did not begin moving until shortly after the peak and their movement only lasted approximately 1 hr. The hydrograph for the 1985 peak is not reproduced because discharges for this year were below the coarse sediment entrainiment threshold, so that a significant transport event did not occur. The transport of a coarse particle every few minutes in 1985, however, indicates that the threshold for large-scale coarse bedload transport was approached.

The slope as transport began was probably the same for both events. Following the 1982 flood, bed slope was measured and found to be two percent. In 1985, water surface slope was measured as well. Although there was no significant motion of material larger than very coarse pebbles in 1985, discharges exceeded those of 1982. Water surface slope approached two percent when discharges approached

TABLE 1.—PARTICLE SIZE OF BED MATERIAL FROM CUMULATIVE SIZE DISTRIBUTIONS FROM SQUAW CREEK.

Size Class (Cum. %)	Size Millimeters				
	#1	#2	#3	#4	#5
d_5	0.35	0.95	0.34	0.33	0.54
d_{16}	0.47	7.60	1.12	3.80	3.20
d_{25}	1.05	10.70	3.50	10.00	18.00
d_{50}	10.30	18.00	13.00	24.00	128.00
d_{75}	33.00	30.00	26.00	39.00	180.00
d_{84}	42.00	36.00	33.00	49.00	200.00
d_{95}	64.00	44.00	58.00	78.00	240.00

#1 76.2-mm Helley-Smith sample at 8.2 m^3/sec on 25 May 1982.
#2 76.2-mm-Helley-Smith samle at 5.3 m^3/sec on 28 May 1983.
#3 Gravel bar upstream of detector.
#4 Gravel bar downstream of detector.
#5 Gravel in exposed channel.

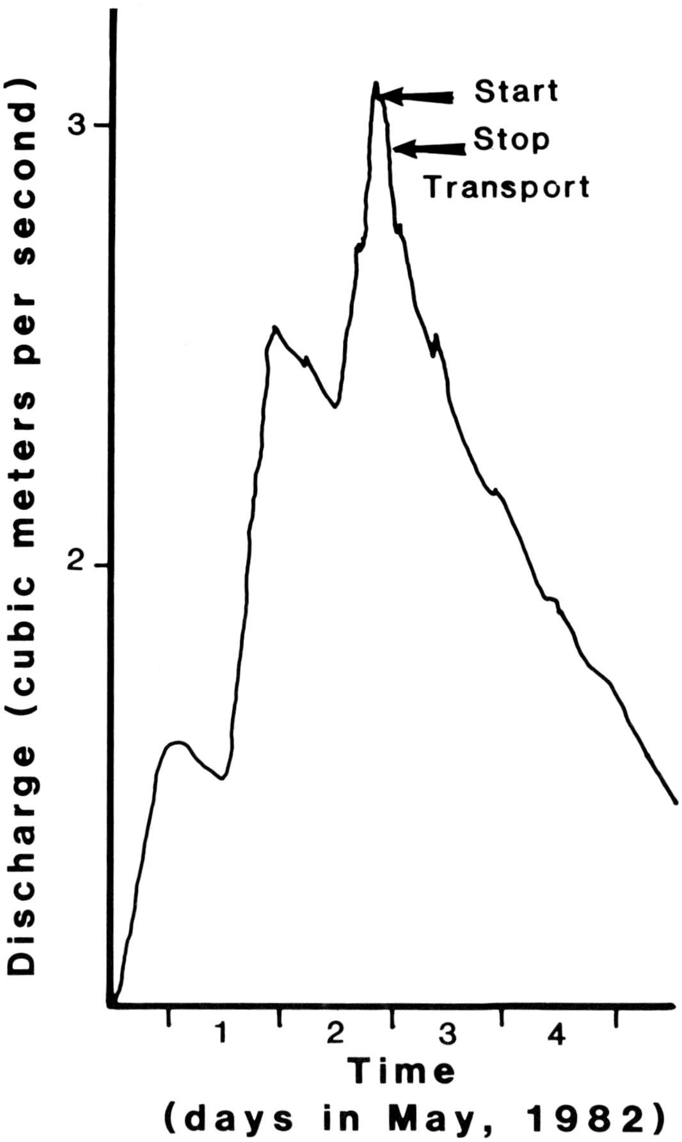

FIG. 2.—Hydrograph of the flood event in May 1982.

the threshold of very coarse pebble transport. Thus, both bed slope and water slope measurements independently indicate a two percent gradient as coarse transport begins. All slope measurements were made in the stream reach directly upstream of the log sill holding the detectors. The pebbles and cobbles at the detector were assumed to have been subjected to this slope, and to the discharge, velocity, and water depth present where the stream was confined at the bridge above the log sill.

RESULTS

Beginning and end of transport.—

The real-time chart of coarse sediment transport provides a graphic record of the nature of the transport event (Fig. 3). This record is better than a plot of averaged counts per second because such averaging obscures the variability of the transport process. The record begins at 22:36.8 hrs on 2 May 1985 and ends at 23:54.8 hrs the same day. The time of day on the chart demonstrates one of the advantages of remote detection of coarse particle transport. The transport event took place at midnight.

Careful examination of Figure 3 reveals that the nature of the onset of pebble and cobble motion differs from the nature of the termination of pebble and cobble motion. Figure 3A shows the beginning. Note that the beginning of coarse material transport is characterized by a few spikes indicating the occasional passage of a particle. At approximately 22:37, short bursts of transport take place lasting approximately 12–24 sec. The bursts do not contain many particles. Each particle spike is still resolvable. At 22:45.6, approximately two thirds of the way across Figure 3A, the transport event begins in earnest. This shows that large particle transport is initiated gradually. In contrast, such transport ends abruptly at 23:55 hrs in Figure 3D. Coarse material transport does not decline gradually; rather, the coarse pebble motion is truncated abruptly, and after 23:55, only a few isolated signals continue into the night. No major activity is seen after the last pulse of transport takes place. More investigation is needed to explain this behavior. Total bedload transport at the detector in 1982 during this event is estimated to be 0.83 kg/s-m using the technique described by Ergenzinger and Custer (1983). No mechanical samples of material were obtained.

The conditions that initiated the transport event were a discharge of approximately 3.4 m^3/sec with a total stream power per unit bed area ($\omega = \rho QS/w$) of approximately 9.7 kg/s-m. (Notation follows that of Bagnold (1977) where ω = total stream power per unit bed area, ρ = mass density of water of 1000 kg/m^3, Q = discharge, and S = slope of the bed surface the bed surface slope in 1982 and the water surface slope in 1985 are both 0.02, and w = stream width at the 7-m-wide bridge where the detectors were installed.) When transport ended, the discharge was approximately 2.7 m^3/sec and the total stream power per unit bed area was approximately 7.7 kg/s-m. If one calculates the mean boundary shear stress ($\tau = \rho dS$) at the start of transport at the bridge where the data were collected, values of the order of 60 kg/m^2 are obtained. (τ = mean boundary shear stress, ρ = mass density of water, S = slope of the bed surface, d = stream depth at the detector.)

Sediment transport waves.—

Another characteristic of the record shown in Figure 3 is the wave-like behavior of the signals from the cobbles and pebbles as they pass. This behavior may be akin to the kinematic waves described by Langbein and Leopold in 1968. The spacing of the waves is variable. The early spacing between the onset of one wave and the onset of the next wave is approximately 1.5 min. Once the event begins in earnest at 22:45.8, however, the time between the onset of each wave ranges from 0.10 to 0.40 min with the coarse-material-transport gaps lasting approximately 0.10 min and the coarse-material-transport waves lasting approximately

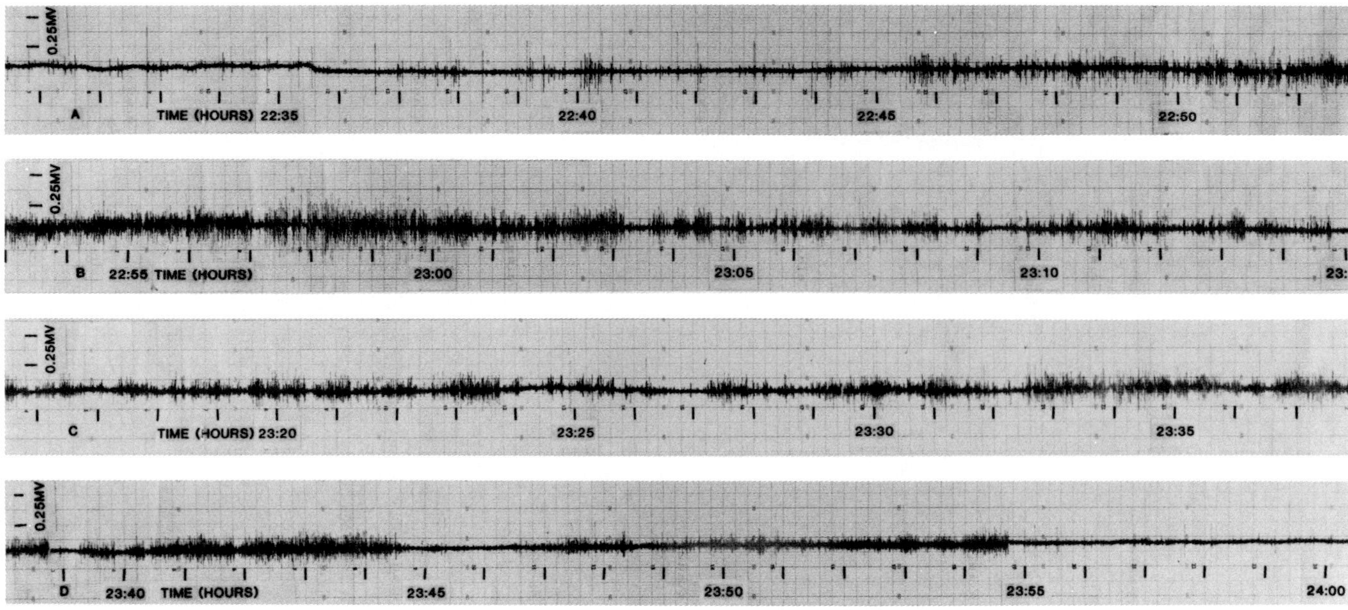

FIG. 3.—Potential-versus-time record of the sediment transport event on 2 May 1985.

0.15 to 0.25 min. If one ignores the short breaks between bursts of activity, the coarse material transport might be characterized as continuous for approximately 15 min. At approximately 23:03 (Fig. 3B), the gaps in coarse material transport become more distinct. The time between the onset of waves is approximately 0.6 min. The waves last from 0.2 to 0.4 min and the gaps last from 0.1 to 0.3 min. As transport continues, the waves are less staccato. By the time coarse material transport ends, sustained waves are apparent (Fig. 3D). Now the waves are spaced approximately 6 min apart with burst duration gradually increasing from 4.8 to 8.2 min and gap duration rising from 0.4 to 2.0 min. These latter bursts of coarse material transport average approximately 6 min in length and are quite similar to the 6-min sediment transport waves from flume research reported by Kuhnle and Southard (1985). The entire coarse-material-transport event was shorter than the 1.7 hr spacings reported by Reid and others (1985).

The fact that researchers have reported several different periodicities for waves indicates that there is probably "wave" behavior or pebble-and-cobble-transport bursts at more than one scale. In a stream such as Squaw Creek, bursts of coarse material transport such as those in Figure 3 may be due to the periodic exposure of part of the bed by movement of a clast in the armor, or response of the bed to interaction between microtopography and cluster bedforms such as those described by Brayshaw (1985). The waves may also be related to kinematic wave behavior of particles in motion, such as that described by Langbein and Leopold (1968). Alternatively, the complex local hydraulic behavior within the stream may influence the nature of sediment transport waves on the bed. This behavior may include bursts and sweeps, eddies, and water waves which are commonly averaged out in most fluvial research. Bursts of sediment transport larger than those displayed in Figure 3 might be caused by sediment supply waves related to stream tributary input, landslides, bank collapse, bursting log or ice jams, or logs or ice scraping bottom. Much research remains to be done to understand these phenomena.

Location of transport.—

If detectors are placed end to end across the stream, the location of coarse sediment transport in the channel should also be identifiable in a real-time way. In 1985, eight detectors 0.3 m long and 0.05 m high with 20,000 windings of wire were placed across the stream to monitor the location of pebble and cobble transport. Despite the fact that large-scale transport did not occur, some information regarding the part of the channel where coarse material transport was beginning was obtained (Fig. 4). Whereas the record is not as graphic as that from 1982, it illustrates the potential of the technique. The three ink traces show three of the detectors simultaneously recording data. Notice that only one recorder channel displayed occasional spikes; the other two channels did not (Fig. 4). Each day, all the detectors were tested with a magnetic cobble tied to a rope which was passed over the detectors to be certain that transport would be detected if it occurred. All the recorder channels depicted in Figure 4 showed a positive response to the test cobble.

The data indicate that minor pebble or cobble transport, much like that before the onset of the main event in 1982, was taking place on the bed of Squaw Creek in 1985. Threshold was approached but was not exceeded. An individual pebble moved every few minutes. No waves of sediment transport occurred. What transport was taking place was present at detector "a" near the center of the channel, 3 m from the left bank (Fig. 4A). At the time this record was obtained, the discharge was 3.74 m^3/sec. The stream

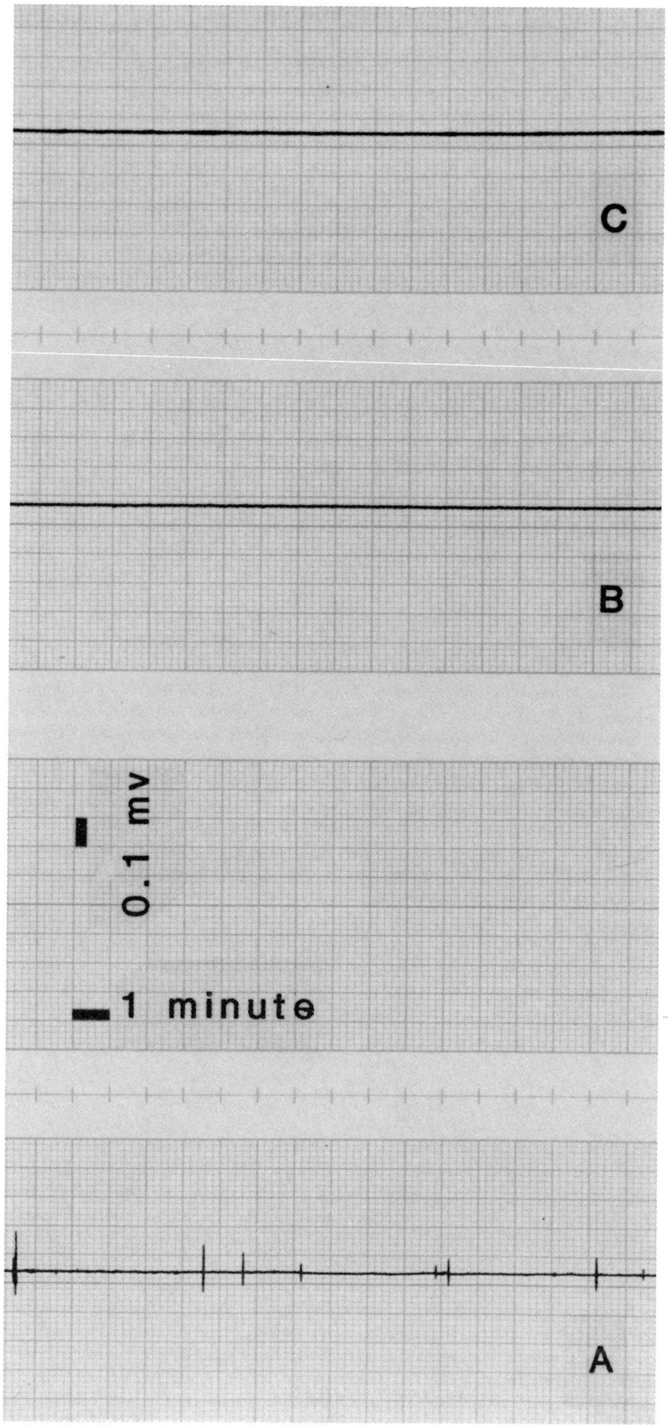

FIG. 4.—Potential-versus-time record for detectors at three locations in Squaw Creek (A = 3 m from left bank; B = 5 m from left bank; C = 1 m from left bank).

power was 10.7 kg/s-m. These values exceeded those which produced a transport event in 1982. Apparently, the bed configuration was more stable in 1985 so that more stream power was needed to initiate transport.

Detector noise.—

While these detectors appear to have promise as data collection devices in gravel-bed streams with naturally magnetic pebbles and cobbles, the devices also have some inherent problems. The most serious problems are maintenance and noise. One of the reasons that no data can be presented for 1983 and 1984 is that ice destroyed electrical connections and noise problems precluded obtaining usable records.

A variety of noise sources have been identified. One of the most serious is detector vibration. If the detectors are moved through the Earth's magnetic field, a voltage larger than that produced by a passing pebble may be recorded. Figure 5A shows a record from a vibrating detector bolted to a log on the bottom of Squaw Creek in 1983. Test pebbles passed over this detector could not be distinguished from the noise generated by vibration. Another source of noise is line noise such as that produced when an electric drill is turned on nearby (Fig. 5B). These pulses are unidirectional as opposed to the bidirectional pulse of a passing cobble. A signal similar to that produced by a drill is produced by lightning strikes in the area. The detector is also quite sensitive to magnetic metal objects. A car pass-

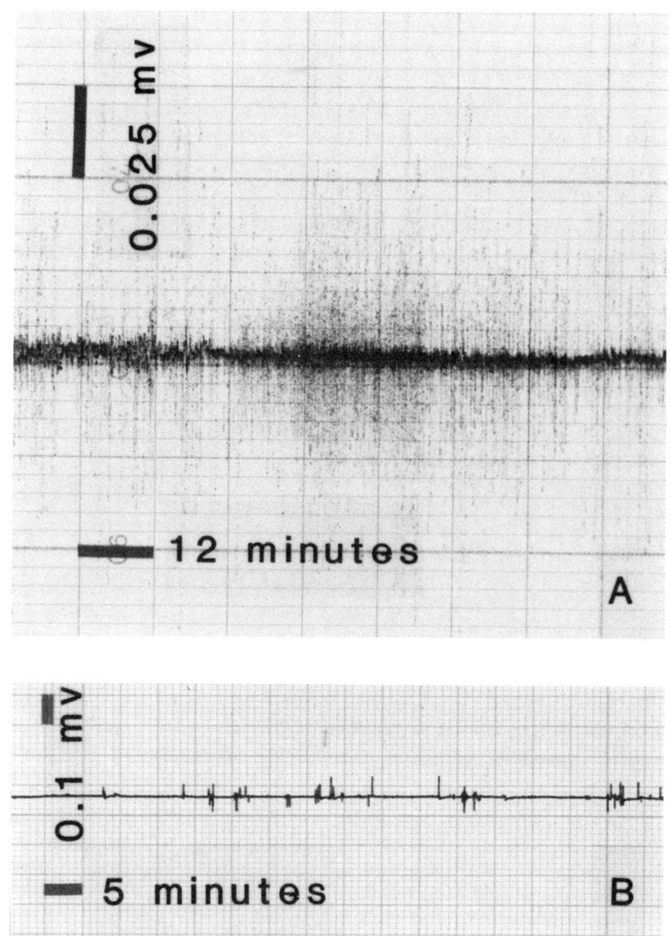

FIG. 5.—Record of noise due to A. detector vibration and B. line noise.

ing on the bridge can produce a signal. Because of this sensitivity, great care must be taken to keep the detector as far away as possible from metal objects, such as nails and other commonly used construction materials, and to insure at the same time that the detector is attached very firmly to a non-vibrating substrate. As can be seen in Figures 3 and 4, the problem of noise can be overcome, but the problem cannot be taken lightly.

APPLICABILITY

This detector should work in any stream where coils can be installed without noise and where there are naturally magnetic cobbles and pebbles. Magnetic cobbles and pebbles are remarkably common in igneous and metamorphic terrane as well as some sedimentary terranes. If a carpenter's stud finder is attracted to some particles, detection of transport is probably possible. A stream with 100 percent detectable cobbles may be less workable for transport detection unless extremely fast recorders are available. Nevertheless, this technique may be applicable in many regions of the world.

CONCLUSION

The records collected by the elecromagnetic detector at Squaw Creek, Montana show that very coarse pebble and cobble transport began gradually but ended abruptly. During the transport event, coarse particle waves of several scales were detected. The wave spacing from onset to onset ranged from 0.6 min to 6 min. The wave spacings suggest that time averaging of coarse-particle-transport data may obscure the short period behavior of an event. The data show that the electromagnetic bedload detection device described by Ergenzinger and Custer (1983) can be used to better characterize coarse-sediment-transport waves and the nature of coarse particle transport when noise problems are overcome. The hypothesis that transport waves are not related to stream hydraulic variations, as proposed by Reid and Frostick (this volume), should be testable in more detail when detector data are combined with real-time high-speed records of hydraulic data.

ACKNOWLEDGMENTS

Funds for the research in 1985 were provided by Deutsche Forschungsgemeinschaft, Bonn-Bad Godesberg, and by the Research Creativity Program sponsored by Montana State University. The Castle Rock Baptist Bible Camp kindly provided electricity, and the Gallatin National Forest approved access. Steve Glasser of the National Forest Service assisted with stream discharge records.

REFERENCES

BAGNOLD, R. A., 1977, Bed load transport by natural rivers: Water Resources Research, v. 13, p. 303–312.

BRAYSHAW, A. C., 1985, Bed microtopography and entrainment thresholds in gravel-bed rivers: Geological Society of America Bulletin, v. 96, p. 218–223.

BUTTLER, P. R., 1977, Movement of cobbles in a gravel-bed stream: Geological Society of America Bulletin, v. 88, p. 1072–1234.

EMMETT, W. W., 1980, A Field Calibration of the Sediment-Trapping Characteristics of the Helley-Smith Bedload Sampler: Washington, United States Government Printing Office, 44 p.

ERGENZINGER, P. J., AND CUSTER, S. G., 1982, A new tracer technique for measuring bedload in natural channels: Catena, v. 9, p. 77–80.

———, AND CUSTER, S. G., 1983, Determination of bedload transport using naturally magnetic tracers: First experiences at Squaw Creek, Gallatin County, Montana: Water Resources Research, v. 19, p. 187–193.

HEY, R. D., BATHURST, J. C., AND THORNE, C. R., 1982, Gravel-Bed Rivers: Chichester, John Wiley & Sons Ltd., 875 p.

HUBBELL, D. W., 1964, Apparatus and techniques for measuring bedload: U.S. Geological Survey Water Supply Paper 1748, 74 p.

———, AND SAYNE, W. W., 1964, Sand transport studies with radioactive tracers: Journal of Hydraulics Division, American Society of Civil Engineers, v. 4, p. 1227–1234.

KLINGEMAN, P. C., AND EMMETT, W. W., 1982, Gravel bedload transport processes, *in* Hey, R. D., Bathurst, J. C., and Thorne, C. R., eds., Gravel-bed Rivers: Chichester, John Wiley & Sons Ltd., p. 141–180.

KUHNLE, R. A., AND SOUTHARD, J. B., 1985, Sediment transport fluctuations in a gravel-bed laboratory channel: Third International Fluvial Sedimentology Conference at Colorado State University, Fort Collins, Colorado, Abstracts, p. 25.

LANGBEIN, W. B., AND LEOPOLD, L. B., 1968, River Channel Bars and Dunes-Theory of Kinematic Waves: U.S. Geological Survey Professional Paper 422-L, p. L1–L20.

LEOPOLD, L. B., AND EMMETT, W. W., 1976, Bedload measurements, East Fork River, Wyoming: Proceedings of the National Academy of Science, v. 73, p. 1000–1004.

OLDFIELD, F., THOMPSON, R., AND DICKSON, D. P. E., 1981, Artificial magnetic enhancement of stream bedload: A hydrological application of superparamagnetism: Physics of the Earth and Planetary Interiors, v. 26, 107–124.

RAEMY, F., AND JAEGGI, M., 1981, Some problems related to sediment transport measurement in steep mountain streams *in* Erosion and Sediment Transport Measurement: Proceedings of the Florence Symposium, 1981, International Association of Hydrologic Science, Publication 133, p. 231–239.

REID, I., BRAYSHAW, A. C., AND FROSTICK, L. E., 1984, An electromagnetic device for automatic detection of bedload motion and its field applications: Sedimentology, v. 31, 269–276.

———, FROSTICK, L. E., AND LAYMAN, J. T., 1985, The incidence and nature of bedload transport during flood flows in coarse-grained alluvial channels: Earth Surface Processes and Landforms, v. 10, p. 33–44.

TYWONIVK, N., AND WARNOCK, R. G., 1973, Acoustic detection of bedload transport: Proceedings of Hydrologic Symposium at University of Alberta, Edmonton, May 8 and 9, 1973: Department of Environment Natural Resource Council, Canada, p. 728–743.

WHITE, W. R., AND DAY, T. J., 1982, Transport of graded gravel bed material, *in* Hey, R. D., Bathurst, J. C., and Thorne, C. R., eds., Gravel-Bed Rivers: Chichester, John Wiley & Sons Ltd., p. 181–223.

FLOW DYNAMICS AT RIVER CHANNEL CONFLUENCES: IMPLICATIONS FOR SEDIMENT TRANSPORT AND BED MORPHOLOGY

JAMES L. BEST
Department of Geology, University of Hull, U.K.

ABSTRACT: Flow dynamics at river channel confluences can be characterized by six major regions of flow stagnation, flow deflection, flow separation, maximum velocity, flow recovery and distinct shear layers. The dominant controls upon the magnitude of these regions are shown to be the junction angle and the ratio of discharges between the confluent channels. Through the combined use of scaled laboratory modelling and an analysis of field evidence, the dynamics of flow are found to produce a confluence morphology which consists of avalanche faces at the mouth of each confluent channel, a deep central scour and a bar within the separation zone. Tracing of sediment in both laboratory and natural channels reveals distinct sediment pathways within the junction which can be explained through the model of flow dynamics. A knowledge of confluence flow dynamics is important when assessing channel design criteria, junction bed morphology and ancient confluence sediments.

INTRODUCTION

River channel confluences are a feature common to all fluvial networks and are the points at which significant contributions of water (Frostick and Reid, 1977; Richards, 1980) and sediment (Miller, 1958; Lodina and Chalov, 1971; Knighton, 1980, 1982) are added to the main channel. It is because of this that major changes in the channel hydraulic geometry can occur at channel confluences, one morphological consequence of these changes being the formation of distinct bed scours at these sites (Komura, 1973; Mosley, 1975, 1976, 1982; Mosley and Schumm, 1977; Kjerfve and others, 1979; Ashmore, 1979, 1982; Ashmore and Parker, 1983; Best, 1986). Channel confluences are also of importance in other areas, for instance, in computations of flood routing through channel networks (Stoker, 1957; Sevuk and Yen, 1973; Sanmuganathan, 1973; Murota and others, 1973; Yen and Akan, 1976; Soliman, 1976; Prohaska, 1978). These morphological and channel management considerations make a knowledge of flow dynamics and sediment transport of interest to both the sedimentologist and engineer. Yet river junctions have received surprisingly little attention from earth scientists, this paucity of research perhaps reflecting the extreme complexity of the situation where two streams join. Studies that have sought to address the problem of junction flow dynamics over the last 40 years include those of Taylor (1944), Greated (1964), Webber and Greated (1966), Itakura (1972), Sevuk and Yen (1973), Soong (1976), Lin and others (1976), Lin and Soong (1979), and Modi and others (1981). Insights into the possible morphological and paleohydraulic significance of confluence sediments have also been given by Mosley (1975, 1976), Frostick and Reid (1977), Mosley and Schumm (1977), Reid and Frostick (1979), Ashmore (1979, 1982), Ashmore and Parker (1983), Kennedy (1984) and Bryant and others (1983).

These studies, however, have not produced a general model to describe the dominant features of flow at channel confluences which would enable a more complete interpretation than is currently possible of sediment transport, bed morphology, depositional facies and design criteria at channel junctions. This paper describes the dynamics of flow at channel confluences and examines its implications for sediment transport and bed morphology. This has been achieved through the complementary use of controlled laboratory experiments and an analysis of field evidence.

EXPERIMENTAL DESIGN

A confluence flume was designed and constructed for this study which can accommodate any one of five junction angles through an interchangeable center section. Confluence angles of 15°, 45°, 70°, 90°, and 105° were chosen to cover both a wide range and reflect those angles most common in the natural environment as determined from map analyses. Only confluences where the post-confluence channel forms a linear extension of the upstream main channel were studied, these corresponding to the non-equal angle junctions of Mosley (1975, 1976) and excluding '*Y*' planform junctions. All the flume channels are rectangular and 0.15 m wide by 0.20 m deep. Water is recirculated within the flume, and the discharge to each channel is controlled by valves to the supply line in each channel. Through calibration of these valves, the discharge ratio, Q_r, between the tributary and mainstream can be accurately controlled, where

$$Q_r = Q_t/Q_m$$

where Q_t and Q_m are the discharges from the tributary and main channels, respectively. The slope of the flume can also be adjusted while water depth is controlled by a tailgate at the end of the flume. For each experimental run, the water depths at the inlet and outlet of the flume were adjusted so as to be equal. The terms used in this article are defined in Figure 1.

The dynamics of flow were investigated in sediment-free, rectangular channels. Fluid velocities within the junction were measured using a miniature current meter and averaged over 1 min to give a representative mean sample time. Indications of turbulence were also gained by continuously recording the output on a potentiometric chart recorder. Besides this direct instrumentation, two methods of flow visualization were employed. First, the video recording of dye injected both into the flow and onto the water surface enabled the detailed observation and measurement of several distinctive features within the junction. A second flow visualization technique used neutrally buoyant tracer particles. The material employed was Pliolite (trade name of Imperial Chemicals Inc.) which was first used as a flow tracer by Nychas and others (1973). Pliolite particles of between 75 and 105 μm diameter were found suitable in this study.

Mobile bed conditions were modelled within the flume using a well sorted sand with a D_{50} diameter of 0.49 mm. In the simulation of mobile bed conditions, equilibrium was

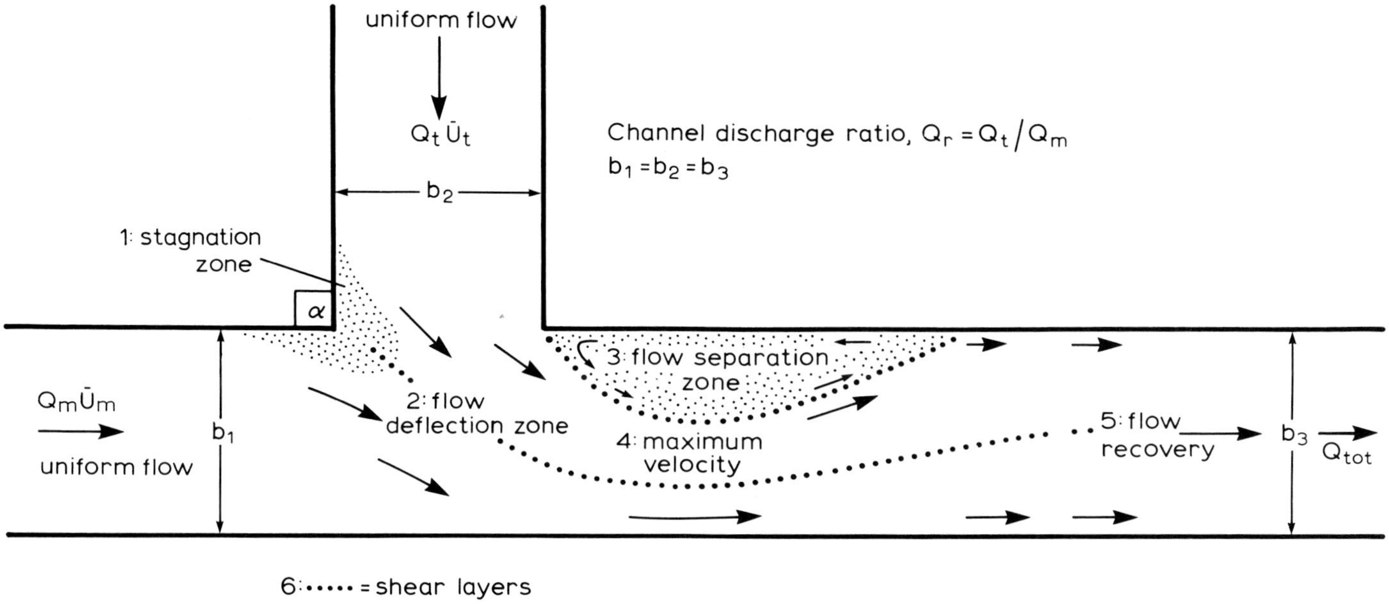

FIG. 1.—A model of flow dynamics at river channel confluences and definition of terms used in the text.

judged to have been attained from four criteria which have been widely used in previous flume studies: 1) when sediment infeed equalled sediment outfall (i.e., when there was no net erosion or deposition within the flume) (Williams, 1970); 2) when water and bed surface slopes became constant (Guy and others, 1966; Hooke, 1974); 3) when flow depths at many localities were constant in time, and, 4) when the bed morphology attained a constant form (Ackers and Charlton, 1970; Williams, 1970; Costello and Southard, 1981). Sediment infeed was accurately controlled through hoppers over each upstream channel and ensured there was no scour to the base of the flume. Sediment transport rates within the flume were measured using a unique tracer technique which allowed particle paths from each channel to be monitored, thus providing data on the total transport rate within the flume and the paths and transport rates from each confluent channel.

In both sets of experiments the appropriate scaling criteria were employed to ensure comparison could be made between flume and natural channels. These criteria included Froude and Reynolds number similarity together with the scaling of grain size, flow depth and width characteristics. The flume experiments followed the approach proposed by Hooke (1968) in a 'similarity of process' modelling criterion. This advocates the similarity of gross scaling relationships, of bed morphology and of the processes operative in order to ensure analogy between the laboratory and natural environment. Details of the experimental procedure, scaling criteria and sediment transport measurement techniques can be found in Best (1985).

FLOW DYNAMICS AT RIVER CHANNEL CONFLUENCES

The dynamics of flow at channel confluences were examined at all five angles and at discharge ratios of approximately 0.50, 1.0 and 1.50 at each junction configuration. From these experiments a model of confluence flow dynamics can be proposed and is illustrated for the 90° junction in Figure 1. Flow within the junction is dominated by six distinct elements: 1) a zone of relative flow stagnation near the upstream junction corner, 2) an area of flow deflection where each stream enters the confluence, 3) a zone of separated flow below the downstream junction corner, 4) an area of maximum velocity, 5) an area of gradual flow recovery downstream from the flow separation zone, and 6) several distinct shear layers with associated vortex generation. The two dominant controls on these zones are the confluence angle and the channel discharge ratio. These regions are now examined and the precise influence of the controlling variables identified.

Zones of flow stagnation and flow deflection.—

Stagnation of flow within each channel is caused by the mutual deflection of flows away from the upstream junction corner. This deflection and the presence of the flow stagnation region are evident in Figure 2, which shows Pliolite visualization of flow at three discharge ratios for the 70° confluence. The reduced velocities within the stagnation zone are also reflected in the near-bed velocity vector maps at three junction angles shown in Figure 3. As junction angle is increased, the confluent streams undergo a progressively greater deflection to flow parallel to the post-confluence channel. Because of this, the relative flow stagnation increases; this trend is illustrated in Figure 4 where the relative near-bed velocity at point *A* in the tributary channel is plotted against junction angle. Relative velocity is expressed as the near-bed velocity at that point, U, divided by the tributary mean near-bed freestream velocity, $\bar{U}_t$. It is evident that at each discharge ratio the level of flow stagnation increases at higher junction angles where the deflection of flows and turning of the tributary stream increase. At the 105° junction, this turning becomes so severe at high discharge ratios that a reverse flow is induced into

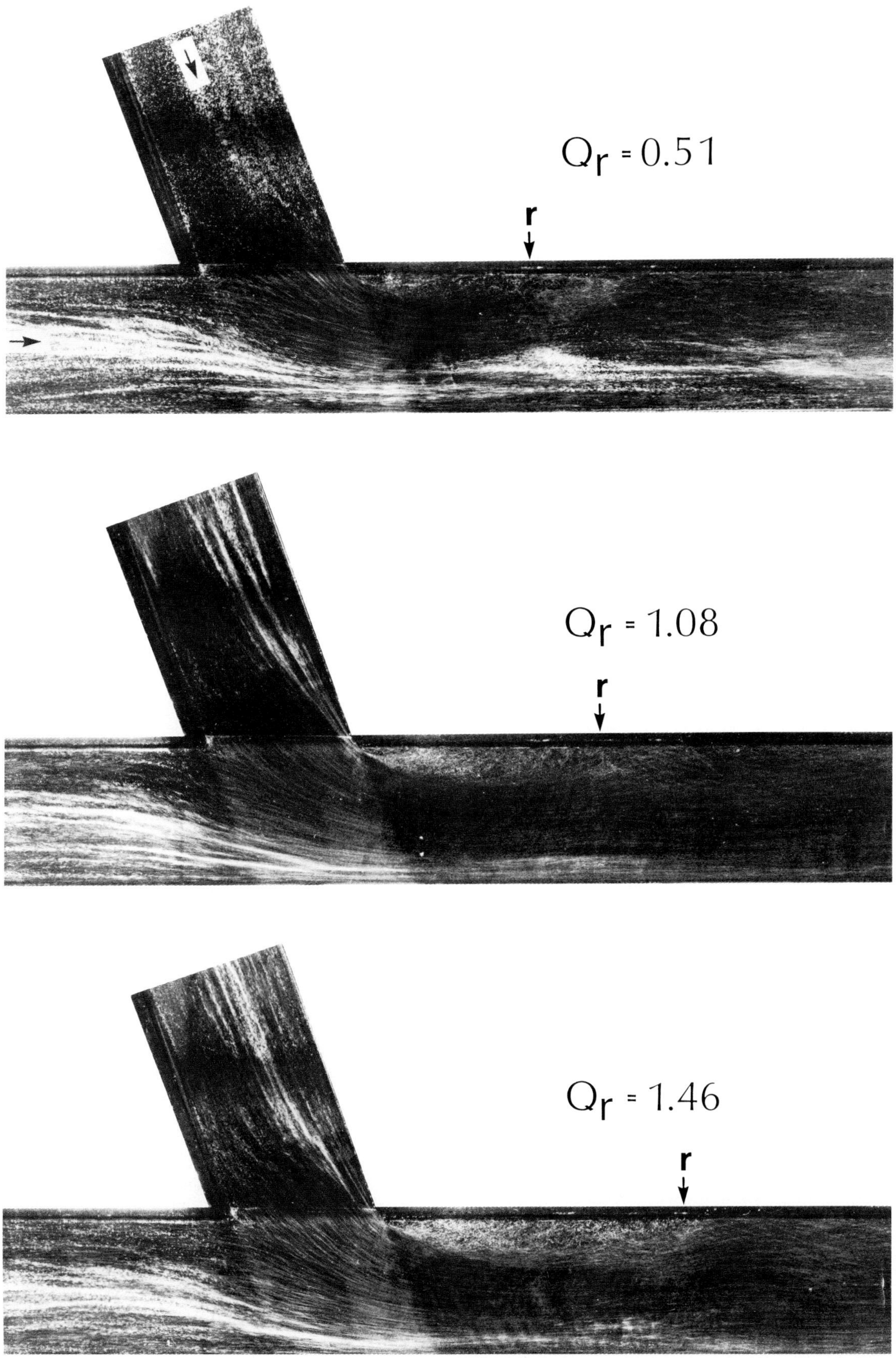

FIG. 2.—Visualization of flow using Pliolite at the 70° junction for three discharge ratios. '*r*' denotes the time-averaged position of the reattachment point of the separated flow.

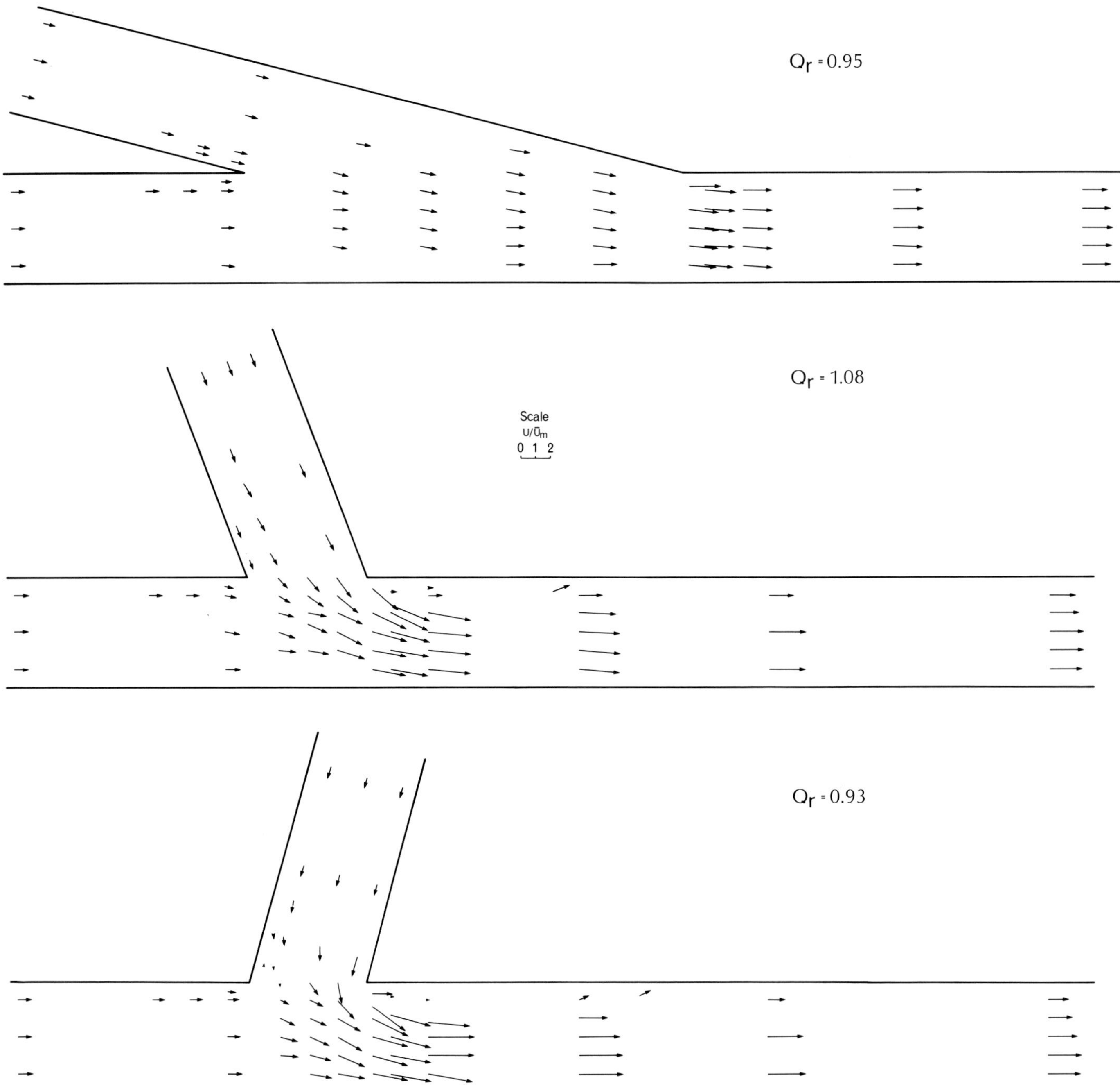

FIG. 3.—Maps of near-bed velocity vectors for junction angles of 15°, 70° and 105° at discharge ratios of near unity. Near-bed velocities were measured at a depth of 0.12 y/Y above the bed, where y is height in the flow and Y is the total flow depth. The length of the arrows expresses velocity in relation to the mean flow velocity in the freestream of the main channel.

the stagnation zone in the tributary from the main channel (Fig. 3). Similar reverse flows into this region in natural channels have been described by Teissyre (1975), and Alam and others (1985). Except for this anomalous oblique angle tributary, the graph illustrates that the degree of flow stagnation in the tributary channel is also greater at lower discharge ratios, where the relative momentum of the tributary flow is less. This trend is again complicated in the 105° junction where the turning that the tributary flow has to undergo at higher discharge ratios is more severe. The nature of flow deflection, levels of velocity and hence shear stress within the stagnation zone can therefore be expected to be instrumental in determining the transport of sediment in this region.

The mutual deflection of flows increases at both higher junction angles and higher discharge ratios, with the 15° junction showing little deflection of either stream (Fig. 3). A feature of great significance in this region is that a shear

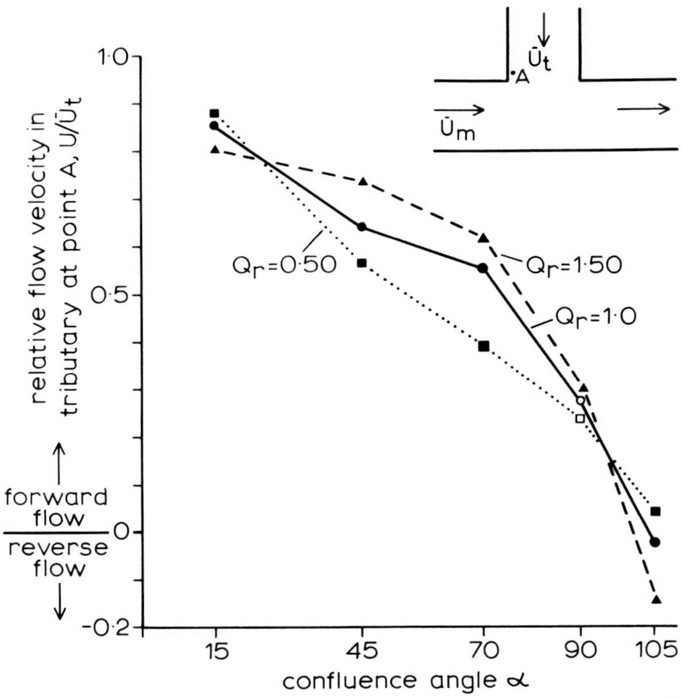

FIG. 4.—Retardation of near-bed flow velocity within the tributary stagnation zone at point A (see inset sketch) and its variation with confluence angle and discharge ratio. Relative flow velocity is expressed as $U/\bar{U}_t$, where U is the near-bed velocity at point A and $\bar{U}_t$ is the mean near-bed velocity in the freestream of the tributary channel.

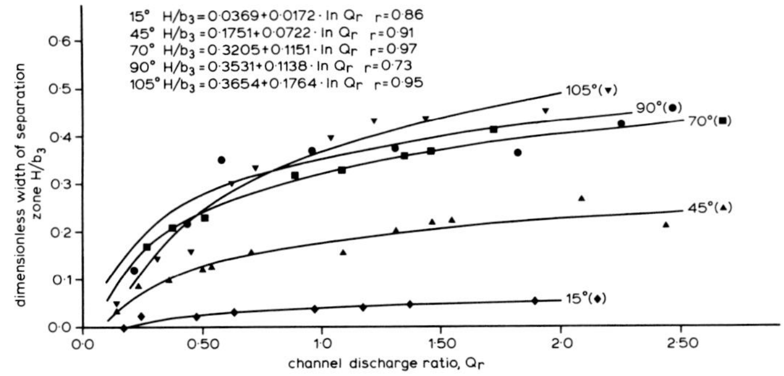

FIG. 5.—Plot of the dimensionless width of the separation zone, H/b_3, against discharge ratio for five junction angles.

layer is created between the two convergent flows along which powerful, vertical vortices are generated. These are responsible for increased bed shear stresses within the junction which, together with the increase in velocity as both flows enter the confluence, are responsible for considerable bed scour.

Zones of flow separation and maximum velocity.—

Perhaps the most distinctive zone within the confluence is the region of separated flow that forms below the downstream junction corner. Flow separation occurs where the tributary flow cannot remain attached to the channel wall as it enters the confluence because of the abrupt change in boundary geometry at the downstream junction corner. The formation of a distinct zone of separated flow at this locality has been documented by several authors (Taylor, 1944; Kanda and Kitada, 1977; Modi and others, 1981), and the dimensions of the zone have been quantified by Itakura (1972) and Best and Reid (1984). The separation zone, which may occupy half the width of the post-confluence channel, grows in both length and width as junction angle and channel discharge ratio are increased. Figure 5 illustrates this by plotting the dimensionless width of the flow separation zone against discharge ratio for five junction angles. At each junction angle, the width of the separation zone increases at higher discharge ratios (see also Fig. 2). It can be seen, however, that the rate of increase declines at higher discharge ratios where the effects of the confining far bank, together with pressure effects within the separation zone (Artt, 1970; Best, 1985), become important. Additionally, Figure 5 demonstrates that at any single discharge ratio the separation zone grows in width at higher junction angles where the change in boundary geometry at the downstream junction corner is more severe. A slight anomaly to this pattern is introduced by the 105° junction which at low discharge ratios has a smaller separation zone than either the 70° or 90° junctions. This complexity may again be attributed to the greater momentum loss suffered by the tributary flow at low discharge ratios at this oblique angle confluence. The effect of this is to lessen the relative penetration of the tributary flow into the confluence.

Flow velocities within the separation zone may be as little as 5 percent of those in the freestream of the combined flows. A recirculating flow is created within the separation zone with upstream flow near the channel wall occurring between the reattachment point and the downstream junction corner. A shear layer is created between the separation zone flow and the combined flows along which vortices are generated. The separation zone is also characterized by both lower fluid pressures and water surface elevations than the rest of flow within the junction (McGuirk and Rodi, 1978; Best, 1985).

An important corollary of the width of the flow separation zone is that it delimits an area of channel through which the combined flows must pass. Because of this, the region between the separation zone and channel wall is the zone of maximum velocity increase within the junction (Figs. 1 and 3). Best and Reid (1984) show that the maximum velocity within the confluence increases by a factor of 1.3 as

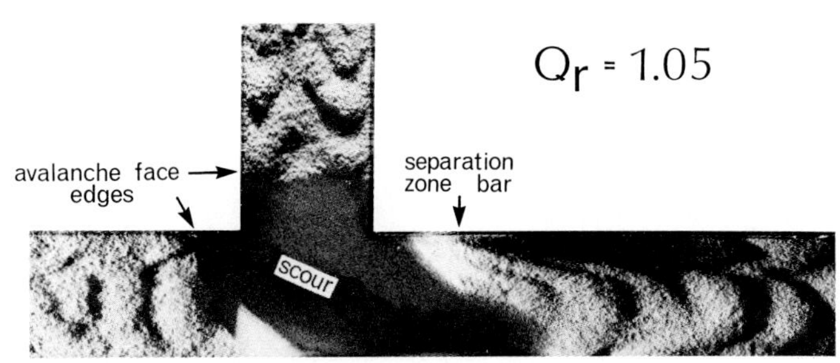

FIG. 6.—Photograph illustrating the three characteristic features of bed morphology developed at the 90° confluence flume at a discharge ratio of 1.05.

confluence angle is raised from 15° to 90° at any one discharge ratio. Maximum flow acceleration therefore responds sympathetically to the growth of the separation zone.

The flow separation zone therefore has considerable importance for sediment erosion, transport and bed morphology within the junction.

Zone of flow recovery.—

Downstream from the flow separation zone, the flow recovers from the influence of the confluence dynamics; the two flows gradually mix across the shear layer between the combining flows and levels of turbulence decline; the asymmetric cross-channel velocity distribution produced by flow acceleration and flow separation at the confluence begins to diminish and return to a more symmetrical pattern with flow parallel to the channel walls. This zone of flow recovery, where the effects of the confluence become less pronounced and the streams mix fully, may extend for many kilometers downstream in the natural environment (Mackay, 1966, 1970).

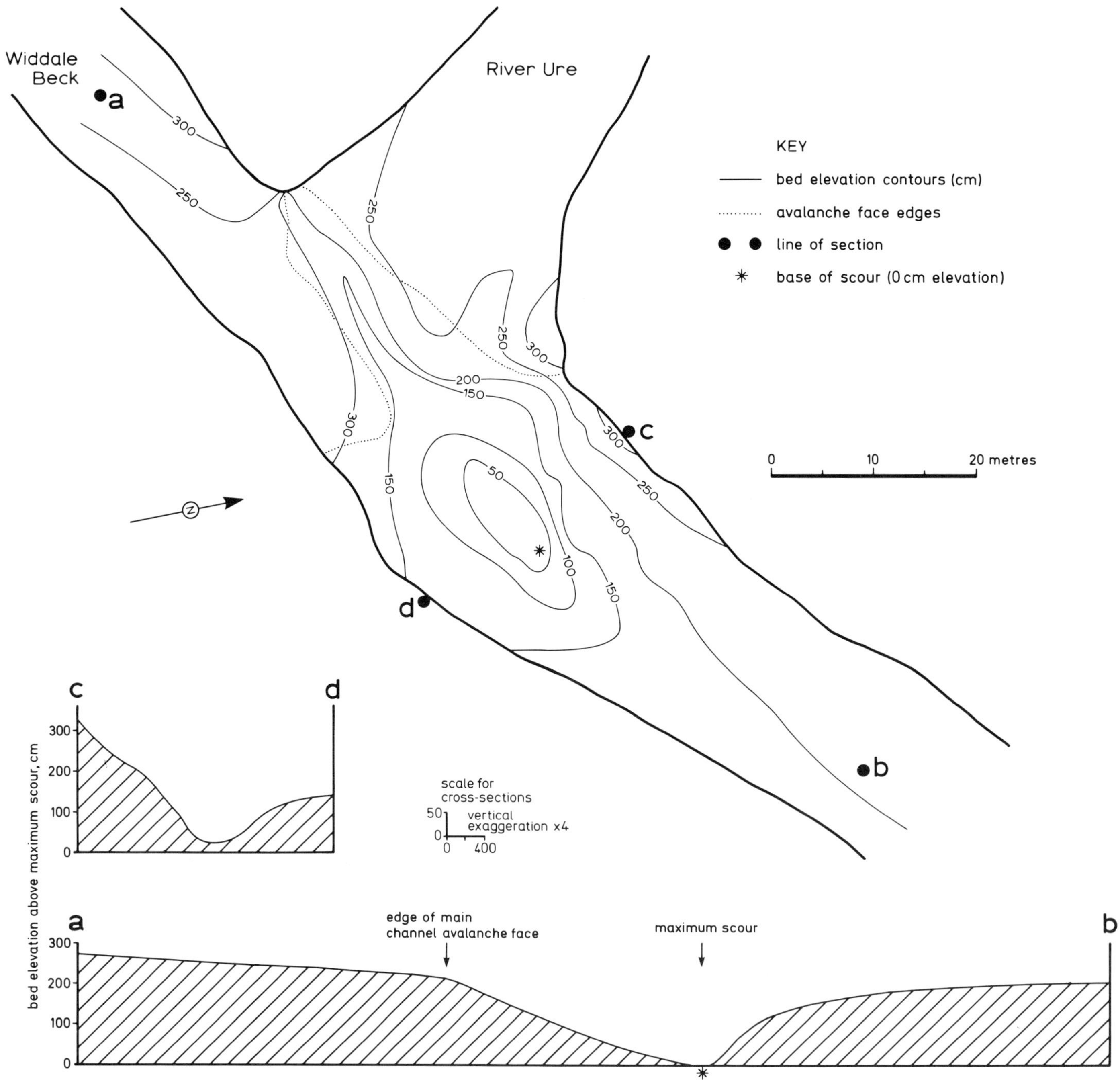

FIG. 7.—Planimetric map and cross sections of bed morphology at the confluence of Widdale Beck and the River Ure at Appersett, North Yorkshire, U.K.

IMPLICATIONS FOR SEDIMENT TRANSPORT AND BED MORPHOLOGY AT CHANNEL CONFLUENCES

The model of flow dynamics outlined above, although developed in sediment-free channels, has many significant implications for the transport of sediment and development of bed morphology at channel confluences. An examination of both sediment transport and bed morphology at channel confluences has been achieved through the simulation of mobile bed conditions within the flume and by examination of a natural channel confluence.

The flume experiments reveal three characteristic elements of bed morphology at channel confluences (Fig. 6): 1) a scour whose orientation approximately bisects the junction angle, 2) a bar formed within the flow separation zone, and, 3) distinct avalanche faces at the mouth of each channel which dip into the scour hollow. These features have also been found at natural channel confluences, as is shown by the isopleth map of bed morphology at the confluence of the River Ure and Widdale Beck at Appersett, North Yorkshire, U.K. (Fig. 7). At this natural channel confluence, the scour originates at the upstream junction corner and reaches a maximum depth 2.5 m lower than the edge of the avalanche faces, which stretch across the mouth of each confluent channel (section a–b in Fig. 7). A bar is also present below the downstream junction corner whose elevation is as much as 3 m above the maximum scour depth (section c–d in Fig. 7). The similarity of morphology between flume and natural channels not only demonstrates the validity of laboratory modelling but also enables the examination of factors that were effectively controlled in the flume, such as channel width adjustment and the effect of non-equilibrium bed conditions.

Flume and natural channel experiments have also allowed tracing of sediment within channel junctions. Maps of total sediment transport distribution and the contributions from each upstream channel for a discharge ratio of 1.05 at the 90° confluence flume are shown in Figure 8A and B. Figure 9 illustrates particle paths traced at the Appersett field site over a period of one month and through four bedload generating hydrographs, all of which had discharge ratios of essentially unity. The nature of bed morphology can be assessed in the light of the flow dynamics model and these sediment transport observations.

The avalanche faces are a product of the severe flow deflection suffered by both streams as they enter the confluence, together with the rapidly increasing velocities and shear stresses within the junction. Both of these deflect sediment as is shown in the laboratory and natural channel experiments. The flume sediment transport distributions, which are expressed as the percentage of the channel bedload passing through a given cross section, show both mainstream and tributary sediment to be deflected across the mouth of each channel (Fig. 8B). The deflection of sediment across the mouth of the River Ure is also evident in Figure 9, with sediment being deflected away from the upstream junction corner or the stagnation zone within the River Ure. Particle mobility within this stagnation zone was also found to be much lower, often only 20 percent, of the average entrainment rate in the upstream reaches of the River Ure. Reverse flows in the stagnation zone of the tributary channel have also been noted in the studies of Teissyre (1975), Baker (1973, 1984) and can produce upstream-dipping foresets within the tributary channel (Alam and others, 1985).

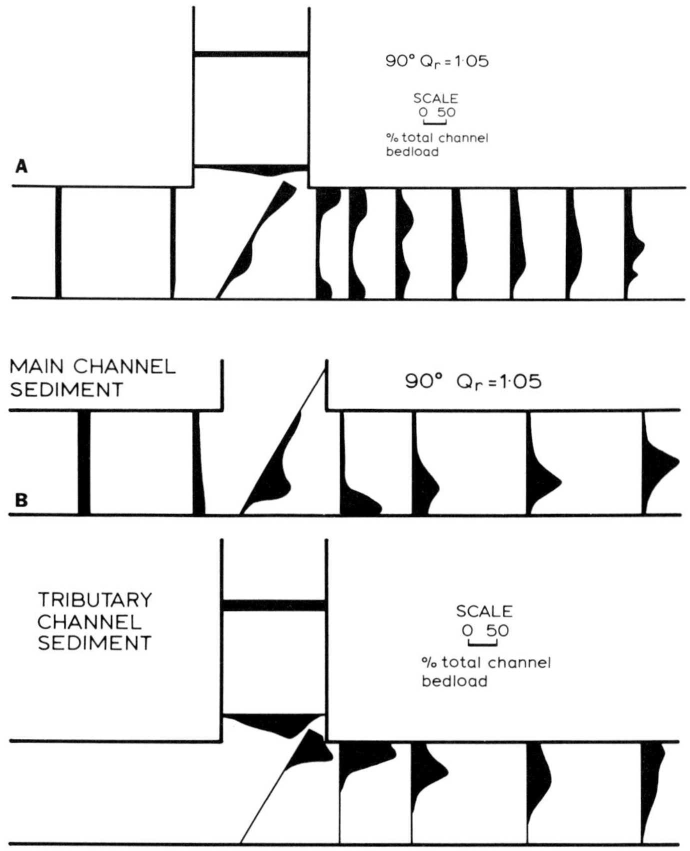

FIG. 8.—A.) Map of total sediment transport distribution within the 90° confluence flume over a bed developed at a discharge ratio of 1.05. Sediment transport rates are expressed as the percentage of bedload passing a given cross section. B.) Maps of sediment transport distribution from each confluent channel at a discharge ratio of 1.05.

The scour present at river confluences is formed by the increased levels of velocity and turbulence within the junction where the two flows combine. In addition, the sediment transport maps indicate that little sediment is transported through the scour, but instead bedload from each confluent channel is transported around the side of the scour (Fig. 8A and B). The segregation of bedload from each channel is a result of both the mutual deflection of flows and the influence of the flow separation zone which entrains material from the tributary channel into this region. The transport paths therefore maintain the existence of the scour, which does not reduce in depth in an attempt to transport the load supplied to any one point as would be necessary if sediment was transported through the scour.

The bar formed below the downstream junction corner owes its origin to the existence of the flow separation zone. Sediment, predominantly from the tributary channel (Fig. 8B), is entrained into this region which initially has both lower fluid velocities and pressure than the flow outside it.

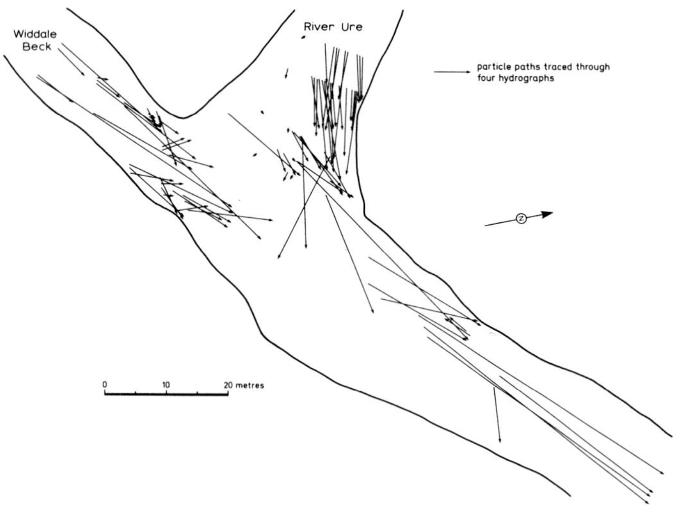

FIG. 9.—Particle movements traced through the Widdale Beck-River Ure confluence through four hydrographs, all of which had a discharge ratio of approximately unity.

Sediment is therefore quickly deposited once entrained into the separation zone and a bar is formed. In the flume this reaches an equilibrium state where flow depth is adjusted to provide the bed shear stresses to transport the sediment load supplied to any one point. In the natural channel, upstream particle paths indicate the presence of reverse flows within the separation zone (Fig. 9). This separation zone bar may be a site for the preferred accumulation of heavy mineral grains (Best and Brayshaw, 1985). Downstream the zone of flow recovery is marked by a gradual return to a bed morphology which has no asymmetry in cross-channel elevation. Also, sediment loads from each confluent channel, which were segregated as they passed through the confluence, begin to mix across the downstream channel.

CONCLUSIONS

A model of flow dynamics at river channel confluences is proposed in which six characteristic flow zones are recognized: regions of flow stagnation, flow deflection, flow separation, maximum velocity, downstream flow recovery and several distinct shear layers. These attributes of flow reflect directly upon the observed patterns of sediment transport and confluence bed morphology. The mutual deflection of flow *and* sediment from each confluent channel, together with marked increases in velocity within the junction, generate avalanche faces at the mouth of each channel. These avalanche faces dip into a pronounced scour hollow which owes its origin to the increased velocities and levels of turbulence within the confluence *and* the paths of sediment transport, which are predominantly around and not through the scour. A bar is formed within the flow separation zone where mostly tributary channel bedload is deposited in this region of reduced flow velocities.

An understanding of both flow dynamics and sediment transport at channel confluences is important in three major areas. Besides aiding channel design criteria in both natural and artificial channels and interpretation of junction bed morphology, these results provide a physical basis from which to interpret confluence sediments in the ancient record. This may be of special significance in braided streams where the many junctions and frequent channel abandonment can be expected to increase greatly the preservation potential of these sediments and to produce a complex but characteristic confluence facies.

ACKNOWLEDGMENTS

I thank the Committee and Sponsors of the Third International Fluvial Sedimentology Conference, and the University of Hull for financial support. I am also grateful to Charlie Bristow and an anonymous referee for their constructive comments and to Paul McSherry and John Garner at Hull for drafting and photographing the figures.

REFERENCES

ACKERS, P., AND CHARLTON, F. G., 1970, The geometry of small meandering streams: Proceedings of the Institute of Civil Engineers, Supplement 12, Paper 732BS, p. 289–317.

ALAM, M. M., CROOK, K. A. W., AND TAYLOR, G., 1985, Fluvial herring-bone cross-stratification in a modern tributary mouth bar, Coonamble, New South Wales, Australia: Sedimentology, v. 32, p. 235–244.

ARTT, D. W., 1970, An experimental investigation of film cooling, with particular reference to the injection region: Unpublished Ph.D. Thesis, Faculty of Applied Science & Technology, Queens University of Belfast, 255 p.

ASHMORE, P. E., 1979, Laboratory modelling of braided streams: Unpublished M.S. Thesis, Department of Geography, University of Alberta, 195 p.

———, 1982, Laboratory modelling of gravel braided stream morphology: Earth Surface Processes, v. 7, p. 201–225.

———, AND PARKER, G., 1983, Confluence scour in coarse braided streams: Water Resources Research, v. 19, p. 392–402.

BAKER, V. R., 1973, Paleohydrology and sedimentology of Lake Missoula flooding in eastern Washington: Geological Society of America Special Paper 144, 79 p.

———, 1984, Flood sedimentation in bedrock fluvial systems, *in* Koster, E. H., and Steel, R. J., eds., Sedimentology of Gravels and Conglomerates: Canadian Society of Petroleum Geologists Memoir 10, p. 87–96.

BEST, J. L., 1985, Flow dynamics and sediment transport at river channel confluences: Unpublished Ph.D. Thesis, Birkbeck College, University of London, 393 p.

———, 1986, The morphology of river channel confluences: Progress in Physical Geography, v. 10, p. 157–174.

———, AND BRAYSHAW, A. C., 1985, Flow separation—a physical process for the concentration of heavy minerals within alluvial channels: Journal of the Geological Society of London, v. 142, p. 747–755.

———, AND REID, I., 1984, Separation zone at open channel junctions: Journal of Hydraulic Engineering, Proceedings, American Society of Civil Engineers, v. 110, p. 1588–1594.

BRYANT, I. D., HOLYOAK, D. T., AND MOSELEY, K. A., 1983, Late Pleistocene deposits at Brimpton, Berkshire, England: Proceedings of the Geologists Association, v. 94, p. 321–343.

COSTELLO, W. R., AND SOUTHARD, J. B., 1981, Flume experiments on lower flow regime bed forms in coarse sand: Journal of Sedimentary Petrology, v. 51, p. 849–864.

FROSTICK, L. E., AND REID I., 1977, The origin of horizontal laminae in ephemeral stream channel fill: Sedimentology, v. 24, p. 1–9.

GREATED, C. A., 1964, A theoretical study of flow behaviour at open channel junctions: Department of Civil Engineering, University of Southampton, U.K., Report No. 6, 26 p.

GUY, H. P., SIMONS, D. B., AND RICHARDSON, E. V., 1966, Summary of alluvial channel data from flume experiments, 1956–1961: U.S. Geological Survey Professional Paper 462-I, 96 p.

HOOKE, R., Le B., 1968, Model geology: prototype and laboratory streams:

Discussion: Geological Society of America Bulletin, v. 79, p. 391–394.

———, 1974, Shear stress and sediment distribution in a meander bend: University of Uppsala, Sweden, Ung Report No. 30, 58 p.

ITAKURA, T., 1972, Study of the mechanism of flow at confluences (in Japanese): Proceedings, 16th Hydraulics Conference, Japanese Society of Civil Engineers, p. 7–12.

KANDA, T., AND KITADA, T., 1977, An implicit method for unsteady flows with lateral inflows in urban rivers: Proceedings, 17th Congress of the International Association for Hydraulic Research, Baden-Baden, Paper A102, p. 213–220.

KENNEDY, B. A., 1984, On Playfair's law of accordant junctions: Earth Surface Processes and Landforms, v. 9, p. 153–173.

KJERFVE, B., SHAO, C. C., AND STAPOR, F. W., 1979, Formation of deep scour holes at the junction of tidal creeks: An hypothesis: Marine Geology, v. 33, p. M9–M14.

KNIGHTON, A. D., 1980, Longitudinal changes in size and sorting of stream-bed material in four English rivers: Geological Society of America Bulletin, v. 91, p. 55–62.

———, 1982, Longitudinal changes in the size and shape of stream bed material: Evidence of variable transport conditions: Catena, v. 9, p. 25–34.

KOMURA, S., 1973, River-bed variations at confluences: Proceedings of the Symposium on River Mechanics, Bangkok, 9–12 January, Paper A66, p. 773–784.

LIN, J. D., SCOTTRON, V. E., AND SOONG, H. K., 1976, Friction effects on spatially varied open channel flows with increasing discharge, *in* Rivers '76, Annual Symposium of the Waterways, Harbors and Coastal Engineering Division of the American Society of Civil Engineers, Fort Collins, Colorado, 10–12th August, p. 1517–1534.

———, AND SOONG, H. K., 1979, Junction losses in open channel flows: Water Resources Research, v. 15, p. 414–418.

LODINA, R. V., AND CHALOV, R. S., 1971, Effect of tributaries on the composition of river sediments and of deformations of the main river channel, *in* Soviet Hydrology, Selected Papers, No. 4, p. 370–374.

MACKAY, J. R., 1966, The mixing of the waters of the Laird and Great Bear Rivers with those of the Mackenzie: Geographical Bulletin of Canada, v. 8, p. 166–173.

———, 1970, Lateral mixing of the Laird and Mackenzie Rivers downstream from their confluence: Canadian Journal of Earth Sciences, v. 7, p. 111–124.

MCGUIRK, J. J., AND RODI, W., 1978, A depth-averaged mathematical model for the near field of side discharges into open channel flow: Journal of Fluid Mechanics, v. 86, p. 761–781.

MILLER, J. P., 1958, High mountain streams: Effects of geology on channel characteristics and bed material: State Bureau of Mines and Mineral Resources, New Mexico Institute of Mining and Technology, Socorro, New Mexico, Memoir No. 4, 42 p.

MODI, P. N., ARIEL, P. D., AND DANDEKAR, M. M., 1981, Conformal mapping for channel junction flow: Journal of Hydraulics Division, American Society of Civil Engineers, v. 107, p. 1713–1733 (and errata, v. 108, p. 1549, December, 1982).

MOSLEY, M. P., 1975, An experimental study of channel confluences: Unpublished Ph.D. Thesis, Colorado State University, 216 p.

———, 1976, An experimental study of channel confluences: Journal of Geology, v. 84, p. 535–562.

———, 1982, Scour depths in branch channel confluences, Ohau River, Otago, New Zealand: Proceedings of New Zealand Institute of Professional Engineers, v. 9, p. 17–24.

———, AND SCHUMM, S. A., 1977, Stream junctions—A probable location for bedrock placers: Economic Geology, v. 72, p. 691–694.

MUROTA, A., KANDA, T., AND ETO, T., 1973, Flood routing for urban river network: International Association for Hydraulic Research Symposium on River Mechanics, Bangkok, Thailand, 9–12 January, v. 3, p. 339–350.

NYCHAS, S. G., HERSHEY, H. C., AND BRODKEY, R. S., 1973, A visual study of turbulent shear flow: Journal of Fluid Mechanics, v. 61, p. 513–540.

PROHASKA, S. J., 1978, Hydrologic flood routing in confluence reaches: Proceedings, International Conference on Water Resources Engineering, Bangkok, Thailand, p. 689–702.

REID, I., AND FROSTICK, L. E., 1979, Semi-arid sedimentology and geomorphology of the Lake Turkana basin, North Kenya: Natural Environment Research Council Report, Grant GR3/2897, 39 p.

RICHARDS, K. S., 1980, A note on the changes in channel geometry at tributary junctions: Water Resources Research, v. 16, p. 241–244.

SANMUGANATHAN, K., 1973, The passage of a flood wave through a junction: International Association for Hydraulic Research Symposium on River Mechanics, Bangkok, Thailand, 9–12 January, v. 3, p. 351–360.

SEVUK, A. S., AND YEN, B. C., 1973, Comparison of four approaches in routing flood wave through junction: Proceedings, 15th Congress of the International Association for Hydraulic Research, Istanbul, v. 5, p. 169–172.

SOLIMAN, M. M., 1976, Flood control arrangements for river junctions, *in* Rivers '76, Third Annual Symposium of the Waterways, Harbors and Coastal Engineering Division of the American Society of Civil Engineers, Fort Collins, Colorado, 10–12th August, v. 3, p. 1057–1070.

SOONG, H. K., 1976, Effect of lateral inflow on steady open channel flows: Unpublished Ph.D. Thesis, University of Connecticut, 104 p.

STOKER, J. J., 1957, Water Waves: Interscience Publishers, Inc., New York, 496 p.

TAYLOR, E. H., JR., 1944, Flow characteristics at open channel junctions: Transactions of the American Society of Civil Engineers, v. 109, Paper 223, p. 893–912.

TEISSYRE, A. K., 1975, Pebble fabric in braided stream deposits: Geologia Sudetica, v. 10, p. 7–46.

WEBBER, N. B., AND GREATED, C.A., 1966, An investigation of flow behavior at the junction of rectangular channel: Proceedings and Papers of the Institute of Civil Engineers, Paper 6901, p. 321–334.

WILLIAMS, G. P., 1970, Flume width and water depth effects in sediment transport experiments: U.S. Geological Survey Professional Paper 562-H, 37 p.

YEN, B. C., AND AKAN, A. O., 1976, Flood routing through river junctions, *in* Rivers '76, Third Annual Symposium of the Waterways, Harbors and Coastal Engineering Division of the American Society of Civil Engineers, Fort Collins, Colorado, August 10–12th, p. 212–231.

LOCATION-DEPENDENT SEDIMENT SORTING IN BEDFORMS UNDER WANING FLOW IN THE RIO GRANDE, CENTRAL NEW MEXICO

DAVID W. LOVE

New Mexico Bureau of Mines and Mineral Resources, Socorro, NM 87801

ALLAN GUTJAHR

Mathematics Department, New Mexico Institute of Mining and Technology, Socorro, NM 87801

AND

SYLVEEN ROBINSON-COOK

Geoscience Department, New Mexico Institute of Mining and Technology, Socorro, NM 87801

ABSTRACT: Grain size sorting in channels of the Rio Grande varies with location within ripples and megaripples and within downstream progressions of bedforms. In the natural sand-dominated floodway channel, plane beds and ripples commonly have log-hyperbolic grain size distributions. Under waning flow with minor suspended sediment, late-forming ripples contain unimodal (0.15 mm) log-hyperbolic distributions. Where these ripples have overtaken and mixed with earlier formed migrating coarser grained ripples, bimodal (0.15 and 0.30 mm modes) ripples result. These mixed ripple distributions may be modelled mathematically by mixing hyperbolic distributions.

In subjacent dunes formed from mixed ripples overtopping the dunes' brinks, upper foresets are slightly bimodal. Toesets are also bimodal, but lower foresets are more poorly sorted and approximate one or more hyperbolic distributions. In a 50-m reach of dunes, upper foresets contain progressively more pebbles downstream.

Waning flow in the artificial Rio Grande conveyance channel left gravel-dominated straight-crested megaripples with entirely different grain size distributions. Gravel armor developed from the troughs to the crests of the 2.8-m-wavelength megaripples. Both troughs and crests have distributions that exhibit broad "plateaus" from 0.25 to about 10 mm, and both have secondary modes at 0.063 mm. Upper slipfaces have roughly log-hyperbolic distributions, but contain irregular amounts of coarse and fine grains. Lower slipfaces have log-hyperbolic distributions with modes ranging from 1.4 mm upstream to 0.71 mm downstream. The slipfaces deposited during waning flow are better sorted and finer grained than the foresets of the megaripple interiors, which are composed of a mixture of upper and lower slipface grains along with even coarser clasts.

INTRODUCTION

A hallmark of streams in arid and semiarid regions is their large variation in discharge. For example, the Rio Grande in central New Mexico was historically an intermittent stream until modern flow control structures were built. Floods (recorded to as much as 1400 m^3/sec) alternated with seasons of no flow. Excellent exposures of Pliocene and Pleistocene deposits of the ancestral Rio Grande indicate that the river was alternately a gravel-dominated and sand-dominated stream. These deposits lead to questions concerning the long-term behavior of the stream and the detectability of intermittent flow conditions in the past. In order to try to quantify conditions of deposition to compare with sediments of the ancestral Rio Grande, modern deposits formed under waning flow (decreasing to no flow) were studied to look for flow indicators and locational indicators within channels beyond sedimentary structures alone. This paper summarizes the results of two studies of location-dependent sorting of modern sediments in bedforms developed during relatively simple waning flow (Love and others, 1985; Love and others, 1986). The results extend some mathematical models for grain size distributions and raise further questions regarding sorting processes within bedforms and progressions of bedforms.

The bedforms studied include small and large ripples, sandy megaripples, and gravelly megaripples. Ripples are small undulations on the surface of noncohesive sand and silt. The studied ripples are unidirectional asymmetric undulatory-to-lingoid forms with a range of wavelengths and amplitudes. Megaripples are ripple-shaped bedforms having wavelengths larger than 60 cm and a range in amplitudes typically larger than ripples. Crests of megaripples can be straight, sinuous or crescentic (commonly the lunate forms are called "dunes"). Megaripples form in gravel- or sand-bedded unidirectional flows.

METHODS

Study sites, flow measurement, and bedform measurement.—

At San Acacia, 25 km north of Socorro, New Mexico, flow in the Rio Grande is directed into three major channels: the Rio Grande floodway, the Rio Grande conveyance channel, and the Socorro Main Canal. The Middle Rio Grande Conservancy District diverts water to the channels depending on seasonal needs and flow conditions. The U.S. Geological Survey records water levels (stage) in all three channels and takes water quality samples periodically. A second set of gauges is located 74 km downstream on the floodway and 73 km downstream on the conveyance channel. Unpublished water level records, charts, and water quality data were obtained from the U.S. Geological Survey after bedforms had been sampled in order to estimate declining flow conditions at the sample sites.

Unfortunately, bedforms in the Rio Grande floodway and conveyance channel were discovered only after flow had waned and after the water had been totally diverted out of each of the channels. Study sites were selected where the bedforms were most fully developed in reaches near Socorro. The sandy ripples and megaripples occurred in the floodway channel 24 km downstream from the floodway gauge, and the gravelly megaripples occurred in the conveyance channel 24 km downstream from the conveyance gauge.

Bedforms in the floodway channel were photographed and sampled and some measurements were taken on January 27, 1984 (11 days after flow had ceased), but a series of windstorms destroyed or buried many of the features before detailed measurements could be made. In the conveyance channel 10 megaripples were surveyed on April 19, 1985, along a 30-m reach with a Leitz TM 20 theodolite and a rod measuring to the nearest hundredth of a foot (about 3 mm).

Details of sampling of bedforms are given in Love and others (1985) and Love and others (1986). In the sand-dominated ripples and dunes, 20–40 g samples were taken from locations along the bedforms (see RESULTS). Only subsurface samples were taken in order to eliminate any eolian contribution at the surface. Generally, whole sections of individual ripples were taken (as much as 2 cm thick). Interiors of sandy megaripples were sampled by trenching perpendicular and parallel to the form and selecting uniform structures from upper (~5 cm deep) and lower (~15 cm deep) parts of the form. Larger samples were taken from locations with gravel. In the gravelly megaripples, 1 kg samples were taken in an area of 200–400 cm^2 in troughs and on crests. Ten megaripples and adjacent downstream troughs were numbered in a downstream progression and four were sampled along troughs, lower slipfaces, upper slipfaces, crests, and interiors (Love and others, 1985).

In the laboratory, the samples were dried and sieved using sieve openings of about a $\sqrt[4]{2}$ progression. All of the coarse fraction (>1 mm) for each sample was sieved in order to have an adequate estimate of large clasts. Samples with abundant portions of sand less than 1 mm in size were split to 50 g or less and sieved, and the proportion of the total was calculated. Duplicate splits of some samples were sieved to test the adequacy of the procedure.

Mass-frequency diagrams.—

Mass-frequency diagrams were plotted following Bagnold's (1941) procedures. Details of the technique are given in Love and others (1986). Both axes of the plots are logarithmic because of the wide ranges in grain size and mass percentages. Irregular intervals between sieves are standardized by dividing the mass percent of each size class by the difference between logarithms of the diameters of the largest and smallest grains in the size class. The diagrams show the mass frequency of different size grains in such a way that proportions of one population may remain similar even though other populations are added or subtracted from the total in other parts of the distribution. Some sieves caught consistently less mass than the sieves adjacent to them because of deviations in aperture size on that particular sieve. These irregularities have not been corrected. Some curves are discontinuous at the coarse end because no clasts were caught on some sieves. For ease of comparison, some of the grain size data presented below are superimposed on the same graph by addition or subtraction of constant amounts to the ordinate. The amounts of offset are given in the figure captions.

Log-hyperbolic distributions and estimation.—

Some log plots of grain size distributions from the Rio Grande have hyperbolic distributions similar to those of Bagnold (1941), Barndorff-Nielsen (1977), and Bagnold and Barndorff-Nielsen (1980), so a technique similar to theirs was used to fit the data. Fitting log-hyperbolic distributions is described below and more details are given in Love and others (1986).

Hyperbolic distributions are generalizations of lognormal distributions and require four parameters. For grain size data, the logarithm of the grain diameter is assumed to have a hyperbolic distribution. By analogy with the lognormal, the grain size is thus said to be a log-hyperbolic random variable.

If $v = \log_e d$ (where d is grain diameter), the probability density function of v with four parameters is written $f(v; \mu, \delta, \phi, \gamma)$, separating the argument v from the parameters by a semicolon. The log-hyperbolic assumption for the probability density function means

$$\log_e f(v;\mu,\delta,\phi,\gamma) = Co - \frac{(\phi + \gamma)}{2}(\delta^2 + (v - \mu)^2)^{1/2} + \frac{(\phi - \gamma)}{2}(v - \mu) \quad (1)$$

In equation (1), Co is a constant related to parameters μ, δ, ϕ and γ via an expression involving K_1, a modified Bessel function of the third kind with index $\nu = 1$. Co is the ordinate of the point where the two asymptotic lines of the hyperbola cross. The parameters are identified as follows: μ locates the abcissa of the point where the asymptotes cross, δ is a scale parameter, and ϕ and γ represent the slopes of the asymptotic lines on the grain size plot. The mode, or maximum value for the probability density function is given by

$$v_m = \mu + \delta(\phi - \gamma)/2\phi\gamma \quad (2)$$

The second and third terms on the right hand side of (1) are a general equation of a hyperbola. If $\delta = 0$ and $\phi = \gamma$, the density $f(v; \mu, 0, \phi, \phi)$ would reduce to a normal probability density. If (1) is used to describe the behavior of $v = \log_e d$, the proportion (i) of grains with d between d_{1i} and d_{2i} can be calculated by letting $v_{1i} = \log_e d_{1i}$, $v_{2i} = \log_e d_{2i}$ and evaluating

$$P_i(\mu,\delta,\phi,\gamma) = \int_{v_{1i}}^{v_{2i}} f(v;\mu,\delta,\phi,\gamma)dv \quad (3)$$

Three other useful parameters are

$$\kappa = (\phi\gamma)^{1/2} \quad (4)$$

$$\omega = (\phi\gamma)/(\phi + \gamma) \quad (5)$$

$$\beta = (\phi - \gamma)/2 \quad (6)$$

Using these parameters, Barndorff-Nielsen (1977) shows that

$$Co = \log_e(\omega/\delta\kappa K_1(\delta\kappa)) \quad (7)$$

In the graphs and figures given later, $\log_{10} d$ is used in place of $\log_e d$. With this conversion the parameters change as indicated below (where primes indicate the values for $\log_{10}$ values)

$$\phi' = \phi \log_e 10 \quad (8)$$

$$\gamma' = \gamma \log_e 10 \tag{9}$$

$$\delta' = \delta \log_{10} e = \delta / \log_e 10 \tag{10}$$

$$\mu' = \mu \log_{10} e = \mu / \log_e 10 \tag{11}$$

Figure 1 shows a log-hyperbolic curve along with the two linear asymptotes for the hyperbola. $P_i/(\log_{10} d_{2i} - \log_{10} d_{1i})$ is graphed at midpoints $(\log_{10} d_{2i} + \log_{10} d_{1i})/2$ along with the density. If $\log_{10} d_{2i} - \log_{10} d_{1i}$ is small, then this ratio will approximate the density.

With a given set of data, the four parameters are estimated by using the method of maximum likelihood which is equivalent to the method of maximum likeness described by Barndorff-Nielsen (1977). The data are probability data and hence the estimation procedure finds μ, δ, ϕ, and γ so that the P_i values are close to the observed proportions. More specifically, if r_i is the observed proportion of grains found in the range d_{1i} to d_{2i}, the estimated parameters μ, δ, ϕ and γ are those values which maximize the following expression:

$$MLE = \sum_{i=1}^{m} r_i \log_e P_i(\mu, \delta, \phi, \gamma) \tag{12}$$

where P_i is the expression given in equation (3). The sum is over all the number of classes or divisions and $d_{1,i+1} = d_{2i}$, $i = 1, 2, \ldots m - 1$.

The maximum likelihood (or likeness) equations (*MLE*) are non-linear and thus an iterative scheme is used to obtain the final estimates. The procedure used is the Rosenbrock hill-climbing method with constraints (Kuester and Mize, 1973) and the integral for P_i is approximated numerically.

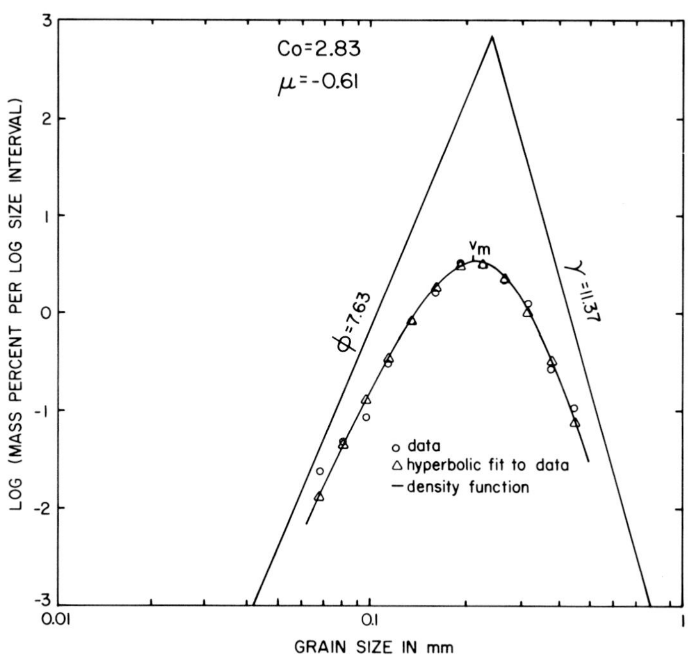

FIG. 1.—Bagnold-style plot of grain size data from a plane bed at the beginning of a dune field in the Rio Grande floodway channel. The curve is the log-hyperbolic density fit to the data. The two straight lines are the two linear asymptotes of the curve, along with ϕ and γ, the slopes of the asymptotes, (μ, $\log_e Co$) the intersection point, and v_m the mode of the distribution.

RESULTS

Flow conditions and bedform measurement.—

Flow in the two channels was determined after sampling and is based on unpublished data from the U.S. Geological Survey. At San Acacia, 24 km upstream from the floodway sample site, a pulse of water was put into the Rio Grande floodway channel between 5:30 pm on January 13, 1984, and 4:30 pm January 15, 1984. The water initially rose from 0 to 21 cm in one-half hour and peaked at 45 cm 6 hrs later. Flow depth ranged from 37 to 45 cm with a mean discharge of 2 m^3/sec until flow was cut off. Water depth dropped 38 cm in less than half an hour. The next gauge, 74 km downriver, did not record the rising flow (float initially stuck in mud?) but showed that declining flow began 22 hrs later than at the first gauge and took 20 hrs to wane completely. Waning flow at the floodway sample site is interpreted to have taken only a couple of hours, with no lingering low flow.

In the floodway channel, bedforms consisted of three-dimensional megaripples (dunes) approximately 15 cm high with wavelengths ranging from 3 to 5 m. Ripples developed from troughs to crests of the dunes with increasing amplitudes and wavelengths to 3 cm and 30 cm respectively (Figs. 2 and 3). These bedforms had not been reworked by extremely shallow flows and the only indication of waning flow was the extent of ripples within troughs and on the crests of the dunes. Only a thin clay drape was present in the troughs. Average channel slope through the reach was about 0.0006.

On April 12 and 13, 1984, flow in the conveyance channel at the gauge 24 km upstream from the sample location averaged a nearly constant 43 m^3/sec (U.S. Geological Survey, unpublished data). On April 14, 1984, water level dropped about 1.6 m in 10.5 hrs and flow declined to 0.9 m^3/sec. The gauge 51 km downstream from the sample site recorded declining flow 9 hrs later. The lower gauge de-

FIG. 2.—Ripple-covered dunes of the floodway channel of the Rio Grande, after waning flow had ceased (photographed January 27, 1984, 11 days after pulse).

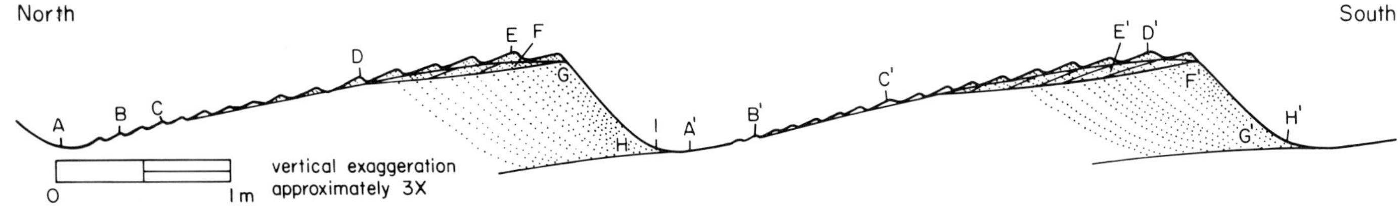

FIG. 3.—Sample locations designated by letters for ripples and portions of dunes in the Rio Grande floodway channel.

creased 1.2 m in 24 hrs and continued to record flow at 8–9 m^3/sec.

In the conveyance channel, bedforms were gravelly megaripples with wavelengths of 2.8 m and amplitudes of 12 cm (Fig. 4). The crests were straight to slightly sinuous. These bedforms were reworked along the thalweg. Average slope along the conveyance channel was 0.016.

Distributions from floodway ripples and dunes.—

In a ripple train progressing from the trough to the crest of the adjacent dune downstream, grain size distributions progress from an essentially unimodal distribution to a bimodal distribution (Fig. 5). In small ripples downstream from the trough, both fine and coarse tails of the distribution contain less mass than in the troughs, and the peak of the distribution at 0.15 mm is more pronounced (Fig. 5A–C). In bigger ripples, a secondary mode at 0.30 mm appears (Fig. 5D) and increases to equal the amount of the finer mode (Figs. 5E, F). The trough-and-ripple train of the next dune downstream contains similar distributions, indicating that the sorting processes repeated in at least two trough-ripple trains (Love and others, 1986).

Sorting in the subjacent dunes is related to position of sampling along the slipfaces of buried foresets (Fig. 5G, H, I). The upper foreset is composed of a mixture similar to the two ripple fractions, dominated by the coarse ripple fraction (Fig. 5G). The lower foreset consists of essentially the coarser mode with even coarser sand and pebbles added (Fig. 5H). The added coarser fraction tends to make the coarse side of the distribution irregular. At the toe of the

FIG. 4.—Gravelly megaripples with wavelengths of 2.8 m in the conveyance channel of the Rio Grande. Low-flow thalweg is reworking sand and modifying megaripples, April 19, 1984 (from Love and others, 1985).

slipface, bimodal distributions occur again (Fig. 5I). Some toes of slipfaces and adjacent troughs contain a few larger pebbles (as much as 10 mm in diameter).

In a downstream progression of five dunes, upper foresets (similar to locations G and F' of Fig. 3) consist of a modal sand fraction and variable amounts of finer and coarser clasts (Fig. 6 A). The size of the mode of the dominant sand fraction shifts from 0.21 to 0.30 mm downstream. Amounts of clasts greater than about 1 mm increase downstream, although the size of the coarsest clast is variable and does not increase downstream.

Size distributions in five lower foresets of dunes (similar to locations H and G' of Fig. 3) are much more variable downstream (Fig. 6 B). In general, only the fractions in the 0.075–0.150 mm range are similar throughout the reach. Commonly the modal sand fraction is coarser than the corresponding modal fraction of the upper foreset, and the coarse side of the modal fraction slopes less and is more irregular in the lower foreset.

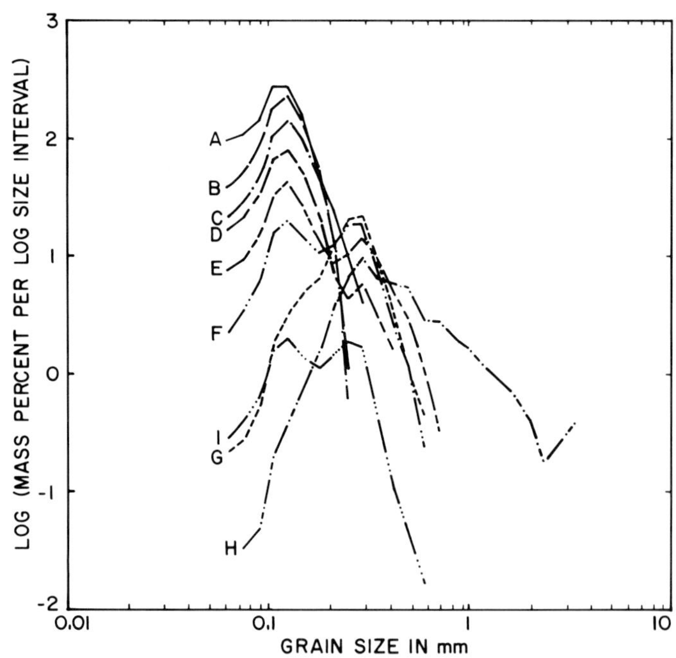

FIG. 5.—Grain size distributions of a progression of samples from trough to toe of dune. Letter designation of each curve refers to sample location on Figure 3 and numbers in parentheses indicate amount added or subtracted from Y-values to separate curves from each other. (A) trough (+0.5); (B) small ripple (+0.2); (C) small ripple (0); (D) medium-size ripple (−0.2); (E) large ripple (−0.4); (F) remains of large ripple buried at crest of dune (−0.6); (G) top of foreset (−1.1); (H) base of slipface (−1.6); (I) distal toe of slipface (−2.0).

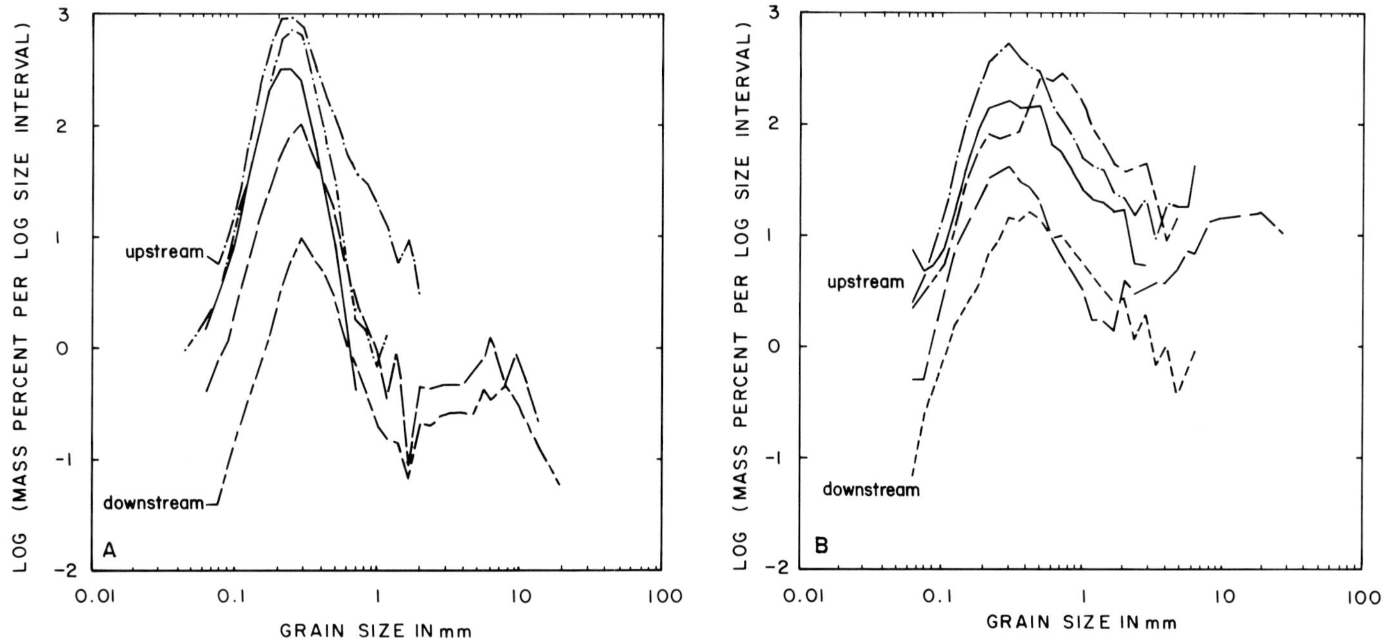

FIG. 6.—A. Grain size distributions of upper foresets in a downstream progression of dunes. Numbers added to or subtracted from *Y* values to separate curves from top to bottom are +0.5, +0.3, 0. −0.5, −1.5, respectively. B. Grain size distributions from lower foresets in same downstream progression of dunes. Numbers added or subtracted to separate curves from top to bottom are 0, −0.5, −1.0, −1.4, −1.8, respectively.

Distributions from gravelly megaripples.—

As shown in Figure 7A and B, troughs and crest armor of four of the 10 megaripples examined have broad, slightly skewed distributions from 0.25 to about 8 mm. The troughs are slightly coarser, are skewed to the fine side, and decrease in amount of coarse grains downstream. The crest-armor distributions are more uniform or are slightly skewed to the coarse end. Both distributions have a secondary mode at about 0.063 mm. Megaripple interiors are even more skewed to the coarse end and have similar secondary modes (Fig. 7C). Both upper and lower slipfaces are essentially hyperbolic distributions, but lower slipfaces are better sorted and decrease in mode downstream from about 1.3 to 0.78 mm (Fig. 7D, E).

Downstream progressions along individual megaripples also reveal trends in the following sequence: 1) trough to 2) megaripple crest to 3) interior to 4) upper slipface to 5) lower slipface to 6) trough. As Figure 7 shows, distributions become better sorted and unimodal in progressions from troughs to lower slipfaces. The coarse fractions (>8 mm) of troughs decrease in crests and interiors and are not present in slipfaces. Secondary modes at 0.063 mm also decrease from troughs to slipfaces. In two of the progressions from lower slipfaces to troughs, the troughs have a deficiency of clasts in the range of the peak of the lower slipfaces.

As Table 1 shows, the mass of clasts larger than 2 mm in troughs decreases downstream, and d_{50} (diameter in mm at 50 percent cumulative mass) decreases downstream, but the mass of all clasts larger than 0.6 mm shows no significant differences downstream. The lower slipfaces contain progressively less mass greater than 0.6 mm downstream, but other megaripple portions do not show significant trends.

DISCUSSION

Comparison of sorting mechanisms in ripples, dunes, and gravelly megaripples.—

Before clasts arrive at a given reach, numerous sorting processes have determined the distribution of sizes in transit. General factors include production of clasts from source rocks, contributions of clasts from tributaries, differential longitudinal and/or lateral transport and storage, and abrasion and dissolution. Group behavior of grains in transit is affected by macro- and micro-scale pulses in bedload and/or suspended load, micro-scale lateral and longitudinal transport among small bedforms, destruction of small bedforms at the brinks and slipfaces of larger bedforms, reverse and lateral transport in the lee of bedforms, and accumulation rates. Within a dune and ripple field the major depositional processes which affect sorting in the bedforms are due to flow separation at the brinks of ripples and dunes (Jopling, 1964; Fig. 8). Sorting occurs as grains move to the brink of the ripple or dune and either accumulate at the top of the slipface and avalanche to the toe, or move with the current beyond the brink, settling through zones of differential turbulence and relative velocity to the slipface, trough, or stoss side of the bedform downstream. Small grains are carried away in suspension.

Grains caught in avalanches down the slipface are sorted by differential shearing and relative dispersive pressure (Jopling, 1964; Bagnold, 1968). Pulses of relatively coarser and finer grains may avalanche down the slipfaces of dunes as coarse-grained troughs and finer grained ripples overtop the brink of the larger bedforms (Smith, 1972).

Subsidiary processes of sorting include overpassing by both larger and smaller grains (Everts, 1973; Allen, 1983),

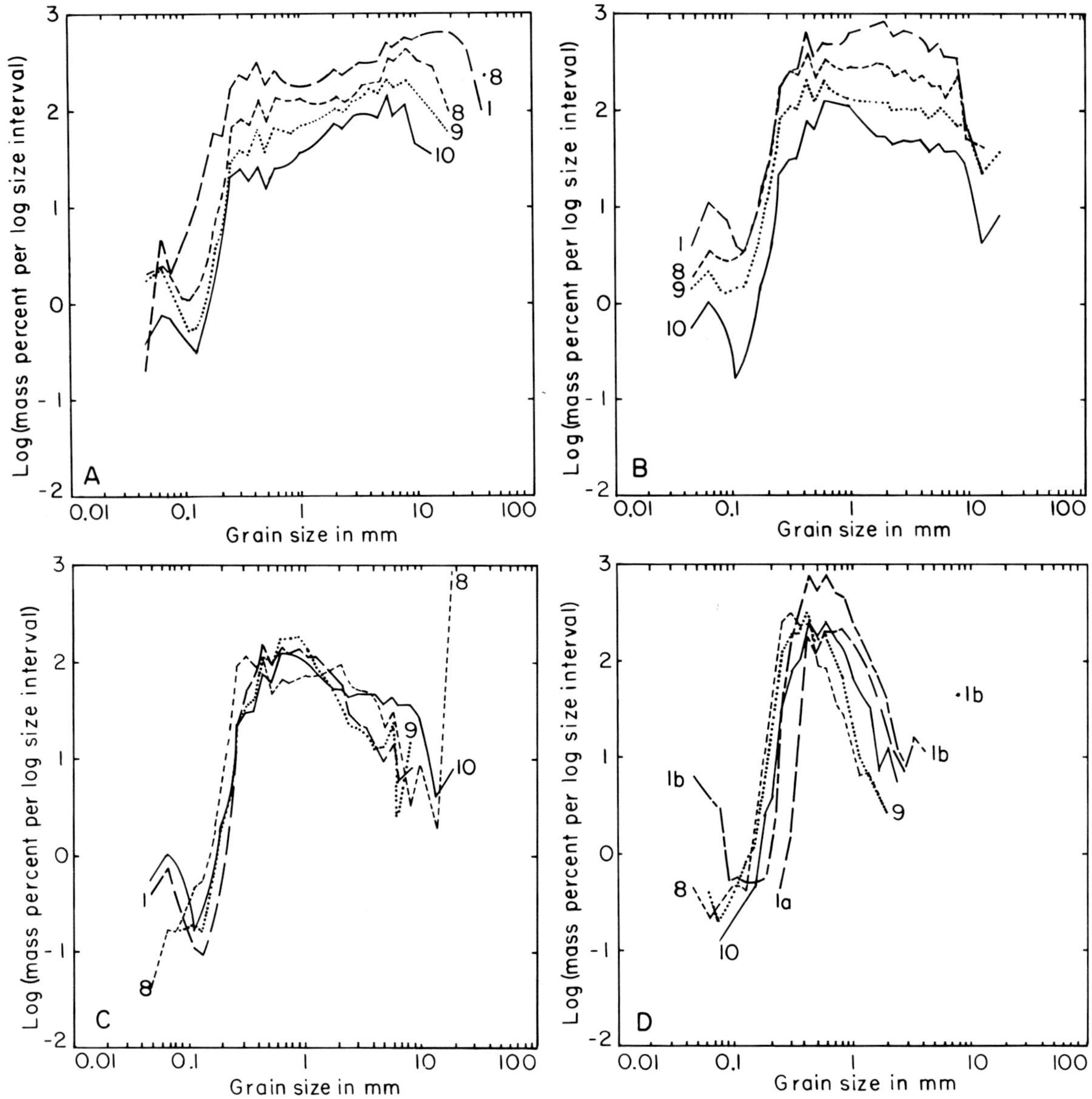

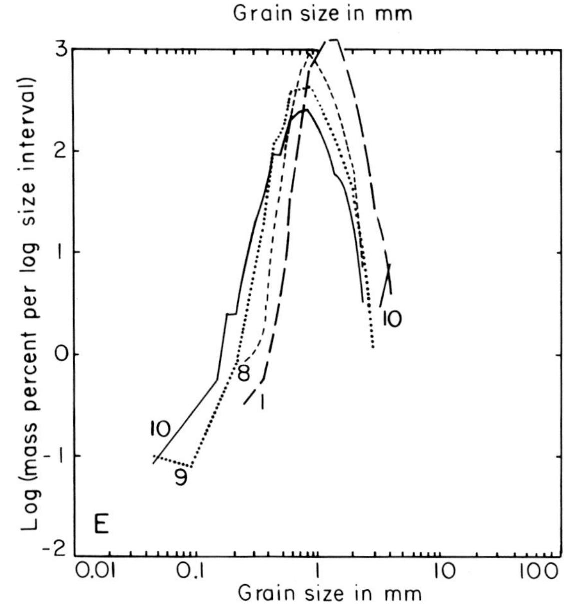

FIG. 7.—Grain size distributions from gravelly megaripples. Modified from Love and others (1985). A. Troughs located downstream from megaripples 1, 8, 9, and 10. For clarity, curves 1, 8, and 9 have been offset vertically by the respective addition of 0.9, 0.6, and 0.3 to their *Y* values. B. Crest armor. For clarity, curves 1, 8, and 9 have been offset vertically by the respective addition of 0.9, 0.6, and 0.3 to their *Y* values. C. Megaripple interiors. No curves have been offset. D. Upper slipfaces. Curves 1a, 8, 9 and 10 are not offset; 1b is offset by 0.5 *Y* units to avoid overlap. 1a is from active slipface in thalweg. 1b is from inactive slipface. E. Lower slipface. Curves 1, 8, and 9 are offset vertically by the respective addition of 0.6, 0.4, and 0.2 to their *Y*-values.

TABLE 1.—LONGITUDINAL RANGES IN MASS PERCENT AND D_{50} OF SELECTED CLAST SIZES FROM PORTIONS OF MEGARIPPLES ALONG THE CONVEYANCE CHANNEL OF THE RIO GRANDE.

Megaripple	Percent >2 mm in Troughs	d_{50} (mm) in Troughs	Percent >0.6 mm in Troughs	Lower Slip	Upper Slip	Armor	Interior
1	75.3	8.10	83.3	99.8	57.3*	83.7	71.4
8	75.1	6.91	91.3	97.7	12.8	75.4	65.6
9	73.6	4.77	91.8	86.4	28.7	76.5	77.2
10	72.2	3.93	91.3	81.9	52.3	83.0**	

*Only lb, inactive slipface, represented here.
**Armor and interior sampled together in megaripple 10.

trapping and/or grain substitution (Parker and Klingeman, 1982; Steidtman, 1982) on stoss sides of ripples and dunes, and sorting due to properties of individual grains (size, density, settling velocity, shape, electrical charge of small grains).

Rates of transport and deposition also determine whether grains are buried within sedimentary structures or are able to move farther downstream. Depending on conditions of waning flow, dunes may undergo progressive modification until they become totally reworked into ripples or even plane beds within the low-flow channel. In the present study, however, dunes were only slightly modified, but ripples are interpreted to have expanded into the troughs and to the crests of dunes before flow ceased.

In mobile gravel-bed streams all grains move episodically as a result of fluid shear stress along the bed and no grains can be considered lag. An armor or pavement at the bed-water interface develops that results in nearly equal transport rates for all sizes of clasts in the bedload (Andrews, 1983; Parker and Klingeman, 1982); winnowing of finer sizes by differential bedload transport apparently does not take place. Similarly, armor is developed from troughs to crests of gravelly megaripples. Andrews (1983) and Parker and Klingeman (1982) imply that grain size distributions of armor should reflect nearly equal amounts of a large range in sizes.

Flow separation at the brink of gravelly megaripples is similar to that for ripples and dunes. Under waning flow, the gravelly mobile bed in the stoss sides of megaripples should come to a stop and only finer sizes (sand-clay) should continue to move. Sand caught in the zone of mixing and reverse flow should be sorted by movement up the lee side of the megaripple by reverse flow.

Grain-size sorting in ripples and dunes.—

According to the above mechanisms of sediment movement over ripples and dunes, some variation in grain size might be expected to result from the range in amplitudes of ripples or dunes, influencing flow conditions and the amount of avalanching of coarse grains on slipfaces into troughs. The change from unimodal small ripples to bimodal larger ripples was not anticipated but may be explained as a consequence of waning flow. The change may be visualized in the following manner.

Dunes apparently formed and migrated before flow waned. Ripples with unimodal coarse grain size distributions formed on the stoss sides of dunes, but may have been sheared out on the crests of the dunes under high discharge conditions (Bridge, 1982). Troughs were relatively large. As flow waned, finer sand came out of suspension, occupied troughs, and formed small ripples. These small ripples continued to form and migrate, mixing with coarser ripples that formed earlier on the stoss sides of dunes. The mixed ripples continued to migrate to the crests of dunes, but amounts of finer sand sizes decreased in amount along the bedforms, probably because of increased transportability of fines along the crests of the dunes and decreased amounts available from the lower stoss sides of dunes. As ripples were destroyed beyond the dunes' brinks, amounts of fine sizes incorporated in upper slipfaces remained noticeable.

Lower slipfaces, on the other hand, accumulated distinctive distributions of coarser grains that avalanched from the brinks. Dispersive pressure between grains did not operate over a distance long enough to sort the grains so the distributions are irregular (Figs. 5H and 6B). The bimodal distributions of the toes of the slipfaces (Fig. 5I) suggest that fine grains from waning flow in troughs and coarser grains from ripples destroyed at the brinks of the dunes were mixed similarly to bimodal ripples, but without the corresponding bedforms.

Grain-size sorting in gravelly megaripples.—

The mechanisms of grain movement in megaripples suggest that the interiors of the bedforms should be made up

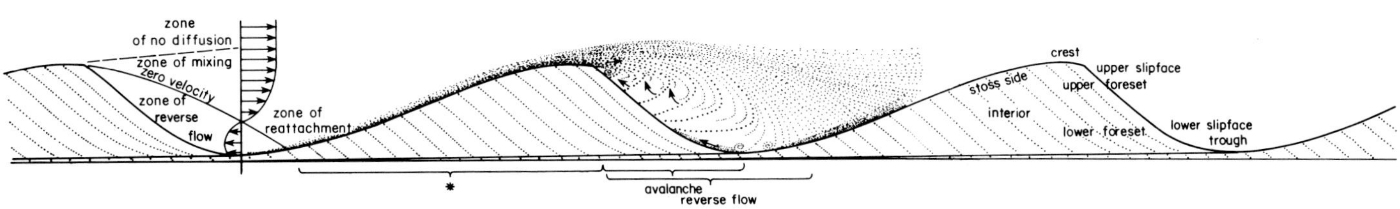

FIG. 8.—Flow, sediment movement, and anatomy of ripples, dunes, and gravelly megaripples. For ripple or sand-dominated subaqueous dunes, * indicates plane-bed conditions on stoss sides of ripples and some dunes and rippled stoss sides on dunes under different flow conditions. For gravelly megaripples, * indicates where mobile-bed armor develops from troughs to crests of the bedforms. Modified from Reineck and Singh (1980).

of upper and lower slipfaces, and that armor on crests and stoss sides of megaripples should be derived in part from reworking megaripple interiors along with the addition of grains that passed over previous bedforms. These implied relations appear to be substantiated by the illustrated grain size distributions.

The size distributions of interiors of megaripples (Fig. 7C) may be simulated by mixing distributions from upper and lower slipfaces (Fig. 7D, E), although some clasts coarser than those of the slipfaces also occur. These coarse clasts in the interior may be due to flow conditions at an earlier time of deposition in contrast to the slipfaces formed under waning flow.

The excellent degree of sorting of the lower slipfaces (Fig. 7E) is unexpected considering the proposed mechanism of intermittent avalanching of grains down the slipface. If avalanching was the chief sorting mechanism, coarser grains should be expected at the base of the slipface. Instead, the shape of the grain size distributions suggests sorting during unidirectional flow similar to that of sand in ripples and

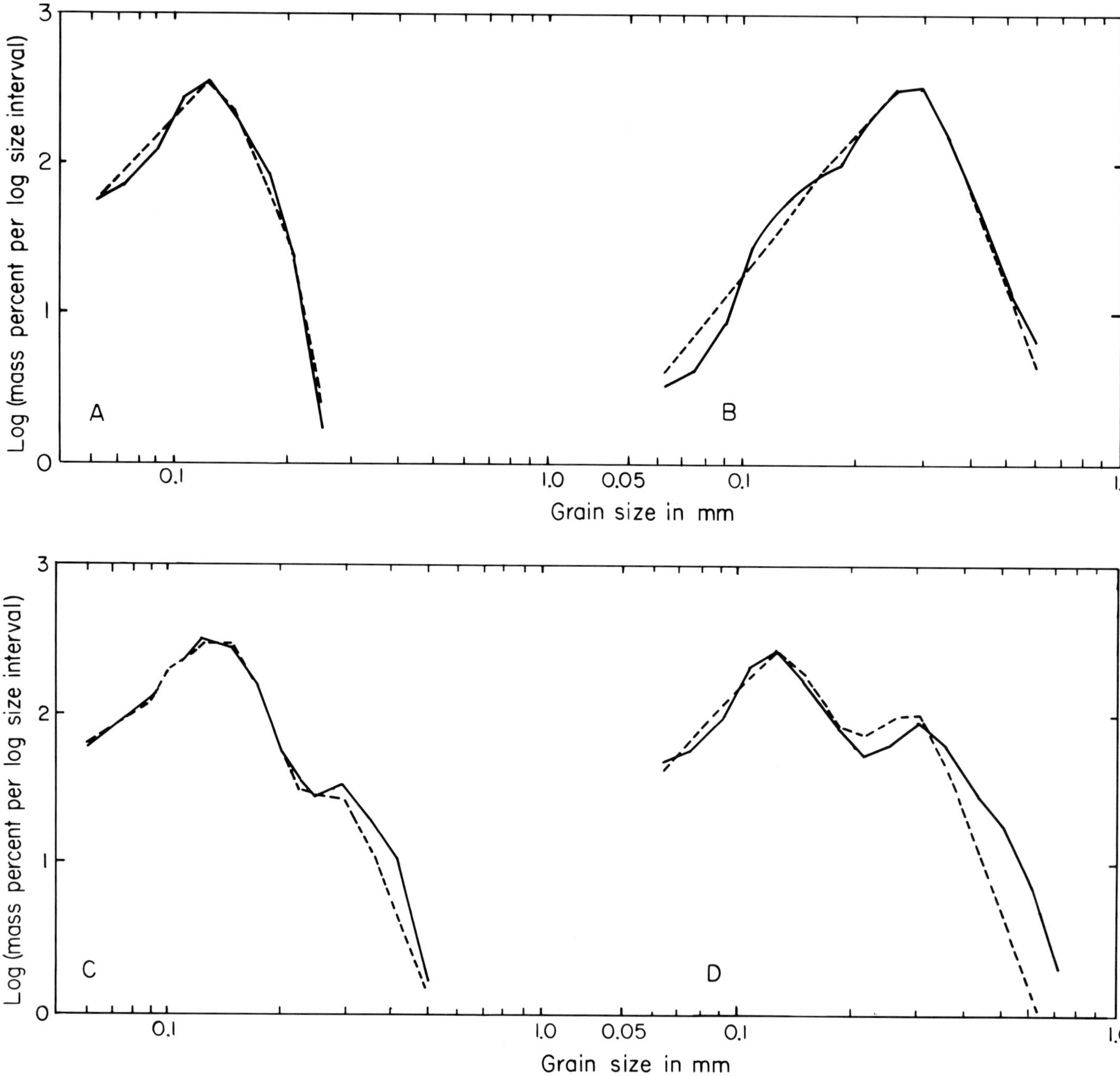

FIG. 9.—Calculated hyperbolic distributions (dot-dash) compared with actual distributions (solid line) for samples from the progression illustrated in Figure 5. A. Small ripple (Fig. 5B); B. upper foreset (5G); C. mixed ripple (5D), mixture of 89 percent small ripple A, 11 percent upper foreset B; D. mixed ripple (5E), mixture of 29 percent small ripple A, 71 percent upper foreset B.

TABLE 2.—PARAMETERS OF LOG-HYPERBOLIC FITS FOR GRAIN SIZE DISTRIBUTIONS FROM RIPPLES, DUNES, AND MEGARIPPLES FROM THE RIO GRANDE.

Sample	ϕ	γ	δ	μ	Mean	Mode
Floodway Channel						
A	4.73	11.44	.0136	−.87	−1.17	−.87
B	6.59	16.03	.0288	−.83	−1.03	−.85
D	6.05	8.61	.0037	−.87	−1.02	−.87
E	6.30	4.59	.0186	−.88	−.90	−.88
F	12.68	32.55	.730	−.33	−.76	−.68
G	7.51	15.55	.044	−.50	−.58	−.52
H	4.98	1.82	.198	−1.22	−.35	−.48
Floodway Channel Tops of Foresets						
1	10.82	7.32	.05	−.61	−.57	−.60
2	12.92	15.22	.09	−.55	−.57	−.56
3	12.32	16.69	.12	−.57	−.61	−.59
4*	9.05	9.31	.05	−.52	−.52	.52
5*	9.76	7.52	.03	−.49	−.42	.49
Conveyance Channel Upper Slipface						
lb	27.05	9.80	.29	−.38	−0.17	−.22
8	31.38	7.62	.11	−.56	−0.39	−.48
9	63.04	16.91	0.54	−0.72	−0.31	−.34
10	34.87	14.41	0.54	−.48	−0.21	−.23
Lower Slipface						
1	21.24	21.08	.27	.14	.14	.14
8	26.69	16.37	.25	−.07	.008	−.009
9	24.51	13.02	.30	−.18	−.06	−.09
10	9.19	8.52	.13	−.08	−.07	−.07

*Only hyperbolic part of distribution less than 2 mm considered.

plane beds. Therefore, the sand of the lower slipface apparently was re-sorted and accumulated by reverse flow in the wake of the megaripples during waning flow.

Armor could have been derived either by reworking megaripple interiors, removing large amounts of finer clasts and accumulating coarse clasts as lag, or by adding coarse clasts in a mobile bed that moved from the troughs to the crests of the megaripples. The nearly flat "plateaus" between 0.25 and 8 mm of the distributions of the armor (Fig. 7B) could indicate either that clasts in that size range were equally mobile because of equal shear stress to move all sizes (Andrews, 1983) or that the grains were differentially exposed to shear stress depending on the size and placement of surrounding grains (Parker and Klingeman, 1982). Lower shear stress in troughs would force accumulation of otherwise mobile large grains, accounting for the slight "hump" in the mass of coarse grains (Fig. 7A).

The abrupt decrease in mass percent of grains less than 0.25 mm in the armor of crests and troughs suggests an efficient sorting process such as suspension of finer grains; however, secondary peaks commonly occur at 0.09–0.11 and 0.06 mm in the armor of the troughs and crests. This sorting of secondary modes takes place despite the fact that no ripples form on beds with average grain size larger than about 0.6 mm. Neither mechanical crushing nor infiltration of fine grains appears to explain these secondary modes (Love and others, 1985). Therefore, some sorting process of fine sand along the bed must be taking place during mobile conditions.

The downstream trends in grain size distributions from analogous portions of megaripples along such a relatively short reach may be explained as a combination of changes in flow conditions, pulses of sediment and relative mobility of grains. Possibly the waning stages of deposition changed progressively downstream so that deposition, particularly along slipfaces, continued downstream after it had ceased upstream. Different pulses of grains may have reached different portions of megaripples at different times during waning flow, providing more coarse grains to upstream megaripples than to downstream megaripples, or providing more grains in the range of 0.85–8 mm to downstream troughs after passing upstream bedforms.

Hyperbolic fits.—

Hyperbolic fits to the unimodal ripples and nearly unimodal upper slipfaces were quite good (Fig. 9A, B). The ϕ, δ, and γ parameters for the two end distributions are similar (Table 2), the major difference being in μ, the location parameter. Note that in Table 2 the results are given for $\log_{10}$ data; thus, the means, modes and other parameters refer to $\log_{10}$ of grain size. It is tempting to try to convert these to millimeters but the only one that converts directly is the mode (the point where the density is maximum). For example, sample A has a mode of $10^{-.87} = .134$ mm. The bimodal ripples were modelled mathematically as mixtures of the finer grained and coarser grained unimodal distributions. As shown in Figure 9C and D, the mixed hyperbolic distributions are reasonably good.

In the progression of dunes downstream, the modal size of the upper slipfaces shifts slightly (from 0.21 to 0.30 mm), but the amount of coarse grains increases downstream. The hyperbolic fits show that the location parameter μ does not change appreciably but the other three parameters change downstream. The causes for these changes appear to be the increase in width and decrease in depth and slope of the reach of dunes as they died out in a riffle downstream.

In Barndorff-Nielsen (1977) the log-hyperbolic distribution is itself represented as an infinite mixture of log-normal distributions in which the mixing involves the variance. Namely, the distribution can be represented as a mixture or combination of log-normal distributions with variances which themselves are random quantities. The analyses presented here suggest that further mixing occurs as one moves from one end member to another, where this second level of mixing essentially involves the means of distributions.

The use of hyperbolic equations for grain size distributions is relatively new (Barndorff-Nielsen, 1977; Bagnold and Barndorff-Nielsen, 1980; Barndorff-Nielsen and others, 1982). For distributions that are reasonably hyperbolic, the advantages of this family of equations are that (1) the distributions can be characterized by four parameters, (2) unlike the moments of normal curves (mean, standard deviation, skewness, and kurtosis), the parameters can be used to generate the density function of the distribution, (3) the slope of each side of the distribution is quasi-independent of the slope of the other side, and (4) weight- or mass-percent data can be used in a statistically valid fashion.

CONCLUSIONS

Grain size distributions from longitudinal sequences of bedforms or portions of bedforms along the Rio Grande suggest that sorting is determined by sediment movements related to formation, migration, and cessation of movement of bedforms under waning flow.

In the sand-dominated floodway channel after waning flow ceased, the grain size of ripples varied with location along the stoss sides of dunes, being unimodal fine sand (mode 0.15 mm) in troughs and lower stoss sides of dunes and being bimodal (modes 0.15 and 0.30 mm) in ripples near the crest. The probable explanation is that the fine sand came out of suspension and formed small ripples in troughs as flow waned and the zone of mixing and reattachment shrunk downstream from the dunes. The small ripples mixed with previously formed ripples on the upper stoss sides of dunes to form bimodal ripples. Ripples that avalanched over the brink of the dunes formed slightly bimodal upper slipfaces. Lower slipfaces reflect incomplete resorting by avalanching. Toes of slipfaces were bimodal from mixing of brink-destroyed ripples and fine sand caught in reverse flow from the troughs.

In gravelly megaripples along the conveyance channel of the Rio Grande under waning flow, samples within a longitudinal sequence of bedforms are typified by certain grain size distributions and some of their derivative parameters. Each part of a megaripple (crest, interior, upper slipface, lower slipface, or trough) has a diagnostic grain size distribution. Along a reach of only 10 megaripples, the crests, lower slipfaces, and troughs of upstream megaripples are coarser than their downstream counterparts. Similarly, the mass percent of clasts larger than 2 mm and the size of d_{50} decrease downstream. In lower slipfaces, the mass of clasts larger than 0.6 mm decreases downstream.

Sorting mechanisms that produce different grain size distributions along gravelly megaripples are related to mobile-bed formation of armor along troughs, stoss sides and crests, and to flow separation, avalanching and reverse flow along slipfaces. Sorting of sandy slipfaces during waning flow apparently is dominated by reverse flow after the gravelly bedforms are no longer mobile.

The grain size distributions illustrated here show the importance of knowing the location of samples with respect to bedforms. The variation in grain size distributions along bedforms demonstrates that many fundamentally different distributions occur in fluvial deposits.

A majority of these sand-dominated grain size distributions may be modelled using log-hyperbolic distributions. The bimodal ripples can be modelled as mixtures of two hyperbolic distributions. For ripples, the location parameter μ changes from fine to coarse modes, but parameters ϕ, δ, and γ do not change appreciably. In upper foresets of dunes, on the other hand, μ does not change very much (slightly coarser downstream), but the other three parameters do change. If log-hyperbolic distributions are considered as infinite mixtures of log-normal distributions, two levels of mixing are represented by the distributions from the Rio Grande. The first-level mixture is related to the variance, while the second is related to the mean. Not all distributions can be modelled as log-hyperbolic distributions, however. Coarsest portions of distributions, in particular, are commonly irregular and may form either separate lines or even separate modes apart from the sand-size hyperbolic distributions.

These grain size data taken from bedforms developed under relatively simple waning flow suggest that it may be possible to relate grain size distributions to flow conditions more quantitatively. If quantification of flow conditions is possible using grain size distributions from similar bedforms, it also may be possible to determine what entire bedforms were like, or their relative longitudinal position within a progression of bedforms, based on the portions that are preserved in geologic sections. Alternatively, more data may demonstrate that several sets of flow conditions can produce similar distributions or that one set of flow conditions can produce dissimilar distributions. Nonetheless, the results of these studies may be useful in looking at fluvial deposits of suspected intermittent-ephemeral character.

ACKNOWLEDGMENTS

We thank John P. Borland of the U.S. Geological Survey for use of unpublished flow and sediment data. We are especially grateful to Garnett P. Williams and J. R. K. Daniel for reviewing the manuscript and suggesting many improvements. Jim Boyle, Ron Broadhead, and Mark Hemingway helped in the field and laboratory. We are grateful to Mickey Wooldridge, Cherie Pelletier, and Irean Rae for drafting the figures, and to Lynne McNeil for typing the manuscript.

REFERENCES

ALLEN, J. R. L., 1983, Gravel overpassing on humpback bars supplied with mixed sediment: Examples from the lower Old Red Sandstone, south Britain: Sedimentology, v. 30, p. 285–294.

ANDREWS, E. D., 1983, Entrainment of gravel from naturally sorted riverbed material: Geological Society of America Bulletin, v. 94, p. 1225–1231.

BAGNOLD, R. A., 1941, The physics of blown sand and desert dunes: Methuen, London, 265 p.

———, 1968, Deposition in the process of hydraulic transport: Sedimentology, v. 10, p. 45–56.

———, AND BARNDORFF-NIELSEN, O., 1980, The pattern of natural size distributions: Sedimentology, v. 27, p. 199–207.

BARNDORFF-NIELSEN, O., 1977, Exponentially decreasing distributions for the logarithm of particle size: Proceedings of the Royal Society of London Series A, v. 353, p. 401–419.

———, DALSGAARD, K., HALGREEN, C., KUHLMAN, H., MOLLER, J. T., AND SCHOU, G., 1982, Variation in particle size distribution over a small dune: Sedimentology, v. 29, p. 53–65.

BRIDGE, J. S., 1982, Reply: Bed shear stress over subaqueous dunes, and the transition to upper-stage plane beds: Sedimentology, v. 29, p. 744–747.

EVERTS, C. H., 1973, Particle overpassing on a flat granular boundary: American Society of Civil Engineers Proceedings, Journal of the Waterways, Harbors and Coastal Engineering Division, v. 99, no. WW4, Paper 10125, p. 425–438.

JOPLING, A. V., 1964, Laboratory study of sorting processes related to flow separation: Journal of Geophysical Research, v. 16, p. 3403–3418.

KUESTER, J. L., AND MIZE, J. H., 1973, Optimization Techniques with FORTRAN: McGraw-Hill, New York, 500 p.

LOVE, D. W., BOYLE, J. T., ROBINSON, S., AND HEMINGWAY, M., 1985, Location-dependent sediment sorting in gravelly megaripples from the Rio Grande, central New Mexico: New Mexico Geology, v. 7, p. 26–30.

LOVE, D. W., GUTJAHR, A., AND ROBINSON-COOK, S., 1986, Sediment sorting in three-dimensional ripple and dune fields in the Rio Grande, central New Mexico: New Mexico Bureau of Mines and Mineral Resources Open-file Report 240, 61 p.

PARKER, G., AND KLINGEMAN, P. C., 1982, On why gravel bed streams are paved: Water Resources Research, v. 18, p. 1409–1423.

REINECK, H. E., AND SINGH, I. B., 1980, Depositional sedimentary environments: Springer-Verlag, New York, 549 p.

SMITH, N. D., 1972, Some sedimentological aspects of planar cross-stratification in a sandy braided river: Journal of Sedimentary Petrology, v. 42, p. 624–634.

STEIDTMAN, J. R., 1982, Size-density sorting of sand-size spheres during deposition from bedload transport and implications concerning hydraulic equivalence: Sedimentology, v. 29, p. 877–883.

PART II
SEDIMENT DEPOSITION IN MODERN CHANNEL, FLOODPLAIN AND FLUVIAL-DELTAIC ENVIRONMENTS

A RECONNAISSANCE OF SEDIMENTATION ON THE KOSI ALLUVIAL FAN OF INDIA

NEIL A. WELLS
AND
JOHN A. DORR, JR.
Department of Geology, Kent State University, Kent, Ohio 44242;
and Department of Geological Sciences, The University of Michigan, Ann Arbor, Michigan 48109

ABSTRACT: Significant attributes of the Kosi fan are its flat surface, fine-grained, highly bioturbated sediments, perennial water saturation, and abundant abandoned channel facies. Pre-existing drainages controlled fan shape, and disparate monsoon and dry season discharges cause some channels to be vegetated and muddy while simultaneously containing sandy point bars. Similar fans should be recognizable by isopach-defined wedges, tectonic position, bowing of grain-size isograds away from the orogen, clustering of large channels, and interfingering with interfan facies.

INTRODUCTION

The Kosi fan has been cited as the archetypal humid alluvial fan (McGowen and Groat, 1971; Schumm, 1977; Kochel and Johnson, 1984) and at least 24 papers since 1974 have cited it in interpreting ancient deposits. Data are scarce, however, so it tends to be used too freely as an analog for any atypical large alluvial fan deposit. Because better facies descriptions are needed, a 1765-km, 9-day, dry season reconnaissance of the Kosi fan was made in December 1981. During this reconnaissance, pits and bank cuts were investigated wherever possible (at 41 sites), surface conditions were logged, and Landsat imagery was assessed using the 1:1,000,000 Operational Navigation Chart H9 map (India restricts detailed maps and aerial photographs).

Setting and dimensions of the fan.—

The Tamar, Arun and Sun Kosi Rivers drain the area around Mount Everest and join to form the Kosi River just before flowing onto the Gangetic plain (Figs. 1, 2). Rainfall in the 59,000 km^2 Kosi catchment basin is about 152 cm/year; 81 percent falls in the summer monsoon, but monsoon rainfall may be 50 percent above or below normal and individual storms can yield excessive rain (Imperial Gazetteer, 1908). Annual runoff is high, at 42.73 million acre-feet (52,700 km^2-m or 5.21 million hectare-m) and the annual sediment load is 96,000 acre-feet (11,700 hectare-m) (Gole and Chitale, 1966). River flow varies greatly and the area is notorious for floods. Rare discharges attain 913,000 cfs (25,850 m^3/sec) (Wood, 1976), 100 times regular low flow (Jones and Hofmann, 1967).

The Kosi River created its broad, lobate or teardrop-shaped fan by sweeping 50° from east to west; it once flowed into the Mahananda and flowed just east of Purnea as indicated by a 1731 survey (Fig. 3). It shifted 113 km in 228 years (0.5 km/yr), mostly in episodes of rapid to gradual opening and abandonment of channels. During one year there was a westward avulsion of 19 km (Wood, 1976).

Deposition begins as the Kosi spreads out beyond its gorge and its slope decreases from 7.08 m/km to 0.89 m/km at the fan head (Gole and Chitale, 1966). The fan is 154 km long and 117 to 147 km wide (depending on choice of boundaries), and it covers as much as 16,500 km^2. Downfan slopes average 0.78 m/km (4 ft/mile or 0.04°), but they decrease abruptly 25 km from Chatra (Fig. 2B). The center of the fan is mostly 3 to 5 m higher than the sides, with as much as 10 m difference in elevation at the north. The onfan slope of the modern bed of the Kosi River decreases from 0.89 m/km (4.7 ft/mile) to 0.06 m/km (0.30 ft/mile or 0.003°) (Gole and Chitale, 1966; Fig. 2E). In short, the slope of the fan is relatively gentle.

RESULTS

Facies distributions.—

Inglis (1967) described the Kosi gorge and fan head. Side valleys above the apparently similar Tista River fan head (Fig. 4A) are choked by slides, small fans, mudflows and streamflood conglomerates, showing tremendous sediment supply. Such deposits were reworked by the Tista and never reach the fan head.

Grain size decreases sharply away from the mountains. Gole and Chitale (1966) reported that "coarser pebbles and shingles" reach 15.5 km downstream from the Kosi fan head, "lighter shingle" essentially disappears by 51.5 km, and "coarse material" (0.2–0.6 mm) is practically absent by 125 km. The bulk of the fan is thus clay- to medium sand-size. The gravel zone, called the *bhaber* (bhabar) zone (Geddes, 1960; Pascoe, 1964), extends all along the Himalayan front. Widest on fans, this gravel zone is 8 to 24 km wide and 9.5–17 m/km in slope. The bhaber is a zone of braided rivers and infiltration of surface waters. Only the largest rivers flow across it at the height of the dry season. Downslope of the bhaber, beyond the spring line, is the *terai* (tarai), which is flatter (0.4 m/km), forested where uncut and marshy with meandering rivers. Here, even ostensibly abandoned parts of the Kosi fan have groundwater-fed rivers which carry off 125 to 180 cm of rain that falls on the fan each year. The plains are by no means affected solely by the large Himalayan rivers.

Figure 3 presents a new map of Kosi fan channels made from dry season, false color Landsat imagery with a camera lucida on a stereomicroscope. Several problems were encountered in making the map. First, the camera lucida produces minor (<10 percent) areal distortion toward the edge of the field of view. Corrections were made by centering the work area and drawing wide overlaps to force fits between map segments. Second, resolution varies with contrast. Bridges over rivers, canals, wet channels, and two lane roads on tree-lined embankments were usually visible, but small dry channels, two lane roads without trees, and

FIG. 1.—Late dry season Landsat image of the Kosi alluvial fan (19 March 1977).

railways were not. Narrow channels thus "disappeared" over some reaches. Third, interpretation depends on magnification. Connecting channel segments not apparent at low magnification were visible under high magnification, and some channels that seemed reasonably connected became disjunct. Nevertheless, the image could be examined directly in color under bright lights and moderate magnification, so the method was superior to either direct tracing or inking on black and white prints, which were affordable alternatives. The map and imagery show that the lower fan is characterized by many meandering channels and abandoned channel segments, whereas the top, the 50-km-long neck of the lobe, shows splayed belts of poorly developed and discontinuous braided channels. Disconnected channel reaches become common farther south and the many subsidiary channels and cutoffs demonstrate many minor shifts between major course changes.

The upper fan was visited at the nearby Kosi barrage on the Birpur-Saharsa road and along the Birpur-Basantpur-Purnea road. The barrage and associated canals and embankments changed sediment loads and discharge, creating marshes to the north and redistributing water to the south. As seen on satellite images, the modern Kosi here consists of a 5- to 7-km-wide belt of 1 to 4 active channels whose width is as much as 0.75 km, with wide vegetated islands and barren flood channel sand belts. Upfan, the islands are very small and compose only 10–20 percent of the braided channel belt, but downfan they increase to 3 by 25 km and 45 percent. Each channel is essentially a separate braided river with smaller scale channels and bars, except for the distinctly meandering small channel in the 165- to 185-km-wide reach (point 4, Fig. 3). Just south of the barrage are flat dry plains and broad marshes (some probably in imperceptible large channels) with reeds and water hyacinths (Fig. 4C, D). Soils include light colored, massive sands (presumably former broad sand bars), micaceous brown sands and clays, and olive-grey to black clays. Red-stained root traces and root mottling are evident locally.

The lower fan, traversed from Purnea to Saharsa, Dalkola, and Kursela, is characterized by meandering streams, abundant abandoned channels and oxbow lakes, and floodplains. The older, eastern, rainier area is slightly wetter, with more channels and oxbow lakes but half as many floodbasin marshes (1 channel per 1.5 km between Purnea and the Mahananda River vs. 1 per 5.3 km between Saharsa and Purnea). West of Saharsa, a short straight stretch of the modern Kosi River neither meanders nor braids, although the Kosi clearly meanders further downstream. The meanders are not very regular, but they have an average wavelength of just under 11 km with amplitudes of 5 km or less.

In order of abundance, the principal environments on the Kosi fan are dry floodplains, wet floodplains (mostly broad marshes of water hyacinth), abandoned channels, active meandering channels, and active braided channels. The overall environment differs little from adjacent parts of the Gangetic plain, so the following facies descriptions will include examples from on and around the fan.

FIG. 2.—Index map (stipples on inset map of India show Indo-Gangetic plain), contour map (A: after Geddes, 1960) and profiles of the Kosi fan (B, C, and D). E shows a longitudinal profile of the entire Arun + Kosi river system, based on the Army Map Service 1:1,000,000 1965 NG-45 and 46 sheets. Note the large vertical exaggeration. Abbreviations: Bh Bhagalpur, Bm Bhimnagar, Br Birpur, Bs Basantpur, Ch Chatra, Cp Chapra, Da Darbhanga, Dj Darjeeling, Dl Dalkhola, Fo Forbesganj, Go Gogri, Hn Hanumannagar, Jo Jogbani, Ka Katihar, Ku Kursela, *Ma* Mahananda River, Mo Monghyr, Mu Muzaffarpur, Ni Nirmali, Pa Patna, Pu Purnea, Sa Saharsa, Si Siliguri, Su Supaul.

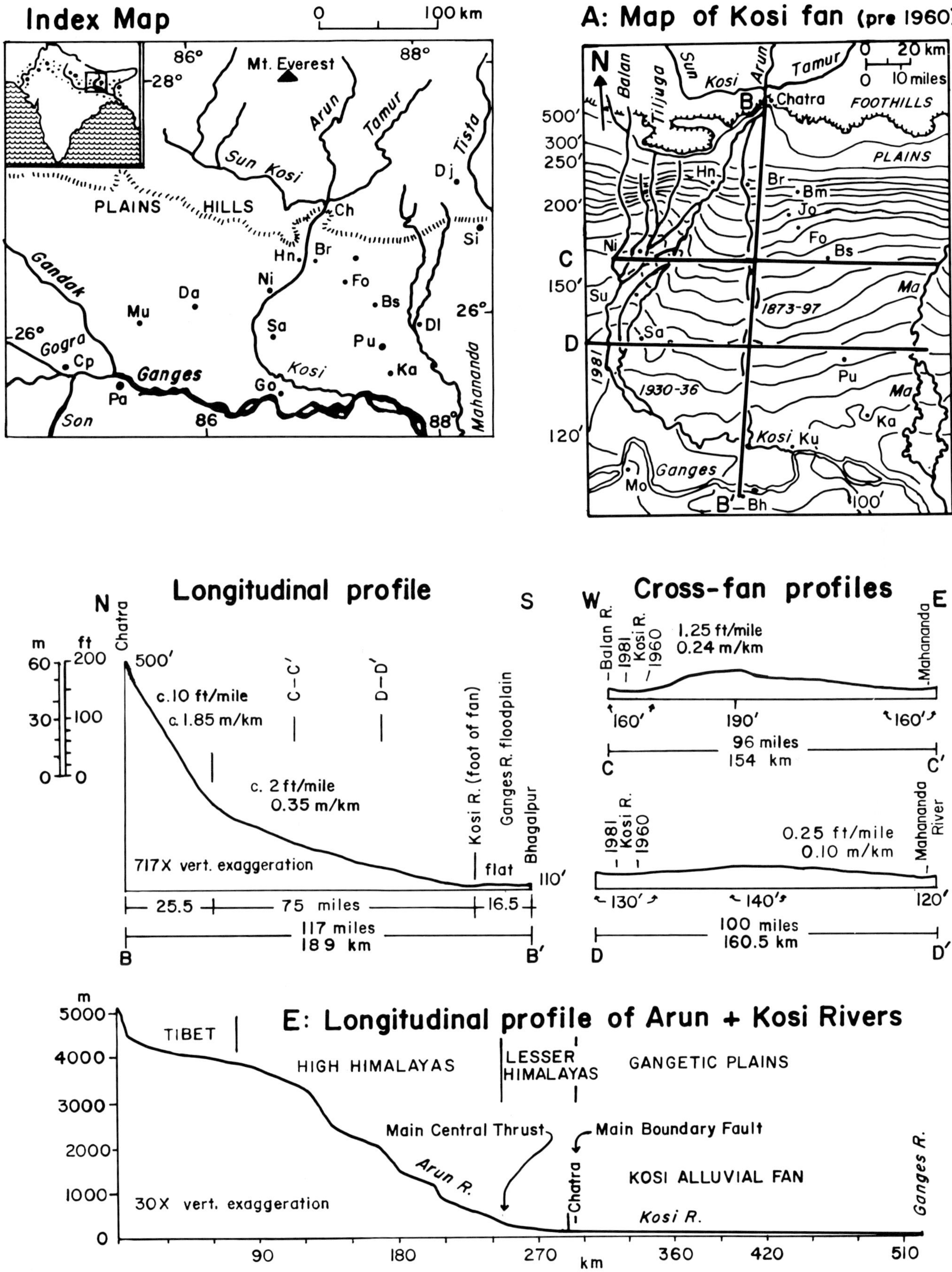

Index Map
0 100 km
86° Mt. Everest 88°
28°
Sun Kosi
Arun
Tamur
Tista
Dj
PLAINS HILLS
Ch
Si
Gandak
Hn. Br
Ni
Fo
Da
Mu
Bs
26°
Sa
Dl
26°
Gogra
Cp
Pu
Ganges
Kosi
Ka
Pa
Go
Son
86
88°
Mahananda
A: Map of Kosi fan (pre 1960)
N
0 20 km
0 10 miles
Balan
Tiljuga
Sun
Kosi
Arun
Tamur
B
Chatra
FOOTHILLS
500′
300′
250′
200′
150′
120′
PLAINS
100′
Hn
Br
Bm
Jo
Fo
Bs
Ma
C
Ni
C′
Su
1873-97
D
Sa
D′
1981
1930-36
Pu
Ka
Kosi
Ku
Ganges
Mo
B′
Bh
Longitudinal profile
N
S
Chatra
m ft
60 200
30 100
0 0
500′
c.10 ft/mile
c. 1.85 m/km
C–C′
D–D′
c. 2 ft/mile
0.35 m/km
Kosi R. (foot of fan)
Ganges R. floodplain
Bhagalpur
717X vert. exaggeration
flat
110′
25.5
75 miles
16.5
117 miles
189 km
B
B′
Cross-fan profiles
W
E
Balan R.
1981
Kosi R.
1960
1.25 ft/mile
0.24 m/km
Mahananda
160′
190′
160′
96 miles
154 km
C
C′
1981
Kosi R.
1960
0.25 ft/mile
0.10 m/km
Mahananda River
130′
140′
120′
100 miles
160.5 km
D
D′
E: Longitudinal profile of Arun + Kosi Rivers
m
5000
4000
3000
2000
1000
0
TIBET
HIGH HIMALAYAS
LESSER HIMALAYAS
GANGETIC PLAINS
Main Central Thrust
Main Boundary Fault
Arun R.
Chatra
KOSI ALLUVIAL FAN
Ganges R.
30X vert. exaggeration
Kosi R.
90
180
270
360
420
510
km

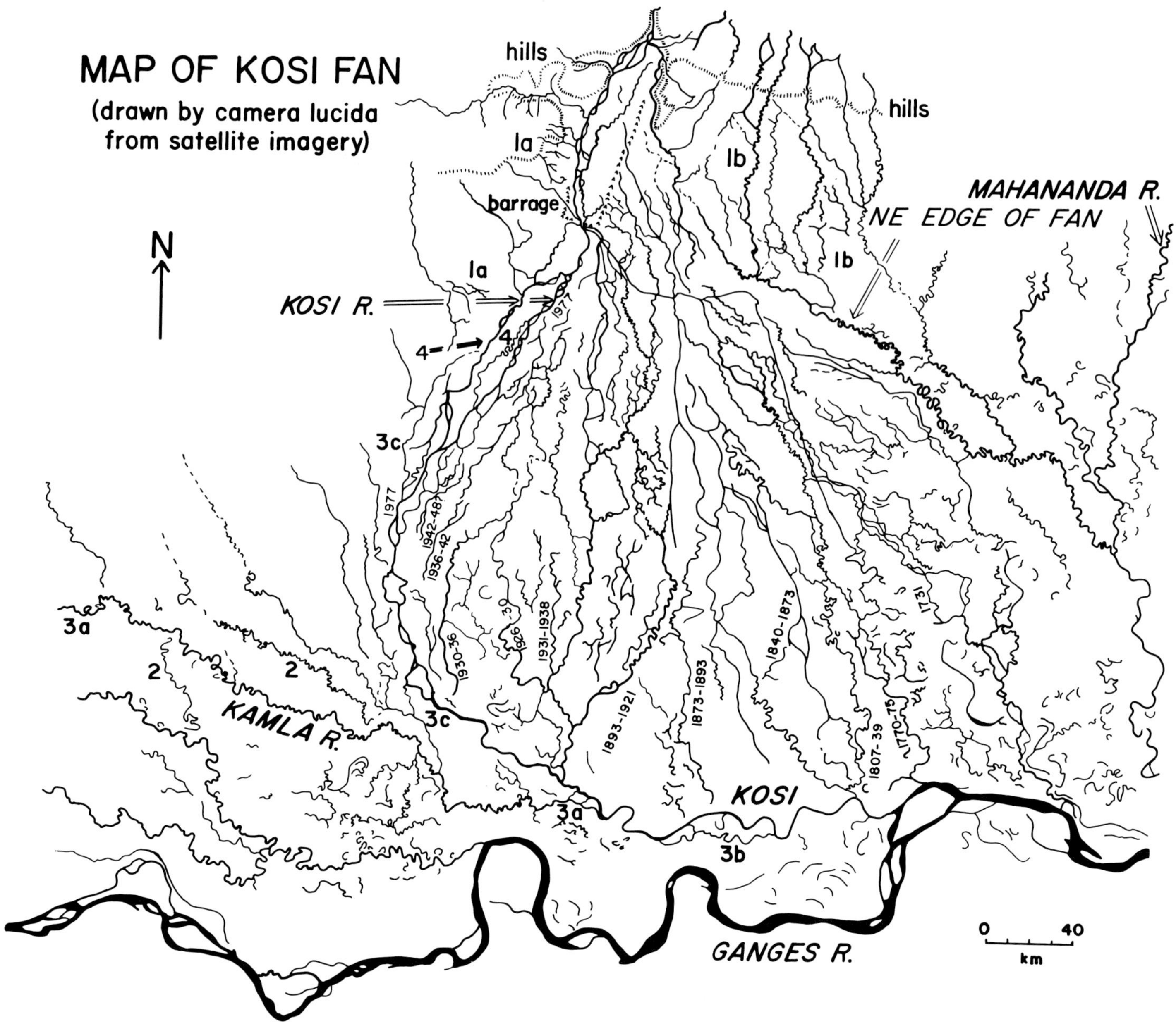

FIG. 3.—New map of Kosi drainage system, constructed from 19 March 1977 false color satellite image. Dates for occupation of channels are from Singh (1971, Fig. 5.16), although his 1931–1938 channel was perhaps active in 1921–1926. Numbered areas are as follows. 1A: Interfan area west of Kosi fan, where dry watercourses feed into small distributary fans and commonly fail to reach the Kosi River. 1B: Interfan east of the fan. Channels contain water near the foothills, but many rivers drain underground while crossing the bhaber gravel zone. 2: Long discontinuous sections of meandering channels show the mobility of drainage patterns on adjacent fans. 3A: Kamla (Kamala) River on Kosi-Gandak interfan. 3B: Former part of Kamla? (adjacent reaches overwhelmed by the Kosi). 3C: Possibly, the next major shift of the Kosi River will be into these channels. 4: Meandering secondary channel in the modern Kosi.

Channels and point bars.—

Except for the braided rivers of the upper fan, most streams are best characterized as meandering, although the distinction was not always easy to make on the ground. It was difficult to assess channel patterns and high water channel shapes, particularly of indistinct abandoned channels, while standing on a plain in the dry season without detailed maps or aerial photos. Most channels are not classically braided or meandering (Jackson, 1978; Schumm, 1981) but are irregularly sinuous with somewhat irregular side bars and some mid-channel bars. The abundant abandoned channel segments range from oxbow lakes, developed best on the lower fan, to elongate and somewhat sinuous water hyacinth bogs, representing nearly filled or drained segments of irregular channels (Figs. 4E, F). Almost all the abandoned channels on the fan (except the 1731 channel in Fig. 4B) are too small to hold the entire Kosi River.

FIG. 4.—A) Head of Tista fan, next fan east of the Kosi fan (foothills in distance, entrenched old fan head surface at mid-distance). B) Large abandoned 1731 channel, locally called the Mara (Dead) Kosi, south of Basantpur. C) View east onto Kosi fan from artificial levees across the Kosi River, near Nirmali below barrage. Note that the upper fan is unlike a typical, steep, mudflow-dominated, semiarid alluvial fan. D) The upper fan due east of C, on Birpur-Supaul road. The fan is mostly flat and muddy, and varies from well drained to open ponds. E and F) Typical mid-fan and lower-fan abandoned channels along the Purnea-Saharsa road. Note that these channels are too small to have held the entire Kosi.

The Burhi Gandak and the Kamla (or Kamala) have relatively wide and deep (>50 m by >4 m), complex, bilevel channels with an upper set of unvegetated to overgrown point bars, chutes, etc. and a lower, inner, unvegetated set. Onfan, the Kilaba (Tilabeh) and the "Old Kosi" 30 km west of Purnea have tiny inner "levees" and low areas. The two levels likely relate to disparate monsoonal and dry season discharges; Gohain and Parkash (1985a) reported well developed two-level channels, each with a fining-upward sequence related to peak, sheet, and post-sheet flows, in the modern Kosi River.

Typical point bars in such rivers include channel floor bars, dunes and ripples, and low unvegetated channel edge bars with cross-top flow or bank side chutes. Above this, the main body of the point bar commonly consists of poorly developed, lateral-accretion (epsilon) sets that dip at 7° to 16° and contain ripples, plane-bedded sand, and/or irregular mud drapes. Accumulation was noted on tops and downstream, upstream, bank, and river sides of bars. Crossbar channels and chutes may be filled with rippled or laminated sands and muds by either lateral or vertical accretion. Vegetation is variable. Boils during floods may have formed parabolic, 3- to 5-m-long, >1-m-wide, 30- to 60-cm-deep, out-of-channel scours that widen and shallow downstream and have upstream-dipping drapes of sand and mudcracked clay. Some demonstrably active onfan point bars are particularly noteworthy because their deposits are coarser than those of the adjacent channel floor (Fig. 5A–D). The channel floors are overgrown by subaquatic plants that decrease flow velocity and trap fine sediment. The point bars are presumably active only at high flow, when rivers deposit sand on the point bar, which intercalates downslope with low-stage channel mud (Fig. 5B). In short, mud drapes and zones disturbed by roots can occur anywhere; both lateral and vertical aggradation can occur; and lateral accretion surfaces dip in any direction. Most deposits are composed of <30-cm-thick units of sand and/or mud, and most sequences fine upward.

Point bars on shallower, sandier, and less constrained interfan tributaries of the Mahananda River are little more than emergent channel floor bars that happen to be attached to the bank, although they do occur as alternating side bars opposite small cut banks along a sinuous river course. These bars are thin and consist of stacked ripple packages and low bar foresets. Channel floor bedforms are 10–40 cm high, overlapping, flat, broad, lobate bars made of sheets of ripples and local small megaripples in 10–50 cm of water. Avalanche faces diverge 30° to 70°. At slightly lower water levels, exposed mid-stream bars undoubtedly cause braiding.

Overall, bars accumulate sediment on any or all sides in packages of ripples, small foresets, and laminated sediment. Viewing restrictions gave a strong bias toward tops of fluvial cycles, but ripples are abundant, even in large sand bodies low in large rivers. Gohain and Parkash (1985b) reported common reworking of large-scale crossbeds in the modern Kosi River.

Floodplains and sediment disturbance.—

Floodplains (floodbasins) make up most of the Kosi fan, much of which is very wet in the monsoon season. Inglis (1967) noted that the Kosi River can be 15 to 30 km wide in the wet season and after snowmelt. Road and railway embankments are built with many dry bridges that are far from stream beds in order to drain widespread flood waters. The dry season water table is shallow (mostly <1 m), and villagers tell of huge floods and high wet season water levels. Fan floodplains include permanently wet areas and more widespread drier areas. Wet areas are marked by permanent open water, reed marshes, and water hyacinth bogs. Their sediments are fine-grained, variable in organic content, olive-grey to black, and imperfectly layered to massive, depending on oxygen concentration, degree of bioturbation, permanence and depth of water, rate of flow, and biotic productivity.

Drier areas show interlayered sands and muds, whose ultimate appearance is controlled by frequency and magnitudes of additions of sediment and degree and kind of bioturbation (Figs. 5E, F). Clays probably settle from stagnant floodwaters, whereas sands, where preserved, are mostly plane-bedded or rippled, suggesting active flood currents. A few sparsely grassed damp areas in low floodplains and abandoned channels had a surficial algal mat but lacked cryptalgal laminations. Local areas of deeply and completely bioturbated sand, as around Saharsa, could be a thoroughly bioturbated but recent Kosi bar or a bypassed remnant of a pre-1900 alluvial surface. Very dry areas elsewhere on the Gangetic plain develop powdery white efflorescences called *reh,* and old surfaces develop calcrete called *kankar* (Mithal and others, 1973; Spate and others, 1967), but neither of these were seen in recent Kosi sediments. Well drained, off-fan Pleistocene soils (under tea plantations on bhaber terraces) show the ultimate result of exposure: deep leaching, reddening and complete homogenization.

Deposits on the Kosi fan are extensively and rapidly bioturbated (Fig. 6). Worms create a thoroughly churned pelletal soil. Termites and their labyrinthine and somewhat flat bottomed tunnels were observed in two sections, but termite mounds were absent. A nonmarine crab and its dry, round, gently sloping burrow were encountered 45 cm above the water level. Trampling by large vertebrates was evident in many wet areas. The large vertebrates today are water buffaloes and cows, but elephants, rhinoceroses, bovids, deer, and other large mammals were abundant there since the Miocene, so trampling has long been important.

Deep mudcracks also destroy depositional fabrics. Low areas along the Ganges have 10-cm-wide and >60-cm- (probably >1.2 m) deep mudcracks arranged in 60- to 100-cm-wide polygons. A side bar of fine sand in the lower Kosi was cut into rectangles by long, deep, narrow and widely spaced cracks aligned parallel and perpendicular to bar sides. Figure 6D shows floodplain sediments above the Bagmati River with 30 cm of grass-covered hard grey clay with snails and >1 m of mottled clay cut by vertical crack-and-fill zones spaced <1 m apart. The cracks evidently open, collect some snails and darker clay, close when rewetted, and later reopen in not quite coincident positions. This soil is therefore capable of self-mulching, although it does not show the curvilinear slickensides typical of vertisols. Gleying is common where the water table is high but fluctuating.

FIG. 5.—Point bar and floodplain deposits. A) Meander with cut bank, point bar, and cutoff channel on Gomati River, east of Saharsa. Arrows mark sections shown in B and C. B) Section through channel edge of point bar (channel to left). Note grass on top of point bar and interfingering wedges of mud from channel and sand from point bar. C) Section along back of point bar in wall of cutoff channel. Body of rippled sands, fining and thinning upward to vegetated top of point bar, flow to right. Trough crossbeds occur under clay cap on floor of cutoff. D) Sursar River, between Birpur and Forbesganj, which is similar to A but has a higher water level. Note grass and sand on point bars and mud and vegetation in channels, and also meandering thalweg (although extensive cutoffs almost create braiding, making this channel intermediate between types 3 and 9 of Schumm, 1981, Fig. 6). E) Distal Kilaba floodplain east of Saharsa: fine sand layers (marked by lines) in silty clay. Note plowed area at top and increased homogenization of sands with age. F) Complex depositional history of floodplain near Nirmali: 1) deposition and bioturbation of clay *a*, 2) deposition of sand *b* as crevasse-splay, 3) lobes of sand layer sink into clay *a*, from trampling or quakes, 4) shallow bioturbation destroys lamination in sand *b* and makes base diffuse, but cannot reach disrupted laminae in lobes, and 5) thick mud *c* deposited after erosion, at end of next flood, but base not affected by subsequent bioturbation.

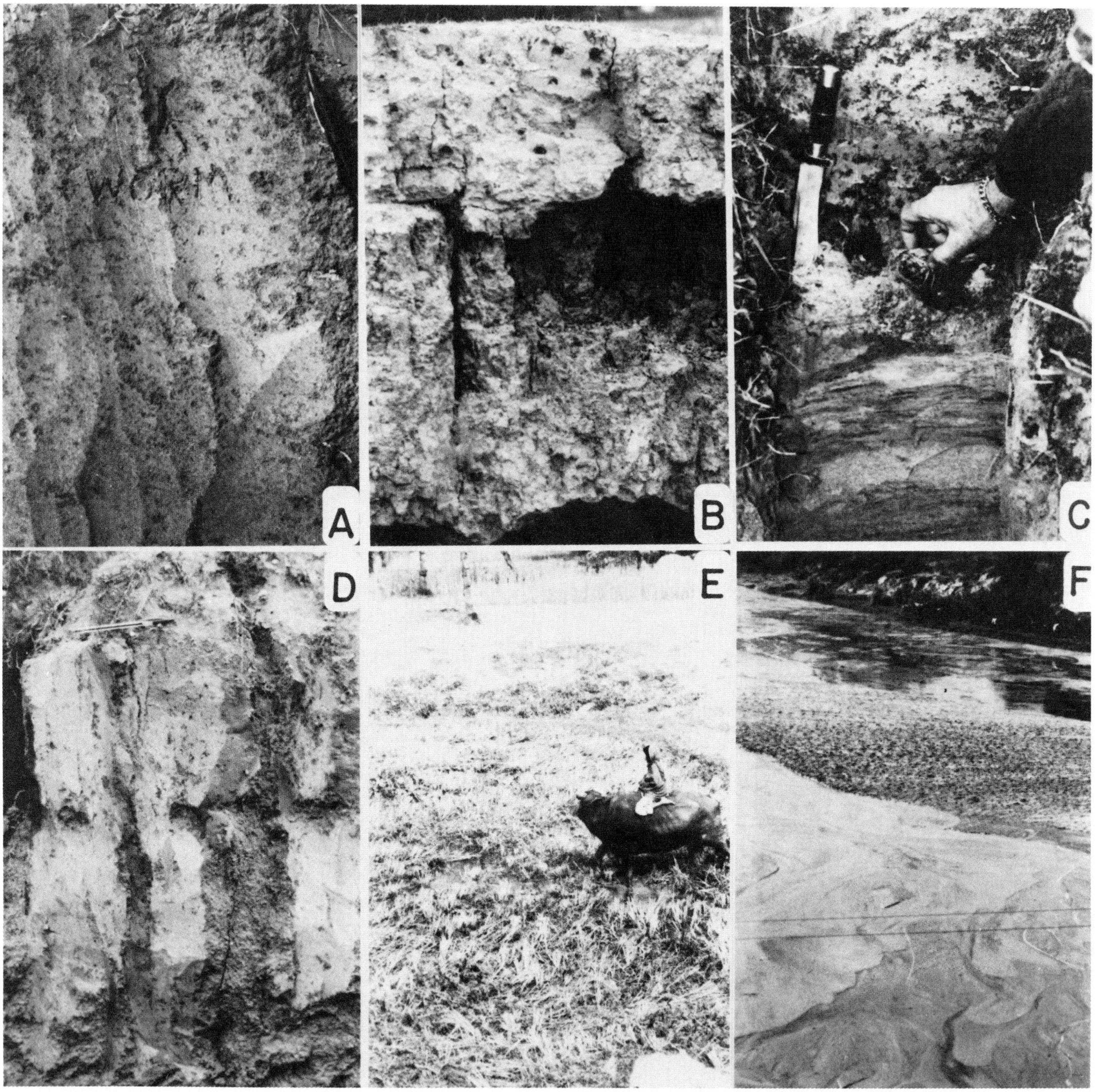

FIG. 6.—Disruption of sediments. A) Burrowing by worms causes mottling and pelletization; NW of Kishanganj. B) Flat-bottomed termite burrows in a low broad mound, Romghat River, west of Kursela. C) Nonmarine crab and its burrow, north of Kursela. D) Deep mudcracks showing multiple generations of opening, partial infilling, and reclosing, and resultant self-mulching tendency, Bagmati River floodplain, east of Muzaffarpur. E) Water buffalo tramping in a mire, near Dharbanga. F) Emergent sand bar at cattle crossing remains trampled, but only slightly lower submergent surfaces are reworked. F is from Madagascar, but illustrates destruction of river banks and bar tops by cattle, as observed on Kosi fan. This could be preserved as an isolated lens of massive sand in a rippled sand body.

Earthquake-generated features (sand volcanoes, pipe structures, clastic dikes) were not identified but are surely present. Seventy-four destructive quakes have occurred under or just north of the plains in the last 200–250 years (Wadia, 1975). Wadia (p. 38) describes results of the 1897 Assam quake: "Wide, gaping earth-fissures opened out in all directions in the alluvial plains, from which issued innumerable jets of water and sand, like fountains, spouting up to 1 or 1–3 m in the air. Beds of rivers, tanks, and even wells were ridged up, or filled, by the outpouring sand, thus greatly disturbing the drainage system of the land and causing extensive flooding." The 1934 Bihar quake, with one of its epicenters at Monghyr just down the Ganges from the mouth of the Kosi, caused 28,500 km^2 of the Ganges basin to be "riddled by fissures and sand-vents which ejected large volumes of water and sand, flooding the cultivated country" (Wadia, 1975, p. 40). The Soil Survey of East Pakistan (Bangladesh) has mapped numerous large sand dikes and blowouts that apparently formed during two major quakes (Coleman, 1969).

DISCUSSION

Is the Kosi an alluvial fan, since it lacks mudflows, sheetflood conglomerates, a steep fan head, and an arid climate? It meets Blissenbach's 1954 (p. 176) definition, "An alluvial fan is a body of detrital sediments, built up by a mountain stream at the base of a mountain front." It seems valid to consider the Kosi fan as a huge, flat and wet end member in a spectrum of alluvial fans, although it is clearly also allied with alluvial plains. Alternative names include alluvial cone, megafan and fan plain. The term wet alluvial fan is useful to emphasize its perennial rivers, in contrast with arid fans, although the climate is better described as monsoonal rather than humid.

Despite low fan head and toe elevations of 150 and 35 m, depths to the base of the Quaternary deposits are 878 and 70 m, respectively (Niyogi, 1981), which indicate subsidence near the orogen as in front of other overthrusts. It should be noted, however, that the lobate shape of the fan is partly fortuitous. Old maps show predetermination of the western flank and toe by early south-flowing west-bowed rivers into which the Kosi has moved, whereas the fan's eastern side has not been encroached upon by major rivers or converted into interfan. The western edge of the fan is thinly superimposed on the Balan-Tiljuga interfan deposits. The latter are probably in turn superimposed on an earlier Kosi fan, and so on for thousands of meters; nevertheless, the latest, westernmost, mid-fan veneer is only 2 m thick, as calculated by Gole and Chitale (1966).

Kosi-like fans must be important in molasse. Most major Himalayan rivers deposit large but less perfectly shaped fans (Geddes, 1960) and limited data show major south-trending clusters of channel sands in the subsurface. Fans have been fixed since the mid-Tertiary, except for progradation of thrusting, because the courses of most trans-Himalayan rivers, like the Arun and Sun Kosi, predate the mountains (Pascoe, 1964; Wadia, 1975). Similar deposits have been identified in the Morrison, Gannett and Wayout Formations of Wyoming and Idaho (Derman and others, 1984; Dorr, 1985) and could also occur in Alpine and Appalachian (Catskill) molasse. Recognition should allow for vagaries of shape, broad interfingering of adjacent systems, and sedimentologic similarities between fans and interfans. Identification will depend on recognition of isopach-defined wedges, clusters of large channels, and outward bowing of grain-size dispersal plots away from the orogen. A final important feature is that the Kosi and other actively migrating Himalayan rivers usually move into pre-existing channels of less active streams, thus possibly leading to multistory sandstones showing sudden changes in discharge and perhaps provenance but not flow direction (Wells and Dorr, in prep.).

Channels are transformed by rapid abandonment which creates elongate wet grasslands, water hyacinth bogs, or oxbow lakes, depending on degree of meandering, drainage, height of water table, and stage of infilling. Gradual abandonment, by slow diversion into other distributaries, can cause anastomosing and underfitting (e.g., the shrunken modern Bagmati in its overgrown channel). Many channels that lose Himalayan runoff remain active, though diminished, because they receive local drainage from subbhaber springs or runoff from the 130–180 cm of rain that falls yearly on the plains, or more recently by use as uncontrolled canals. In short, extensive reworking is possible; large channels may be reformed into lakes, marshes or smaller meander belts.

The distance from the fan head to the point of transition from braided to meandering channels should be sensitive to tectonism and environmental change. The modern Kosi starts to meander at 147 km from the fan head, but progressively older major channels (now sub-Kosi-size) meander at 117, 121, 124, and 118 km from the fan head. Either the transition zone has been stable until recently, or low discharge anastomosing subchannels and post-abandonment rivers begin meandering at a distance of about 120 km. Lesser channels are more variable. Recent changes include reduced discharge by removal of water for irrigation and increased sediment loads due to deforestation in Nepal (Gole and Chitale, 1966; Inglis, 1967). Both might extend braiding downstream, although a decrease in the variability of discharge might promote meandering. Completion of the Kosi barrage has complicated the assessment of sediment loads since 1963 (Inglis, 1967).

The Kosi area has undergone important changes at the hands of man. Besides the barrage and its 9365 km^2 canal system (Biggs, 1976), old river courses have been used as uncontrolled canals. Extensive deforestation has occurred on the plains as well as in the mountains. Records from Alexander the Great, Mahmud of Ghazni, the Moghul emperors, and early British traders show that as recently as 200–300 years ago, the Indo-Gangetic plains were extensively forested and inhabited by elephants, buffaloes, rhinoceroses, and tigers (Imperial Gazetteer, 1908; Spate and others, 1967; Fairley, 1975). Deforestation from overpopulation has been an increasing problem in Nepal and has long been recognized as the cause of increased erosion and sediment loads (Inglis, 1967). Long, large earthen embankments have been built for roads, railways, and flood control. Paddy-field agriculture and removal of clay for bricks are extensive practices that go hand in hand in significantly flattening and lowering the landscape. The population, however, has not always been large and expanding. In the 1700's, when records were kept but before it was possible to alleviate natural disasters, mortality rates from diseases, floods, and droughts locally reached nearly 50 percent (Wood, 1976), and it was not until World War II that the population of the Purnea District exceeded approximately 165 people per km^2 (Wood and Biggs, 1976).

Lastly, there are larger aspects to the Kosi story that are

not discussed here. The progressive westward shifting of the Kosi River and climatic and tectonic controls will be discussed elsewhere. More significantly, the 3500-km-long, 100- to 250-km-wide, 5000- to 10,000-m-thick Indo-Gangetic molasse, of which the Kosi fan is a small part, provides an exceptional and underutilized opportunity to study controls on sedimentation. Paleomagnetic work has yielded remarkable results in linking local tectonics and sedimentation (e.g., Visser and Johnson, 1978; Burbank and Reynolds, 1984). The molasse belt encompasses very wet and dry areas (rainfall varies east-west from 1080 to 9 cm/yr). Geology of source areas is reasonably consistent for considerable distances along strike, and debouchements of major rivers have been fixed during orogenesis. The desert-traversing and tributaryless lower Indus, the huge and unstable Brahmaputra, and the traversely draining and craton-hugging Ganges provide diverse fluvial styles. Historical records could provide excellent data but need much more attention. More subsurface information is needed.

CONCLUSIONS

Reconnaissance of the Kosi fan, though limited to the surface in the dry season, nevertheless improves the fan as a model for wet alluvial fans. It is large, flat, wet, relatively fine-grained, and bioturbated. Abandoned channels abound. Probably, rapid abandonment creates oxbows and water hyacinth bogs, depending on rates of drainage and infilling, whereas slow abandonment allows underfitting or shrinkage of channels. Important aspects include a wet and dry monsoonal climate and greatly fluctuating discharge; the partly braided, partly meandering, partly anastomosed character of many streams; and the broad interfingering of fans and interfans. The fan's identifying characteristics include its wedge shape, tectonic position, clustering of channels, and relatively distant transport of sediment. Basic vertical successions represent braided streams with complex laterally interwoven channels and bars in the north and meandering channels, point bars, and homogenized floodplain sediments in the south.

ACKNOWLEDGMENTS

We thank Mr. Arijeet Sengupta for help in the field and Champlin Oil Company for principal funding. We also thank the Smithsonian Institution Foreign Currency Program for travel funds for NAW (via a grant to Dr. Philip D. Gingerich of The University of Michigan); the Scott Turner Fund of the University of Michigan's Geological Sciences Department, for satellite imagery; and Mr. John Potter for image processing. Mr. Ram Nad drove us safely through the region. Martin Dawson and an anonymous reviewer made many helpful comments.

REFERENCES

BIGGS, S. D., 1976, The Kosi area development project, Chapter 8, *in* Joy, J. L., and Everitt, E., eds., The Kosi Symposium: Institute of Development Studies, p. 151–158.

BLISSENBACH, ERICH, 1954, Geology of alluvial fans in semiarid mountains: Geological Society of America Bulletin, v. 65, p. 175–190.

BURBANK, D. W., AND REYNOLDS, R. G. H., 1984, Sequential late Cenozoic structural disruption of the northern Himalayan foredeep: Nature, v. 311, p. 114–118.

COLEMAN, J. M., 1969, Brahmaputra River: Channel processes and sedimentation: Sedimentary Geology, v. 3, p. 129–239.

DERMAN, A. S., WILKINSON, B. H., AND DORR, J. A., JR., 1984, Jurassic-Cretaceous nonmarine foreland basin sedimentation in western United States: American Association of Petroleum Geologists Bulletin, v. 68, p. 470–471.

DORR, J. A., JR., 1985, Newfound Early Cretaceous dinosaurs and other fossils in southeastern Idaho and westernmost Wyoming: Contributions from the Museum of Paleontology of the University of Michigan, v. 27, p. 73–85.

FAIRLEY, JEAN, 1975, The Lion River—the Indus: Lahore, Pakistan, Gillani, 290 p.

GEDDES, ARTHUR, 1960, The alluvial morphology of the Ingo-Gangetic plain: Its mapping and geographical significance: Institute of British Geographers, Transactions and Papers, Publication 28, p. 253–276.

GOHAIN, KRISHNAN, AND PARKASH, B., 1985a, Effects of two distinct discharges on the Kosi river bed and its fan, north Bihar, India and Nepal: Third International Fluvial Sedimentology Conference, Abstracts Volume, p. 19–20.

———, AND ———, 1985b, Lithofacies of the Kosi river bed, north Bihar, India and Nepal: Third International Fluvial Sedimentology Conference, Abstracts Volume, p. 20.

GOLE, C. V., AND CHITALE, S. V., 1966, Inland delta building activity of the Kosi River: Proceedings of the American Society of Civil Engineers, Journal of the Hydraulics Division, v. 92, p. 111–126.

IMPERIAL GAZETTEER, 1908, The Imperial Gazetteer of India: Clarendon, Oxford.

INGLIS, C. C., 1967, Inland delta building activity of Kosi River (discussion): Proceedings of the American Society of Civil Engineers, Journal of the Hydraulics Division, v. 93, p. 93–100.

JACKSON, R. G., 1978, Preliminary evaluation of lithofacies models for meandering alluvial streams, *in* Miall, A. D., ed., Fluvial Sedimentology: Canadian Society of Petroleum Geologists Memoir No. 5, p. 543–576.

JONES, P. H., AND HOFMANN, W., 1967, Water resources investigation program for upper Gangetic plain, India: U.S. Geological Survey Open-File Report GPO 974-423, 111 p.

KOCHEL, R. C., AND JOHNSON, R. A., 1984, Geomorphology and sedimentology of humid temperate alluvial fans, central Virginia, *in* Koster, E. H., and Steel, R. J., eds., Sedimentology of Gravels and Conglomerates: Canadian Society of Petroleum Engineers Memoir No. 10, p. 109–122.

MCGOWEN, J. H., AND GROAT, C. G., 1971, Van Horn Sandstone, West Texas: An alluvial fan model for mineral exploration: Bureau of Economic Geology, University of Texas at Austin, Report of Investigations No. 72, 57 p.

MITHAL, R. S., SINGHAL, B. B. S., AND BAJPAI, I. P., 1973, Groundwater conditions in the Gangetic alluvium of western Uttar Pradesh: International Symposium on Development of Groundwater Resources, Madras, India, Nov. 26–29, 1973, Proceedings, v. 3, p. 53–61.

NIYOGI, D., 1981, Geomorphology and morphotectonics of the Neogene-Quaternary sediments in India, *in* Sastry, M. V. A., Kurien, T. K., Dutta, A. K., and Biswas, S., eds., Neogene-Quaternary Boundary Field Conference, India, 1979, Proceedings: Geological Survey of India, Calcutta, p. 131–142.

PASCOE, E. H., 1964, A manual of the geology of India and Burma, v. 3: Geological Survey of India, Delhi, 2030 p.

SCHUMM, S. A., 1977, The Fluvial System: Wiley, New York, 338 p.

———, 1981, Evolution and response of the fluvial system, sedimentological implications, *in* Ethridge, F. G., and Flores, R. M., eds., Recent and Ancient Nonmarine Depositional Environments: Models for Exploration: Society of Economic Paleontologists and Mineralogists, Special Publication 31, p. 19–29.

SINGH, R. L., ed., 1971, India—a regional geography: National Geographical Society of India, Varanasi, India, 992 p.

SPATE, O. H. K., LEARMONTH, A. T. A., AND LEARMONTH, A. M., 1967, India and Pakistan, A General and Regional Geography: Methuen, London, 877 p.

VISSER, C. F., AND JOHNSON, G. D., 1978, Tectonic control of Late Pliocene molasse sedimentation in a portion of the Jhelum re-entrant, Pakistan: Geologische Rundschau, v. 67, p. 15–37.

WADIA, D. N., 1975, Geology of India: Tata McGraw-Hill, New Delhi, 508 p.

WOOD, G., 1976, The legacy of the past: The agrarian structure, Chapter 5, *in* Joy, J. L., and Everitt, E., eds., The Kosi Symposium: Institute of Development Studies, p. 93–112.

———, AND BIGGS, S. D., 1976, The growing pressure of population, Chapter 3, *in* Joy, J. L., and Everitt, E., eds., The Kosi Symposium: Institute of Development Studies, p. 63–76.

BRAHMAPUTRA RIVER: CHANNEL MIGRATION AND DEPOSITION

CHARLES S. BRISTOW
Department of Earth Sciences, University of Leeds, Leeds LS2 9JT, U.K.

ABSTRACT: This study follows the pioneering work of Coleman (1969) and presents the results of a study of a 200-km-long reach of the Brahmaputra River in Bangladesh. Channel pattern and migration has been monitored using LANDSAT imagery and historic maps. Channel pattern in the Brahmaputra is varied. The river is mostly braided but some reaches are anastomosed or meandering. Channel movement is dominated by lateral migration with some minor channel switching and one major avulsion in the last 200 years. There is a hierarchy of channels in the Brahmaputra. First-order channels encompass the whole river and may comprise several second-order channels which, in turn, have third-order channels within them. The channels divide and rejoin around bars which scale with the bankfull width and depth of each channel. Bar types include lateral (point) bars, diagonal bars, medial bars and tributary bars. Bars sometimes coalesce to form semi-permanent bar assemblages. Channel cross sections observed by echo-sounding show a high degree of channel asymmetry and extremely complex cross sections which can be related to bar development. The pattern of deposition is divided into four styles: new mid-channel bars (13 percent), channel abandonment (15 percent), lateral accretion to bank (19 percent) and addition to bars (53 percent). The latter is the most important and is subdivided into three elements of deposition: upstream accretion (14 percent) downstream accretion (29 percent) and flank accretion (57 percent).

INTRODUCTION AND PREVIOUS WORK

Current models for sand-bed braided river deposits depend heavily upon studies of relatively small rivers (Smith, 1970; Williams and Rust, 1969; Collinson, 1970; Cant and Walker, 1978; Blodgett and Stanley, 1980), yet many ancient fluvial sand bodies were deposited by braided rivers of probably much larger magnitude (Campbell, 1976; Conaghan and Jones, 1975; Kirk, 1983). The earliest study of the Brahmaputra was made by Rennell (1781) who first described the growth and formation of bars around a point of nucleation. A more detailed study was made by Coleman (1969) which revealed very large bedforms with rapid rates of migration and high rates of channel movement. His studies of bedform patterns and migration revealed a recurring sequence of events during single flood cycles. The most significant modifications occur during falling stage when sediment is deposited within the channels, accompanied by rapid changes in local flow directions and migration of the thalweg. Coleman also used aerial photographs and maps to show braid bar migration in response to channel movements and large-scale river migration (from geomorphic mapping). Coleman studied paleocurrents from crossbedding measurements on exposed bars and described some of the internal structures. Other work on the Brahmaputra includes engineering studies (Latif, 1969; J.I.C.A., 1976) and a study of suspended sediment transport and basin aggradation (Goswami, 1983).

This study provides additional information on rates of channel migration and deposition and their patterns to gain a more comprehensive understanding of the deposits of a large sand-bed braided river. Channel movements have been recorded from historical maps and LANDSAT images. These reveal large and smaller scale patterns of channel migration and deposition. Field work has included some studies of deposits at low flow stage and the collection of 400 km of echo-sounder traces over a 2-month period during and immediately following the major 1984 flood. This paper presents the conclusions of this first phase, viz. investigation of the patterns of channel morphology and migration. Future reports will deal in more detail with bedform distribution, flow dynamics and floodplain sedimentation.

RIVER HYDROLOGY

The Brahmaputra River is 2840 km long and drains an area of some 380,000 km^2 from Tibet to the Bay of Bengal. The section of the river described in this paper is a 200-km-long reach in Bangladesh immediately upstream from the confluence with the River Ganges (Fig. 1). This is similar to the area studied by Coleman (1969). The study reach is a multichannel system with a braided appearance in plan view. The maximum discharge recorded in the river is 76,500 m^3/sec in 1970, and the maximum discharge recorded during this study was 75,700 m^3/sec (18/9/84) (data from Bangladesh Water Development Board gauging station at Bahadrabad, Fig. 2). This discharge is in excess of the predicted 10-year flood and gave rise to widespread flooding and overbank deposition. The minimum discharge in 1984 was 3,950 m^3/sec (20/2/84). The large variation in discharge results from the seasonal monsoon rains.

During the present field study, the maximum measured surface velocity in the Brahmaputra was 3 ms^{-1}. Mean flow depth was 5 m and the maximum recorded depth was 35 m. The width of the river varies from 3 km to 18 km but the average width is about 10 km. Width/depth ratios for individual channels vary from 50:1 to 500:1. The gradient of the Brahmaputra in Bangladesh is 0.000077 decreasing to 0.00005 near the confluence with the Ganges.

GRAIN SIZE AND SEDIMENT TRANSPORT

The Brahmaputra River has a high suspended load of up to 4,544 ppm (Latif, 1969). The suspended load is dominantly silt with some very fine sand (Goswami, 1983). Total suspended load discharge has been estimated at 725 million tons per annum (Latif, 1969), which is close to the maximum 696 millions tons reported by Coleman (1969) and somewhat more than his calculated mean annual suspended sediment discharge of 478 million tons. It has not been possible to measure accurately total sediment transport in the Brahmaputra River, but some estimates have been made based on measurements of suspended sediment discharge. Goswami (1983) used a series of different formulas to calculate bedload and total load discharge. Goswami

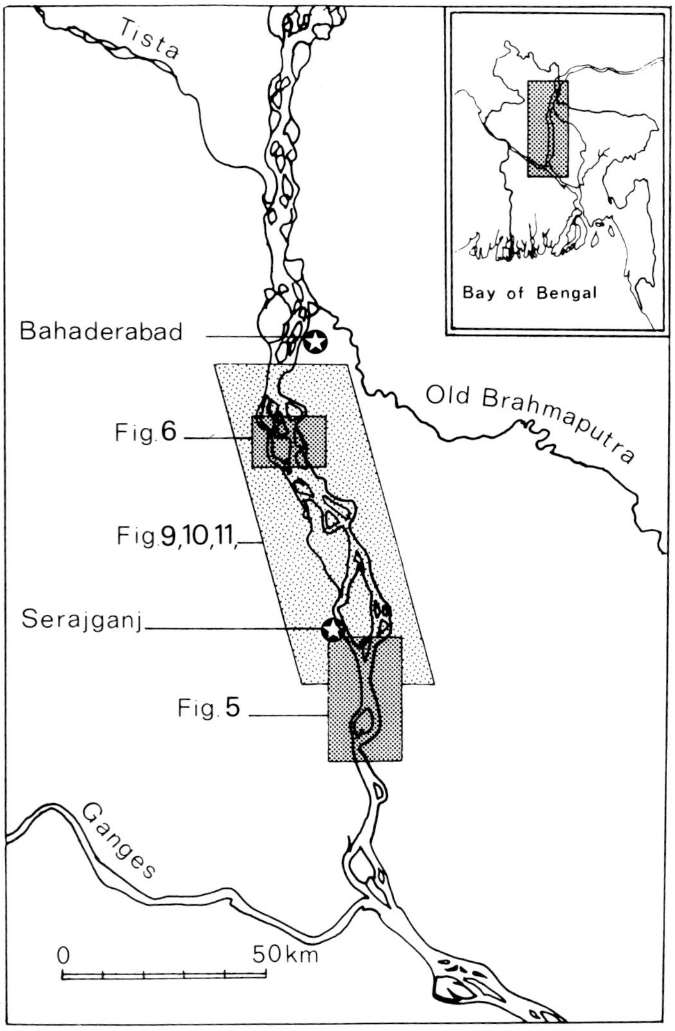

FIG. 1.—Location of the study area within Bangladesh. Shaded areas indicate locations of Figures 5, 6, 9, 10 and 11. Old Brahmaputra River is the former course of the Brahmaputra before the avulsion which occurred around 1787.

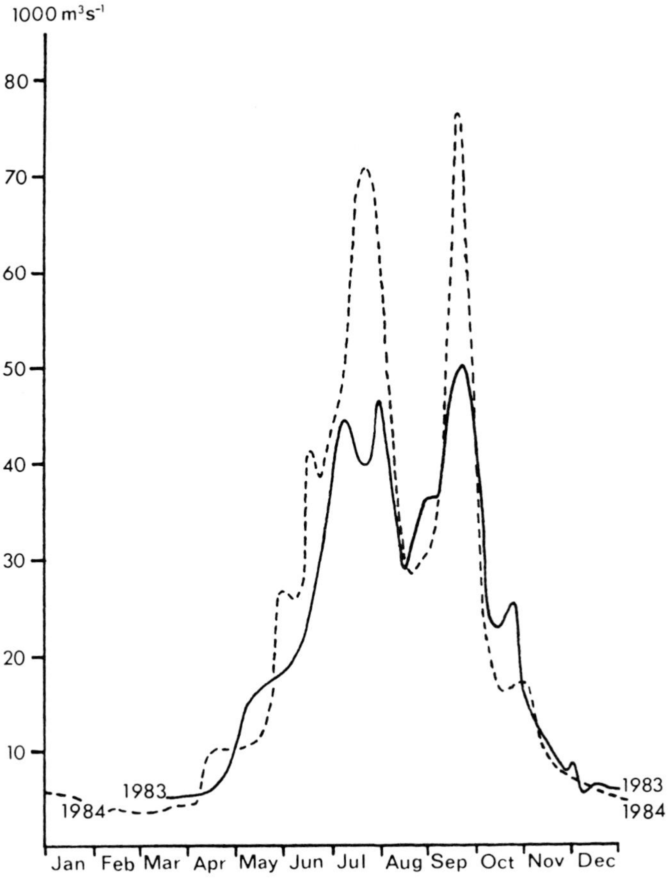

FIG. 2.—Brahmaputra River hydrograph for 1983 and 1984. Discharge measurements made by Bangladesh Water Development Board at Bahaderabad. Peak discharge of 75,700m^3/sec on September 19th, 1984 was in excess of the predicted 10-year flood and caused extensive overbank flooding.

preferred the estimate of 15 percent bedload based on the formula of Laursen (1958). This gives a total annual sediment discharge of 4 million tons, two orders of magnitude less than the estimates of Coleman (1969) and Latif (1969). The difference is due to an anomalous measured suspended load of 3,520,631 tons (Goswami, 1983). Coleman (1969) estimated bed load discharge at 25 percent but later (in Curray and Moore, 1971) suggested that the percentage may be as high as 50 percent.

Bed load grain size.—

Bed load grain size changes spatially and temporarily in response to changes in sediment supply and discharge. Measurements of bed load grain size in Assam (Goswami, 1983) do not show a progressive downstream decrease in grain size (Fig. 3). This is due to the large number of tributaries contributing sediment of varying quantity and calibre along the Brahmaputra. Temporal changes in bed load grain size have been observed in Bangladesh (J.I.C.A., 1976) by sampling bed load each month at two localities (Fig. 3). Bed load is finest during low flow, increases during peak flow and again during falling stage. Coarsening of bed load during falling stage may be a winnowing phenomenon. Alternatively, the coarsening of grain size may come from coarser sediment supplied by one or a group of tributaries late in the monsoon season. More observations of bed load supply and transport are needed to resolve the alternatives.

THE HIERARCHY OF CHANNELS WITHIN THE BRAHMAPUTRA RIVER

A hierarchy of channels exists in the Brahmaputra River (Fig. 4). The largest first-order channel is the entire river, comprising several second-order channels, which in turn have smaller channels within them (Figs. 5, 6). All the channels are active at bankfull stage but some of the third-order channels become abandoned at low stage. Each channel appears to have a pattern that is formed in response to its slope, discharge and sediment supply. Each channel has slightly different characteristics and as a result each channel behaves differently.

The hierarchy of channels in the Brahmaputra is very different from the one described in the Donjek River (Williams and Rust, 1969) in both scale and process (Fig. 4). In the channel hierarchy of Williams and Rust (1969), each

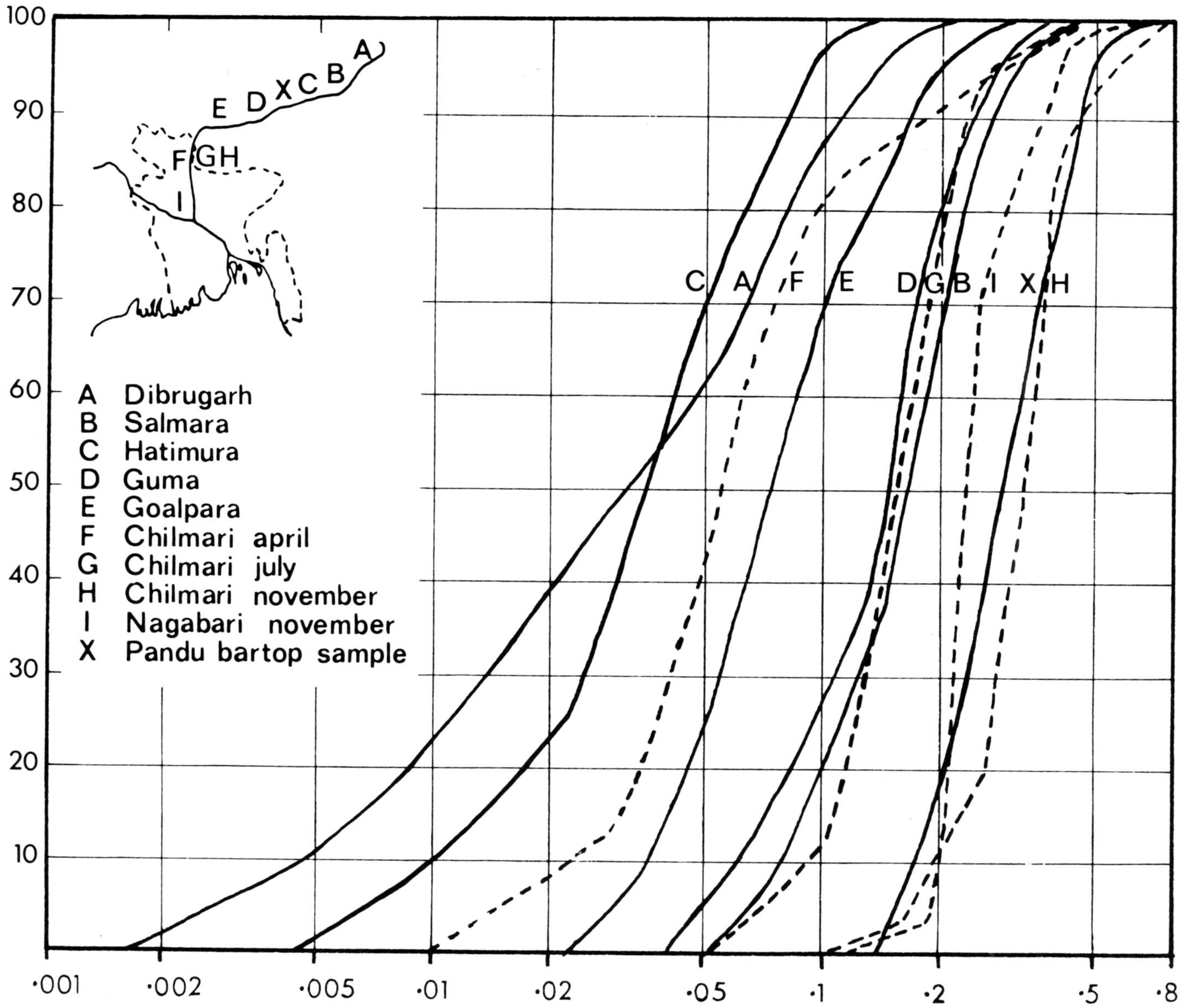

FIG. 3.—Bed load grain size distribution in the Brahmaputra (A-E after Goswani, 1983) shows no downstream decrease in grain size. Samples F-H show seasonal changes in bed load grain size (after J.I.C.A., 1976), which is finest during low flow in April, coarsens during peak flow in July and coarsens again during falling stage in November.

channel scaled with individual bars within the river and its form is dependent on the level of flow around the bar. Thus, the river changes from a single channel at high stage to a multiple channel at low stage. In the Brahmaputra the second-order channels maintain their characteristics and pattern at all stages. Some of the third-order channels may be produced during falling stage, but others appear to maintain their identity at bankfull stage.

First-order channels.—

The first-order channel includes the entire river (Fig. 4), the channel margin being defined by the outermost river banks. The channel changes its pattern along its length from braided in the north to a more sinuous course in the south. This may be in response to the decrease in gradient. There are also several narrow straight reaches along the river called node points. Coleman (1969) suggested that the position of these was controlled by more cohesive banks at abandoned channel localities. Historic maps show that node points are not fixed but are transient on a scale of tens of years. One node point identified by Coleman is located just south of Sirajganj. Historic maps (Figs. 7, 8) show that the river has previously been wider here and is not constricted by more cohesive banks but by deposition downstream from a confluence.

Second-order channels.—

Second-order channels show a variety of patterns. In the area north of Sirajganj (Fig. 7) a large megaform or bar assemblage has developed from a number of medial bars

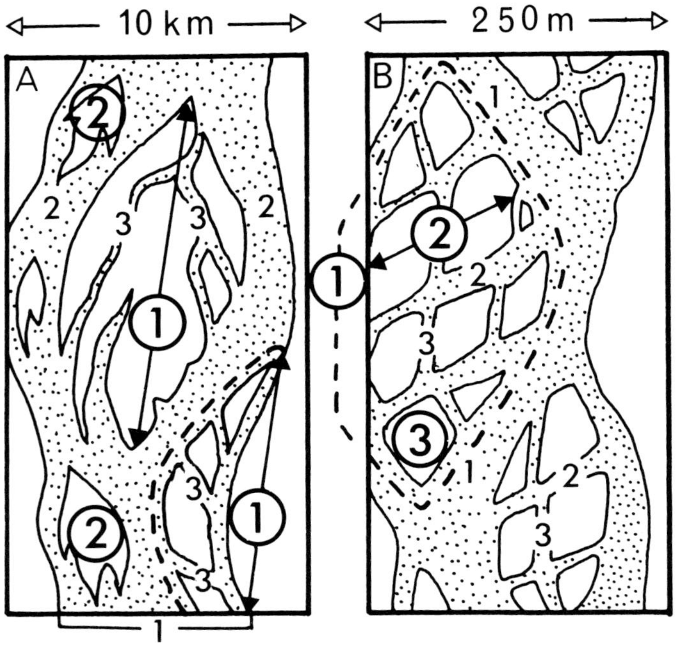

FIG. 4.—Channel hierarchies in the Brahmaputra River (A) and the Donjek River (B) (after Williams and Rust, 1969). Numbers in circles refer to bars, other numbers refer to channels. In the Brahmaputra River the first-order channel comprises the whole river, which includes several second-order channels. Bars scale with the channels in which they occur. Third-order channels modify higher order bars but still have bars within them which are not possible to illustrate at this scale.

since 1954. The western and easternmost second-order channels now have braided and meandering courses, respectively. These channels have maintained fairly constant patterns over the last 10 years (Fig. 8). They appear to be in equilibrium with the local discharge slope and sediment supply. They also migrate independently within the channel belt and could be considered as an anastomosing reach within the braided system. The different behavior of the two channels is probably due to different discharges, assuming other factors to be constant, the braided channel having a greater discharge than the meandering channel.

Third-order channels.—

Third-order channels have been observed with meandering and braided courses. Some occur on the top of first- or second-order bars and are produced in response to a hydraulic gradient across bar tops flowing from the main channel into the chute channel. Some third-order channels have been observed by echo-sounding at high stage, however. These may be produced by multiple flow cells in channels with a very high width/depth ratio. There is insufficient data to distinguish the form of these channels or to say if channels exposed at low stage are similar to those observed at high stage.

MACROFORMS OR BARS

The type of bars and their distribution within the Brahmaputra have been identified using LANDSAT images, hydrographic charts, echo-sounder surveys, maps and field examination. A limited number of bar types (macroforms using the terminology of Jackson, 1975, Church and Jones, 1982, and Bridge, 1985) can be recognized. These include lateral bars (including point bars), medial (braid) bars, diagonal bars, scroll bars and tributary (chute) bars. The bars scale with the bankfull channel dimensions. Bar height is between one-half and just less than total bankfull depth. Bar length from crossover to crossover is between 3 and 5 times channel width. Individual bars vary in size because of the hierarchy of different channel sizes. Bars have a time span which is greater than or equal to the passing of a single flood event, which in Bangladesh can be assumed to equal 1 year.

At low flow stage, water level within the Brahmaputra falls as much as 8 m and large areas of bar top are exposed. These emergent areas range in length from a few tens of meters to 15 km. The size and shape is the result of complex sequences of erosion and deposition. Great care has to be taken in distinguishing partially exposed areas of bars in the field from modified mesoforms (dunes or megaripples) (cf. Church and Jones, 1982; Bridge, 1985). Hydrographic charts which show the full three-dimensional form of bars are the most useful data source in this respect.

Lateral (point) bars (Fig. 5D, E; Fig. 6C, D) occur on the inside of channel bends and in straight reaches. They are attached to the bank at their upstream ends and generally have subsidiary scroll bars and chute channels (Fig. 5E). Lateral bars are commonly modified by third-order channels which flow over the bar from first- or second-order channels into the chute channel, at which point third-order tributary bars have been observed. One sinuous chute channel observed at low flow stage had point bars which scale with bankfull width of the chute channel. Medial (braid) bars (Fig. 5C; Fig. 6A, B) are detached forms which are deposited in areas of flow divergence. With further flow divergence and channel widening, a more complex medial bar may be formed with scroll bars attached to one or both sides and with third-order channels dissecting it. Tributary bars (Fig. 5A, B; Fig. 6E, F) occur at junctions of channels of the same or different order. Figure 5A shows a tributary bar at the junction of two second-order channels; Figure 5B shows a tributary bar deposited where a third-order chute channel enters a second-order channel at the downstream margin of a medial bar which may be evolving into a diagonal bar. Diagonal bars (Fig. 6G) are attached to the bank at both ends and occur at crossovers where flow diverges as the main current moves from one bank to another.

Bar assemblages or megaforms.—

In several places bars become coalesced to form bar assemblages (the sedimentation zones of Church and Jones, 1982, or flood plain segments of Romashin in Church and Jones). The length of bar assemblages from crossover to crossover is around 30 km. This does not scale with first-order channel width (3 km at node points). Bar assemblages are associated with Channel anastomosis in the Brahmaputra (Fig. 7). Church and Jones interpreted bar assemblages as the product of locally increased sediment deposition. Relatively rapid deposition is also associated with channel anastomosis in other rivers (Smith and Smith, 1980).

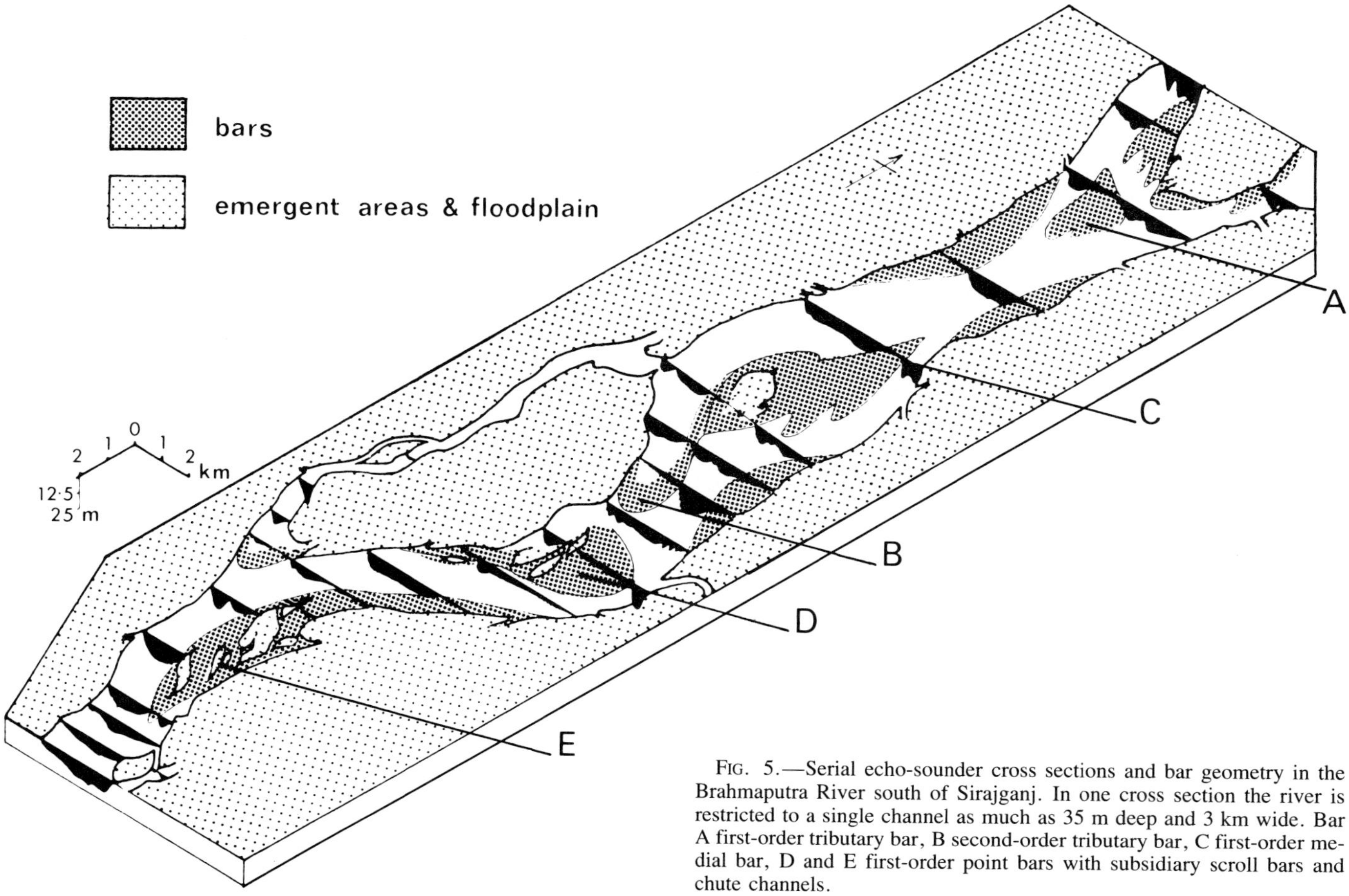

FIG. 5.—Serial echo-sounder cross sections and bar geometry in the Brahmaputra River south of Sirajganj. In one cross section the river is restricted to a single channel as much as 35 m deep and 3 km wide. Bar A first-order tributary bar, B second-order tributary bar, C first-order medial bar, D and E first-order point bars with subsidiary scroll bars and chute channels.

The time span of these megaforms is on the order of tens of years (Fig. 7); several of these megaforms are shown in Figure 9.

CHANNEL MOVEMENT

Channel displacement in the Brahmaputra River occurs at a variety of magnitudes. First-order channel migration and avulsion (switching of the channel belt) occurs on a larger time scale than second- and third-order channel movements. Coleman (1969) described two distinct types of channel movement: gradual and continuous transverse movement (lateral migration), and sudden shifts (channel switching within the channel belt). The different methods of channel movement produce different patterns of deposition. Lateral migration produces lateral accretion, whereas switching produces channel enlargement in one area and channel fill in another. Both processes have been observed at all levels of the channel hierarchy, and the deposits of the Brahmaputra contain examples of lateral accretion and cut-and-fill at different scales.

First-order channel movements.—

The movements of the first-order channels have been recorded from historical maps. Rennell (1781) shows the Brahmaputra River running along a course now occupied by the Old Brahmaputra River (Fig. 1). Around 1787 the Brahmaputra avulsed into its present course, capturing the Tista River. There is some dispute in the literature as to the nature and trigger for the avulsion (Morgan and McIntire, 1959). Coleman (1969) concluded that the change in course took place gradually, due to increased flood discharge, faulting, or a combination of both. Latif (1969) refers only to a devastating flood in 1787. The avulsion resulted in a maximum channel displacement of 100 km.

The avulsion occurred gradually over 50 years. The new course of the Brahmaputra was initially sinuous, as shown in Wilcox's 1833 survey (Fig. 8). By 1867 the river had become braided. This change in pattern was probably due to increased discharge as less water diverted along the Old Brahmaputra. On Rennell's (1781) map the Old Brahmaputra was braided; since the avulsion, this river has assumed a meandering course in response to decreased discharge. The Brahmaputra avulsed into an unnamed tributary of the Konaie River, not the Jenni as reported by Coleman (1969). Maps from 1830 to the present (Fig. 8) show that the river has migrated back and forth within a channel belt approximately 15 km wide. Coleman recognized a net westward migration of as much as 10 km over this 150-year period, a rate of approximately 70 m per annum. This net migration is most apparent in the upstream reaches. There has been less first-order migration downstream and the confluence with the Ganges has remained almost static.

Coleman concluded that net migration westward was due

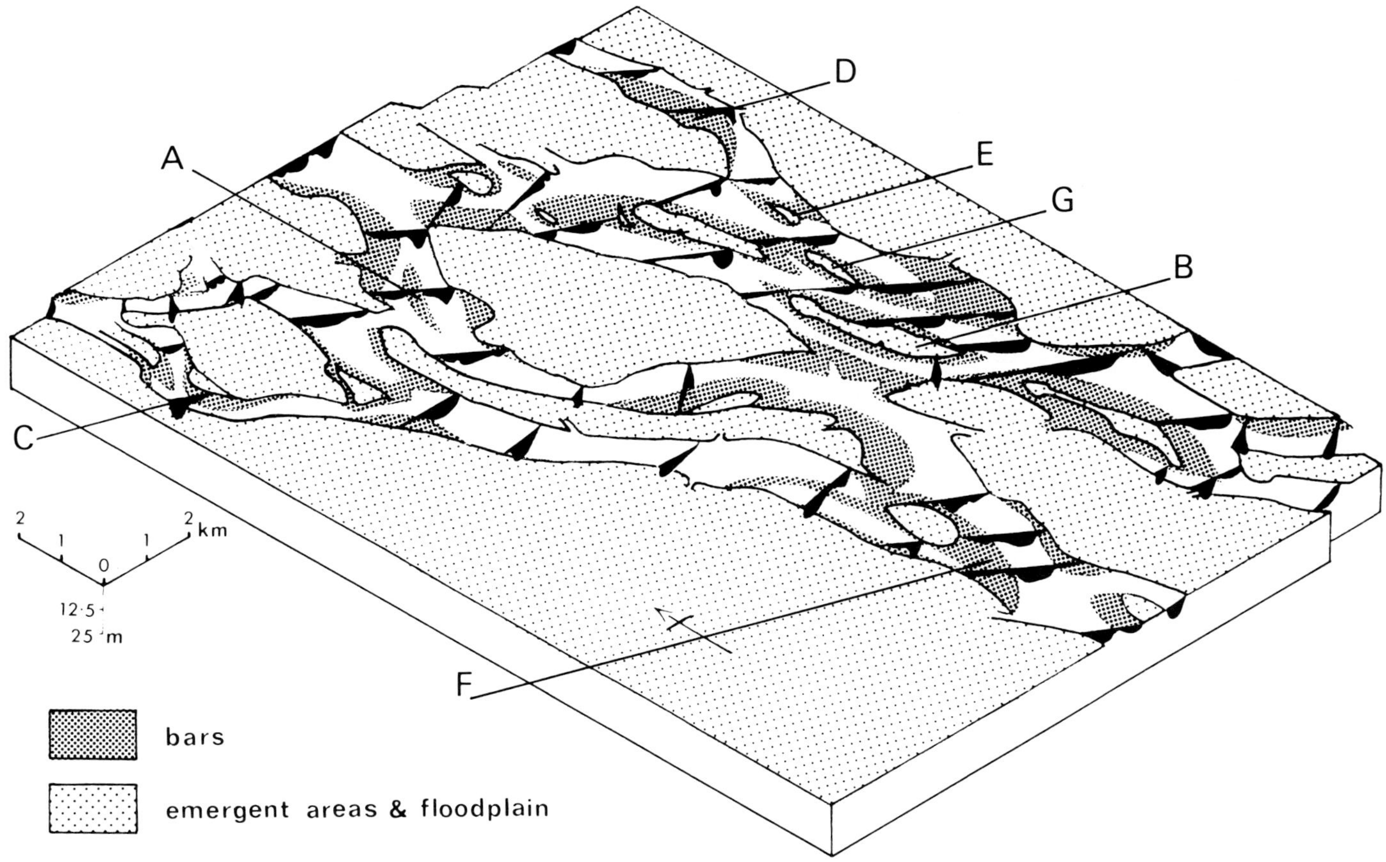

FIG. 6.—Multichannel reach of the Brahmaputra River between Shareakhandi and Maderganj with echo-sounder cross sections and bar geometry. Bars A and B are elongate second-order medial bars, C and D are second-order point bars on the sides of medial bars with strongly asymmetric channel cross sections. E and F are second-order tributary bars, G is a diagonal bar with a series of small third-order channels superimposed.

to tectonic tilting, whereas different rates of migration along the river were substrate controlled. Westward migration is against the regional tectonic slope, but the river course is probably influenced by intrabasinal faults (Morgan and McIntire, 1959). The changing rate of migration may also be due to changes in the pattern of the river downstream and the reduced gradient near the confluence with the Ganges.

Second-order channel movements.—

Studies of LANDSAT images from 1972–1978 show second-order channel switching and lateral migration (Fig. 9). Second-order channel migration is comparatively rapid; migration of 1 km per annum is not uncommon. This rate of movement exceeds the average figure of 76 m per annum for first-order channel migration and suggests that second-order channels will totally rework any first-order accretion surfaces. Most of the movement of second-order channels is by lateral migration, which has a major effect on the internal sand body geometry.

When a second-order channel erodes one bank, there is generally an equal amount of deposition on the opposite bank (Figs. 9–11). This is contrary to Coleman's (1969, p. 159) observation that "in a braided channel . . . this type of cut and fill does not occur." The difference lies in Coleman's use of bankline studies which were restricted to observations of the first-order migration without considering the second- and third-order channels, which are fundamental component parts of a multichannel river. Using the channel hierarchy concept, it is easier to explain channel migration and resulting patterns of deposition (see below). It is hoped that the same concept can be used in interpreting ancient sand bodies deposited by multichannel rivers. The movements of the first-order channel control the regional distribution of sand and sand body shape. Second-order channel migration controls the internal stratigraphy.

Third-order channel movements.—

Third-order channels are too small to be clearly defined on LANDSAT images which have a resolution of 80 m on the ground. Studies are restricted to observations at low flow and echo-sounder traces. Coleman (1969) described a third-order channel moving within a second-order channel and changing its behavior in response to changes in discharge. During the rising stage, movement is gradual lateral migration of as much as 1 km. During peak flood, there is little movement of the third-order channels, but during falling stage the channels switch in a sudden and irregular fashion. Coleman attributed this to deposition of material during falling stage producing changes in local flow direction. Thus, in a period of bed aggradation, channel switching is greatest, many new third-order channels are formed

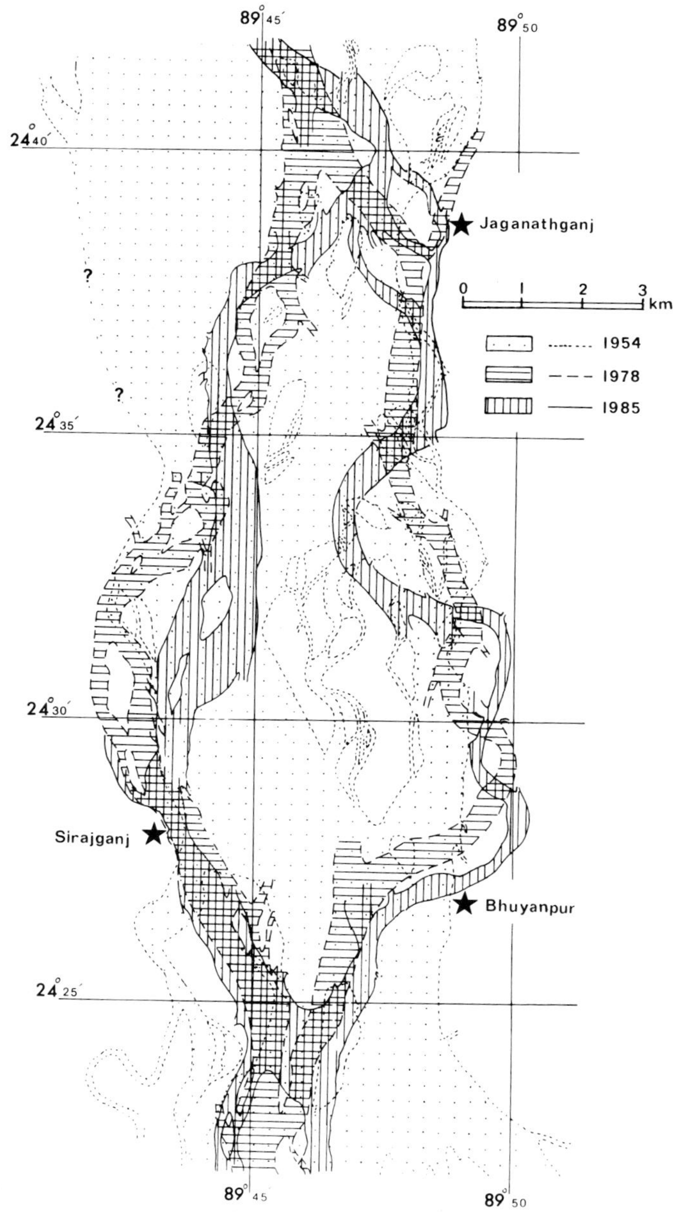

FIG. 7.—An anastomosed reach of the Brahmaputra which has developed following the coalescence of medial bars into a bar assemblage. The division of discharge has produced two contrasting channel patterns in the anastomosed reach. The western channel is braided, the eastern channel is meandering. The bar assemblage megaform has remained stable over the past 15 years, while the second-order channels have migrated independently along either side.

and function for a short time and are then abandoned. The rapid and contrasting changes in movement of third-order channels will produce complex internal erosion surfaces and sequences of deposition (see below). The third-order channels, which are low stage features draining off higher order bars, do not migrate very much due to the infrequent discharge and high cohesion of the substrate on bar tops. Drainage channels may be enlarged during floods and then become preferential sites for higher order channel avulsions, or abandoned and filled with mud plugs at the top of the sand body.

PATTERNS OF CHANNEL MIGRATION AND DEPOSITION

Maps showing channel migration and deposition (Figs. 9, 10, 11) have been compiled from LANDSAT imagery with resolution of 80 m on the ground; this is sufficient to monitor migration of channels and bars in a large river such as the Brahmaputra. The images used were all selected for the dry season periods (1972–1978) because at this time of year, there is very little cloud cover and low water level exposes the bars. During flood stage, the high sediment load in the water unfortunately prevents any use of LANDSAT imagery for bathymetric or bedform observations.

The images reveal many changes in channel position (Fig. 9). Areas of deposition have been analysed to show temporal variations (Fig. 10). The style of deposition has been interpreted from these successive images (Fig. 11). The areas of deposition shown in Figures 10 and 11 are areas of net deposition; some of these areas were subsequently eroded so that the gross pattern of deposition is likely to be more complex than that illustrated. The erosion of some bars explains the apparent misnomer of some deposits classified as addition to bars (Fig. 11) when there is no pre-existing bar indicated due to its subsequent erosion.

The styles of deposition recorded here refer only to the deposits resulting from migration of the first- and second-order channels. With LANDSAT imagery, it is not possible to record the movements of third-order channels and bars. Thus, the geometry of sand bodies inferred from the styles of deposition refer only to the large-scale first- and second-order channel deposits. The third-order elements may show a different architecture within this framework. Smaller scale "deposition" patterns from third-order channels and bedforms have been measured from echo-sounder cross sections, aerial photographs, and field observations.

Temporal changes in deposition.—

Figure 10 shows the net pattern of deposition from six annual floods on the Brahmaputra River. The amount of erosion, sediment transport and deposition varies from year to year. These are probably controlled by discharge. Very high discharges, or prolonged high discharges, produce greater erosion, channel migration, sediment transport and deposition. The period December 1972 to January 1976 includes the major 1973 flood when overbank flooding was maintained for 2 months. This period shows the greatest amount of change.

STYLE OF DEPOSITION

The areas of deposition identified in Figure 10 have been subdivided into four categories: addition to bars, lateral accretion to the bank, new mid-channel bars and channel abandonment (Fig. 11). The relative proportions of these four styles of deposition are shown in Figure 11. From this compilation it is apparent that the most important component of deposition in the Brahmaputra River is addition to pre-existing bars. This category accounts for 53 percent of the area of deposition, whereas lateral accretion to the bank, new mid-channel bars and channel abandonment account for 19, 13 and 15 percent, respectively.

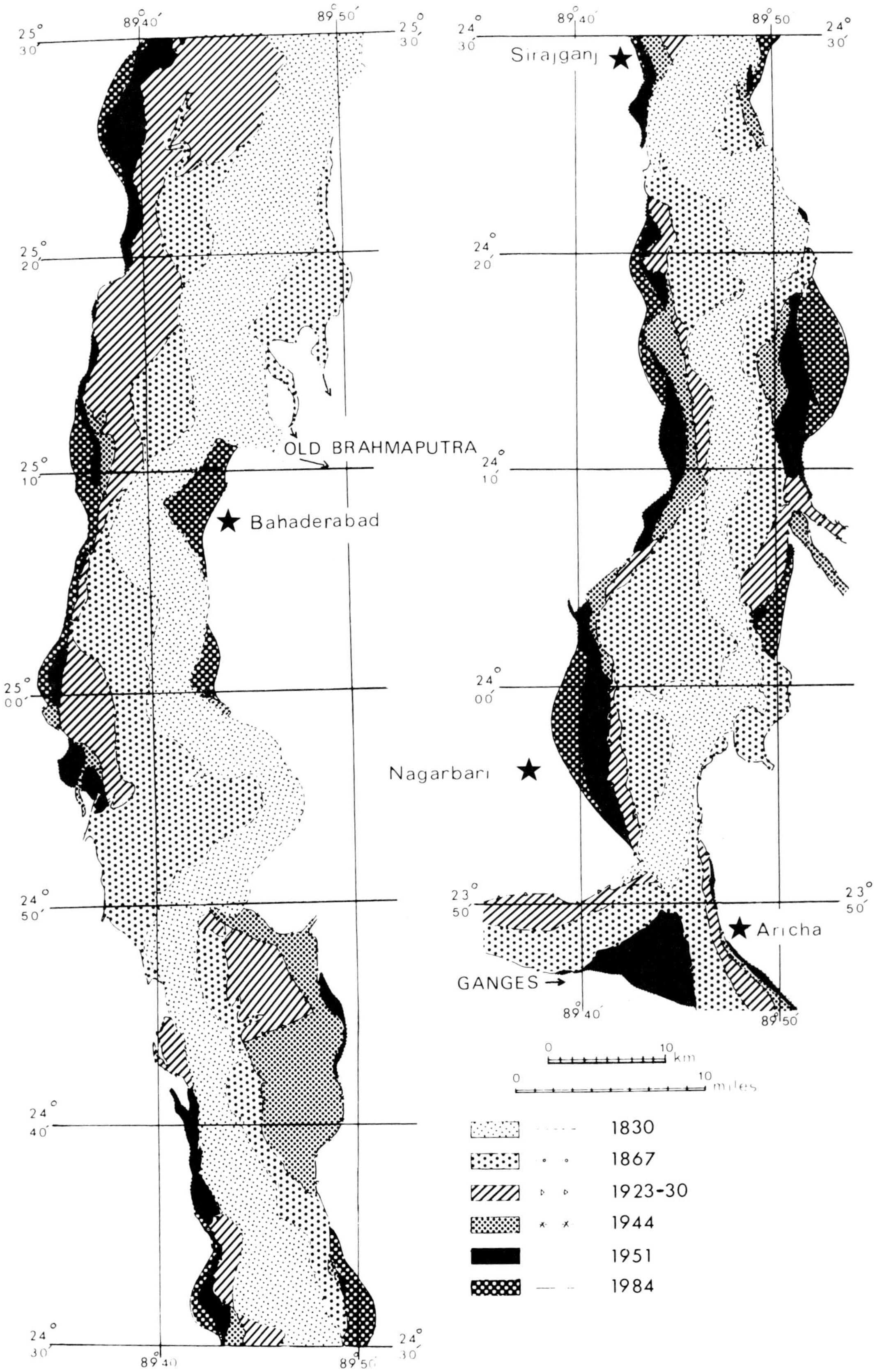

FIG. 8.—Bank line changes over the past 150 years. The river has changed from a sinuous course in 1830 to a braided course by 1867. The channels have migrated back and forth within a 15-km-wide belt which shows a minor amount of net migration to the west.

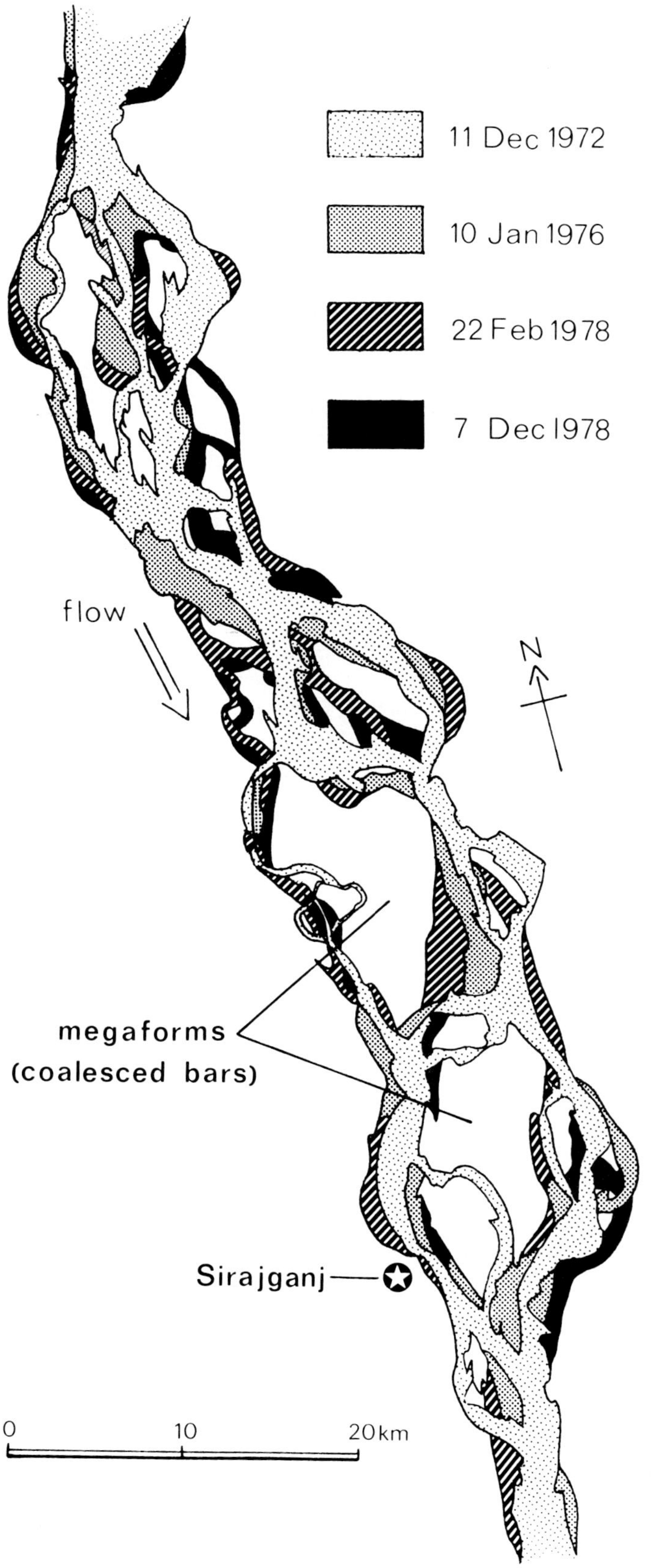

FIG. 9.—Second-order channel locations recorded from LANDSAT images, 1972 to 1978. Most channel movement occurs by lateral migration at rates of as much as 1 km per annum. The megaforms have remained stable over this period.

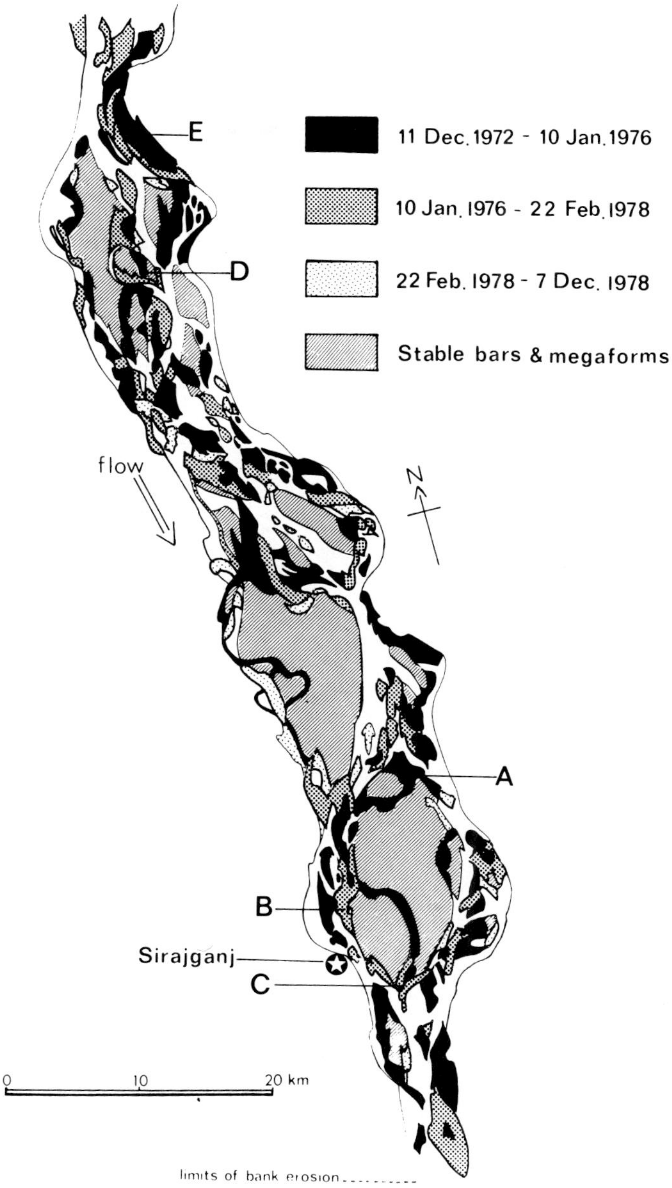

FIG. 10.—Areas of deposition from the map of successive channel locations; most deposition occurs on the margins of the megaforms. A, upstream accretion; B, Flank accretion; C, Downstream accretion; D, shows the development of a complex medial bar from an initial new mid-channel bar, and E is a large point bar accreted on the left bank. Blank areas within the limits of bank erosion are areas characterized by erosion or nondeposition.

Addition to bars.—

The addition to bars can be subdivided into three elements of deposition: upstream accretion, flank accretion and downstream accretion. All three elements combine different proportions of vertical and lateral accretion as the bar is enlarged. Upstream accretion (14 percent) may be accomplished by megaripples and scroll bars wrapping around the upstream margin of a bar in a manner similar to that described for the formation of sandflats in the South Saskatchewan River (Cant and Walker, 1978). Alternatively, deposition may occur in a zone of comparatively slack water

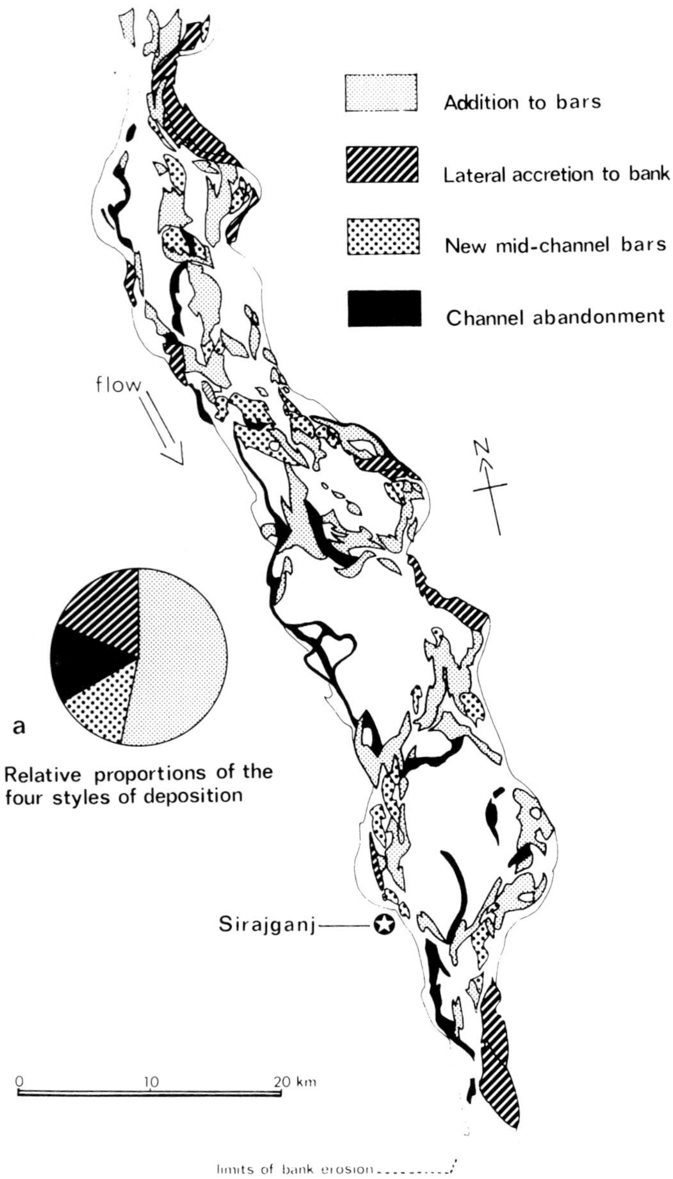

FIG. 11.—Styles of deposition, 1972 to 1978. The areas of deposition identified in Figure 10 are classified according to the area in which they were deposited. This figure represents gross deposition; subsequent erosion in many areas complicates the pattern of net deposition.

upstream from a bar where the current diverges. There is not necessarily a scour pool upstream of every bar as described by Coleman (1969).

Flank accretion is the most important element in medial bar growth (57 percent). Flank accretion is produced by megaripples and scroll bars migrating around and up the flanks of convex bar margins in a manner similar to that described from point bars in meandering rivers. Allen (1968) suggested that braid bars could form as double point bars and Bridge (1985) proposed a similar model. Recognizing this lateral accretion in the deposits is very difficult due to the extreme width/depth ratios and very low angle of lateral accretion surfaces in the Brahmaputra.

The downstream accretion element (29 percent) includes a considerable amount of deposition from suspension. During falling stage, large flow separation zones exist downstream from bars. In these areas current velocity is reduced and sometimes moves upstream. Deposits are dominated by climbing ripple lamination and mud drapes up to 2 m thick. Megaripples and scroll bars also migrate around the downstream margin of medial bars.

These three elements can be distinguished on the maps of deposition (Figs. 10, 11). The dominance of lateral accretion produces ridge-and-swale topography on bars (Fig. 12) which records successive downstream and lateral growth of the bar. Superimposed on the ridges and swales are natural levees. Thus, the bar is highest at its margins and lowest in the middle. Similar accretion topography has been observed on bars in the low sinuosity Calamus River (N. D. Smith, and J. S. Bridge pers. commun.).

Lateral accretion to bank.—

Lateral accretion to the bank comprises 19 percent of the deposition, most of which is in the form of large point bars as much as 10 km long and 3 km wide (Fig. 8E). In some localities the lateral accretion may be onto a bar margin which has been incorporated into the floodplain following channel abandonment. The largest areas of lateral accretion are all on the east bank of the river (Fig. 8), which correlates well with Coleman's observations that the Brah-

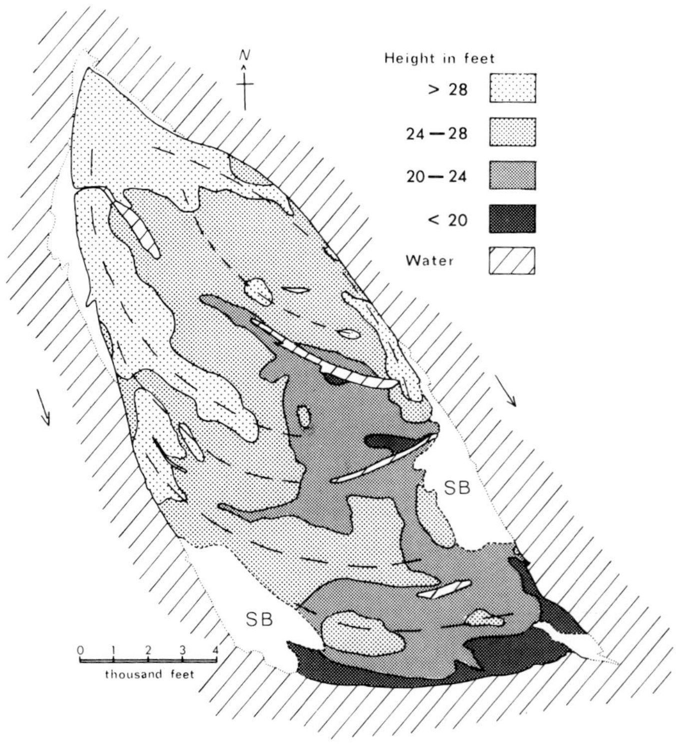

FIG. 12.—Bar topography (compiled from maps with 1-foot contours) shows ridge and swale topography from lateral and downstream accretion. Lateral accretion is accomplished by scroll bars (SB) wrapping around the downstream margin of the bar. The highest areas of the bar are along the upstream and lateral margins which have been raised as natural levees. The lowest part of the bar is along the center and downstream margins, where flood waters are ponded and sediment is deposited from suspension.

maputra River is migrating westwards. It is interesting to note that Coleman interpreted all the recent deposits of the Brahmaputra as those of a meandering river with very broad low-curvature meander loops. This interpretation is supported by the 1830 map, but the large component of lateral accretion observed in bar growth (57 percent) produces a topography similar to ridges and swales on point bars (Fig. 12). This presents problems for the interpretation of recent and ancient deposits of a large braided river based on the conventional association of lateral accretion with meandering rivers.

New mid-channel bars.—

New mid-channel bars form the smallest component of the styles of deposition 13 percent. This is due to the small size of new bars. New mid-channel bars generally have an elongate diamond or petaloid form. If the channel widens, they may enlarge into a more complex medial bar.

Channel abandonment.—

The channel abandonment comprises 15 percent of the deposition recorded, which is considerably less than expected from previous models of braided river sedimentation (Bridge, 1985). Furthermore, the channel fill deposits generally preserve the minor second- and third-order channels (Fig. 11). Coleman's observations suggest that the smaller third-order channels are more susceptible to avulsion than the larger second-order channels. Using these abandoned channel-fill sequences for paleodischarge measurements should be treated with caution, because the abandoned channels may not be representative of the true magnitude of the fluvial system.

DISCUSSION

Recent sedimentation in the Brahmaputra River is predicted to produce a sand body of a sheet-like nature approximately 40 m thick and 20 km wide. The dimensions of the sand body and total sediment distribution are controlled by first-order channel movements. The most important of these are the 1787 avulsion and the subsequent westward migration. This movement occurs on a time scale of 10^2–10^3 years and is probably controlled by tectonic movements within the Bengal basin (Morgan and McIntire, 1959).

The internal stratigraphy of this sand body is controlled by second-order channel movements. This is dominated by lateral migration at rates of as much as 1 km per annum. Due to the extremely high width/depth ratios of channels in the Brahmaputra (as much as 500:1), the resulting lateral accretion surfaces will be very difficult to recognize in the geologic record. More asymmetric sinuous channels (Fig. 6C), however, have width depth ratios of 45:1. Continual reworking within the channel belt and occasional channel switching will produce complex internal erosion and deposition surfaces in a multistory sand body. During falling stage, large amounts of suspended sediment are deposited within the channel, particularly on the downstream margin of medial and lateral bars. These silt drapes may preserve some record of second-order bar and channel form. Low stage reworking and the entrenching or abandonment of third-order channels will further complicate the internal geometry of the sand body.

CONCLUSIONS

The Brahmaputra River is a multichannel river. Channels of many different sizes (from hundreds of meters to kilometers wide) and different patterns (braided, meandering and anastomosed) occur in Bangladesh.

The presence of three different channel patterns within the Brahmaputra River shows that there is a continuum between braided, meandering and anastomosed rivers. The main factor controlling channel pattern in the Brahmaputra is channel discharge. Other factors, sediment supply, grain size and slope, are locally similar. The change in channel pattern from braided in the north to a more sinuous course in the south is probably caused by the downstream reduction in valley gradient from 0.000077 to 0.00005. There is also a possible downstream reduction in peak discharge due to diversion of flood waters along the Old Brahmaputra and Dharleswari Rivers and into the Atrai Gur flood basin. These changes in slope and discharge may affect the rate of channel migration which also decreases downstream.

The channels are organized in an hierarchy. The first-order channel comprises the entire river. This has an average width of 10 km and maximum depth of 45 m. The first-order channel has made one avulsion in the last 200 years. Since then, it has migrated within a channel belt as much as 20 km wide. First-order channel migration is an average 70 m per annum. Second-order channels have maximum widths of 5 km and depths to as much as 40 m. They migrate at rates of as much as 1 km per annum. Most channel movement occurs by lateral migration, although some channel switching has been observed. Third-order channels occur within second-order channels. They have widths of hundreds of meters and migrate in response to local hydraulic gradients within second-order channels. Third-order channels dissect and modify higher order bars at falling stage. Some may also be active at high stage when a high degree of switching has been observed.

Five types of bar are recognized in the Brahmaputra: lateral bars, medial bars, tributary bars, diagonal bars and scroll bars. Of these the first three are dominant. All of these bars scale with the bankfull width and depth of the channel. Bar length from crossover to crossover is approximately 3 to 4 times channel width, and bar height is greater than one-half channel depth. There is a wide range in bar size due to the variety of channel magnitudes. Thus, there are first-, second- and third-order bars which scale with first-, second- and third-order channels, respectively. Bars are stable over several floods and are modified by succeeding depositional and erosional events which produce complex bar forms.

The most important component of deposition on bars is lateral accretion (57 percent), followed by downstream accretion (29 percent) and upstream accretion (14 percent). Together, these form 53 percent of all deposition within the river. Other sites of deposition are lateral accretion to the bank 19 percent, channel abandonment 15 percent and new mid-channel bars 13 percent.

In some localities bars coalesce to form bar assemblages which are as much as 20 km long. At these points, the river is anastomosed and each second-order channel migrates independently within the channel belt. These megaforms are stable over tens of years.

Recent deposits of the Brahmaputra form a 20-km-wide multistory and multilateral sand body which is predicted to have a complex internal geometry. This will be marked by a hierarchy of internal erosion and deposition surfaces, scaling with the hierarchy of channels. Due to the high width/depth ratios, lateral accretion surfaces will be extremely low-angle. Second-order bounding surfaces are expected to dominate the internal stratigraphy due to their high rates of migration and deposition within the channel belt.

ACKNOWLEDGMENTS

This work forms part of Ph.D. research at the University of Leeds and was funded entirely by Texaco.

I thank my supervisor, Dr. M. R. Leeder, and Dr. J. S. Bridge and Professor N. D. Smith for critically reading this paper; the personnel of Bangladesh Inland Waterways Transport Authority, particularly the river surveyor Mr. F. R. Khandokar, and the crew of the survey boats "Sayeeda" and "Alert," without whose cooperation this work would not have been possible. I am indebted to Mr. Ken Cross of Rendel Palmer Tritton for looking after me during my convalescence and introducing me to the B.A.G.H.A. club, and to all the other people in Bangladesh who were so hospitable and made my visit a happy one.

REFERENCES

ALLEN, J. R. L., 1968, The nature and origin of bedform hierarchies: Sedimentology, v. 10, p. 161–182.

BLODGETT, R. H., AND STANLEY, K. O., 1980, Stratification bedforms and discharge relations of the Platte braided river system, Nebraska: Journal of Sedimentary Petrology, v. 50, p. 139–148.

BRIDGE, J. S., 1985, Palaeochannel patterns inferred from alluvial deposits: A critical evaluation: Journal of Sedimentary Petrology, v. 35, p. 579–589.

CAMPBELL, C. V., 1976, Reservoir geometry of a fluvial sheet sandstone: American Association of Petroleum Geologists Bulletin, v. 60, p. 1009–1020.

CANT, D. J., AND WALKER, R. G., 1976, Development of a braided-fluvial facies model for the Devonian Batter Point Sandstone, Quebec: Canadian Journal of Earth Sciences, v. 13, p. 102–119.

———, AND ———, 1978, Fluvial processes and facies sequences in the sandy braided South Saskatchewan River, Canada: Sedimentology, v. 25, p. 625–648.

CHURCH, M., AND JONES. D., 1982, Channel bars in gravel-bed rivers, *in* Hey, R. D., Bathurst, J. C., and Thorne, C. R., eds., Gravel-Bed Rivers: John Wiley & Sons, Ltd., p. 291–338.

COLEMAN, J. M., 1969, Brahmaputra River channel processes and sedimentation: Sedimentary Geology, v. 3, p. 129–239.

COLLINSON, J. D., 1970, Bedforms of the Tana River, Norway: Geografiska Annaler, v. 52, p. 31–56.

CONAGHAN. P. J., AND JONES, J. G., 1975, The Hawkesbury sandstone and the Brahmaputra: Journal of the Geological Society of Australia, v. 22, p. 275–283.

CURRAY, J. R., AND MOORE, D. G., 1971, Growth of the Bengal deep-sea fan and denudation in the Himalayas: Geological Society of America Bulletin, v. 82, p. 563–572.

GOSWAMI, D. C., 1983, Brahmaputra River Assam (India) suspended sediment transport, valley aggradation and basin denudation: Unpublished Ph.D. Thesis, Johns Hopkins University, Maryland, 199 p.

JACKSON, R. G., 1975, Hierarchial attributes and a unifying model of bedforms composed of cohesionless material produced by shearing flow: Geological Society of America Bulletin, v. 86, p. 1523–1533.

JAPANESE INTERNATIONAL COOPERATION AGENCY, 1976, Feasibility study for Jamuna River bridge construction project, 8 volumes, 960 p.

KIRK, M., 1983, Bar development in a fluvial sandstone (Westphalian 'A'), Scotland: Sedimentology, v. 30, p. 727–742.

LATIF, A., 1969, Investigation of the Brahmaputra River: Journal of the Hydraulics Division, American Society of Civil Engineers, v. 95, HY5, p. 1687–1698.

LAURSEN, E. M., 1958, The total sediment load of streams: Journal of the Hydraulics Division, American Society of Civil Engineers, v. 54, HY1, p. 1–36.

MORGAN, J. P., AND MCINTIRE, W. G., 1959, Quaternary geology of the Bengal Basin, East Pakistan and India: Geological Society of America Bulletin, v. 70, p. 319–342.

RENNELL, J. J., 1781, The Ganges and Brahmaputra Rivers: Royal Society of London, Philosophical Transactions, LXXI, p. 91–103.

———, 1793, An account of the Ganges and Burrampooter rivers, appendix to memoir of a map of Hindoostan or the Mogul Empire; W. Bulmer and Company, London, p. 336–364.

SMITH, D. G., AND SMITH, N. D., 1980, Sedimentation in anastomosed river systems: Examples from alluvial valleys near Banff, Alberta: Journal of Sedimentary Petrology, v. 50, p. 157–164.

SMITH, N. D., 1970, The braided stream depositional environments: Comparison of the Platte river with some Silurian clastic rocks, North-Central Appalachians: Geological Society of America Bulletin, v. 81, p. 2993–3014.

WILCOX, R., 1833, Copy of the map prepared by Captain R. Wilcox, season 1830–31 Sheets No. 1–7 inclusive. Office of the Surveyor General of India, Calcutta, 7 p.

WILLIAMS, P. F., AND RUST, B. R., 1969, The sedimentology of a braided river: Journal of Sedimentary Petrology, v. 39, p. 649–679.

GRAVEL COUNTERPOINT BARS: EXAMPLES FROM THE RIVER TYWI, SOUTH WALES

SIMON A. SMITH

Sedimentology Research Laboratory, Department of Geology, Reading University, Whiteknights, Reading RG6 2AB, U.K.

ABSTRACT: Accretion in meandering rivers does not exclusively occur in point bars. This is demonstrated by the River Tywi, a meandering gravel-bed stream in South Wales, in which deposition occurs at the concave sides of meander bends. These deposits, termed counterpoint bars in this paper, form at sites commonly assumed to be ones of erosion. There are two types of counterpoint bar in the Tywi; fine-grained examples, similar to the concave bank benches of Page and Nanson (1982), and large gravel bars. Gravel counterpoint bars are a little known type. They are bulbous to elongate in plan view, are as much as 30–40 m wide and 60–70 m long. They consist of a bar platform, secondary channel and riffle.

Gravel counterpoint bars appear to have a relatively high preservation potential in the Tywi valley. This is suggested by historical maps and the presence of several abandoned examples on the floodplain. These form elongate and bulbous assemblages of two or more bars. Morphology and sedimentary structures are distinct from the more familiar fine-grained examples described from other rivers; however, gravel counterpoint bars may be more difficult to distinguish from gravel point bars if preserved in the geologic record.

INTRODUCTION

Although facies models for meandering rivers have become increasingly detailed in recent years, many studies concentrate exclusively on point bars and their sediments. As a result, bars at other positions within the river have been neglected. This paper describes examples of sedimentation along the concave banks of a meandering river, a site commonly assumed to be one of erosion. The examples come from the River Tywi of South Wales which contains two types of concave bank deposit. Bars in similar positions have been described from several other rivers (Carey, 1969; Taylor and Woodyer, 1978; Hickin, 1979) and were termed concave bank benches by Taylor and Woodyer (1978). They consist of fine sand and mud, have a bench-like shape and are separated from the outer bank by a narrow secondary channel. The bars are deposited in a slack water zone created by flow separation about a tightly curved meander bend (Page and Nanson, 1982). As Lewin (1982, 1983) has pointed out, however, gravel can also accrete against the concave sides of meander loops. Although he gave no detailed descriptions of these bars, he noted that they were elongate to bulbous in plan and were not bench-like. Accordingly, Lewin suggested the more general term counterpoint bar for all concave bank deposits, a terminology that is used in this paper.

Both fine-grained and gravel counterpoint bars occur in the River Tywi. Hence, the river provides an excellent opportunity to study gravel counterpoint bars, a type which is poorly known, and to compare them with the more familiar fine-grained examples. This paper concentrates on the morphology, growth and formation of gravel counterpoint bars. The significance of these deposits is assessed by comparing them with fine-grained examples, estimating their preservation potential and outlining their possible appearance in the geologic record.

BACKGROUND AND METHODS

The River Tywi drains 1090 km^2 of the Welsh plateau (Fig. 1) and is a substantial gravel-bed river with a mean annual discharge of 38 m^3 s^{-1}, and mean annual flood discharge of 370 m^3 s^{-1} (N.E.R.C., 1975). In the study reach, the Tywi flows within a relatively narrow (1–2 km) rock-bound valley, which is locally constricted by railway embankments built in the late 1850s. The Tywi has a history of limited aggradation and degradation (Lewin and Manton, 1975) and gravel and silt are currently being deposited on Pleistocene tills and clays. Gravel and fine-grained counterpoint bars form in actively migrating bends in the middle reaches of the river. Here the channel width is about 35 m, and average bankfull depth is 3–4 m. Other bar types include lateral bars and several types of gravel point bar. The channel pattern is rather irregular and consists of alternating high sinuousity and low sinuousity reaches. The sinuous reaches contain between one and five actively migrating meanders which are asymmetrical or box-like. Low sinuosity reaches are stable over a time scale of at least 50 years and range in length from 0.4–2 km.

Counterpoint bar morphology and structures were studied by field mapping and levelling of the seven gravel and three fine-grained examples present in the Tywi. Observations were taken at a variety of flood stages. An estimate of maximum grain size was obtained by measuring the "*b*" axes of the 20 largest clasts in a 4 m^2 area (mean maximum clast size), following a procedure similar to that of Bluck (1982). Flow over the bars was studied directly by estimating the direction of surface flow with a compass and indirectly by measuring clast imbrication, transverse rib and chute orientations. "Hiller" and "Russian" augers were used to sample the fine-grained bars. Historical sources, air photographs taken in 1971 and 1946, and 6-inch Ordnance Survey maps dated 1907 and 1886 were used to study the long-term evolution of counterpoint bars.

THE MORPHOLOGY OF GRAVEL COUNTERPOINT BARS

Gravel counterpoint bars occur against the concave sides of meander bends, just upstream from the associated point bar and bend apex. They are associated with a range of meander planforms, but the upstream limb of the meander is typically long and relatively straight. The mean sinuosity of eight such limbs calculated over 250 m is 1.02. Gravel counterpoint bars come in various shapes and sizes in the River Tywi. Typically, they are ogive-shaped, the convex side of the bar facing the concave bank. Other examples are more elongate. Bar areas range from a few square meters to 23,000 m^2. The position of some counterpoint bars in the channel and their range of shape and size are shown in Figure 2.

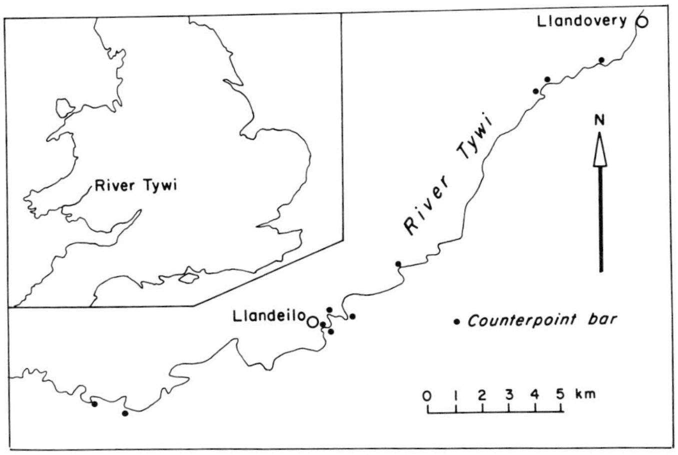

FIG. 1.—Location of the study reach.

Gravel counterpoint bars normally consist of three components: (1) the bar platform, (2) a narrow secondary channel that separates the bar from the floodplain, and (3) a riffle that joins the counterpoint bar to the point bar on the convex side of the channel. These features are defined in Figure 3 and will be described in turn. The descriptions are generalized and represent the morphology of most gravel counterpoint bars in the Tywi.

In most cases, the bar platform is bounded on the inner side by the main channel and on the outer by a secondary channel. The inner margins of the platform are rather steep, whereas the outer flank dips at about 15°. The bar surface is inclined at about 2° toward the concave bank, so that the bar has an asymmetrical cross section. The elevation of the bar decreases upstream toward the bar head, which is the lowest part of the bar. There are several chutes at this point. In some examples, the bar head is joined to the floodplain by a gently rising grass-covered ramp. This is inundated during floods, which may deposit small sheets of imbricated gravel. The maximum relief of the bar is about 1.5–2 m, and maximum bar widths are about 30–40 m. These features are illustrated by the example in Figures 4 and 5.

FIG. 2.—Gravel counterpoint bars (arrows) in the River Tywi. Flow is from right to left, bar scale is 200 m.

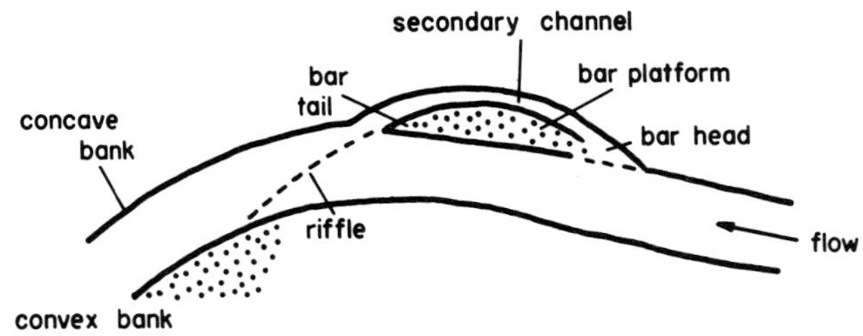

FIG. 3.—Definition sketch.

The bars consist of poorly sorted, imbricated sandy gravel, which fines toward the bar tail and outer flank. The gravel is clast-supported with a closed framework of coarse sand and granules.

The bar surface has few sedimentary structures. The tail is featureless apart from patches of rippled sand and falling-stage marks in sand, which occur around the margin of the bar. Small chutes as much as 2 m wide, however, are an important feature at the bar head. These chutes, presumed to have formed under supercritical flow, are incised into the lowest part of the bar and run obliquely across it toward the secondary channel where they end in small gravel deltas. These deltas build obliquely into the secondary channel and are elongated downstream. Transverse ribs—ridges of gravel 1–2 m long and 1–2 grain diameters wide—are common on the bar head and are presumably formed by supercritical flow (Shaw and Kellerhals, 1977; Koster, 1978) over the bar head.

An arcuate secondary channel separates the outer bank from the bar. The channel floor consists of poorly sorted sandy gravel with a thin mantle of sand, silt, and leaf detritus. The adjacent floodplain is commonly littered with mounds of tree trunks and branches (individual trunks may be as much as 4 m long), which are stranded as floodwaters expand over the bank.

The final component of gravel counterpoint bars is the riffle face which runs obliquely across the main channel. This joins the counterpoint bar with the point bar. Riffles are the most poorly known part of the counterpoint bars because they are submerged for most of the time. Enough observations have been made to show that riffles are highly variable, in particular, there is a wide range in the size of the riffle relative to the associated counterpoint bar. The latter ranges from a mere bulbous projection to 2–3 times the size of the riffle. Some riffles are simple in form and have a gently curved avalanche face, whereas others are more complicated. These have linguoid or irregular avalanche faces which are dissected by small chutes and deltas. Some examples have a "humpbacked" profile consisting of an avalanche face, a gently downstream-dipping segment and a gently upstream-sloping segment which passes into a deep scour pool. Some riffles are dissected by a large chute which may occur on either its inner or outer margin. The riffles are always the coarsest grained part of the complex, a relationship that they often share with other bar types (Hirsch and Abrahams, 1981; Milne, 1982). Maximum clast size decreases toward the margins of the riffle, whereas aprons and sheets of sand may occur in the lee of the riffle.

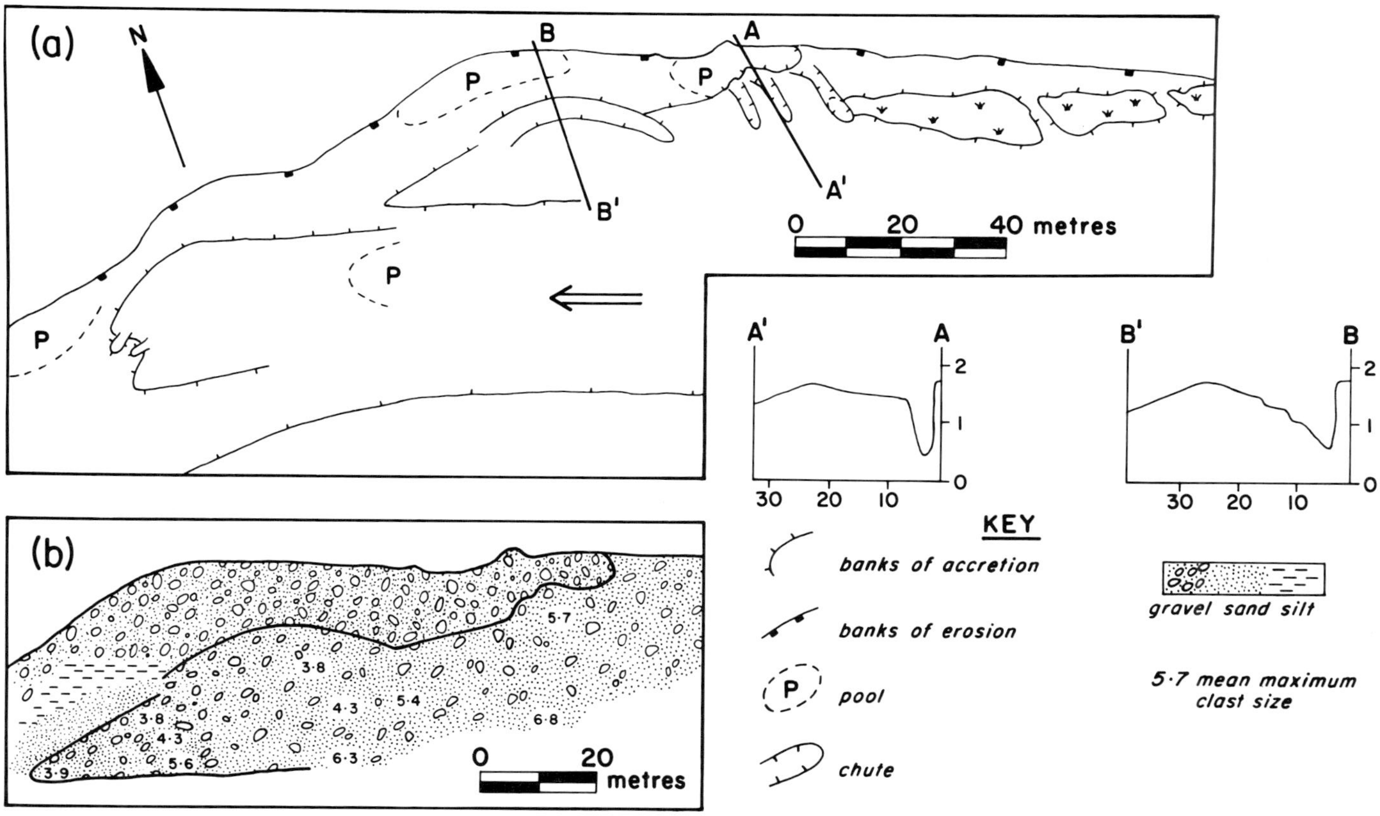

FIG. 4.—Field map of a gravel counterpoint bar. (a) Morphology, levelled sections in meters, (b) grain size distribution.

FLOW OVER GRAVEL COUNTERPOINT BARS

Some idea of flow patterns over the bars is given by measured surface flow directions and the orientation of sedimentary structures. Two examples are shown in Figure 6, which are based on readings taken during and shortly after a bankfull flow.

FIG. 5.—Field photograph of a gravel counterpoint bar (the example shown in Fig. 4). Flow is toward the observer, field of view is about 50 m.

During low river stages, the flow is confined to the main channel and the bars are exposed. As the river rises, the bar head is inundated, and then the tail. As flow enters the meander bend, it is deflected across the counterpoint bar toward the concave bank. At the bar tail, the flood waters move over the bar as a relatively shallow sheet, travelling almost parallel to the bar. At the bar head, however, the deeper flood waters are abruptly deflected across the bar where much of the flow is concentrated into the chutes. These direct jets of water toward the concave bank, where turbulent eddies are formed against the bank. The eddies scour deep pools immediately opposite the chutes and then move down the secondary channel. This flow undermines the bank and encourages erosion.

THE EVOLUTION OF GRAVEL COUNTERPOINT BARS

The growth of gravel counterpoint bars can be studied from two sources. First, the earliest stages of bar growth are highlighted by comparing bars of different sizes and examining the bars over a 4-year period. Second, sequential air photographs (taken in 1946 and 1971) and 6-inch Ordnance Survey maps (1886 and 1907) show bar evolution over a longer time span. Unfortunately, these maps can only give a rough indication of bar growth and channel migration because of the large time intervals between sources (average gap of 29 years).

Field observations show that the gravel counterpoint bars start as a bulbous deposit on the upstream edge of a riffle. These mounds of gravel are only a few square meters in

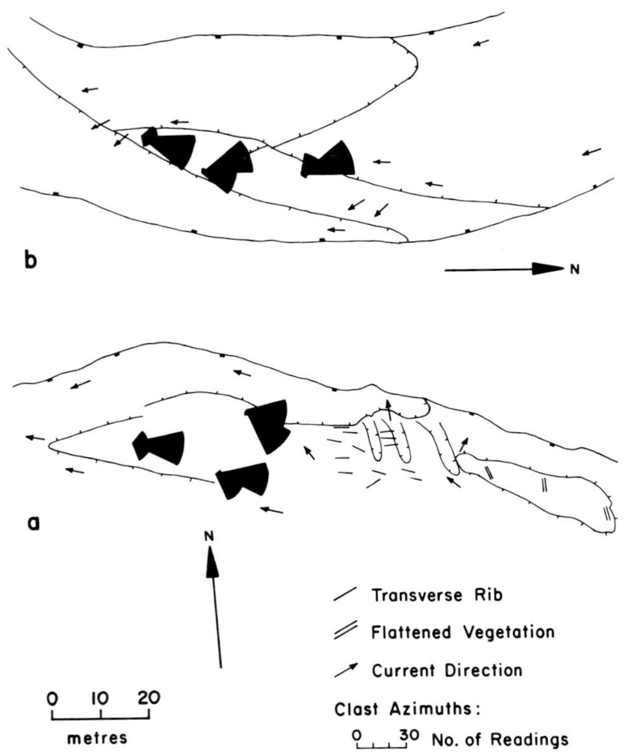

FIG. 6.—Flow patterns over gravel counterpoint bars. Rose diagrams represent the dip directions of the 50 largest clasts. Example (a) is also shown in Figure 4 and 5.

size, but have a relief of about 0.5 m (Fig. 7). There is a narrow deep "bite" in the adjacent bank, which suggests that bar formation is associated with localized and intense erosion. Over a few years, the gravel mound extends downstream along the edge of the riffle, forming a secondary channel between the bar and bank. At this stage, the deposit is a full fledged counterpoint bar.

The longer term evolution of a representative gravel counterpoint bar is shown in Figure 8a. A large bar was present in 1886. The position of this bar had not changed by 1907, although the associated point bar had migrated slightly downstream. By 1946 the point bar had translated even further downstream, but the counterpoint bar had been abandoned and incorporated into the floodplain. The outer margin of this bar is marked by a long, narrow and gently curved depression which reflects the abandoned secondary channel (Fig. 8b). This feature can be distinguished from meander cutoffs which are wider (above 10 m), have greater relief and are more tightly curved. The 1971 air photograph shows further downstream translation of the point bar and initiation of a new counterpoint bar. Both air photographs show two gently curved depressions just to the west of the railway embankment (Fig. 8b). These have a geometry similar to that of the bar abandoned between 1907–1946 and probably formed in the same way. Similar floodplain topography is seen elsewhere along the Tywi valley, where it is associated with active gravel counterpoint bars (Fig. 8c).

Field observations and historical records therefore suggest:

(1) Gravel counterpoint bars are associated with translation of the parent meander bend.

(2) Translation may eventually cause the bar to be abandoned and incorporated into the floodplain. This is because the bar becomes ensconced in the concave bank while the point bar remains free to migrate downstream. This is similar to the process of "mobile bar cutoff" described from other Welsh rivers (Lewis and Lewin, 1983).

(3) More than one sequence of bar initiation, growth and abandonment may occur during loop migration. The possible sequence of bar evolution is summarized in Figure 9.

THE FORMATION OF GRAVEL COUNTERPOINT BARS

The accounts of bar morphology, evolution and associated flow patterns can be combined to suggest a model of bar formation. Unfortunately, this is provisional because of

FIG. 7a,b.— The first stage in the formation of gravel counterpoint bars; small gravel mounds (arrows) on the upstream edge of a riffle. In both cases, flow is toward the observer and the channel is about 30 m wide. Note revetment of the retreating bank in (b).

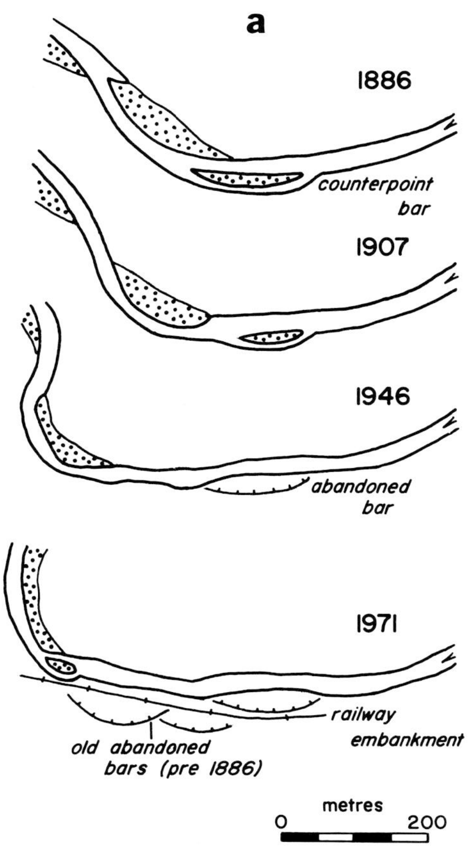

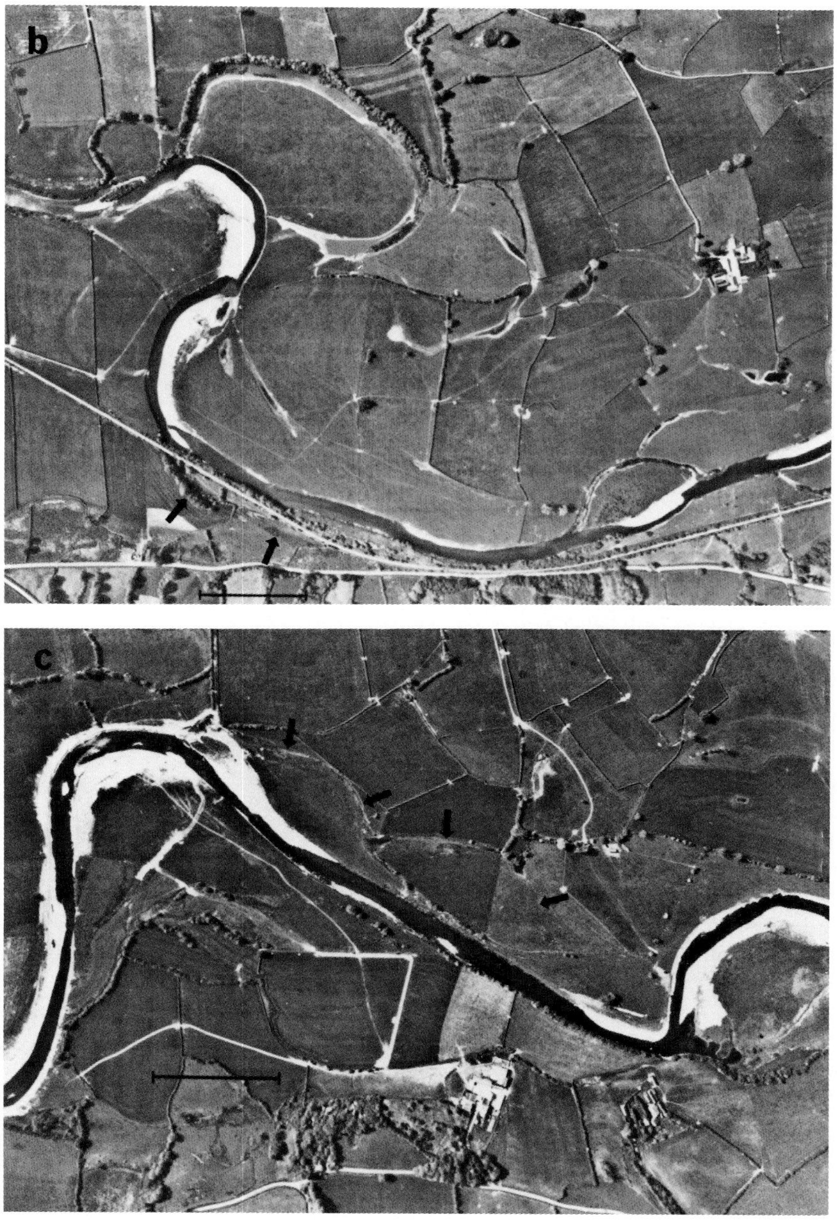

FIG. 8.—(a) The evolution of a gravel counterpoint bar, (b) air photograph of the same reach, taken in 1971, showing older (pre-1886) abandoned counterpoint bars (arrows). The bar which was abandoned between 1907 and 1946 can be seen to the right of the railway embankment. Scale is 200 m. (c) Abandoned counterpoint bars (marked), flow is from right to left, bar scale is 200 m.

the absence of detailed flow data. Gravel counterpoint bars appear to be initiated on the upstream edge of a diagonal riffle. Localized bank erosion occurs opposite this site, while gravel deposition occurs in the space provided by bank retreat. This pattern of erosion and deposition strongly suggests a component of cross-stream bedload and fluid transport. Such a feature has recently been recognized in some sand-bed streams (Dietrich) and others, 1979; Dietrich and Smith, 1983, 1984; Thorne and others, 1985). According to these authors, cross-stream movement of bedload is caused by three processes; flow acceleration induced by shoaling about the point bar head, flow deflection by obliquely oriented bedforms, and avalanching on a slope at the head of the point bar. Shoaling and flow deflection across an obliquely oriented gravel riffle are possible causes of cross-stream sediment transport in the Tywi.

Cross-stream movement must be especially effective if a gravel counterpoint bar is to form. This may be ensured by the presence of a long and relatively straight reach immediately upstream of the counterpoint bar. There are several field studies of flow leaving a long straight reach and entering a meander bend (Bhowmilk, 1982; Geldoff and DeVriend, 1983; Van Alpen and others, 1984). These studies show that maximum flow velocities and sediment transport rates do not occur along the convex upstream part of the meander bend, as they may when two bends are immediately adjacent (Jackson, 1976; Bridge and Jarvis, 1976, 1982). Instead, maximum velocities may occur along the concave bank, a condition which may encourage the formation of counterpoint bars.

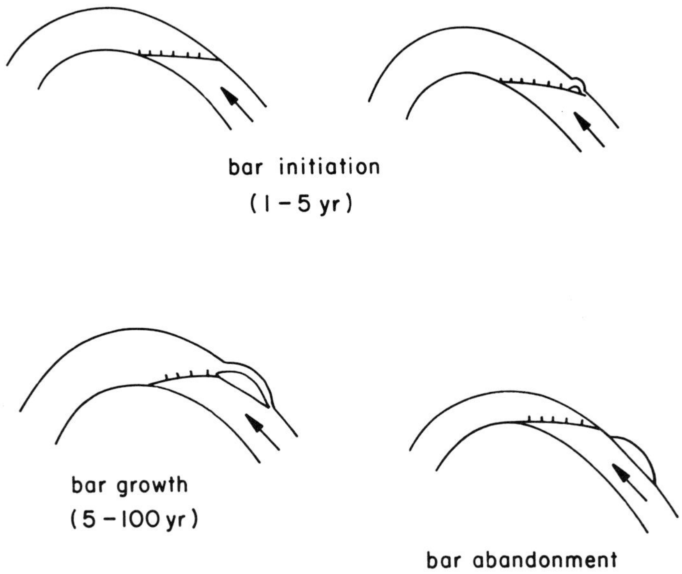

FIG. 9.—A summary of gravel counterpoint bar evolution, with possible time scale in years. Flow from right to left.

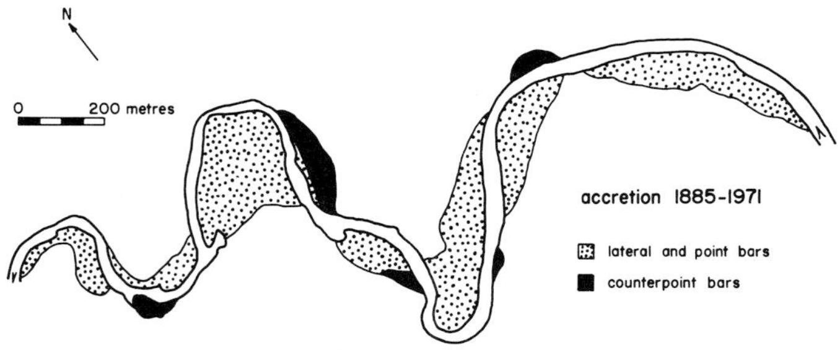

FIG. 10.—Floodplain composition near Llandeilo. Map shows bars accreted between 1885 and 1971, flow is from right to left.

SIGNIFICANCE AND CHARACTERISTICS

Gravel counterpoint bars are an important component of sedimentation in the River Tywi. To be of wider significance, however, they must occur in other gravel-bed rivers and have some likelihood of preservation. According to Lewin (1982, 1983), gravel couterpoint bars do indeed occur in many other Welsh rivers. Similar bars have not yet been described from other regions; however, this may reflect the comparative neglect of meandering gravel-bed rivers and the fact that existing studies concentrate on point bars.

The short-term (10–100 years) preservation potential of bars in the River Tywi can be estimated from historical sources. Gravel counterpoint bars can be recognized in air photographs and Ordnance Survey maps which go back about 100 years. If channel and bar positions are plotted on a common base, it is possible to estimate the area covered by counterpoint bar deposits during this period. This has been done for a typical reach of the Tywi (Fig. 10). Gravel counterpoint bars make up about 14 percent, by area, of the bars deposited since 1886, a significant proportion. As noted in a previous section, these bars have a distinctive topographic expression on the floodplain. As a result, older (pre-1886) abandoned counterpoint bars are easily recognized. Such abandoned bars are common at several sites along the Tywi valley, as suggested by Figures 8b and 8c. Hence, historical records and floodplain topography suggest that gravel counterpoint bars have a high preservation potential in the River Tywi. This may reflect the behavior of the parent meander bend; the counterpoint bar escapes immediate reworking as it is stranded upstream of a downstream migrating meander.

The sedimentological characteristics of gravel counterpoint bars are now compared with those of fine-grained counterpoint bars and gravel point bars. Such a comparison may be useful to sedimentologists attempting to recognize gravel counterpoint bars in the geologic record. In the River Tywi, fine-grained counterpoint bars consist of sandy silt and sand, have a bench-like shape and are separated from the outer bank by a narrow secondary channel. The deposit fines from the bar tail to the bar head. Augering shows that the bar tail consists of alternating laminae of coarse sand, silty clay and leaves. The sequence as a whole does not fine upward, in contrast to examples described from other rivers. Bar head sediments are sandy silts with many dispersed organic fragments but no discrete leaf layers. Fine-grained counterpoint bars, described by Page and Nanson (1982) and Nanson and Page (1983), and found in the Tywi, are compared with gravel counterpoint bars in Table 1. Clearly, channel position is the only shared characteristic.

Differences between gravel counterpoint bars and gravel point bars are less marked, however. The characteristics of the two types of bar are compared in Table 2, which is based on my own observations of gravel point bars in the Tywi and on accounts of modern and ancient examples by Lewin (1978), Blacknell (1980, 1982), Ori (1982) and Ricci-Lucchi and others (1981). The main differences are the elongate geometry of gravel counterpoint bars and the pres-

TABLE 1.—A COMPARISON OF FINE-GRAINED AND GRAVEL COUNTERPOINT BARS, BASED ON EXAMPLES IN THE RIVER TYWI AND THE ACCOUNTS OF PAGE AND NANSON (1982) AND NANSON AND PAGE (1983).

	Fine-Grained	Gravel
Morphology	Elongate and bench-like, steep side faces outer bank, narrow secondary channel, located upstream of bend apex, associated with confined bends	Elongate to bulbous, steep side faces inner bank, not always found with confined bends
Grain Size, Sedimentary Structures	Medium sand- to mud-size, finer than other within-channel deposits, overall fining upward, may fine upstream. Massive, basal planar cross bedding, organic detritus	Gravel and sand, coarsens upstream and toward outer bank, coarsens upward (?). *Gm* and low-angle *Gp*
Geometry	Ribbon-like bodies formed in the wake of the meander bend (?)	Ribbon-like bodies composed of offlapping wedges
Formation	Flow separation about a tightly curved meander bend. Associated with downstream migration of the meander bend and channel widening	Cross-stream fluid and bedload movement (?), induced by shoaling about the point bar head

TABLE 2.—A COMPARISON OF GRAVEL POINT BARS AND COUNTERPOINT BARS, MAINLY BASED ON THE RIVER TYWI, BUT ALSO ON THE ACCOUNTS OF BLACKNELL (1980, 1982), ORI (1982) AND RICCI LUCCHI AND OTHERS (1981).

	Point Bar	Counterpoint Bar
Morphology	Morphologically variable; simple and gently dipping, or well developed bar platform with a complicated and mosaic-like bar supra-platform, or may have a well developed avalanche face at the bar tail	Elongate to bulbous
Grain Size Patterns	Head-to-tail fining, may also fine from the channel to bar top. Bar tail of sand, silt and leaf layers	Head-to-tail fining, but coarsens toward the bar top (toward the main channel). No fine-grained tail
Sedimentary Structures	Mainly laterally extensive and low-angle *Gp*. Some basal high-angle *Gp*, capped by *Gm*. Gravel is laterally accreted	*Gm* with subordinate low-angle *Gp*. Gravel is laterally accreted
Geometry	Forms both ribbon and sheet bodies	Ribbon-like bodies composed of offlapping wedges. May have abrupt contact with channel-fill deposits (counterpoint bar deposits will wedge out from this contact)

ence of extensive, laterally accreted and low-angle cross-stratification in gravel point bars.

CONCLUSION

Sedimentation is important along the concave banks of a meandering gravel-bed river in South Wales. Deposition occurs as gravel bars, a poorly known type, as well as the more familiar concave bank bench. Gravel counterpoint bars seem to have a high preservation potential and may be a useful addition to the facies models for meandering rivers.

ACKNOWLEDGMENTS

This paper was written while the author was in receipt of a N.E.R.C. post-graduate studentship, for which the author is grateful. Professor J.R.L. Allen, Dr. T. Astin, and fellow graduate students suggested improvements to earlier versions of the manuscript. Mr. A. Cross drafted the figures and Mr. J. Watkins gave photographic assistance. N. Lewis is thanked for moral support and copious reserves of patience.

REFERENCES

BHOWMIK, N. G., 1982, Shear stress distribution and secondary currents in straight channels, *in* Hey, R. D., Bathurst, J. C., and Thorne, C. R., eds., Gravel-Bed Rivers: Fluvial Processes, Engineering, and Management: Chichester, Wiley, p. 31–55.

BLACKNELL, C., 1980, Point bar formation in Welsh rivers: Unpublished Ph.D. Thesis, University College of Wales, Aberystwth, 385 p.

———, 1982, Morphology and surface sedimentary features of point bars in Welsh gravel-bed rivers: Geologic Magazine, v. 119, p. 181–192.

BLUCK, B. J., 1982, Texture of gravel bars in braided streams, *in* Hey, R. D., Bathurst, J. C., and Thorne, C. R., eds., Gravel-Bed Rivers: Fluvial Processes, Engineering and Management: Chichester, Wiley, p. 339–355.

BRIDGE, J. S., AND JARVIS, J., 1976, Flow and sedimentary processes in the meandering river South Esk, Glen Cova, Scotland: Earth Surface Processes and Landforms, v. 1, p. 303–336.

———, AND ———, 1982, The dynamics of a river bend: A study of flow and sedimentary processes: Sedimentology, v. 29, p. 449–542.

CAREY, W. C., 1969, Formation of floodplain lands: Journal of the Hydraulics Division, American Society of Civil Engineers, v. 95, p. 981–994.

DIETRICH, W. E., SMITH, D. J., AND DUNNE, T., 1979, Flow and sediment transport in a sand-bedded meander: Journal of Geology, v. 87, p. 305–315.

———, AND SMITH, D. J., 1983, Influence of the point bar on flow through a curved channel: Water Resources Research, v. 19, p. 1173–1192.

———, AND ———, 1984, Bedload transport in a river meander: Water Resources Research, v. 20, p. 1355–1380.

GELDOF, H. J., AND DEVRIEND, H. J., 1983, Distribution of main flow velocity in alternating river bends, *in* Collinson, J. D., and Lewin, J., eds., Modern and Ancient Fluvial Systems: International Association of Sedimentologists Special Publication 6, p. 85–96.

HICKIN, E. J., 1979, Concave bank benches in the Squamish River, British Columbia: Canadian Journal of Earth Sciences, v. 16, p. 200–203.

HIRSCH, P. J., AND ABRAHAMS, A. D., 1981, The properties of bed sediments in pools and riffles: Journal of Sedimentary Petrology, v. 51, p. 757–760.

JACKSON, R. G., 1976, Depositional model of point bars in the lower Wabash River: Journal of Sedimentary Petrology, v. 46, p. 579–594.

KOSTER, E. H., 1978, Transverse ribs: Origin and paleohydraulic significance, *in* Miall, A. D., ed., Fluvial Sedimentology: Canadian Society of Petroleum Geologists Memoir 5, p. 161–186.

LEWIN, J., 1978, Meander development and floodplain sedimentation: A case study from mid-Wales: Geological Journal, v. 13, p. 25–36.

———, 1982, British floodplains, *in* Adams, B. H., Fenn, C. R., and Morris, L., eds. Papers in Earth Sciences, Norwich, Geobooks, p. 21–37.

———, 1983, Changes in channel pattern and floodplain, *in* Gregory, K. J., ed., Background to Palaeohydrology, Chichester, Wiley, p. 303–319.

———, AND MANTON, M. M. M., 1975, Welsh floodplain studies: The nature of floodplain geometry: Journal of Hydrology, v. 25, p. 37–50.

LEWIS, G. W., AND LEWIN, J., 1983, Alluvial cutoffs from Wales and the Borderlands, *in* Collinson, J. D., and Lewin, J., eds., Modern and Ancient Fluvial Systems: International Association of Sedimentologists Special Publication 6, p. 145–154.

MILNE, J. A., 1982, Bed-material size and the riffle-pool sequence: Sedimentology, v. 29, p. 267–278.

NANSON, G. C., AND PAGE, K. P., 1983, Lateral accretion of fine-grained concave bank benches in meandering rivers, *in* Collinson, J. D., and Lewin, J., eds., Modern and Ancient Fluvial Systems: International Association of Sedimentologists Special Publication 6, p. 133–143.

NATURAL ENVIRONMENT RESEARCH COUNCIL, 1975, Flood studies Report, Her Majesty's Stationary Office, p. 136–138.

ORI, G. G., 1982, Braided to meandering channel pattern in humid-region alluvial fan deposits, River Reno, Po plain, (Northern Italy): Sedimentary Geology, v. 31, p. 231–248.

PAGE, K., AND NANSON, G., 1982, Concave bank formation and associated floodplain formation: Earth Surface Processes and Landforms, v. 7, p. 529–543.

RICCI LUCCHI, F., COLELLA, A., ORI, G. G., AND OGLIANI, F., 1981, Pliocene fan deltas of the Intra-Apenninic basin, Bologna: Excursion Guidebook, Second European Regional Meeting, International Association of Sedimentologists, p. 81–162.

SHAW, J. AND KELLERHALS, R., 1977, Paleohydraulic interpretation of antidune bedforms with application to antidunes in gravel: Journal of Sedimentary Petrology, v. 47, p. 257–266.

TAYLOR, G., AND WOODYER, K. D., 1978, Bank deposition in suspended load streams, *in* Miall, A. D., ed., Fluvial Sedimentology: Canadian Society of Petroleum Geologists Memoir 5, p. 257–275.

THORNE, C. R., ZEVENBERGEN, L. W., PITLICK, J. C., RAIS, J. B., BRADLEY, J. B., AND JULIEN, P. Y., 1985, Direct measurements of secondary currents in a sand-bed river: Nature, v. 316, p. 746–747.

VAN ALPEN, J. S. L. J., BLOKS, P. M., AND HOEKSTRA, P., 1984, Flow and grain size patterns in a sharply curved river bend: Earth Surface Processes and Landforms, v. 9, p. 513–522.

MEANDERING RIVER POINT BAR LITHOFACIES MODELS: MODERN AND ANCIENT EXAMPLES COMPARED

DERALD G. SMITH

Department of Geography, University of Calgary, Calgary, Alberta T2N 1N4, Canada

ABSTRACT: Modern meandering river point bar deposits formed in fluvial and tidally-influenced environments were investigated to explain why large-scale epsilon cross-stratification (ECS) is common in ancient fluvial rocks but appears to be absent in modern deposits. To resolve this problem several modern meandering systems were studied; the Athabasca upper delta plain in northeast Alberta, Canada; the mesotidally-influenced reach of the Willapa River, southwest Washington State, U.S.A.; and the lower Daule and Babahovo Rivers, Ecuador, which have micro- and mesotidally-influenced depositional conditions. As well, exposed point bar facies in 18-m-high cutbanks were examined in the fluvial, lower Liard River (Middle Holocene), NWT, Canada, and tidally-influenced Willapa River (Late Pleistocene), Washington State.

Based on sedimentologic results obtained from these areas, a threefold lithofacies classification of point bar deposits is proposed: (1) fluvial sandy point bar facies, (2) low-energy fluvial and microtidally-influenced (upper estuary) point bar facies, (3) mesotidally-influenced point bar facies deposited in upper and middle estuary settings. The latter two facies are very similar to many reported ancient meandering river point bar rocks. The three lithofacies models are compared with four ancient examples of point bar rocks selected from Alberta, Canada, the Lower Cretaceous middle McMurray Formation (Athabasca Oil Sands), the Upper Cretaceous Judith River and Horseshoe Canyon Formations and the Paleocene Paskapoo Formation.

INTRODUCTION

In 1977, at the First International Fluvial Sedimentology Symposium, two important papers about stratigraphic lithologies in meandering river point bar deposits indicated that there exists a considerable difference between modern and ancient deposits, rather than the expected similarity. Jackson (1978) pointed out that most documented examples of ancient point bar deposits contain alternating sandstone and shale sequences (couplets) as epsilon cross-stratification (ECS), whereas modern point bar examples are dominated by sand, lacking mud interbeds. Supporting that observation, Mossop (1978) reported that as much as 25 m of ECS with sand-mud couplets, averaging 12° inclination, occur in the Lower Cretaceous sediments of the Athabasca Oil Sands.

In spite of the absence of mud interbeds in modern sandy point bar deposits, there has been an increasing tendency by geologists to assume that sand and mud sequences in ancient ECS are associated with fluvial meandering river deposits (Cotter, 1971; Leeder, 1974; Elliott, 1976; Stewart, 1981; Flach and Mossop, 1985, to name a few).

While investigating modern, low-gradient (energy), fluvial meanders, Bluck (1971), Bridge and Jarvis (1976, 1982) and Taylor and Woodyer (1978) reported ECS in the upper parts of point bars; however, Jackson (1981) and Calverley (1984) observed the presence of mud interbeds of irregular thickness at irregular stratigraphic intervals in sand in both upper and lower parts of point bars. Jackson worked on two small streams (mean discharge 2.5 cms) with channel gradients of 22 cm/km and point bar slopes averaging 12°. Calverley worked on the Embarras distributary channel of the Athabasca upper delta plain, Canada. The Embarras and Athabasca Channels together carry 2600 cms at bankfull discharge, and point bar slopes average 12.5°.

From modern mesotidal creeks (2–4 m tidal stage range) on intertidal flats in Europe, van Straaten (1954), Reineck (1958), Oomkems and Terwindt (1960), Bridges and Leeder (1976) and de Mowbray (1983) documented the presence of sand-mud couplets in ECS. While these point bar deposits rarely exceeded 1 m in thickness, they did contain decimeter-thick sand-mud couplets.

Results from studies of the modern tidal creeks and low-energy fluvial meander deposits suggest that the ECS of the Lower Cretaceous McMurray Formation, described by Mossop and Flach (1983) and Flach and Mossop (1985), was probably deposited in a reach of a meandering river-estuary with a micro- and mesotidal influence. To verify this hypothesis, two relatively undisturbed river systems were selected for study, the Willapa River in southwest Washington State, U.S.A. and the Daule-Babahovo River system in Ecuador, S. A. Estuarine conditions are present in the lower reaches of all three rivers. An upper estuary-river reach is defined here as the part of a lower river reach influenced by tidal fluctuation. An estuary ends upriver where tidal influence becomes inactive.

In the course of studying the point bar deposits in tidally-influenced meanders, it became clear why previous researchers of modern point bar deposits may have avoided fluvial-tidal systems, because of the difficult muddy field conditions, making trenching and coring nearly impossible. The lack of detailed sedimentologic data from point bars in tidally-influenced rivers suggested that an important area of sedimentology had been avoided. By studying point bar deposits in several different depositional environments, it became clear that several different lithofacies trends exist for meandering rivers.

The prime objective of this paper is to report on and classify the different lithostratigraphic trends observed in vibracores taken from point bars in the different modern fluvial and tidally-influenced systems. A second objective is to compare the three commonly observed lithofacies trends (models) from modern deposits with several ancient meandering river deposits located in Alberta, Canada.

FIELD METHODS

Stratigraphic logs taken from vibracores and trenches included the following information: mean grain size of sediment (grain size was determined by comparing sand grains with an Amstrat grain size chart with the aid of a 30-power field microscope) from each lithologic unit; presence of mud blocks, mudballs and mud chips; slope angle of point bar surfaces; mud interbeds in sands; sedimentary structures; presence of organic debris (type of material, ground-up litter, logs, etc.); presence of bioturbation; type of lithologic transition (erosional or gradational). Locations of each trench or core on a point bar were surveyed.

A geophysical bottom profiler was used to determine the underwater morphology of both point bar slopes and channel bedforms. To provide horizontal scale, a continuous readout current meter to measure boat speed was used in conjunction with the profiler. On a given point bar, 5 to 10 channel cross section profiles were recorded, whereas channel bedforms in the thalweg were measured along one or two longitudinal profiles parallel to flow.

Laboratory work consisted of detailed logging of vibracores from the Willapa River. Grain size analysis of sand and mud samples was carried out using sieving and hydrometer methods.

MODERN MEANDERING RIVER DEPOSITS

Modern (Holocene and Late Pleistocene) meandering river point bar deposits were studied in four different field sites, located in Canada, the United States and Ecuador, South America (Fig. 1). The areas were selected because either they represented prime examples of a specific meandering river deposition environment with minor disturbance by man, or I had previously worked on each area.

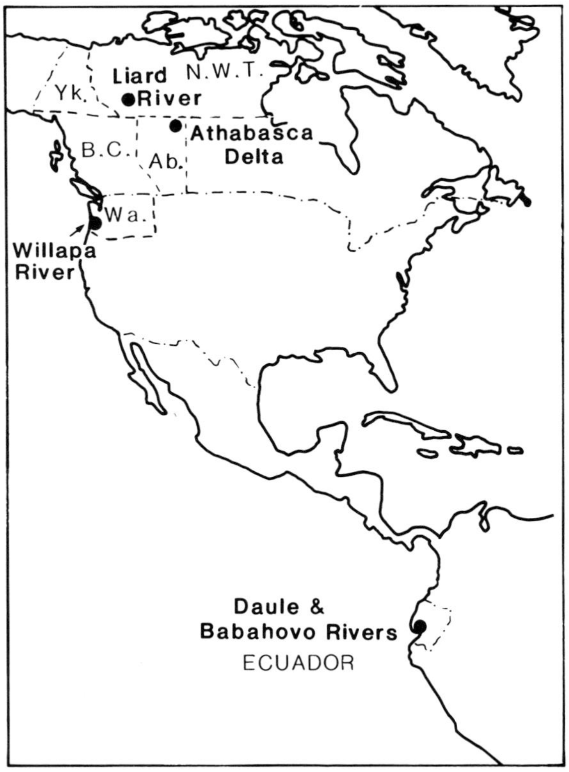

FIG. 1.—Index map showing the locations of the four modern meandering river point bar deposits that were investigated.

Liard River, Canada, fluvial sandy point bars.—

The Liard River study area consists of a 45-km meandering channel reach, located in the southwest corner of the Northwest Territories, Canada, at 61° latitude (Fig. 1). The study reach straddles the confluence of the Nahanni River near the town of Nahanni Butte. Over the past 5,750 years (C-14 date, GSC-4103), the Liard has downcut 18 m, exposing point bar facies in cutbanks of the Middle Holocene Liard River terrace. Aerial photographs show patterns of oxbow lake fills and scroll bars on the terrace surface, which verify the deposits as point bars. Seven stratigraphic logs were taken at seven different cutbanks.

Underlying the point bar deposits, channel bed facies consist of pebbly sand to sandy gravel 0.5–2 m thick. Point bar facies of medium, trough crossbedded, clean sand with a few pebbles extend to as much as 16 m thick. The upper 2 to 3 m of point bar is dominated by fine to medium rippled sand, with overbank silt 0.5 m thick capping the sequence. Similar point bar sequences have been reported by Jackson (1976), Nanson (1980), and Bridge and Jarvis (1982). Such sand-dominant point bar deposits are classified in this paper as fluvial sandy point bar lithofacies.

Athabasca Delta, Canada, low-energy fluvial point bars.—

Located in the northeast corner of Alberta, the Athabasca River Delta contains two meandering distributaries, the Athabasca and Embarras Channels, in a 25-km reach of the upper delta plain (Fig. 1). The Embarras Channel point bar facies were studied primarily by trenching and taking vibracores to depths of 8.5 m, following the method of Calverley (1984). The channel has a sinuosity index of 1.75, whereas point bar slopes average 12.5°. Point bar surfaces are draped with mud as much as 30 cm thick which causes considerable difficulty when vibracoring.

Point bar facies of the Embarras Channel are incomplete because the vibracorer was unable to sample the complete point bar and channel-bed sediment sequence (Fig. 2). In spite of incomplete cores, Calverley (1984) felt that at least 70 percent of the sequence was penetrated based on adjacent channel depths. Five stratigraphic logs from vibracores taken from point bars down the Embarras Channel show two important trends (Fig. 2). First, fining upward of sediment grain size (coarse to fine sand) is prominent in all of the cores. Second, in each core mud beds of irregular thickness at irregularly spaced intervals are present. Mud beds are inferred to be parallel to the point bar surface which averages 12.5° slope. Only Jackson (1981) has reported similar lithofacies for several low-gradient streams in the American middle west.

Based on the low channel gradient coupled with the presence of ECS with irregularly thick and variably spaced mud beds, I classify the Embarras Channel point bars as low-energy fluvial point bar lithofacies.

Willapa River, U.S.A., mesotidally-influenced point bars.—

The Willapa River is located in southwest Washington, U.S.A., with the study reach located on the east side of

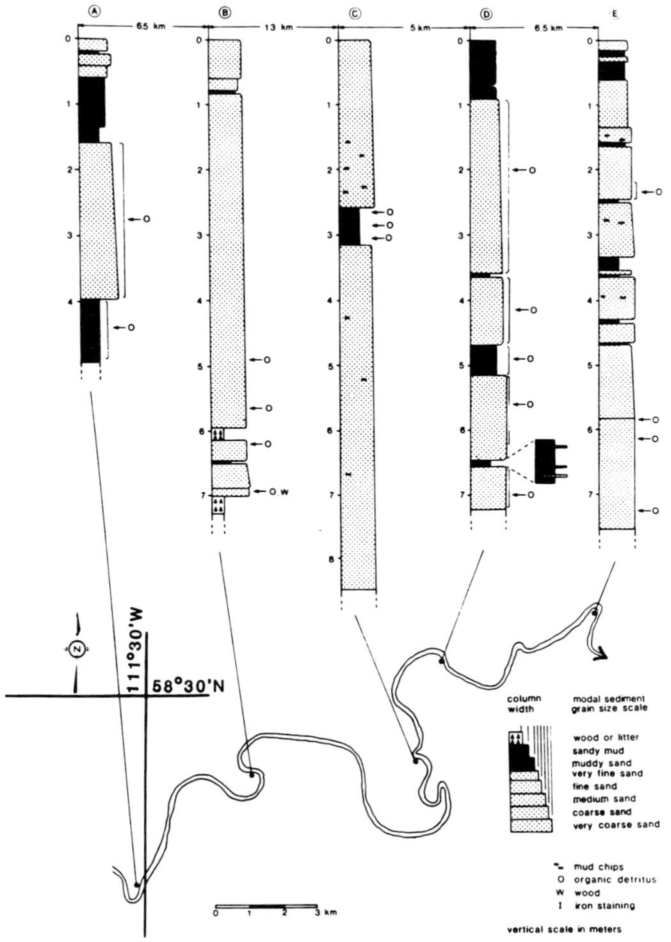

FIG. 2.—Five stratigraphic logs from vibracores taken from the meandering Embarras distributary channel of the Athabasca upper delta plain, northeast Alberta (Calverley, 1984). Datum is the river surface 2 m below bankfull stage.

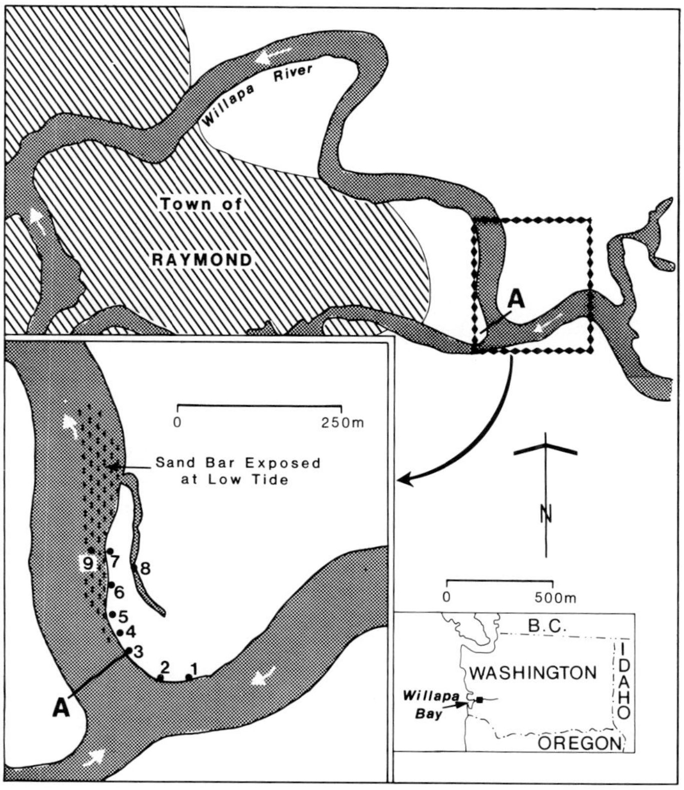

FIG. 3.—A detailed location map of the vibracore sites from the Willapa River-upper estuary study area near Raymond, southwest Washington State, U.S.A.

the town of Raymond (Figs. 1, 3). The Willapa River experiences tidal flow reversals and a tidal range of 3.4 m (spr. tide), which is mesotidal (2–4 m range) in Hayes' (1976) classification. Channel gradient is not measurable and point bar surfaces dip at 18° to 9.5° (upstream to downstream), averaging 12.5°. Overbank tidal marsh mud deposits form steep blocky cliffs 1–3 m high at the top of the point bars.

Point bar facies are dominated by ECS of sand-mud couplets extending the full length of the point bar (inferred from similar looking Late Pleistocene point bar deposits exposed in a nearby sea cliff). Gravel and mudballs rest in erosional contact with underlying mud at the channel base. Upward, inclined sand-mud couplets contain mudballs, mud chips, mud blocks and organic litter. Bioturbation is most prevalent in the upper point bar sediments.

Several lithologic trends are present in the Willapa point bars. First, mean grain size fines both upsection and downstream around a point bar (upstream to downstream) (Fig. 4). Second, and most important, point bar facies are dominated by a rhythmic style of alternating sand and mud beds (Fig. 4).

Similar mesotidally-influenced point bar lithofacies of Late Pleistocene age are exposed in sea cliff outcrops at Bruce Park, located 10 km west of Raymond. Here, very similar stratigraphic sequences dominate the outcrop. The only differences are the ECS unit is 3 m thicker than the present day Willapa point bars and organic-rich mud is rare.

Based on the location in an upper estuary with a tidal range of 3.4 m, I classify these point bars with sand-mud couplets as meso-tidally-influenced.

Daule and Babahovo Rivers, Ecuador, fluvial and micro- and mesotidally-influenced point bars.—

In Ecuador, South America, 2° latitude south of the equator, the south-flowing Daule and Babahovo Rivers join near the port city of Guayaquil to form the Guayas River, an estuary 5 km wide by 70 km long before entering the Gulf of Guayaquil (Figs. 1, 5). At Guayaquil the tidal range is 4.5 m (spr. tide). Tidal fluctuations affect the Daule and Babahovo 65 km upriver from Guayaquil. Measurable salanities extend 20 km north of Guayaquil. The Daule is 300 m wide by 12 m deep and the lower Babahovo is 800 m wide by 15 m deep at bankfull flows. Bottom profiles of both rivers indicate underwater point bar slopes averaging 12°.

Due to the overbank tidal marsh mud, varying between 4 and 7 m thick, vibracoring with 6-m-long core barrels was only successful by working over the side of a boat at low to middle tide. Thus, taking cores as much as 2 m long

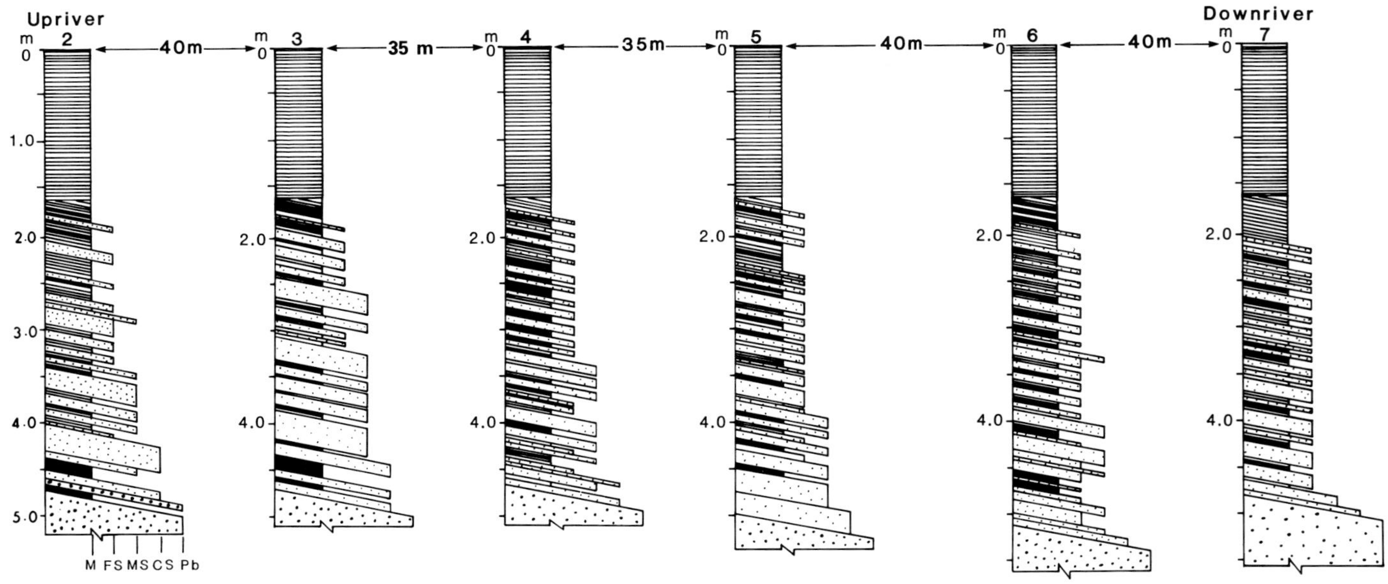

FIG. 4.—Stratigraphic logs 2 to 7 are from point bar A (Fig. 3) in the Willapa River. Note the ECS sand-mud couplets in each log. The upper 1.6 m of sediment is tidal marsh mud. Both vertical and horizontal distances are in meters. Datum is bankfull stage.

was the only means of obtaining sedimentologic information of point bar deposits.

Six stratigraphic logs from 13 vibracores taken from Daule point bars show a downriver sequential change of point bar lithofacies styles (Fig. 5). A very similar sequence is present in the point bars of the Babahovo River. The lithofacies trend in logs 5 and 6 is very similar to the rhythmic sand-mud couplets present in the Willapa River (Fig. 4). Mudballs, organic litter and bioturbation are present, though not shown (Fig. 5). Note the similarity between logs 1 and 2 and point bar facies in the Liard and descriptions by Jackson (1978), and the similarity between logs 3 and 4 and lithofacies data from the Athabasca Delta by Calverley (1984).

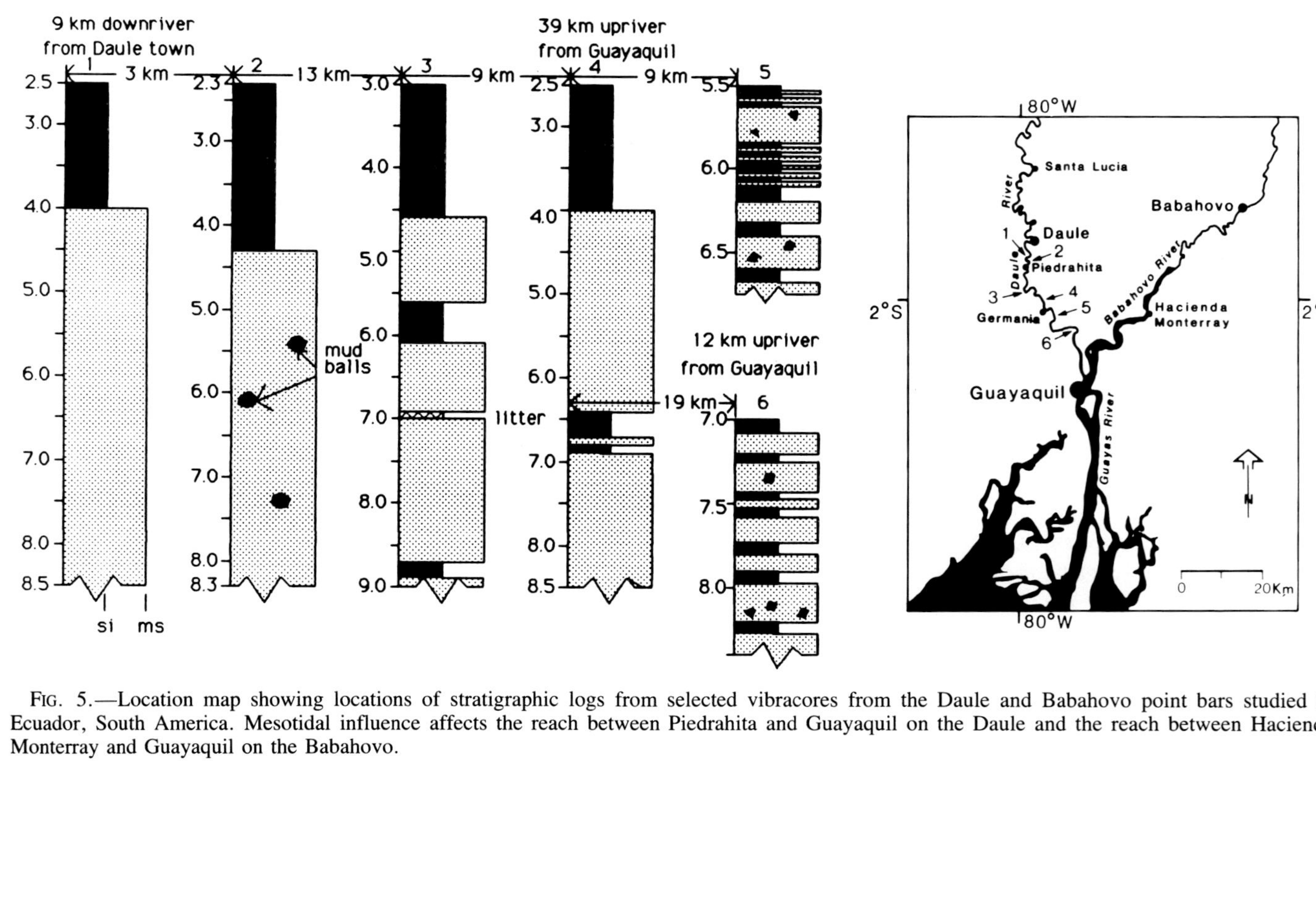

FIG. 5.—Location map showing locations of stratigraphic logs from selected vibracores from the Daule and Babahovo point bars studied in Ecuador, South America. Mesotidal influence affects the reach between Piedrahita and Guayaquil on the Daule and the reach between Hacienda Monterray and Guayaquil on the Babahovo.

Based on the variety of lithofacies styles and ranges of tidal influence, all of the aforementioned lithofacies classes are present in the Daule River and estuary.

DISCUSSION OF POINT BAR LITHOFACIES MODELS

A history of meandering river lithofacies models began with Allen (1970). His model incorporated several earlier ideas: 1) that low-angle lateral accretion bedding of ECS could be used to identify ancient meandering river rocks (Wright, 1959), later termed by Allen (1963) as epsilon cross-stratification (ECS), 2) that meandering river deposits contain an upward-fining trend in sediment grain size (Allen, 1963), and 3) that ripple and trough cross-stratification current directions form at right angles to the dip direction of epsilon cross-stratification (Allen, 1965). These early ideas set the stage for minor refinements to the model (Jackson, 1976, 1978, 1981; Collinson and Thompson, 1982; Galloway and Hobday, 1983; Walker and Cant, 1984). Besides the commonly cited sandy point bar facies, Jackson (1981) and Calverley (1984) noted the presence of variably thick mud interbeds in sand at irregular stratigraphic intervals while working on modern low-energy fluvial point bars.

Vibracore results from point bars in the Willapa, Daule and Babahovo Rivers strongly suggest that an additional lithofacies trend occurs in modern meandering tidally-influenced rivers in estuarine settings (Figs. 4, 5). The three rivers which contain ECS of rhythmic sand-mud couplets are located in reaches with a 2 to 4 m tidal range (mesotidal). Jouanneau and Latouche (1981) refer to such areas as upper estuary or riverine estuary in the Gironde Estuary of France. Mesotidally-influenced point bar facies are characterized by ECS with rhythmic sand-mud couplets (Figs. 4, 5), a facies common in many ancient meandering rock sequences. This facies is most similar to Jackson's (1978) lithofacies class number 1, formed in a muddy fine-grained stream.

By combining the often cited sand-only lithofacies model (Allen, 1970; Walker and Cant, 1984), plus the lithofacies trends reported by Jackson (1981) and Calverley (1984), coupled with the present results, three lithofacies models are proposed for different meandering river conditions (Fig. 6). The downriver sequential change implies that all three point bar styles can occur in one fluvial-estuarine system, beginning on the left with fluvial sandy deposits, then downriver to either low-energy fluvial or micro-tidally-influenced deposits, and ending with mesotidally-influenced deposits (Fig. 6). Point bar data from the Daule River is very similar to the proposed threefold classification scheme for point bar deposits (Figs. 5, 6).

While it might be argued that the tidally-influenced point bar facies located in upper and middle estuarine settings should not be classified as fluvial, two reasons can be presented for a fluvial classification. First, sand in the ECS is thought to be deposited by fluvial floods coupled with ebb tide, though mud is thought to be deposited during low stage fluvial-ebb tidal and flood tidal flows (Mossop and Flach, 1983; Flach and Mossop, 1985). Second, most ECS with rhythmic sand-mud couplets in ancient rock sequences have been interpreted as fluvial in origin, though they may have been of micro- and mesotidally-influenced river/upper to middle estuary origin (Stewart, 1981a, b; Cotter, 1971; Leeder, 1974, to name a few).

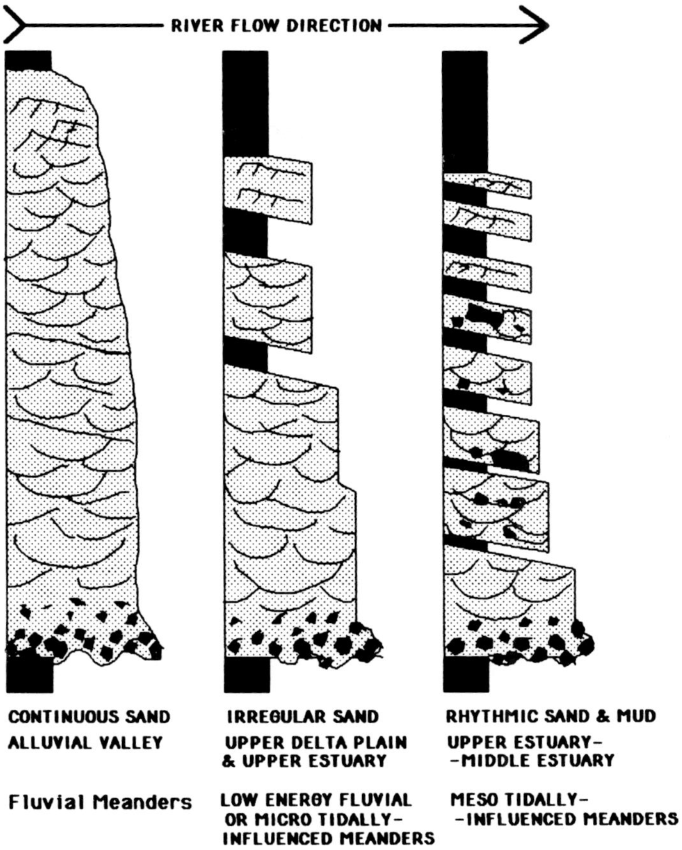

FIG. 6.—A threefold lithofacies classification for meandering river-estuarine point bar deposits based on modern point bar deposits from previous research and results from this study.

It must be remembered that transitions from one facies trend to another must exist and should be expected in studies of point bars in either modern deposits or ancient rocks. Moreover, Calverley (1984) noted that considerable variation of lithofacies style can occur within one point bar. She observed that sand-only facies can occur in the upstream part of a point bar, while irregular thicknesses of mud at irregular intervals in sand can occur in the downstream part of the same point bar. The same variation in a single point bar sequence was noted by Wood (1985) in Upper Cretaceous point bar rocks in Alberta. The three lithofacies models presented here are not intended to represent all possible trends present in point bar deposits, but instead are an expansion of a concept, which, if applied to the interpretation of a longitudinal trend of ECS in an ancient rock unit, can prove useful to predict both upchannel and downchannel facies trends.

MODERN AND ANCIENT EXAMPLES COMPARED

Four published examples of ancient rocks interpreted as point bar sequences have been selected from Alberta, Canada, for comparison with the three lithofacies models pre-

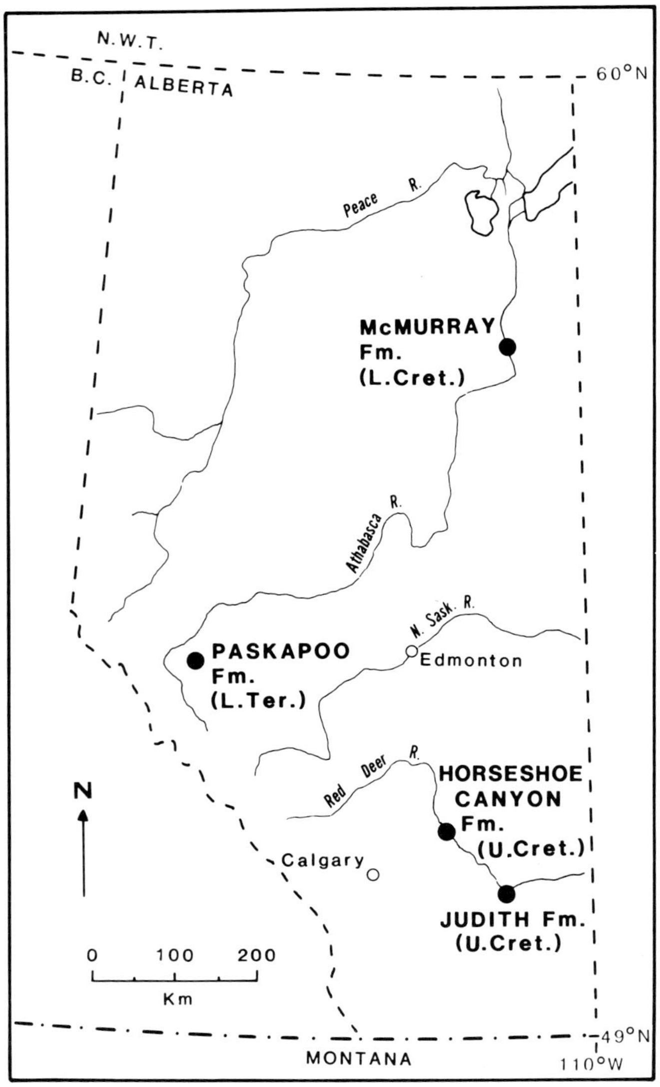

FIG. 7.—Index map showing the locations of four ancient interpreted point bar deposits in Alberta, Canada, which are compared with similar facies from modern point bars.

sented. The ancient examples include the Lower Cretaceous middle McMurray Formation, northeast Alberta, the Upper Cretaceous Judith River and Horseshoe Canyon Formations located along the Red River in southeast Alberta, and the Lower Tertiary Paskapoo Formation located in the foothills of west central Alberta (Fig. 7). Extensive outcrops of all four formations are readily accessible.

Middle McMurray Formation.—

The Lower Cretaceous middle McMurray Formation consists of fluvial-estuarine deposits, which contain bitumen-rich sand in channel meander belts oriented northwest-southeast (Fig. 7) (Stewart and MacCallum, 1978). The meandering channel deposits have been interpreted as fluvial point bar deposits by Mossop and Flach (1983) and Flach and Mossop (1985). They describe ECS with rhythmic sand-mud couplets in point bar deposits as much as 25 m thick (Fig. 8). The upper point bar sediments and overbank mud have been removed by a transgression. Decimeter-thick sand and mud beds are inclined at an average of 12° with fine sand and silt dominating the grain size distribution. Sand beds have mudballs and chips, whereas mud beds commonly have sand grains and sometimes very thin (one grain-diameter-thick) sand interbeds (not reported in the literature but observed in core). Bioturbation of laminated mud and sand beds is common in the upper point bar but decreases with depth. There is an absence of shells and shell fragments throughout. Thin beds or organic litter of coffee ground-size particles to woody fragments several centimeters in diameter are common, particularly upsection. In the upper point bar, sand beds contain cosets of ripple structures, whereas lower in the section trough cross-stratification is more common. Paleocurrent directions from sedimentary structures are at right angles and unidirectional to the dip direction of the ECS (Mossop and Flach, 1983). In outcrops the continuity of mud interbeds is rarely broken along dip and strike, often extending for tens of meters.

A comparison of the ECS deposits in the McMurray with the proposed lithofacies models indicates a similar facies trend to the mesotidally-influenced meanders located in upper to middle estuarine environments of the Willapa, Daule and Babahovo Rivers (Figs. 6, 8). The only notable difference is that the McMurray point bar sequences are much thicker than those studied in the modern.

Judith River Formation, Dinosaur Park.—

The Upper Cretaceous Judith River Formation, exposed in badland topography at Dinosaur Provincial Park along the lower Red Deer River, stratigraphically underlies the Horseshoe Canyon Formation (Fig. 7). The outcrops contain as much as 90 m of vertical exposure, sometimes exhibiting several complete ECS units in stacked vertical sequences (Koster, 1983; Wood, 1985).

Point bar deposits of the Judith River Formation consist of both sand-only and sand with irregularly thick and variably spaced mud beds (Fig. 9) (Wood, 1985). From outcrops Wood shows 23 lithostratigraphic logs, some with complete point bar sequences. The average thickness of point bar ECS sequences is 3.8 m, but the sequences can reach thicknesses of 20 m, with bedding planes of sand-mud couplets dipping at an average of 7° (Koster, 1983). Sand grain size is very fine to fine with ripple cross-stratification. In the lower point bars sand beds are dominated by trough cross-stratification.

Wood (1985) interprets the ECS sequences as low-energy fluvial point bar meanders with no tidal influence. Koster (pers. commun.), however, argues for a tidal estuarine origin with fluvial and both micro- and meso-tidally-influenced point bars, based on the transgressive marine sediments which immediately overlie the formation and the presence of marine trace fossils. It is quite probable that all three point bars are present in the area and both Koster (1983) and Wood (1985) are correct in their interpretations.

Horseshoe Canyon Formation.—

Point bar facies in the Upper Cretaceous Horseshoe Canyon Formation are best exposed in outcrops along Willow

FIG. 8.—(A) An outcrop of the Lower Cretaceous middle McMurray Formation (inclined unit) near the Steepbank River in northeast Alberta, showing ECS as alternating sand and shale sequences dipping to the right (Flach and Mossop, 1985). (B) The decimeter-thick bitumen-saturated sands are dark colored, and the shale interbeds are light toned. The sediments are interpreted in this paper as mesotidally-influenced point bar deposits. (C) A closeup view of ECS in outcrop. The pen for scale is 15 cm long.

Creek near Drumheller, Alberta (Fig. 7). Most of the recent research in this area has been carried out by Rahmani (1983), who interpreted the complete stratigraphic section as part of a marine regressive sequence in a tidally-influenced estuarine setting. He interpreted the point bar ECS deposits exposed along Willow Creek (Fig. 10) as middle to upper estuarine, influenced by mesotidal conditions.

The ECS deposit at Willow Creek consists of as much as 14 m of stacked sandstone-shale couplets. Thickness of individual beds varies, rarely exceeding 40 cm and averaging 20 cm. Sandstone makes up 65 to 75 percent of the facies. Rip-up shale clasts as well as millimeters-thick litter beds (appearance of coffee grounds) are common (Rahmani, 1983). Large slide blocks of overbank mud are present midway within the ECS; these blocks consist of rhythmic laminated silt-dominant mud. Similar blocks were observed in the modern point bars of the Willapa River. The dip angle of the ECS averages 12° and is very similar to the Willapa.

Based on the sedimentology as compared to the lithofacies models in Figure 6, I would concur with Rahmani's (1983) interpretation of meso-tidally-influenced point bars in an upper estuarine setting.

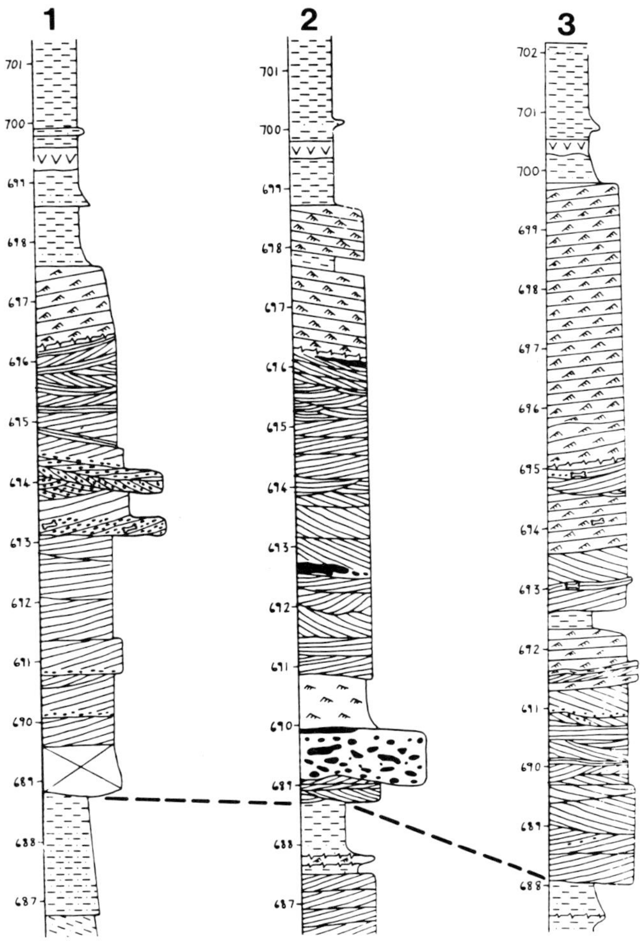

FIG. 9.—In the Upper Cretaceous Judith River Formation, Dinosaur Provincial Park, Alberta, three logs of the same channel sandstone unit show the basal erosional contact with underlying shale, followed by sandy point bar deposits, then capped by overbank shale. Many point bar units, though not in this figure, show shale bedding of irregular thickness at various stratigraphic intervals, suggesting either a low-energy fluvial or micro-tidally-influenced meandering river point bar depositional environment (Wood, 1985). Datum is sea level and the scale is in meters. Dominant grain size is fine sand. Mud chips and chunks are shown as darkened areas.

FIG. 10.—ECS is exposed in the lower badland slopes in the Upper Cretaceous Horseshoe Canyon Formation, located near Drumheller, Alberta. The point bar unit of sand-mud couplets is 12 m thick. The sandstone is light and shale is dark colored. The depositional environment is interpreted to have been a meso-energy, tidally-influenced meandering river in an upper to middle estuarine setting.

Paskapoo Formation.—

The Lower Tertiary (Paleocene) Paskapoo Formation with point bar meandering river deposits is best observed in abandoned open pit coal mines and road cuts in the vicinity of the Coal Valley area of the west central Alberta Foothills (Fig. 7). Jerzykiewicz and McLean (1980) have interpreted laterally extensive, thick (as much as 11 m) sandstone facies as fluvial sandy point bars (Fig. 11). The lower sandstone contains trough cross-stratification, whereas the upper sandstone exhibits ripple structures. Often a pebbly channel lag occurs at the base. In general, sedimentary structures are poorly preserved, making interpretations difficult.

Based on stratigraphic logs coupled with the great distance to the nearest marine shoreline at the time of deposition (at least 300 km to the northeast; R. McLean, pers. commun.), I concur with Jerzykiewicz and McLean (1980) who interpreted the deposits as fluvial sandy point bars. A comparison of Figure 11 with the lithofacies logs in Figure 6 indicates a close similarity with fluvial point bars.

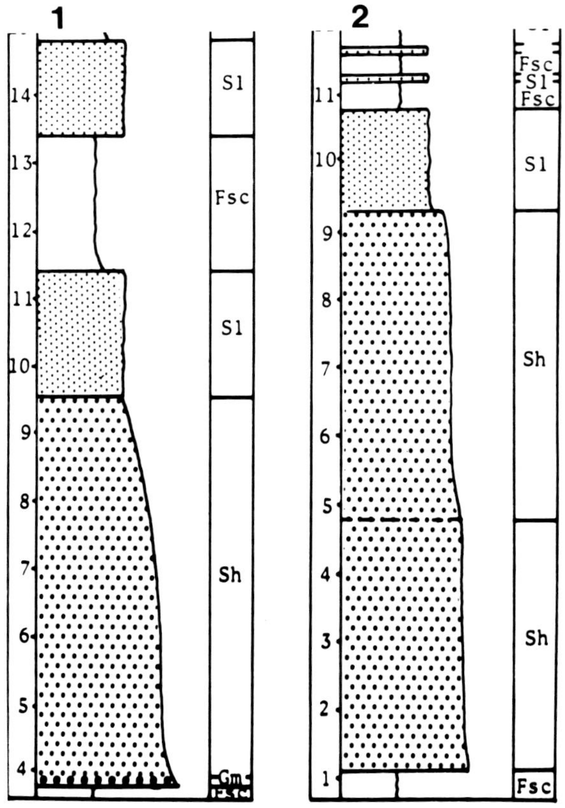

FIG. 11.—Two stratigraphic logs of the Lower Tertiary (Paleocene) Paskapoo Formation show sandstone-only point bar deposits, interpreted to be of fluvial origin (Jerzykiewicz and McLean, 1980). Datum is the base of the outcrop and the scale is in meters. Grain size of the sand is medium to coarse.

SUMMARY AND CONCLUSIONS

1. The paradox of mud interbeds being absent in modern sandy point bar deposits but present in ancient point bar rocks (Jackson, 1978; Collinson, 1978) is explained by the fact that modern point bar deposits with mud beds in low-energy fluvial and micro- and mesotidally-influenced settings have been avoided by previous researchers. Avoidance was probably due to the very difficult field working conditions and the absence of light weight, portable, deep vibracoring devices.
2. Proposed is a threefold classification of meandering river point bar deposits based on lithofacies style and depositional environments.
3. Determination of lithologic style of point bars in ancient meandering river rock sequences is important when using the proposed meandering river models to reconstruct depositional environments.
4. For the deposits studied, the sedimentologic and stratigraphic trends of ECS with mud interbeds include the following:
   a) mean grain size in sand beds fine upsection (from base to top),
   b) intensity of bioturbation increases upsection,
   c) thickness of mud beds increases upsection,
   d) thickness of sand beds decreases upsection,
   e) no shells or shell fragments were found in cores or exposures, in spite of micro- and mesotidally-influenced conditions,
   f) the presence and amount of wood fragments, organic litter and organic-rich mud increases upward,
   g) mud interbeds often contain a small percentage of sand grains,
   h) mesotidally-influenced point bars are overlain by 2–7 m of overbank marsh mud, whereas overbank mud on fluvial meanders is minimal,
   i) mean grain size of sand beds decreases downstream, from upstream to downstream, around point bars, whereas the number and thickness of mud beds increase.

ACKNOWLEDGMENTS

Special thanks are extended to R. Jackson and A. Calverley for their research on modern river deposits and to R. McLean, T. Jerzykiewicz, P. Flach, G. Mossop, R. Rahmani, E. Koster, R. Thomas, J. Wood, and J. Visser, who carried out research on ancient fluvial point bar rocks in Alberta. Field assistance, laboratory work, drafting and photography by L. Morrison, M. Woofter, T. Bartkiewicz, S. Vanderburgh, W. Halase, K. McMullen, J. Millen, M. Styk and R. Wheate are most appreciated. R. Davies, R. Rahmani and P. Warwick are thanked for providing many suggestions which improved the manuscript. Generous financial support from the Canadian National Science and Engineering Research Council, the Alberta Research Council and the Alberta Oil Sands Technology Research Authority is gratefully acknowledged.

REFERENCES

ALLEN, J. R. L., 1963, The classification of cross-stratified units, with notes on their origin: Sedimentology, v. 2, p. 93–114.

———, 1965, Fining upward cycles in alluvial successions: Geological Journal, v. 4, p. 229–246.

———, 1970, Studies in fluviatile sedimentation: A comparison of fining-upward cyclothems with special reference to coarse-member composition and interpretation: Journal of Sedimentary Petrology, v. 40, p. 298–323.

BLUCK, B. J., 1971, Sedimentation in the meandering River Endrick: Scottish Journal of Geology, v. 7, p. 93–138.

BRIDGE, J. S. AND JARVIS, J., 1976, Flow and sedimentary processes in the meandering river South Esk, Glen Clova, Scotland: Earth Surface Processes, v. 1, p. 303–336.

———, AND ———, 1982, The dynamics of a river bend: A study in flow and sedimentary processes: Sedimentology, v. 29, p. 499–540.

BRIDGES, P. H. AND LEEDER, M. R., 1976, Sedimentary model for intertidal mudflat channels, with examples from the Solway Firth, Scotland: Sedimentology, v. 23, p. 533–552.

CALVERLEY, A., 1984, Sedimentology and geomorphology of the modern epsilon cross-stratified point bar deposits in the Athabasca upper delta plain: Unpublished M.S. Thesis, University of Calgary, Alberta, 116 p.

COLLINSON, J. D., 1978, Alluvial sediments, *in* Reading, H. G., ed., Sedimentary Environments and Facies: Elsevier, New York, p. 15–59.

———, AND THOMPSON, D. B., 1982, Sedimentary Structures: George Allen and Unwin Press, Boston, 194 p.

COTTER, E., 1971, Sedimentary structures and the interpretation of paleoflow characteristics of the Ferron Sandstone (Upper Cretaceous), Utah: Journal of Sedimentary Petrology, v. 41, p. 129–138.

DE MOWBRAY, T., 1983, The genesis of lateral accretion deposits in recent intertidal mudflat channels, Solway Firth, Scotland: Sedimentology, v. 30, p. 425–435.

ELLIOTT, T., 1976, The morphology, magnitude and regime of a Carboniferous fluvial-distributary channel: Journal of Sedimentary Petrology, v. 46, p. 70–76.

FLACH, P. D. AND MOSSOP, G. D., 1985, Depositional environments of Lower Cretaceous McMurray Formation, Athabasca Oil Sands, Alberta: American Association of Petroleum Geologists Bulletin, v. 69, p. 1195–1207.

GALLOWAY, W. E. AND HOBDAY, D. K., 1983, Terrigenous clastic depositional systems; Applications to petroleum, coal and uranium exploration: Springer-Verlag, New York, 423 p.

HAYES, M. O., 1976, Morphology of sand accumulations in estuaries, *in* Cronin, L. E., ed., Estuarine Research: Academic Press, London, p. 3–22.

JACKSON, R. G., 1976, Depositional model of point bars in the Lower Wabash River: Journal of Sedimentary Geology, v. 46, p. 579–594.

———, 1978, Preliminary evaluation of lithofacies models for meandering alluvial streams, *in* Miall, A. D., ed., Fluvial Sedimentology: Canadian Society of Petroleum Geologists Memoir 5, p. 543–576.

———, 1981, Sedimentology of muddy fine-grained channel deposits in meandering streams of the American Middle West: Journal Sedimentary Petrology, v. 51, p. 1169–1192.

JERZYKIEWICZ, T. AND MCLEAN, J. R., 1980, Lithostratigraphical and sedimentological framework of coal-bearing Upper Cretaceous and Lower Tertiary strata, Coal Valley area, central Alberta Foothills: Geological Survey of Canada Paper 79-12, 47 p.

JOUANNEAU, J. M. AND LATOUCHE, C., 1981, The Gironde Estuary: Stuttgart, E. Schweizerbartsche Verlagsbuchhandlung, 115 p.

KOSTER, E. H., 1983, Sedimentology of the Upper Cretaceous Judith River (Belly River) Formation, Dinosaur Provincial Park, Alberta: Canadian Society Petroleum Geologists Conference, The Mesozoic of Middle North America, May 8–11, 1983, Calgary, Field Trip Guidebook No. 1, 121 p.

LEEDER, M. R., 1974, Lower Border Group (Tournaision) fluvio-deltaic sedimentation and paleaeogeography of the Northumberland Basin: Yorks, Geological Society Proceedings, v. 40, p. 129–180.

MOSSOP, G. D., 1978, Epsilon cross-strata in the Athabasca Oil Sands, (abs.), *in* Miall, A. D., ed., Fluvial Sedimentology: Canadian Society of Petroleum Geologists Memoir 5, p. 854–855.

———, AND FLACH, P. D., 1983, Deep channel sedimentation in the Lower Cretaceous McMurray Formation, Athabasca Oil Sands, Alberta: Sedimentology, v. 30, p. 493–509.

NANSON, G. C., 1980, Point bar and floodplain formation of the meandering Beatton River, northeastern British Columbia, Canada: Sedimentology, v. 27, p. 3–29.

OOMKEMS, E. AND TERWINDT, J. H. J., 1960, Inshore estuarine sediments in the Haringvliet, Netherlands: Geologie en Mijnbouw, v. 39, p. 701–710.

RAHMANI, R. A., 1983, Facies relationships and paleoenvironment of a Late Cretaceous tide-dominated delta, Drumheller, Alberta: Unpublished Field Guidebook, 33 p.

REINECK, H. E., 1958, Longitudinale schragschicht in Watt: Geologische Rundschau, v. 47, p. 73–82.

STEWART, D. J., 1981a, A meander-belt sandstone of the Lower Cretaceous of Southern England: Sedimentology, v. 28, p. 1–20.

———, 1981b, A field guide to the Wealden Group of the Hastings and the Isle of Wright, *in* Elliott, T., ed., Field Guides to Modern and Ancient Fluvial Systems in Britain and Spain: Second International Fluvial Sedimentology Symposium, University of Keele, U.K. Sept. 21–25, p. 3.1–3.31.

STEWART, G. A. AND MACCALLUM, G. T., 1978, Athabasca Oil Sands Guidebook: Canadian Society of Petroleum Geologists, Calgary, Alberta, Canada, 33 p.

TAYLOR, G. AND WOODYER, K. D., 1978, Bank deposition in suspended-load streams, *in* Miall, A. D., ed., Fluvial Sedimentology: Canadian Society of Petroleum Geologists Memoir 5, p. 257–276.

VAN STRAATEN, L. M. J. U., 1954, Composition and structure of recent marine sediments in the Netherlands: Leidse Geolkogische Mededelingen, v. 19, p. 1–110.

WALKER, R. G. AND CANT, D. J., 1984, Sandy fluvial systems, *in* Walker, R. G., ed., Facies Models, second edition: Geoscience Canada, Reprint Series 1, Geology Association of Canada, Toronto, p. 71–89.

WOOD, J., 1985, Sedimentology of the Late Cretaceous Judith River Formation "Cathedral Area," Dinosaur Provincial Park, Alberta: Unpublished M.S. Thesis, University of Calgary, Alberta, 215 p.

WRIGHT, M. D., 1959, The formation of crossbedding by a meandering or braided stream: Journal of Sedimentary Petrology, v. 29, p. 610–615.

LATERAL ACCRETION SURFACES IN EPHEMERAL-STREAM POINT BARS, RIO PUERCO, NEW MEXICO

RUSSELL G. SHEPHERD
Department of Geological Sciences, University of Kentucky, Lexington, Kentucky, 40506

ABSTRACT: Lateral accretion surfaces (LAS) are rarely seen in modern point-bar sequences. In point bars in the arroyo of the Rio Puerco west of Albuquerque, they are clearly seen because 1) beds of texturally and structurally distinct wind-blown sand are deposited on point bars during no-flow periods and not totally removed during floods, and 2) point bars are excised soon after they form because of rapid channel sweep and swing. LAS are readily identified texturally because eolian units have a marked paucity of coarse silt, apparent in number-frequency but not weight-frequency cumulative size distributions. The surfaces characteristically are rugged (notably like staircases) in plan and section, and separate overlapping tabular and wedge-shaped units locally truncated by scour in chute channels.

INTRODUCTION

Lateral accretion surfaces (LAS) are important criteria for recognizing fluvial point-bar deposits and for distinguishing them from other types of fluvial deposits (Bridge, 1985; Allen, 1982; Gardner, 1983; Miall, 1982; Ethridge and Schumm, 1978; Reading, 1986). As bounding surfaces in epsilon cross-stratified units (Allen, 1963), they generally represent the topographic surface of a point bar at a point in time; thus, they may be surfaces of erosion, deposition, or a combination of the two. Numerous investigators have noted, and some have emphasized, the paucity of observations of LAS in modern fluvial point bars (Jackson, 1978; Smith, D.G., this volume). Indeed, Reading (1986) stated that they are not known from modern point bars.

In the Rio Puerco alluvial valley west of Albuquerque, New Mexico, lateral accretion surfaces are commonly observed in modern point-bar sequences (Fig. 1). In this dry climate, wind ripples and dunes are commonly found in ephemeral channels (Shepherd, 1976; Shepherd and Macke, 1978; Love, 1979). They occur both along the thalweg (which may be dry more than three-fourths of the year), and higher up on the point bars of the sinuous, entrenched channel (Fig. 2). Furthermore, wind-blown sediments are extensively preserved with almost unaltered textures and sedimentary structures within point bars when floods inundate them and rapidly deposit large amounts of fine sand and silt from suspension (Fig. 3). The eolian beds are deposited on the LAS and may protect them from erosion during runoff events.

PURPOSE AND LOCATION

The purpose of this paper is to describe some lateral accretion surfaces in the Rio Puerco and to discuss their origin and significance. Although LAS are well exposed in many excised Rio Puerco point bars, a readily accessible bar with several LAS near Albuquerque is exemplary (Fig. 1) and was selected for detailed study. Meander migration has almost exactly bisected the point bar parallel to the direction of lateral accretion, and a complete two-dimensional vertical sequence parallel to that direction is well exposed. The point bar is south of the junction of I-40 and the Rio Puerco in Sec. 15, T. 9 N., R. 1 W., Bernalillo County (Fig. 1).

GENERAL DESCRIPTION

In the Rio Puerco, sheets and wedges of eolian sediment are deposited on the surfaces of fluvial beds (Fig. 2) and are easily distinguished from them. Eolian deposits are more friable, contain less clay, are more evenly laminated, and commonly have inversely graded laminae. They are, consequently, more easily eroded, compared to adjacent fluvial units. In addition, the fluvial units have sedimentary structures, textures, and sequences indicative of floods in dry regions (McKee and others, 1967; Denny, 1965; Picard and High, 1973). Both fluvial and eolian sediments can be observed and studied as they are deposited in the Rio Puerco.

The complete point-bar sequence of Figure 2 is approximately 5 m high. The central part of this exposure contains several wedge-shaped units of eolian sand (Figs. 2, 4, 5) which are texturally and structurally different from adjacent fluvial units. The eolian beds are poorly cohesive and evenly laminated compared to adjacent units. The unit is truncated where it adjoins a scarp formed in rippled fluvial sediment, and it contains small angular clasts of the scarp-forming material.

TEXTURAL IDENTIFICATION OF LAS

Many lateral accretion surfaces in Rio Puerco point-bar sequences can be readily observed from the distance because they are overlain by thin beds of eolian sand and silt that are preferentially eroded by sandblasting during common windy periods (Figs. 2, 3). Others are less distinctive but can be recognized through the identification of an eolian signature by textural analysis using number-frequency methods. This is illustrated by textural analysis of fluvial and eolian beds on the surface of the modern bar and in two vertical sequences through it (Figs. 4–7).

A sample from wind ripples was collected from ripples that were migrating while the wind was blowing, similar to those shown in Figure 2. Size-frequency distributions were obtained for the sample using both sieves and number-frequency measurements (Fig. 5). The texture of this sample was compared to the eolian beds and the rippled fluvial beds in the point-bar sequence (Figs. 6, 7).

Both weight-frequency (sieve) and number-frequency analyses were performed for all samples. Standard sieve analyses were performed using quarter-phi intervals and 8-inch screens. To obtain number-frequency distributions, the

FIG. 1.—Location and photographs of Rio Puerco point bar. Oblique aerial photo, lower right, shows that the point bar is being excised almost parallel to the direction of lateral accretion. Arrows indicate location of section shown at top. Lateral accretion topography and surfaces are apparent in plan view and vertical section (top). Note 4-m rod and man for scale.

lengths of four grain diameters were measured at 45° rotations and averaged for each of approximately 500 grains in a sample. A Quantimet electronic automatic image analyzer (AIA) coupled with a computer was employed. A discussion of the method and initial results was provided by Shepherd and Macke (1978).

IDENTIFICATION OF EOLIAN BEDS

Sieve analysis results and the AIA number-frequency results are markedly different for the eolian deposit (Fig. 5). Mean sizes differ, but in addition, a prominent plateau, principally occupying the coarse-silt and the 55th-percentile to 60th-percentile range, exists in the AIA cumulative distribution. Indeed, only 8 of 497 grains measured were coarse silt. This plateau on the number frequency curve was not observed in the sieve analysis data (Fig. 5).

To compare the two methods of measurement, the AIA number-frequency data were converted to weight-frequency data. Grains of quartz (specific gravity = 2.65) of spherical shape were assumed. The weight of each grain was calculated from its volume, which was calculated using the average of the four measured diameters. The resulting weight-frequency curve has a shape similar to the sieve curve, but

FIG. 2.—Photograph of eolian sand shadow and wind ripples typical of Rio Puerco point bars. Preservation of some of this deposit will result in a sand wedge.

FIG. 3.—Photograph of eolian wedge (E) preserved in point-bar sequence.

FIG. 4.—Photograph showing vertical sequence located at rod in Figure 1. Note sedimentary structures, chiefly flat laminae, climbing ripples, and deformed laminae, and eroded nature of the poorly indurated eolian bed, demarcated naturally by sand blasting during periods of high winds.

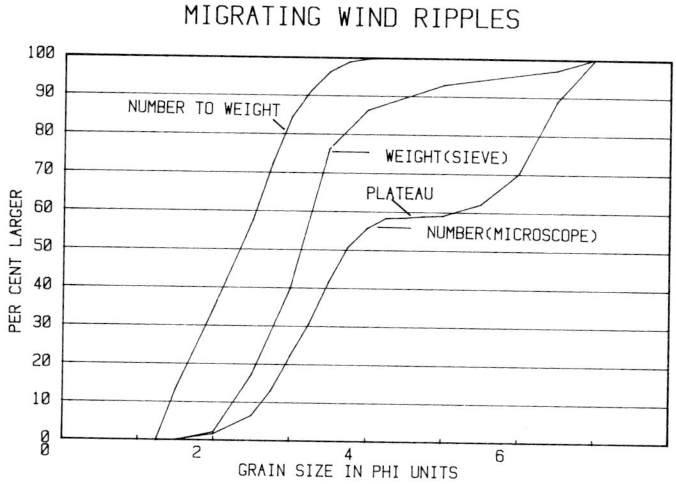

FIG. 5.—Cumulative size frequency curves showing sieve and image analysis results from the same sample, and calculation of the weight-frequency distribution from the number-frequency distribution. Note the prominent plateau displayed in the number-frequency data.

even more notable is that the plateau of the number-frequency curve is not apparent. This is simply because the weight of a grain is proportional to the cube of its diameter. Thus, silt and clay grains constitute little of the total weight of the sample, though they are many times more abundant than sand grains. For the eolian sample of Figure 6, only 58 percent (the coarsest fraction) of the grains of the sample constitute almost 90 percent of the weight.

COMPARISON OF EOLIAN AND FLUVIAL BEDS

Two suites totaling 19 samples from two vertical sequences through the point bar were collected. Each unit with distinctly different sedimentary structures and/or textures was sampled, including the eolian deposits on the lateral accretion surfaces and the fluvial deposits above and below them (Figs. 6, 7).

The eolian units on the LAS have mean sizes significantly different from the other units for both sieve and AIA methods. Considering all the samples from the two sequences, two distinct types of number-frequency size-distribution curves can be recognized (Figs. 5, 6, 7). The eolian units all have size distribution curves with marked plateaus above the 40th percentile on the cumulative frequency curve and also a notable paucity of coarse silt grains. Accordingly, eolian curves indicate higher proportions of sand (numerically, at least 40 to 50 percent). In contrast, the fluvial distributions commonly have a lower percentage of sand (<20 percent) and no plateau or a plateau below the 30th percentile (Figs. 6, 7).

The dearth of coarse-silt grains in eolian ripples was documented many years ago by Udden (1898; 1914). He noted this paucity in relation to the common abundance of very fine sand and coarse silt in loess. Udden (1898) also obtained number-frequency distributions of fine sand and mud by using counts of loose grains under the microscope. In the Rio Puerco, coarse silt grains are apparently removed from fluvial sediment by the wind, resulting in a lack of coarse-silt particles like that Udden observed. Nevertheless, texture generally has not been as useful as originally indicated for recognizing eolian sediments (Bigarella, 1972; Ahlbrandt, 1975).

Textural analysis alone makes it possible to readily distinguish lateral accretion surfaces in the Rio Puerco point bar. Although textural analysis is not necessary to recognize eolian units in the Rio Puerco, the method may have application in recognition of eolian units in ancient fluvial point-bar sequences. It must be emphasized, however, that other depositional environments may have laterally moving channels, for example braided (Bridge, 1985) and tidal (Reading, 1986) channels, and thus also lateral accretion surfaces.

CONFIGURATION OF LAS

LAS in the Rio Puerco tend to be rugged in cross section. This is chiefly because bars that are deposited during floods

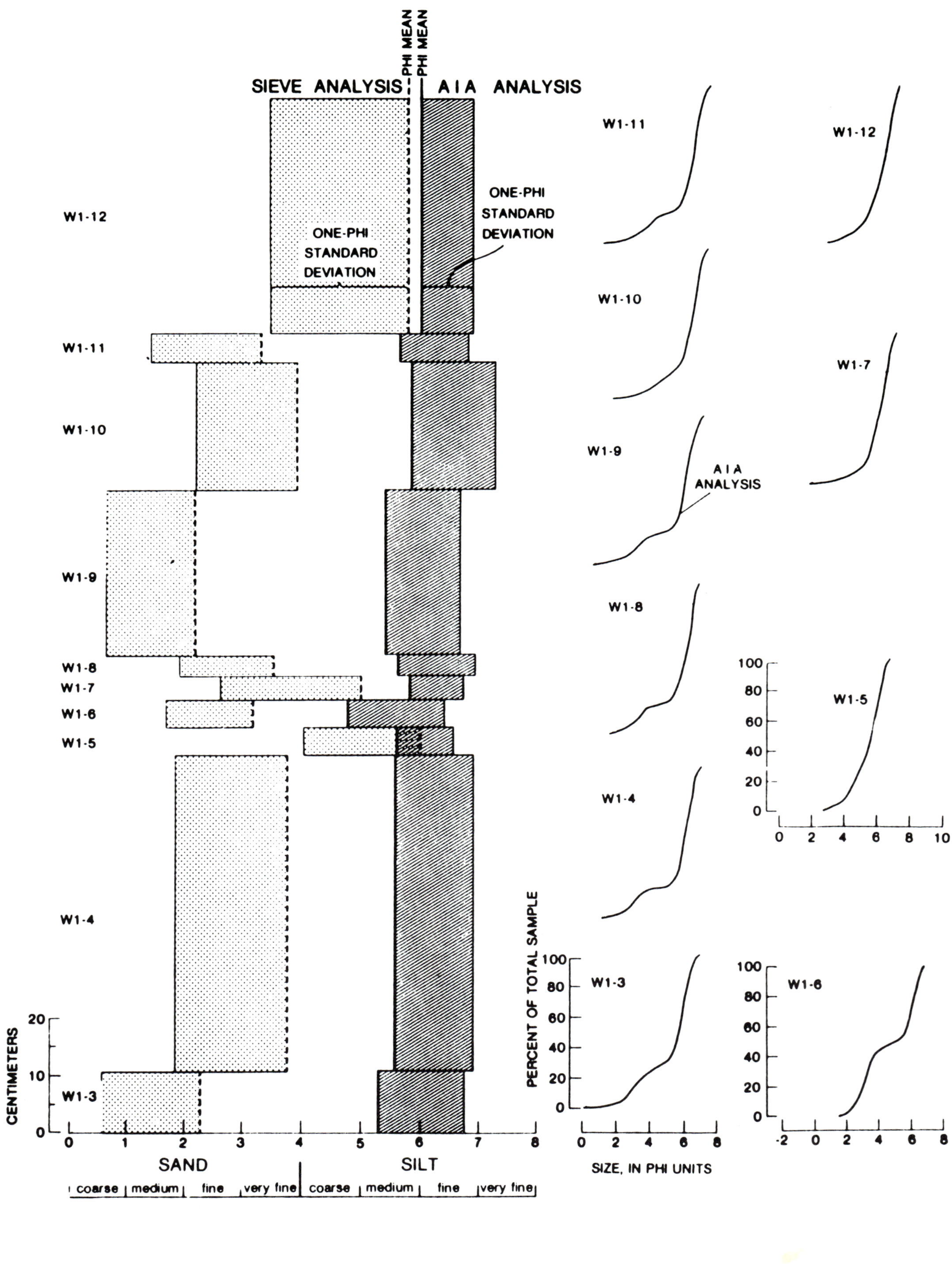
SIEVE ANALYSIS
PHI MEAN
PHI MEAN
A I A ANALYSIS
ONE-PHI STANDARD DEVIATION
ONE-PHI STANDARD DEVIATION
W1-12
W1-11
W1-10
W1-9
W1-8
W1-7
W1-6
W1-5
W1-4
W1-3
CENTIMETERS
0
10
20
0 1 2 3 4 5 6 7 8
SAND
SILT
coarse medium fine very fine coarse medium fine very fine
A I A ANALYSIS
PERCENT OF TOTAL SAMPLE
SIZE, IN PHI UNITS
0 20 40 60 80 100
0 2 4 6 8 10
-2 0 2 4 6 8

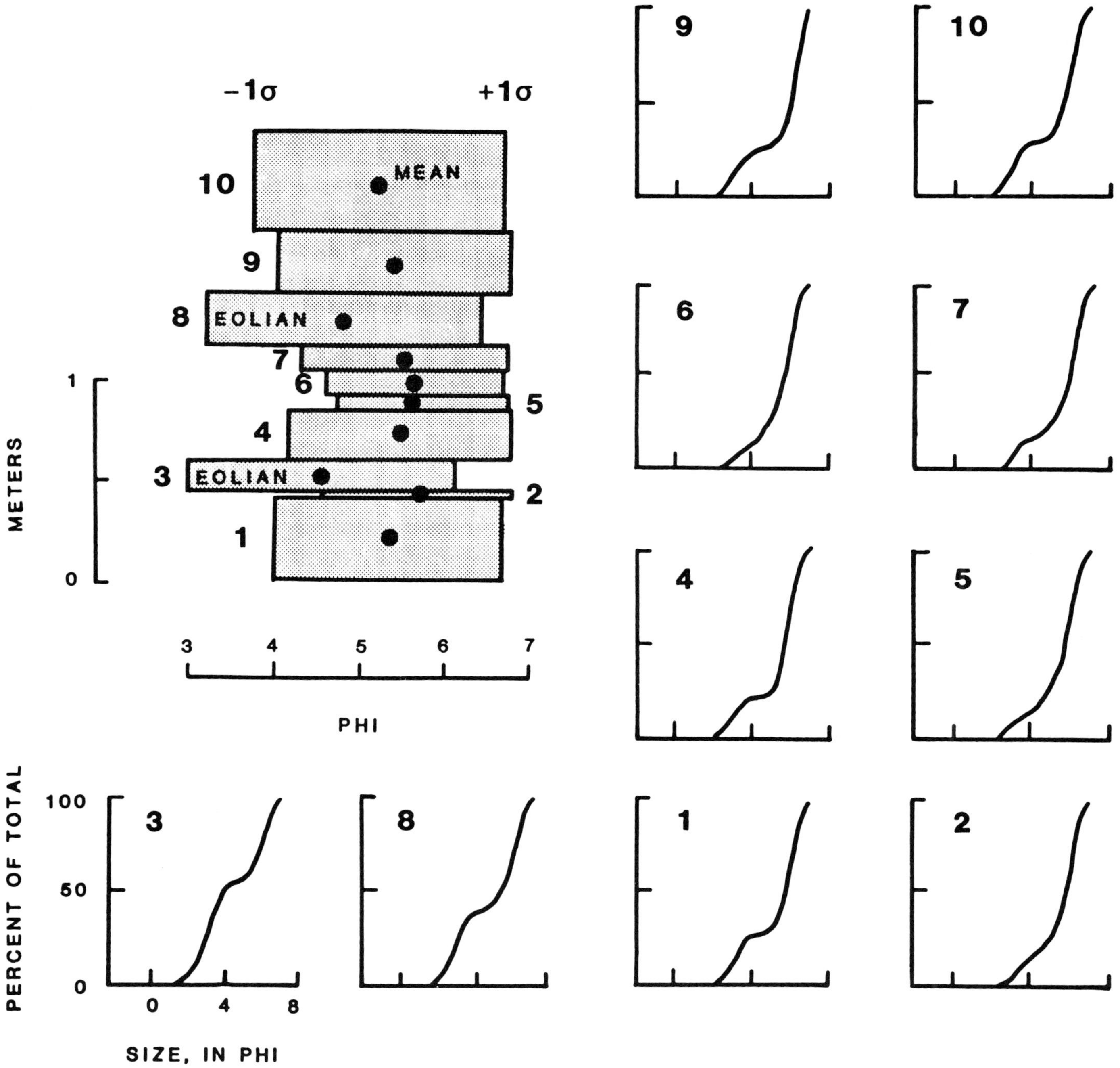

FIG. 7.—Textural properties of second section through the point bar showing two lateral accretion surfaces, with eolian units #3 and #8 on them. Note differences in mean sizes and plateaus of eolian units, shown by number-frequency analysis.

are dissected by later floods, leaving numerous prominent erosional scarps like that shown in Figure 2. Eolian sand shadows form downwind of and prevent complete scour and erosion of the scarps and adjacent bar surfaces during subsequent flood events. Thus, the configuration of a single LAS is similar to that of a flight of stairs, with relatively long trends and short risers (Fig. 1). In addition, chute channels are abundant, and these commonly have sharp scour bases that truncate bar surfaces and may become depositories of eolian sand. Chute channels thus add to the ruggedness of LAS.

In recent years, numerous investigators have emphasized

←

FIG. 6.—Textural properties of vertical point-bar sequence shown in Figure 4, illustrating differences in results obtained from weight frequency (sieve) and number frequency (automatic image analyzer) of the same samples. The eolian unit (W1–6) is markedly different in mean size due to a paucity of coarse silt.

the importance of recognizing the three-dimensional nature of processes and sedimentary attributes of point bars (Jackson, 1978; Bridge, 1985; Miall, 1982; Reading, 1986). Point bars in the Rio Puerco are actively migrating, and the shapes of lateral accretion surfaces vary markedly depending on the relationship between orientation of the vertical section observed and direction of migration of the bar.

The complexity of the surfaces can be easily understood by cutting one end of a deck of cards to form the subrounded shape of a nose. Individually crumple the cards and then stack them. Fan the cards out as you would when inspecting a Missouri Pitch hand, with the noses at the top, representing the meander bend. Each card represents the LAS formed during one major flood event, with fluvial beds between. Some flood events will produce thicker beds and some thinner; some will scour much of the underlying card while others will scour little. Some cards will protrude out from the rest, and some will be truncated at the bottom. Lateral accretion surfaces are similarly highly variable in configuration, but wedge-shaped units and chute-channel scours are common. The high sinuosity of the channel and the crosscutting chutes produce a wide variance of current direction indicators.

A section cut parallel to the downvalley direction through a Rio Puerco point bar that is relatively swinging but not sweeping may not display prominent lateral accretion surfaces. Rather, continuous, subparallel beds, laterally uniform in thickness and representing sheet bars and chute-channel deposits, are observed. Well developed LAS are observed perpendicular to swing direction and are stepped. Sections oriented between the swing and sweep directions display a trend, from prominent steps, to wedges thinning towards the thalweg, to parallel beds. Identification of orientation of a section with respect to sweep and swing directions is possible using these attributes of the configuration of lateral accretion surfaces. Additional evidence is provided by current direction indicators such as ripple cross-stratification, parting lineations, clast orientation, chute channel orientation and by changing thicknesses of beds.

ACKNOWLEDGMENTS

David Macke assisted with the early phases of this work, which were supported by the U.S. Geological Survey through Ed Clifton. Macke, Robert F. Gantnier, and Carla Potter performed size analyses and computations.

REFERENCES

AHLBRANDT, T. S., 1975, Comparison of textures and structures to distinguish eolian environments, Killpecker dune field, Wyoming: Mountain Geologist, v. 12, p. 61–73.

ALLEN, J. R. L., 1963, The classification of cross-stratified units, with notes on their origin: Sedimentology, v. 2, p. 93–114.

———, 1982, Sedimentary structures: Their character and physical basis, two volumes: Elsevier, Amsterdam, 594 p.

BIGARELLA, J. J., 1972, Eolian environments: Their characteristics, recognition and importance, *in* Rigby, J. K., and Hamblin, W. R., eds., Recognition of Ancient Sedimentary Environments: Society of Economic Paleontologists and Mineralogists Special Publication 16, p. 12–62.

BRIDGE, J. S., 1985, Paleochannel patterns inferred from alluvial deposits: A critical evaluation: Journal of Sedimentary Petrology, v. 55, p. 579–589.

DENNY, C. S., 1965, Alluvial fans in the Death Valley region, California and Nevada: U.S. Geological Survey Professional Paper 466, 62 p.

ETHRIDGE, F. G., AND SCHUMM, S. A., 1978, Reconstructing paleochannel morphologic and flow characteristics: Methodology, limitations, and assessment, *in* Miall, A. D., ed., Fluvial Sedimentology: Canadian Society of Petroleum Geologists Memoir 5, p. 703–722.

GARDNER, T. W., 1983, Paleohydrology and paleomorphology of a Carboniferous, meandering, fluvial sandstone: Journal of Sedimentary Petrology, v. 53, p. 991–1006.

JACKSON, R. G., II, 1978, Preliminary evaluation of lithofacies models for meandering alluvial streams, *in* Miall, A. D., ed., Fluvial Sedimentology: Canadian Society of Petroleum Geologists Memoir 5, p. 543–576.

LOVE, D. W., 1979, Quaternary fluvial geomorphic adjustments in Chaco Canyon, New Mexico, *in* Rhodes, D. D., and Williams, G. P., eds., Adjustments of the Fluvial System: Kendall-Hunt, Dubuque, Iowa, p. 277–280.

MCKEE, E. D., CROSBY, E. J., AND BERRYHILL, H. L., 1967, Flood deposits, Bijou Creek, Colorado, June, 1965; Journal of Sedimentary Petrology, v. 37, p. 829–851.

MIALL, A. D., 1982, Analysis of fluvial depositional systems: American Association of Petroleum Geologists Education Course Note Series No. 20, 75 p.

PICARD, M. D., AND HIGH, R. L., Jr., 1973, Sedimentary structures of ephemeral streams, No. 17 of Developments in Sedimentology: Elsevier Publishing Co., Amsterdam, 222 p.

READING, H. G., ed., 1986, Sedimentary Environments and Facies: Blackwell Scientific, 2nd edition, 615 p.

SHEPHERD, R. G., 1976, Sedimentary processes and structures of ephemeral-stream point bars, Rio Puerco west of Albuquerque, New Mexico: Geological Society of America Abstracts with Programs, v. 8, no. 6, p. 1103.

———, AND MACKE, D. L., 1978, Discrimination of fluvial and eolian deposits by number-frequency analysis of sediment of sand through silt size from a point bar, Rio Puerco, New Mexico: Journal of Research, U.S. Geological Survey, v. 6, no. 4, p. 499–504.

UDDEN, J. A., 1898, The mechanical composition of wind deposits: Augustana Library Publication 1, 69 p.

———, 1914, Mechanical composition of clastic sediments: Geological Society of America Bulletin, v. 25, p. 655–744.

CHANNEL AND FLOODPLAIN FACIES IN A WANDERING GRAVEL-BED RIVER

JOSEPH R. DESLOGES† AND MICHAEL CHURCH
Department of Geography, The University of British Columbia, Vancouver, B.C., V6T 1W5

ABSTRACT: Bella Coola River drains 5050 km^2 of glacierized mountains on the central coast of British Columbia. Spring snowmelt and autumn rainstorm floods occur up to 1000 m^3s^{-1}. The valley-fill is composed of Quaternary glacial and fluvial sediments, with modern alluvium representing less than 3 percent of the total volume. Proportions of light and heavy minerals in alluvial deposits indicate that headwater volcanic terrain is the dominant sediment source; however, distal tributaries draining plutonic rocks are locally important.

Sedimentation within the modern channel yields medial, lateral and point bars and vegetated channel islands. Avulsions within 'sedimentation zones' result in a network of slough channels. Floodplain development occurs in three ways: infilling of sloughs after channel avulsions, lateral accretion of channel bars, and overbank deposition. The most prominent facies assemblage is horizontally and trough cross-stratified sands over massive gravels. Intervening stable reaches are cobble-paved zones through which sediment is transported directly. Facies within these reaches comprise horizontally stratified sands and silts, indicating that floodplain development also occurs by vertical accretion. This association of channel zone and floodplain facies is widespread in mountain valley rivers in the northern Cordillera today. They are not aggrading and would not be prominent in the stratigraphic record.

INTRODUCTION

Sediments which accumulate within a river valley reflect both short- and long-term alluvial processes. Deposition of alluvium is controlled by the volume of material delivered by transporting events of a given frequency and duration, and by the spatial distribution of sediment sources. In the long term, river slope and sediment texture become adjusted to the prevailing sediment supply and discharge regimes. Channel morphologies and patterns of sediment deposition reflect these adjustments. In this study, the alluvial portion of a postglacial valley-fill is examined to determine the influence of sediment supply, discharge regime and valley bottom vegetation on the development of channel zone and floodplain facies within a wandering gravel-bed river.

This river type, defined by Neill (1973) and by Church (1983), exhibits an irregularly sinuous channel, sometimes split about channel islands and in some places braided; seasonal or perennial 'side' channels—subordinate anabranches of the river—are common. The alluvial sediments, not previously characterized for this channel type, consist of 0–3 m of sandy channel-fill or overbank silty sands deposited above several meters of channel cobble-gravel. Similar sequences have been described for sinuous gravel channels by Bluck (1971), Jackson (1978) and Forbes (1983). Lateral accretion is the dominant mode of deposition. The type is sufficiently widespread in stable mountain valley and foreland settings to merit more detailed attention (cf., for example Kellerhals and others, 1972).

Our objectives here are (1) to characterize the sedimentary facies for a wandering gravel-bed river, and (2) to illustrate the way in which sediment supply, hydrologic regime, and pattern of river instability control development of the deposits.

Bella Coola River drains 5050 km^2 of Triassic to Tertiary volcanic and plutonic terrain in the central Coast Mountains of British Columbia (Fig. 1). Two principal tributaries join to form this river. Talchako River, draining 1300 km^2 of the heavily glacierized southern catchment, contributes the majority of discharge and sediment to Bella Coola River. Atnarko River drains the largest area of any tributary (2030 km^2) but contributes less than 25 percent of the discharge. The Atnarko headwater is an uplifted plateau containing numerous lakes along principal drainage lines. The high storage capacity of the lakes severely limits sediment delivery to the Bella Coola River which, in comparison with the Talchako River, might be no more than 5 percent of the total load at their confluence. The southern boundary of the basin includes some of the highest areas in the Coast Mountains (to +3035 m). The entire basin was heavily glaciated in Pleistocene time and a late neoglacial resurgence of alpine glaciers ended at the turn of the twentieth century. The Talchako basin remains heavily glacierized over 25 percent of its area. Upland surficial materials in Bella Coola basin consist of till, kames, colluvium and local pockets of alluvium. Approximately 22 percent of the basin is exposed bedrock.

Orographic effects over the Coast Mountains produce a strong precipitation gradient within the basin. In the western section of the catchment annual precipitation exceeds 2000 mm and declines to less than 500 mm in the east. Runoff to Bella Coola River is dominated by two processes: spring snowmelt and fall rainstorms. The largest flows (maximum mean daily $Q > 1000\ m^3s^{-1}$) occur in the autumn with the incidence of heavy rains and melting of recently fallen snow. Spring snowmelt floods are usually smaller.

Bella Coola River is a cobble-gravel stream with gradients between 0.0010 and 0.0031. Single thread and multiple channels occur in both sinuous and non-sinuous reaches of the river. Contemporary alluvial processes along the Bella Coola are controlled by sediments which have accumulated within the valley during the late glacial and Holocene epochs, and by the distribution of upland sediment sources. A brief description is provided of the origin and significance of the older sediments, and the influence of upland sources is assessed. Details of sediment deposition, facies development and floodplain evolution are then given.

SEDIMENTOLOGICAL CONTEXT

Valley-fill sediments.—

Contemporary alluvium in valleys of coastal British Columbia caps a sequence of sediments associated with the final Pleistocene deglaciation. Most of what is known about these sediments is derived from limited borehole informa-

†Present address: Department of Geography, University of Toronto, Toronto, Ontario, Canada M5S1A1

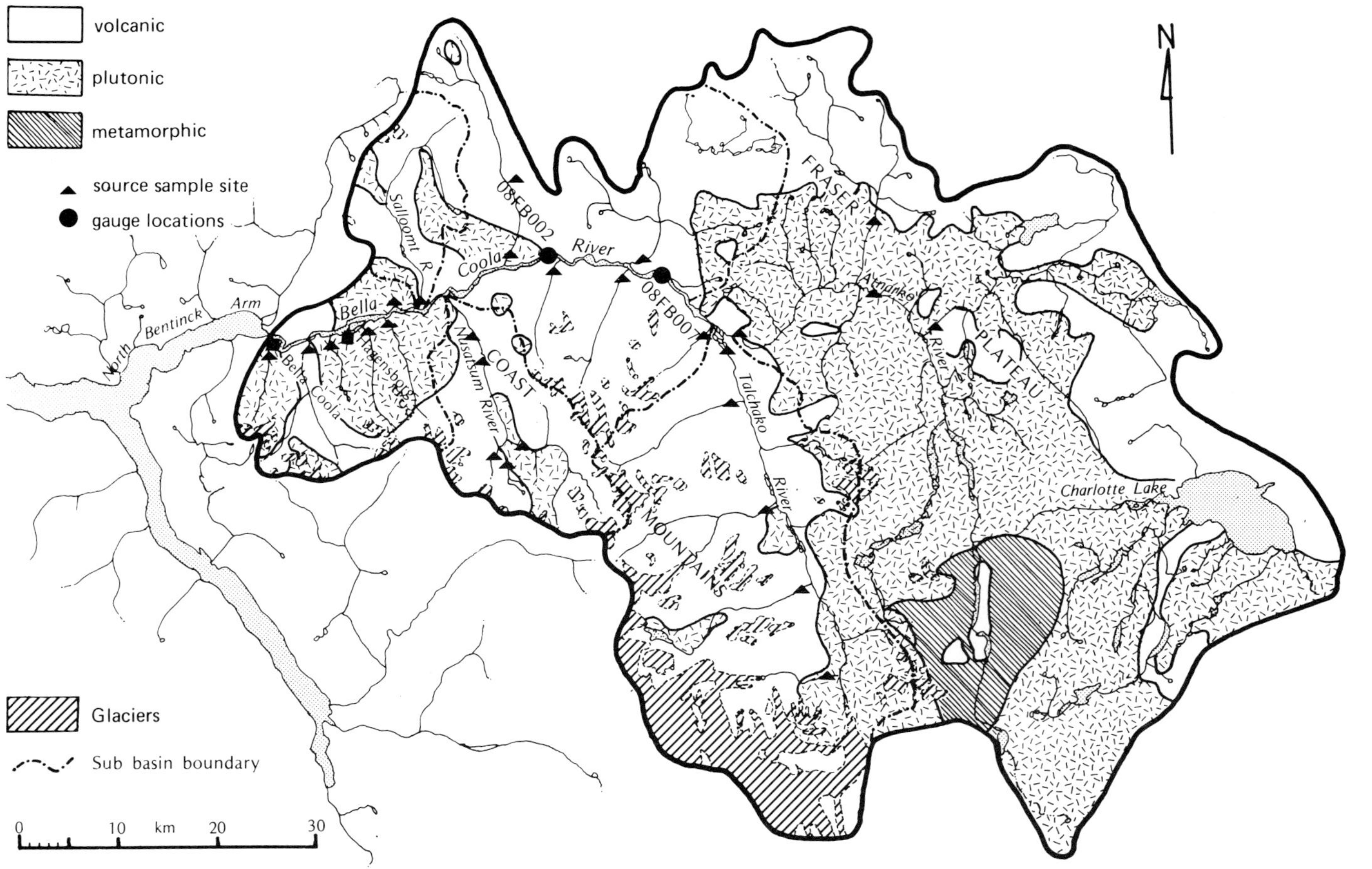

FIG. 1.—Location and generalized geology. Bella Coola River straddles the boundary between the Coast Plutonic Complex and Intermontane Volcanic Belt and thus exhibits rock types which range from coarse-grained quartz monzonite and granodiorite to fine-grained greenstone, andesite and basalt (Baer, 1973; Tipper, 1979). Triangles show sampling sites for mineral characterization of tributary creeks (see text).

tion (Armstrong, 1981; Howes, 1983; Clague, 1985). The character of the deposits depended on the position of sea-level, controlled by isostatic and eustatic responses of the land and ocean as Cordilleran ice melted. Sea levels were 100–150 m higher until approximately 9000 yrs B.P. (Andrews and Retherford, 1978; Clague and others, 1982) and ice fronts terminated at sea level. Similar conditions are found today in Glacier Bay, Alaska (Powell, 1983). Sediment deposition in proglacial fiords produced a stratigraphic record of these changing environmental conditions. The Bella Coola valley-fill sequence is given in Figure 2.

Overall, the Quaternary valley-fill represents facies from glacial, glaciomarine, marine and alluvial processes. The prograding delta of the Bella Coola River is composed of thick deposits of sand and gravel which coarsen upward. Concurrently, alluvial fans along the valley margin prograded across the valley depositing sandy gravels and cobbles within the deltaic sediments. Contemporary terrestrial deposits consist of coarse sandy gravel and a thin layer of sandy floodplain material derived from the Bella Coola River and continued alluvial fan development.

Alluvial sediment sources.—

The principal sources for alluvial sediments deposited along a river reach are as follows:

a) headwater tributaries introducing sediment to the active channel;
b) downstream tributaries entering along the river reach contributing sediment directly to the floodplain surface or to the active channel;
c) bank erosion and redeposition of previously deposited alluvial material;
d) fluvial erosion of non-alluvial deposits resident within the zone of channel activity.

Sources (a) and (b) are directly connected with the uplands. Sources (c) and (d) represent remobilization of sediment previously introduced into the valley. Since less than 13 percent of the channel of Bella Coola River is in contact with non-alluvial deposits (the majority of which are either well-vegetated and stable colluvial slopes, bedrock or deposits of glaciomarine silt and clay), they are not important sources for the sand and gravel sediments that dominate the alluvial valley-fill.

Klages and Hsieh (1975) demonstrated that the mineralogy of floodplain deposits may reflect provenance in a fashion similar to the more widely studied case of glacial tills (cf., Slatt, 1970; Shilts, 1973; Clague, 1975). A major difference between the two processes is that fluvial transport leads to a higher degree and more rapid mixing of sed-

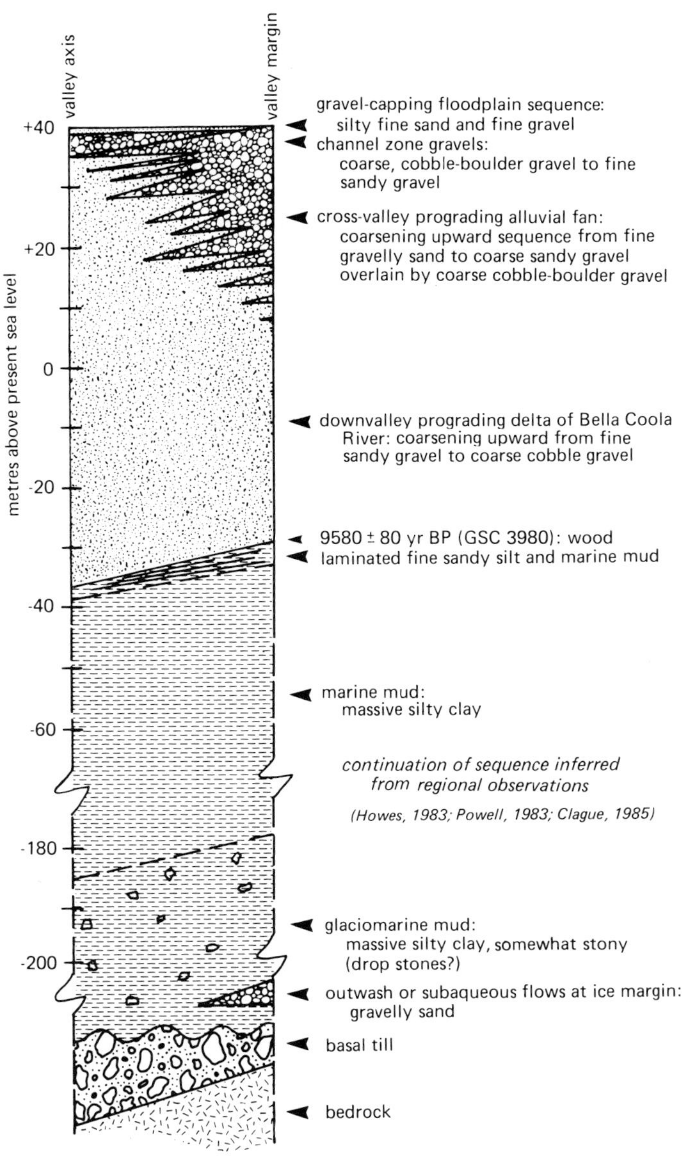

FIG. 2.—Valley-fill seidments of the Bella Coola River valley. Note that below about −60 m, the sequence is conjectural.

iments away from respective source areas than do glacial processes.

The requirements for using deposit mineralogy as an index for sediment source are that source areas yield sediments dominated by distinctive lithologies, characteristic proportions of certain minerals, or containing unique minerals not found in other source areas; and that these properties are not appreciably attenuated by mixing during transport to a particular depositional site.

To characterize source area materials, subsurface, bulk gravel samples were obtained at the mouths of 11 major tributaries of Bella Coola and Talchako Rivers, and coarse sand samples were taken from an additional five tributaries representing, in total, 95 percent of the sediment source area for the floodplain. Sand samples were also collected from 11 sub-basins within some of the larger tributaries (Fig. 1).

Bulk gravel samples also were extracted from 17 sites along 52 km of the river channel and clast lithologies were identified for 100 grid-sampled, bar-surface particles at each site. The texture of the channel gravels shows no systematic downstream trend; maximum sizes approach -7.5ϕ (180 mm) everywhere. Floodplain sediments were extracted from 106 auger boreholes positioned along 18 cross-valley transects. Coarse sands were sampled at the contact between underlying gravels (i.e., auger refusal depth) and the overlying floodplain material; this procedure was adopted as likely to yield the most consistent suite of samples from the boreholes with respect to depositional environment.

Megascopic identifications of quartz, feldspar, mafic and accessory minerals were made along with chlorite-rich groundmass particles for four groups of 100 grains in the 0.50–0.71 mm size fraction (the largest in which non-aggregate minerals or homogeneous groundmass particles were visible) of the sand samples. Microscopic examination of some samples and examination of other size ranges confirmed the consistency of the results. Clast lithologies were classified using standard petrologic guidelines (Ehlers and Blatt, 1981).

The source areas (tributary samples) can be divided into three dichotomous groups that reflect bedrock geology (Fig. 3). Comparison of the valley alluvial samples with these groups reveals the dominance of volcanic source areas. All 17 sand samples from the active channel were classified as having volcanic affinity (Fig. 3). Paired t-tests between upstream and downstream samples for percentages of quartz, feldspar and mafic minerals reveal significantly higher quartz and feldspar and lower mafic content in the downstream group ($\alpha = 0.05$). In the channel gravels, 61 percent of the clasts are volcanic and 31 percent plutonic, a proportion comparable with the ratio of bedrock types in the basin. This ratio does not change significantly when the sample is split into upstream and downstream subsets. The plutonic fraction from downstream samples, however, exhibits an increasing percentage of quartz monzonite, a lithology found from Nusatsum River west. Shale clasts are present in sampling sites immediately downstream from Nusatsum and Salloomt Rivers, the only major sources for this particular rock type.

Of the 106 floodplain sand samples, 51 percent have volcanic, 9 percent plutonic and 40 percent mixed affinity (Fig. 3). All sites with plutonic affinity are located downstream adjacent to alluvial fans, tributary creeks or rock walls comprised primarily of quartz monzonite. Samples of volcanic origin are located close to the active channel or active slough channels. Mixed sediments are mostly found in the downstream floodplain within a transition zone between single source sites or downstream from the Nusatsum alluvial fan.

These results indicate that fine sediment (sands) along the active channel and in the floodplain derives mainly from the glacierized volcanic terrain of upper Talchako River. Downstream, there is in the channel a detectable admixture of sands from lateral tributary basins with plutonic sources. Sands dominated by lateral sources, however, are restricted to the vicinity of the tributary junction. Mixed sediments originate from tributaries which produce material that truly

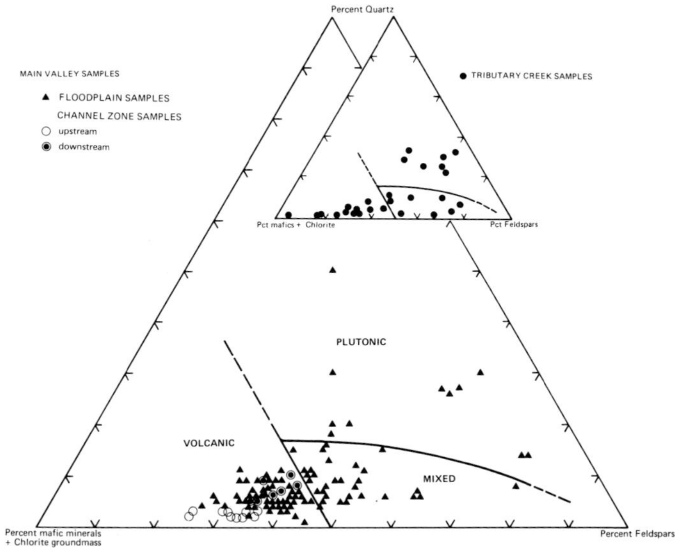

FIG. 3.—Ternary diagram showing mineral composition of channel and floodplain sand samples. Boundaries separating the three sediment source groups are based on discriminant functions derived from grouped mineral data of tributary creek sand samples (inset). The discriminating functions correctly classify more than 80 percent of the tributary creek samples.

TABLE 1.—LITHOFACIES AND SEDIMENTARY STRUCTURES OF CHANNEL ZONE AND FLOODPLAIN DEPOSITS, BELLA COOLA RIVER+

Facies Code	Lithofacies	Sedimentary Structure	Interpretation
Gm	gravel, including coarse cobbles and boulders	imbricated, massive	point bars, medial bars, lateral bars
Gp	gravel, weak to well developed stratification	imbricated, crude planar crossbeds	gravel deltas on point bars, edge of riffles
St	sand, medium to coarse	grouped (pi) trough crossbeds	dunes, lower flow regime
Sp	sand, medium to coarse	planar crossbeds	sand waves
Sr	medium to fine sand	small-scale cross-strata	ripples lower flow regime
Sh	sand, very fine to very coarse	horizontal bedding	planar bed (mostly u. flow regime)
Se	coarse sand, may be pebbly	weak cross-stratification	discontinuous scour/fills
Ss	medium to coarse sand, may be pebbly	horizontal to low-angle planar cross-strata	continuous, broad, shallow scour/fills
Fsc	silt, mud, organics	massive	back channel or elevated floodplain

+ after Miall (1978)

reflects a mixed bedrock source (e.g., Nusatsum River) and from mixing of plutonic and volcanic sediments on the floodplain. Larger clasts appear to reflect more uniformly the overall distribution of source terrain in the basin. About this several things need be said: certain plutonic lithologies—though by no means all of those in the basin—resist attrition and may become over represented in lag deposits; plutonic clasts may be derived disproportionately from eroded Quaternary deposits; finally, size-related effects may influence the results (size was not controlled in the surface clast sample).

Sediments are delivered from volcanic headwater sources to the downstream channel and floodplain during basin-wide high flows, and during summer glacial runoff. Sediments originating from plutonic terrain are delivered to the margin of the floodplain during basin-wide floods, and also during storms in which the contribution of runoff from the lower basin is much greater than from the headwater areas. This is not an unusual occurrence because of the eastward decline of freezing levels during late autumn storms, and because of the marked gradient in precipitation across the basin.

ALLUVIAL DEPOSITS: CHANNEL ZONE FACIES

Storage of sediment in the contemporary channel zone varies along the river. In several reaches, large volumes of resident gravel influence channel pattern producing multiple channels and associated gravel bar development. These reaches have been referred to as 'sedimentation zones' (Church, 1983). Intervening reaches have greater lateral stability and occur where the valley width is restricted by bedrock or tributary alluvial fans. Hence, sediment accumulation and resulting facies associations within the active channel zone can be divided into those within unstable and those within stable reaches. In this and following sections, stratigraphic codes follow Miall (1978) and are summarized in Table 1.

Unstable channel reaches.—

Medial bars and islands dominate the channel within these zones of high sediment storage. Medial bar complexes are composed of several mid-channel bars forming clusters upstream from confluences with major tributary channels. Medial bar surfaces consist of cobbles and gravel (facies Gm and Gp). Sand occurs as thin plane-bed sheets (10–20 cm thick: facies Ss and Sh) behind logs or around stable bar vegetation. Bar surfaces and gravel exposed in cut banks are 1 to 1.5 m above the mean summer stage (which corresponds to a flow of approximately 200 m^3s^{-1}), although relative relief within the bars is not great (0.5–1.0 m). Barhead gravels exhibit massive structure where exposed; however, planar cross-stratification probably is formed along the outer margin of the bar head where flows cross over the leading edge of the riffle (Fig. 4a).

Islands typically are wooded and gravel is exposed only on the upstream side of the island. Channel shifting occurs by avulsion. The islands represent old floodplain surface which has been isolated by avulsion. After such an episode, some channels begin to fill with gravel and sand. The process is slow, however, because events capable of completing the infilling to near bankfull level have a return period of more than 5 years (Fig. 5).

Approximately 40 percent (22 km) of Bella Coola River is dominated by channel macroforms typical of laterally unstable channels (Fig. 6). Sequential air photographs taken over the last decade indicate that the downstream migration rate of some medial bars is in the order of 10 m per year.

Stable channel reaches.—

These sections of the channel have a relatively high channel sinuosity (1.2–1.5) and consist of a series of bank-at-

FIG. 4.—Cobble-gravel bar forms in Bella Coola River (see Table 1 for explanation of facies codes). A) Medial bars in an unstable reach of Bella Coola River. Massive (Gm) and planar cross-stratified (Gp) gravels dominate these features. B) Sinuous course of a stable reach of lower Bella Coola River showing a gravel-dominated point bar (a) and lateral bar (b).

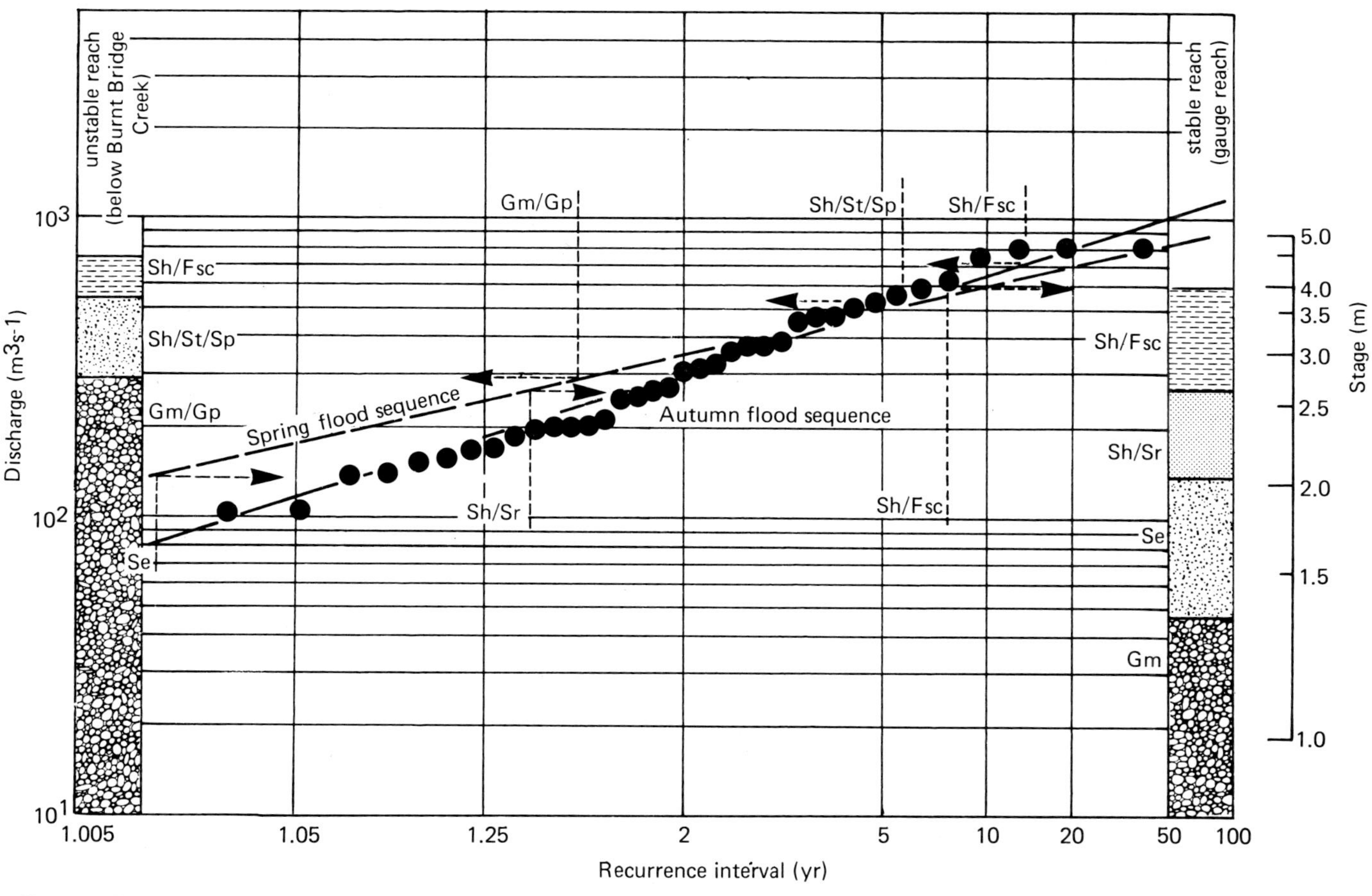

FIG. 5.—Flood frequency and vertical zonation of facies types in stable and unstable reaches of the Bella Coola River. Data of the autumn flood sequence only, which yields the most extreme floods, are plotted. Elevations of facies types in stable reaches (stratigraphic column on right side of plot; see also figs. 4B and 7B) are represented by data from the gauging station, where the stage-discharge relation is well known. The recurrence interval at the upper limit of each facies type is indicated by the vertical dashed lines below the flood frequency plot, and the stage scale on the extreme right refers to water elevations at the gauge. Representative upper facies limits for an unstable reach (stratigraphic column on left; see also figs. 4A and 7A) are given above the flood frequency plot. Overbank flows are associated with floods that have a recurrence interval of approximately 2–5 years in stable river reaches and 5–10 years in unstable reaches.

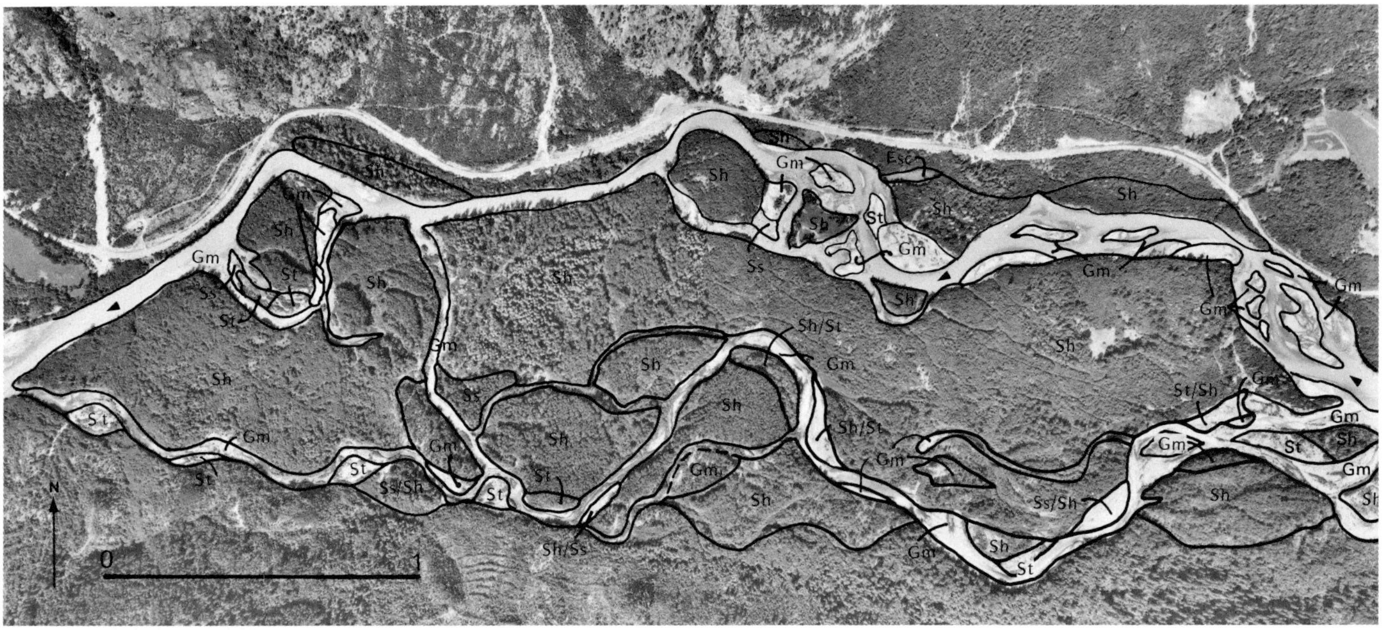

FIG. 6.—Morphology and distribution of surficial facies in an unstable reach of Bella Coola River. See Table 1 for facies code. Photo BC82040/071 taken 18 August 1982; Q = 128 m^3s^{-1}.

tached lateral bars or point bars (Fig. 4b). The lateral bars commonly exhibit a surface of boulders, cobbles and coarse gravel (32 < D_{50} < 150 mm: facies Gm) extending downstream from the bar-head region along the outer (streamside) margin. A low-flow back channel or chute active only at high flow is almost always present containing a cobble-gravel bed (probably crudely stratified) and structureless sand levees on both flanks. A wedge of thicker sands in the bar tail (facies St/Se) exhibits some planar and trough cross-stratification particularly in the low-flow back channel. A drape of silt over sands was also observed; however, these finer sediments are preserved only until the next high water event. Bar vegetation clearly demonstrates the sequence of lateral bar growth.

Six major point bar complexes occur within the more sinuous reaches of Bella Coola River, with well connected low-flow and high-flow slough channels. The outer bar-head region typically contains massive, loosely packed cobbles and gravels (32 < D_{50} < 128 mm: facies Gm) forming ridges or terraces along the upstream edge. Chute channels immediately behind the gravel ridge are downfilled with horizontal to cross-stratified sands (facies Sh and St) and side-filled by planar to massive gravels. Back bar regions contain poorly structured sands (facies Sh or Ss: 10–50 cm in thickness) over gravel. In comparison with lateral bars, the direction of point bar migration is more orthogonal to the main channel.

In straight river reaches basal gravels are exposed only during winter low flows. The vertical zonation of primary facies found in these bars contrasts sharply with those found in unstable river reaches (Fig. 5). In stable reaches, gravels are restricted to low elevations and sands are mobilized from the stream bank by flows that recur every year. In unstable reaches, where abundant channel deposits occur, gravel on the bar approaches the stage of mean annual flood (the comparison in Fig. 5 remains approximate because there is no guage record from an unstable reach). The stage at which the floodplain becomes covered with water is also higher for unstable reaches, probably because of the substantial secondary channel capacity there. The difference is not so great as that of the top elevation of gravel between the two reach types; the accumulation of sand on the floodplain is substantially thicker adjacent to stable reaches.

ALLUVIAL DEPOSITS: FLOODPLAIN FACIES

Spatial variations in the character of the sandy 'floodplain' sediments were examined using data collected from auger boreholes, excavations and cut-bank sections along 22 cross-valley transects oriented orthogonally to the active channel and primary sloughs. This allowed examination of sediments along the principal axis of channel migration where the most complete range of alluvial deposits is likely to exist.

The overall importance of the sandy facies in terms of contemporary alluvial sediment volumes can be determined by assuming that the depth of scour (thalweg depth) in the active channel represents the bottom of the gravel basement or 'bar platform.' This assumption can be made because of the apparent vertical stability of the river (Church, 1983). Thicknesses of the sandy deposits range from a few centimeters to over 3 m of coarse gravelly sand to fine sandy silt and represent from 17 to 39 percent of the bar platform sediments by volume (that is, 14 to 28 percent of the total 'active' sediment pile). Nowhere does clay constitute more than 15 percent of any sample; most frequently it is much less. In addition to the bar platform gravels, five dominant facies can be recognized within the floodplain portion of the alluvial valley-fill (Fig. 7) comprising bar top, channel fill and true overbank sediments.

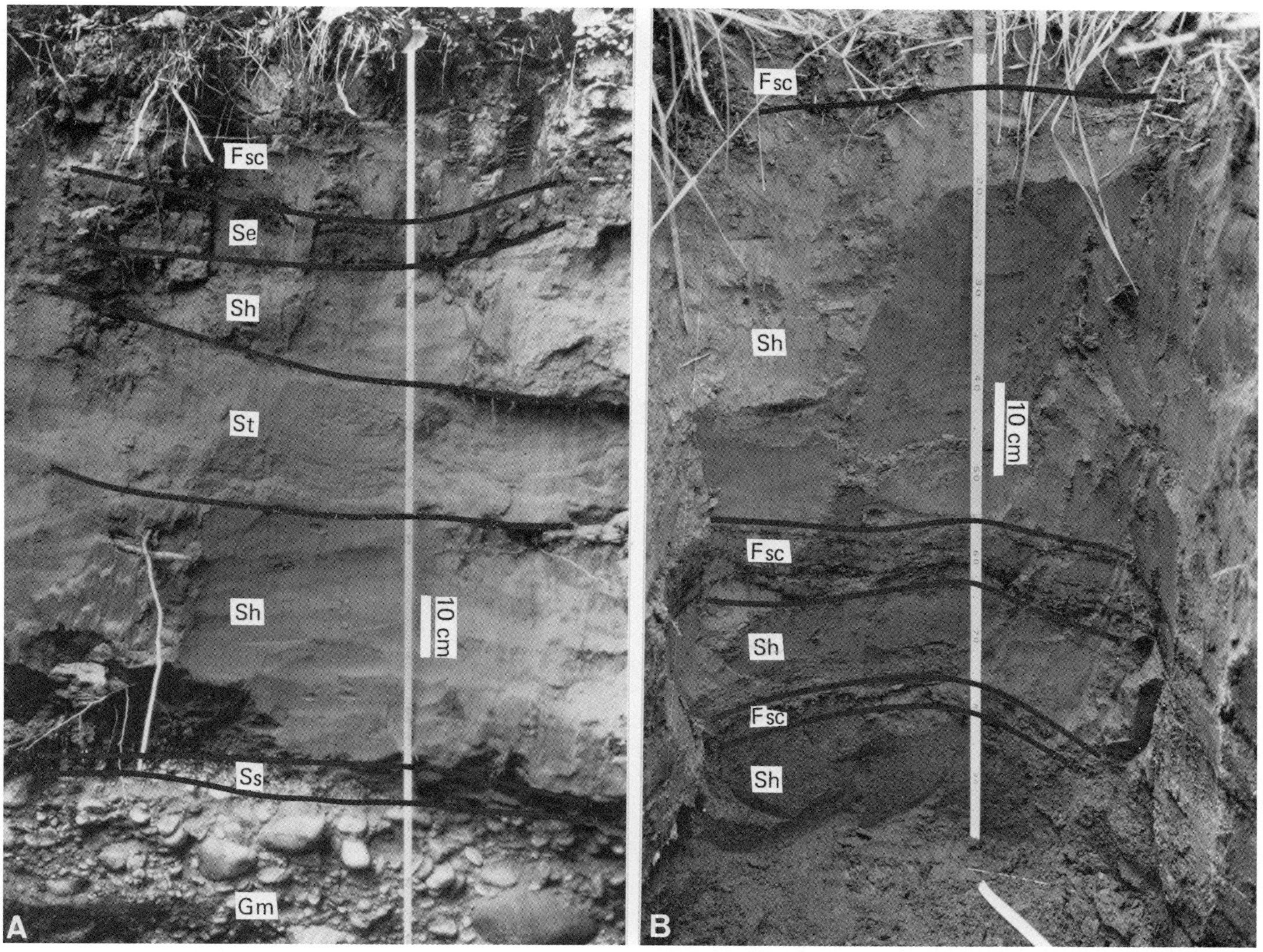

FIG. 7.—A) Facies Gm/Ss/Sh/St/Sh(Se)/Fsc of an unstable reach of Bella Coola River. Medium sands (D_{50} = 0.42 mm) form trough cross-stratified units, whereas horizontally bedded units are comprised of fine sand (D_{50} = 0.12 mm). B) Facies Gm/Sh/Fsc of a stable reach of Bella Coola River. Organic layering above the Sh unit represents multiple inundations of quiet flood waters. Fsc sediments are primarily silts (D_{50} = 0.025 mm).

Facies Ss.—

Thin, discontinuous sheets of pebbly coarse sand or granules ($1 < D_{50} < 5$ mm) overlie many of the basement gravel bars. Thicknesses rarely exceed 20 cm and are typically less than 10 cm. The Ss facies is most likely a scour/fill sediment of finer gravel laid down during the waning stages of flood flow. The thickness of this unit appears to increase in the direction of older channel fills, suggesting that it may have been part of thin channel deposits similar to some of the levees found along the margin of the contemporary active channel.

Facies Sh.—

Horizontally bedded, fine-to-coarse sands (0.13 mm $< D_{50} <$ 0.71 mm) varying in thickness from 10 cm to over 2 m characterize this facies. Thicker deposits often are found in stable reaches without Ss. The most probable origin is plane-bed sands of the upper flow regime laid down in stable sections of the river valley as channel fill or overbank sediments. Velocities in shallow floodplain depressions or swales adjacent to the main channel are known to exceed 1 m/sec during peak floods. In other instances, such as along back channel margins, it is probable these sediments aggrade as lower regime plane-bed sands. Occasional planar cross-stratification (Sp) is found but rarely exceeds 10 to 15 percent of this unit.

Facies Se.—

Discontinuous scour/fill deposits of coarse sand (0.5 mm $< D_{50} <$ 1.0 mm) (Se) within finer horizontally bedded sands (Sh) are found frequently in floodplain sections where old slough channels have been filled. The coarse sand forms sediment wedges near the surface of the sections examined and may represent the final infilling of the channel just prior to abandonment. Crude crossbedding is visible only when the scour/fill deposit is laterally continuous for several meters.

Facies St.—

Trough cross-stratified medium sands ($0.25 < D_{50} < 0.71$ mm) between 20 and 50 cm in thickness disconformably overlie finer planebed sands. This facies is derived from the migration and aggradation of large-scale lunate ripples seen forming today within the active channel or on inner, accretionary concave benches of unstable river reaches. St sediments are less common than Sh because the conditions for lower flow-regime bedforms in this high-energy river are transitory; that is, quickly changing flow depths and velocities discourage the preservation of these bedforms.

Facies Fsc.—

Facies Fsc is silt and minor fine sand (0.012 mm $< D_{50} < 0.063$ mm) combined with organics formed on the floodplain far removed from the active channel. No cross-stratification is visible, and organic-rich (wood and charcoal) layers may be present. Thicknesses of individual layers range from 5 to 15 cm, although a sequence of layers can exceed 1.5 m. The higher elevation and distance of this unit from the active channel suggest that these sediments were laid down as backwater deposits. Preservation of aggradational layers indicates little or no erosion between or during flood events.

The two most common associations of the facies units described above are illustrated in Figure 7. Facies assemblages Gm/Ss, Gm/Ss/Sh/St and Gm/Ss/Sh/St/Se are common in floodplain areas adjacent to a laterally unstable active channel (Fig. 7a). Sediments accumulate by lateral accretion of the main channel or by infilling of abandoned sloughs following channel avulsion. Shallow, high velocity flows are common in many areas of the bar and thus sedimentation units are thin and not laterally continuous for large distances. Gm/Sh and Gm/Sh/Fsc assemblages dominate in reaches of long-term stability (Fig. 7b). Overbank flows are common, and standing water occurs frequently in depressions and along the outer margin of the floodplain.

FLOODPLAIN EVOLUTION

Channel and point bar development is accompanied by the cross-valley migration of the channel, resulting in the formation of a bar platform and, in addition to the main channel, several active slough channels or a single, poorly connected back channel. Once established, vegetation type and age provide a chronology for channel migration (Fig. 8). Alders (*Alnus rubra*) colonize recently deposited barhead sediments, often in the first growing season following a flood. Alders and black cottonwood (*Populus trichocarpa*) dominate the central portions of the bar and recently developed floodplain surfaces. On older surfaces, decadent black cottonwood (100+ years), western red cedar (*Thuja plicata*) and sitka spruce (*Picea sitchensis*) prevail. This succession indicates the mean rate of channel migration to be in the range of 3.0 to 3.5 m/yr.

Sediment volumes (see above) indicate that lateral accretion of gravels accounts for more than three-quarters of sedimentation during floodplain construction. As the vegetation develops, sands are trapped effectively on bar tops and overbank, so the floodplain rises by vertical accretion to its final level. Channel migration rates in sinuous reaches indicate that the replacement or turnover rate for sediment along a unit length of the 0.5-km-wide floodplain would be between 145 and 165 years. Once deposited there, floodplain sediments remain undisturbed for periods of the order one to a few centuries. Sands are very mobile in this high-energy, cobble-gravel river, however, moving predominantly in suspension. Once re-entrained, the sand may be flushed from the system or redeposited a substantial distance downstream. Much of the sand probably moves through the system in only one or two stages and remains immobile during residence in the floodplain. Hence, there is little opportunity for mixing, and the mineral characteristics of the source areas are well preserved, with the proximal glacierized volcanics clearly dominant. In comparison, step lengths for gravel are relatively short, so that there is ample opportunity for mixing of materials from different sources in a much longer total time scale.

The actual modes of channel migration in unstable reaches are progressive erosion on the cut-bank opposite a growing channel bar, and rapid channel avulsion, or reduction of flow in the channel around one side of a medial bar, leaving behind a series of active, "subordinate" sloughs (cf. Fig. 6). Certain sloughs along the inner margins of point bars remain active, sediment transporting channels for periods in excess of 50–80 years following subordination. Infilling is likely to occur only if the slough entrance becomes blocked by large organic debris. In sinuous reaches of lower gradient, sloughs become poorly connected at most flow stages and a back channel develops with deposition of fine sands, silts and organic material. Infilling may never be complete as some of the older sloughs and back channels are preferentially occupied during high magnitude flows. As a result, channel avulsions frequently lead to the occupation of older channels rather than the development of new ones. Although avulsion produces abrupt realignment of the channel, progressive lateral accretion and erosion may be substantially more significant in floodplain development and sediment exchange.

Progressive erosion of the channel banks topples mature trees into the river. Hence, large organic debris, including complete trees, logs, root wads, and forest refuse, is an important component in bar formation and channel diversion. Medial bar heads or leading edges of cross-stream riffles frequently are cluttered with organic debris deposited during flood flows. Intervening flows are diverted either to the main channel—promoting the attachment of a mid-channel bar—or to an existing chute which is subsequently widened or deepened. Estimates of the volume of large organic debris measured at four sites, including bar-heads and entrances to slough channels, range from 0.5 to 3.5 m^3 per square meter of bar surface; that is, locally comparable with sandy facies volumes. Considering the large quantities present on the surface, organic debris is a surprisingly insignificant component of sand and gravel sedimentation units. It most likely is transported out of the system during floods or destroyed relatively rapidly.

The sedimentary sequences associated with stable and unstable reaches of the river reflect the integrated processes

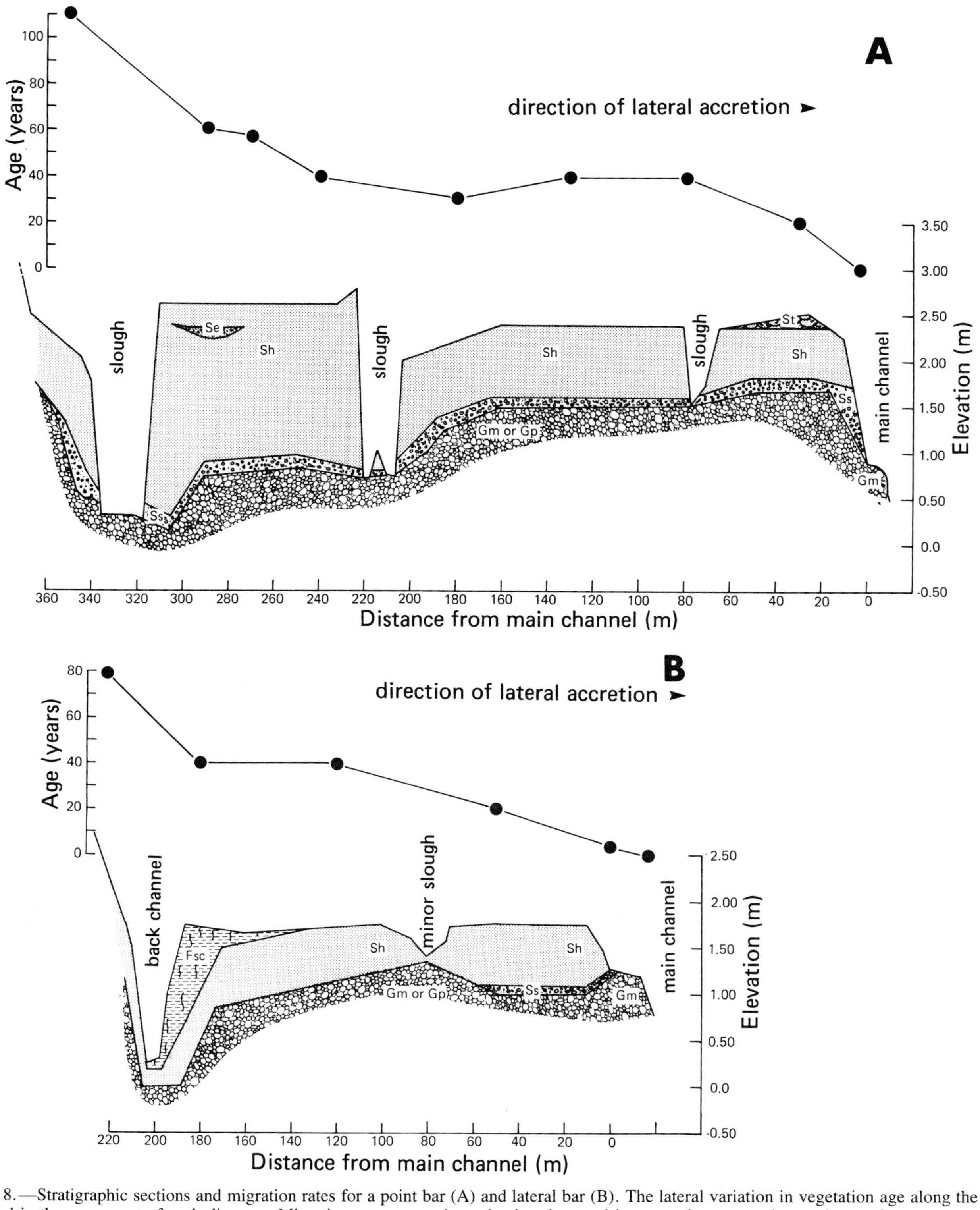

FIG. 8.—Stratigraphic sections and migration rates for a point bar (A) and lateral bar (B). The lateral variation in vegetation age along the section is plotted in the upper part of each diagram. Migration rates are estimated using the trend in vegetation succession on bar surfaces.

of floodplain development over time. In unstable reaches of the river, rapid channel avulsions, the occurrence of multiple channels with varying capacities for sediment and water discharge, and more variable flow conditions result in a diversified set of facies units within the channel and on the vegetated floodplain surface. With increasing stability, the spatial variation in flow conditions declines and the probability of transitions between facies units becomes lower than in unstable reaches of the river. Progressive lateral accretion and the vertical stability of the river appear to favor the simpler facies sequence (Gm/Sh/Fsc), although infilled back channels may still exhibit the wider variety of sedimentary structures. The greater incidence of overbank flooding in stable reaches (Fig. 5) produces vertical accretion of the floodplain here and development of the Fsc facies. Lateral accretion deposits, however, dominate the overall sequence in just the manner described by Wolman and Leopold (1957). The relatively short turnover time for the alluvial valley-fill suggests frequent modification of the sedimentary sequence at any given site. Reach-averaged proportions of dominant facies sequences may remain constant for long periods, however.

SUMMARY AND CONCLUSIONS

Sediments within the Bella Coola Valley represent a variety of processes operating during the late glacial and Holocene periods. Late Pleistocene glaciomarine and deltaic sediments comprise the major proportion of the valley-fill. Volumetrically, the alluvial channel and floodplain sediments constitute less than 3 percent of the total sediment pile, yet they represent 60 percent or more of postglacial time, during which the alluvium has been reworked and exchanged repeatedly. Mineral composition of these sediments indicates that glacierized volcanic terrain in the headwaters is the dominant sediment source; however, sediment contributions from downstream tributaries are locally notable. Channel gravels represent a mixture of plutonic and volcanic rock lithologies derived from different tributaries, whereas most overbank sands retain distinctive source area characteristics. Non-alluvial deposits (colluvium and glaciomarine sediments) are found adjacent to only 13 percent of the channel length and are not important sources for alluvial sediments.

Channel zone deposits of the contemporary river are dominated by massive to poorly stratified gravels and cobbles forming medial bars in unstable reaches and lateral and point bars in stable reaches. Horizontally stratified sand sheets and trough cross-stratified sands are minor components of the active channel sediments. The gravels, which are formed by lateral accretion during migration of the active channel, constitute about three-quarters of the Holocene alluvial valley-fill.

Floodplain sediments over the gravels can be divided into several principal facies types. Floodplain areas adjacent to a stable active channel are dominated by horizontally stratified sands and fine overbank silts mixed with organics. Reaches characterized by an unstable active channel have floodplain sediments ranging from gravels to horizontally bedded and trough cross-stratified sands. These latter deposits are usually thinner but of a greater lateral extent than those found in stable reaches. Vegetation successions on point bars and lateral bars indicate a mean annual channel migration rate in excess of 3 m/yr, so the floodplain may be reworked in about 150 years. Large organic debris introduced into the river by bank undercutting of stable mid-channel islands strongly influences channel diversions and the development of gravel bars.

The spatial variability in channel zone and floodplain facies of Bella Coola River precludes the classification of these assemblages into any single previously described facies model. Similarities in facies types occur between this river and the gravel-dominated systems described by Bluck (1976) and Forbes (1983), although there is an absence of clays in overbank and backwater deposits here. Sedimentation on bars and floodplain surfaces is closely linked to adjacent channel morphologies, which vary according to the volume of stored sediment derived from upstream and from downstream tributaries. Alternating reaches of laterally stable and unstable active channel respond differently to flooding and this results in different styles of sedimentation.

The long-term vertical stability of this river indicates that sediment is being passed through the system at a rate commensurate with the supply. Hence, although it represents an active sedimentary environment that has persisted for some time (during most of the current non-glacial interval), it is apt to be poorly represented in the sedimentary record. If aggradation does occur, it is likely that only the channel gravel facies will generally be preserved since the sands are easily moved during flood. The ultimate sedimentary record could not then be interpreted correctly as deriving from this river regime type.

ACKNOWLEDGMENTS

This study is supported by a grant (MC) and scholarship (JRD) from the Natural Sciences and Engineering Research Council of Canada. M. Cantwell, L. Davies, P. Desloges, T. Millard and J. Weninger assisted in the field. The paper has benefited substantially from reviews by Dr. L. E. Frostick and J. Pitlick.

REFERENCES

ANDREWS, J. T., AND RETHERFORD, R. M., 1978, A reconnaissance survey of late Quaternary sea levels, Bella Bella/Bella Coola region, central British Columbia coast: Canadian Journal of Earth Sciences, v. 15, p. 341–350.

ARMSTRONG, J. E., 1981, Post-Vashon Wisconsin glaciation, Fraser Lowland, British Columbia: Geological Survey of Canada Bulletin 322, 32 p.

BAER, A. J., 1973, Geology of the Laredo Sound/Bella Coola map area, British Columbia: Geological Survey of Canada Memoir 372, 122 p.

BLUCK, B. J., 1971, Sedimentation in the meandering River Endrick: Scottish Journal of Geology, v. 7, p. 93–183.

———, 1976, Sedimentation in some Scottish rivers of low sinuosity: Transactions, Royal Society of Edinburgh, v. 69, p. 425–456.

CHURCH, M., 1983, Pattern of instability in a wandering gravel bed channel, *in* Collinson, J. D., and Lewin, J., eds., Modern and Ancient Fluvial Systems: International Association of Sedimentologists Special Publication 6, Blackwell Scientific, p. 169–180.

CLAGUE, J. J., 1975, Glacier flow patterns and the origin of late Wisconsinan till in the southern Rocky Mountain Trench, British Columbia: Geological Society of America Bulletin, v. 86, p. 721–731.

———, 1985, Deglaciation of the Prince Rupert—Kitimat area, British Columbia: Canadian Journal of Earth Sciences, v. 22, p. 256–265.

———, HARPER, J. R., HEBDA, R. J., AND HOWES, D. E., 1982, Late Quaternary sea-levels and crustal movements, coastal British Columbia: Canadian Journal of Earth Sciences, v. 19, p. 597–618.

EHLERS, E. G., AND BLATT, H., 1981, Petrology: W. H. Freeman and Company, San Francisco, 732 p.

FORBES, D. L., 1983, Morphology and sedimentology of a sinuous gravelbed channel system: Lower Babbage River, Yukon coastal plain, Canada, *in* Collinson, J. D., and Lewin, J., eds., Modern and Ancient Fluvial Systems: International Association of Sedimentologists Special Publication 6, Blackwell Scientific, p. 195–206.

Howes, D. E., 1983, Late Quaternary sediments and geomorphic history of northern Vancouver Island, British Columbia: Canadian Journal of Earth Sciences, v. 20, p. 57–65.

Jackson, R. G., II, 1978, Preliminary evaluation of lithofacies models for meandering alluvial streams, *in* Miall, A. D., ed., Fluvial Sedimentology: Canadian Society of Petroleum Geologists Memoir 5, p. 543–576.

Kellerhals, R., Neill, C. R., and Bray, D. I., 1972, Hydraulic and geomorphic characteristics of rivers in Alberta: Research Council of Alberta, River Engineering and Surface Hydrology Report 72-1, 55 p.

Klages, M. G., and Hsieh, Y. P., 1975, Suspended solids carried by the Gallatin River of southwestern Montana: II. Using mineralogy for inferring sources: Journal of Environmental Quality, v. 4, p. 68–77.

Miall, A. D., 1978, Lithofacies types and vertical profile models in braided river deposits: A summary, *in* Miall, A. D., ed., Fluvial Sedimentology: Canadian Society of Petroleum Geologists Memoir 5, p. 597–604.

Neill, C. R., 1973, Hydraulic and morphologic characteristics of Athabasca River near Fort Assiniboine: Alberta Research Council, Edmonton. Highway and River Engineering Division Report REH/73/7, 23 p.

Powell, R. D., 1983, Glacial-marine sedimentation processes and lithofacies of temperate tidewater glaciers, Glacier Bay Alaska, *in* Molina, B. F., ed., Glacial Marine Sedimentation: Plenum Press, London, p. 185–232.

Shilts, W. W., 1973, Glacial dispersal of rocks, minerals and trace elements in Wisconsinan till, southeastern Quebec: Geological Society of America Memoir 136, p. 189–219.

Slatt, R. M., 1970, Texture and composition of till derived from parent rocks of contrasting texture: Southeastern Newfoundland: Sedimentary Geology, v. 7, p. 283–290.

Tipper, H. W., 1979, Geology of the Parsnip River sheet: Geological Survey of Canada, Map 1826A.

Wolman, M. G., and Leopold, L. B., 1957, River floodplains: some observations on their formation: U.S. Geological Survey Professional Paper, v. 282-C, p. 83–109.

SEDIMENTOLOGY AND FACIES ARCHITECTURE OF OVERBANK DEPOSITS OF THE MISSISSIPPI RIVER, FALSE RIVER REGION, LOUISIANA

KATHLEEN M. FARRELL

Department of Geology, Louisiana State University, Baton Rouge, 70803

ABSTRACT: This detailed study of the sedimentology and facies architecture of three overbank subenvironments which occur marginal to a meander belt in the lower Mississippi River Valley leads to the following conclusions: 1) backswamp, levee and splay deposits can be subdivided into units related to the establishment of the associated channel belt; 2) cycles related to avulsion, levee progradation, splay progradation and abandonment, and sheet-flood events are preserved in flood basin deposits; and 3) flood basin facies prograde basinward as the levee builds upward over time along a channel belt margin.

The establishment of the channel belt is divided into four phases: 1) a pre-avulsion stage, 2) an avulsion stage, 3) an early channel belt, and 4) a late channel belt stage. In the study area, the flood basin sequence affiliated with the avulsion and establishment of the channel belt (30 m) has a maximum thickness of about 10 m. The avulsion event is probably recorded in the sedimentary sequence as the lithologic change from blue clay with detrital organic debris deposited in standing water to sheet-flood silts and sands related to levee and splay progradation during incipient formation of the channel belt. Surficial levee deposits (silty sand unit) exhibit an overall coarsening-upward sequence (2.5 m) that reflects meander bend migration toward the sampling site during the late stage of channel belt development. Surficial splay deposits (silty sand unit) initially coarsen upward during the progradational phase and then fine upward as the splay is abandoned (3 m). Individual flood cycles (mm–cm) occur as small-scale fining-upward rhythmites with poorly preserved stratification in all sub-environments.

INTRODUCTION

Overbank deposits have been traditionally neglected by fluvial sedimentologists in favor of studies of the stratigraphic sequence emplaced in channel belts during point bar migration (Jackson, 1978). In flood basins, however, sedimentary sequences can potentially record the avulsion history of the basin (channel belt establishment), crevassing events, progradational and abandonment phases of levee and splay development, and individual flood events (Elliott, 1974; Bridge, 1984). The purpose of this research is to examine the sedimentology and stratigraphy of overbank subenvironments which occur along a meander belt margin in the lower Mississippi River Valley and to interpret the preserved sequences in light of the processes operating in the basin.

Sedimentation processes in overbank regions and general characteristics of the resultant deposits have been discussed by Bridge (1984), Smith (1983), Hughes and Lewin (1982), Elliott (1974), and Allen (1965). A number of studies describe sedimentary structures and sequences in specific modern overbank subenvironments of fluvial systems including the natural levee (Smith, 1983; Singh, 1972; Coleman, 1969), slays (Cross and Smith, 1985; Smith, 1983; Coleman, 1969) and backswamp and lacustrine deposits (Cross, 1985; Krinitzsky and Smith, 1969; Coleman, 1966). Upper point bar deposits are also considered overbank deposits because they are the result of post-depositional modification by overbank flooding as well as lateral accretion in channels. The stratification of the upper point bar has been described by Jackson (1976), Ray (1976), Singh (1972), and Davies (1966). A review of stratification resulting from lateral accretion of modern lower point bars in channels will not be included here.

Most of the modern studies of stratification mentioned above were based on descriptions from shallow trenches, low scarps and cutbanks (<3 m of section) preserved near channel margins and which mainly represent channel belt deposits. Several studies, however, have used cores through overbank deposits to document the succession and spatial relationships of lithofacies in flood basins (Smith, 1985; Cross and Smith, 1985; Smith, 1983; Krinitzsky and Smith, 1969; Coleman, 1966).

In this study, the approach used by Ray (1976) of examining a specific overbank subenvironment (upper point bar) for flood-related sequences is applied to levee, splay and backswamp deposits. These subenvironments were sampled with a large number of closely spaced cores taken across the margin of the present meander belt in the lower Mississippi Valley. The cores were first subdivided into a number of units based on gross lithology. Then, rhythmites, which represent individual flood events, were identified in the coarsest units. Lastly, an attempt was made to identify larger scale sedimentation cycles attributed to channel belt establishment, crevassing, progradation of levees and splays into flood basins, and splay abandonment. This paper presents the stratification types and sequences observed in these overbank deposits.

GEOLOGY OF THE LOWER MISSISSIPPI RIVER VALLEY

The study was conducted in the lower Mississippi Valley about 35 km northwest of Baton Rouge, Louisiana (Fig. 1). It was centered about False River, an oxbow lake with excellent relict topography, which is located along the margin of the presently occupied channel belt (Fig. 2). The difference in elevation between the highest point on the levee (11.6 m above M.S.L.) and the lowest point (2.5 m above M.S.L.) in the backswamp in the adjacent flood basin to the west is 9.1 m, making the levees at False River the most highly developed anywhere along the Mississippi River (Fisk, 1947). This could be an artifact of sedimentation on the downthrown block of an east-west-trending fault zone (Fig. 2). Faulting could also have controlled the sites of levee crevassing. This difference in elevation is caused by progradation of overbank sediment into the backswamp during the establishment and evolution of the meander belt.

The segment of the modern channel belt which extends through the study area has been occupied for approximately 3,000 years (Saucier, 1974). Before then, the active chan-

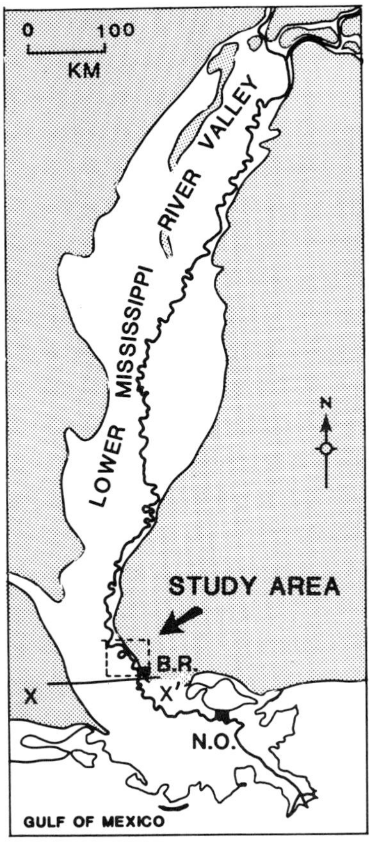

FIG. 1.—Location map showing the position of the False River region in the lower Mississippi Valley. Cross section X-X′ is shown in Figure 3.

nel belt was positioned along the western wall of the lower Mississippi Valley and a poorly developed drainage network existed in the vicinity of the present channel belt (Fisk, 1944). An average sedimentation rate of 0.3 cm/yr is calculated as the rate of levee accretion, based on the maximum thickness of the levee and the age of the channel belt.

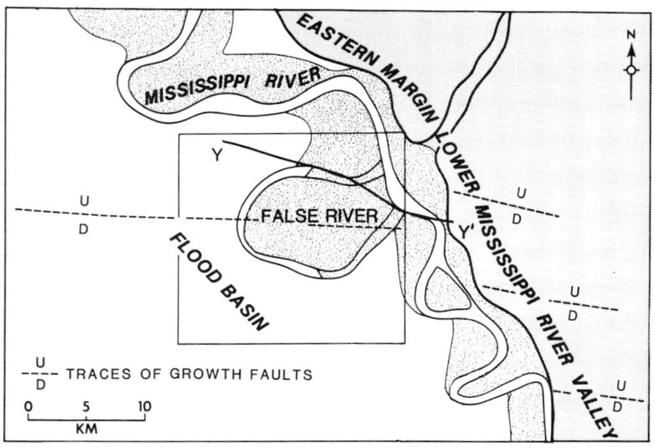

FIG. 2.—Location of the study area (box) with respect to the channel belt (stippled pattern), the flood basin (non-stippled area) and the eastern wall of the Mississippi Valley (gray pattern). Cross section Y-Y′ is shown in Figure 4. Growth fault data is from Saucier (1969).

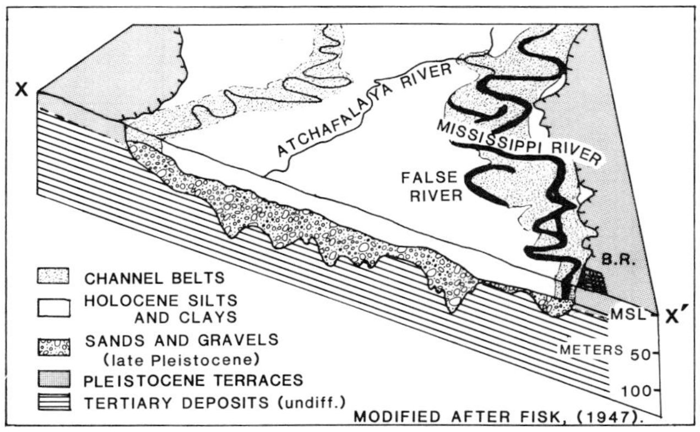

FIG. 3.—Geologic cross section through the lower Mississippi Valley near Baton Rouge, Louisiana. The location of cross section X-X′ is shown in Figure 1.

This figure is awaiting confirmation from radiocarbon dates.

A cross section (X-X′) through the flood plain near Baton Rouge shows that approximately 30 m of Holocene overbank clays and silts overlie an undifferentiated sand and gravel unit of late Pleistocene age (Figs. 1, 3). The Holocene section is divisible into a stack of alternating poorly drained swamp, well drained swamp and lacustrine facies (Krinitzsky and Smith, 1969; Coleman, 1966). The overbank subenvironments studied here include the upper 10 m of these Holocene deposits.

Sharp facies contacts between channel belt sands and flood basin clays are ubiquitous along channel belt margins in the lower Mississippi Valley (Fisk, 1947, Plate 5; Allen, 1965; Jackson, 1978). These high angle contacts are principally the result of cutbank erosion along the margin of the channel belt due to the downstream translation of the meandering thalweg. In Figure 4, Saucier's (1969) cross section through the channel belt margin in the study area has been modified to show the vertical extent of the flood basin-channel belt contact and the discrete positions of upper point bar, lower point bar and levee subenvironments.

Fisk (1944) inferred a meandering history for the Mississippi River in the presently occupied channel belt based on orientation of ridges and swales on preserved point bars

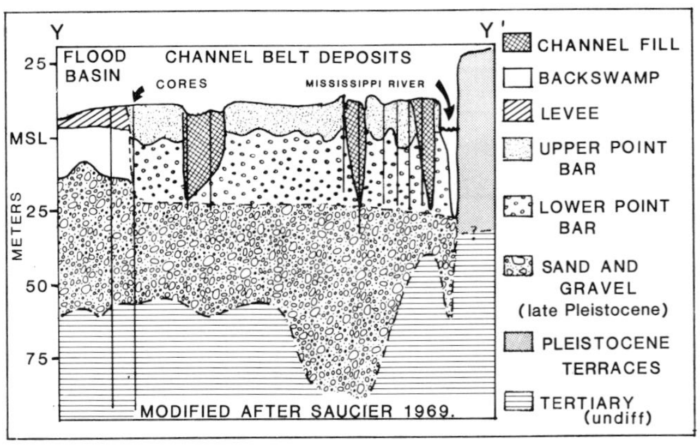

FIG. 4.—Geologic cross section through the False River region showing distribution of several overbank subenvironments. The location of cross section Y-Y′ is shown in Figure 2.

and the assumption that point bars migrate at a constant rate. In the study area, a meander loop and its associated point bar migrated gradually southwestward between 1200 and 400 years B.P. (Fig. 5A, 5B). Historic records indicate that in approximately 1700 A.D., a neck cutoff resulted in the formation of False River, an oxbow lake (Fig. 5B) (Sternberg, 1956). Since that time, the active channel thalweg has remained close to the eastern valley wall.

OVERBANK SUB-ENVIRONMENTS IN THE FALSE RIVER AREA

Overbank subenvironments in the study area (Fig. 6) include well developed ridge-and-swale topography of the upper point bar on the convex side of the oxbow lake and a natural levee along the concave side of the relict meander

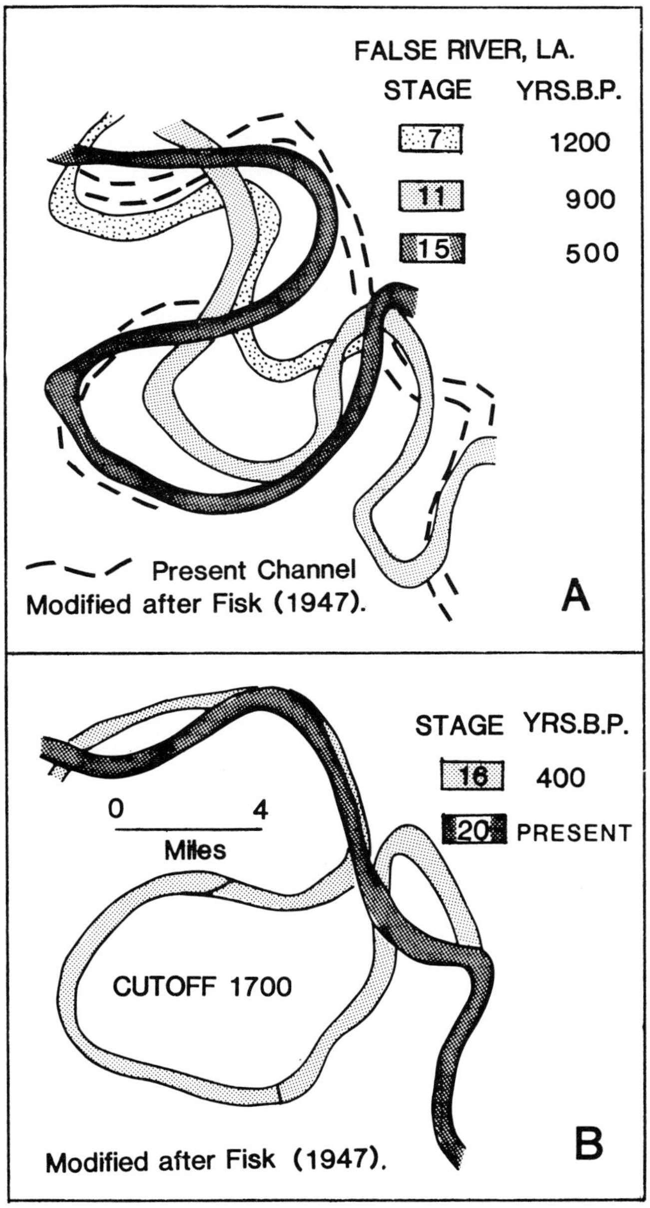

FIG. 5.—The meandering history of the Mississippi River: A) between 1200 and 500 years B.P.; B) between 400 years B.P. and the present with a neck cutoff occurring in 1700 A.D.

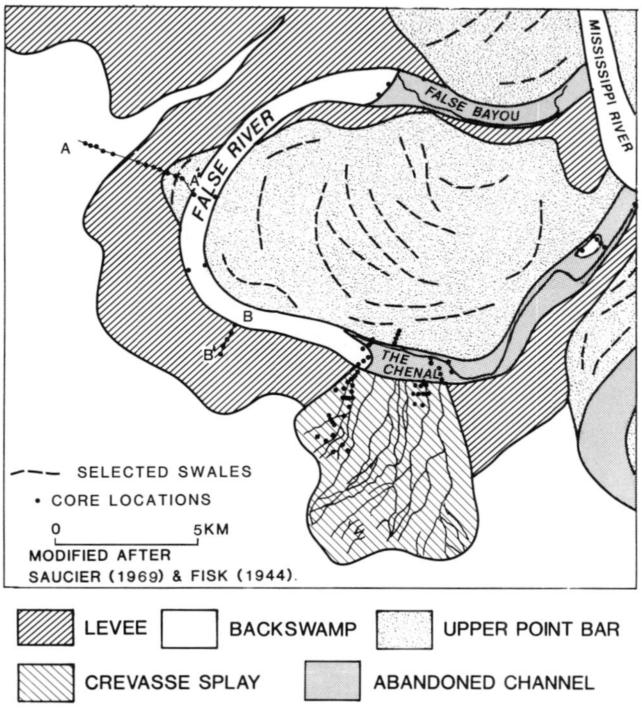

FIG. 6.—The distribution of overbank subenvironments around False River with core locations.

bend. A region of levee crevassing and splay development is located along the downstream bend of the meander. Backswamp deposits lie in the adjacent flood basin. The crevasse topography (Fisk, 1947), including seven main crevasses and the associated splay deposits, covers an area of approximately 50 km^2, and extends approximately 8 km out into the flood basin. Crevasse channels are deeply incised (to a maximum depth of about 5 m) into the splay and levee deposits and for the most part are still open and unfilled. The crevasses are anastomosing and flow into distal drainage networks which parallel the slope of the flood basin floor.

SAMPLING

Approximately 100 cores were collected from the overbank environments associated with the channel belt margin including the backswamp, levee, crevasse and splay deposits (Fig. 6). In subaerial environments the cores were taken in a shelby tube (9-cm-diameter × 1.1 m in length) lined with PVC tubing and attached to a Giddings soil probe. This type of coring proceeds incrementally to a maximum penetration depth of 9 m as the shelby tube is hydraulically pushed into the ground. The upper four to five segments of core were generally in excellent condition, but at depths greater than 5.5 m scrapings from the core hole walls usually caused compaction and disruption of sedimentary structures in the upper 60 cm of the core tube. Vibracoring was restricted to subaqueous subenvironments such as the crevasse channels, and penetration depth was shallow (<3.5 m) because of the tightly packed backswamp deposits underlying the channel fill. In the laboratory the cores were split, slabbed for X-radiography, described and photographed.

SEDIMENTOLOGY OF THE OVERBANK SUB-ENVIRONMENTS

Backswamp

The four major units in the backswamp cores are: 1) a basal blue clay, 2) a silty sand unit, 3) mudball conglomerate beds, and 4) concretion-rich rooted clays (Fig. 7). The occurrence and early diagenesis of nodules and precipitants in backswamp deposits are discussed in detail by Ho and Coleman (1969) and Coleman (1966).

Blue Clays.—The deepest sediments cored, which are at an elevation of about 5 m below M.S.L. (Fig. 7), consist mostly of bluish-gray clays with intercalations of silt and macerated organic debris. Organic-rich zones occur at intervals approximately 1 to 1.5 cm apart and consist of several leaf layers with minor woody material interlaminated with clay. The silt layers which appear as patchy, slightly less dense material in the X-radiographs have been disrupted into diffuse layers by burrowing organisms. This unit also contains vivianite nodules and tiny rootlets with pyrite druses. Upsection, the clays are overlain by a narrow zone (10 cm thick) of rooted clay/silt laminates with iron oxides as root linings. These are in turn overlain by the silty sand unit.

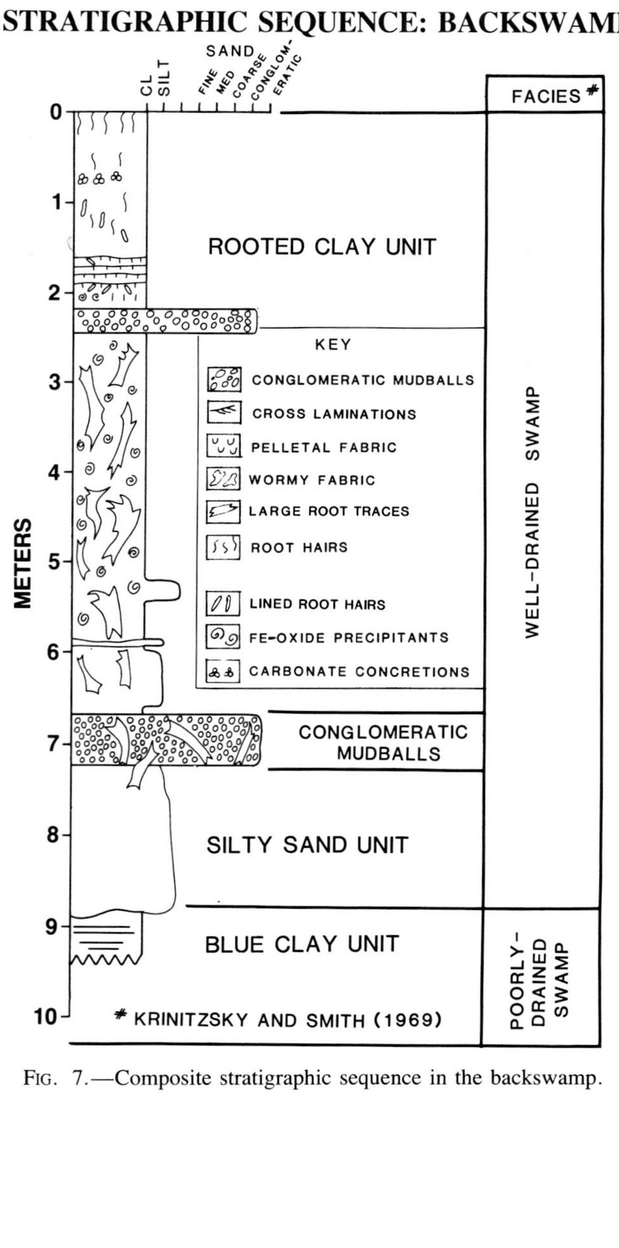

FIG. 7.—Composite stratigraphic sequence in the backswamp.

Silty Sand Unit.—Above these clays are pale green very fine silty sands and sandy silts which comprise the silty sand unit. The base of this silty sand unit contains gray clay-rich material which either represents rip-up clasts or infillings of root burrows. The coring method at this depth (5–8 m) often causes disruption of the upper halves of the core tubes. In the lower half of each core tube, however, faintly laminated silt and more rarely, single cross-sets of very fine sand are visible where bioturbation has not disturbed the primary fabric. Large burrows (3–4 cm in diameter) with partial clay linings and filled with illuviated debris such as clay balls, loose silt, and partially cemented silt clasts cross cut the original bedding features. A common fabric consists of silt with clay-filled root traces and clay films (argillans) segregated by illuvial processes.

Rooted Conglomerate.—Above the silty sand unit occurs a bed of rip-up clasts (2 cm maximum diameter) in a matrix of silt and smaller rip-up clasts. These clasts are angular to rounded and mainly consist of a variety of different colored clays. Upsection, the basal bed of rip-up clasts merges indistinctly with a zone of highly plastic, but jumbled, clay lithologies and then into the overlying root-mottled clay. The jumbled zone may be interpreted as: 1) extremely large rip-up clasts, or 2) slumped blocks into large vacated root burrows, or 3) an artifact of coring. It is difficult, if not impossible, to differentiate the conglomerate from slumped zones and the overlying rooted clay in this unit because of extensive root mottling.

Rooted Clays.—Above the jumbled zone are 3–5 m of clay which ranges in color from green to gray to brownish-gray with decreasing depth. This unit is gradational with the underlying unit. The main characteristics of this clay are: 1) extensive root mottles, 2) carbonate nodules as growths around fragmentary plant remains (leaf fragments and small root hairs), 3) iron oxide precipitants as coatings on small root hairs and in silty 'mottles,' 4) vertically oriented root hairs, and 5) sporadic silty beds. Root hairs are about 1 mm in diameter and exhibit a wide range of lengths.

These clays were deposited from suspension after a levee overtopping event. The original bedding probably consisted of very thin rhythmites (mm) with thin silty bases fining upward into clay. Remnants of this stratification are present sporadically; however, they are mostly obliterated by large tree roots, small root hairs, burrowing organisms and even dense nodule formation. Burrowing organisms disrupt many rhythmites into small (1 mm) pockets of silt surrounded by clay (pelletal fabric described below). In X-radiographs, faint precipitants of iron oxide are visible in these pockets. Other rarer rhythmites consist of clay layers with intervening horizons of organic debris. Original organic material is not commonly preserved here; usually leaves and root hairs are replaced or lined with carbonate or iron oxides.

The mottling caused by root penetration is very distinct at the base of this unit. Elongate, irregularly-shaped gray clay zones penetrate through a green clay discolored by orange stains from iron oxide precipitants in silt- or sand-rich mottles. The gray zones are reduced areas which include fragmentary plant remains. Slickensides and/or fragmented green clay may occur along mottle edges. Upsection, however, as the clay changes in color from green to gray, the root mottling is still present but less obvious.

Retallack (1983a, 1983b) presented several hypotheses to explain the origin of these zones of reduced sediment surrounding plant remains. Here, they are probably the result of both the anaerobic decay of roots in ground water after burial and illuviation of clays into vacated root burrows. Brecciation of the green clay along mottle edges is probably caused by a combination of shrinking and swelling of clay, penetration by root hairs which are attached to tree roots and displacement of clays during penetration by tree roots.

Distinct beds of rip-up clasts are interstratified with the rooted clays (Fig. 7) in upper parts of two of the cores. Rounded mudballs as much as 1.5 cm in diameter are present and the matrix appears to be clay or smaller mudballs.

Interpretation.—Krinitzsky and Smith (1969) provided the following descriptions for the three major facies in flood basins. The lacustrine facies consists primarily of dark colored reduced clay with minor amounts of sand and silt. Typically it is well bedded, with shell beds, detrital organics and carbonate nodules. Locally it is extensively burrowed. The poorly drained swamp facies also consists mainly of dark colored reduced clay. Primary stratification is extensively disturbed by roots, and detrital organic layers and peats may be locally present. Concretionary material present includes carbonates, sulfides and vivianite. The well drained swamp facies consists of iron oxide-mottled clay exhibiting a variety of colors including light gray, light yellowish-brown and dark brown depending on the type and amount of iron oxides present. Primary stratification is generally obliterated by penetrating roots, and organic fragments are rare because of the oxidizing environment.

The units observed here can be reclassified according to Krinitzsky and Smith's (1969) facies descriptions (Fig. 7). The blue clay unit is interpreted as either the lacustrine or poorly drained swamp facies because it consists of dark colored, reduced clay, is well bedded, contains detrital organic laminations, pyrite druses and vivianite nodules. The units above the blue clay are extensively root-mottled and similar to the well drained swamp facies. Although most of the clays are reduced, they are patchily oxidized with abundant iron oxide precipitants and rarely contain preserved organic material.

The sequence of four units preserved beneath the backswamp (Fig. 7) indicates four periods of deposition from the bottom to top. These are: 1) a period of standing water deposition (blue clay unit) in a poorly drained swamp or lake environment; 2) incursions of probable splay or lacustrine-delta sands (silty sand unit) into the standing body of water; 3) a major flood event resulting in the deposition of the bed of rip-up clasts (mudball conglomerate bed); 4) progradation of overbank sediment from the channel belt margin into the flood basin. This resulted in the upbuilding of relief and the establishment of a well drained swamp which trapped fine-grained suspended sediment from distal sheet floods caused by levee overtopping.

Levee and Splay Deposits

Levee and splay sediments are similar lithologically because both are deposited as rhythmites from decelerating sheet floods that overtop natural levees and crevasse channels. The levee and splay cores include four major units: 1) a basal blue clay, 2) a lower sandy silt unit, 3) a rooted clay unit, and 4) an upper silty sand unit (Fig. 8). The blue clay occurred only in the deepest splay cores taken at low elevations in crevasse channels and presumably marks the base of the entire splay/levee complex associated with this channel belt. It is similar lithologically to the blue clay intercepted at the base of the backswamp cores and is also interpreted as lake or poorly drained swamp deposits. The rooted clays are backswamp deposits that interfinger with the levee and splay deposits and will not be redescribed here.

The sandy silt and silty sand units (m) represent periods of levee and/or splay progradation into the flood basin. In the following paragraphs, the origin of stratification types that occur in these two units are discussed: sandy rhythmites, silty rhythmites and pelletal silt.

Rhythmites.—Rhythmites are alternating beds or laminations of contrasting lithologies. Each rhythmite (cm) is deposited during a single sheet flood. Sandy rhythmites consist of well sorted, very fine ripple-laminated sand which

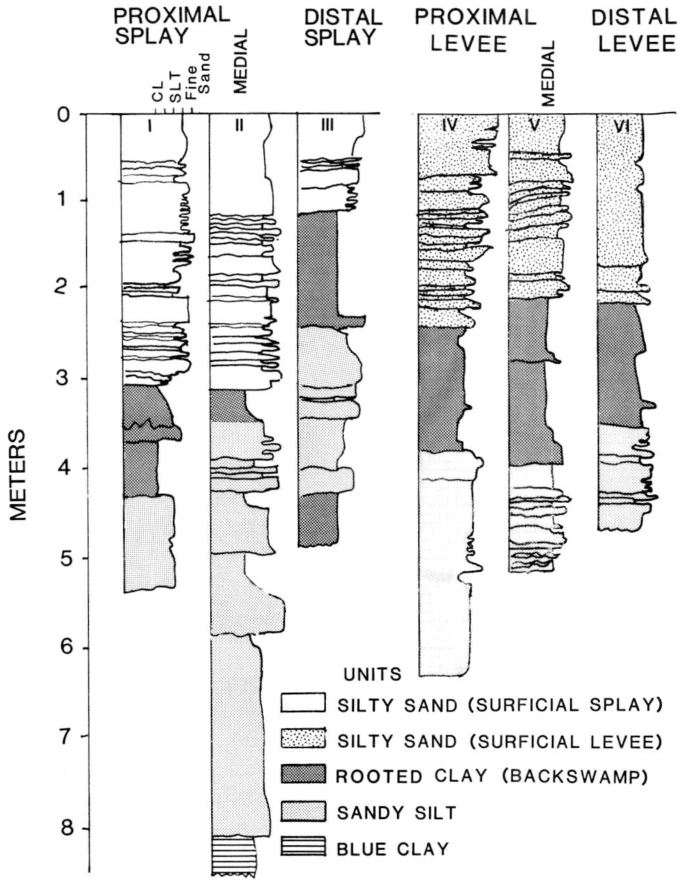

FIG. 8.—Stratigraphic sequences in levee and splay deposits.

fines upward into clay, mud or muddy sand. Silty rhythmites are similar with laminated silts or very small ripple forms in silt as the coarser part of the bedset. Amalgamated sand beds are rare. Well stratified sandy rhythmites are common in surficial levee deposits immediately adjacent or proximal to False River and in smaller scale levees which occur along crevasses (Plate 1D). Primary stratification in the rhythmites, however, may be partially to completely obliterated because of biogenic and pedogenic overprinting which causes the development of a pelletal (Plate 1F) or wormy fabric (Plate 1E). These poorly stratified deposits occur surficially in splay lobes distal to the crevasse channels and in levee deposits distal to False River.

Origin of Wormy Fabric.—A common type of burrow that disrupts the rhythmites is a worm-shaped burrow with a circular cross section about 0.8 cm in diameter. These burrows are horizontal to sub-vertical in orientation and are concentrated in the muddy part of the rhythmite (Plate 1B, at k). Close inspection of burrow cross sections indicates that the burrow fill consists of radially arranged mud-rich pellets (mm) in a coarser matrix. They appear to have been formed by organisms that mostly crawl laterally through substrate and pelletize the sediment while doing so. A wormy fabric may result when individual burrows extend downward as worm-shaped protrusions of mud from densely burrowed layers into the sands and silts below disrupting primary stratification (Plate 1B, D).

Origin of Pelletal Fabric.—Pelletal fabric results when the burrows described above coalesce so that individual burrow boundaries are lost and all that is visible is a mass of burrow fill (Plate 1C, F, H). Pelletal clay, pelletal mud, pelletal silt and even pelletal very fine sand beds can result.

Pelletal silt, a very common bedding type in levee and splay deposits, consists of small pockets (mm) of clay or silt or mud in a contrasting matrix of silt, clayey silt or sandy silt (Plate 1F). Iron oxides are also present as linings around microroot traces (mm in diameter) and as precipitants in coarser grained pockets of sediment in the pelletal fabric. The pelletal silt beds occur in the surficial levee sediments distal to the oxbow lake and in splay lobes distal to crevasses.

Summary of Bedding Type in Levees and Splays.—Figure 9 shows the condition of rhythmite beds as a function of progressive disturbance by these burrowers. The ideal rhythmite consists of well stratified sand or silt fining upward into mud or clay (Fig. 9A). If suspension layers at the top of the cycle are very thin, then the rhythmites may consist of ripple-laminated sand beds separated by wormy clay drapes (Fig. 9B). Extensive burrowing in thicker muds at the top of rhythmites results in zones of well developed wormy fabric separating disrupted crossbedded sands with only traces of the original bedding being present (Fig. 9C). Sedimentary structures in densely burrowed rhythmites are completely obliterated to form rather homogeneous appearing pelletal silt (Fig. 9D).

Other Bioturbation Structures.—Superimposed on and cross-cutting the stratification types described above are additional biogenic structures which even further modify the rhythmites. Large dwelling burrows(?) or root traces which are as much as 10 cm in diameter may cross-cut both primary stratification and secondary fabrics (Plate 1A, I). Cross sections through segments of these structures are irregular in shape and may be partially lined by clay (Plate 1I). Internal sediment appears to consist of either mud pellets or intraclasts in a matrix of very fine sand and silt. In some vertical sections the internal sediment appears to be horizontally layered and may indicate either geopetal infilling or active backfilling by a burrower (Plate 1A). Other burrows which are present but rare include simple vertical burrows with iron oxide linings and backfilled vertical burrows (not shown).

Cyclic Sedimentation in Levee Deposits.—The levee cores include a lower sandy silt unit, a rooted clay and an upper silty sand unit (Fig. 8). This sequence indicates that multiple episodes of levee and splay progradation separated by intervening periods of backswamp-style deposition are preserved in the subsurface along the margin of this channel belt. The sandy silt unit consists of pelletal silts and silty rhythmites and resembles distal surficial levee and splay lobe deposits. It can be interpreted as 1) sheet flood deposits associated with the initial establishment of the channel belt, or 2) a progradational package related to the passage of an early meander bend down the channel belt, or 3) progradation of a lacustrine delta. Deposition of backswamp-like rooted clays indicates a distal thalweg or a distal origin for flood waters filling up the flood basin.

The surficial silty sand unit consists mainly of a coarsening-upward package of well stratified sandy rhythmites. Upsection, the rhythmites thicken, coarsen slightly and fine into muddy sand instead of clay or mud. This sequence is observed best in the proximal levee core (Fig. 8, Core IV) positioned closest to the oxbow lake (at B, Fig. 6). This sequence is caused by the progradation of the natural levee associated with documented meandering of the paleochannel toward the core site between 1200 and 400 years B.P.

Cyclic Sedimentation Due to Splay Development.—Cores taken in crevasse levees and close to the oxbow lake in the splays (the zones of coarsest sediment; Fig. 8, Core I) exhibit a surficial splay sequence about 3 m thick. At the base of the sequence very thin silty rhythmites are present which represent incipient crevassing (Fig. 10). These rhythmites consist of single layers of small ripples in coarse silt with wormy mud layers separating adjacent rhythmites. Upsection, sandy rhythmites overlie the silty rhythmites. As the splay built into its fully developed phase, the sandy part of the rhythmites coarsens upsection and thickens with stratification becoming better preserved. During the abandonment phase which follows, the sandy part of the rhythmites fines slightly and the muddy part of the rhythmites thins. A stronger biogenic overprint is superimposed on the original bedding. Thus, during incipient crevassing and initial progradation of a splay into a backswamp or lake, a coarsening-upward sequence is observed with the coarsest sediment deposited when the splay is fully developed. This would be followed by a fining-upward sequence as the splay is progressively abandoned. Distal lobe deposits also coarsen upward slightly (Fig. 8, Core III) but the main bedding type is pelletal silt.

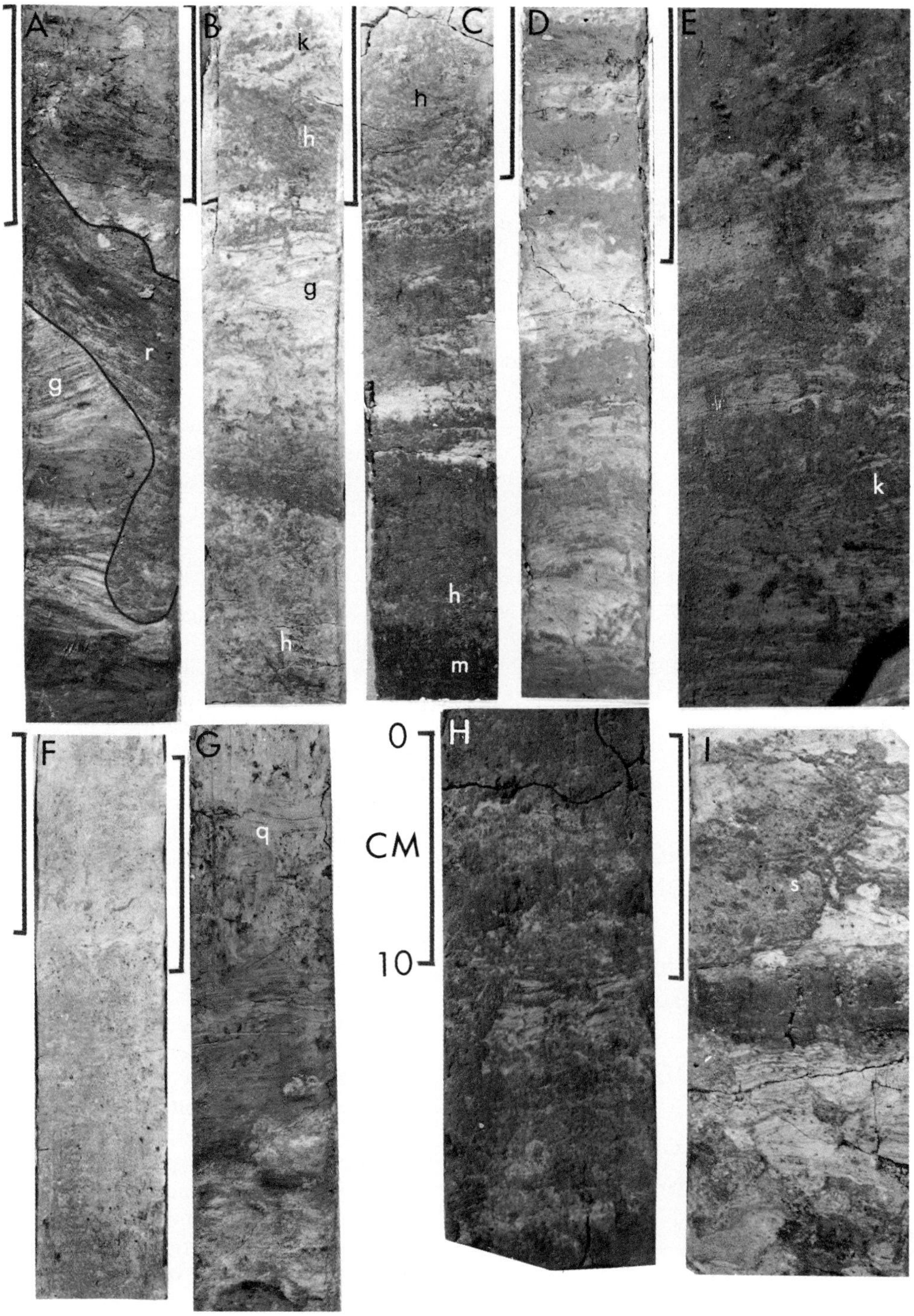

PLATE 1.—Stratification in levees and splays. Scale bars 10 cm, to left of photographs. A. Sandy rhythmites with well preserved cross-laminations (g) which are truncated by a large backfilled burrow (r), outlined in black. Levee Core 86, Profile A, depth 2.35–2.7 m, silty sand unit. B. Silty rhythmites with poorly preserved cross-laminations (g) and pelletal fabric in mud and silt (h). An individual burrow with its pelleted infill is shown at k. Levee Core 74, Profile B, depth 1.6–2.0 m, silty sand unit. C. Dark brown, organic-rich mud at base (soil horizon?) overlain by nearly obliterated silty rhythmites with pelletal fabric (h). Levee Core 74, Profile B, depth 1.6–2.0 m, silty sand unit. D. Stacks of rhythmites with wormy mud intervals and partial obliteration of primary sedimentary structures. Levee Core 87, Profile A, depth 0.45–0.75 m, silty sand unit. E. Closeup of well developed wormy fabric composed of coalescing burrows (k) which are partially obliterating laminations in rhythmites. Splay Core 44, depth 2.2–3.2 m, silty sand unit. F. Pelletal fabric in the lower silty sand unit. Dark dots are root traces with coatings of iron oxide. Levee Core 75, Profile B, depth 5.5–5.9 m, sandy silt unit. G. Typical gray rooted clay of the well drained swamp facies which is interfingering with levee and splay deposits. Levee Core 86, Profile A, depth 1.25–1.6 m, rooted clay unit. H. A combination of pelletal and wormy fabric overprinting on faintly cross-laminated silty rhythmites. Splay Core 54, depth 4.5 m, sandy silt unit. I. Cross section through a large clay-lined burrow (s) or root trace which truncates cross-laminated rhythmites and has an intraclast infilling. Splay Core 54, sandy silt unit.

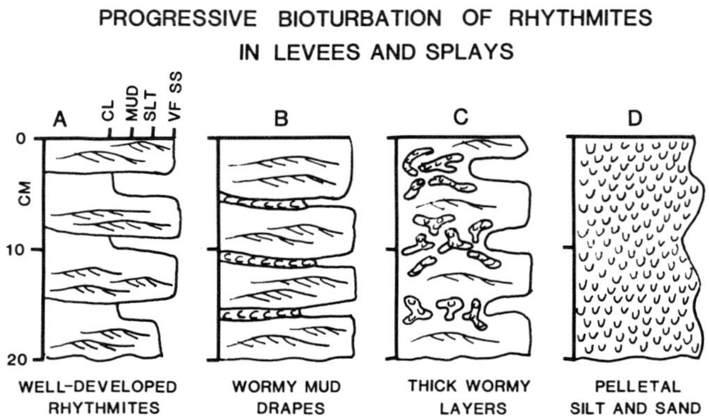

FIG. 9.—Progressive bioturbation of rhythmites in levee and splay deposits.

THE EVOLUTION OF A CHANNEL BELT AS RECORDED IN THE FLOOD BASIN

Facies Architecture

Here, the term facies or alluvial architecture refers to the geometry, spatial arrangement and stacking of sand bodies in a stratigraphic sequence. The two main types of sand bodies that exist in flood plain sequences of the lower Mississippi Valley are the ribbon or shoestring sands of channel belts and crevasse-splay sheet sands. Here, ribbon sands (channel belts) are large sand bodies (30 m × 5 km × 100 km). The 'sheet sands' that are the result of splay progradation are much smaller features (2.5 m × 2 km × 8 km) and are a complex of interlaminated sand and mud (rhythmites) with poorly preserved stratification. In this section, the spatial relationships between these sand bodies and flood basin facies will be discussed with respect to the data observed in the study area and the establishment of the channel belt at this location of the flood basin.

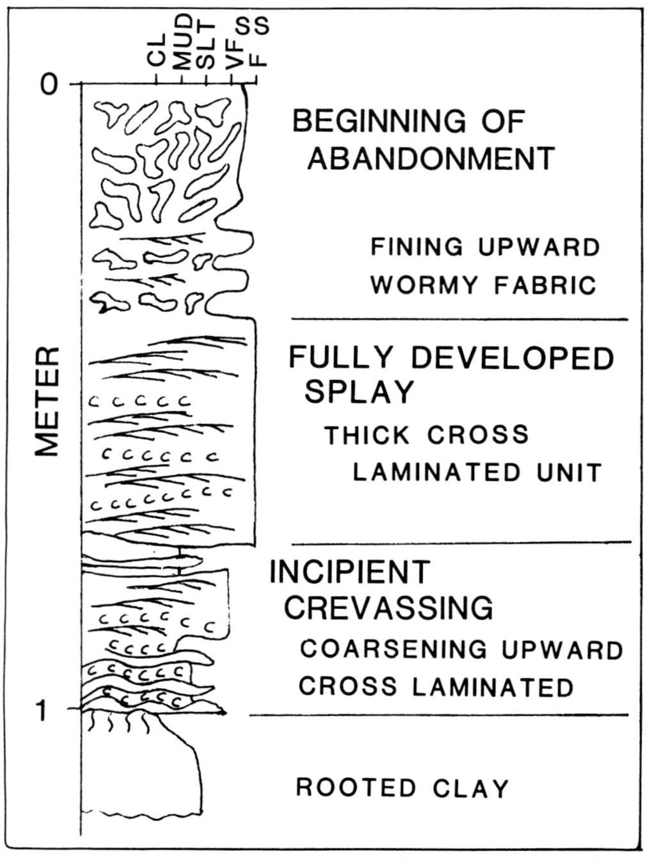

FIG. 10.—Stratigraphic section in a proximal splay deposit which shows incipient crevassing, stabilization of the splay and the beginning of abandonment.

Recognition of the Avulsion Event

The avulsion event related to the establishment of the channel belt is recorded in the flood basin section as a lithology change from blue clay (poorly drained swamp facies) to the overlying silty sands (well drained swamp facies) (Fig. 7). The mudball conglomerate unit above the silty sand unit in backswamp cores probably represents a major flood event associated with the establishment of the channel belt.

Stages in the Evolution of the Channel Belt

The establishment and subsequent evolution of a channel belt and its associated subenvironments can be shown in four block diagrams which are based on the lithostratigraphy observed in the flood basin cores (Fig. 11).

Before the channel belt was established, a poorly drained swamp with probable localized lakes existed in the study area in the pre-avulsion stage (Fig. 11A). During the avulsion stage (Fig. 11B), a channel came to exist in this low lying area of the flood basin, possibly due to the extension of a crevasse channel associated with the then active channel belt. During large floods, this channel captured a significant volume of the main channel flow, causing incisement of the channel into the flood basin silts and clays and deposition of sheet-flood silts and sands along its margin. The levee was nonexistent at this time so flow was completely unconfined. The channel belt began to form as the stream began to meander through the flood basin clays and silts (Fig. 11C). Initially it was narrower and shallower than the modern channel belt. As stream flow continued to be captured, the thalweg degraded while the levee aggraded. This was a period of rapid levee upbuilding because flow was still unconfined and levee overtopping was common due to its low elevation.

With a full development of the channel belt, the levees achieved their highest elevation (Fig. 11D). The depositional relief caused by the progradation of overbank deposits into the flood basin causes a partitioning of the flood basin margin into levee facies, well drained swamp and poorly drained swamp facies. As the channel belt grew in width and the levee became more confining, deposition on the levee (principally silts and sands) became restricted to the convex side of meander bends in the flood basin. Elsewhere, the clays associated with well drained swamps were deposited along the channel belt margin. Splay deposits, of course, carried sand farther into the backswamp. Levee deposits are interstratified with well drained backswamp clays because of the passage downstream of meanders and their

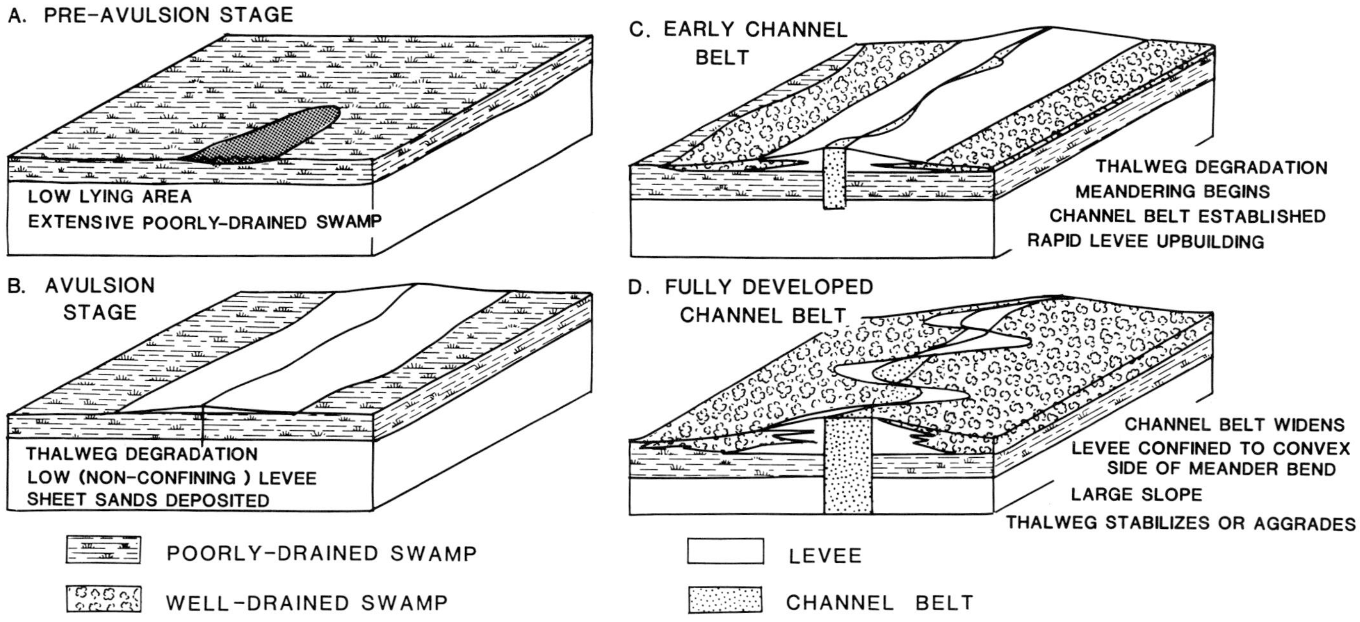

FIG. 11.—Four stages in channel belt evolution as recorded in overbank sequences.

associated levee deposits on the concave side of the meander.

CONCLUSIONS

This detailed study of the sedimentology and facies architecture of three overbank subenvironments which occur marginal to a meander belt in the lower Mississippi River Valley leads to the following conclusions: 1) backswamp, levee and splay deposits can be subdivided into units related to the establishment of the associated channel belt; 2) cycles related to avulsion, levee progradation, splay progradation and abandonment, and sheet-flood events are preserved in flood basin deposits; and 3) flood basin facies prograde basinward as the levee builds upward over time along a channel belt margin.

The deepest unit encountered in the backswamp, levee, and splay cores is a blue clay with leaf layers, interpreted as lacustrine or poorly drained swamp in origin. The units occurring above this are related to the progradation of the overbank levee, splay and backswamp deposits into the flood basin during the establishment of the channel belt.

In backswamp cores, the blue clay is succeeded upsection by 1) a silty sand unit, 2) mudball conglomerate, and 3) rooted clays. The silty sand unit represents the progradation of incipient splay, levee or lacustrine-delta deposits. Above this, at the base of a thick clay sequence, is a mudball conglomerate zone indicating a major flood event. After the silty sand unit and mudball conglomerate were deposited, a well drained swamp with trees became established which trapped clays (rooted clay unit).

In levee and splay cores, the basal blue clays are overlain in succession by 1) a sandy silt unit, 2) rooted clays, and 3) a silty sand unit. The rooted clays are backswamp deposits which interfinger with levee and splay progradational deposits (the silty sand and sandy silt units). Rhythmites are the main bedding type in these units and consist of small-scale sequences (cm) which fine upward from cross-laminated sand or laminated silt into mud or clay. Normally, primary stratification in the rhythmites is partially or completely obliterated by pedogenic and biogenic overprinting which results in the formation of pelletal and wormy fabrics.

The establishment of the channel belt is divided into four phases: 1) a pre-avulsion stage, 2) an avulsion stage, 3) an early channel belt, and 4) a late channel belt stage. In the study area, the flood basin sequence affiliated with the avulsion and establishment of the channel belt (30 m) has a maximum thickness of about 10 m. The avulsion event is probably recorded in the sedimentary sequence as the lithologic change from blue clay with detrital organic debris deposited in standing water to sheet-flood silts and sands related to levee and splay progradation during incipient formation of the channel belt. Surficial levee deposits (silty sand unit) exhibit an overall coarsening-upward sequence (2.5 m) that reflects meander bend migration toward the sampling site during the late stage of channel belt development. Surficial splay deposits (silty sand unit) initially coarsen upward during the progradational phase and then fine upward as the splay is abandoned (3 m). Individual flood cycles (mm–cm) occur as small-scale fining-upward rhythmites with poorly preserved stratification in all subenvironments.

ACKNOWLEDGMENTS

Special thanks are extended to Dr. Robert Miller of the Louisiana State University Agronomy Department for allowing me to use his department's drilling rig. Bruce Lindsay was my principal field assistant. In addition, thanks to all sources of funding for this research including Sohio, Arco, Shell Foundation, Geological Society of America,

Gulf Coast Section of the SEPM and Gulf Coast Association of Geological Societies (GCAGS). Drs. Indra Singh and Dag Nummedal reviewed the manuscript.

REFERENCES

ALLEN, J. R. L., 1965, A review of the origin and characteristics of Recent alluvial sediments: Sedimentology, v. 5, 89–191.

BRIDGE, J. S., 1984, Large-scale facies sequences in alluvial overbank environments: Journal of Sedimentary Petrology, v. 54, p. 583–588.

COLEMAN, J. M., 1966, Ecological changes in a massive fresh-water clay sequence: Transactions, Gulf Coast Association of Geological Societies, v. XVI, p. 159–174.

———, 1969, Brahmaputra River—channel processes and sedimentation: Sedimentary Geology, v. 3, p. 129–239.

CROSS, T. A., AND SMITH, N. D., 1985, Comparative anatomy of contemporary and ancient crevasse splay and anastomosed crevasse channel complexes: Third International Fluvial Sedimentology Conference, Abstracts Volume, Fort Collins, Colorado, August 7–9, 1985, p. 14.

DAVIES, D. K., 1966, Sedimentary structures and sub-facies of a Mississippi River point bar: Journal of Geology, v. 74, p. 234–239.

ELLIOTT, T., 1974, Interdistributary bay sequences and their genesis: Sedimentology, v. 21, p. 611–622.

FISK, H. N., 1944, Geological investigation of the alluvial valley of the lower Mississippi River: Mississippi River Commission, Vicksburg, Mississippi, 78 p.

———, 1947, Fine-grained alluvial deposits and their effects on Mississippi River activity: Mississippi River Commission, Vicksburg, Mississippi, 82 p.

HO, CLARA, AND COLEMAN, J. M., 1969, Consolidation and cementation of Recent sediment in the Atchafalaya Basin: Geological Society of America Bulletin, v. 80, p. 183–191.

HUGHES, D. A., AND LEWIN, JOHN, 1982, A small scale flood plain: Sedimentology, v. 29, p. 891–895.

JACKSON, R. G., II, 1976, Depositional model of point bars in the lower Wabash River: Journal of Sedimentary Petrology, v. 46, p. 579–594.

———, 1978, Preliminary evaluation of lithofacies models for meandering alluvial streams, *in* Miall, A. D., ed., Fluvial Sedimentology: Canadian Society of Petroleum Geologists Memoir 5, Calgary, Alberta, Canada, p. 543–576.

KRINITZSKY, E. L., AND SMITH, F. L., 1969, Geology of backswamp deposits in the Atchafalaya Basin, Louisiana: U.S. Army Corps of Engineers, Waterways Experiment Station, Technical Report S-69-8, Vicksburg, Mississippi, 26 pp.

RAY, PULAK, 1976, Structure and sedimentological history of the overbank deposits of a Mississippi River point bar: Journal of Sedimentary Petrology, v. 46, p. 788–801.

RETALLACK, G. J., 1983a, A paleopedological approach to the interpretation of terrestrial sedimentary rocks: The mid-Tertiary fossil soils of Badlands National Park, South Dakota: Geological Society of America Bulletin, v. 94, p. 823–840.

———, 1983b, Late Eocene and Oligocene paleosols from Badlands National Park, South Dakota: Geological Society of America, Special Paper 193, Boulder, Colorado, 82 p.

SAUCIER, R. T., 1969, Geological Investigation of the Mississippi River area, Artonish to Donaldsonville, LA: U.S. Army Corps of Engineers, Waterways Experiment Station, Technical Report S-69-4, Vicksburg, Mississippi, 20 Plates.

———, 1974, Quaternary Geology of the lower Mississippi Valley: Arkansas Archeological Survey, Publications on Archeology, Research Series No. 6, 26 p.

SINGH, I. B., 1972, On the bedding in the natural levee and the point-bar deposits of the Gomti River, Uttar Pradesh, India: Sedimentary Geology, v. 7, p. 309–317.

SMITH, D. G., 1983, Anastomosed fluvial deposits: Modern examples from Western Canada, *in* Collinson, J. D., and Lewin, J., eds., Modern and Ancient Fluvial Systems: International Association of Sedimentologists Special Publication No. 6, Blackwell Scientific Publications, Oxford, England, p. 155–168.

SMITH, N. D., 1985, Avulsion-controlled fluvial evolution: The Cumberland Marshes, east-central Saskatchewan: Third International Fluvial Sedimentology Conference, Abstracts Volume, Fort Collins, Colorado, August 7–9, 1985, p. 36.

STERNBERG, H. O. R., 1956, A contribution to the geomorphology of the False River area, Louisiana: Unpublished Ph.D. Dissertation, Louisiana State University, Baton Rouge, LA, 169 pp.

DELTAIC DEPOSITIONAL PROCESSES IN A GLACIER-FED LAKE: A MODEL FOR THE FLUVIAL/LACUSTRINE INTERFACE

JIM BOGEN

Norwegian Water Resources and Energy Administration, Glacier Division, P.O. Box 5091, Maj. 0301 OSLO 3

ABSTRACT: The sediment supply, deltaic deposits and processes of sedimentation in the glacier-fed lake downstream from Tunsbergdalsbre in western Norway have been investigated. The depositional processes were analyzed by a computer model which is structurally different but of the same type as that of Bonham-Carter and Sutherland (1968). Particle trajectories are computed by numerical integration through the lake flow field and the zones of sedimentation of different particle sizes in the bottomset layers determined. The effect of turbulence was represented by the introduction of a stochastic element causing the particles to move in a random walk. Various test runs showed that the turbulent spread of particle trajectories increased with decreasing particle fall velocity.

Application of the model to Lake Tunsbergdalsvatn involved investigations of the lake flow field during different sediment and water discharge situations. Calculations of the extension of the zones of sedimentation during this latter situation tallied with the results obtained from analysis of grain size distributions of bottomset beds.

A process response model of the long-term delta development is presented. The present lake is a remnant of an 8-km-long and 100-m-deep fjord valley lake that existed after deglaciation 9000 BP. A major part of the volume of the delta sediments was deposited by flow divergence processes at the delta front. The deltaic deposits are regarded as the results of a series of different phases of delta development through time. During each phase, the magnitude and composition of sediment supply, and the basin flow fields, were decisive for delta formation processes. The morphology of former basin conditions, however, influences the flow field in a feedback loop, and future development is dependent upon past history.

INTRODUCTION

This paper analyzes the relationship between deltaic morphology and sedimentology and the depositional processes in the proglacial system in terms of a quantitative model. A delta may be defined as a partly subaerial deposit formed when a river enters a large permanent basin. The deltaic processes and the associated deposits together comprise the deltaic system. Deltas have been the subject of a great number of investigations, though as Sundborg (1964) and Thakur and McKay (1973) emphasized, most have been of a descriptive nature. Very few studies have analyzed the morphological and sedimentological features in terms of the dynamic processes that take place. The first one was Bates (1953), who suggested that various types of jet flow resulted in different delta morphologies. Axelsson (1967), however, stressed the decisive role of sediment load in the development of separate delta types. He showed that the morphology at the river mouth and the processes of transportation and deposition of bed load and suspended load should be taken into consideration. Sundborg (1956, 1958) derived a sedimentation formula that was applied to calculate delta development in the Tabqua reservoir (Sundborg, 1964).

Laboratory experiments carried out by Allen (1965, 1968) and Jopling (1963, 1964, 1967) on deposition in the lee of ripples and small sand deltas indicate that these systems are subject to the interactions of a multitude of variables. Church and Gilbert (1975) discussed the deltaic depositional process with special reference to the glaciofluvial system.

Bonham-Carter and Sutherland (1968) developed a computer-based model for simulation of the development of a marine delta. The purpose of the model was to illustrate the multidimensional structure of the system. A sediment-loaded river flows into a tideless and currentless marine basin. Input parameters included river width, depth, water discharge and sediment transport, and the bottom topography of the marine basin. The spatial distribution of sediments and the development of bars and submerged levees in the basin were determined by a computation of trajectories of nominal particles.

Some of the discrepancies which have arisen in the interpretation of deltaic systems are probably due to insufficient consideration of changes through time. For the duration of a short time increment, the processes of sedimentation beyond the delta front are mainly governed by the flow field in the basin. Delta development and growth over a long time span, however, are closely dependent on the sediment yield of the catchment area. In this connection the extent to which the magnitude and intensity of sediment transport are reflected in the sedimentary deposits should be assessed.

The glacier-fed Tunsbergdal delta in western Norway was selected as a test area for the present study. Possessing a high yield of sediment load across a broad range of grain size fractions and highly variable water discharge throughout the whole season, the proglacial environment offers especially favorable conditions for field experiments.

MODEL DEVELOPMENT

The formulation of a quantitative model is a means to explore the structure of a system. A model that is of practical use in the analysis of any natural system would be even more convenient. Thus, a model of the same type as that of Bonham-Carter and Sutherland and others was developed. The model was, however, structurally different, limited to two dimensions and aiming at an analysis of size distributions. The essential features of the model are shown in Figures 1 and 2.

A river jet expands into a larger basin of depth H. Sediment particles are introduced into the basin at a level h_L in the river with depth h. The sedimentation zone X_L of each particle size of diameter d and equivalent fall velocity V_S is calculated by numeric integration of particle trajectories through the lake flow field.

The particle movement is described in a basin coordinate system XY. Integration of particle trajectories presupposes

FIG. 1.—Definition sketch of depositional model.

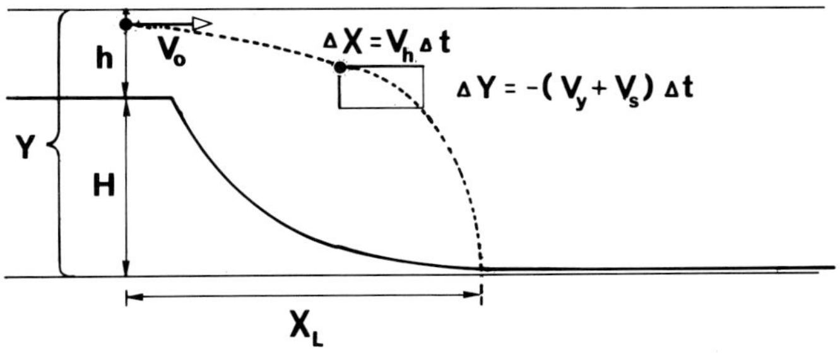

that a velocity V = $V(X,Y)$ may be assigned to each point in the basin.

In symmetrical simple delta systems, the velocity field in the vicinity of the river mouth resembles a jet expansion. The velocity decay along the axis of a jet expansion may be expressed by a relation:

$$\frac{V_m}{V_0} = a\left(\frac{X}{D}\right)^{-b} \tag{1}$$

where a and b are parameters which have to be determined by experiments. D is the diameter of the orifice, and V_0 the mean velocity of the river. Parameter values for some common cases are given in Table 1.

Away from the jet axis, the velocity V at a coordinate (X,Y) is expressed by a Gaussian function:

$$\frac{V}{V_m} = e^{-k(Y/X)^2} \tag{2}$$

where V_m is the velocity at the axis and k is a parameter which has to be determined experimentally.

The presence of solid boundaries increases the values of the parameter a in table 1 (Allen, 1965). Thus, in a shallow basin the flow is faster than in a deeper one.

Density differences due to differences in content of suspended particles or dissolved solids between river and basin water cause the jet axis to curve downward and the spread parameters to vary. Parameter values may be computed from equations given by Abraham (1963). If density differences are large, a continuous turbidity current may be generated.

Mixing conditions at the river mouth influence density gradients. Turbidity currents are probably more likely to occur with low Reynolds numbers and a limited flow separation.

Experiments performed by Rao and Carstens (1971) show that a jet carrying particles in suspension behaves differently from a comparable jet with dissolved matter, due to the independent motion of the suspended particles. Sediments in suspension initially cause a faster flow, owing to excess energy. The progressive fallout of particles, however, renders the jet weightless. Thus, the axial velocity decay rate is higher in the case of suspension matter jets than in dissolved matter jets or momentum jets.

In the case of a river flowing into a marine basin, the salt water may underlie the fresh water in a two-layer stratification. Thus, flow divergence is confined to two dimensions only. The limited mixing causes only a slow velocity decay compared to the other cases (Table 1).

Flow chart.—

A flow chart illustrating the operation of a FORTRAN computer program is shown in Figure 2. The main program is divided into five separate modules. Physical constants, sediment data and characteristics of the actual delta system are read into the main program in Module 1. Settling velocities are calculated in subroutine SETVEL according to a formula given by Rubey (1933). In Module 2 the basin is divided into a number of discrete cells and coordinates are assigned to each cell.

The water stage valid for a time increment ΔT is read in Module 3. All instructions between Module 3 and Module 5 are executed for each water stage read. Time-varying variables such as velocity, sediment concentration and the velocity field of the basin are assumed constant during the time increment. Module 3 illustrates the hydrodynamic process of sorting. The equations describing the flow field of the basin are in subroutine JET. The trajectories are computed through a numerical integration in subroutine TRAJEC, where JET is called at each step of the integration. Thus, computations in a time-varying velocity field modelling the turbulence is possible. A random number gen-

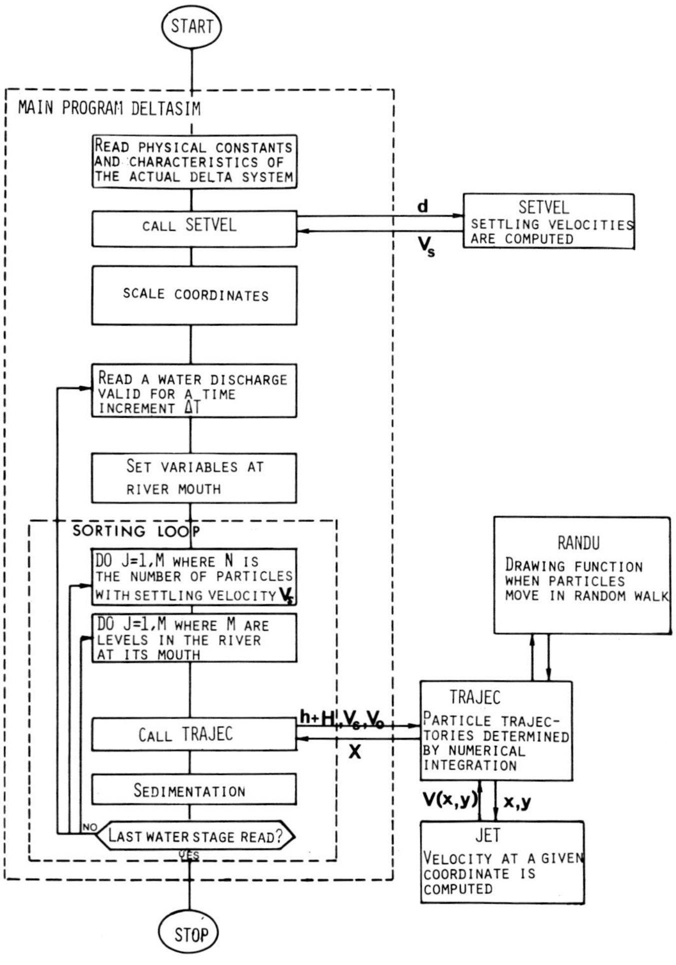

FIG. 2.—Flow chart illustrating the operation of the computer model.

TABLE 1.—PARAMETERS OF JET EXPANSIONS.

Parameter	Free jet with dissolved solids	Two dimensional free jet	Free jet carrying particles in suspension	River jet, Lake Tunsbergdalsvatn Q=35–45 m/s
a	6,2	2,28	45,66	54,11
b	1,0	0,5	1,47	1,51

erator is called from subroutine TRAJEC and causes the particles to move in a random walk. Application of this structure does not limit the model to a particular flow field. Thus, the structure of Module 3 is the major difference from the Bonham-Carter and Sutherland model.

Computation of the trajectories.—

Consider a particle, moving through the basin (Fig. 1). Now assume that we are able to assign a velocity vector $V = V(X,Y)$ to each point in the system, with components V_h and V_v and that the horizontal velocity of the particle equals V_h. Thus, no relative motion between particle and liquid in the horizontal direction is assumed.

The equations in Table 1 represent time-meaned velocity fields. In addition, the turbulence creates fluctuating velocity components v_x' and v_y'. According to Allen (1968), the momentary velocity may be written as a sum of the time meaned and fluctuating velocity:

$$V_h = V_x \pm v_x'$$
$$V_v = V_y \pm v_y' \tag{3}$$

During a time increment Δt, the particle moves a distance ΔX in the horizontal direction and $-\Delta Y$ in the vertical. If $V = V(X,Y)$ is assumed constant across the increment ΔX, the coordinates are:

$$\Delta Y = -(V_v + V_s)\Delta t$$
$$\Delta X = V_h \Delta t \tag{4}$$

V_s is the particle settling velocity. Elimination of Δt gives:

$$\Delta Y = \frac{-(V_v + V_s)\Delta X}{V_h}$$

$$\Delta Y = \frac{-((V_y + v_y') + V_s)\Delta X}{V_x + v_x'} \tag{5}$$

In a time-meaned velocity field with no vertical components, the expression reduces to:

$$\Delta Y = \frac{-V_s}{V_x}\Delta X \tag{6}$$

The trajectories of the particles may now be computed by a simple numerical integration. At the river mouth, the velocity of the particle is given by the mean velocity of the river water, V_0.

Assume then that the X coordinate increases with a constant step length, e.g., $X = 0.1$ m. The Y coordinate may now be computed by calculating V_X from the equations describing the flow field and substituting into equation (6).

In the n-th step the coordinates of the particles are:

$$X = \Delta X_1 + \Delta X_2 + \ldots + \Delta X_n$$
$$Y = H + h + \Delta Y_1 + \Delta Y_2 + \ldots + \Delta Y_n \tag{7}$$

The summation is repeated until the particle has reached the bottom, i.e., the negative Y coordinate equals $H + h$.

Model experiments—the effect of turbulence.—

In any turbulent natural system, the velocity vector fluctuates and a real particle moves in a complex path line through the basin. Regarding the motion of discrete sediment particles through the turbulent region as a "random walk," it is possible to investigate the effect of the fluctuating velocities when the model is subject to different parameter settings. When the particles move in a random walk, the coordinates are computed by relation (7), using (5) for a calculation of the Y-increment. The magnitude and direction of the fluctuating velocities are drawn during each step in the computation of the trajectory. The drawing was performed by a subroutine generating random numbers. The magnitude of the fluctuating velocities is dependent on the velocity at the river mouth and the type of flow field. Referring to experiments carried out by several authors, Allen (1968) takes $v_x' = 0.25V_0$ and $v_y' = 0.125V_0$ as characteristic values for the fluctuating velocities in the mixing region of a jet subject to solid boundaries. Thus, it seems reasonable that v_x' and v_y' assume values in the range of $0.0–0.25V_0$ and $0.0–0.125V_0$, respectively, and allow another computer-generated drawing to decide the direction of the velocity.

Using the equation of a sediment-loaded jet as a velocity field (Table 1), several experiments with various combinations of input parameters were carried out. During each run, a hundred particles were introduced at the surface of the river mouth. Turbulent fluctuations outside the turbulent region were neglected. In the first experiment with this model design, the fluctuating velocities were allowed to assume both positive and negative values. The resultant spread of the particles seemed quite unrealistic, however. They were distributed throughout a large region on the bottomset, apparently without any uniform structure. Presumably this implies that reverse motion of the particles occurs less frequently than was postulated. In the next experiment v_x was allowed to assume only positive values, whereas v_y' assumed both positive and negative values as in the first case. The results of two computer runs with the velocity at the river mouth $V_0 = 0.2$ m/s and $V_0 = 1.2$ m/s are shown in Figure 3. The particles are distributed in small "heaps" on the bottom. The shape of each individual distribution seemed to vary with the ratio V_0/V_s. When V_s assumes

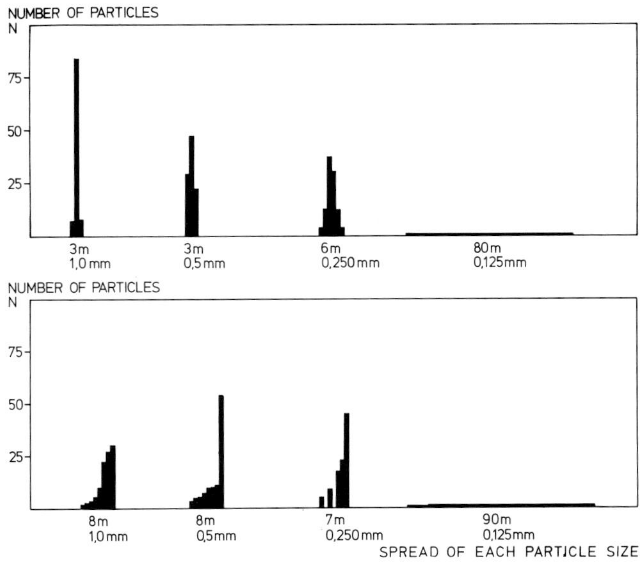

FIG. 3.—Result of test runs when particles are moving in a random walk. Velocity at river mouth $V_0 = 0.2$ m/sec and $V_0 = 1.2$ m/sec.

large values compared to V_0, the particle distribution is sharply peaked and symmetrically distributed around the mean. As V_s decreases, the distribution becomes skewed. Smaller values of V_s cause a spread of the particles throughout a larger region. These experiments illustrate some properties of the particle motion similar to those Allen (1968) found in laboratory experiments with a related ripple system. The response effect of this probabilistic model is therefore assumed to be real.

Applying the principle of conservation of mass-to-sediment discharge through a stream tube, the accumulation of suspended sediments in the bottomset may be computed. The sediment discharge passing between elevations h and $h + \Delta h$ at the inlet accumulates in the bottomset cells between X_L and $X_L + \Delta X_L$. This method was utilized by Jopling (1967) in computing grain size distributions of sedimented sand grades in a laboratory delta, and by Bonham-Carter and Sutherland (1968) in simulating a marine delta.

In most natural systems, a considerable amount of silt and clay are present. The undetermined character of the trajectories of these fractions invalidates the application of this method to such systems. The test runs, however, indicate that the model is capable of predicting the zones of sedimentation of the coarser particles with the spread expected on account of the stocastic element. Thus, the model may be utilized in a discussion of the relative influence of various parameter settings, and in this way be a powerful tool when interpreting observed characteristics of deposits in a real system. In principle, the effect of any observed velocity field may be studied. As an example, the equations of a free, two-dimensional jet may be introduced into the model, thus simulating the conditions in a marine system.

If contents of dissolved solids and concentrations of suspended sediments are moderate, the flow field may be taken from Table 1 and relation (2). In the absence of tides and wind-driven currents, the decisive parameters of the flow are the width of the inflowing river and the vertical distance down to the halocline.

The model response to variation in mean velocity at the river mouth for different grain sizes is shown in Figure 4. The curves are valid for particles situated at the surface of the inlet, thus indicating a maximum extension X_L of the zone of sedimentation of a certain grain size possessing a settling velocity V_s and an initial velocity V_0 at the inlet. The mean temperature of the river water is assumed to be 4° C. X_L increases almost linearly with V_0, and the range of the zones decreases with increasing grain size. It is evident that the process of sedimentation is very sensible to changes in grain settling velocity V_s.

THE TUNSBERGDAL DELTA

The Tunsbergdal delta and its drainage basin are situated in the mountains of western Norway, just north of Sognefjord. The drainage area comprises in total 136.8 km^2, of which 45 percent is covered by glaciers (Fig. 5). The 18-km-long Tunsbergdalsbre is an outlet glacier from the Jostedalsbreen ice cap. Tunsbergdal and its tributary Tverrdal are U-shaped glacial valleys that are sharply incised into the paleic surface that existed prior to the onset of the ice age. The entire drainage basin is situated in rocks of the Precambrian basal gneiss complex.

Water discharge and sediment transport were measured at the glacier front, at the river mouth in Lake Tunsbergdalsvatn and at the lake outlet (Bogen, 1976, 1982). In 1972 the transport of suspended sediments amounted to 44×10^6 kg/yr. The supply of bed load to the delta in Lake Tunsbergdalsvatn was calculated by the Meyer-Peter and Müller (1948) equation as 9.6×10^6 kg/yr. On average, this corresponds to about 11 percent of the total sediment discharge through this section. The subaerial part of the delta has a length of about 6 km and slopes southward with a mean gradient of 1–2°. The recent advances of Tunsbergdal glacier have deformed the northern part of the delta. In this area the river breaks through an area with a number of neoglacial moraines.

The channel pattern within the moraine area is controlled by the moraine topography. In front of some of the moraines, small sandur plains are present, the largest one lying

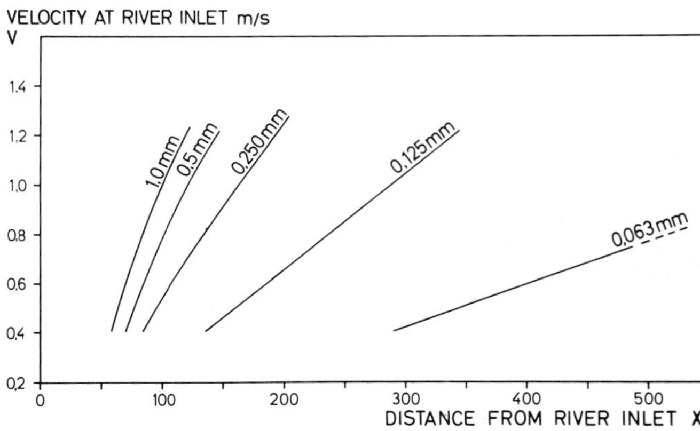

FIG. 4.—Calculated zones of sedimentation for different grain sizes in a marine basin.

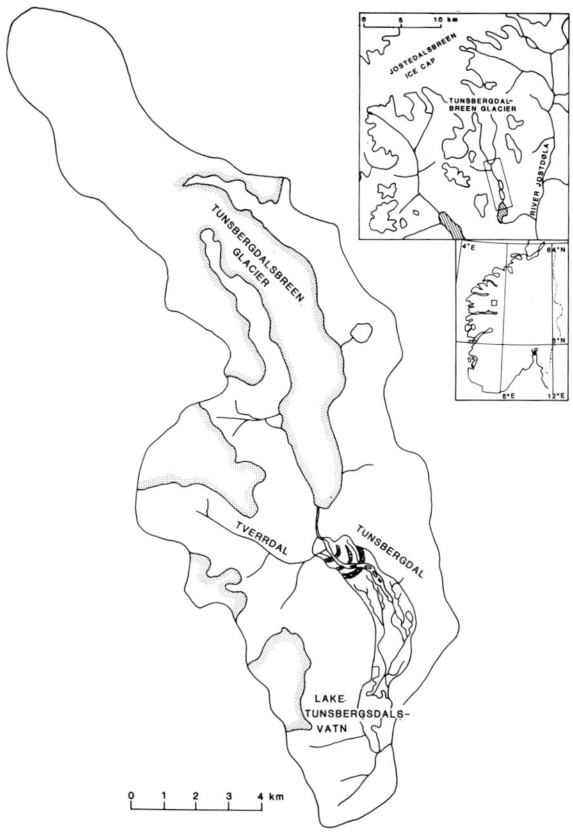

FIG. 5.—Index map of drainage basin of the Tunsbergdal delta.

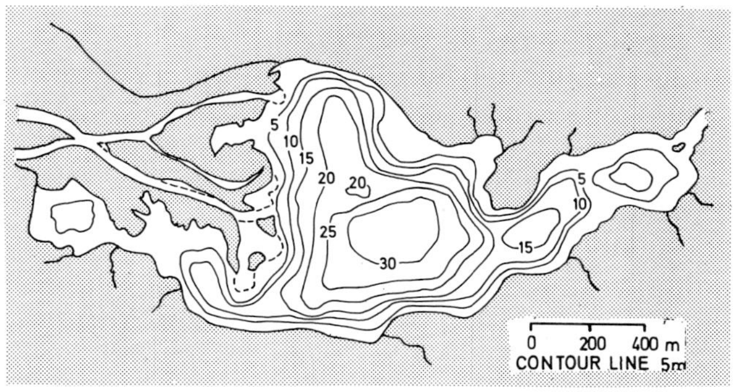

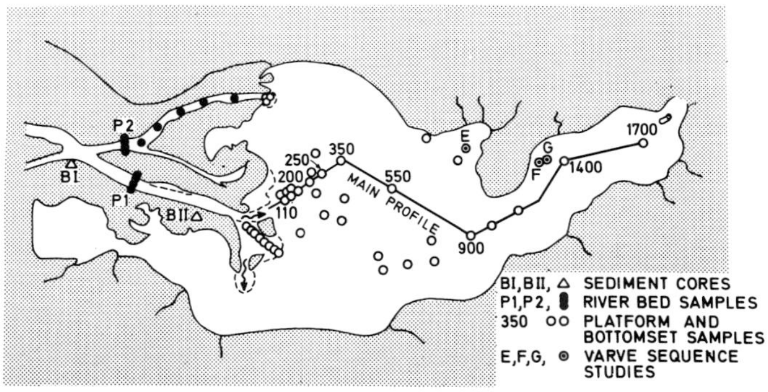

FIG. 6.—Bathymetry and sampling locations in Lake Tunsbergdalsvatn.

outside the terminal moraine. At present the river cuts through the sandur and branches into a braided river system about 1 km downstream from the outermost moraine. In this area the morphology of the original delta landscape is preserved, with active and abandoned channels, levees and interlevee basins. The grain size of the river bed decreases slowly in a downstream direction.

The conditions at the river mouth do not seem to favor the development of a subaquatic mouth bar. This is probably due to the deep basin. Despite this, bifurcations are present at the mouth of the main river. The broad shape of the Tunsbergdal delta indicates that bifurcations have often occurred during postglacial time. In Tunsbergdal, the delta platform sediments are associated with a two-dimensional jet expanding laterally across the platform during discharges in excess of 40 m^3/sec. Parts of the water discharge, however, follow the smaller distributaries and expand in three-dimensional jets.

The large flood which occurred in August 1972 accumulated a marked layer on the delta platform. Samples from this layer were collected along a profile assumed to coincide with the axis of the flow field, as shown in Figure 6. At the distal part of the profile, a thin layer composed of fine fractions had accumulated on the top of the coarser layer during the falling water stage. The peak discharge was 60 m^3/sec, corresponding to a mean velocity of 1.3 m/sec. Most of the sediments were probably transported as bed load. The grain size decreases considerably across the platform. It appears that all of the coarse fractions attain a maximum before they are sorted out. Some fractions attain a more peaked maximum than others. The three-dimensional graph in Figure 7 shows the downstream change in grain size frequency. Close to the river mouth, the coarser fractions dominate. Downstream, the maximum shifts to smaller grain sizes. The shifting is not monotonic.

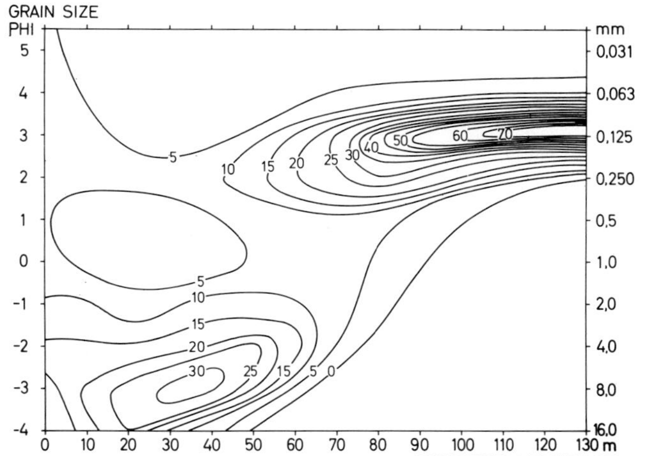

FIG. 7.—Variations in relative amounts of individual size fractions as a function of downstream distance on the river mouth platform. Numbered contour lines in lower diagram are percentages.

A saddle point occurs around 1–4 mm. These fractions never seem to dominate and are less frequent than any of the others along the whole profile. The line connecting points of maximum values seems to follow the line connecting the points where the fractions disappear. The positions of these lines is probably determined by the spread parameters of the flow divergence process.

The flow field of Lake Tunsbergdalsvatn.—

Current measurements in Lake Tunsbergdalsvatn were carried out with an ultrasonic CMI-instrument (Bogen, 1976). The flow field of the lake appeared to vary with the water discharge and sediment concentration of the inflowing river. The axis of maximum flow velocity maintained a constant position, however. The general features of the flow are shown in Figure 8.

In a situation representative for low-water discharges, 20–25 m^3/sec, the flow seems to spread along the bottom. At distances of 350–650 m, the largest velocities were measured along the bottom. The observed differences are small, however, with no abrupt transition and only a slight increase in velocity with depth. No distinct plunge point was observed at the surface.

At intermediate discharges, the velocity field seems to be of interflow type. The maximum velocities, at 500–600 m from the mouth, are observed at intermediate depths. The concentration of suspended sediments increases slightly toward greater depth, but the temperature and conductivity are not subject to any significant change.

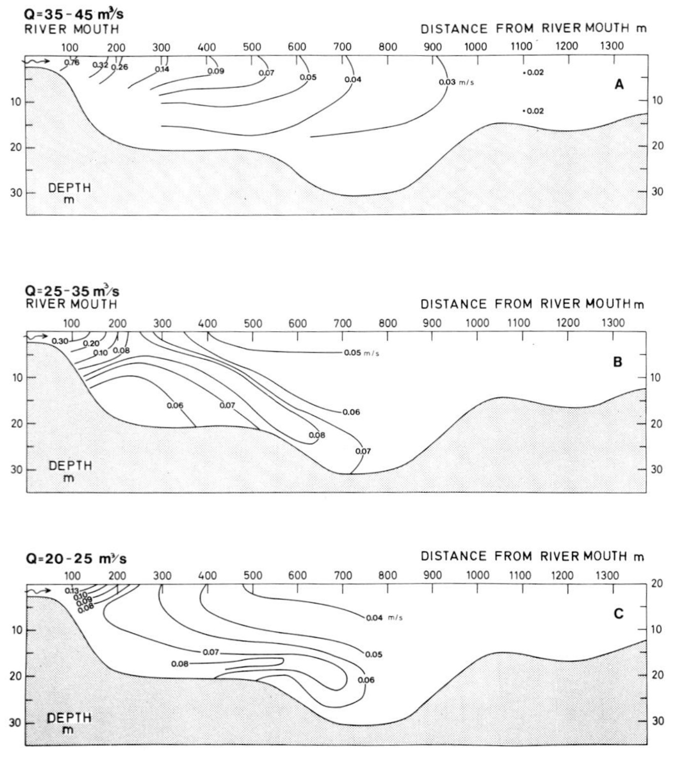

FIG. 8.—Flow field of Lake Tunsbergdalsvatn during different discharges Q, in the inflowing river. A;Q = 20–25 m^3/s, B;Q = 25–35 m^3/s, C;Q = 35–45 m^3/s.

The flow field at high discharges of 35–45 m^3/sec is more regular than at lower discharges. There is no increase in velocity with depth. Along the main axis of flow, the velocity field resembles that of a free jet (Bogen, 1982).

Turbidity currents were not observed in Lake Tunsbergdalsvatn. The low concentrations of sediments and the low density difference between the inflowing river and the small lake volume are the main causes. At the maximum a concentration of 320 mg/l was measured in the inflowing river. The relatively steep foreset slope also favors mixing. Several glacier-fed lakes have been investigated in Norway (Bogen, 1982; Raubakken, 1985). Turbidity current have only been observed in large lakes during floods, or in lakes where mixing conditions are limited by a low-angle (4–5°) foreset slope as in Lake Vågåvatn (Karlsen and Stene, 1978).

Grain size distributions of bottomset deposits.—

The bathymetry of Lake Tunsbergdalsvatn is irregular with several small basins (Fig. 8). The initial topography has been smoothed out by sedimentation so that the deeper parts of the lake are flat. Steeply inclined delta front slopes of a real foreset character are only found in front of the river mouth. Apart from this area, bottomset sediments supplied from the distributaries dominate everywhere in the lake.

The dip angle of the steeper part of the foreset slope is only 12°. This low inclination may be the result of the transport being dominated by suspended sediments and the large supply of sand fractions. As a large amount of the bed load accumulates on the platform and only a small part is dumped on the foreset slope, the effective bed load-suspended load ratio becomes small.

The distribution of sediments with respect to their median diameter shows that the largest grain sizes occur along the main axis of flow where the largest velocities are present. The relative amounts of the different fractions deposited along the main profile are also shown in Figure 9. There is a systematic variation in grain size along the profile. The curves are, however, complex and may be compared with those of Figure 7. All the sand fractions, 0.5–0.250 mm, 0.250–0.125 mm and 0.125–0.063 mm, attain a maximum before they are sorted out at a certain distance from the river mouth. The 0.063–0.031 mm fraction attains a maximum around 350 m but has a less peaked curve than the sand fractions. The fractions finer than 0.031 mm all constitute a small share of the sediments close to the river mouth but become more dominant at greater distances.

A three-dimensional representation of the downstream variation in grain size composition is also shown in Figure 9. It is seen that the largest percent values are in the region of the coarsest sand fractions close to the inlet and shift to silt fractions at increasing distance. A local minimum, a saddle point, occurs about 200 m from the river mouth due to a lower content of 0.125–0.063 mm material.

The line connecting points of maximum values seems to follow the line connecting the points where the fractions disappear. The latter represents the zones of sedimentation during the highest discharges and velocities of the river in 1971. These zones may be predicted with the computer model when the velocity field is known.

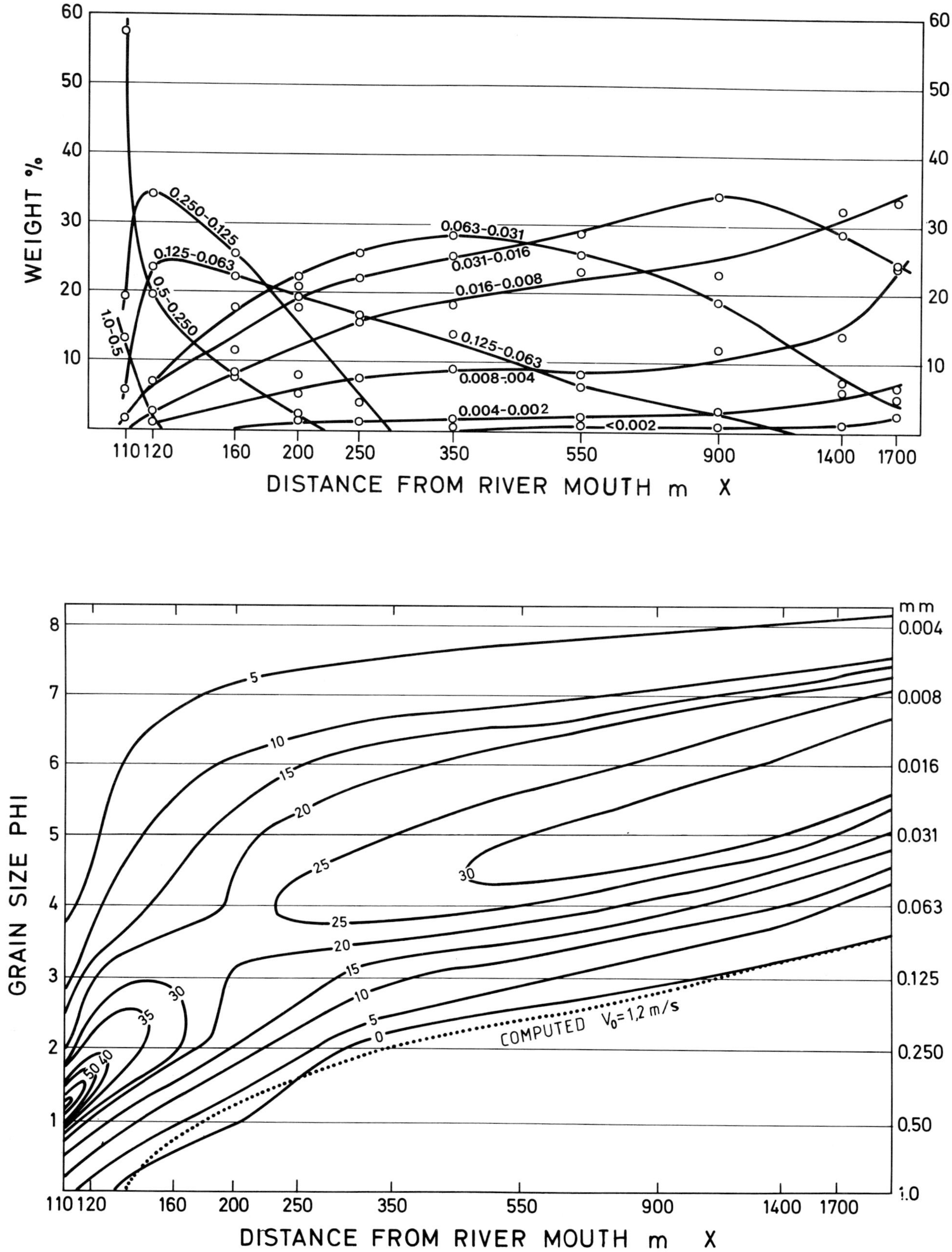

FIG. 9.—Variations in relative amounts of size fractions with distance along the main profile in the bottomset deposits of Lake Tunsbergdalsvatn.

The spread parameters of the river jet were determined by curve fitting (Bogen, 1982 and Table 1). By supplying the computer model with this relation as velocity field and approximating the vertical velocities by Gaussian curves [relation (2)], the theoretical curve in Figure 9 was obtained. The deviation of the computed curve is no greater than would be expected from the experiments with the stochastic model. Thus, the model seems to be able to represent the system for these maximum conditions.

Long-term delta development.—

Sediment cores and the seismic measurements reveal certain structural features of the Tunsbergdal delta and its growth and development during postglacial time (Bogen, 1976). The major part of the sediment in the Tunsbergdal basin was deposited after the last known glacial advance, dated at 9100 BP by Vorren (1973). During the final downwasting of the ice, large amounts of silt and sand were deposited. The final retreat seems to have been comparatively rapid. According to radiocarbon dates performed by Mottershead and others (1974), the ice receded from the Tunsbergdal basin before 8000 BP. A regeneration with a new advance took place after 3850 BP. At this time, the whole valley was probably occupied by a large lake with a water level close to the present one, 450 m above sea level. During the final deglaciation, large amounts of sediment were transported and deposited in the inner part of the Tunsbergdal basin. These deposits were later covered up by debris during the glacial advance in the 18th century.

The next known stage of the deltaic development is indicated by a core dated at 3370 BP (Fig. 10). The effective basin depth when the delta front passed this location was probably close to the condition in the present Lake Tunsbergdalsvatn. In the course of the next 2000 years, the whole of the inner basin was filled with sediments. A large part of their volume is bottomset sediments. When the delta front advanced across the rock bar, the basin there was only 4 m deep. An estimate of the linear velocity based on the present sediment supply suggests the delta front was advancing in this area at about 2 m/yr. During the subsequent advance into deeper basin conditions, the linear growth decreased considerably. The amount of sediment deposited is, however, consistent with a transport intensity of the same order of magnitude as the present one. The amount of bedload-derived material which accumulated downstream of the rock bar is estimated at about $5.6'10^9$ kg. Assuming that the delta front crossed the rock bar about 1500 BP, the mean value of the rate of bedload transport was $3.7'10^6$ kg/yr during the last 1500 years.

The general outline of the delta development given above is merely a sketch of some events in the delta's history. It does, however, show that the material deposited by flow divergence processes similar to those studied at the delta front accounts for the major part of the delta's volume. Knowledge about the effect of this type of process is of fundamental importance to a better understanding of delta formation.

A delta may be interpreted as a dynamic system operating through time. Such a system is denoted by Chorley and Kennedy (1971) as a process-response system. As a result of the operation of the system, the morphological output from each time increment influences the subsequent one through a negative feedback loop (Fig. 11). Deltaic growth may thus be looked upon as a sequence of separate system states. During small time increments, the solid boundaries of the basin may be considered stationary. As the basin may shoal during a long time span, the flow field will change and the sedimentation rate decrease. The sedimentary processes in the basin are governed by the conditions in the considered time increment. Each state is the result of the preceding states of the system. Knowledge about the influence of flow field on the sedimentary processes, when subject to various basin conditions and sediment load, is thus decisive for an understanding of the deltaic system as a whole.

Early workers related different delta types to particular flow fields (e.g., Bates, 1953). A certain type of flow field, however, is a mere event in the deposition of a deltaic sedimentary sequence. The process-response model put forward here accounts for the interaction between the amount and character of the sediment supply, the flow field and the sedimentary deposits.

GRAIN SIZE DISTRIBUTIONS AND DEPOSITIONAL PROCESSES

A modal diagram of all samples from the deltaic deposits in Tunsbergdalsvatn shows a pronounced multimodality (Fig. 12). The effect of the deltaic processes on textural features of the deposits was analyzed in a series of scatter plots. Statistical analysis of the data is based on the grain size parameters suggested by Folk and Ward (1957). In Figure 13 mean size M_z is plotted against standard deviation, σ_I. The bulk of the samples cluster around a sinusoidal or M-shaped trend, well known from fluvial deposits. The ma-

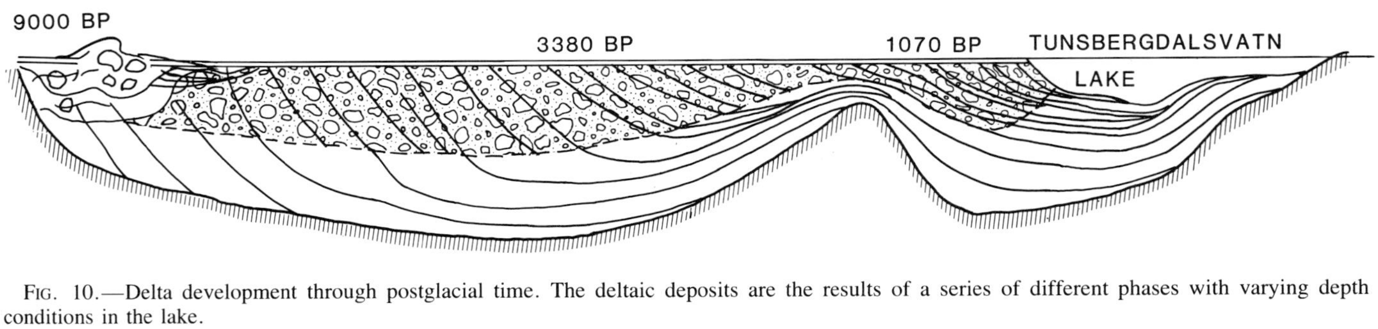

FIG. 10.—Delta development through postglacial time. The deltaic deposits are the results of a series of different phases with varying depth conditions in the lake.

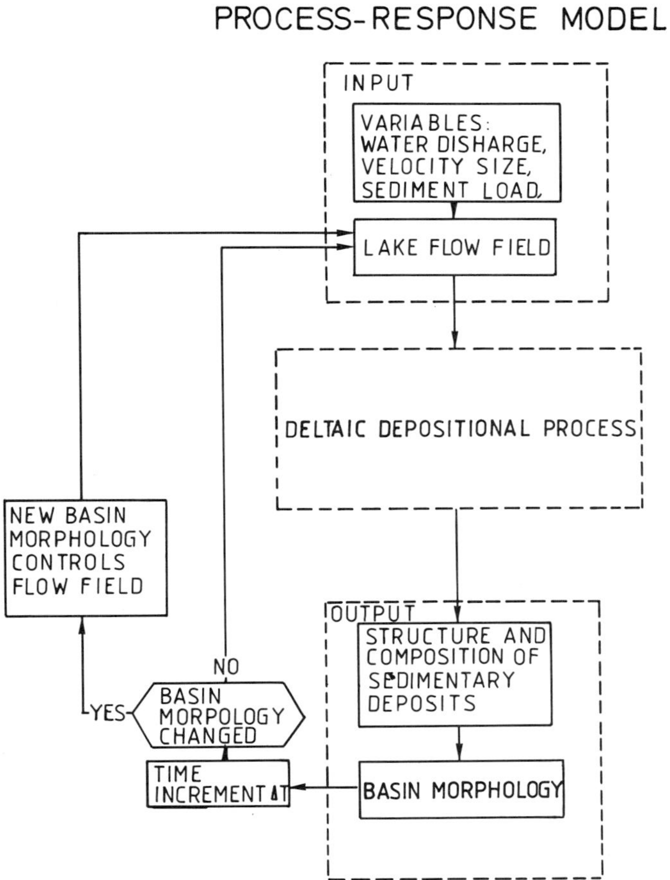

FIG. 11.—Process-response model of delta development through time.

jority of samples is mainly grouped inside area A. These samples reveal poor sorting from 1.5–3. The sorting is at its best for the samples with a large mean grain diameter and decreases with decreasing diameter. According to Folk and Ward (1957), the best sorting coincides with the modes in the sediments, and the poorest sorting correspond to grain

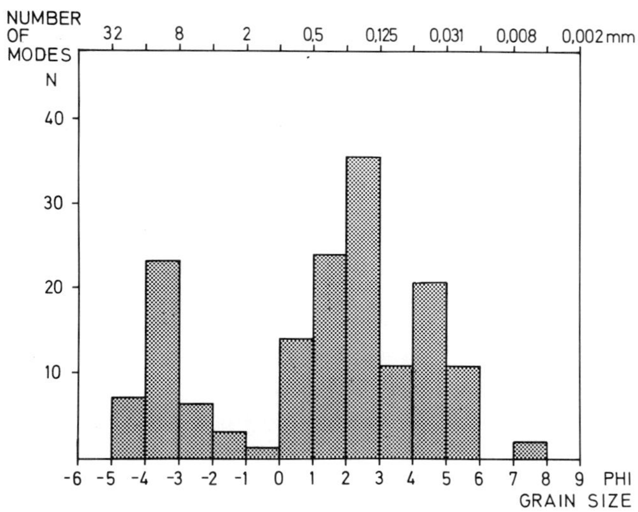

FIG. 12.—Modal diagram of grain size distribution curve from the deltaic deposits.

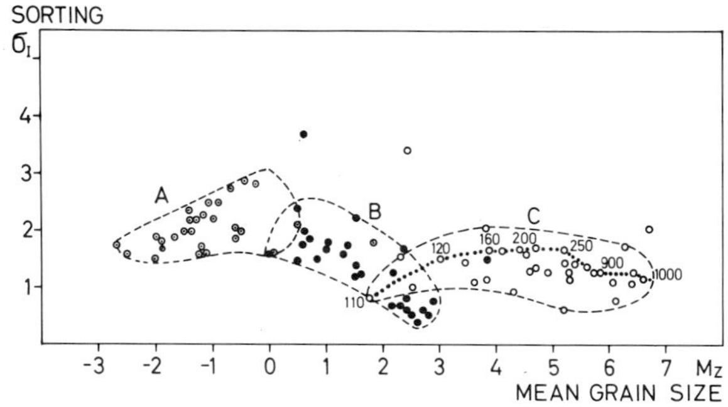

FIG. 13.—Sorting versus mean grain size. A. Channel deposits. B. Platform deposits. C. Bottomset deposits. Samples along main profile connected with dotted line. Distance from river mouth indicated.

sizes midway between modal diameters. In the river channels of the Tunsbergdal delta, the gravel fractions are subject to a gradual downstream deposition. The poorest sorting occurs when the gravel mode and the sand mode are equal. Flow conditions may vary locally, however, and sediments transported in suspension are also subject to deposition in the channels. Consequently, there is a large spread in the samples, and some plot outside area A.

The flow separation at the delta front manifests itself as a more efficient sorting process. The samples inside areas B and C may be related to this type of process. Area B contains samples from the recent river mouth platform and from sediment cores from the older foreset deposits. The best sorting, σ_i = 0.26, occurs close to the main mode at 2.6 Phi. Area C includes material deposited in bottomsets. It appears that the dynamics of the sorting processes outside the mouth of the main river lead to the poorest sorting of the sand fractions. Samples positioned far from the main profile possess a poorer sorting than samples situated along the profile.

Skewness is plotted against mean diameter in Figure 14. The skewness indicates to what extent a grain size distribution deviates from the normal distribution. The majority of the river-bed deposits possesses negative skewness, whereas the well sorted samples associated with stream flow divergence processes at the river mouth have a skewness close to a normal distribution. The majority of bottomset samples possesses positive skewness, skewed toward finer grain sizes.

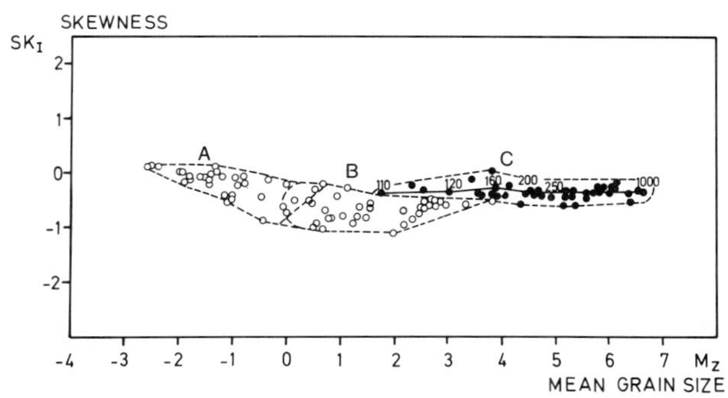

FIG. 14.—Skewness versus mean grain size.

In Figure 15 skewness is plotted versus sorting for samples with mean grain size larger than 0.125 mm. River-bed deposits and delta front deposits tend to group inside two different areas with comparatively few samples in between. Folk and Ward (1957) showed that a mixing of sediments from two modes in gravel and sand resulted in an elliptic or near circular trend in the SK_I-σ_I diagram, if all possible combinations are found. A corresponding theoretical curve may be constructed from the delta sediments (Fig. 15). It appears that the sedimentary processes associated with the deltaic system do not provide all the theoretically possible combinations. Positively skewed distributions with good sorting, in the lower right part of the circle, and symmetrical distributions with poor sorting, in the upper left part, seem to be less frequent.

The first type concerns deposits where a coarser mode dominates a finer one. In the investigated part of the delta, this type of distribution only occurs in combinations of sediments from the sand and silt modes. As shown in Figure 9, this combination is only observed outside the river mouth. The other type of distribution, symmetrical with poor sorting, has been found in a small area on the delta platform. The reason for this may be inferred from the experiments with the computer model (Fig. 4). There are hydrodynamic restrictions as to how many fractions may be deposited at the same location at the delta front. Thus, a certain type of distribution tends to dominate the deposit, whereas others are less frequent.

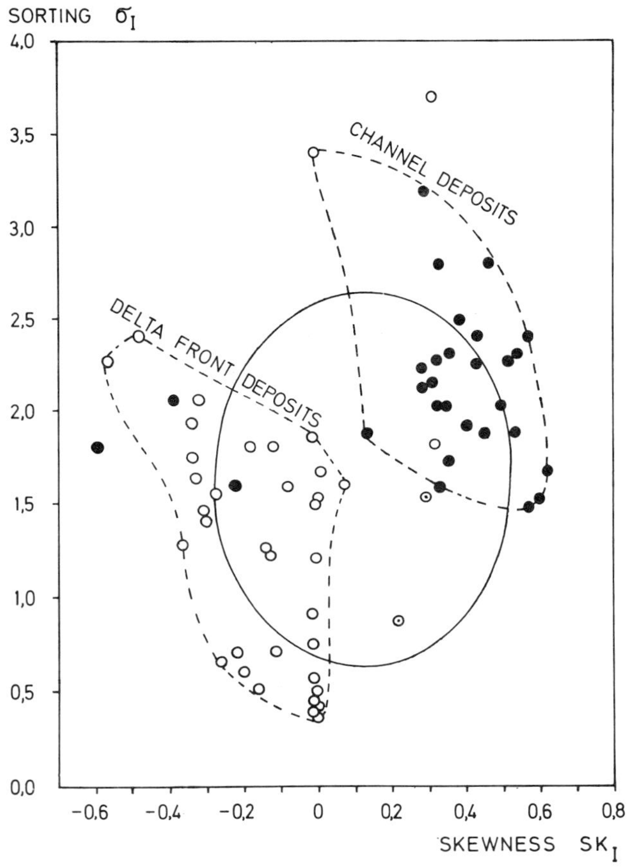

FIG. 15.—Scatter plot of skewness versus sorting of channel deposits, (closed circles) and delta front deposits, (open circles). Bottomset samples not included. Open circles with dots denote samples from the foreset slopes.

DISCUSSION

The proglacial environment of deposition is characterized by a relatively large supply of sediment, where a large share of the sediment budget is comprised of cobble and gravel fractions transported as bed load. Deltas developed in this type of environment rarely pass through any destructive phase as modifying littoral processes progress only slowly when compared to the intensity of the fluvial ones.

The general mode of deltaic depositional processes put forward in this paper is best suited to such a system. The question now is to what extent the grain size distribution features found to be characteristic of the delta deposits in Tunsbergdal are generally valid in other proglacial systems.

Very probably, they express merely the effect of the depositional processes subject to the physical conditions and the source material present in Tunsbergdal. Conditions in similar systems, however, may be discussed in terms of the general model established.

The major processes that control the deposition of sediments may be divided into three types:

1. the three-dimensional flow divergence that takes place at the delta front,
2. the two-dimensional flow divergence that takes place across the delta platform, and
3. the depositional processes in the river channel and on the river plain.

The grain size distribution of the supplied sediments is a decisive parameter in the process of sedimentation. Thus, every proglacial deltaic system is characterized by the sediment-producing glacier. When the sediments are deposited, the mixing of material from the different modes is dependent upon the interaction between the dynamics of the depositional processes and the available material.

The distance from a glacier to the basin of sedimentation, and the hydrological regime of the catchment area, are the decisive factors determining the amount and diameter of sediment fractions transported all the way down to the river mouth, and possibly also sorting of the sediments during transport.

Expressed in another way, the glacial erosion and sediment production give rise to a range of possible grain size distributions. The number of the potential distributions that is actually produced in significant amounts is dependent, however, on the specific properties of a certain catchment area and the dynamics of the depositional processes as predicted by the model. Even if the supplied sediments possess a dominant sand mode, their composition, as depicted for instance in the $M_z - \sigma_I$ diagram (Fig. 15), is dependent upon the efficiency of sedimentary processes with regard to sorting. The relatively distinct curves that arise when the composition of the sediments is studied along representative profiles (Figs. 9, 13) indicate that it is possible to define a relationship between the composition of the sediments and their genesis by a specification of the parameters of the depositional processes.

This approach seems to be of special interest when applied to deltaic deposits. Information may, in a wider sense, be regarded as a measure of the amount of organization in a given system. The flow separation process in certain types of deltas possesses a high level of organization, and each individual state is depicted in the deposits and preserved as the system is operated through time. Even though it has only been possible to quantify part of the system in the present work, it has demonstrated that deltaic processes result in deposits with a higher degree of information about earlier system states than is the case, for example, with river plain deposits.

ACKNOWLEDGMENTS

Financial support for this work was provided by the Norwegian Research Council for Science and the Humanities (NAVF) and "Nansenfondet og de derved forbundne fond."

REFERENCES

ABRAHAM, B., 1963, Jet diffusion in stagnant ambient fluid: Delft Hydraulics Laboratory, Publication No. 29, 183 p.

ALLEN, J. R. L., 1965, Sedimentation in the lee of small underwater sand waves: An experimental study: Journal of Geology, v. 73, p. 95–116.

———, 1968, Current Ripples: Their Relation to Patterns of Water and Sediment Motion: North-Holland, Amsterdam, 433 p.

AXELSSON, V., 1967, The Laitaure delta: Geografiska Annaler, v. 49A, p. 1–127.

BATES, C. C., 1953, Rational theory of delta formation: American Association of Petroleum Geologists Bulletin, v. 37, p. 2119–2162.

BOGEN, J., 1976, Sedimentasjonsprosessens dynamikk i et deltasystem: Unpublished Ph.D. Dissertation, Department of Geography, University of Oslo, 103 p.

———, 1982, Morphology and sedimentology of deltas in fjord and fjord valley lakes: Sedimentary Geology, v. 36, p. 245–267.

BONHAM-CARTER, G. F., AND SUTHERLAND, A. J., 1968, Mathematical model and FORTRAN IV program for computer simulation of deltaic sediments: State Geological Survey, The University of Kansas, Lawrence, Computer Contribution 24, 56 p.

CHORLEY, R. J., AND KENNEDY, B. A., 1971, Physical Geography: A System Approach: Prentice-Hall, London, 370 p.

CHURCH, M., AND GILBERT, R., 1975, Proglacial fluvial and lacustrine environments, *in* Jopling, A. V., and McDonald, B. C., eds., Glaciofluvial and Glaciolacustrine Sedimentation: Society of Economic Paleontologists and Mineralogists Special Publication 23, p. 22–100.

FOLK, R. L., AND WARD, W. C., 1957, Brazos river bar: A study in the significance of grain size parameters: Journal of Sedimentary Petrology, v. 27, p. 3–26.

JOPLING, A. V., 1963, Hydraulic studies on the origin of bedding: Sedimentology, v. 2, p. 115–121.

———, 1964, Laboratory study of sorting processes related to flow separation: Journal of Geophysical Research, v. 1969, p. 3403–3418.

———, 1967, Some applications of theory and experiments to the study of bedding genesis: Sedimentology, v. 7, p. 71–102.

KARLSEN, O. G., AND STENE, R. N., 1978, Bøvra i Jotunheimen, En fluvialgeomorfologisk undersøkelse, Prosjektledere: J. Gjessing & K. Nordseth, Kontaktutvalget for Vassdragsreguleringer, University i Oslo rap. 78/02, 53 p.

MEYER-PETER, E., AND MULLER, R., 1948, Formulas for bedload transport: International Association of Hydraulic Structures Research, Stockholm, App. 2, p. 39–64.

MOTTERSHEAD, D., COLLIN, R. L., AND WHITE, I. D., 1974, Two radiocarbon dates from Tunsbergdal: Norsk Geologisk Tidsskrift, v. 54, p. 131–134.

RAO, U. M., AND CARSTENS, T., 1971, Sediment-laden submerged horizontal jet: Proceedings, 14th Congress of the International Association for Hydraulic Research, p. 135–142.

RUBEY, W. W., 1933, Settling velocities of gravel, sand, and silt particles: American Journal of Science, v. 225, p. 325–338.

RAUBAKKEN, V., 1985, Suspensjonstransport og sedimentasjon i Erdalsvassdraget, Stryn: Unpublished Ph.D. Dissertation, Department of Geography, University of Oslo, 108 p.

SUNDBORG, ÅKE, 1956, The river Klaraelven: A study of fluvial processes: Geografiska Annaler, v. 38, p. 127–316.

———, 1958, A method for estimating the sedimentation of suspended material: International Association of Science Hydrology Publication, v. 43, p. 249–259.

———, 1964, The importance of the sediment problem in the technical and economic development of river basins: Kungliga vetenskapssamhaellet i Årbok 8, p. 33–53.

———, 1967, Some aspects on fluvial sediment and fluvial morphology: Geografiska Annaler, v. 49A, no. 2–4, p. 333–343.

THAKUR, T. K., AND MCKAY, D. K., 1973, Delta processes, *in* Fluvial Processes and Sedimentation: Proceedings of Hydrology Symposium, University of Alberta, Edmonton, p. 509–530.

VORREN, T. O., 1973, Glacial geology of the area between Jostedalsbreen and Jotunheimen, South Norway: Norges Geologiske Undersøkelse, v. 291, p. 1–46.

PART III
PALEOHYDROLOGY AND VERTEBRATE TAHPONOMY

LARGE-SCALE FLOOD DEPOSITS ASSOCIATED WITH THE FORMATION OF COARSE-GRAINED, BRAIDED TERRACE SEQUENCES

JUDITH K. MAIZELS

Department of Geography, University of Aberdeen, St. Mary's, High Street, Old Aberdeen, AB9 2UF, Scotland, U.K.

ABSTRACT: Paleohydraulic data from outwash terrace sequences in New Zealand and Iceland suggest that they were formed by glacially controlled, high magnitude, incising flood events. Relic incision channels cut into the terrace surfaces provide evidence of steeper gradients, coarser sediment and thus greater bed-shear stresses, velocities, Froude numbers, and discharges than those of the braided channel systems on the terrace surfaces.

The paper proposes that channel incision and terrace formation can be generated by catastrophic floods which exceed the thresholds for channel entrenchment. These thresholds are defined by the flow conditions predicted by paleohydraulic modeling of the relic channels.

INTRODUCTION

Many outwash terrace sequences exhibit low relief, low-gradient, coarse-grained terrace surfaces, often characterized by extensive braided paleochannel systems. These surfaces, however, are frequently cut by deeply incised, steep, narrow channels often containing very coarse cobble or boulder deposits and left 'hanging' at the terrace edge or grading downward to the next terrace surface. This paper assesses the significance of these channels and associated deposits in terms of the nature of the flow event(s) that created them and their role in subsequent terrace formation.

SELECTION OF FIELD AREAS

Evidence for former high-magnitude flood events was investigated in the Tekapo Valley in South Island, New Zealand and at Skogasandur in Iceland. In each of these areas, sequences of deglaciation outwash terraces have been cut by deeply incised channels. The paleochannel systems in these areas developed under glacial meltwater flow regimes, characterized not only by marked diurnal and seasonal discharge fluctuations, but also by episodic high-magnitude flood events such as 'jökulhlaups' caused by 'glacier bursts' (Arnborg, 1955; Church, 1972; Björnsson, 1974; Jonsson, 1982). These large-scale flood events can occur at any stage during deglaciation, and hence are superimposed on an overall trend of discharge change during deglaciation (Maizels, 1983b,c).

During deglaciation, many sandur deposits are periodically incised to form a terrace sequence. The precise nature of the flow events that act to incise these deposits cannot be determined directly from geomorphic or sedimentologic evidence, since little or no such evidence remains of these particular channel-cutting events. Hence, the absence of direct evidence for the former terrace-forming channels means that one cannot directly determine the nature of the flow event(s) that cut the terraces.

One approach for indirectly determining the origin of the terrace-forming channels is through comparison of paleohydrologic data for the deeply incised channel systems with those for the braided channel systems that occupied the adjacent terrace surfaces. Hence, the field areas selected were those that exhibited both types of channel system within their terrace sequence.

METHODOLOGY

The paleohydraulic analyses presented here estimate a range of former flow parameters apparently associated with two discrete populations of flood events. These populations are, first, the relatively regular overbank flow events which occurred during progressive aggradation or reworking of sediments in a proglacial braided channel environment, and second, the episodic high-magnitude glacier-burst flood events associated with the incision of these deposits and the transport and subsequent deposition of large boulder and cobble beds along these deep, steep channels. These two environments are distinguished by cross-sectional area, depth, gradient, sedimentology and particularly by relative roughness of the channels. The methodology therefore distinguishes between (1) models derived from the hydraulics of channels exhibiting low gradients and small-scale roughness, which will be applied to the braided terrace sequence, and (2) those derived from steep, intermediate-scale, boulder-bed channels, which will be applied to the incised channels.

SOURCES OF ERROR FROM PALEOHYDRAULIC MODELS

Each stage in the determination of paleoflow parameters relies on a number of assumptions which, if not met, give rise to numerous sources of error which reduce the reliability and accuracy of the computed estimates. The paleoflow models only provide approximate 'order of magnitude' estimates of former flow conditions, since one cannot readily demonstrate the validity of many of the basic assumptions. The degrees of confidence need to be estimated wherever possible based on calculations of an 'envelope' of probable values for each paleoflow parameter.

The present approach for determining former flow conditions is based on initial estimates of the critical shear stress at which bed particles begin to move. This computation uses Shields' (1936) function (see details in Maizels, 1983a) and allows subsequent calculation of former flow depth using the Du Boys equation, and resistance to flow and flow velocity using Manning- and Darcy-Weisbach-type equations.

Steady Uniform Flow

The presence of steady uniform flow is the most fundamental assumption underlying the application of many theoretical flow equations in paleohydraulic analysis. This assumption is particularly significant in paleohydraulic analysis, as it means that (i) the measured channel bedslope may be considered to parallel the slope of the former energy gradient; (ii) the flow continuity equation

$$Q = w\bar{d}\bar{V} \tag{1}$$

(where Q = discharge, w = flow width, d = mean flow depth, and V = mean velocity) may be applied to paleodischarge determinations; and (iii) paleohydraulic flow conditions may be reliably predicted from a number of well established theoretically-derived hydraulic models based on steady uniform flow conditions. This assumption is also the most questionable in paleoflow analysis, however, particularly in relation to former steep gradient, rough-bed meltwater streams, which would have been subject to rapid fluctuations in velocity and discharge, and in channel boundary conditions. Flow in such channels is often characterized by large-scale turbulence and flow funneling which promote particle lifting, episodic transport of pulses of sediment, and expansion and contraction energy losses, processes not accounted for in the steady uniform flow models based on estimates of bed-shear stress. Flow resistance is then due not only to skin resistance but to resistance associated with non-uniform, unsteady flow conditions. As a result, paleovelocities may be significantly overestimated.

Shields' Function

Recent work has demonstrated that Shields' coefficient is highly variable and may range from as low as 0.01 to as much as 0.25 (Williams, 1983) depending on the degree of sediment sorting, imbrication, packing and armoring, as well as on particle shape. Hence, the linear function proposed for the relationship between critical shear stress and particle diameter cannot necessarily be accepted, and critical shear stresses may be underpredicted by as much as 250 percent (Ashida and Bayazit, 1973).

Roughness Elements

These elements are assumed to be uniformly distributed on the bed such that the initiation of particle motion may be regarded as a random probabilistic or stochastic process (A.S.C.E., 1966; Church, 1978). Gravel and boulder beds frequently exhibit local sorting of surface materials, however, reflected in the formation of transverse ribs, imbricated pebble-and-boulder trains, longitudinal sorted sediment 'streams,' and areas of differential compactness and armoring. Hence, initiation of particle motion may be strongly controlled by local patterns of grain size, shape, packing, and resistance.

Sediment Availability

The approach assumes that all sizes of sediments were available for transport, implying that the maximum particle sizes present in the deposit reflect former stream competence rather than the lack of supply of larger particles. In most proglacial environments this assumption is likely to be acceptable on a catchment scale but may not be on a local within-channel scale, especially where bed armoring has occurred.

Transport Mechanism

The sediment particles are assumed to have been transported by competent Newtonian fluid flows (Costa, 1984) rather than by debris flows and have not reached the site by ice-rafting, bank slumping, or landslides. It is also assumed that the sediments do not represent the non-transportation of older, coarse lag deposits that may have been exposed during scour of the channel bed.

Channel-Forming Flow Events

The events recorded on the terrace surfaces are likely to represent a series of penultimate flood events that affected each surface prior to the occurrence of floods, possibly of higher magnitude, which actually incised that surface to form the deep, steep 'flood channels.' This interpretation implies that flood events represented on successive terrace surfaces are directly comparable in terms of the relative magnitude-frequency relations that they represent. This assumption still needs to be tested in present outwash environments, particularly since many flood channels may represent multiple periods of sudden flood activity (Lord and Kehew, 1984).

DERIVATION OF INPUT PARAMETERS

Further sources of error are associated with the field determination of each of the input parameters required by the paleohydraulic models.

Former Energy Gradients

Former energy gradients of the channel bed need to be established as a primary control on former bed-shear stress, flow depth and mean velocity. Measurement errors arise, however, in the selection of the length and orientation of the slope measurement in relation to local flow directions, main thalweg directions and valley axis orientation. Even during high flow events, former water surface slopes would have exhibited some local variability and may well have differed significantly from the overall energy slopes (Lord and Kehew, 1984; Costa, 1985). Errors also arise in accurately determining, and correcting for, any differential tectonic disturbance. In this study, gradients were determined by levelling over distances of as much as 200 m in the downvalley direction; the Tekapo terrace gradient values were adjusted for tectonic uplift.

Particle Size Characteristics

Particle size characteristics of the channel deposits provide a means of estimating former skin resistance, and hence former bed-shear stress, flow depth and mean velocity. Since the objective was to determine the maximum discharge conditions associated with the formation of each terrace deposit and each flood channel, the coarsest clasts present were considered to be of maximum hydraulic significance (c.f., Nolan and Donnelly-Nolan, 1984). Major errors occur, however, in trying to determine the coarse particle size representative of the former flow environment. These errors arise from:

(a) Selection of sample sites. The location of sample sites is frequently controlled by the distribution of suitable exposures, and hence may not be fully representative of the deposit. In this study nearly 1150 sample sites at Tekapo and over 400 sites at Skogasandur were selected at 20–25 m intervals along a series of traverses across the exposed terrace surfaces, along terrace bluffs, and along the flood channel courses.

(b) Sample size. The largest size fraction on the bed is considered to have primary control on grain resistance, and hence on eddy intensity and particle entrainment conditions (Limerinos, 1970; Church and Gilbert, 1975; Church and Kellerhals, 1978; Day, 1978). A number of workers have measured only the largest clasts (Malde, 1968; Scott and Gravlee, 1968; Clague, 1975; Church, 1978; Maizels, 1983a,b,c), thereby obviating the need to ensure that a sufficiently large bulk sample has been collected, a procedure that creates severe practical problems when handling coarse cobble and boulder deposits. The fraction of the total population that these coarse clasts represent, however, remains uncertain. In this study the 10 largest clasts at each site were sampled.

(c) Clast measurement procedures. The intermediate or hydraulic axis of each clast was measured, with operator error being minimized by the use of a single operator throughout (Hey and Thorne, 1983). Since intermediate clast diameters are used in the calculation of relative bed roughness rather than the short clast axis (Limerinos, 1970), however, relative roughness values are overestimated.

Channel and Flow Dimensions

Depth of flow was computed from shear stress equations (see below), allowing the width of flow to be estimated by extrapolation of the calculated flow depth across the levelled cross-profile on the terrace surface, or across the flood channel. Major sources of error can arise in: (a) determining the degree of contemporaneity of individual channels in the system, and (b) extrapolating the proportionate flow width on preserved terrace fragments to the whole former extent of the valley floor.

Channel Planform Characteristics

Channel pattern, sinuosity, degree of braiding, channel extent and continuity were largely determined from aerial photograph analysis, combined with some field levelling. Problems arise in the interpretation of channel planform characteristics where only terrace fragments have been preserved and in assessing the degree to which the observed braided channels were active contemporaneously.

PALEOHYDRAULIC COMPUTATIONS

In addition to errors associated with application of the paleohydraulic models to former meltwater flood deposits, and with field determination of the required input parameters, a number of inaccuracies can arise in the manipulation, precision, and standardization of units of measurement during paleohydraulic computations (Williams, 1984). Each successive stage of analysis is, therefore, accompanied by a cumulative expansion in error values (Gardner, 1983).

Critical Tractive Force

Both theoretical and empirical methods of estimating the critical tractive force are available. The theoretical approach is based on the Shields' function, where the critical tractive force (τ_c, in N m^{-3}) is regarded as a linear function of particle size (D, in m). The value of this function depends on the value of Shields' coefficient, ϕ, in:

$$\tau_c = \phi(\gamma_s - \gamma)D \tag{2}$$

where γ = the specific weight of water, taken as 9810 N m^{-3}, and γ_s = the specific weight of the sediment, taken as 26000 N m^{-3}, assuming clear, sediment-free water. Since the outwash deposits exhibited strong imbricated fabrics, ϕ values of 0.04 and 0.07 have been employed to represent the probable range of former bed conditions.

Williams (1983) has proposed that the upper and lower limits of critical tractive force can be predicted from empirical envelope equations that define the range within which particles have been observed to move. The upper and lower limits are defined by linear functions:

$$\tau_c = C1\ D \tag{3}$$

where $C1 = \phi(\gamma_s - \gamma)$, and ranges between 170 and 3900 N m^{-3}, thus representing a much wider range of possible Shields' coefficients (0.01 to 0.24).

Critical Flow Depth

Mean critical flow depth, d_c, can be estimated by substituting computed values of τ_c and measured slope values, (S), into the Bu Boys equation for boundary shear, such that:

$$d_c = \tau_c/\gamma S \tag{4}$$

where d_c replaces the hydraulic radius, R. If flow depths considerably exceed the equivalent hydraulic radius, however, O'Brien and Rindlaub (1934) claimed that shear forces may be overestimated by as much as 20 percent, and therefore that former flow depths may also be overestimated by as much as 7.5 percent. Substitution of equation (3) into equation (4) gives:

$$d_c = C2\ D\ S^{-1} \tag{5}$$

where $C2 = C1\ \gamma^{-1}$, which ranges between 0.066 and 0.115 for ϕ = 0.04 and 0.07, respectively.

Resistance to Flow

The accurate determination of mean paleovelocity is largely dependent on obtaining a valid and reliable measure of flow resistance. In low sinuosity, coarse-grained, high width-depth ratio channels, such as those of the terraced braided channel networks, the dominant source of flow resistance is from bed particles. This skin resistance can be readily determined using well established friction coefficients for small-scale roughness channels, such as Manning's n and the Darcy-Weisbach f.

The value of Manning's n for paleoflood channels is most accurately determined by computation rather than by empirical methods (Maizels, 1983a) and here three different methods of calculating Manning's n are employed. Strickler (1923) derived a simple functional relationship between n and the representative particle diameter such that:

$$n = 0.039\, D_{50}^{0.167} \qquad (6)$$

where D_{50} = 50th percentile of the sediment size distribution. Limerinos (1970), however, demonstrated that n was best determined by the use of a logarithmic function relating the resistance coefficient to a measure of relative roughness (defined as the ratio of d to D):

$$n = \frac{0.113\, d^{0.167}}{1.16 + 2.0 \log (d/D_{84})} \qquad (7)$$

where D_{84} = 84th percentile of sediment size distribution. More recently, Jarrett (1984a) has derived a predictive equation for n based on multiple regression:

$$n = 0.32\, S^{0.38}\, d^{-0.16} \qquad (8)$$

for $0.002 < S < 0.034$, $0.14 < d < 2.01$ m, $0.028 < n < 0.159$, $0.09 < D < 0.79$ m, and where the standard error = 28 percent.

The Darcy-Weisbach friction factor (f) has, however, a more sound theoretical basis, as it is related to the von Karman-Prandtl velocity 'law' and is also dimensionally correct. Values of f may be determined by using the approach proposed by Hey (1979):

$$\frac{1}{\sqrt{f}} = C3 \log \left(\frac{aR}{K_s}\right) \qquad (9)$$

where $C3 = 2.3/(k/\sqrt{8})$, k = the von Karman coefficient, and Ks = the representative grain size (m). The value of 'a' is a function of channel cross-sectional shape and spacing of roughness elements (Silberman and others, 1963) and may be determined either graphically (Hey, 1979) or from:

$$a = 11.1\, (R/d\text{max})^{-0.314} \qquad (10)$$

where $d$max = maximum flow depth (m) (Thorne and Zevenbergen, 1985). Where the hydraulic radius (R) and/or maximum flow depth are unknown, however, f may be estimated from relative roughness alone, according to Henderson's (1966) model:

$$f = 0.113\, (d/D)^{0.33} \qquad (11)$$

For channels exhibiting large-scale roughness, i.e., where $D/d > 0.3$, there are no wholly satisfactory resistance equations. Thompson and Campbell (1979) proposed that calculation of the Darcy-Weisbach friction factor for boulder channels should be modified to take into account the large 'blocking' effects on flow produced by high relative roughness conditions:

$$\frac{1}{\sqrt{f}} = (1 - 0.1\, K_s/R) \cdot 2 \log_{10} (12R/K_s) \qquad (12)$$

where Ks was found to be best represented by $4.5D$. Bathurst (1978, 1985) and Bathurst and others (1981) have developed some theoretical equations for resistance determination in boulder-bed channels, but these are complex, cumbersome and inappropriate for paleohydraulic analysis. At present, resistance in such channels appears to be most accurately estimated from modified small-scale roughness equations, such as that of Hey (1979):

$$\left(\frac{8}{f}\right)^{1/2} = 5.62 \log \left(\frac{aR}{3.5\, D_{84}}\right) \qquad (13)$$

and Bathurst's (1985) adjusted version of the Hey equation:

$$\left(\frac{8}{f}\right)^{1/2} = 5.62 \log \left(\frac{aR}{3.5\, D_{84}}\right) + 4 \qquad (14)$$

Jarrett's (1984a) empirical equation for prediction of n [equation (8) above], may also be valid for rough-bed channels.

Mean Flow Velocity

Mean critical flow velocity can be computed from either the Manning equation:

$$\bar{V}_c = \frac{d^{2/3}\, S^{1/2}}{n} \qquad (15)$$

substituting in values of n determined from equations (6) to (8), or the Colebrook-White equation:

$$\bar{V}_c = \left[\frac{8gdS}{f}\right]^{1/2} \qquad (16)$$

substituting in values of f determined from equations (9), (11) to (13). These equations provide seven different estimates of critical paleoflow velocity for each Shields' coefficient selected during computation of the critical tractive force and mean flow depth [equations (2) to (5); Table 1]. Further estimates of mean flow velocity can be determined from an equation derived by Costa (1983) based on the mean of four theoretical and empirical functions:

$$\bar{V} = 0.18\, (D.\, 10^3)^{0.487} \qquad (17)$$

In addition, the minimum and maximum probable paleovelocities may be estimated from Williams' (1983) envelope equation:

$$\bar{V}_c = C4\, D^{1/2} \qquad (18)$$

where $C4$ min = 2.05 and $C4$ max = 14.53 (D in m).

TABLE 1.—INPUT PARAMETERS AND ESTIMATES OF SELECTED PALEOFLOW PARAMETERS FOR INDIVIDUAL RELIC FLOOD CHANNELS

A. TEKAPO VALLEY, NEW ZEALAND

Flow Parameter	Wolds Flood Channel		Balmoral F.C.		Mt John F.C. 1		Mt John F.C. 2		Mt John F.C. 3		Tekapo F.C.	
Mean gradient (m m^{-1})	0.0194		0.0117		0.0194		0.0206		0.0221		0.0208	
Mean clast size (m)	0.215		0.248		0.371		0.4182		0.7728		0.5250	
Mean channel width (m)	180.0		130.0		360.0		440.0		350.0		240.0	
Mean channel sinuosity	1.040		1.019		1.094		1.088		1.004		1.099	
Shields coefficient	0.04	0.07	0.04	0.07	0.04	0.07	0.04	0.07	0.04	0.07	0.04	0.07
Tractive force (N m^{-2})	139.49	244.11	160.60	281.06	240.13	420.23	270.83	473.95	500.47	875.81	339.99	594.98
Flow depth (m)	0.73	1.28	1.41	2.46	1.27	2.21	1.34	2.35	2.31	4.05	1.67	2.92
Resistance coefficients*												
n_s (6)	0.030	0.030	0.031	0.031	0.033	0.033	0.034	0.034	0.037	0.037	0.035	0.035
n_L (7)	0.048	0.044	0.045	0.042	0.053	0.048	0.055	0.049	0.062	0.055	0.057	0.051
n_J (8)	0.075	0.069	0.056	0.051	0.069	0.063	0.070	0.064	0.066	0.060	0.068	0.062
f_{Hey} (9)	0.097	0.073	0.074	0.058	0.097	0.073	0.100	0.075	0.104	0.077	0.101	0.075
f_{Hend} (11)	0.170	0.205	0.201	0.243	0.170	0.205	0.167	0.201	0.163	0.196	0.166	0.200
$f_{T\&C}$ (12)	0.362	0.203	0.212	0.136	0.361	0.202	0.389	0.214	0.425	0.228	0.395	0.216
$f_{T\&Z}$ (13)	0.030	0.019	0.020	0.014	0.030	0.019	0.031	0.020	0.033	0.021	0.031	0.020
f_{Bath} (14)	0.010	0.008	0.008	0.006	0.010	0.008	0.011	0.008	0.011	0.008	0.011	0.008
OVERALL MEAN	0.103	0.081	0.081	0.073	0.103	0.081	0.107	0.083	0.113	0.085	0.108	0.083
Mean flow velocities (m s^{-1})												
$V n_s$ (6 + 15)	3.75	5.45	4.38	6.37	4.92	7.15	5.18	7.52	6.96	10.10	5.79	8.41
$V n_L$ (7 + 15)	2.35	3.78	3.02	4.72	3.08	4.97	3.20	5.17	4.22	6.87	3.56	5.77
$V n_J$ (8 + 15)	1.51	2.39	2.43	3.85	2.36	3.76	2.50	3.98	3.96	6.29	3.00	4.76
$V_{f\ Hey}$ (9 + 16)	3.39	5.18	4.16	6.25	4.45	6.80	4.65	7.12	6.20	9.52	5.20	7.96
$V_{f\ Hend}$ (11 + 16)	2.56	3.09	2.53	3.04	3.36	4.05	3.61	4.35	4.96	5.98	4.05	4.88
$V_{f\ T\&C}$ (12 + 16)	2.92	6.89	5.34	11.06	3.84	9.06	3.79	9.11	4.70	11.60	4.18	10.10
$V_{f\ T\&Z}$ (13 + 16)	6.15	10.05	8.02	12.68	8.07	13.19	8.35	13.72	11.03	18.23	9.32	15.34
$V_{f\ Bath}$ (14 + 16)	10.37	15.64	12.55	18.68	13.61	20.52	14.24	21.51	19.02	28.82	15.92	24.06
V_{Cos} (17)	2.46	2.46	2.64	2.64	3.21	3.21	3.40	3.40	4.59	4.59	3.80	3.80
V_{Wmin} (18)	0.95	0.95	1.03	1.02	1.25	1.25	1.33	1.33	1.80	1.80	1.49	1.49
V_{Wmax} (19)	6.74	6.74	7.24	7.24	8.85	8.85	9.40	9.40	12.77	12.77	10.53	10.53
OVERALL MEAN (excl. eq. (18))	3.94	6.10	5.01	7.70	5.21	8.08	5.44	8.43	7.29	11.33	6.09	9.45
Mean Froude numbers from *dc* and *V*												
Fr V_{nS}	1.96	2.36	1.39	1.68	1.95	2.36	2.03	2.45	2.13	2.57	2.05	2.47
Fr V_{nL}	0.77	1.14	0.66	0.92	0.77	1.14	0.78	1.16	0.79	1.19	0.78	1.17
Fr V_{nJ}	0.32	0.45	0.43	0.62	0.45	0.65	0.48	0.69	0.69	0.99	0.55	0.79
Fr V_{fHey}	1.60	2.13	1.25	1.62	1.60	2.13	1.64	2.20	1.69	2.28	1.63	2.22
Fr V_{fHend}	0.91	0.76	0.46	0.38	0.91	0.76	0.99	0.82	1.08	0.90	1.00	0.83
Fr $V_{fT\&C}$	1.09	1.94	1.44	2.25	1.09	1.94	1.04	1.90	0.99	1.84	1.03	1.89
Fr $V_{fT\&Z}$	5.25	8.02	4.67	6.67	5.25	8.01	5.29	8.16	5.36	8.37	5.32	8.22
Fr V_{fBath}	14.96	19.43	11.43	14.47	14.93	19.40	15.39	20.06	15.95	20.91	15.51	20.24
Fr V_{Cos}	0.92	0.69	0.71	0.54	1.10	0.69	0.94	0.71	0.96	0.73	0.94	0.71
Fr V_{Wmin}	0.13	0.07	0.08	0.04	0.13	0.07	0.13	0.08	0.14	0.08	0.14	0.08
Fr V_{Wmax}	6.33	3.61	3.80	2.17	6.31	3.60	6.70	3.83	7.19	4.11	6.78	3.87
OVERALL MEAN**	3.00(0.97)	4.03(1.26)	2.49(0.91)	3.24(1.14)	3.12(1.12)	4.12(1.38)	3.18(1.13)	4.24(1.42)	3.29(1.19)	2.36(1.50)	3.20(1.14)	4.28(1.44)
Relative roughness (*d/D*)	3.40	5.96	5.67	9.92	3.41	5.97	3.21	5.62	2.99	5.24	3.17	5.55
Maximum flow discharge (m^3 s^{-1}) based on (*w.d.V*) for ϕ = 0.04 only (excl. eq 18)												
$Q\ V_{ns}$	494.96		800.74		2242.44		3060.16		5632.12		2314.91	
$Q\ V_{nL}$	309.61		551.63		1403.50		1889.20		3419.13		1425.08	
$Q\ V_{nJ}$	198.69		443.33		1076.66		1479.34		3203.98		1197.66	
$Q\ V_{fHey}$	447.43		759.09		2027.44		2748.18		5021.48		2078.66	
$Q\ V_{fHend}$	338.15		461.49		1530.69		2131.20		4015.68		1618.71	
$Q\ V_{f\ T\&C}$	385.13		976.25		1749.56		2239.19		3808.71		1671.61	
$Q\ V_{f\ T\&Z}$	810.77		1465.31		3674.85		4936.89		8927.60		3728.24	
$Q\ V_{f\ Bath}$	1368.28		2293.62		6199.63		8417.15		15408.68		6366.22	
$Q\ V_{Cos}$	325.05		482.04		1461.61		2011.62		3716.39		1520.29	
$Q\ V_{Wmin}$	125.53		186.51		568.48		783.63		1459.27		593.97	
$Q\ V_{Wmax}$	889.74		1321.92		4029.29		5554.12		10343.23		4210.02	
OVERALL MEAN	519.79		914.83		2374.04		3212.55		5905.97		2435.64	

*Equation number given in brackets (see text)

**Figures in brackets exclude Fr from eqns. (13) and (14)

TABLE 1.—(continued)

B. SKOGASANDUR, S. ICELAND		Channel S1		Channel S2		Channel S3	
Flow parameter							
Mean gradient (m m^{-1})		0.0254		0.0315		0.0349	
Mean clast size (m)		1.095		0.670		0.304	
Mean channel width (m)		350.0		370.0		290.0	
Mean channel sinuosity		1.011		1.004		1.014	
Shields coefficient		0.04	0.07	0.04	0.07	0.04	0.07
Tractive force (N m^{-2})		709.12	1240.90	433.89	759.31	196.87	344.52
Flow depth (m)		2.85	4.98	1.40	2.46	0.58	1.01
Resistance coefficient							
n_s	(6)	0.040	0.040	0.037	0.037	0.032	0.032
n_L	(7)	0.068	0.060	0.066	0.057	0.061	0.051
n_J	(8)	0.067	0.061	0.082	0.075	0.098	0.089
f_{Hey}	(9)	0.113	0.083	0.129	0.093	0.138	0.099
f_{Hend}	(11)	0.155	0.187	0.145	0.174	0.140	0.168
$f_{T\&C}$	(12)	0.517	0.262	0.726	0.331	0.872	0.374
$f_{T\&Z}$	(13)	0,037	0.023	0.047	0.028	0.052	0.030
f_{Bath}	(14)	0.012	0.009	0.013	0.010	0.014	0.011
OVERALL MEAN		0.126	0.091	0.156	0.101	0.176	0.107
Mean flow velocities (m s^{-1})							
$V\, n_s$	(6 + 15)	8.09	11.74	6.10	8.86	4.04	5.87
$V\, n_L$	(7 + 15)	4.73	7.79	3.36	5.64	2.15	3.65
$V\, n_J$	(8 + 15)	4.78	7.59	2.73	4.34	1.32	2.10
$V_{f\ Hey}$	(9 + 16)	7.09	10.95	5.18	8.08	3.38	5.29
$V_{f\ Hend}$	(11 + 16)	6.04	7.28	4.90	5.90	3.36	4.05
$V_{f\ T\&C}$	(12 + 16)	4.60	12.02	2.57+	7.44	1.44	4.44
$V_{f\ T\&Z}$	(13 + 16)	12.31	20.63	8.63	14.80	5.50	9.54
$V_{f\ Bath}$	(14 + 16)	21.84	33.23	16.08	24.66	10.52	16.18
V_{Cos}	(17)	5.44	5.44	4.28	4.28	2.91	2.91
V_{Wmin}	(18)	2.15	2.14	1.68	1.68	1.13	1.13
V_{Wmax}	(19)	15.20	15.20	11.89	11.89	8.01	8.01
OVERALL MEAN		8.32	12.96	5.98	9.33	3.85	6.00
Mean Froude numbers from *dc* and *V*							
Fr V_{nS}		2.34	2.82	2.70	3.26	2.89	3.49
Fr V_{nL}		0.80	1.24	0.82	1.32	0.82	1.35
Fr V_{nJ}		0.82	1.18	0.54	0.78	0.31	0.45
Fr V_{fHey}		1.80	2.45	1.95	2.71	2.02	2.83
Fr V_{fHend}		1.31	1.09	1.74	1.45	2.00	1.66
Fr $V_{fT\&C}$		0.87	1.72	0.69	1.52	0.61	1.41
Fr $V_{fT\&Z}$		5.43	8.71	5.41	9.09	5.35	9.23
Fr V_{fBath}		17.08	22.60	18.78	25.23	19.60	26.53
Fr V_{Cos}		1.03	0.78	1.15	0.87	1.22	0.92
Fr V_{Wmin}		0.16	0.09	0.20	0.12	0.22	0.13
Fr V_{Wmax}		8.28	4.73	10.27	5.87	11.38	6.50
OVERALL MEAN (excl. eqs 13,14)		3.50(1.28)	4.73(1.61)	3.75(1.37)	5.14(1.70)	3.87(1.41)	5.32(1.73)
Relative roughness (*d/D*)		2.10	3.67	2.60	4.55	1.89	3.31
Maximum flow discharge (m^3 s^{-1}) for ϕ = 0.04 only							
$Q\ V_{ns}$		8054.18		3169.94		673.72	
$Q\ V_{nL}$		4715.27		1742.98		358.23	
$Q\ V_{nJ}$		4757.21		1419.53		220.46	
$Q\ V_{fHey}$		7059.11		2692.81		562.96	
$Q\ V_{fHend}$		6019.92		2545.63		559.85	
$Q\ V_{f\ T\&C}$		4586.38		1333.29		240.12	
$Q\ V_{f\ T\&Z}$		12260.66		4484.63		916.42	
$Q\ V_{f\ Bath}$		21750.37		8356.28		1753.49	
$Q\ V_{Cos}$		5417.00		2224.21		485.84	
$Q\ V_{Wmin}$		2136.76		971.75		188.48	
$Q\ V_{Wmax}$		15144.67		6178.77		1335.88	
OVERALL MEAN		8291.12		3107.70		641.23	

Discharge

Determination of paleodischarge from coarse channel deposits can be based on the flow continuity equation [equation (1)]. This somewhat simplified approach, however, needs to be modified for braided channel systems where extensive overbank flooding has occurred. The two components of channel flow and overbank flow are associated with different conditions of relative roughness, bed resistance, critical mean depth, and velocity. Discharges therefore need to be computed separately, and then summed, for each individual flow environment.

PALEOHYDRAULICS OF LARGE-SCALE FLOOD DEPOSITS

Tekapo Valley, South Island, New Zealand

Flood channels and the terrace sequence.—

The Tekapo valley is marked by a sequence of eight main terrace levels, extending as much as 20 km downstream from terminal moraine complexes. The oldest, highest outwash deposits appear to date from the Wolds and Balmoral glaciations, probably dating from about 100,000 to 50,000 years BP. The more recent terrace sequences date from the Mt. John and Tekapo glacial phases, probably occurring from between 16,000 and 14,000 years BP (Speight, 1961; Gair, 1967; Mansergh, 1973). The surface of each of these terrace deposits is marked by the courses of former meltwater channels, the older ones exhibiting subdued dendritic and braided paleochannel systems partially masked by loess infill, whereas the two younger terrace sequences exhibit distinctive and extensive braided paleochannel networks. Each of these deposits is, in turn, cut by at least one deeply incised channel normally extending from a former ice-marginal zone and grading down to the next terrace surface. These include the Wolds (Wc), the Balmoral (Bc), and four Mt. John/Tekapo (MJ1, MJ2, MJ3, T1) flood channels (Fig. 1).

Channel morphology and gradients.—

The older channels (Wolds and Balmoral) exhibit both low width-depth ratios and sinuosities (Table 1); their cross-sectional morphology remains relatively constant downchannel, suggesting that flow was probably near uniform. The upper reaches of the Mt. John and Tekapo channels, however, were broader, shallower and more sinuous (Fig. 1; Table 1). Width-depth ratios gradually increase downstream as the channels meander through successive moraine ridges, suggesting that flow was less likely to have been uniform and that resistance may have been underestimated from measures of bed roughness alone.

The flood channel gradients, corrected for differential tilting (Mathews, 1972), are all high, ranging between 0.0117 and 0.0221 m m^{-1} (Fig. 2; Table 1). These gradients are of a similar order of magnitude to those of present mountain rivers in which the high relative roughness flow resistance equations adopted here have been tested (Jarrett, 1984a, b; Bathurst, 1985; Thorne and Zevenbergen, 1985). The flood channel gradients at Tekapo exceed those of the adjacent terrace surfaces by as much as about 60 percent (e.g., MJ3).

Boulder sizes and critical tractive force.—

The flood channels are mantled with large cobbles and boulders as much as 0.77 m in diameter (Table 1A) and with sizes exceeding those of the adjacent terrace surfaces by as much as about 30 percent. Estimates of the critical tractive force are hence relatively high but compare closely with the range of values estimated, for example, for the largest recorded U.S. rainfall-runoff floods (Costa, 1985).

Critical flow depth and relative roughness.—

Estimates of critical flow depth are proportional to the critical tractive force present over a given channel gradient [equation (5)], and since both gradients and clast sizes are large, critical flow depths emerge as relatively modest. Maximum flood flow depths rarely exceeded 2.3 m (for $\phi = 0.04$; or 4.1 m for $\phi = 0.07$; Table 1A). These depths compare closely with maximum depths of between 0.6 and 5.5 m recorded for recent large-scale floods in steep boulder-bed channels (Bradley and Mears, 1980; Jarrett, 1984a; Bathurst, 1985; Costa, 1985; Thorne and Zevenbergen, 1985). Hence, although the Tekapo flood channels are relatively deeply incised, flood waters did not fill the whole channel, suggesting that the channels were progressively incised, possibly during successive floods. Relative roughness values for the Tekapo flood channels all lie close to the boundary condition for small-scale and intermediate-scale roughness (Table 1A). Since relative roughness values tend to be overestimates due to use of the intermediate clast axis (see above), however, these flood events were probably characterized by small-scale roughness conditions.

Resistance to flow.—

Estimates of resistance to flow compare closely with those that were obtained during recent maximum flood events (Bradley and Mears, 1980; Costa, 1985), although the range of values differs according to the model used. Thus, the lowest n values have been predicted by Strickler's equation, whereas the Limerinos model predicts intermediate values and Jarrett's equation predicts relatively high values (Table 1A; Fig. 3A). Even higher n values have recently been reported from steep mountain rivers by Jarrett (1984a) and Thorne and Zevenbergen (1985), however.

Maximum values of the Darcy-Weisbach friction factor have been estimated from Thompson and Campbell's equation and compare closely with those estimated in Colorado mountain streams (Thorne and Zevenbergen, 1985). Hey's equations have produced significantly lower estimates of flow resistance than those predicted by the large-scale roughness model of Thompson and Campbell (Table 1A; Fig. 3A).

Critical mean flow velocity.—

Paleovelocities have been estimated using the eight different resistance equations, together with Costa's (1985) equation and Williams' (1983) envelope equations for empirically observed flow velocity relations. The predicted paleoflow velocities exhibit a wide range of values, with the majority of velocity estimates ranging between 1.5 and 7.0 m s^{-1} (for $\phi = 0.04$; for $\phi = 0.07$, velocity estimates are about 60 percent higher; Table 1A; Fig. 3B). These values lie both within Williams' (1983) predicted minimum and maximum limits, and within the lower range of values estimated by Costa (1985) for U.S. floods. The two modified Hey equations [equations (13) and (14)] predict excessively high velocities. When the latter are excluded, the predicted Froude numbers for the flood flow events conform closely to those estimated for other large flood events (0.43 to 2.49; Costa, 1985). The Froude numbers (Table 1A) suggest that both the paleoflow depth and velocity estimates for the Tekapo channels are likely to represent the correct order of magnitude.

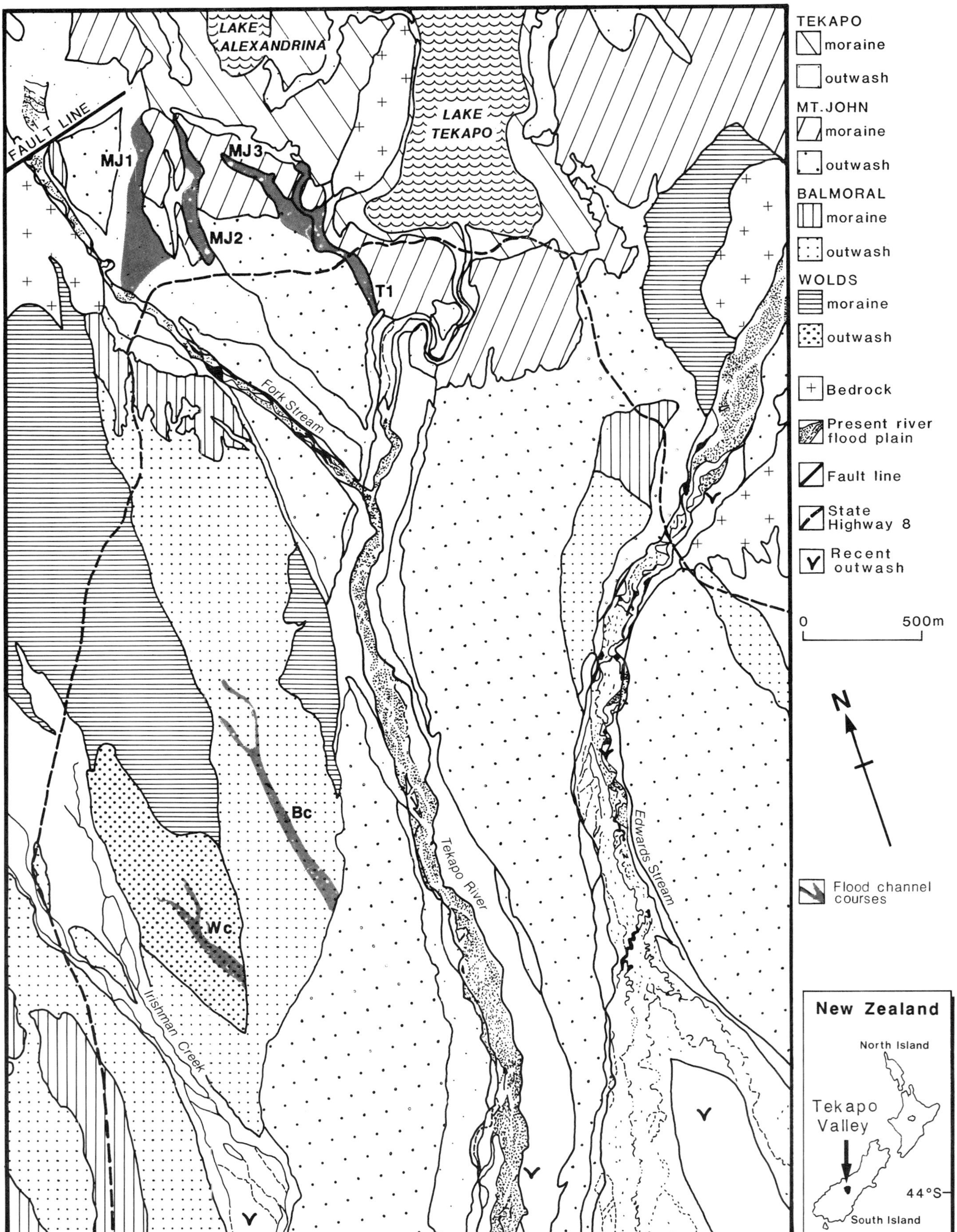
LAKE ALEXANDRINA
LAKE TEKAPO
FAULT LINE
MJ1
MJ2
MJ3
T1
Fork Stream
Bc
Wc
Irishman Creek
Tekapo River
Edwards Stream
TEKAPO
moraine
outwash
MT. JOHN
moraine
outwash
BALMORAL
moraine
outwash
WOLDS
moraine
outwash
Bedrock
Present river flood plain
Fault line
State Highway 8
Recent outwash
0
500m
N
Flood channel courses
New Zealand
North Island
Tekapo Valley
44°S
South Island
170°E

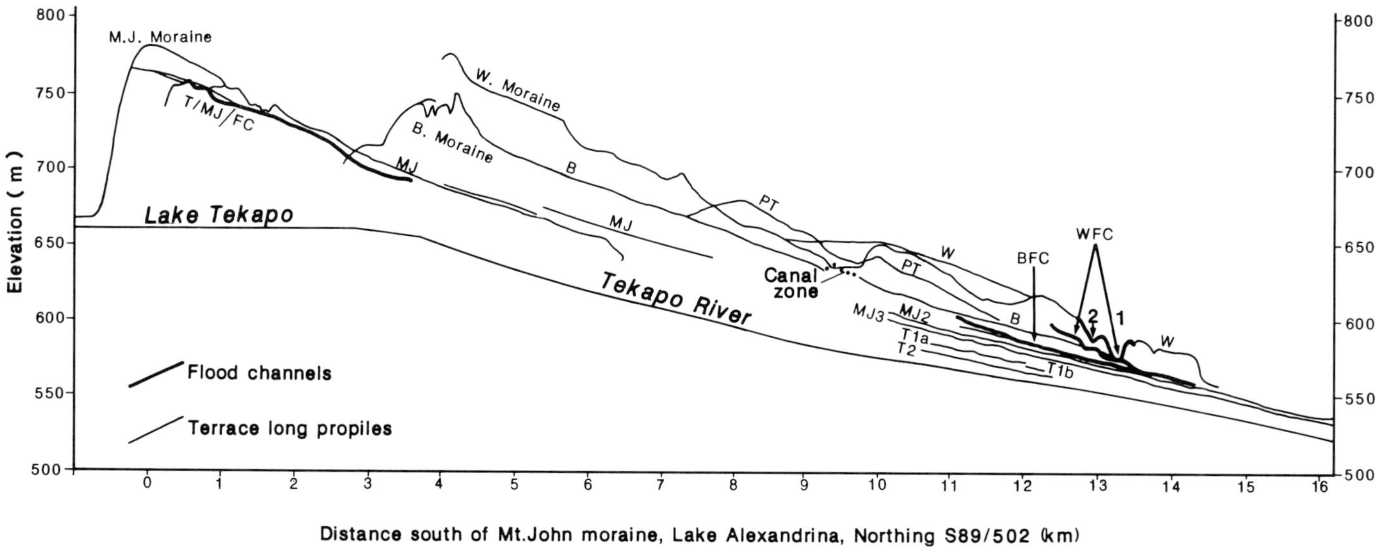

FIG. 2.—Longitudinal profiles of terraces and flood channels, Tekapo valley, South Island, New Zealand.

Maximum discharge.—

The estimated maximum discharges of the flood channels ranged between about 350 and 4200 m^3 s^{-1} (for $\phi = 0.04$; Table 1A; Fig. 3C). The Tekapo flood channels appear to have accommodated almost 30 times the discharge volume of the adjacent terrace surface channel systems.

Skogasandur Channels, Solheimàjökull, Southern Iceland

Flood channels and the terrace sequence.—

Three distinctive flood channels have been incised into a pumice gravel fan deposit dating from the seventh to tenth centuries A.D. and lying about 3 km south of Solheimàjökull, a southern outlet glacier of the Myrdalsjökull ice cap in southern Iceland (Fig. 4). The channels appear to have been cut during tenth century flood or jökulhlaup events relating to eruptions of the subglacial volcano Katla (Maizels and Dugmore, 1985). Subsequent jökulhlaups have dissected these deposits to form a sequence of terrace levels over an elevation range of about 35 m.

Channel morphology and gradients.—

The three channels exhibit similar morphologies, with width and depth initially constant but then both increasing downchannel as flow fanned out onto the former outwash plain (Fig. 4). Distal bedslopes exhibit large-scale stepped or wave-forms with 'wavelengths' averaging 200 to 400 m. Hence, uniform flow was unlikely in the downchannel reaches, so that analysis has concentrated on the proximal flow conditions. Channel gradients were relatively steep with overall slopes exceeding 0.025 m, and hence were considerably greater than those of the Tekapo channels (Fig. 5; Table 1B).

Boulder sizes.—

Boulder sizes varied considerably between each channel with the smallest boulders occurring in the steepest channel (Table 1B) and with many boulders exhibiting a distinctive upstream imbrication. Critical tractive forces computed from boulder sizes were of a similar magnitude to those of the Tekapo channels and to those estimated for other flood events (Costa, 1985).

Critical flow depths and mean velocities.—

Critical flow depths were also of a similar magnitude to those estimated for the Tekapo flood events, ranging between about 0.6 and 2.9 m (Table 1B). Hence, intermediate-scale roughness conditions prevailed during these flood events. The most accurate estimates of mean flow velocity for the rough-bed channels were provided by Thompson and Campbell's (1979) model, ranging between about 1.4 and 4.6 m s^{-1} (Table 1B; Fig. 3B). These values also provide plausible Froude numbers.

Maximum discharges.—

If the Froude numbers, and hence the velocity and resistance estimates, at least for Thompson and Campbell's model, are accepted, maximum discharges averaged between about 240 and 4600 m^3 s^{-1} in the Skogasandur flood channels (Table 1B; Fig. 3C). These values are very similar to those of the Tekapo flood channels. The channel S1 flow event represents, as does the MJ3 channel at Tekapo, an increased discharge compared with the adjacent terrace channel systems of almost 40 times.

FIG. 1.—Location of flood channels, Tekapo valley, South Island, New Zealand.

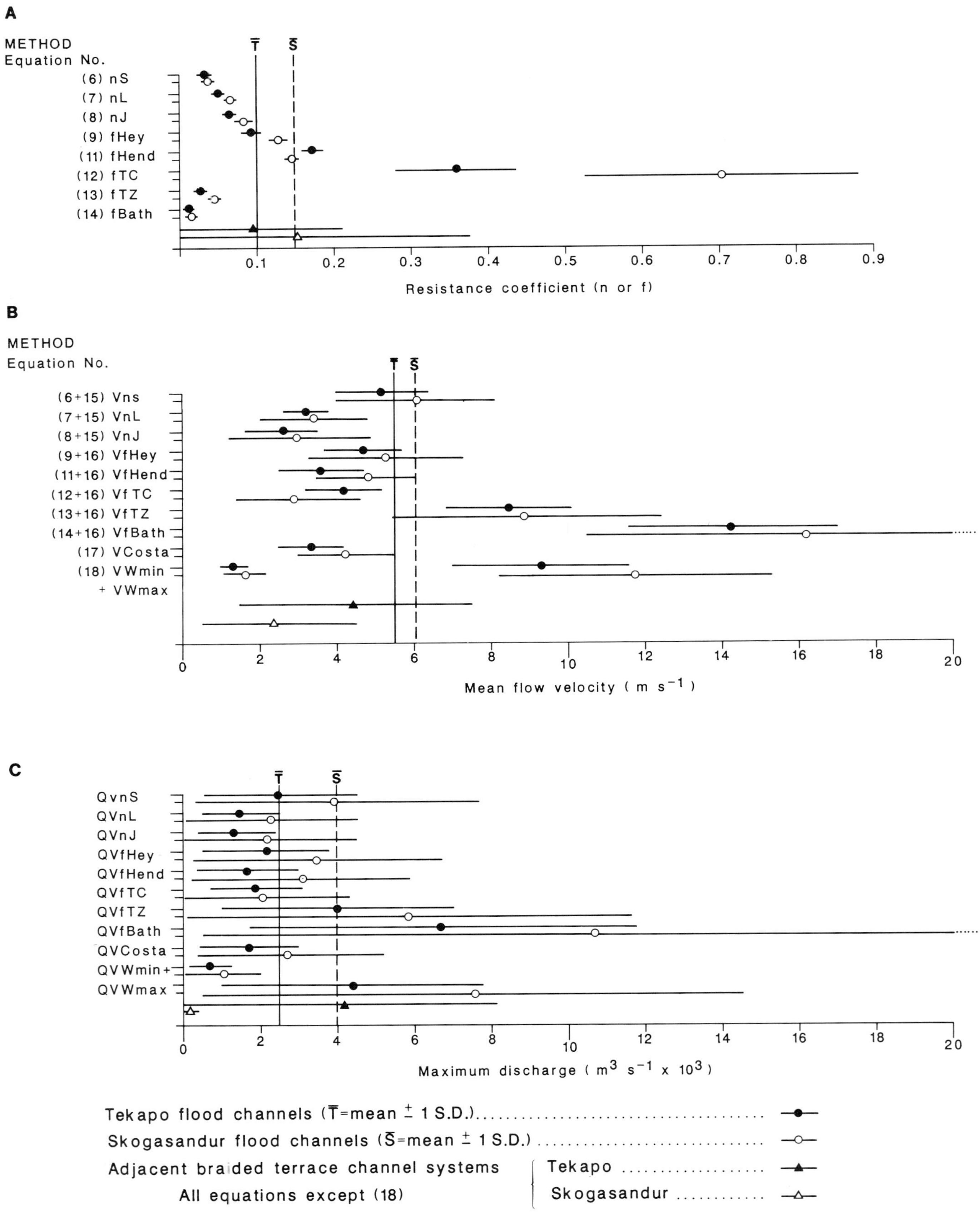

FIG. 3.—Estimates of mean paleoflow resistance, velocity and discharge in relic flood channels, Tekapo valley, South Island, New Zealand, and Skogasandur, southern Iceland: A. Estimates of mean paleoflow resistance. B. Estimates of mean paleoflow velocity. C. Estimates of mean paleoflow discharge. Note: The estimates of paleoflow parameters are based on the series of hydraulic models discussed in the text and represent the mean flow conditions for the six Tekapo flood channels and the three Skogasandur channels, respectively.

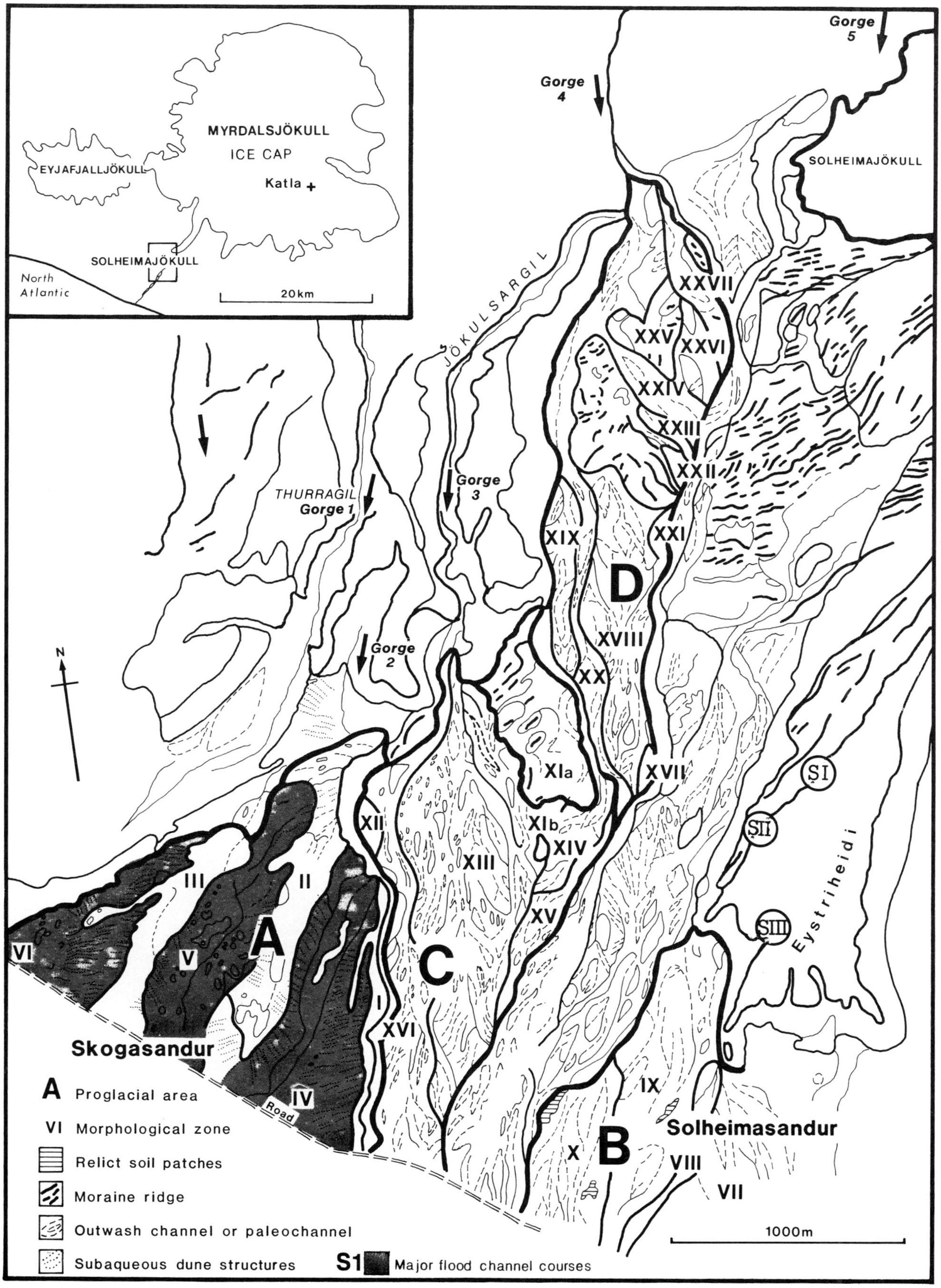

FIG. 4.—Location of flood channels, Skogasandur, southern Iceland.

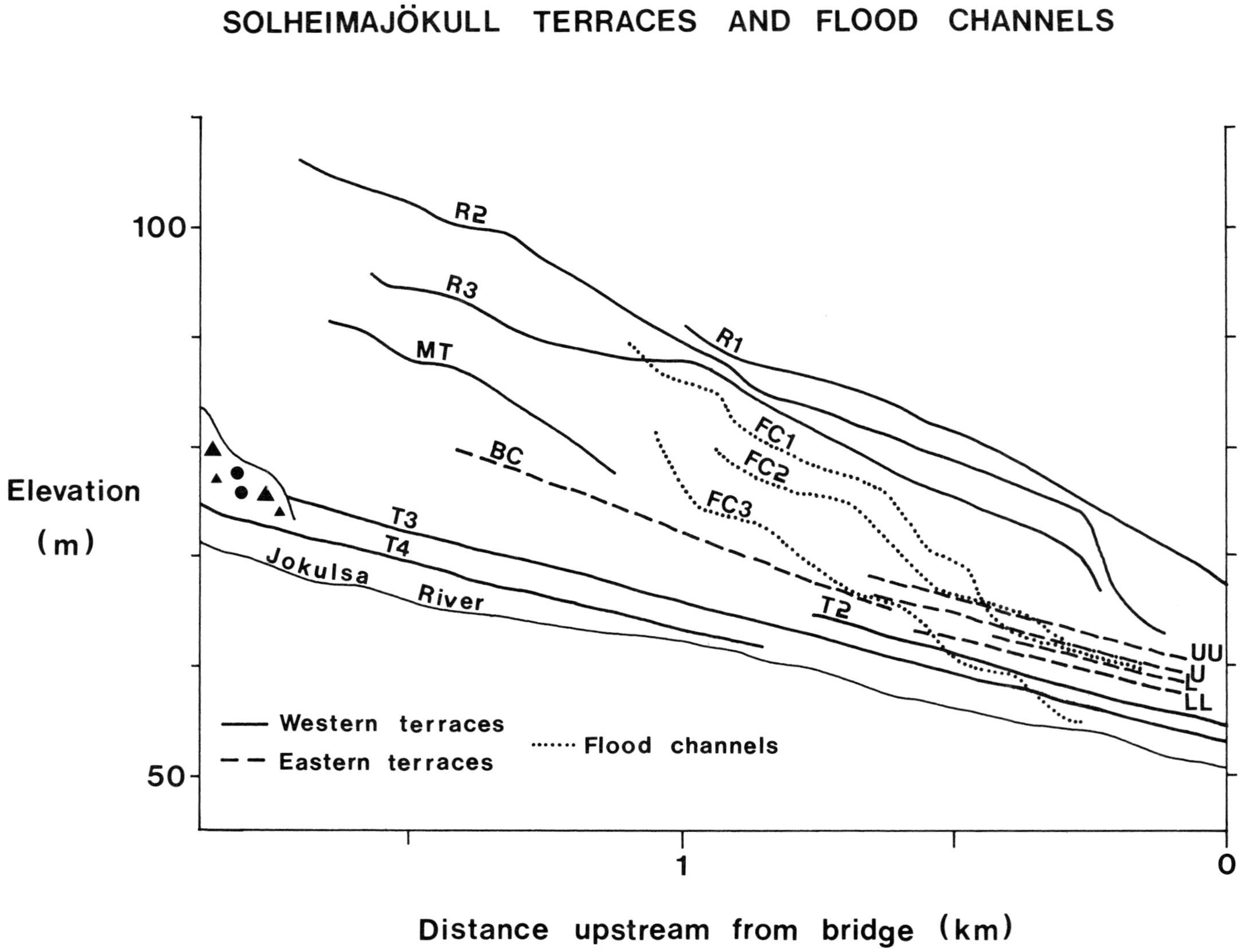

FIG. 5.—Longitudinal profiles of terraces and flood channels, Skogasandur, southern Iceland.

DISCUSSION AND CONCLUSIONS

Nature of Flow Events Associated with Deeply Incised Channels

The deeply incised channels clearly accommodated high-magnitude flood events in which the thresholds for channel change across the pre-existing sandur surface were exceeded. Channel incision was hydrologically controlled, with bed-shear stresses and velocities reaching sufficiently high values to entrain and transport coarse bed and bank sediments for several kilometers over steep downstream slopes. These ice-marginal flood events were probably short-lived (c.f., Missoula flood, Baker, 1973), although the Tekapo flood channels in particular may have been repeatedly occupied and overdeepened.

The frequency of these high-magnitude flow events is unknown, the time resolution associated with both the Tekapo and Skogasandur channels being too crude, particularly in the absence of precise dates, to allow any magnitude-frequency analysis to be attempted (Kochel and Baker, 1982; Knox, 1985). Compared with present flow regimes, however, the peak floods estimated for the Tekapo channels represent discharges 6.5 times greater than the 2500-year flood reported in 1878 (Ministry of Works and Development, Wellington, N.Z., pers. commun.).

Flood Channel Role in Terrace Formation

Terrace sequences have been attributed to a range of possible mechanisms that might trigger regional entrenchment, such as tectonic uplift, base-level fall, and complex response (Schumm, 1983; Begin and Schumm, 1984). The presence of the Tekapo and Skogasandur flood channels supports the hypothesis that these terrace sequences were formed, however, by single or successive flood events associated with hydrologic controls of the source glaciers. The observed relic flood channels were probably abandoned soon after their formation as the flood event ceased. Other flood channels, however, continued to be occupied and deepened, finally developing into the present incised meltwater stream courses. The paleodischarges estimated for the formation of the abandoned relic channels probably represent the minimum flow conditions required for incision of each terrace surface and which continued to be exceeded in the flood channels antecedent to the present river courses.

Implications for Non-Glacial Terrace Sequences

Since major flood events probably acted as the main incising agent in the formation of glacial terrace sequences, it seems likely that non-glacial terrace sequences were instigated by the floods exceeding the threshold for channel entrenchment during major flood events. Indeed, there are many well documented examples of channel entrenchment during large flood events (Gupta, 1983; Knox, 1983). Many non-glacial terrace deposits exhibit deeply incised channels associated with coarse sediment infills, suggesting that terrace formation in general may result from single flood events that exceed local geomorphic thresholds for channel stability. The actual conditions for incision may have varied through the channel system, particularly where different channel reaches exhibited different degrees of sensitivity to input changes or complex response (Begin and Schumm, 1984; Patton and Schumm, 1981; Schumm, 1983). Despite this within-channel variability, the minimum threshold flow conditions for terrace formation can be sought in the abandoned relic channels that remain on the preserved terrace surfaces.

Implications for Future Paleohydraulic Research

The paleohydraulic modeling techniques used are based on a number of questionable and unvalidated assumptions, and the models rely on limited empirical data and theoretical analysis of flow and sediment entrainment processes within a restricted range of coarse-grained fluvial environments. Hence, none of the methods can be expected to provide very accurate results, but the estimates based on the many different approaches do provide an envelope of the probable correct order of magnitude of the various paleoflow parameters.

This study highlights the areas of research that would benefit future paleohydraulic analysis. First, the models require further testing, particularly in relation to high-magnitude flow events in present steep, coarse-grained, and boulder-bed channels. Second, more precise estimates of error sources in the prediction of each paleoflow parameter are required, in relation to both errors in field and map determination of input parameters and errors associated with the individual models and their associated assumptions. Finally, the magnitude-frequency relations of these flow events need to be determined more precisely. This can be achieved both by extending these paleohydraulic studies to the sedimentary sequences underlying the terrace surfaces and by examining flood channel development within more precisely dated terrace sequences.

REFERENCES

American Association of Civil Engineers, 1966, Sediment transport mechanisms: Initiation of motion: Proceedings of the American Society of Civil Engineers, Journal of the Hydraulics Division, v. 92, p. 291–314.

Arnborg, L., 1955, Hydrology of the glacial River Austerfljot: Geografiska Annaler, v. 37, p. 185–201.

Ashida, K. and Bayazit, M., 1973, Initiation of motion and roughness of flows in steep channels: Proceedings, 15th Congress, International Association for Hydraulic Research, Instanbul, v. 1, p. B1–10.

Baker, V. R., 1973, Paleohydrology and sedimentology of Lake Missoula flooding in eastern Washington: Geological Society of America Special Paper 144, 79 p.

Bathurst, J. C., 1978, Flow resistance of large-scale roughness: Proceedings of the American Society of Civil Engineers, Journal of the Hydraulics Division, v. 104, p. 1587–1603.

———, 1985, Flow resistance estimation in mountain rivers: Proceedings of the American Society of Civil Engineers, Journal of the Hydraulics Division, v. 111, p. 625–643.

———, Li, R. M., and Simons, D. B., 1981, Resistance equation for large-scale roughness: Proceedings of the American Society of Civil Engineers, Journal of the Hydraulics Division, v. 107, p. 1593–1613.

Begin, Z. B., and Schumm, S. A., 1984, Gradational thresholds and landform singularity: Significance for Quaternary studies: Quaternary Research, v. 21, p. 267–274.

Björnsson, H., 1974, Explanation of jökulhlaups from Grimsvötn, Vatnajökull, Iceland: Jökull, v. 24, p. 1–24.

Bradley, W. C., and Mears, A. I., 1980, Calculations of flows needed to transport coarse fraction of Boulder Creek alluvium at Boulder, Colorado: Geological Society of America Bulletin, v. 91, p. 1057–1090.

Church, M., 1972, Baffin Island sandurs: A study of Arctic fluvial processes: Geological Survey of Canada Bulletin, v. 216, 208 p.

———, 1978, Palaeohydrological reconstructions from a Holocene valley fill, *in* Miall, A. D., ed., Fluvial Sedimentology: Canadian Society of Petroleum Geologists Memoir 5, p. 743–772.

———, and Gilbert, R., 1975, Proglacial fluvial and lacustrine environments, *in* Jopling, A. V., and McDonald, B. C., eds., Glaciofluvial and Glaciolacustrine Sedimentation: Society of Economic Paleontologists and Mineralogists Special Publication 23, p. 22–100.

———, and Kellerhals, R., 1978, On the statistics of grain size variation along a gravel river: Canadian Journal of Earth Sciences, v. 15, p. 1151–1160.

Clague, J. J., 1975, Sedimentology and paleohydrology of Lake Wisconsin outwash, Rocky Mountain Trench, southeastern British Columbia, *in* Jopling, A. V., and McDonald, B. C., eds., Glaciofluvial and Glaciolacustrine Sedimentation: Society of Economic Paleontologists and Mineralogists Special Publication 23, p. 223–237.

Costa, J. E., 1983, Paleohydraulic reconstruction of flash-flood peaks from boulder deposits in the Colorado Front Range: Geological Society of America Bulletin, v. 94, p. 986–1004.

———, 1984, Fluvial paleoflood hydrology (abs.): EOS Transactions, American Geophysical Union, v. 65, p. 890.

———, 1985, Interpretations of the largest rainfall-runoff floods measured by indirect methods on small drainage basins in the conterminous United States: US-PRC Bilateral Symposium on the Analysis of Extraordinary Flood Events, Nanjing, China. 42 p.

Day, T. J., 1978, Aspects of flow resistance in steep channels having particulate beds, *in* Davidson-Arnott, R., and Nickling, W., eds., Research in Fluvial Geomorphology: Geo Abstracts Limited, p. 45–58.

Gair, H. S., 1967, Geological map of New Zealand, Sheet 20-Mt. Cook: New Zealand Geological Survey, 1:250,000.

Gardner, T. W., 1983, Paleohydrology and paleomorphology of a Carboniferous, meandering fluvial sandstone: Journal of Sedimentary Petrology, v. 53, p. 991–1005.

Gupta, A., 1983, High-magnitude floods and steam channel response, *in* Collinson, J. D., and Lewin, J., eds., Modern and Ancient Fluvial Systems: International Association of Sedimentologists Special Publication 6, p. 219–227.

Henderson, F. M., 1966, Open Channel Flow: MacMillan, 522 p.

Hey, R. D., 1979, Flow resistance in gravel-bed rivers: Proceedings of the American Society of Civil Engineers, Journal of the Hydraulics Division, v. 105, p. 365–379.

———, and Thorne, C. R., 1983, Accuracy of surface samples from gravel bed material: American Society of Civil Engineers, Journal of Hydraulic Engineering, v. 109, p. 842–851.

Jarrett, R. D., 1984a, Hydraulics of high-gradient streams: American Society of Civil Engineers, Journal of Hydraulic Engineering, v. 110, p. 1519–1539.

———, 1984b, Evaluation of methods of estimating paleofloods of high-gradient streams (abs.): EOS Transactions, American Geophysical Union, v. 65, p. 893.

JONSSON, J., 1982, Notes on the Katla volcanological debris flows: Jokull, v. 32, p. 61–68.

KNOX, J., 1983, Response of river systems to Holocene climates, *in* Wright, H. E., Jr., ed., Late-Quaternary Environments of the United States: University of Minnesota Press, Minneapolis, v. 2, p. 26–41.

———, 1985, Responses of floods to Holocene climatic change in the upper Mississippi Valley: Quaternary Research, v. 23, p. 287–300.

KOCHEL, R. C., AND BAKER, V. R., 1982, Paleoflood hydrology: Science, v. 215, p. 353–360.

LIMERINOS, J. T., 1970, Determination of the Manning coefficient from measured bed roughness in natural channels: U.S. Geological Survey Water Supply Paper 1898-B, 47 p.

LORD, M. L., AND KEHEW, A. E., 1984, Paleohydraulic characterization of sudden proglacial lake discharges in the northern Great Plains (abs.): EOS Transactions, American Geophysical Union, v. 65, p. 893.

MAIZELS, J. K., 1983a, Palaeovelocity and palaeodischarge determination for coarse gravel deposits, *in* Gregory, K. J., ed., Background to Palaeohydrology: John Wiley and Sons Limited, p. 101–139.

———, 1983b, Channel changes, paleohydrology and deglaciation: Evidence from some late glacial sandur deposits of northeast Scotland: Quaternary Studies in Poland, v. 4, p. 171–187.

———, 1983c, Proglacial channel systems: Change and thresholds for change over long, intermediate and short time-scales, *in* Collinson, J. D., and Lewin, J., eds., Modern and Ancient Fluvial Systems: International Association of Sedimentologists Special Publication 6, p. 251–266.

———, AND DUGMORE, A. J., 1985, Lichenometric dating and tephrochronology of sandur deposits, Solheimajokull area, southern Iceland: Jokull, v. 35, p. 71–79.

MALDE, H. E., 1968, The catastrophic Late Pleistocene Bonneville flood in the Snake River Plain, Idaho: U.S. Geological Survey Professional Paper 596, 52 p.

MANSERGH, G., 1973, Quaternary of the MacKenzie basin: International Quaternary Association Guidebook, Section I, p. 102–111.

MATHEWS, W. H., 1972, Profiles of Late Pleistocene glaciers in New Zealand: New Zealand Journal of Geology and Geophysics, v. 10, p. 146–163.

NOLAN, K. M., AND DONNELLY-NOLAN, J. M., 1984, Catastrophic flooding related to a sub-glacial eruption at Medicine Lake Volcano, northeastern California (abs.): EOS Transactions, American Geophysical Union, v. 65, p. 893.

O'BRIEN, M. P., AND RINDLAUB, B. D., 1934, The transportation of bedload by streams: American Geophysics Union Transactions, v. 15, p. 593–602.

PATTON, P. C., AND SCHUMM, S. A., 1981, Ephemeral-stream processes: Implications for studies of Quaternary fills: Quaternary Research, v. 15, p. 24–43.

SCHUMM, S. A., 1983, River morphology and behaviour: Problems of extrapolation, *in* Elliot, C. M., ed., River Meandering: Waterway, Port, Coastal, and Ocean Division, American Society of Civil Engineers, New Orleans, p. 16–29.

SCOTT, K. M., AND GRAVLEE, G. C., 1968, Flood surge on the Rubicon River, California—Hydrology, hydraulics and boulder transport: U.S. Geological Survey Professional Paper 422-M, 38 p.

SHIELDS, A., 1936, Anwendung der Ahnlichkeitsmechanik und der Turbulenzforschung auf die Geshiebebewegung. (Application of similarity principles and turbulence research to bedload movement, Translated by Ott, W. P. and Van Uchelen, J. C.): U.S. Department of Agriculture, Soil Conservation Service Cooperative Laboratory, California Institute of Technology, 70 p.

SILBERMAN, E., CARTER, R. W., EINSTEIN, H. A., HINDS, J., AND POWELL, R. W., 1963, Friction factors in open channels: Progress Report Task Force on friction factors in open channels of the Committee on Hydromechanics of the Hydromechanics Division: Proceedings of the American Society of Civil Engineers, Journal of the Hydraulics Division, v. 69, p. 97–143.

SPEIGHT, J. G., 1961, Late Pleistocene historical geomorphology of the Lake Pukaki area, New Zealand: New Zealand Journal of Geology and Geophysics, v. 6, p. 160–188.

STRICKLER, A., 1923, Beitrage zur Frage der Geschwindigkeitsformel und der Rauhigkeitsamen fur Strome, Kanale und geschlossene Leitungen: Mitteilungen des Amts für Wasserwirtschaft der Schweiz Bern No. 16, 77 p.

THOMPSON, S. M. AND CAMPBELL, P. L., 1979, Hydraulics of a large channel paved with boulders: Journal of Hydraulics Research, v. 17, p. 341–354.

THORNE, C. R., AND ZEVENBERGEN, L. W., 1985, Estimating mean velocity in mountain rivers: American Society of Civil Engineers, Journal of Hydraulic Engineerinig, v. 111, p. 612–624.

WILLIAMS, G. P., 1983, Paleohydrological methods and some examples from Swedish fluvial environments, I—Cobble and boulder deposits: Geografiska Annaler, v. 65A, p. 227–243.

———, 1984, Paleohydrologic equations for rivers, *in* Costa, J. E., and Fleisher, P. J., eds., Developments and Applications of Geomorphology: Springer-Verlag, Berlin, p. 343–367.

PALEOHYDROLOGY OF SOME OGALLALA (NEOGENE) STREAMS IN THE SOUTHERN PANHANDLE OF NEBRASKA

R. G. GOODWIN
Department of Geology, University of Nebraska-Lincoln 68588-0340
AND
R. F. DIFFENDAL, JR.
Conservation and Survey Division, IANR, University of Nebraska-Lincoln 68588-0517

ABSTRACT: Stratification and estimated paleoflow conditions for valley-fill deposits suggest that Ogallala Group streams in western Nebraska were similar to modern streams of south-central Alberta.

Ogallala stratification includes medium-scale (0.5 to 2.0 m thick) trough crossbedded sand and gravel, tabular indistinctly horizontally bedded and imbricated gravel, and horizontally bedded sand and pebbly sand. Valley fills are 15 to 55 m thick and 800 to 1800 m wide at the top. Some are in bedrock-floored channels resembling the "inner channels" of Shepherd and Schumm (1974).

Gradients for three well exposed paleovalley floors range from 0.0014 to 0.0020 (m/m) after tectonic correction. This compares with 0.00135 for the modern North Platte River Valley in Nebraska. The average intermediate diameters of the 10 largest clasts from tabular gravel beds found at 17 sites varied between 0.077 and 0.15 m. The average median intermediate diameter for gravel from four well exposed tabular gravel beds is 0.024 m.

Consistent paleodepth estimates of about 2 m correlate well with the scale of cross-stratification observed in the valley fills. Paleovelocities are estimated at about 2 m/sec, and Froude numbers of about 0.4 are consistent with a lower flow regime in the stability field of dunes. Two-dimensional specific in-channel paleodischarges were 3 to 4 m^2/sec. Total paleodischarge estimates based on slope-discharge relationships for gravel-bed rivers range from 340 to 1240 m^3/sec and are comparable to average annual peak discharges on the North Platte River reported 80 to 90 years ago.

Ogallala streams were probably dominated by macroforms similar to the "crescent-shaped bars" of the North Saskatchewan River. Deposition also took place on longitudinal bars in deeper channels. Shallow upper-flow regime transport and deposition is recorded by horizontally bedded sand and pebbly sand.

INTRODUCTION

The Ogallala Group (Neogene) is a continental deposit of markedly variable thickness that covers most of the southern third of the Nebraska panhandle (Fig. 1). The sediments and sedimentary rocks of the group are dominantly epiclastic and alluvial. Airfall volcanic ash, lacustrine diatomaceous beds, and pedogenic calcareous and siliceous horizons occur within the sequence. Source areas for the epiclastic component were primarily in the Southern Rocky Mountains of southern Wyoming and northern Colorado, but some sediments were derived locally in Nebraska from stream erosion of older Tertiary deposits.

The purposes of this paper are to review briefly the stratigraphic and sedimentologic work done on the Ogallala Group in western Nebraska since 1970, to provide site-specific data on the sediments of individual valley fills and the paleovalleys they occupy where possible, and to provide insight into the nature and behavior of the fluvial systems that produced the group.

PREVIOUS WORK

Robinson (1970) described the Ogallala Group in general terms for parts of Wyoming, Colorado and western Nebraska and demonstrated its complexity. Since his work, regional syntheses have confirmed his general outline of Ogallala geology. Izett (1975) and Scott (1975) have discussed the geology of Cenozoic deposits including the Ogallala and its equivalents in Colorado and adjacent areas. Blackstone (1975) prepared a similar treatment for the Cenozoic rocks of the Laramie Basin of Wyoming. Swinehart and others (1985) described the Cenozoic paleogeography of western Nebraska including the Ogallala.

Historical reviews have been prepared by Schultz (1977) for the Ogallala in general and by Diffendal (1984) for the Ogallala primarily in Nebraska. These works contain references to most of the pre-1970 reports done on the unit across the Great Plains and descriptions of the concepts used earlier to explain its nature and mode of deposition.

General descriptions and site-specific studies of Ogallala paleovalleys and valley fills of Nebraska and adjacent areas have been published by Skinner and others (1977), Swinehart (1979), Scott (1982), Diffendal (1982, 1983, 1985), and Diffendal and others (1985).

Alluvial sediment dispersal patterns, stream drainage patterns and types, and general variations in stream competence through time for Ogallala and younger deposits in western Nebraska have been the subjects of a few studies. Stanley (1971a, b) and Stanley and Wayne (1972) concluded that source areas and dispersal paths for sediments carried by Ogallala streams were different from those of Early Pleistocene streams and that Ogallala streams were not as competent as their Early Pleistocene counterparts. Stanley (1976) studied variations in sandstone petrofacies for the Ogallala and other units in southeastern Wyoming and adjacent parts of Colorado and western Nebraska. Swinehart (1979) and Diffendal (1982) used Stanley's techniques to separate most Ogallala gravels from younger gravels and to trace individual paleodrainages. Breyer (1975) used Markov-chain analysis to conclude that Ogallala streams were braided.

Most Ogallala Group sediments have undergone little diagenesis since deposition and are slightly cemented. Incipient diagenetic changes can be attributed to ". . . pedogenic processes, evaporative concentration of groundwater in the capillary zone, or by groundwater reaction with sediment" (Stanley and Benson, 1979, p. 401).

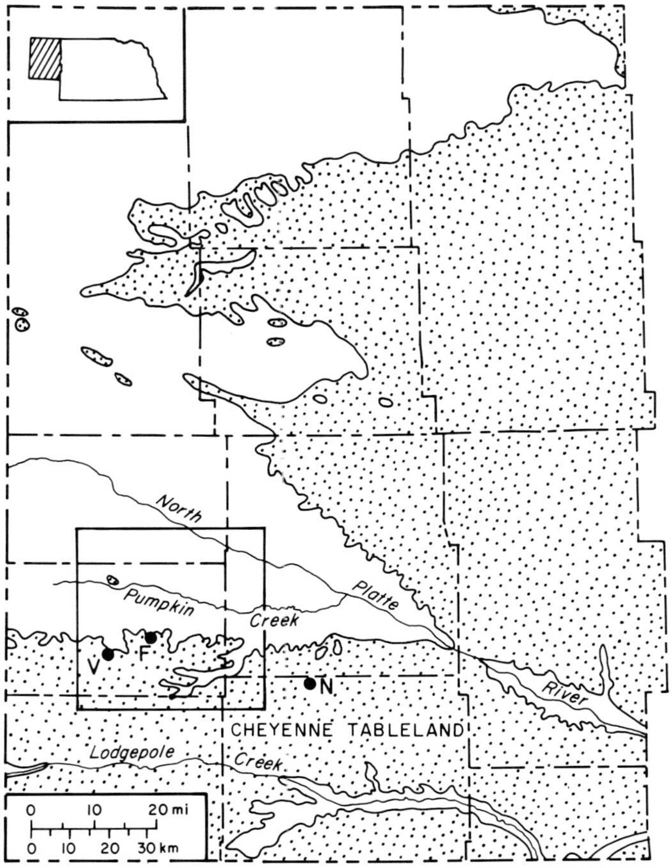

FIG. 1.—Extent of Ogallala Group (stippled) in the Nebraska panhandle. Modified after Swinehart and others (1985) with permission of the Rocky Mountain Section of the Society of Economic Paleontologists and Mineralogists. Black rectangle is the area covered by the satellite image in Figure 2. F is Faden paleovalley, sites 2 and 22; N is Nelson paleovalley, site 6; V is Van Pelt paleovalley, site 20. Site locations are listed in Table 1.

CHARACTERISTICS OF OGALLALA FLUVIAL SYSTEMS

Quaternary erosion of Neogene deposits of the Nebraska panhandle has revealed the nature of both Ogallala paleovalleys and valley-fill deposits. Diffendal and others (1985) have described two paleovalleys filled with Ogallala sediments. The valley wall slopes are a function of the composition of the walls. Valleys cut into siltstones of the Paleogene Brule Formation have slopes of at least 55°. Valleys with coarser grained walls have gentler slopes. Tributary valleys have preserved depths as great as 50 m and widths of more than 1 km (Diffendal, 1982).

Ogallala paleovalley and paleovalley-fill sites examined in this study are listed in Table 1. Sites 1 through 17 were sampled to find the 10 largest distantly derived clasts. Two of these sites and two additional sites were sampled to obtain a measure of the median grain size of exposed massive gravel beds (sites 18 through 21, Table 1). Site 22 was studied solely to examine stratification.

Paleovalley characteristics.—

Ogallala paleovalleys are either alluvial or non-alluvial (Table 1). Prestegaard (1984) has emphasized the importance of non-alluvial controls on stream channel morphology. Carling (1983) has shown that resistance exerted by the walls of narrow bedrock valleys will influence sediment transport. The paleovalley at site 2 divides into two narrow subvalleys, each between 100 and 120 m wide, around a bedrock island that is capped by Ogallala sediments. This arrangement of valley elements is similar to the "inner channels" described by Shepherd and Schumm (1974). At all other locations, estimates of the widths of paleovalleys are based on the widths of preserved valley-fill deposits. Three of the paleovalleys (sites 2, 6, and 20) have floors exposed over distances of from 1.3 to 18 km. The slopes of these floors were used in paleohydrologic analyses and are listed with valley parameters for streams presently in the study region (Table 2).

TABLE 1.—LOCATION OF STUDY SITES

Site	Decription		Type
1	NE 1/4 SE 1/4 NW 1/4	Sec 33, T18N, R50W	NA
2	NW 1/4 SW 1/4 SW 1/4	Sec 8, T18N, R54W	NA
3	SE 1/4 NW 1/4 SW 1/4	Sec 27, T18N, R48W	A
4	SE 1/4 NE 1/4 NW 1/4	Sec 33, T18N, R48W	A
5	C NE 1/4 SE 1/4	Sec 32, T17N, R50W	A
6	SW 1/4 NE 1/4 SW 1/4	Sec 21, T17N, R50W	A
7	SW 1/4 SE 1/4	Sec 27, T17N, R51W	NA
8	SW 1/4 SE 1/4	Sec 19, T18N, R51W	NA
9	NW 1/4 NE 1/4	Sec 36, T18N, R52W	NA
10	NE 1/4 NW 1/4	Sec 31, T17N, R53W	NA
11	C S 1/2	Sec 30, T17N, R53W	NA
12	SW 1/4	Sec 25, T17N, R54W	NA
13	SW 1/4 NW 1/4 SW 1/4	Sec 35, T17N, R54W	NA
14	SE 1/4 NE 1/4	Sec 5, T16N, R54W	NA
15	NE 1/4 NW 1/4	Sec 9, T18N, R54W	NA
16	NE 1/4 SW 1/4 NW 1/4	Sec 15, T18N, R55W	NA
17	N 1/2 NW 1/4 NE 1/4	Sec 13, T18N, R55W	NA
18	NW 1/4 SW 1/4 SW 1/4	Sec 8, T18N, R54W	NA
19	SW 1/4 NE 1/4 SW 1/4	Sec 21, T17N, R50W	A
20	NW 1/4 NE 1/4 NW 1/4	Sec 25, T18N, R56W	NA
21	SW 1/4 NW 1/4 SW 1/4	Sec 19, T14N, R46W	A
22	NE 1/4 NE 1/4 NE 1/4	Sec 18, T18N, R54W	NA

NA = Nonalluvial valley; A = Alluvial valley

Paleovalley-fill characteristics.—

Individual valley-fill deposits, seen as distinctive erosional remnants (Fig. 2), vary between 800 and 1800 m in width, between 15 and 55 m in preserved thickness, and occupy non-alluvial paleovalleys. The relatively great width of these paleovalleys probably rendered non-alluvial control of sediment transport and deposition important only near steep valley walls. Similarities in grain size and stratification observed at alluvial and non-alluvial sites and summarized below suggest this is true.

TABLE 2.—COMPARISON OF MODERN LAND SURFACE AND STREAMS TO PALEOVALLEYS AND FILLS

Topographic Feature	Slope (m/m)	Width (m)	Depth (m)
LODGEPOLE CREEK VALLEY	0.0027	7000	60
PUMPKIN CREEK VALLEY	0.0046	16000	200
NORTH PLATTE VALLEY	0.00135	25000	180
CHEYENNE TABLELAND	0.0027		
FADEN PALEOVALLEY	0.0041	1800	55
VAN PELT PALEOVALLEY	0.0043	1100	25
NELSON PALEOVALLEY	0.0047	800	12

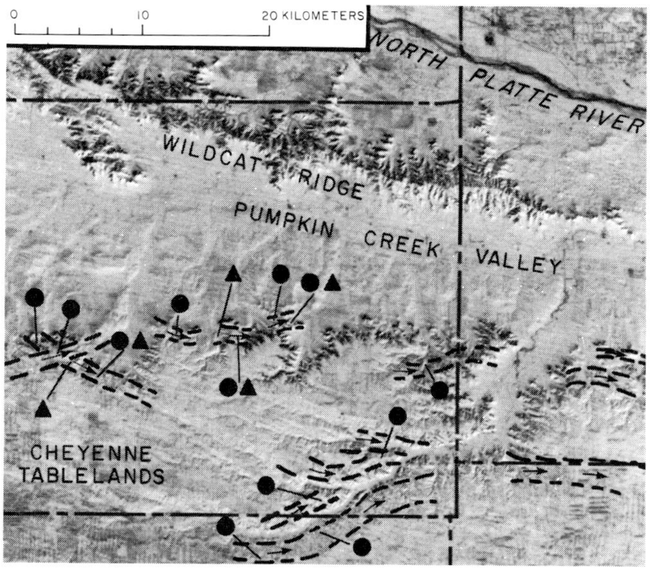

FIG. 2.—Satellite image of the area noted in Figure 1. Dashed lines mark sides of Ogallala valley fills. Arrows point in the direction of flow. Triangles are locations of locally derived megaclast blocks; circles are megaclast boulders.

Locally derived megaclasts are present in Ogallala valley-fill deposits. Rounded clasts as much as 2 m in largest exposed dimension were observed in the troughs of crossbedded sands and gravels (Fig. 3). They are similar to the large intraclasts described by Gibling and Rust (1984) from the Pennsylvanian Morien Group of Nova Scotia. Megaclast rounding, similar to the shape modification of some Morien intraclasts, may have been produced by undercutting and overturning of each clast while the megaclast surface was "sandblasted" by fluvial sand and gravel. The megaclasts were excluded from subsequent paleohydrologic analyses because of the uncertainties about their transport histories and the mechanisms responsible for their transport (see Miall, 1976, for a paleohydraulic treatment of similar clasts).

Granitic, metamorphic, and volcanic pebbles from the Front, Laramie, and possibly the Medicine Bow and Park Ranges of Colorado and southern Wyoming are common in Ogallala gravels. Rhyolite-bearing feldspathic sands and plagioclase sands comprise most Ogallala sandstones (Stanley, 1971a, b, 1976; Swinehart, 1979; Diffendal, 1982).

Bedding and sedimentary structures were impossible to observe at many sites where Ogallala deposits are either very poorly cemented or uncemented. Well indurated sedimentary rocks with a very fine-grained calcite cement occur at sites 2, 22, and 20 (Fig. 3A, B, C). All three of these valley fills are close to bedrock valley walls. Sites 6 and 21 (Fig. 3D, E) are not, and they are very poorly cemented or non-cemented. Restricted groundwater circulation in rock-walled valleys may have affected cement development.

Locally derived sediment covers the floors of many of the Ogallala paleovalleys. Fine, massive gravel composed of locally derived and fluvially reworked calcareous concretions from the underlying Brule Formation occurs at site 20 (Fig. 3C). Similar sediments cover floors of other Cenozoic paleovalleys in the area (Diffendal and others, 1985). This suggests that valley-cutting occurred prior to incorporation of valleys into regional drainage systems.

Several bedding styles and sedimentary structures were observed in Ogallala valley fills (Fig. 3). Fines are very rare in these deposits, but a bed of laminated fine sand and silt 1 m thick was observed at site 21 (Fig. 3E). This unit was probably deformed by vertebrates before the emplacement of overlying gravels (Loope, pers. commun.).

Horizontally bedded sands and fine gravels with a few intercalated scour-and-fill structures predominate at site 20 (Fig. 3C). These sediments were probably deposited from shallow upper flow regimes and are similar to sandy proximal braided stream deposits (Rust, 1978). Horizontally bedded sand was the principal bedding type formed by the 1965 Bijou Creek flood (McKee and others, 1967). The two horizontally bedded sand and gravel units separated by the erosional bounding surface shown in the lower part of Figure 3C have maximum preserved thicknesses of 1.5 to 2 m. Typical thicknesses for horizontally bedded sands deposited during the Bijou Creek flood were just under 1 m, but thicknesses of nearly 4 m were recorded (McKee and others, 1967).

Ten- to 40-cm-thick trough crossbeds of sand and fine gravel at sites 2 and 20 (Figs. 3A, 3C) also occur in many fluvial depositional systems and are very common in braided stream environments (Ore, 1964; Miall, 1977; Rust, 1978).

Beds of massive to imbricated and indistinctly normally graded, medium to coarse pebbles are present at site 2 (Fig. 3A). Each bed is nearly 1 m thick, and similar beds were observed at sites 19, 20, and 21. Foreset bedding was not observed in any of these units, and we believe that the beds were deposited as longitudinal bars. The largest distantly derived clasts always occurred in massive beds.

Medium-scale trough crossbedded sands and gravels with preserved trough-fill sequences as much as 1.5 m thick occur at sites 6, 21, and 22 (Figs. 3D, 3E, 3B). The lower unit of Figure 3A appears to consist of planar crossbedding in the section available, but nearby exposures revealed features of similar scale to be trough crossbedded. The very large locally derived megaclasts of siltstone or sandstone discussed previously are associated with many of the troughs and massive gravels.

Troughs exposed at gravel pits at sites 6 and 21 (Figs. 3D, 3E) reveal the internal structure of large Ogallala troughfills. Coarse and medium pebbles occur both as lag deposits along the base of individual troughs and as discrete graded crossbeds 5 to 10 cm thick that have either openwork textures or a medium sand matrix. Gravel crossbeds are separated by single or stacked, 2- to 5-cm-thick sandy graded crossbeds. Granule-size clasts are generally concentrated at the base of these crossbeds, and grain size fines upward to medium sand at the top. Floating pebbles are common in the sandy crossbeds.

Trough crossbedding composed exclusively of coarse gravel- to cobble-size clasts was observed in exposures of indurated Ogallala valley-fills (Figs. 3A, 3B). Set thicknesses are as great as 1.5 to 2 m. Individual crossbeds are

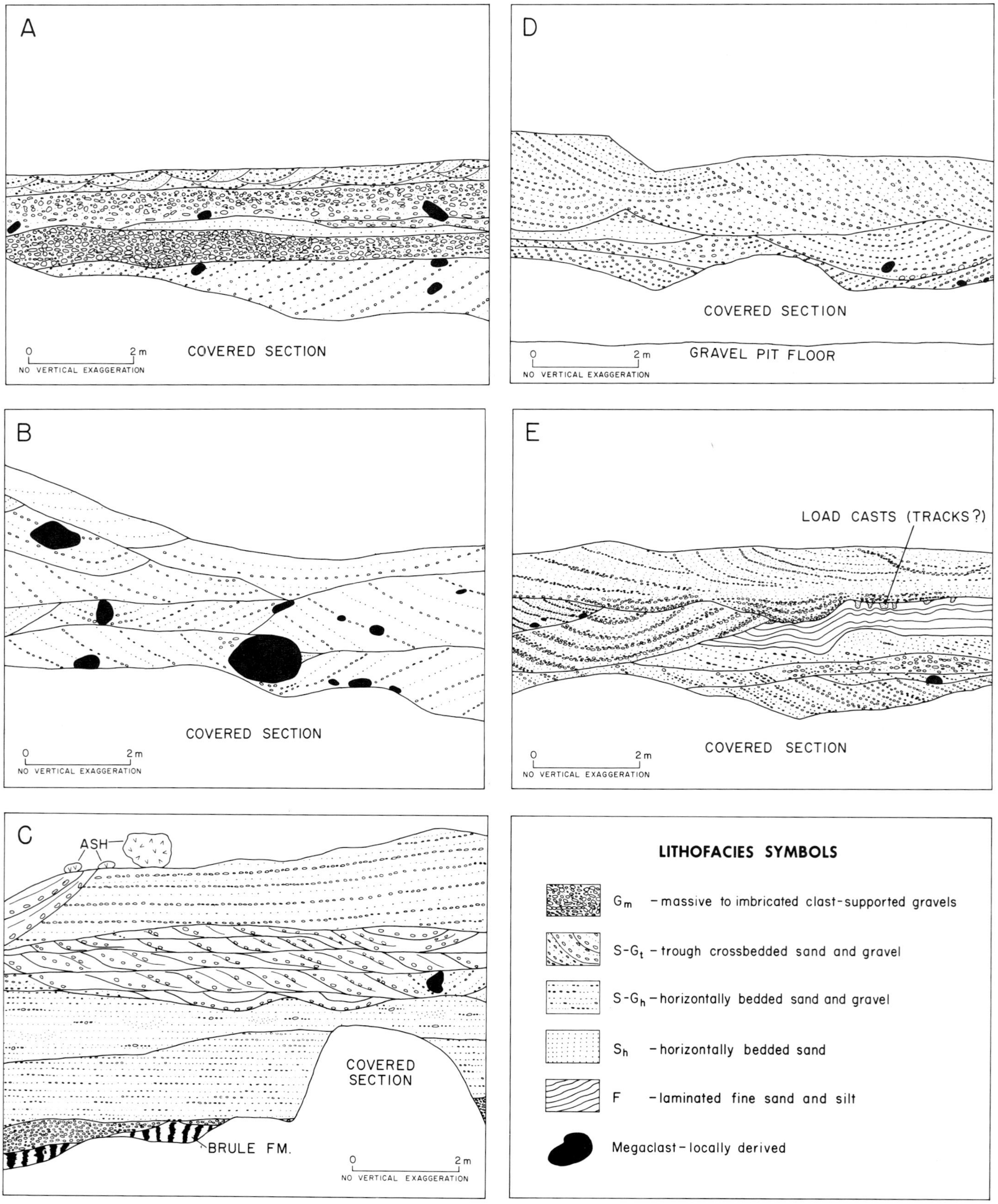

FIG. 3.—Scale outcrop maps of lithified fluvial sequences in rock-walled valley (A-C, sites 2, 22, and 20) and unconsolidated sediments in alluvial valleys (D and E, sites 6 and 21). See Table 1 for site locations.

very poorly sorted and appear massive or indistinctly normally graded. Differences in overall grain size between adjacent crossbeds allow them to be identified.

The stratification types present in the Ogallala suggest deposition in braided stream systems. Kraus (1984) described a G_t assemblage for her lithosome *F* of the Willwood Formation of Wyoming that is very similar to the collection of lithofacies described above for the Ogallala. She compared the G_t assemblage to the upper mid-fan sediments of the Scott outwash fan (Boothroyd and Ashley, 1975). The Ogallala sediments also resemble those of the Donjek-type braided stream (Williams and Rust, 1969; Rust, 1972; Miall, 1977, 1978). This is consistent with the paleogeographic position of Ogallala streams in western Nebraska. They were beyond proximal longitudinal bar-dominated braided streams that may have existed near the Southern Rocky Mountain front. Ogallala streams of western Nebraska were also quite different from the modern South Platte and Platte River systems which are dominated by deposition of sandy planar-tabular crossbeds and small-scale sandy trough crossbeds (Smith, 1971; Blodgett and Stanley, 1980; Crowley, 1983).

PALEOHYDROLOGY

Paleoflow parameters were estimated for sampled sites in Table 1. Valley slopes were corrected for tectonic tilting and were used as estimates of the energy grade lines for Ogallala streams. Values of D_{10x}, the average intermediate diameter of the 10 largest clasts, were used to infer flow magnitudes.

Methods and results.—

A number of methods are available for the estimation of paleoflow conditions for coarse alluvial sediments (Baker and Ritter, 1975; Bradley and Mears, 1980; Maizels, 1983; Costa, 1983; Williams, 1983; Baker, 1984; Steer and Abbott, 1984). Most are based on estimation of the tractive force necessary to move the largest clasts transported by a stream-flow event.

The methods we employ are based on threshold of motion estimates. Results must be regarded as qualitative (Maizels, 1983; Church, pers. commun.), and they apply, strictly speaking, only to gravel-floored riffle stretches of Ogallala streams.

Several authors have investigated the conditions necessary to entrain single coarse clasts from both natural and artificial stream beds. Parker and others (1982) and Andrews (1983) have studied this problem for bedload transport in natural flows having mixed particle-size beds. In each of these works, the Shield's parameter (dimensionless critical shear stress parameter) is found using a power relationship of the form $\tau^*_{ci} = a(D_i/D_{50})^b$, where D_i = diameter of the *i*th percentile of the bed surface material and D_{50} = the diameter of the median subsurface material, both in meters. Both papers present values of the constant, *a*, and the coefficient, *b*, that are similar. We used the relationship of Andrews (1983), $\tau^*_{ci} = 0.0834\ (D_i/D_{50})^{-0.872}$, because it was derived from large data sets from three rivers with the western United States.

D_{50} was evaluated by a modified grid-by-number sampling technique (Kellerhals and Bray, 1971). At sites 18–21 (Table 1), a horizontal line was placed along the center of a well exposed massive gravel bed. The exposed intermediate diameters of 50 pebbles located at 10-cm intervals were measured and ranked by size. The diameter of the 25th pebble in the ordered set was taken as a measure of D_{50} for the site. The average for all four sites was 0.024 ± 0.002 m where the plus-minus indicates the range in values. Using this average and values of D_{10x} for D_i in the equation of Andrews (1983), an average value for the dimensionless critical shear-stress parameter of 0.021 ± .006 was derived for sites 1–17. A shear-stress value of 0.02 was used for these sites in all subsequent calculations. Unique values rather than an average were used for sites 18–21.

Paleodepth estimates were calculated from the relationship $d_c = \tau^*_{ci}[(\gamma_s - \gamma)/\gamma]D/S$, where d_c = critical flow depth required to transport a clast of diameter, *D* (D_{10x} was used for this value); γ_s = the specific weight of sediment, taken as 2650 kg/m^3 and γ = the specific weight of water, taken as 1000 kg/m^3 for clear water and 1250 kg/m^3 for a suspended sediment concentration of 10 percent; and *S* = the slope of the energy grade line, taken as valley slope. The use of two values for the specific weight of water follows the methods of Maizels (1983) to account for uncertainties in the suspended sediment content of flows in Ogallala streams.

We have also used two values for the valley slope in estimating flow depths at each site. One of us (RFD) has noted that many Ogallala valley fills in western Nebraska are erosionally truncated to the west and buried by younger sediments to the east. This suggests that tectonic tilting of the area has taken place since deposition of Ogallala sediments. In order to correct for this post-depositional tilting, we constructed an uplift curve (Fig. 4) based on a map of crustal movement during the past 10 million years in the High Plains and Southern Rocky Mountain regions (Gable and Hatton, 1983). Post-depositional tilting of the study area in the direction of paleovalley elongation is estimated at 0.0027 (m/m). By subtracting this from the present paleovalley slopes (Table 2), estimates of slopes at the time of deposition of 0.0014, 0.0017, and 0.0020 (m/m) were calculated. These are comparable to a slope of 0.00135 for the modern North Platte River Valley in western Nebraska. Both the high and low values were used in the estimation of Ogallala stream-flow conditions. Each site provided four estimates of depth, one each for the different combinations of valley slope and specific weight of water. Flow depths in Ogallala streams at the time of re-entrainment of large distantly derived clasts from riffle segments were between 1.6 and 2.6 m, with the most likely depths near 2 m (Table 3).

Paleovelocity was estimated using equations of Bray (1979), who used data gathered from 67 gravel-bed rivers in Alberta to empirically derive statistically significant relationships between water surface slope and both Manning's *n* (r^2 = 0.298) and $1/\sqrt{f}$ (r^2 = 0.473), where *f* is the widely used dimensionless friction factor. Substitution of these expressions into the original Manning equation and the Darcy-Weisbach friction equation yields power equa-

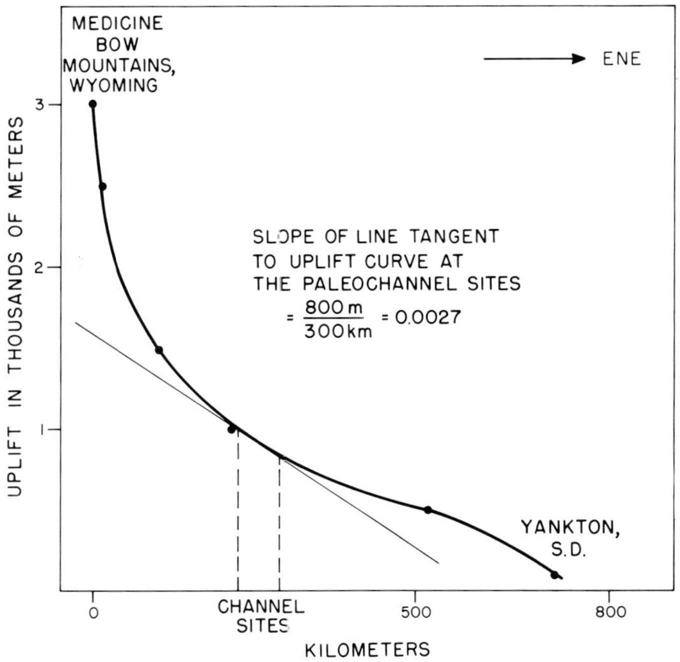

FIG. 4.—Uplift curve based on data from Gable and Hatton (1983). Line tangent to the curve at the channel sites has a slope of 0.0027 (m/m). See text for explanation.

tions similar to the Lacey equation, where $V = ad^bS^c$. An expression for the best fit of the Lacey equation to the Alberta data [Bray, 1979, see equation (41)] was also used. Richards (1982) has commented that such power equations are a limited representation of the general logarithmic resistance equation and are useful only within confined grain size conditions. We feel justified in using Bray's expressions for Ogallala valley fills because of the similarities in both valley slope and grain size between the Alberta rivers and the paleovalley fills.

The following equations were used to estimate paleovelocity:

$$1.\ V_1 = 9.62\, d^{0.667}S^{0.323} \text{ (Manning equation)} \quad (1)$$

$$2.\ V_2 = 6.16\, d^{0.5}S^{0.244} \text{ (Darcy-Weisbach equation)} \quad (2)$$

$$3.\ V_3 = 8.0\, d^{0.60}S^{0.29} \text{ (Lacey equation),} \quad (3)$$

where d = flow depth and S = water surface slope. Each of the two slope estimates was matched with the four depth estimates for the 21 sites (Table 3). Flow velocities were in the range of 1.5 to 2.3 m/sec with the most likely velocities slightly less than 2 m/sec.

Estimates of specific in-channel discharge, a sort of two-dimensional discharge (units of m^2/sec; Maizels, 1983), were calculated for each site by multiplying the average paleodepth estimate by the most commonly occurring average paleovelocity estimate (Table 3). In cases where channel width can be estimated, it can be multiplied by specific in-channel discharge to obtain total in-channel discharge. Specific in-channel discharges for Ogallala streams are between 2.1 and 5.7 m^2/sec with the most likely values between 3 and 4 m^2/sec (Table 3).

Estimates of total paleodischarge for Ogallala streams are difficult to make. From the limited outcrops available, it is impossible to reconstruct the total across-valley area of the Ogallala paleovalleys that may have been occupied by riffled channels at any given time. Even if this information were available, there is no way of calculating paleodischarge across the three-dimensional bedforms that must have been present outside the channels and that produced the trough crossbeds preserved in the valley fills (Fig. 3).

Steer and Abbott (1984) presented several equations to estimate bankfull or dominant (channel-forming) discharge for gravel-bed streams. All are power relationships based on slope. Application of equations (32), (34), and (36) of Steer and Abbott (1984, p. 38) to the slopes 0.0014 and 0.0020 (m/m) results in paleodischarge estimates of 340 to 1240 m^3/sec. Williams (1975) suggested that average an-

TABLE 3.—RESULTS OF THE PALEOHYDROLOGIC ANALYSES

Site	D_{10x}(m)	d_c(m)	V_1(m/s)	V_2(m/s)	V_3(m/s)	Specific Discharge (m^2/s)	Froude Number
1	0.113 ± 0.015	2.0 ± 0.6	1.8 ± 0.2	1.9 ± 0.3	1.9 ± 0.2	3.8 ± 1.8	0.41
2	0.124 ± 0.011	2.2 ± 0.6	1.9 ± 0.2	2.1 ± 0.3	2.0 ± 0.3	4.4 ± 1.8	0.43
3	0.087 ± 0.006	1.6 ± 0.4	1.6 ± 0.2	1.6 ± 0.2	1.6 ± 0.2	2.6 ± 0.8	0.40
4	0.115 ± 0.010	2.0 ± 0.6	1.8 ± 0.2	2.0 ± 0.2	1.9 ± 0.3	3.8 ± 1.2	0.43
5	0.116 ± 0.007	2.0 ± 0.6	1.9 ± 0.2	2.0 ± 0.2	1.9 ± 0.2	3.8 ± 1.2	0.43
6	0.077 ± 0.010	1.4 ± 0.4	1.5 ± 0.2	1.5 ± 0.2	1.5 ± 0.2	2.1 ± 0.6	0.40
7	0.109 ± 0.016	2.0 ± 0.6	1.8 ± 0.2	1.9 ± 0.3	1.9 ± 0.2	3.8 ± 1.8	0.43
8	0.099 ± 0.009	1.8 ± 0.6	1.7 ± 0.2	1.8 ± 0.2	1.8 ± 0.2	3.2 ± 1.2	0.43
9	0.100 ± 0.008	1.8 ± 0.6	1.7 ± 0.2	1.8 ± 0.2	1.8 ± 0.2	3.2 ± 1.2	0.43
10	0.114 ± 0.009	2.0 ± 0.6	1.8 ± 0.2	1.9 ± 0.3	1.9 ± 0.3	3.8 ± 1.8	0.43
11	0.127 ± 0.014	2.2 ± 0.6	1.9 ± 0.2	2.1 ± 0.3	2.0 ± 0.3	4.6 ± 1.8	0.43
12	0.117 ± 0.018	2.0 ± 0.6	1.9 ± 0.2	2.0 ± 0.3	2.0 ± 0.2	4.0 ± 1.8	0.45
13	0.118 ± 0.022	2.2 ± 0.6	1.9 ± 0.2	2.0 ± 0.3	2.0 ± 0.2	4.4 ± 1.8	0.43
14	0.121 ± 0.016	2.2 ± 0.6	1.9 ± 0.2	2.0 ± 0.3	2.0 ± 0.2	4.4 ± 1.8	0.43
15	0.107 ± 0.005	2.0 ± 0.6	1.8 ± 0.2	1.9 ± 0.2	1.9 ± 0.2	3.8 ± 1.2	0.43
16	0.135 ± 0.014	2.4 ± 0.6	2.0 ± 0.2	2.2 ± 0.3	2.1 ± 0.3	5.0 ± 1.8	0.43
17	0.150 ± 0.036	2.6 ± 0.8	2.1 ± 0.2	2.3 ± 0.3	2.2 ± 0.3	5.7 ± 2.4	0.44
18	0.104 ± 0.015	1.9 ± 0.5	1.7 ± 0.2	1.8 ± 0.3	1.8 ± 0.2	3.4 ± 1.5	0.42
19	0.053 ± 0.008	1.9 ± 0.5	1.8 ± 0.2	1.9 ± 0.2	1.8 ± 0.2	3.4 ± 1.0	0.42
20	0.056 ± 0.008	2.0 ± 0.5	1.8 ± 0.2	1.9 ± 0.3	1.9 ± 0.2	3.8 ± 1.5	0.43
21	0.049 ± 0.004	1.7 ± 0.5	1.7 ± 0.2	1.8 ± 0.2	1.8 ± 0.2	3.1 ± 1.0	0.44

Expressions for d_c, V_1, V_2, V_3, specific discharge and Froude number are found in the text.

nual peak flows of the North Platte River at North Platte, Nebraska were between 450 and 650 m^3/sec near the turn of this century. These historic discharges were probably subject to upstream control of flow (Bentall, 1982), and pre-control peak annual flows may have been greater than the values cited by Williams (1975).

Average depth and velocity estimates were also used to calculate Froude numbers for in-channel flow over riffles (Table 3). All sites have Froude numbers between 0.40 and 0.45. These suggest that flows remained in the stability field of dunes in the lower flow regime (Guy and others, 1966). This result agrees well with field observation of trough crossbedding in Ogallala valley fills.

DISCUSSION

The high and low ranges for the values of the flow conditions estimated for sites 1 through 17 (Table 3) are for sites with D_{10x} significantly higher or lower than the average for the 17 sites. The sites that produced high and low estimates have estimated dimensionless critical shear-stress parameters that are different than 0.02. Paleoflow estimates calculated for sites 18 through 21, where a unique dimensionless critical shear-stress parameter was established for each site, are very similar despite differences in D_{10x} between the sites. Clearly, the use of a single dimensionless critical shear-stress parameter for all sites is unsatisfactory. Our estimates of paleodepth, paleovelocity, and specific in-channel paleodischarge fall into a relatively narrow range only because of the relative uniformity in D_{10x} between sites (note that we have assumed a narrow range for D_{50}, as well).

In order to explore the possibility that tectonic tilting of the study site has not occurred, paleoflow estimates were made for site 5 using slope values of 0.0041 and 0.0047. All other factors were kept the same. This scenario assumes that erosion of Ogallala paleovalleys to the west and their burial to the east represent modification of a Miocene depositional surface strictly by erosion and deposition. Estimated depth was 0.8 ± 0.1 m, velocity was 1.4 ± 0.2 m/sec, and specific in-channel discharge was 1.1 ± 0.2 m^2/sec. The flow depth seems too shallow to accommodate the large crossbeds described above. The calculated Froude number was 0.50, which suggests transition to a plane-bed stage rather than a stable dune-stage bed.

Estimation of bankfull stage or dominant paleodischarge using the methods of Steer and Abbott (1984) and paleovalley slopes of 0.0041 and 0.0047 gives results of 42 to 84 m^3/sec. It seems likely that the Ogallala valley-fill sediments were transported and deposited by streams of greater size.

SUMMARY AND CONCLUSIONS

Paleoflow depths estimated for riffled channels of Ogallala streams using valley slopes of 0.0014 and 0.0020 (m/m) are close to 2 m. This compares favorably with the observation that maximum preserved thicknesses of individual trough-scour fills (Fig. 3) are 1.5 to 2 m. Paleodepths calculated using uncorrected valley slopes are 0.8 m, about one-half the value suggested by stratification. Velocities slightly less than 2 m/sec, estimated using the corrected slopes, produce Froude numbers that are consistent with the trough cross-stratification in Ogallala valley-fill deposits (Fig. 3), but this result is misleading since the largest Ogallala trough crossbeds were probably made by migrating macroforms, as discussed below.

Paleodischarge is impossible to reconstruct from Ogallala deposits, but estimates based on valley slopes give quantities that are the same order of magnitude as average annual peak flows on the North Platte River for the period of 80–90 years ago. Estimates of paleodischarge based on uncorrected valley slopes are an order of magnitude smaller.

Control of the North Platte since before its earliest gauging measurements render strict comparisons between it and Ogallala streams difficult. The best modern models for flow conditions in Ogallala streams probably are among the 67 gravel-bed rivers from Alberta (Bray, 1979). The Pembina River near Entwistle, the North Saskatchewan River near Rocky Mountain House, and the Red Deer River at Red Deer have bankfull discharges estimated at between 200 and 700 m^3/sec (Bray, 1979). All have sources near the Continental Divide of the Canadian Rockies, and the three sites mentioned above are 100 to 150 km from the divide. In these respects they are probably similar to the Ogallala streams that had sources near the Continental Divide in the Rocky Mountains of Colorado 200 to 250 km from our study area.

Flow depths for the three Canadian streams listed above are given as 1.7 to 1.8 m. In addition, Galay and Neill (1967) discussed large crescent-shaped bars (probably macroforms in the terminology of Crowley, 1983) from the North Saskatchewan River near Drayton Valley, Alberta. These are 150 to 300 m in length, 1.8 to 3 m high, are composed of gravel with a mean size of about 5 cm, and form in flow depths of about 3 m. The bars are shaped much like sandy subaerial barchan dunes, and deposition on the curved slipfaces of these bars would presumably produce large trough crossbeds. The North Saskatchewan River flows in a straight reach about 240 to 370 m wide and appears to have one sinuous channel or thalweg 100 m wide (Galay and Neill, 1967). Such a channel in an Ogallala stream could have carried an estimated discharge of 300 to 400 m^3/sec during bankfull conditions.

Macroforms described from the modern South Platte and Platte Rivers of Colorado and Nebraska have maximum vertical heights and preserved slipfaces that are similar to the flow depths that produced them. Flow acceleration over a Platte River macroform may cause upper flow regimes to develop (Crowley, 1983). The sandy planar-tabular crossbedding produced by migration of these forms is in marked contrast to the massive gravels and trough crossbedding of the Ogallala deposits. This suggests that the Ogallala streams were dominated by a different "North Saskatchewan"—type macroform. To the authors' knowledge, such macroforms have not been described for the North Platte River. In this respect, the Ogallala valley fills may be unique in the western Nebraska panhandle. It is also possible that control of the flow on the North Platte River during the past 100 years has changed the bedforms present in the river.

ACKNOWLEDGMENTS

We acknowledge the assistance of Mr. Jack Faden, Mr. Richard Van Pelt, Mr. Kermit Nelson, Mr. Walter Dykman, and other land owners in the study area who provided access to the sites that we investigated. We also thank Dr. Michael Church, who gave us valuable suggestions that led to improvement of the paleohydrologic analyses, and Dr. Paul Komar, who supplied personal insight and unpublished reports about the physical process of coarse particle entrainment from a gravel bed. Tom Gustavson, Neil Wells, and K. Nibbelink reviewed an early version of this manuscript.

REFERENCES

Andrews, E. D., 1983, Entrainment of gravel from naturally sorted riverbed material: Geological Society of America Bulletin, v. 94, p. 1225–1231.

Baker, V. R., 1984, Flood sedimentation in bedrock fluvial systems, *in* Koster, E. M., and Steel, R. J., eds., Sedimentology of Gravels and Conglomerates: Canadian Society of Petroleum Geologists Memoir 10, p. 87–98.

———, and Ritter, D. F., 1975, Competence of rivers to transport coarse bedload material: Geological Society of America Bulletin, v. 86, p. 975–978.

Bentall, R., 1982, Nebraska's Platte River: A graphic analysis of flows: University of Nebraska-Lincoln, Conservation and Survey Division, Water Supply Paper 53, 47 p.

Blackstone, D. L., Jr., 1975, Late Cretaceous and Cenozoic history of Laramie basin region, southeast Wyoming, *in* Curtis, B. F., ed., Cenozoic History of the Southern Rocky Mountains: Geological Society of America Memoir 144, p. 249–279.

Blodgett, R. H., and Stanley, K. O., 1980, Stratification, bedforms, and discharge relations of the Platte braided river system, Nebraska: Journal of Sedimentary Petrology, v. 50, p. 139–148.

Boothroyd, J. C., and Ashley, G. M., 1975, Processes, bar morphology, and sedimentary structures on braided outwash fans, northeastern Gulf of Alaska, *in* Jopling, A. V., and McDonald, B. C., eds., Glaciofluvial and Glaciolacustrine Sedimentation: Society of Economic Paleontologists and Mineralogists Special Publication 23, p. 193–222.

Bradley, W. C., and Mears, A. I., 1980, Calculations of flows needed to transport coarse fraction of Boulder Creek alluvium at Boulder, Colorado: Geological Society of America Bulletin, v. 91, Part II, p. 1057–1090.

Bray, D. I., 1979, Estimating average velocity in gravel-bed rivers: Journal of the Hydraulics Division, American Society of Civil Engineers, v. 105, p. 1103–1122.

Breyer, J., 1975, The classification of Ogallala sediments in western Nebraska: University of Michigan Museum of Paleontology Papers on Paleontology, v. 3, n. 12, p. 1–8.

Carling, P. A., 1983, Threshold of coarse sediment transport in broad and narrow natural streams: Earth Surface Processes, v. 8, p. 1-18.

Costa, J. E., 1983, Paleohydraulic reconstruction of flash-flood peaks from boulder deposits in the Colorado Front Range: Geological Society of America Bulletin, v. 94, p. 986–1004.

Crowley, K. D., 1983, Large-scale bed configurations (macroforms), Platte River basin, Colorado and Nebraska: Primary sedimentary structures and formative processes: Geological Society of America Bulletin, v. 94, p. 117–133.

Diffendal, R. F., Jr., 1982, Regional implications of the geology of the Ogallala Group (upper Tertiary) of southwestern Morrill County, Nebraska, and adjacent areas: Geological Society of America Bulletin, v. 93, p. 964–976.

———, 1983, Megaclasts in alluvial fills from the Ogallala Group (Miocene), Banner, Kimball, and Morrill counties, Nebraska: University of Wyoming Contributions to Geology, v. 22, p. 109–115.

———, 1984, Comments on the geologic history of the Ogallala Formation in the southern panhandle of Nebraska, *in* Whetstone, G. A., ed., Proceedings of the Ogallala Aquifer Symposium II, p. 194–216.

———, 1985, The inapplicability of the concept of the "Sidney Gravel" to the Ogallala Group (late Tertiary) in part of southern Banner County, Nebraska: TER-QUA Symposium Series, v. 1, p. 185–191.

———, Swinehart, J. B., and Gottula, J. J., 1985, Characteristics, age relationships, and regional importance of some Cenozoic paleovalleys, southern Nebraska panhandle: TER-QUA Symposium Series, v. 1, p. 21–32.

Gable, D. J., and Hatton, T., 1983, Maps of vertical crustal movements in the conterminous United States over the last 10 million years: U.S. Geological Survey Miscellaneous Investigations Series, Map I-1315.

Galay, V. J., and Neill, C. R., 1967, Nomenclature for bed forms in alluvial channels: Discussion: Journal of the Hydraulics Division, American Society of Civil Engineers, v. 93, p. 130–133.

Gibling, M. R., and Rust, B. R., 1984, Channel margins in a Pennsylvanian braided, fluvial deposit: The Morien Group near Sydney, Nova Scotia, Canada: Journal of Sedimentary Petrology, v. 54, p. 773–782.

Guy, H. P., Simons, D. B., and Richardson, E. V., 1966, Summary of alluvial channel data from flume experiments: U.S. Geological Survey Professional Paper 462-I, 96 p.

Izett, G. A., 1975, Late Cenozoic sedimentation and deformation in northern Colorado and adjoining areas, *in* Curtis, B. F., ed., Cenozoic History of the Southern Rocky Mountains: Geological Society of America Memoir 144, p. 179–209.

Kellerhals, R., and Bray, D. I., 1971, Sampling procedures for coarse fluvial sediments: Journal of the Hydraulic Division, American Society of Civil Engineers, v. 97, p. 1165–1180.

Kraus, M. J., 1984, Sedimentology and tectonic setting of early Tertiary quartzite conglomerates, northwest Wyoming, *in* Koster, E. H., and Steel, R. J., eds., Sedimentology of Gravels and Conglomerates: Canadian Society of Petroleum Geologists Memoir 10, p. 203–216.

Maizels, J. K., 1983, Palaeovelocity and palaeodischarge determination for coarse gravel deposits, *in* Gregory, K. J., ed., Background to Palaeohydrology, a Perspective: John Wiley & Sons, New York, p. 101–140.

McKee, E. D., Crosby, E. J., and Berryhill, H. L., Jr., 1967, Flood deposits, Bijou Creek, Colorado, June, 1965: Journal of Sedimentary Petrology, v. 37, p. 829–851.

Miall, A. D., 1976, Palaeocurrent and palaeohydrologic analysis of some vertical profiles through a Cretaceous braided stream deposit, Banks Island, Arctic Canada: Sedimentology, v. 23, p. 459–483.

———, 1977, A review of the braided-river depositional environment: Earth Science Reviews, v. 13, p. 1–62.

———, 1978, Lithofacies types and vertical profile models in braided river deposits: A summary, *in* Miall, A. D., ed., Fluvial Sedimentology: Canadian Society of Petroleum Geologists Memoir 5, p. 597–604.

Ore, H. T., 1964, Some criteria for recognition of braided stream deposits: University of Wyoming Contributions to Geology, v. 3, p. 1–14.

Parker, G., Klingeman, P. C., and McLean, D. G., 1982, Bedload and size distributions in paved gravel-bed streams: Journal of the Hydraulics Division, American Society of Civil Engineers, v. 108, p. 544–571.

Prestegaard, K. L., 1984, Progress in understanding fluvial processes: Journal of Geological Education, v. 32, p. 254–260.

Richards, K., 1982, Rivers: Form and Process in Alluvial Channels: Methuen, New York, 358 p.

Robinson, P., 1970, The Tertiary deposits of the Rocky Mountains—A summary and discussion of unsolved problems: University of Wyoming Contributions to Geology, v. 9, p. 86–96.

Rust, B. R., 1972, Structure and process in a braided river: Sedimentology, v. 18, p. 221–245.

———, 1978, Depositional models for braided alluvium, *in* Miall, A. D., ed., Fluvial Sedimentology: Canadian Society of Petroleum Geologists Memoir 5, p. 605–625.

Schultz, G. E., 1977, Field conference on late Cenozoic biostratigraphy of the Texas Panhandle and adjacent Oklahoma: Kilgore Research Center, West Texas State University, Special Publication 1, 165 p.

Scott, G. R., 1975, Cenozoic surfaces and deposits in the Southern Rocky Mountains, *in* Curtis, B. F., ed., Cenozoic History of the Southern Rocky Mountains: Geological Society of America Memoir 144, p. 227–248.

———, 1982, Paleovalley and geologic map of northeastern Colorado:

U.S. Geological Survey Miscellaneous Investigations Series, Map I-1378 and accompanying text.

SHEPHERD, R. G., AND SCHUMM, S. A., 1974, Experimental study of river incision: Geological Society of America Bulletin, v. 85, p. 257–268.

SKINNER, M. F., SKINNER, S. M., AND GOORIS, R. J., 1977, Stratigraphy and biostratigraphy of late Cenozoic deposits in central Sioux County, western Nebraska: American Museum of Natural History Bulletin, v. 158, p. 263–371.

SMITH, N. D., 1971, Transverse bars and braiding in the lower Platte River, Nebraska: Geological Society of America Bulletin, v. 82, p. 3407–3420.

STANLEY, K. O., 1971a, Tectonic implications of Tertiary sediment dispersal on the Great Plains east of the Laramie Range: Wyoming Geological Association Guidebook, 23rd Annual Field Conference, p. 65–70.

———, 1971b, Areal and temporal differences in Plio-Pleistocene gravel composition, Nebraska: Nebraska Geological Survey, Guidebook to the Late Pliocene and Early Pleistocene of Nebraska, p. 22–27.

———, 1976, Sandstone petrofacies in the Cenozoic High Plains sequence, eastern Wyoming and Nebraska: Geological Society of America Bulletin, v. 87, p. 297–309.

———, AND BENSON, L. V., 1979, Early diagenesis of High Plains Tertiary vitric and arkosic sandstone, Wyoming and Nebraska, *in* Scholle, P. A., and Schluger, P. R., eds., Aspects of Diagenesis: Society of Economic Paleontologists and Mineralogists Special Publication 26, p. 401–424.

———, AND WAYNE, W. J., 1972, Epeirogenic and climatic controls of early Pleistocene fluvial sediment dispersal in Nebraska: Geological Society of America Bulletin, v. 83, p. 3675–3690.

STEER, B. L., AND ABBOTT, P. L., 1984, Paleohydrology of the Eocene Ballena Gravels, San Diego County, California: Sedimentary Geology, v. 38, p. 181–216.

SWINEHART, J. B., 1979, Cenozoic geology of the North Platte River valley, Morrill and Garden Counties, Nebraska: Unpublished M.S. Thesis, University of Nebraska, Lincoln, 127 p.

———, SOUDERS, V. L., DEGRAW, H. M., AND DIFFENDAL, R. F., JR., 1985, Cenozoic paleography of western Nebraska, *in* Flores, R. M., and Kaplan, S. S., eds., Paleogeography of the West-Central United States: Society of Economic Paleontologists and Mineralogists, Rocky Mountain Section, p. 209–229.

WILLIAMS, G. P., 1975, The case of the shrinking channels—The North Platte and Platte Rivers in Nebraska: U.S. Geological Survey Circular 781, 48 p.

———, 1983, Paleohydrological methods and some examples from Swedish fluvial environments I. Cobble and boulder deposits: Geografiska Annaler, v. 65A, p. 227–243.

WILLIAMS, P. F., AND RUST, B. R., 1969, The sedimentology of a braided river: Journal of Sedimentary Petrology, v. 39, p. 649–679.

VERTEBRATE TAPHONOMY APPLIED TO THE ANALYSIS OF ANCIENT FLUVIAL SYSTEMS

EMLYN H. KOSTER[1]

Alberta Geological Survey, Alberta Research Council, 4445 Calgary Trail South, Edmonton, Alberta T6H 5R7, Canada

ABSTRACT: Taphonomy is concerned with reconstruction of biological and sedimentary processes leading to preservation of faunal remains in the stratigraphic record; traditionally, it has been viewed as a specialized branch of paleontology. Recognition of the value of taphonomic work on skeletal remains in fluvial sediments is a fairly new development. The fact that many units of Late Paleozoic to Tertiary alluvium apparently lack vertebrate remains implies either surficial or diagenetic destruction, adverse paleoecological conditions, or oversight in the course of sedimentological studies.

In geologic terms, mortality, postmortem developments and burial are stages that develop quickly within channel or overbank settings. Reworking of skeletal debris into younger stratigraphic intervals also occurs, particularly in multistory channel sequences. Remains are classified in terms of their degree of articulation and concentration. Each mode of preservation has a fairly characteristic set of background conditions relating to the cause and location of mortality in the context of concurrent sedimentary activity. Taphonomic studies of well exposed units should adopt a standard scheme of site notation, record locations precisely, and make a comprehensive log of paleontological and sedimentary features.

Depositional systems near base level with lowland taxa, i.e., meander plains and estuarine complexes, are the main target for taphonomic studies. Sedimentary regimes particularly conducive to vertebrate preservation include pedogenesis, standing-water bodies, bankfull and above-bankfull discharge, channel-base lags, and amalgamation surfaces in multistory sequences. As well as assisting in facies and architectural studies of fluvial systems, skeletal remains can provide valuable insight into the hydrologic conditions of paleochannels.

BACKGROUND

The fossil record of pre-Quaternary land vertebrates, comprising aquatic, semi-aquatic and terrestrial taxa, is dominantly found in alluvial sequences near base level (Dodson, 1971; Dodson and others, 1980; Smith, 1980; Bown and Kraus, 1981; Stewart, 1981a, b; Webb and others, 1981; Behrensmeyer and Tauxe, 1982; Koster, 1984; Koster and Currie, in press). Although the record is biased toward lowland habitats, 'contamination' by reworked representatives of more inland settings and 'loss' to marine environments are also known to have occurred (Frey and others, 1975; Horner, 1979).

Preservation mostly occurs as individual skeletal and exoskeletal parts, teeth, and articulated skeletal remains (Bown and Kraus, 1981; Dodson, 1971, 1983; Dodson and others, 1980; Lawton, 1977; Lucas, 1981; Smith, 1980; Webb and others, 1981). Unusual types of preservation include skin impressions (Dodson, 1983), stomach contents (Stokes, 1963), casts of internal organs (Broughton, 1981), coprolites (Broughton and others, 1978) which may contain undigested remains of vertebrates (Mellett, 1974; Dodson and Wexlar, 1979; Fisher, 1981) and microflora (Waldman and Hopkins, 1970), and eggshell (Horner and Makela, 1979; Horner, 1982, 1984). Footprints are also of considerable interest from paleobiological and paleoenvironmental standpoints (Currie and Sarjeant, 1979; Laporte and Behrensmeyer, 1980; Mossman and Sarjeant, 1983; Lockley, 1986), and burrows of small vertebrates have been documented by Bown (1982).

Taphonomy (Greek derivation:—*taphos* for burial, *nomos* for law) is concerned with the reconstruction of biological and sedimentary events occurring between mortality and burial (Efremov, 1940). As such, it builds a valuable interdisciplinary link between the conventional domains of paleontology and sedimentology. Although almost half a century has elapsed since the establishment of this field, it was not until the late 1960's that taphonomy emerged as the major direction of research in vertebrate paleoecology. This trend has continued (Gifford, 1981; Shipman, 1981; Olson, 1985), but a broader stratigraphic perspective is increasingly evident (Behrensmeyer, 1975, 1984; Béland and Russell, 1978; Schindel, 1980; Behrensmeyer and Kidwell, 1985). Papers dealing with the late Tertiary Siwalik Group of Pakistan by Badgley and Behrensmeyer (1980), Behrensmeyer and Tauxe (1982), Badgley (1986), and Behrensmeyer (this volume), which have approached paleoecological questions by examining the lithofacies context of vertebrate remains, are a good illustration of this latter trend. The important issue of distinguishing evolutionary from taphonomic causes of variable fossil assemblages through alluvial sequences is discussed by Schindel (1980) and Behrensmeyer (1982).

The sedimentological literature on Late Paleozoic to Tertiary alluvial units conveys the impression that skeletal remains are comparatively rare. In some cases, this may signify a sparsely inhabited paleoenvironment due to adverse ecological conditions. More commonly perhaps, remains were subject to surficial or diagenetic destruction (Behrensmeyer, 1978; Gordon and Buikstra, 1981). A lack of interest or oversight is a third possibility; as an example, Honey and Robinson (1985) have described a diverse assemblage from Paleocene alluvium in a part of Montana without any previous reports of remains. The taphonomic literature on vertebrate-bearing fluvial deposits has tended to focus on a relatively small number of rich sites, such as the above mentioned late Tertiary of Pakistan, and thus creates a different impression as to overall stratigraphic abundance. The Campanian coastal plain sequence along the North American Western Interior provides an additional example of major interdisciplinary site studies, principally at Dinosaur Provincial Park 180 km east of Calgary in Alberta (Dodson, 1971, 1983, 1985; Farlow, 1976; Béland and Russell, 1978; Fox, 1979; Currie and Russell, 1982; Currie and Dodson, 1984; Currie, 1985; Koster and Currie, in press, in Montana (Sahni, 1972; Horner, 1985), and in New Mexico (Lucas, 1981). The Alberta locality, referred to below, is a UNESCO World Heritage Site in recognition of its un-

[1]New address: Tyrrell Museum of Palaeontology, P.O. Box 7500, Drumheller, Alberta T0J 0Y0, Canada.

TABLE 1.—FAUNAL GROUPS, WITH CURRENT NUMBERS OF SPECIES, REPRESENTED IN CAMPANIAN COASTAL PLAIN SEDIMENTS AT DINOSAUR PROVINCIAL PARK, ALBERTA: FROM KOSTER AND CURRIE (IN PRESS)

Subaerial	Freshwater
Herbivorous Dinosaurs	Mesoreptiles
Hadrosaurids - 8	Turtles - 11
Ceratopsids - 7	Crocodiles - 4
Pachycephalosaurids - 4	Champsosaurs - 1
Nodosaurids - 2	Fish - 14
Ankylosaurids - 1	Salamanders - 7
Thescelosaurids - 1	Frogs - 3
Carnivorous Dinosaurs	Molluscs - 4
Dromeosaurids - 4	
Tyrannosaurids -3	**Marine**
Ornithomimids - 3	
Caenagnathids - 2	Fish - 2
Elmisaurids - 1	Plesiosaurs - 2
Troodontids - 1	? Unreworked Dinoflagellates and Acritarchs
Mammals - 20	
Lizards - 11	
Flying Reptiles - 2	
Birds - 1	

rivalled quality and abundance of dinosaurian and other faunal remains (Table 1).

PURPOSE

Formerly, remains of land vertebrates were often used simply to confirm a continental environment for the host deposits (Vonhof, 1965). Brief reference in Miall (1983, 1984) shows that the advent of taphonomic thinking into the mainstream of fluvial sedimentology is now occurring. In a parallel development, Brett and Baird (1984) have recommended taphonomic study of shelly invertebrates as a tool for paleoenvironmental analysis in marine sequences.

An increased awareness of vertebrate taphonomy by fluvial sedimentologists is desirable because:—preservation potential of local and/or regional components of a prevailing ecosystem is ultimately determined by sedimentary and stratigraphic architecture; vertebrate remains provide novel evidence as to the nature and timing of sedimentary events and may assist in basin-scale analysis of paleodrainage systems; and new fossil localities discovered in the course of sedimentological field studies will have a better chance of being brought to the paleontologist's attention for comprehensive study.

As an aid to comprehensive sedimentological analysis of fluvial sequences that contain hard-part remains of vertebrates, this paper examines the following aspects of taphonomy:—(1) mortality, transport, reworking, and ultimate burial of remains; (2) classification of preservation modes; (3) acquisition and synthesis of field data; (4) preservation potential in each principal type of depositional system. Attention is drawn to key literature, much of which has been published in journals not typically consulted by sedimentologists.

Because hominid influences on taphonomy are a separate consideration (Morlan, 1984), this review only applies to pre-Quaternary deposits.

FROM MORTALITY TO BURIAL

"The passage from the biosphere to lithosphere" (Efremov, 1940, p. 85) of vertebrate material begins with environmental conditions surrounding animal death and ends with those leading to permanent burial of at least one element of the carcass.

The relative locations of mortality and burial (Table II in Smith, 1980) is one of the most significant considerations in vertebrate taphonomy, both for paleoecology and sedimentological aspects of interaction between channel and interchannel areas. Mortality and burial may occur in the same subenvironment with only minor lateral movement (Table 2). Examples of concentrated large remains that preserve a biocoenosis in place due to 'catastrophic' environmental change are provided by Voorhies (1981) and Conybeare and Haynes (1984). Recognizing the complexity of postmortem processes, Voorhies (1969) and Grayson (1978) have shown that Shotwell's (1955) original method for separating representatives of proximal versus distal communities in channel assemblages has doubtful utility. As one aspect of this complexity, Behrensmeyer and others (1979) have shown that the fossil record under-represents small animals because their bones have greater susceptibility to destruction by mastication, trampling, and weathering.

Each principal stage in the 'passage' between mortality and burial is outlined below using descriptors such as location (Table 2), process(es), duration, and the nature of concurrent sedimentary activity. Ability to reconstruct these stages from remains is jointly determined by the mode of preservation and prevailing levels of sedimentological and paleontological understanding for the host sequence. One conclusion of the writer's studies at Dinosaur Park is that large skeletal remains may provide clues for the nature of paleohydrologic conditions beyond those conventionally derived from channel sediments. These taphonomic 'clues' are mentioned at various points in this section with a view to stimulating further investigation by sedimentologists.

Mortality Events

Biologists and paleontologists classify natural mortality patterns as either attritional or catastrophic. In her study of

TABLE 2.—DISTINGUISHING TAPHONOMIC FEATURES OF CHANNEL AND OVERBANK ENVIRONMENTS

Environment	Notation	Comments on Mortality and Probability of Preservation
Overbank: mostly dry with pedogenesis	Od	• subaerial mortality • low potential for movement of intact carcass • biological destruction and dispersal dominant • remains may not outlast next depositional event.
Overbank: areas of standing water	Ow	• aqueous mortality • low-moderate potential for carcass movement • biological and minor mechanical destruction probable • remains have moderate probability of burial
Active channel	C	• aqueous mortality • high potential for major movement of carcass • moderate-high probability for mechanical disarticulation and destruction • remains have moderate probability of burial

Plio-Pleistocene vertebrate assemblages in Kenya, Behrensmeyer (1975) concluded that mode of death is the main factor in determining the preservation potential of different skeletal elements.

Catastrophic events involve mass mortality due to a sudden change in ecological conditions. Floods, drought, epidemics and fallout of volcanic ash are examples of non-attritional causes of mortality. Intact catastrophic death assemblages provide a fairly representative sample of age distribution (Voorhies, 1969; Currie and Dodson, 1984), although sedimentary processes or biological agencies may disperse or eliminate a mass death assemblage in terms of the stratigraphic record. From a geologic perspective, most 'catastrophic' events are more appropriately described as episodic (Dott, 1983). For example, mass mortality in flash floods and at waterholes is virtually an annual occurrence for migratory herds of wildebeest (*Connochaetes*) in the Serengeti area of Tanzania (Schaller, 1972, 1973). The tendency for vertebrates to form large social groups increases the likelihood of mass mortality.

Attritional refers to 'normal' events such as juvenile weakness, non-epidemic disease, predation and old age. The rates of birth and attritional mortality are balanced for a species whose population is in equilibrium within the regional ecosystem. Distinguishing features of attritional events are a bias towards juvenile, old or sick individuals (Voorhies, 1969; Schaller, 1973) and a random distribution in space and time. As discussed in later sections, however, pedogenic or sedimentary processes may lead to spatial and/or stratigraphic concentrations of remains, possibly creating an impression of non-attritional mortality events (Bown and Kraus, 1981; Behrensmeyer, 1982; Webb and others, 1981). Statistical aspects of the connection between attritional assemblages and original community composition are the subject of papers by Behrensmeyer and others (1979), Holtzman (1979), and Damuth (1982).

If mortality occurs in a dry overbank (Od) setting, the subaerially exposed carcass undergoes progressive putrefaction and disarticulation. Predator and/or scavenging action results in rapid dismemberment, dispersal and possible mastication of skeletal components (Schaller, 1973; Mellett, 1974; Korth, 1979). A well trampled substrate may surround a large exposed carcass (Laporte and Behrensmeyer, 1980). Destruction by carnivores is commonly selective (Schaller, 1973; Hill, 1979, 1980) and discarded bone may show chew marks (Haynes, 1980), spiral fractures (Haynes, 1983) or surface damage through trampling (Behrensmeyer and others, 1986). De-fleshing through insect activity is also an important process (Payne, 1965; Behrensmeyer, 1975), especially in warm humid weather.

In the absence of carnivores, a carcass exposed in an Od setting undergoes a two-stage sequence of progressive *in situ* destruction (Table 3). The time required for complete de-fleshing partly depends on the size of the animal, but commonly occurs within a few weeks or months following death (Korth, 1979). For buffalo (*Syncerus caffer*) in Rwenzori National Park in Uganda, Hill (1980) documented an extreme case where two cervical vertebrae were still attached 4 years after death. For mammal bone near Lake Amboseli in Kenya, Behrensmeyer (1978) noted complete decomposition within 10–15 years.

TABLE 3.—PROGRESSIVE STAGES OF DE-FLESHING DUE TO INSECT ACTIVITY AND THEN BONE WEATHERING OF EXPOSED CARCASSES, BASED ON OBSERVATIONS OF MAMMALS BY PAYNE (1965) AND BEHRENSMEYER (1978), RESPECTIVELY; SEE ALSO DISCUSSION BY KORTH (1979).

1. Fresh	No evidence of decomposition or insect activity.
DE-FLESHING:	
2. Bloated	Local insect activity, bloating.
3. Initial Decay	Fully infested, skin broken in stomach area.
4. Advanced Decay	Still infested, most skin and flesh removed.
5. Dry	Bones isolated, dry skin locally present.
BONE WEATHERING:	
6. Cracking	Mostly parallel to fiber structure on longitudinal surfaces and in mosaic patterns on articular surfaces.
7. Flaking	Affects outermost concentric layers near cracks.
8. Exfoliation	Removal of outermost layers with rough patches developing on internal fibrous bone.
9. Advanced	Open cracks with splintered margins and internal weathering becoming penetrative.
10. Disintegration	Fragile, extensively cracked with exposure of innermost bone.

Mortality in aqueous (Ow or C) settings first involves flotation (assuming that water depth exceeds body size), then sinking and finally renewed buoyancy due to early decompositional processes in the body chamber. In transit, some mechanical disarticulation or destruction of skeletal components by aquatic scavengers may occur (Behrensmeyer, 1975).

Processes Leading to Burial

Interest here focusses on skeletal remains that survived postmortem conditions to the point of burial. The duration of this stage spans a wide range from hours, in the case of a flashflood event, to possibly as much as a decade or century for remains to reach their terminal point of down-channel transport.

In the model of Behrensmeyer (1982), there are three 'pathways' for input of attritional remains to fluvial sediments. Pathways A and B both initially involve overbank settings: in A, local burial occurs but is followed by exhumation at eroding channel banks, and in B, above-bankfull flow transports remains before burial into a nearby channel. In Pathway C, remains originate directly in a channel setting. Given the multistage nature of taphonomic processes and the fact that surficial destruction of remains occurs far more commonly than burial, unravelling taphonomic processes with primary reference to subenvironments (Table 2) instead of complete pathways may be more practical from the sedimentological standpoint.

Dry Overbank Areas.—As Bown and Kraus (1981) point out, attention in vertebrate taphonomy has been mostly given to hydraulic assemblages in channel settings. This partly reflects the traditionally unbalanced interest in interchannel versus channel facies (Allen, 1978). Mainly, however, it relates to the lower potential for preservation in overbank settings due to subaerial exposure (Behrensmeyer, 1978) and commonly destructive pH levels during early burial (Gordon and Buikstra, 1981). Behrensmeyer (1975) noted that vertebrate remains on floodplains have the best chance of escaping destruction when flood-deposited units of new sediment have a greater thickness than subsequent soil profiles.

Attritional mortality above paleosol surfaces led to autochthonous assemblages in Lower Eocene overbank sequences of Wyoming (Bown and Kraus, 1981), although Maas (1985) favored a scatological origin for an apparently similar occurrence of later Eocene age in an adjacent basin. In general, one expects that fossil remains in interchannel sediments are broadly autochthonous because carcass or bone transport out of channels is improbable (Dodson and others, 1980) and because interchannel areas are the main habitat for terrestrial vertebrates. In Permian alluvial deposits of the Karoo, Smith (1980) found a decreasing abundance and quality of reptilian remains away from channel margins to distal floodplain subenvironments; crevasse-splay deposits were found to be virtually devoid of remains. By contrast, Stewart (1981a, p. 3.7) interpreted thin beds comprising intraclasts, macrofloral debris and dinosaur remains in Lower Cretaceous overbank mudstone sequences in southern England as indicating "violent flooding to low lying areas of the flood plain".

Standing Water Bodies.—Ecologically, abandoned channels and sloughs function as water holes and as such provide favorable habitats and focal areas for predation (Behrensmeyer, this volume). Therefore, the Ow subenvironment (Table 2) has strong potential to generate and preserve autochthonous attritional assemblages. The Jurassic in the Colorado-Wyoming region (Dodson and others, 1980) and Miocene of Pakistan (Badgley, 1986; Behrensmeyer, this volume) are well documented examples. Sediments peripheral to these water bodies may either preserve trackways or be thoroughly trampled (Laporte and Behrensmeyer, 1980). Fragmented remains around waterholes commonly attain high concentrations as shown by Conybeare and Haynes (1984); Hay's (1976) study points to a similar potential for larger, less ephemeral water bodies. Voorhies (1981) described a rare example of a catastrophic death assemblage in an Ow setting. Excavation of this Miocene water hole in Nebraska, which was engulfed by volcanic ash, revealed such finds as an unborn rhino calf inside an adult skeleton, grass seeds in a rhino's throat cavity, and feather impressions left from scavenging birds.

Although Holocene peat swamps are known to contain perfectly preserved fossil remains (Glob, 1954), ancient examples are unlikely because of the destructive effects of coalification. Rotted tree stumps within coal seams, however, may contain vertebrate remains as shown by the Pennsylvanian Joggins section in Nova Scotia (Carroll, 1967).

Channels.—Active channels are the primary routes for major transport of vertebrate remains and provide the optimum conditions for burial (Table 2). Material enters either directly from mortality associated with flowing water or is reworked from eroding banks (Behrensmeyer, 1975, 1982; see next section).

Preservation of intact skeletons indicates a single episode of carcass flotation, stranding and rapid burial before ligament decay. Stranding occurs due to falling stage and/or movement onto higher relief parts of the channel. For example, Smith (1980) interpreted flood discharge events as the cause for abundant articulated reptile fossils in Permian levee sequences in South Africa. According to Sternberg (1970), the configuration of partial hadrosaur skeletons in Upper Cretaceous strata at Dinosaur Provincial Park (Fig. 14 in Dodson, 1971) indicates that the bloated carcasses may have floated upside down with the head dragging beneath the shoulders. Sternberg also documents a specimen where differential quality in preservation between the lower and upper halves of a stranded *Lambeosaurus* carcass indicates a two-phase burial history. Articulated remains of large vertebrates may provide direct evidence for minimum water depth during flotation.

Transport of individual skeletal elements is a much more complex phenomenon, involving consideration of their size, three-dimensional form, and density (Behrensmeyer, 1975). Each parameter shows wide inter- and intra-specific variation among modern vertebrates (Behrensmeyer, 1975). Attempts to reconstruct the hydraulic significance of fossil bone elements are further complicated by variability in the pre-transport weathering state (Table 3), abrasion and pore saturation once in transport (Behrensmeyer, 1975), and increased density associated with permineralization. Bone transport has been investigated in flumes (Voorhies, 1969; Dodson, 1973; Boaz and Behrensmeyer, 1976; Hanson, 1980), in settling tanks (Behrensmeyer, 1975; Korth, 1979) and in natural channels (Frey and others, 1975; work by C.B. Hanson described in Lawton, 1977; Behrensmeyer, 1982). A principal conclusion of these studies is that certain skeletal elements of large vertebrates are diagnostic of 'normal' and flood velocity levels. Table 4 outlines the basic data from Voorhies (1969), subsequently analyzed in greater detail by Behrensmeyer (1975). Similar relationships have been noted for microvertebrate remains by Dodson (1973) and Korth (1979), although Dodson noted that their low threshold velocities (i.e., 6–35 cm/sec) preclude their usefulness in paleoecological studies. The empirical nomograms for reconstruction of hydraulic parameters from skeletal remains in Behrensmeyer (1975) and Korth (1979) have yet to be applied to the stratigraphic record. Most of this experimental work by paleontologists implicitly assumes that the calibre of skeletal remains in channel sediments is in equilibrium with prevailing competence levels. The 1965 flood of Bijou Creek, Colorado (McKee and others, 1967), however, which transported bridge debris over a sandbed, was a clear indication that this equilibrium may not exist.

TABLE 4.—MAIN RESULTS OF VOORHIES' (1969) FLUME TRANSPORT EXPERIMENTS USING ELEMENTS FROM FRESH COYOTE (*CANIS*) AND SHEEP (*OVIS*) SKELETONS, BASED ON 15 TRIALS, WITH FLOW VELOCITIES AS MUCH AS 150 CM/SEC AND A BED OF FINE-MEDIUM SAND. ONLY THOSE ELEMENTS WITH A SINGLE ASSIGNMENT IN THE THREE 'DISPERSAL GROUPS' ARE LISTED.

Group I	Low threshold velocity; mainly by saltation.	Ribs, vertebra, sacrum* and sternum*.
Group II	Intermediate threshold velocity; mainly by traction.	Femur, tibia, humerous, metapodia, pelvis and radius.
Group III	High threshold velocity; probable lag constituents.	Skull and mandible.

*Elements that float until waterlogged (Behrensmeyer, 1975).

Reworking of Remains

Lateral migration or avulsion of channels may bring about the exhumation of vertebrate remains buried during older alluvial episodes. The original burial site may either have been in overbank sediments (Pathway A in Behrensmeyer, 1982) or channel sediments and may be re-exposed from an eroding bank or degrading bed.

Behrensmeyer (1975) noted that for carcass remains beneath the watertable around Lake Rudolf, Kenya, ligaments decomposed within a year or so from the time of burial. In view of the much longer periods associated with channel reoccupation of a given part of the alluvial surface, a re-exposed carcass will become a point source for a large quantity of unabraded skeletal debris. This material is then dispersed in accordance with prevailing flow conditions, some of which may become included with lag deposits downstream (Stewart, 1981b). Long bone elements, such as ribs or those comprising limbs, typically show good transverse alignment with respect to current direction.

Exhumation of already disarticulated remains from channel sediments introduces a second cycle of abrasion and fragmentation. Their potential for survival to a new burial site depends, to some extent, on the degree to which bone lost organic material and/or was permineralized during initial burial (Table 3; Behrensmeyer, 1975). The resulting loss or gain in bone density will cause differences in transport behavior compared to that of fresh bone. High concentrations of bone included with conglomeratic amalgamation surfaces in multistory channel sequences (Honey and Robinson, 1985; Koster and Currie, in press; Fig. 1) strongly imply filtering from reworked intervals.

Diagenetic Aspects

Data on the destruction of buried skeletal remains are largely limited to the early, shallow stages of burial. Reference has already been made to the conclusions of Behrensmeyer (1975) concerning the destructive effect of pedogenic processes. Further destruction below paleosols within overbank mudstone sequences jointly depends on the acidity of ground water and size of skeletal remains (Gordon and Buikstra, 1981), as well as the physical effects of compaction (Behrensmeyer, 1975). Skeletal remains buried in channel sands are less vulnerable to destruction, and they may serve as preferred nuclei for concretions (Behrensmeyer, 1975). Petrographic study of pore fillings in abraded bone may yield valuable information about the cementation history of the host formation (R.G. Thomas, pers. commun.).

FIG. 1.—Sectional view of a conglomeratic lag deposit marking an amalgamation surface within a multistory channel sequence; Dinosaur Provincial Park. Constituents include pebble-size mudstone intraclasts, a cobble-size ironstone concretion (center right) reworked from overbank mudstones, and an unabraded dinosaur limb bone in horizontal orientation immediately above the erosion surface.

MODES OF PRESERVATION

In general, preserved vertebrate remains can be viewed against a simple classification based on the degrees of articulation and concentration (Table 5). Each of these criteria is recognized as having two 'states.'

'Articulation' refers to skeletal elements of the same individual maintained in their correct relative positions. The term 'association' is used for *in situ* disarticulated assemblages of the same individual. By definition, the number of articulated elements ranges from two to the maximum associated with a perfectly preserved specimen; the skeleton of a modern topi (*Damaliscus korrigum*), for example, comprises about 132 elements (Hill, 1979). In real terms, articulated elements commonly exceed 20 and represent a virtually intact section of the skeleton such as a vertebral column, manus, limb or skull (Fig. 2). Complete preservation of a vertebrate skeleton requires *in situ* burial before soft parts have completely decomposed (Voorhies, 1981). Disarticulated elements, which may occur singly (Figs. 3, 4) or in high concentration, commonly show signs of abrasion and/or breakage. Teeth and scales, however, typically show remarkable preservation (Fig. 5).

'Concentration' refers to the degree of spatial clustering of remains between or within a single sedimentation unit (Fig. 6) and depends jointly on the scale and proximity of the mortality event(s) as well as the rate and nature of deposition. The time connotation varies widely from effectively instantaneous (Voorhies, 1981) to one annual weather cycle (Conybeare and Haynes, 1984) to unspecified longer periods (Currie and Dodson, 1984).

TABLE 5.—CLASSIFICATION SCHEME FOR MODES OF VERTEBRATE PRESERVATION. NUMBERS IN PARENTHESES INDICATE OVERALL RELATIVE ABUNDANCE, 1 BEING THE HIGHEST, IN ANCIENT FLUVIAL DEPOSITS.

		DEGREE OF CONCENTRATION			
		Isolated		Concentrated	
DEGREE OF ARTICULATION	Disarticulated	ID (1)	BONEBEDS	CD (2)	Monogeneric / Multigeneric
	Articulated	IA (3)		CA (4)	Monogeneric

FIG. 2.—Articulated tail section of a lambeosaurine hadrosaur undergoing excavation; Dinosaur Provincial Park. The carcass became stranded on its right side and was quickly buried in the basal part of a channel sequence.

FIG. 3.—Plan view of a degrading pediment surface showing an isolated, tooth-bearing dentary fragment from a lambeosaurine dinosaur in the lower level of a channel sandstone unit; Dinosaur Provincial Park. The breakage and majority of longitudinal cracking are probably original from pre-transport weathering (Table 3), although modern exposure has undoubtedly accentuated the stage of weathering. The surrounding pebble veneer is derived from a nearby Pleistocene till outcrop.

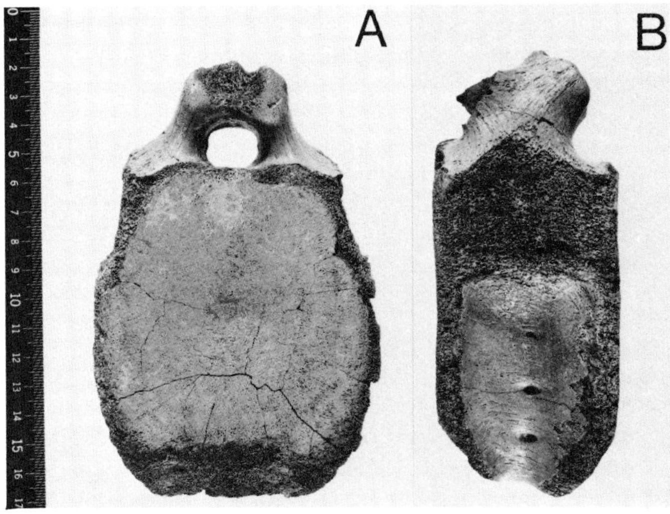

FIG. 4.—Posterior (A) and lateral (B) views of a caudal vertebra from a hadrosaur; Dinosaur Provincial Park. Scale bar in cm. Following disarticulation and decomposition of cartilaginous material in the parent skeleton, traction movement of this vertebra along a channel thalweg caused abrasion of the convex margins to the centrum leading to exposure of inner porous bone. Protected concave surfaces, including the neural canal, are well preserved. Whether breakage causing loss of caudal ribs (upper left and right in A) and the neural spine (top in A and B) occurred before or during initial transport is uncertain.

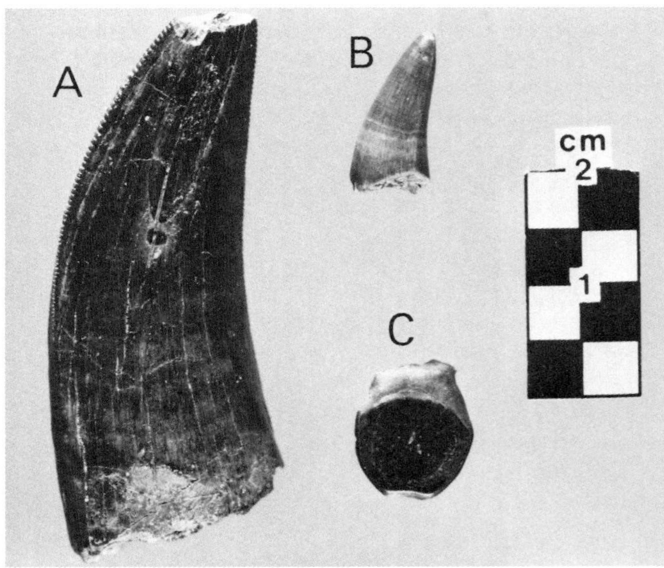

FIG. 5.—Perfect preservation of discarded teeth from a tyrannosaurid dinosaur (A) and from a crocodile (B), as well as a ganoid-type scale from a garpike (C); Dinosaur Provincial Park. Durability relates to their enamelled surfaces. In A, note the preservation of a serrated edge and the upper bevelled surface from contact with an opposing tooth. All three specimens were retrieved from channel deposits.

FIG. 6.—Wide angle view of the *Centrosaurus* Bonebed described by Currie and Dodson (1984); Dinosaur Provincial Park. Excavation of this mass mortality event, which became preserved within a point-bar sequence (Koster, 1984), has revealed that the dominant ceratopsian component consists mostly of ribs, vertebrae and skull parts. An example of the relatively rare CA mode of vertebrate preservation (Table 6).

These two criteria lead to four possible preservation modes (Table 6) abbreviated as IA, CA, ID and CD, where I denotes isolated, C concentrated, A articulated, and D disarticulated. Articulated remains are commonly isolated occurrences (Dodson, 1971, 1983), whereas disarticulated remains either tend toward high concentrations (Lawton, 1977) or scattered occurrences (Stewart, 1981b). As noted earlier, the CA mode requires rapid *in situ* burial of a mass mortality event (Voorhies, 1981), and therefore occurs rarely.

FIELD DATA

The full potential of taphonomic investigations is realized when paleontologists and sedimentologists work collaboratively on the same section. In the more common case where field studies are not concurrent, the paleontologist should record "the manner in which . . . specimens were buried . . . and their position within the sediments" (Sternberg, 1970, p. 9). Such records may provide valuable data to later sedimentological work.

Although an isolated fossil occurrence in a small outcrop is amenable to taphonomic consideration, samples of data that span relatively large intervals of space and time are much more meaningful. The two following sections apply to studies of well exposed stratigraphic units, such as the extensive tract of badlands at Dinosaur Provincial Park (Dodson, 1971; Béland and Russell, 1978; Koster, 1984; Koster and Currie, in press) or the terrain in northern Pakistan (Behrensmeyer and Tauxe, 1982; Badgley, 1986; Behrensmeyer, this volume).

TABLE 6.—EXPECTED GENERAL LEVELS OF PROBABILITY FOR EACH MODE OF PRESERVATION (TABLE 5) IN RELATION TO CHANNEL AND OVERBANK ENVIRONMENTS (TABLE 2) DENSELY INHABITED BY A VERTEBRATE FAUNA.

		Environments		
		Od	Ow	C
Modes of Preservation	(1) ID	Moderate	Moderate	High
	(2) CD	Low	Moderate	High
	(3) IA	Improbable	Low	Moderate
	(4) CA	Improbable	Low	Low

Collection

Notation scheme.—A consistent alphanumeric system that refers both to the mode of preservation (i.e., IA, CA, ID, CD) and sequential number of the locality is recommended. Articulated remains are typically quarried for laboratory research or museum display; their removal from outcrop can be indicated by additional notation.

Location.—Exact positioning of important remains with respect to geographic coordinates and stratigraphic position is essential. Therefore, permanent records on large-scale aerial photographs, detailed metric coordinates and staking of quarry locations are appropriate measures (Sternberg, 1950).

Fossil Contents.—Ideally, the paleontologist should compile a comprehensive written and photographic record of identifiable taxa, their dimensions/age distribution, configuration, mode of preservation, and completeness of remains. Particular attention is needed for representative sampling of microvertebrate remains (Voorhies, 1969; Currie, 1985; Dodson, 1985).

Stratigraphic Context.—Similarly, the sedimentologist should compile a comprehensive written and photographic record of the host facies and position of remains with respect to facies contacts, as well as attempting to decipher sedimentary conditions surrounding stranding and burial.

Consideration of Bias

In taphonomic work based on data spanning a large time-space interval of strata, a useful goal is to distill the typical sedimentological association for observed modes of preservation (Smith, 1980; Badgley, 1986; Behrensmeyer, this volume; Koster and Currie, in press). Data sets of this kind, however, are complicated by two sources of bias in the three-dimensional distribution of data—namely, the hypsometric properties of the outcrop area (Strahler, 1957) and unrepresentative coverage of available outcrop in the course of paleontological prospecting. As an example, the outcrop at Dinosaur Provincial Park extends over 90 km^2 with maximum vertical relief of 140 m; detailed hypsometric calculations reveal that only 20 percent of the badlands occupy the uppermost 70 m. Sets of outcrop-derived frequency data therefore require weighting in terms of number per unit area. Assessing whether there has been any significant bias towards particular areas or elevation intervals in the present tally of over 300 articulated skeletons (Dodson, 1983) is considerably more difficult. A new effort toward systematic prospecting by staff of the Tyrrell Museum of Palaeontology (P. J. Currie, pers. commun.), however, will help to eliminate any bias in the earlier collections (Sternberg, 1950) which formed the basis of work by Dodson (1971) and Béland and Russell (1978).

TAPHONOMIC POTENTIAL OF FLUVIAL DEPOSITIONAL SYSTEMS

The present, relatively small body of taphonomic literature with a sedimentological orientation enables some preliminary comments as to the potential for preservation of skeletal remains in various types of fluvial depositional system.

Fans and Braidplains

In general, these appear to have low potential because of adverse sedimentary and ecological factors. Characteristic features are a high-energy level of transport events, coarse grade of bed sediment, frequent surficial reworking due to lateral switches in channel activity, and a scattered vegetation cover. Therefore, these systems are not typically conducive to populous vertebrate communities or favorable conditions for preservation. Ephemeral hydrologic regimes under an arid climate further reduce the chance of any skeletal remains surviving a period of exposure. Flash floods, however, may bring about quick burial of carcass remains (Smith, 1980) and mass mortality of migratory herds (Schaller, 1972, 1973).

Meander Plains

In contrast, the typical ecological and sedimentary features of low-gradient, single-channel fluvial systems are conducive to preservation of skeletal remains (Dodson and others, 1980; Badgley, 1986; Behrensmeyer, this volume). Under most climatic conditions, interchannel areas are well vegetated and support a varied community of vertebrate fauna. Béland and Russell (1978) cite evidence that alluvial soils enriched by volcanic ashfalls in areas with annual rainfall in excess of 1000 mm are capable of supporting large herbivore biomasses of 20 t/km^2. This, for example, is a plausible set of conditions for the Late Cretaceous coastal plain in the Western Interior. Soil surfaces subject to periodic inundation may accumulate attritional assemblages from carnivore activity, and standing-water bodies have strong potential for a varied autochthonous assemblage. Flood flow in channels may cause point bar accretion over bone-bearing lag deposits and release of previously buried remains from eroding banks. Predation along channel banks or drowning events may lead to flotation of carcasses, which then have a moderate chance of burial as articulated remains. Of 66 bones recorded within a 3.9 km meandering reach of the East Fork River, Wyoming over a 6-year period by Behrensmeyer (1982), 56 percent were found on point bars which comprised only 15 percent of the study area.

Estuaries

The seaward margins of coastal plains, where hinterland rivers become estuarine below high tide limits, are considered to have high potential for studies in vertebrate taphonomy. For example, Frey and others (1975) discovered that Georgia estuaries contain reworked remains from vertebrate assemblages in inland late Tertiary and Quaternary deposits. Bioerosion by invertebrates after their recycling (Frey and others, 1975) is a potentially useful line of evidence for work in the stratigraphic record. An apparently similar estuarine situation was reported by Webb and others (1981) from the late Miocene in northern Florida. At the Dinosaur Park locality in Alberta, Dodson (1971) found that of 84 quarries containing articulated dinosaur skeletons 80 percent were located in a sand-dominated channel facies. According to Koster (1984) and Koster and Currie (in press), this paleochannel type originated in the lower reaches of estuarine systems. From their study in Georgia, Frey and others (1975, p. 257) concluded that ". . . estuarine mixing of faunas is a widespread, normal process that probably has ancient analogs in numerous coastal plain settings; evaluation of possible stratigraphic leakage should be a prerequisite to the biostratigraphic and paleoecologic analysis of ancient estuarine sequences".

ACKNOWLEDGMENTS

Personal interest in vertebrate taphonomy arose from a request from the Provincial Museum of Alberta for an updated sedimentological study at Dinosaur Park. In this regard, I particularly thank Philip J. Currie (Tyrrell Museum of Palaeontology, Drumheller) for his strong involvement in all facets of the study. The stimulus of many others, especially Peter Dodson (University of Pennsylvania, Philadelphia) and Dale A. Russell (National Museum of Natural Sciences, Ottawa), has also been much appreciated. These colleagues as well as Catherine E. Badgley (University of Michigan, Ann Arbor), Larry T. Middleton (Northern Arizona University, Flagstaff) and Anna K. Behrensmeyer (Smithsonian Institution) provided valuable comments on the paper. Linda Jones and Leslie Lorinczi (Alberta Research Council) kindly assisted with word processing and photography.

REFERENCES

ALLEN, J. R. L., 1978, Achievements and prospects in fluvial sedimentology, *in* Miall, A. D., ed., Fluvial Sedimentology: Canadian Society of Petroleum Geologists Memoir 5, p. 849.

BADGLEY, CATHERINE, 1986, Taphonomy of mammalian fossil remains from Siwalik rocks of Pakistan: Paleobiology, v. 12, p. 119–142.

———, AND BEHRENSMEYER, A. K., 1980, Paleoecology of Middle Siwalik sediments and faunas, northern Pakistan: Palaeogeography, Palaeoclimatology, Palaeoecology, v. 30, p. 133–155.

BEHRENSMEYER, A. K., 1975, The taphonomy and paleoecology of Plio-Pleistocene vertebrate assemblages east of Lake Rudolf, Kenya: Bulletin of the Museum of Comparative Zoology, Harvard University, v. 146, p. 474–578.

———, 1978, Taphonomic and ecologic information from bone weathering: Paleobiology, v. 4, p. 150–162.

———, 1982, Time resolution in fluvial vertebrate assemblages: Paleobiology, v. 8, p. 211–227.

———, 1984, Taphonomy and the fossil record: American Scientist, v. 72, p. 558–566.

———, AND KIDWELL, S. M., 1985, Taphonomy's contributions to paleobiology: Paleobiology, v. 11, p. 105–119.

———, AND TAUXE, LISA, 1982, Isochronous fluvial systems in Miocene deposits of northern Pakistan: Sedimentology, v. 29, p. 331–352.

———, GORDON, K. D., AND YANAGI, G. T., 1986, Trampling as a cause of bone surface damage and pseudo-cutmarks: Nature, v. 319, p. 768–771.

———, WESTERN, D., AND BOAZ, D. A. D., 1979, New perspectives in vertebrate paleoecology from a Recent bone assemblage: Paleobiology, v. 5, p. 12–21.

BÉLAND, P., AND RUSSELL, D. A., 1978, Paleoecology of Dinosaur Provincial Park (Cretaceous), Alberta, interpreted from the distribution of articulated vertebrate remains: Canadian Journal of Earth Sciences, v. 15, p. 1012–1024.

Boaz, N. T., and Behrensmeyer, A. K., 1976, Hominid taphonomy: Transport of human skeletal parts in an artificial fluvial environment: American Journal of Physical Anthropology, v. 45, p. 53–60.

Bown, T. M., 1982, Ichnofossils and rhizoliths of the nearshore fluvial Jebel Qatrani Formation (Oligocene), Fayum Province, Egypt: Palaeogeography, Palaeoclimatology, Palaeoecology, v. 40, p. 255–309.

———, and Kraus, M. J., 1981, Vertebrate fossil-bearing paleosol units (Willwood Formation, Lower Eocene, northwest Wyoming, U.S.A.): Implications for taphonomy, biostratigraphy, and assemblage analysis: Palaeogeography, Palaeoclimatology, Palaeoecology, v. 34, p. 31–56.

Brett, C. E., and Baird, G. C., 1984, Comparative taphonomy: A key to interpretation of sedimentary sequences using fossil preservation: Geological Society of America, Abstracts with Programs, v. 16, no. 3, p. 126.

Broughton, P. L., 1981, Casts of vertebrate internal organs from the Upper Cretaceous of western Canada: Journal of Geology, v. 89, p. 741–749.

———, Simpson, Frank, and Whitaker, S. H., 1978, Late Cretaceous coprolites from western Canada: Palaeontology, v. 21, p. 443–453.

Carroll, R. L., 1967, Labyrinthodonts from the Joggins Formation: Journal of Paleontology, v. 41, p. 111–142.

Conybeare, A., and Haynes, Gary, 1984, Observations of elephant mortality and bones in water holes: Quaternary Research, v. 22, p. 189–200.

Currie, P. J., 1985, Small theropods of Dinosaur Provincial Park, Alberta (abs.): Geological Society of America, Abstracts with Programs, v. 17, no. 4, p. 215.

———, and Dodson, Peter, 1984, Mass death of a herd of ceratopsian dinosaurs, *in* Reif, W. E., and Westphal, F., eds., Third Symposium on Mesozoic Terrestrial Ecosystems, Tubingen: Short Papers, p. 61–66.

———, and Russell, D. A., 1982, A giant pterosaur (Reptilia: Archosauria) from the Judith River (Oldman) Formation of Alberta: Canadian Journal of Earth Sciences, v. 19, p. 894–897.

———, and Sarjeant, W. A. S., 1979, Lower Cretaceous dinosaur footprints from the Peace River Canyon, British Columbia: Palaeogeography, Palaeoclimatology, Palaeoecology, v. 28, p. 103–115.

Damuth, John, 1982, Analysis of the preservation of community structure in assemblages of fossil mammals: Paleobiology, v. 8, p. 434–446.

Dodson, Peter, 1971, Sedimentology and taphonomy of the Oldman Formation, Dinosaur Provincial Park, Alberta (Canada): Palaeogeography, Palaeoclimatology, Palaeoecology, v. 10, p. 21–74.

———, 1973, The significance of small bones in paleoecological interpretation: University of Wyoming, Contributions to Geology, v. 12, no. 1, p. 15–19.

———, 1983, A faunal review of the Judith River (Oldman) Formation, Dinosaur Provincial Park, Alberta: The Mosasaur, v. 1, p. 89–118.

———, 1985, Studies of dinosaur paleoecology by repeated sampling in Campanian sediments, southern Alberta (abs.): Geological Society of America, Abstracts with Programs, v. 17, no. 4, p. 216.

———, Behrensmeyer, A. K., Bakker, R. T., and McIntosh, J. S., 1980, Taphonomy and paleoecology of the dinosaur beds of the Jurassic Morrison Formation: Paleobiology, v. 6, p. 208–232.

———, and Wexlar, Diane, 1979, Taphonomic investigations of owl pellets: Paleobiology, v. 5, p. 275–284.

Dott, R. H., Jr., 1983, Episodic sedimentation—How normal is average? How rare is rare? Does it matter?: Journal of Sedimentary Petrology, v. 53, p. 5–23.

Efremov, J. A., 1940, Taphonomy: New branch of paleontology: Pan-American Geologist, v. 74, no. 2, p. 81–93.

Farlow, J. O., 1976, A consideration of trophic dynamics of a Late Cretaceous large-dinosaur community (Oldman Formation): Ecology, v. 57, p. 841–857.

Fisher, D. C., 1981, Crocodilian scatology, microvertebrate concentrations, and enamel-less teeth: Paleobiology, v. 7, p. 262–275.

Fox, R. C., 1979, Mammals from the Upper Cretaceous Oldman Formation, Alberta. I. Alphadon Simpson (Marsupialia): Canadian Journal of Earth Sciences, v. 16, p. 91–102.

Frey, R. W., Voorhies, M. R., and Howard, J. D., 1975, Fossil and Recent skeletal remains in Georgia estuaries: Senckenbergiana Maritima, v. 7, p. 257–295.

Gifford, D. P., 1981, Taphonomy and paleoecology: A critical review of archeology's sister disciplines, *in* Schiffer, M. B., ed., Advances in Archeological Method and Theory, v. 4: Academic Press, New York, p. 365–438.

Glob, P. V., 1954, Lifelike man preserved 2000 years in peat: National Geographic, v. 105, p. 419–430.

Gordon, C. C., and Buikstra, J. E., 1981, Soil pH, bone preservation, and sampling bias at mortuary sites: American Antiquity, v. 46, p. 566–571.

Grayson, D. K., 1978, Reconstructing mammalian communities: A discussion of Shotwell's method of paleoecological analysis: Paleobiology, v. 4, p. 77–81.

Hanson, C. B., 1980, Fluvial taphonomic processes: Models and experiments, *in* Behrensmeyer, A. K., and Hill, A., eds., Fossils in the Making: University of Chicago Press, p. 156–181.

Hay, R. L., 1976, Geology of the Olduvai Gorge: A Study of Sedimentation in a Semiarid Basin: University of California Press, Berkeley and Los Angeles, 203 p.

Haynes, Gary, 1980, Evidence of carnivore gnawing on Pleistocene and Recent mammalian bones: Paleobiology, v. 6, p. 341–351.

———, 1983, Frequencies of spiral and green-bone fractures on ungulate limb bones in modern surface assemblages: American Antiquity v. 48, p. 102–114.

Hill, Andrew, 1979, Disarticulation and scattering of mammal skeletons: Paleobiology, v. 3, p. 261–274.

———, 1980, Early postmortem damage to the remains of some contemporary East African mammals, *in* Behrensmeyer, A. K., and Hill, A., eds., Fossils in the Making: University of Chicago Press, p. 131–152.

Holtzman, R. C., 1979, Maximum likelihood estimation of fossil assemblage composition: Paleobiology, v. 5, p. 77–89.

Honey, J. G., and Robinson, L. N., 1985, New locality of Paleocene mammals in south-central Montana (abs.): Society of Economic Mineralogists and Paleontologists, Annual Midyear Meeting, Abstracts, v. II, p. 42.

Horner, J. R., 1979, Upper Cretaceous dinosaurs from the Bearpaw Shale (marine) with a checklist of Upper Cretaceous dinosaur remains from marine sediments in North America: Journal of Paleontology, v. 53, p. 566–577.

———, 1982, Evidence of colonial nesting and 'site fidelity' among ornithischian dinosaurs: Nature, v. 297, p. 675–676.

———, 1984, The nesting behaviour of dinosaurs: Scientific American, v. 250, no. 4, p. 130–137.

———, 1985, Dinosaur faunal relationships of the Late Cretaceous (Campanian) Judith River Formation of Montana (abs.): Geological Society of America, Abstracts with Programs, v. 17, no. 4, p. 225.

———, and Makela, Robert, 1979, Nest of juveniles provides evidence of family structure among dinosaurs: Nature, v. 282, p. 296–298.

Korth, W. W., 1979, Taphonomy of microvertebrate fossil assemblages: Annals of the Carnegie Museum of Natural History, v. 48, Article 15, p. 235–285.

Koster, E. H., 1984, Sedimentology of a foreland coastal plain: Upper Cretaceous Judith River Formation at Dinosaur Provincial Park: Field Trip Guidebook, Canadian Society of Petroleum Geologists, 115 p.

———, and Currie, P. J., in press, Upper Cretaceous coastal plain sediments at Dinosaur Provincial Park, Southeast Alberta: Geological Society of America Centennial Field Guide, Rocky Mountain Section.

Laporte, L. F., and Behrensmeyer, A. K., 1980, Tracks and substrate reworking by terrestrial vertebrates in Quaternary sediments of Kenya: Journal of Sedimentary Petrology, v. 50, p. 1337–1346.

Lawton, Rebecca, 1977, Taphonomy of the dinosaur quarry, Dinosaur National Monument: University of Wyoming, Contributions to Geology, v. 15, no. 2, p. 119–126.

Lockley, M. G., 1986, The paleobiological and paleoenvironmental importance of dinosaur footprints: Palaios, v. 1, p. 37–47.

Lucas, S. G., 1981, Dinosaur communities of the San Juan Basin: A Case for lateral variations in the composition of Late Cretaceous dinosaur communities, *in* Lucas, S. G., Rigby, J. K., Jr., and Kues, B. S., eds., Advances in San Juan Basin Paleontology: University of New Mexico Press, p. 337–393.

MAAS, M. C., 1985, Taphonomy of a Late Eocene microvertebrate locality, Wind River Basin, Wyoming (U.S.A.): Palaeogeography, Palaeoclimatology, Palaeoecology, v. 52, p. 123–142.

MCKEE, E. D., CROSBY, E. J., AND BERRYHILL, H. L., 1967, Flood deposits, Bijou Creek, Colorado, June 1965: Journal of Sedimentary Petrology, v. 37, p. 829–852.

MELLETT, J. S., 1974, Scatological origin of microvertebrate fossil accumulations: Science, v. 185, p. 349–350.

MIALL, A. D., 1983, Basin analysis of fluvial sediments, *in* Collinson, J. D., and Lewin, J., eds., Modern and Ancient Fluvial Systems: International Association of Sedimentologists Special Publication 6, p. 279–286.

———, 1984, Principles of Sedimentary Basin Analysis: Springer-Verlag, New York, 490 p.

MORLAN, R. E., 1984, Toward the definition of criteria for the recognition of artificial bone alterations: Quaternary Research, v. 22, p. 160–171.

MOSSMAN, D. J., AND SARJEANT, W. A. S., 1983, The footprints of extinct animals: Scientific American, v. 248, no. 1, p. 74–85.

OLSON, E. C., 1985, Vertebrate paleoecology: A current perspective: Palaeogeography, Palaeoclimatology, Palaeoecology, v. 50, p. 83–106.

PAYNE, J. E., 1965, Summer carrion study of the baby pig *Sus scrofa:* Ecology, v. 46, p. 592–602.

SAHNI, ASHOK, 1972, The vertebrate fauna of the Judith River Formation, Montana: Bulletin of the American Museum of Natural History, v. 147, Article 6, 412 p.

SCHALLER, G. B., 1972, The Serengeti Lion: University of Chicago Press, Chicago, 480 p.

———, 1973, Golden Shadows, Flying Hooves: Alfred A. Knopf, New York, 287 p.

SCHINDEL, D. E., 1980, Microstratigraphic sampling and the limits of paleontologic resolution: Paleobiology, v. 6, p. 408–426.

SHIPMAN, PAT, 1981, Life History of a Fossil—An Introduction to Taphonomy and Paleoecology: Harvard University Press, Cambridge, Massachusetts, 222 p.

SHOTWELL, J. A., 1955, An approach to the paleoecology of mammals: Ecology, v. 39, p. 327–337.

SMITH, R. M. H., 1980, The lithology, sedimentology and taphonomy of flood-plain deposits of the Lower Beaufort (Adelaide Subgroup) strata near Beaufort West: Transactions, Geological Society of South Africa, v. 83, p. 399–413.

STERNBERG, C. M., 1950, Map 969A—Steveville, with Notes on the Dinosaur Quarries: Geological Survey of Canada.

———, 1970, Comments on dinosaurian preservation in the Cretaceous of Alberta and Wyoming: National Museum of Canada, Publications in Paleontology, No. 4, p. 1–9.

STEWART, D. J., 1981a, A field guide to the Wealden Group of the Hastings area and the Isle of Wight, *in* Elliot, T., ed., Field Guides to Modern and Ancient Fluvial Systems in Britain and Spain: Chapter 3, Proceedings, Second International Conference on Fluvial Sedimentology, Keele, England, September 1981.

———, 1981b, A meander-belt sandstone of the Lower Cretaceous of southern England: Sedimentology, v. 28, p. 1–20.

STOKES, W. L., 1963, Fossilized stomach content of a sauropod dinosaur: Science, v. 143, p. 567–577.

STRAHLER, A. N., 1957, Quantitative analysis of watershed geomorphology: Transactions of the American Geophysical Union, v. 38, p. 913–920.

VONHOF, J. A., 1965, The Cypress Hills Formation and its reworked deposits in southwestern Saskatchewan, *in* 15th Annual Field Conference Guidebook, Part 1, Cypress Hills Plateau: Alberta Society of Petroleum Geologists, p. 142–161.

VOORHIES, M. R., 1969, Taphonomy and population dynamics of an early Pliocene vertebrate fauna, Knox County, Nebraska: University of Wyoming, Contributions to Geology, Special Paper 1, 69 p.

———, 1981, Ancient ashfall creates a Pompeii of prehistoric animals: National Geographic, v. 159, p. 66–75.

WALDMAN, M., AND HOPKINS, W. S. JR., 1970, Coprolites from the Upper Cretaceous of Alberta, Canada, with a description of their microflora: Canadian Journal of Earth Sciences, v. 7, p. 1295–1303.

WEBB, S. D., MACFADDEN, B. J., AND BASKIN, J. A., 1981, Geology and paleontology of the Love Bone Bed from the Late Miocene of Florida: American Journal of Science, v. 281, p. 513–544.

MIOCENE FLUVIAL FACIES AND VERTEBRATE TAPHONOMY IN NORTHERN PAKISTAN

ANNA K. BEHRENSMEYER

Paleobiology Department, NHB-E207 MRC 121 Smithsonian Institution, Washington, D.C. 20560

ABSTRACT: The deposits of the Potwar Plateau in northern Pakistan show large-scale changes in fluvial deposition during the period between 14 and 6 my BP in three superimposed formations (Chinji, Nagri, Dhok Pathan) approximately 2700 m in total thickness. Five cross sections spaced throughout this sequence document intervals 30–80 m thick over lateral distances of 3–5 km. These show the architectural relationships of sand bodies, fine-grained facies, pedogenic horizons, and vertebrate fossil occurrences within floodplain deposits between major channel belt sandstones. Vertebrate fossil localities are particularly frequent within the fine-grained fill of different scales of abandoned floodplain channels (crevasse-splay and/or tributary channels). Such channels are larger and more common in the lower part of the sequence where average rates of sediment accumulation were 0.14–0.32 m/1000 yrs. as determined from the Siwalik magnetostratigraphic time framework. Higher average accumulation rates of 0.46–0.48 m/1000 yrs for the upper part of the sequence correspond to fewer fine-grained channel fills and lower fossil abundance. Changes in type and frequency of floodplain channels are attributed to a combination of differences in the fluvial regimen and in sediment accumulation rates. The quality of the vertebrate record is linked to the frequency and/or magnitude of abandoned channels on the sub-Himalayan alluvial plain and thus to large-scale tectonic and climatic controls on the fluvial systems.

INTRODUCTION

The Siwalik deposits of the Potwar Plateau in northern Pakistan have been subjected to intensive study and fossil collecting since 1973 by members of the Joint Geological Survey of Pakistan—Harvard University Research Project and other affiliated groups. This has resulted in well established lithostratigraphy and magnetostratigraphy for large areas of Siwalik exposures in the Soan Synclinorium, which lies between the Salt Range and the Himalayan foothills (Barry and others, 1980; Tauxe and Opdyke, 1982; Behrensmeyer and Tauxe, 1982; Raza, 1983; Johnson and others, 1985). The non-marine deposits span Miocene through Pleistocene time and are well over 3000 m thick. Fossil vertebrates occur throughout and provide a record of evolution, immigration and extinction in the Indian subcontinent over the past 18 my (Barry and others, 1982; Barry and others, 1985).

The vertebrate record varies in the degree of concentration of fossils in different lithofacies and stratigraphic levels. The sediments also vary through time in average rates of accumulation, ratios of sandstone to mudstone, and lateral relationships of the course and fine-grained units. The focus of this paper is how occurrences of vertebrate fossils are related to changing fluvial environments within the sub-Himalayan deposits. A more general goal is to understand how conditions leading to fossil preservation may be linked to large-scale controls on fluvial sedimentation.

OVERALL SEDIMENTARY CONTEXT AND METHOD OF STUDY

The Siwalik formations used for this study are distinguished primarily by gross sandstone percentages, with the Chinji Formation <50 percent sand, the Nagri >50 percent sand, and the Dhok Pathan <50 percent sand (in ascending stratigraphic order) (Fatmi, 1973; Pilbeam and others, 1979). The deposits consist of alternating sandstones and finer grained sediments which represent large-scale "cycles" of channel and floodplain deposition. The major sand bodies are tabular units 10–30 m thick which extend tens of kilometers along strike in directions oblique or roughly parallel to the dominantly east-southeast current direction. The floodplain deposits are primarily silt but include significant amounts of clay, marl, pedogenic carbonate, sand, and intraformational conglomerates. They are internally complex, including channel, levee, and overbank facies that represent a variety of smaller scale depositional systems (crevasse-splays, tributaries) which were active on the floodplain(s) of the major rivers.

Since nearly all of the vertebrate fossils occur in the floodplain deposits, these have been the focus for study of the relationships between sediments and fossils. Previous work has been done at a fine level of resolution on specific intervals within the Dhok Pathan Formation (Badgley, 1982) and the Chinji Formation (Raza, 1983). The goal of this study is to gain an overview of large scale patterns of vertebrate preservation throughout the Chinji, Nagri and Dhok Pathan Formations.

Lateral facies studies were done at five different levels within these formations, which total about 2700 m of thickness in the correlated sections from the northern and southern limbs of the Soan Synclinorium (Fig. 1). The exposures are nearly continuous, and documentation consisted of vertical lithologic logs spaced 250 to 700 m apart and detailed mapping of all contacts and sedimentary units that could be traced between the logs. This resulted in five panel sections which show sedimentary relationships within 30- to 80-m-thick intervals over distances of 3.1 to 4.2 km (Fig. 2). Current indicators (large-scale trough cross-stratification, directions of channel axes) were measured wherever possible in the sandstone bodies, and pedogenic horizons (identified by color zonation, ped structure, cutans, carbonate and/or limonite glaebules, rhyzoliths, lack of primary bedding) were documented within both fine- and coarse-grained units. The sedimentary context of all vertebrate fossils was also recorded.

The sections shown in Figure 2 represent sand bodies and finer grained units which are typical of the fossil-bearing floodplain deposits for each level through the sequence of formations. The sand bodies vary from sheet to ribbon geometry (width/depth ratios <50 for ribbon sands); some are quite large (e.g., 10 m thick, >1 km wide normal to current direction) but still distinct from the major channel belt sands. In the following discussion, these will be referred to as secondary versus major sand bodies. The pack-

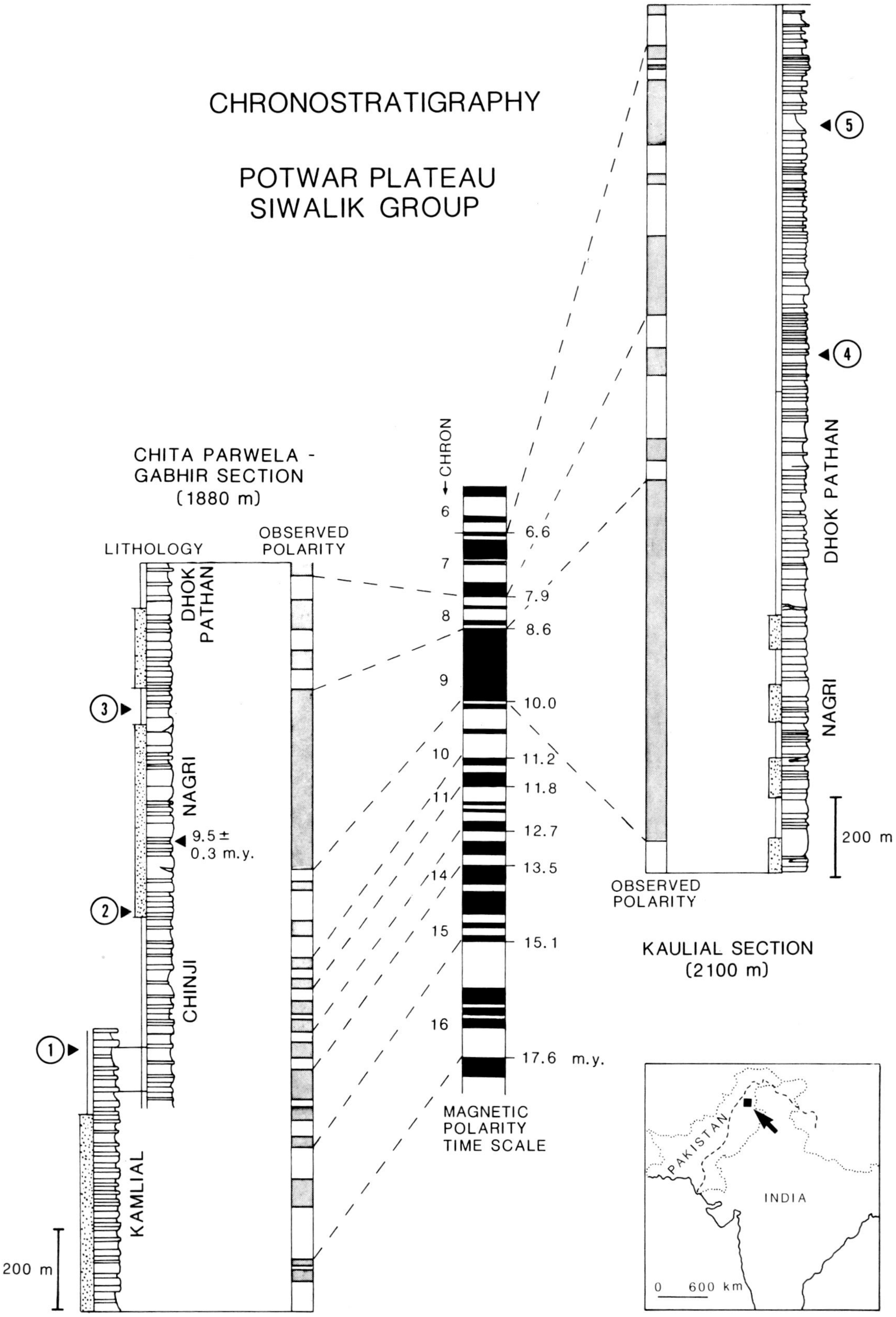

FIG. 1.—Chronostratigraphic framework for the Potwar Plateau Miocene deposits, showing positions of the five levels used for the lateral facies study (circled numbers). Left column represents the section on the southern limb of the Soan Synclinorium, right column the section on the northern limb (distance approximately 50 km). (Compiled from Tauxe and Opdyke, 1982 and Johnson and others, 1985.) Stippled bars to left of the lithologic columns indicate parts of the section with greater than 50 percent sandstone. Inset map shows location of the Potwar Plateau (Box) in northern Pakistan.

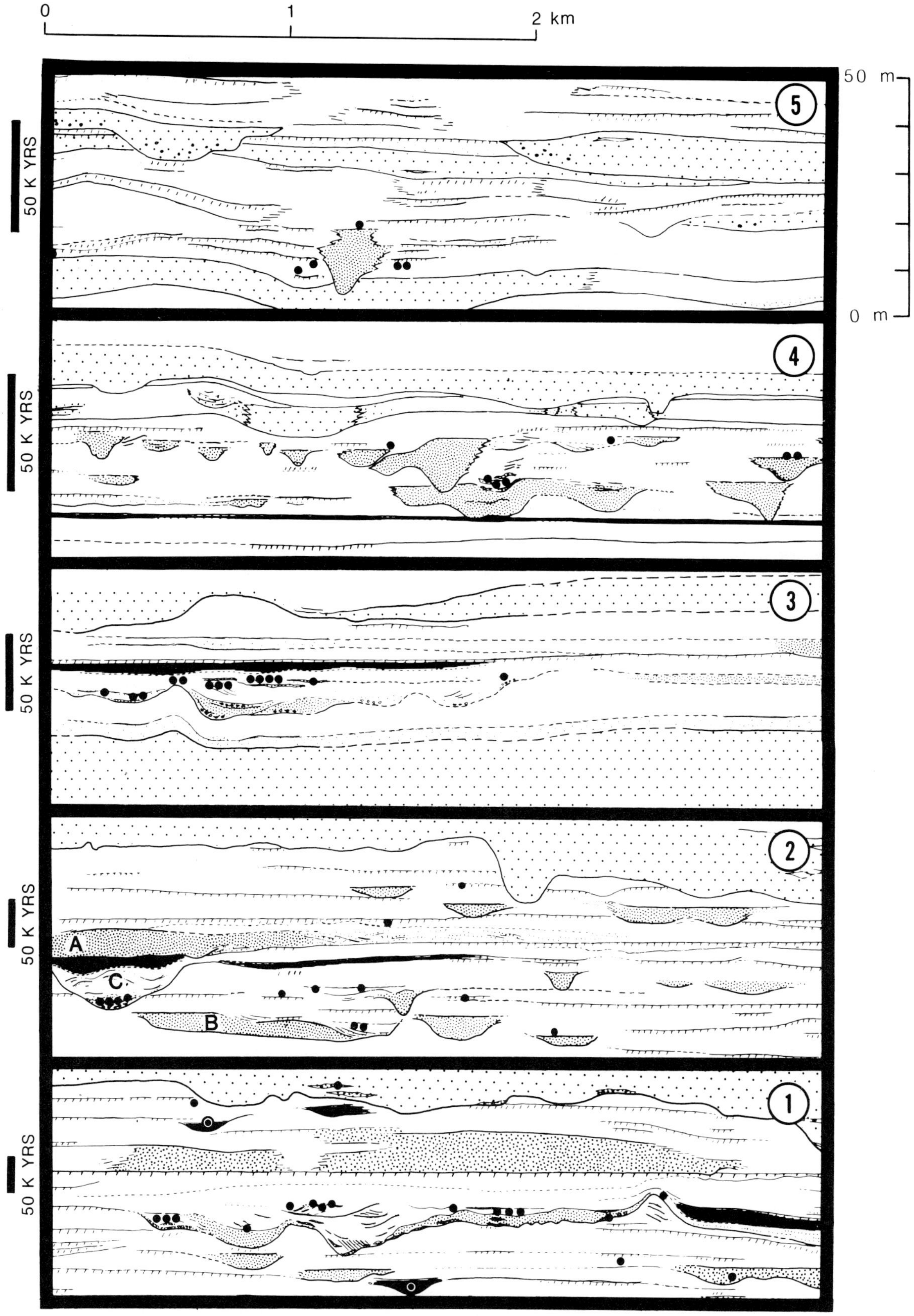

FIG. 2.—The five lateral facies sections, drawn to the same scale with 20× vertical exaggeration. For ease of comparison, only similar-width portions of the overall sections are shown. Vertical bars to the left show the thickness equivalent to 50,000 years, as calculated from data on Figure 1 for 300-m segments centered on each study level. Symbol key: coarse stippling = major channel belt sandstones, fine stippling = secondary channel sands within the floodplain deposits, dark shading = silty clay, blank = silts and marls, hatched lines = pedogenic horizons. The black dots show vertebrate localities, with 1 dot = 1–10 fossils, 2 dots = 11–100, 3 dots = 101–1000, 4 dots = 1000. The three-dimensional outcrop pattern was abstracted somewhat to make the two-dimensional diagrams. Laterally continuous pedogenic horizons were used as horizontal datums.

ages of fine-grained sediment between major sand bodies are "major floodplain deposits" and the distinct siltstones, sand bodies, marls, and clays within them are "secondary floodplain facies."

LITHOFACIES PATTERNS AND TAPHONOMY

Level 1.—The lower Chinji (13.1 my) section covers 3.3 km with nine logs documenting as much as 65 m in thickness. The major floodplain unit lies below a sheet sand 10–15 m thick and contains numerous sand bodies representing channels of varying geometries and sizes. There are also erosional troughs with fine-grained channel fill. Paleosol horizons occur throughout the overbank facies, in some cases directly lateral to channel margins.

The section includes a continuous irregular erosional feature with discontinuous sand lenses along its base. The deposits above the erosion surface fine upward to an extensive clayey-silt paleosol that was used as a horizontal datum. The deepest point relative to this datum is filled with silt/fine sand couplets with large-scale foreset bedding that dips steeply into the erosional contact. Above these inclined couplets are interbedded fine sands, intraformational conglomerates and silts with variable dips showing multiple cross-cutting stages of channel fill that fine upward to the capping paleosol. The lower units above the basal erosion surface have little sign of bioturbation, but middle to upper parts of the fill are root-mottled and locally burrowed.

The erosional feature is interpreted as an abandoned channel, cut by the same river system that laid down the major sheet sands. This feature is so large (1.5 km across normal to current direction) that it might be termed a "failed avulsion," where erosion took place during one or more major flood events, and subsequent fill occurred during sporadic (and waning) flow when the active channel failed to occupy this position on the floodplain. Sedimentary features of the fine-grained channel fill indicate rapid (perhaps 1.5–15 m/1000 yrs) deposition relative to the average overall rate of 0.14/1000 yrs (Table 1). The abandoned channel is regarded as distinct from "paleogully systems" of Kraus and Middleton (1984) and Kraus and others (1985) because the lower part of the fill indicates rapid sedimention in deep (5 m) water, because of the uniqueness of the feature (there are no signs of other erosional features lateral to it), and because laterally associated paleosols do not indicate prolonged periods of land surface stability. Further discussion of this and other sedimentary features will be given elsewhere (Behrensmeyer, in prep.).

TABLE 1.—RELATIONSHIP OF TIME COMPONENTS IN THE LATERAL FACIES SECTIONS TO VERTEBRATE FOSSIL ABUNDANCE

Facies Level:	Older → 1	2	3	4	→ Younger 5
SEDIMENT ACCUM. RATE: m/1000 YRS	0.14	0.19	0.32	0.48	0.46
# DISTINCT PALEOSOLS (IN 20 m)	6.4	8.0	2.6	5.2	7.6
AVERAGE MAXIMUM YRS PER SOIL	22,300	13,200	24,000	8,000	5,700
# FLOODPLAIN CHANNELS (IN 20 m × 1 km)	1.1	2.4	0.6	7.0	0.8
# VERTEBRATE LOCALITIES (IN 20 m × 1 km)	2.3	2.0	1.7	1.4	.9

Note: Numbers are based on chronostratigraphic data in Figure 1 and lithologic data shown in Figure 2, with average rates of accumulation calculated from Figure 1 for a 300-m interval bracketing each level. Numbers of soil horizons are standardized for 20 m of thickness; maximum years per soil = time represented in 20 m/# paleosols in this interval. Numbers of floodplain channels and vertebrate localities are standardized for a 20 m × 1 km portion of each lateral facies section.

The highest density of vertebrate occurrences lies within the middle to upper part of the abandoned channel fill in the crosscutting lenses of silts, conglomerates, and fine sands. The vertebrate remains include associated and articulated body parts, and most are unabraded. A wide range of body sizes and taxa are represented. Bones are fragmentary and whole, and many show evidence of carnivore activity. They occur as dense local patches dispersed through the silt/sand lenses but not as a continuous "bonebed." The evidence points to autochthonous, attritional accumulation and burial, with some local effects of water transport and sorting as sporadic flow filled the abandoned channel.

Low density occurrences of vertebrate remains are associated with some of the other, small-scale channel fills and with levee and overbank facies. A few very abraded fossils were also documented in a conglomerate lens within the major channel sand body.

Level 2.—The upper Chinji (10.5 my) section covers 2.5 km with 9 logs measuring as much as 80 m in thickness. A major sand body caps floodplain deposits which include a number of secondary sand units. Many of these appear to be sheetlike but are actually broad channel sand bodies occupying shallow depressions eroded into the mudstone facies (Fig. 2:2). One (A) passes laterally into well developed levee facies that can be followed more than 1.3 km normal to the channel edge. Another (B) passes from left to right into lateral aggradation surfaces and a fine-grained channel fill over a distance of 225 m normal to current direction.

The largest feature in this section is an erosional trough (Fig. 2:2, C) 10 m in depth which is floored by a conglomerate of algal oncolites and soil glaebules. This is overlain by a thick deposit of interbedded silt and silt/sand couplets with highly variable depositional dips, followed by a clay with deep, sandfilled desiccation cracks. This sequence is interpreted as a channel fill in which active flow was abruptly reduced after deposition of the basal conglomerate.

The richest vertebrate locality occurs in this abandoned channel fill, with thousands of bones concentrated immediately above the basal oncolitic conglomerate in the fine-sand/silt facies. The taphonomic nature of the bone assemblage is much the same as in the major abandoned channel in Level 1, with somewhat more variability in abrasion and fewer associated remains. A second channel in Level 2 has approximately 30 identifiable specimens associated with fine-grained sediments in an abandoned channel lateral to a point-bar deposit.

Other isolated or low density bone occurrences are found in the levee facies and in pedogenic horizons formed on the mudstone facies.

Level 3.—The Nagri (8.8 my) level includes approximately 25 m of floodplain deposits which rest upon a 76-m-thick multistoried sandstone in the sand-dominated Nagri Fm. Exposures are somewhat poorer than at other levels, and six logs document 4.2 km in this section. Secondary sand bodies include sheets with conformable basal contacts and a complex of lenses associated with a major erosional feature within the mudstone facies (left half, Fig. 2:3).

This erosional feature has an irregular lower contact overlain by highly variable units of sand, inclined silt/sand couplets, *in situ* stromatolites as much as 0.6 m in diameter, oncolitic conglomerates, carbonate and iron oxide glaebule conglomerates, and bioturbated silty to sandy clays. These deposits are a maximum of 12 m thick and fine upward to a capping paleosol. The erosional feature and its overlying deposits represent a large-scale channel which bears many similarities to the abandoned channel at Level 1. The channel fill of Level 3 is more complex and indicates highly variable flow before abandonment. The large stromatolites in growth positions also indicate extended periods of relatively quiet (non-turbid) water within the channel.

Fossil vertebrates occur in great abundance within the middle to upper levels of the abandoned channel fill. A number of different taphonomic histories are evident, and assemblages include weathered, abraded bones mixed with fresh, associated elements. Low numbers of vertebrae indicate winnowing, and surface scratches, breakage and vertical orientations of bones in excavated samples document trampling prior to burial (Behrensmeyer and others, 1986). Carnivore activity is also evident from tooth marks and breakage patterns. The unabraded components of the assemblages probably represent attritional accumulation around waterholes within the abandoned channel, with minor redistribution due to sporadic flow events. The fauna represents the entire spectrum of Nagri taxa including the hominoid *Sivapithecus*.

Other fossil occurrences have not been found outside of the context of the abandoned channel. The floodplain facies appear to be more homogeneous than at other levels, lacking small-scale channels and laterally variable paleosol horizons.

Level 4.—The "Sub-U" level (8.2 my) in the Dhok Pathan Formation is characterized by 4- to 10-m-thick sheet sandstones with associated gray silts overlying 15- to 20-m-thick floodplain deposits. Thirteen vertical logs document approximately 3.2 km of nearly continuous exposures. The floodplain deposits include numerous small-scale ribbon sand bodies of a buff sand. A previous study of facies relationships at this level (Behrensmeyer and Tauxe, 1982) demonstrated that the sheet sand bodies and gray silts represent the "blue-gray" fluvial system, which is distinct from the "buff system" that deposited ribbon sands and associated red-brown floodplain deposits. The buff sands are attributed to one or more tributary systems that fed the major river represented by the blue-gray sheet sands (Behrensmeyer and Tauxe, 1982).

The buff sand bodies are more laterally restricted, with lower width/depth ratios, than ribbon sands in Levels 1 and 2. They are also more complex internally, with evidence of mutistory deposition. Fine-grained fills are uncommon and restricted to the uppermost portions of these channels. In their overall geometry and lateral associations with fine-grained facies and other sand bodies, they resemble ribbon sands in Oligocene-Miocene fluvial deposits in the Spanish Ebro Basin (Friend and others, 1979).

Abundant vertebrate fossils occur at the top of one of the larger buff channels. They are associated with dipping lenses of carbonate glaebule and mudclast conglomerate which alternate with fine sands and silts. These fill a small depression cut into the top of the underlying sand body and represent deposition during temporary reactivation of flow in a nearly abandoned channel. The fossils include a mixture of fresh and abraded material representing both local and distant sources of vertebrate remains.

Other vertebrate occurrences are associated with the tops and margins of the buff channels, as is typical of this part of the Dhok Pathan Formation (Badgley and Behrensmeyer, 1980; Badgley, 1982). None were found in the fine grained floodplain sediments or paleosol horizons at this level.

Level 5.—The upper part of the Dhok Pathan Formation is markedly different from the other levels, as shown in the fifth section (6.9 my). Twelve logs document as much as 70 m of thickness over a distance of 3.1 km. The major channel belt sands are as much as 8 m thick and can be traced laterally into well bedded levee facies and paleosols developed on overbank facies.

The floodplain deposits are homogeneous with evenly bedded fine sands and silts and thick silty clays with pedogenic features. There are very few secondary sand bodies. One ribbon sand more than 10 m thick shows evidence of vertical aggradation without lateral migration after the initial erosional event. In contrast to Level 4, there is no difference in lithology between the major sheet sands and the ribbon sand, and they appear to belong to the same ("buff") fluvial system.

Intensive searching for fossils at this level resulted in a cluster of low to moderate density localities in the levee and floodplain paleosols developed on mudstone facies lateral to and above the ribbon sand body.

DISCUSSION

Sixty percent of the 45 vertebrate localities in the five study levels are associated with secondary channels in the major floodplain deposits (Fig. 3), and those with more than 100 specimens occur only in this context. Fossils also tend to occur in the middle to top of relatively fine-grained fills of these channels, although they may be associated with thin beds of coarser sediment. This is true over a wide range of channel widths from 50 m to 1.5 km and is a recurring pattern throughout the vertical sequence of Siwalik formations, even though the overall nature of the fluvial systems changed significantly through time. The richest fossil-bearing units are associated with the large-scale abandoned channels in the Chinji and Nagri Formations (Levels 1–3).

Fossils that are present in other contexts, such as paleosols and levees, are typically few in number, and portions of the Siwalik sequence which have few secondary channels (e.g., Level 5) have relatively poor fossil records.

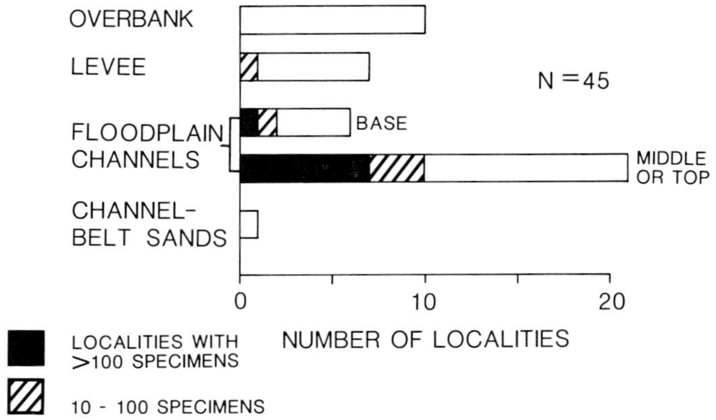

FIG. 3.—Summary of the sedimentary context of the 45 vertebrate localities documented in the five lateral facies studies. (Note: not all localities are shown in the representative portions of the sections in Fig. 2.)

Documentation of patterns of extinction, immigration, evolution, and overall faunal change therefore depend to a considerable degree on the record preserved in secondary channels on the Miocene floodplains.

Some of these channels probably were caused by crevassing events in which various scales of erosion and deposition took place during periods of overbank flow from a major active channel. In avulsion-dominated systems, such as the lower Indus River of recent times, major channel-switching events are well documented (Holmes, 1968) and provide a realistic model for the large-scale abandoned channels in Levels 1 and 3. Smaller scale ribbon sand bodies in floodplain deposits in other fluvial sequences have been attributed to anastamosed channels in rapidly aggrading systems (e.g., Smith and Putnam, 1980; Smith, 1986) and to the evolution of splay channels following major avulsion (Smith and Cross, 1985). Similar causes may account for some of the secondary channels in the Siwalik floodplain deposits. Tributaries that drained the floodplain or local uplands provide an alternative mechanism for these channels, and further work will be needed to distinguish between the two possible causes in the absence of obvious lithologic differences, such as in the buff versus blue-gray sands in Level 4.

From the standpoint of the vertebrate record, distinguishing the cause of floodplain channels is less important at this stage of analysis than establishing the pattern of association of fossils with such channels, particularly with various types of abandoned channel fill. An abandoned channel context for vertebrate fossil concentrations has been recognized by many workers, notable recent examples being bone concentrations in the Cretaceous Wealden Group of Great Britain (Stewart, 1981), the Permian Cutler Formation of New Mexico (Berman and others, 1985), and Carboniferous channels in Ohio (Hook and Ferm, 1985). Such occurrences have been recorded as typical of particular localities or formations rather than as a recurring pattern through a long stratigraphic interval which is documented here for the Siwalik sequence.

Given the litho- and chronostratigraphic framework available for the Siwalik formations, we can address the larger question of possible controls on the frequency of secondary channels and their associated vertebrate record. In Table 1, sediment accumulation rate is shown in relation to some of the measurable characteristics of the five study levels. Paleosol horizons are taken as rough indicators of hiatuses in the section, and these appear to reflect generally shorter time breaks as the rate of accumulation increases upward. The number of secondary channels is highly variable through time, reflecting different styles of floodplain aggradation, but the frequency of abandoned channels with fine-grained fill decreases in the upper two levels.

If rates of floodplain channel formation were constant through time, then we might expect that slower sediment accumulation and longer hiatuses would result in a higher probability of floodplain channel sands occurring in any given outcrop section (particularly if they were tributary channels). With more time represented in any given hiatus, there might also be a greater chance of recording a "rare event" such as a large-scale crevasse channel or "failed avulsion." Thus, one control on the frequency of floodplain channels may be rate of sediment accumulation.

Another control on the number and size of floodplain channels is obviously the nature of the fluvial systems themselves. In the lower three study levels, secondary channels and particularly the large-scale abandoned channels may reflect an avulsion-dominated fluvial regimen enhanced by relatively low average rates of accumulation. A different fluvial system (the "buff" system) was responsible for most of the floodplain facies in the Dhok Pathan Formation on the north side of the Soan Synclinorium. This system was characterized by a different style of floodplain deposition, operating in the context of more rapid average accumulation rates, and it generated fewer secondary channels conducive to vertebrate preservation in the upper two study levels.

Other factors that might have influenced floodplain channel frequency in the study levels include local climate (i.e., as a control on the amount of vegetation on the floodplain) and the proximity of a major channel (e.g., more levee facies with fewer channels proximal to the channel belt) (Behrensmeyer, in prep.).

Since relatively rapid permanent burial is a prerequisite for bone preservation (Behrensmeyer, 1978), the inverse relationship between number of localities and average rate of sediment accumulation in Table 1 is initially puzzling. This can be attributed, however, to the combination of fluvial regimen and a low overall accumulation rate which led to numbers of abandoned channels in the lower study levels. The rate and mode of filling of these channels was optimal for vertebrate preservation, in contrast to sedimentation patterns in other subenvironments on the floodplains. In the upper two study levels, fossils still tend to occur in the tops of margins of secondary channels, but a change in fluvial regimen resulted in fewer opportunities for vertebrate preservation in this context. Other favorable environments for preservation apparently were also uncommon in this fluvial setting.

Floodplain channels represent a complex of processes that could work together to promote bone preservation. The presence of the topographic lows and relatively rapid infilling of the channel troughs are undoubtedly two of the critical factors. Abandoned channels are also places where surface and ground water are concentrated, and they provide favorable habitats for plant growths, animal activity and predation (such as at waterholes) which would contribute attritional vertebrate remains. Faunas from this context should represent localized, time-averaged samples of the original vertebrate communities inhabiting areas proximal to the floodplain channels. Such channels would have important effects on the ecology of the floodplain floras and faunas as well as on their preservation.

CONCLUSION

The taphonomic pattern of fossil preservation in association with abandoned channels in the Siwalik sequence is supported by other studies of the Chinji and Dhok Pathan Formations (Badgley and Behrensmeyer, 1980; Badgley, 1982; Raza, 1983). The pattern seems to hold regardless of the scale of channel or the major fluvial system to which it belongs. Thus, the quality of the vertebrate record is linked to the frequency and scale of abandoned channels in the Siwalik sequence. The occurrence of such channels in fluvial systems should relate to climatic and tectonic controls on the fluvial regimen, as expressed in discharge, flood periodicity, sediment type, net rate of sediment accumulation, and floodplain vegetation. It seems likely that fluvial systems characterized by a greater number and variety of abandoned channel segments may preserve a larger proportion of vertebrate remains than many other types of systems throughout the stratigraphic record. This would be an example of a "taphonomic megabias" (Behrensmeyer and Kidwell, 1985) in which the large-scale tectonic and climatic controls could be approached through study of the causes of abandoned floodplain channels.

The presence of vertebrate fossils in such contexts may hold important information concerning rates and mode of channel fill which would be of value to sedimentologists. Finally, comparative taphonomic studies of preservational modes in the Siwalik Group and other fluvial sequences, such as the Willwood Formation of Wyoming, where vertebrate fossils are most comon in paleosols (Bown and Kraus, 1981), could provide new insights on the relationships between large-scale controls on fluvial sedimentation and different contexts for vertebrate preservation.

ACKNOWLEDGMENTS

I thank David Pilbeam and Ibrihim Shah for their support of sedimentological research as part of the Joint GSP-Harvard Project in Pakistan. Catherine Badgley, John Barry, Mahmood Raza, and Lisa Tauxe have provided much help and inspiration throughout the study. I have greatly appreciated stoic and at times heroic assistance in section measuring from Asif Shah, Khalid Sheikh, Imran Khan, Majeed, Philip Dormitzer, Dana Coyle, Fred Grady, Bakh Jimal, Mark Soloman and Julie Cormack. I thank Tom Bown and Emlyn Koster for their helpful critical reviews of the initial draft of this paper, and Bob Hook and Scott Wing for comments on the final draft.

REFERENCES

BADGLEY, C. E., 1982, Community Reconstruction of a Siwalik Mammalian Assemblage: Unpublished Ph.D. Dissertation, Yale University, New Haven, Connecticut, 364 p.

———, AND BEHRENSMEYER, A. K., 1980, Paleoecology of Middle Siwalik sediment and faunas of the Potwar Plateau: Palaeogeography, Palaeoclimatology, Palaeoecology, v. 30, p. 133–155.

BARRY, J. C., BEHRENSMEYER, A. K., AND MONAGHAN, MARC, 1980, A geologic and biostratigraphic framework for Miocene sediments near Khaur Village, northern Pakistan: Postilla, No. 183, Pts. 1 and 2, 19 p.

———, LINDSAY, E. H., AND JACOBS, L. L., 1982, A biostratigraphic zonation of the middle and upper Siwaliks of the Potwar Plateau of northern Pakistan: Palaeogeography, Palaeoclimatolology, Palaeoecology, v. 37, p. 95–130.

———, JOHNSON, N. M., RAZA, S. M., AND JACOBS, L. L., 1985, Mammalian faunal change in the Neogene of southern Asia and its relation to global climatic and tectonic events: Geology, v. 13, p. 95–130.

BEHRENSMEYER, A. K., 1978, Taphonomic and ecologic information from bone weathering: Paleobiology, v. 4, p. 150–162.

———, GORDON, K. D., AND YANAGI, G. T., 1986, Trampling as a cause of bone surface damage and pseudo-cutmarks: Nature, v. 319, p. 768–771.

———, AND KIDWELL, S. M., 1985, Taphonomy's contributions to paleobiology: Paleobiology, v. 11, p. 105–119.

———, AND TAUXE, L., 1982, Isochronous fluvial systems in Miocene deposits of northern Pakistan: Sedimentology, v. 29, p. 331–352.

BERMAN, D. S., REISZ, R. R., AND EBERTH, D. A., 1985, *Ecolsonia cutlerensis,* an Early Permian dissorophid amphibian from the Cutler Formation of north-central New Mexico: New Mexico Bureau of Mines and Mineral Resources Circular 191, p. 1–31.

BOWN, T. M., AND KRAUS, M. J., 1981, Vertebrate fossil-bearing alluvial paleosols (Willwood Formation, north-west Wyoming, U.S.A.): Implications for taphonomy, biostratigraphy, and assemblage analysis: Palaeogeography, Palaeoclimatology, Palaeoecology, v. 43, p. 95–128.

FATMI, A. N., 1973, Lithostratigraphic units of the Kohat-Potwar Plateau Province, Indus Basin Pakistan: Memoirs of the Geological Survey of Pakistan, No. 10, p. 1–80.

FRIEND, P. F., SLATER, M. J., AND WILLIAMS, R. C., 1979, Vertical and lateral building of river sandstone bodies, Ebro Basin, Spain: Journal of the Geological Society of London, v. 136, p. 39–46.

HOLMES, D. A., 1968, The recent history of the Indus: The Geographical Journal of the Royal Geographical Society of London, v. 134, p. 367–382.

HOOK, R. W., AND FERM, J. C., 1985, A depositional model for the Linton tetrapod assemblage (Westphalian D, Upper Carboniferous) and its palaeoenvironmental significance: Philosophical Transactions of the Royal Society of London, v. B311, p. 101–109.

KRAUS, M. J., AND MIDDLETON, L. T., 1984, Dissected paleotopography and baselevel fluctuations in an ancient fluvial sequence: Geological Society of America, Abstracts with Program, v. 16, p. 227.

———, BOWN, T. M., KVALE, E. P., AND VONDRA, C. F., 1985, Upper Jurassic/Lower Cretaceous and Paleogene alluvial sediments of the Bighorn Basin, northwest Wyoming, *in* Flores, R. M., and Harvey, Michael, eds., Field Guidebook to Modern and Ancient Fluvial Systems in the United States: Third International Fluvial Sedimentology Conference, Fort Collins, Colorado, p. 19–44.

JOHNSON, N. M., STIX, JOHN, TAUXE, LISA, CERVENY, P. F., AND TAHIRKHELI, R. A. K., 1985, Paleomagnetic chronology, fluvial processes, and tectonic implications of the Siwalik deposits near Chinji Village, Pakistan: Journal of Geology, v. 93, p. 27–40.

PILBEAM, D. R., BEHRENSMEYER, A. K., BARRY, J. C., AND SHAH, S. M. I., 1979, Miocene sediments and faunas of Pakistan: Postilla, No. 179, p. 1–45.

RAZA, S. M., 1983, Taphonomy and paleoecology of Middle Miocene vertebrate assemblages, southern Potwar Plateau, Pakistan: Unpublished Ph.D. Dissertation, Yale University, New Haven, Connecticut, 414 p.

SMITH, D. G., 1986, Anastamosing river deposits, sedimentation rates and basin subsidence, Magdalena River, northwestern Columbia, South America: Sedimentary Geology, v. 46, p. 177–196.

———, AND PUTNAM, P. E., 1980, Anastamosed river deposits: modern and ancient examples in Alberta, Canada: Canadian Journal of Earth Sciences, v. 17, p. 1396–1406.

SMITH, N. D., AND CROSS, T. A., 1985, Avulsion-controlled fluvial evolution: The Cumberland Marshes, east-central Saskatchewan: Abstracts Volume, Third International Fluvial Sedimentology Conference, Fort Collins, Colorado, p. 36.

STEWART, D. J., 1981, A field guide to the Wealden Group of the Hastings Area and Isle of Wight, *in* Elliot, T., ed, Field Guides to Modern and Ancient Fluvial Systems in Britain and Spain: Proceedings, Second International Conference on Fluvial Sedimentology, University of Keele, U. K., p. 3:1–30.

TAUXE, LISA, AND OPDYKE, N. D., 1982, A time framework based on magnetostratigraphy for the Siwalik sediments of the Khaur area, northern Pakistan: Palaeogeography, Palaeoclimatology, Palaeoecology, v. 37, p. 43–61.

PART IV
FACIES AND ARCHITECTURE

ANATOMY OF A GRAVELLY MEANDER LOBE IN THE SASKATCHEWAN RIVER, NEAR NIPAWIN, CANADA

JANET E. CAMPBELL
Saskatchewan Research Council, 15 Innovation Blvd., Saskatoon, Sask., S7N 2X8
AND
HUGH E. HENDRY
Department of Geological Sciences, University of Saskachewan, Saskatoon, Sask., S7N 0W0

ABSTRACT: Foxe's Flat, on the south bank of the Saskatchewan River, is underlain by the gravel, sand and silt of a meander lobe. It was produced by northeasterly migration of the river when it was flowing about 5 to 10 m higher than its present level.

The 10 to 12 m of deposits in Foxe's Flat comprise four facies. Facies 1, at the base, consists of crudely stratified pebble and cobble gravels and minor sands with lateral accretion bedding dipping toward the paleochannel at less than 5°. It is overlain by, and interbedded with, discontinuous lenses of sands and gravelly sands (facies 2) which contain trough and planar crossbedding. Deposits of facies 3 are interbedded sand, silt and clay with inclined bedding dipping toward the paleochannel at angles of as much as 12°; the beds of facies 3 may interfinger downdip with either facies 1 or facies 2. The sequence is capped by horizontally layered clayey silt and fine sand of facies 4.

The four facies represent deposition in the channel and on a point bar surface (facies 1), on the top of the gravelly platform of the point bar (facies 2) in chute and slough channels on the point bar (facies 2), on lower parts of the point bar during falling stage (facies 2), on an inner accretionary bank (facies 3), and on a floodplain (facies 4). Individual accretion surfaces are not generally traceable across all four facies but are prominent in both the upper and lower parts of the sequence. Preservation of lateral accretion bedding in the gravels may have been favored by transport of the gravels in low-relief sheets. The more steeply dipping bedding of the inner accretionary bank was scoured repeatedly by strong flows. Wide variations in stream discharge transported gravels onto the flood plain and deposited mud drapes low on the point bar. Some boulders were deposited from rafted ice. Vertical sequences in Foxe's Flat are similar to those described from other gravelly meandering stream deposits and from some braided stream deposits. The lateral transition from lateral accretion bedding in the gravels (facies 1) through lenticular sand deposits (facies 2) to the lateral accretion bedding of silts and sands (facies 3) indicates systematic migration of a stream with only one major channel.

INTRODUCTION

Deposition and development of point bars in meandering rivers has been a major theme of sedimentation studies. The traditional lithofacies model for point bars, which arose from work done on both ancient and modern sediments, incorporates a fining upward sequence (Fisk, 1944, 1947; Sundborg, 1956; Allen, 1963, 1965, 1970a,b; Bernard and others, 1970). In recent years, studies have demonstrated that many modern meandering streams with coarse-grained sediment load conform poorly to this model (McGowen and Garner, 1970; Bluck, 1971, 1976; Jackson, 1976; Nanson, 1980). Yet another example is provided in the postglacial alluvium of the Saskatchewan River Valley, Saskatchewan (Fig. 1). The purpose of this study is to document the facies and the vertical and lateral relationships of the sediments of a meander lobe produced by lateral migration of the graveliferous Saskatchewan River.

LOCATION AND GEOLOGICAL SETTING

The meander lobe (Foxe's Flat) examined in this study is one of a series which forms the lower terraces in the Saskatchewan River Valley. It lies on the south side of the Saskatchewan River approximately 20 km upstream from the town of Nipawin in east-central Saskatchewan. The top of Foxe's Flat was 10–12 m above the river level when the field work was done, but in 1985 it was flooded as part of the Nipawin hydroelectric project.

The Saskatchewan River Valley west of Nipawin is 1 km wide and 50 to 70 m deep. It is cut into Quaternary deltaic sands and silts interbedded with lacustrine silts and clays (Fig. 2). During the last glaciation of the Nipawin area (beginning approximately 11,500 yrs BP), the river eroded through these sediments and into the underlying till to form the present valley (Christiansen, pers. commun.) The upper terraces (Fig. 2) are remnants of a gravel alluvial plain deposited by the river during the initial incision.

The lower terraces within the valley have been interpreted as abandoned point bar deposits of a confined 1-km-wide meander belt. The meander lobe is underlain by coarse gravels deposited during the final stages of deglaciation. The meander belt began to form during the early postglacial period (approximately 7,000 yrs BP) when the discharge of the river was significantly greater than at present. The channel does not appear to be migrating today.

METHOD OF FIELD INVESTIGATION

In the summer of 1984, gravel quarrying of Foxe's Flat exposed a 1-km-long section of point bar deposits. The section is in the upstream portion of a meander bend in the present river (Fig. 3). A temporary road at river level formed the base of the exposure, and the quarry face was from 3.5 to 11.5 m high. The cross section of the exposure (Fig. 4) was constructed by measuring 30 detailed vertical sections along the outcrop, making sketches between sections, and photographing the entire face. From FF-1 to FF-24 the section cuts directly across the ridges and swales of the meander lobe, that is, parallel to the direction of channel migration and accretion of the convex bank. Between FF-24 and FF-29 the section is oblique to the direction of lateral accretion (Fig. 3). The surface topography was determined from a 1-m-contour interval map. The road and vertical sections were surveyed, using the top of the point bar gravel unit as the datum.

FIG. 1.—Location map for the Foxe's Flat section.

SEDIMENTARY FACIES

General.—

The section exposed in Foxe's Flat is from 10 to 12 m thick. It is underlain by poorly sorted cobble and boulder gravel deposited when the river was a meltwater channel. The glaciofluvial gravels have been exposed in pits but their contact with the meander lobe gravels has not been seen. The sequence includes four facies representing deposition on the point bar (gravel and gravelly sand, facies 1 and 2), inner accretionary bank (sand, silt, and clay, facies 3), and flood plain (sand, silt, and clay, facies 4). The top of the sequence is marked by the surface of Foxe's Flat.

Facies 1: gravel - description.—

The lowermost unit consists of 3 to 4 m of crudely stratified, moderately well sorted, pebble- to cobble-grade gravel with thin beds of coarse sand and gravelly sand. Clast sizes are generally smaller than 120 mm, and boulders in the sediment probably were deposited from melting ice. The gravels are mainly clast-supported, typically contain clean sand matrix, and grade through matrix-supported deposits to gravelly sands. In some of the coarse-grained beds the matrix contains small pebbles. All of the clasts are well rounded and local imbrication of platy clasts indicates paleocurrent flow over a wide range of directions, but generally to the northeast, east and southeast (Fig. 4). Clast size is variable both along and across the bedding but shows no consistent trend. Thin strata and sandy interbeds tend to be more common in the stratigraphically higher portions of the gravels but the reverse is true in FF-23 which shows upward coarsening in the gravels. Bedding within the gravels dips at less than 5° to the north-northwest (Fig. 5); it is difficult to discern in outcrop, but is revealed by careful section measurement (Fig. 4). The bedding was produced by lateral accretion of the point bar, and individual strata within it are differentiated principally by changes in the texture of the gravels, and in some places by the orientation of sand layers or of mud drapes. Strata in the gravels are generally as much as 50 cm thick and have sharp, planar bases and flat or uneven top surfaces.

Bedding planes may descend to the base of the exposure but commonly they pinch out, merge with, or are truncated by, the sandy beds of the overlying unit. Though the gravels are confined principally to the lower half of the section, thin pebble layers do occur high in the section (Fig. 4, FF-11 and FF-12) and gravels have been found at the surface on the flood plain.

Facies 1 gravel – interpretation.—

The gravels clearly were deposited on the point bar and in the adjacent channel (Fig. 6). Their internal structure shows that they accumulated as thin gravel sheets that did not generally have steep faces on their lee sides, but the cross stratification in some of the sandier layers may have formed when clasts were swept over the downcurrent edges of the sheets. Continuity of individual layers dipping toward the paleochannel shows that the gravelly surface of the point bar was rarely scoured deeply by flows. Size sorting of gravel clasts across the point bar was not well developed, and deposition of certain sizes of clasts was probably controlled mainly by river stage rather than by the pattern of flow around the bend. Thin mud layers within the gravels and gravel layers on the flood plain indicate wide fluctuations in river stage. The gravels must have accumulated in much the same way as the bar-head gravels of Bluck (1971, 1976), the bar-apex gravels of Levey (1978), or the bar-body deposits of Ori (1979, 1982) and Forbes (1983). The point bar gravels described by Arche (1983) and some of those described by Gustavson (1978) are also very similar.

Facies 2: trough cross-stratified sand and gravelly sand – description.—

Lenses of well sorted medium to coarse sand and gravelly sand are as much as 70 m wide (in the plane of the section) and as much as 1 m thick (Fig. 4). They have flat, gently inclined or concave upward bases and tops. They rest on top of the gravel unit, on individual strata within it, or on scour surfaces cut into the gravel. In places (at FF-18, for example) they interfinger with the underlying gravels. Some lenses lie on thin layers of pebbles or mud which

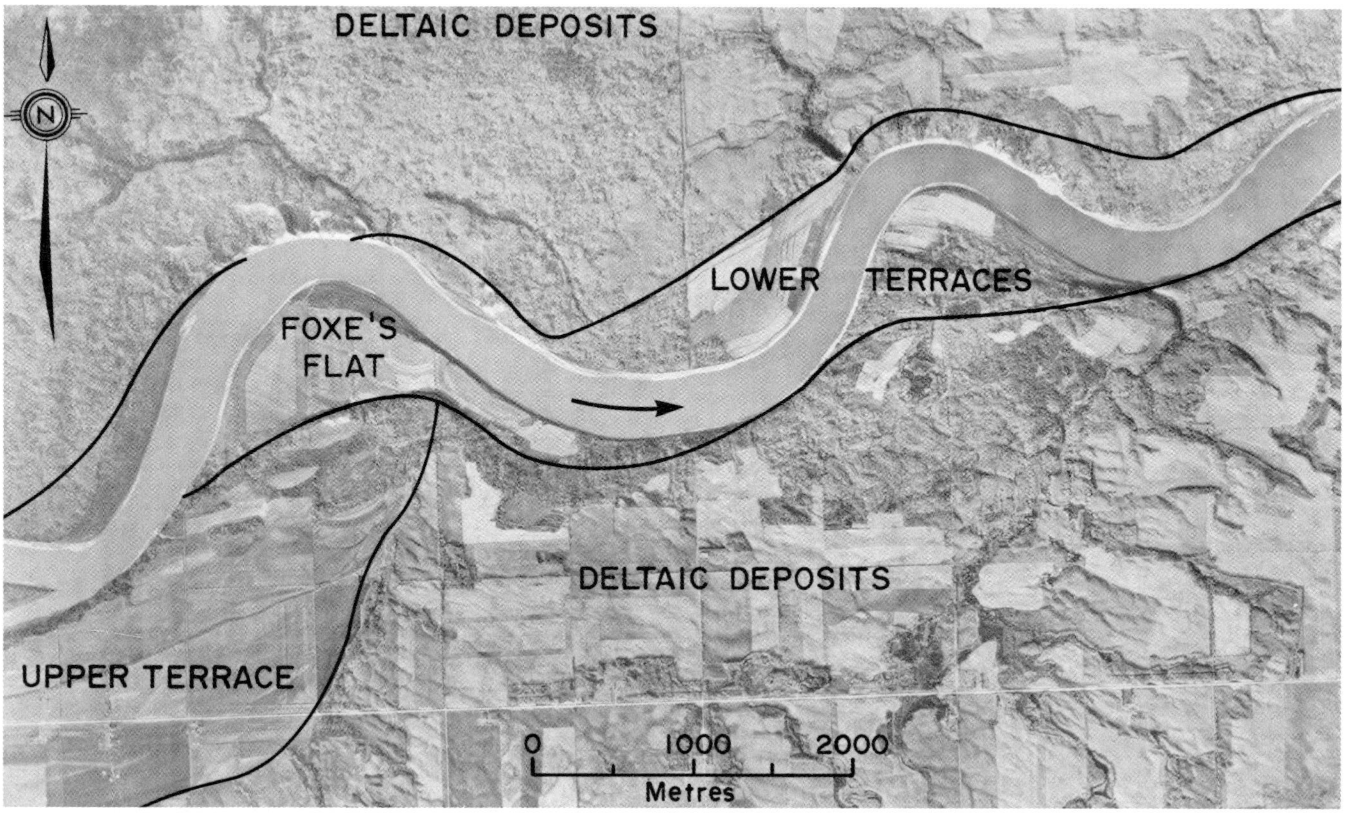

FIG. 2.—Saskatchewan River Valley west of Nipawin. Foxe's Flat, the site of the investigation, lies on the lowermost terrace of the valley. River flows to the east.

cover a scoured surface. The tops of many lenses are erosionally truncated. Lenses of facies 2 interfinger with, or abut against, the deposits of facies 3, which form the inner accretionary bank (Fig. 6). The sands contain mainly trough crossbedding with some planar crossbedding; thickness of individual sets is rarely more than 30 cm, and in some lenses (see FF-17) the grain size and thickness of the sets decrease upward. Paleocurrents from the sands were along, or obliquely up or down, the bedding surfaces in the gravels. Though deposits of this facies are found primarily close to the top of the gravel unit, there are lenses of crossbedded sand near the base of the section (sections FF-1, FF-2, FF-3, and between FF-21 and FF-22) and some beds that extend up to the level of the flood plain (between FF-17 and FF-19).

Facies 2: trough cross-stratified sand and gravelly sand – interpretation.—

The sediments of facies 2 were deposited mainly on the upper part of the point bar close to the inner bank, but on the downstream part of the point bar they were interbedded with gravels low on the bar (Fig. 6, Nos. 3, 4, 7). The sands are most common in the southwestern part of the section, but at any one time in the evolution of the point bar there must have been patches or discontinuous sheets of sand over much of the downstream half of the bar, especially at high levels and close to the inner bank. Most of facies 2 was deposited as megarippled sand sheets or sand waves close to the inner bank (facies 3) on top of the gravel facies, or in depressions that were separated from the main channel (at sections FF-13, FF-16, and FF-19, for example), probably slough or chute channels (Fig. 6, No. 4). Such channels were filled by megarippled sand sheets or by microdeltas (?chute bars), and some show evidence of a series of scour-and-fill episodes separated by deposition of clay drapes (FF-13). The lapping of sands onto the inner bank and the interfingering of facies 2 and 3 resulted when high stage flows swept sand toward the bank, or when flows expanded at the mouths of shallow chutes (Fig. 6, Nos. 3, 4). Fining upward and thinning upward cosets were probably deposited during single floods.

Sand lenses *within* the gravel unit require a different explanation. The sections FF-1 to FF-3 represent deposits that accumulated at the downstream end of a point bar, and the interbedding of sands and gravels is likely a result of the common downstream fining of grain sizes reported by many authors (Bluck, 1971; Jackson, 1976; Arche, 1983). The downstream parts of the point bars probably were the locus of deposition of sand except during very high stages when gravel was swept over the whole bar. Deposition on lower levels of the point bar near the active channel on central and upstream parts of the bend probably took place at low stage and during the falling stage of a flood. At FF-14, between FF-21 and FF-22, and at FF-24, crossbedded sands were deposited by flows parallel to the channel or by flows that were draining down the point bar toward the channel

FIG. 3.—Foxe's Flat, showing locations of a number of selected measured stratigraphic sections along the excavated cross section shown in Figure 4. Arrow indicates direction of flow in channel.

(Fig. 6, No. 9), the so-called "runoff" of Nijman and Puigdefabregas (1978).

Facies 3: interstratified sand, silt, and clay – description.—

This facies is from 2 to 4 m thick and consists of coarse-grained and fine-grained units that dip at angles of as much as 12° toward the paleochannel (Fig. 4). The coarse-grained units consist of medium- to fine-grained sand, in beds from 5 to 15 cm thick with trough cross-lamination, ripple drift cross-lamination, or parallel lamination. They are broadly lenticular (as much as 85 m wide in a direction perpendicular to the paleochannel axis) and pass updip into silts of the flood plain; downdip they rest on gravels or pass laterally into stratified gravelly sands. The fine-grained units consist of beds as much as 30 cm thick of interlaminated fine sand, silt, and clay with fine-grained organic material. Some of the fine sand layers are clean and cross-laminated, others contain ripple drift cross-lamination. The silt and clayey layers typically contain ripple drift cross-lamination or wavy lamination. Clay is common within the ripple drift cross-laminae and as drapes on sand beds. The fine-grained units lose their stratification updip and merge with the more massive silts of the floodplain.

Facies 3: interstratified sand, silt, and clay – interpretation.—

Given the stratigraphic position of facies 3 and the inclined bedding dipping toward the paleochannel, it is best interpreted as a deposit of the inner accretionary bank (Fig. 6, No. 1). Such deposits, first named by Bluck (1971), and described by Sundborg (1956), Jackson (1978), Ori (1979), and Bridge and Jarvis (1982), are built up partly during floods when fine-grained suspension deposits blanket the inner bank of the meander bend, and during lower stages when bedload sands are swept up the bank, especially in the downstream portions of bends. The resulting deposit contains interlaminated sands and muds with layers of leaves and other organic material. Layers very rich in leaves are not present in facies 3 of the Foxe's Flat exposures, most likely because the area was probably grassland with few deciduous trees when the sediments were being deposited (Burly and Meyer, 1982).

Facies 4: bedded silt and sand – description.—

Stratigraphically the highest unit in the section, facies 4 consists of 2 to 4 m of hard, massive, horizontal or gently inclined clayey silt and fine sand with numerous dark colored organic layers. Some of the beds are sandy and some have a faint parallel lamination. The dark carbonaceous layers, which contain plant debris, abundant pollen, *in situ* ash, and charcoal, are probably paleosols; as many as 10 such layers are present over a 3-m depth, but each layer can be correlated over only a very short distance. The base of this facies grades into the upper part of facies 3; the top of the facies is marked by the surface of Foxe's Flat.

Facies 4: bedded silt and sand – interpretation.—

Facies 4 is a true overbank deposit (Fig. 6, No. 10). Sediment that settled from suspension during floods was modified by wind action and soil formation to produce indurated silts and sands. The lack of correlation of paleosols across Foxe's Flat is best explained by irregular construction of the flood plain. The surface of the flat is very uneven so sedimentation during many floods would have been patchy and new soils would not have extended over the whole flat.

LATERAL ACCRETION DEPOSITS

The most striking aspect of the Foxe's Flat cross section is the well developed lateral accretion bedding which dips toward the paleochannel (Figs. 4, 5). There are two types of lateral accretion bedding: that produced in the gravels of facies 1 by the migration of the point bar, and that produced in the silts and sands of facies 3 by the migration of the inner accretionary bank. Jackson (1978) made an explicit distinction between epsilon cross-stratification (ECS), as defined by Allen (1963), and the inclined bedding produced on the inner accretionary bank, as defined by Bluck (1971) from the River Endrick. ECS was originally a purely descriptive term applied to sandy and muddy sediments, but it has more recently been given strong genetic connotations (see review by Jackson, 1978; Smith, D. G., this volume); it has also been applied to gravelly meandering stream deposits (Ricci-Lucchi, 1975, 1978; Ori, 1979, 1982; Ori and Ricci-Lucchi, 1981; Ricci-Lucchi and others, 1981; Ori and Penney, 1982; Arche, 1983). Lateral accretion surfaces have recently been reported from conglomerates interpreted as braided stream deposits (Bryant, 1983; Ramos and Sopena, 1983; Billi and others, this volume), and Bridge (1985) has emphasized that lateral accretion bedding in a deposit is not an indication of channel pattern. Given the original definition of the term and its current genetic implications, it is probably best not to use ECS to describe lateral accretion bedding in gravelly fluvial deposits, and it will not be used here. It is important, however, to maintain the distinction between the two types of lateral accretion bedding (LAB) in the Foxe's Flat exposure. Lateral accretion bedding in the gravels will be described as LAB 1, and that on the inner accretionary bank will be described as LAB 2.

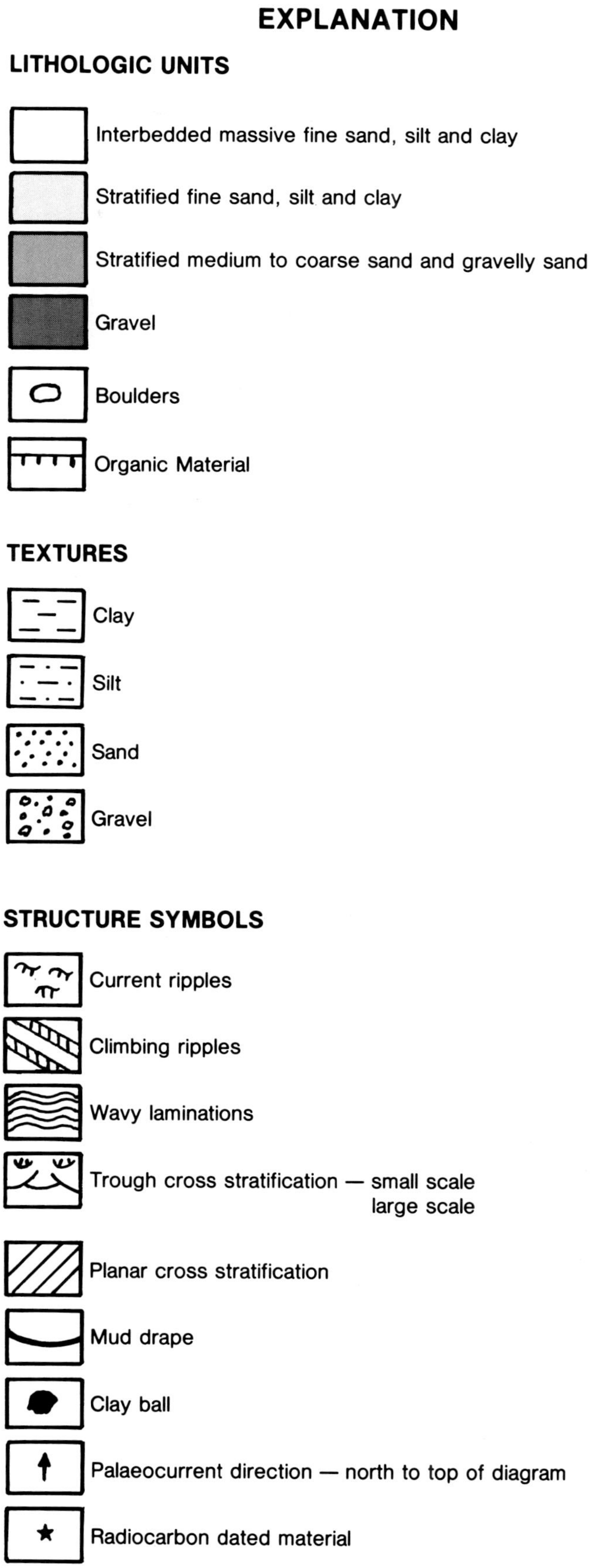

FIG. 4A.—Legend for Figure 4B.

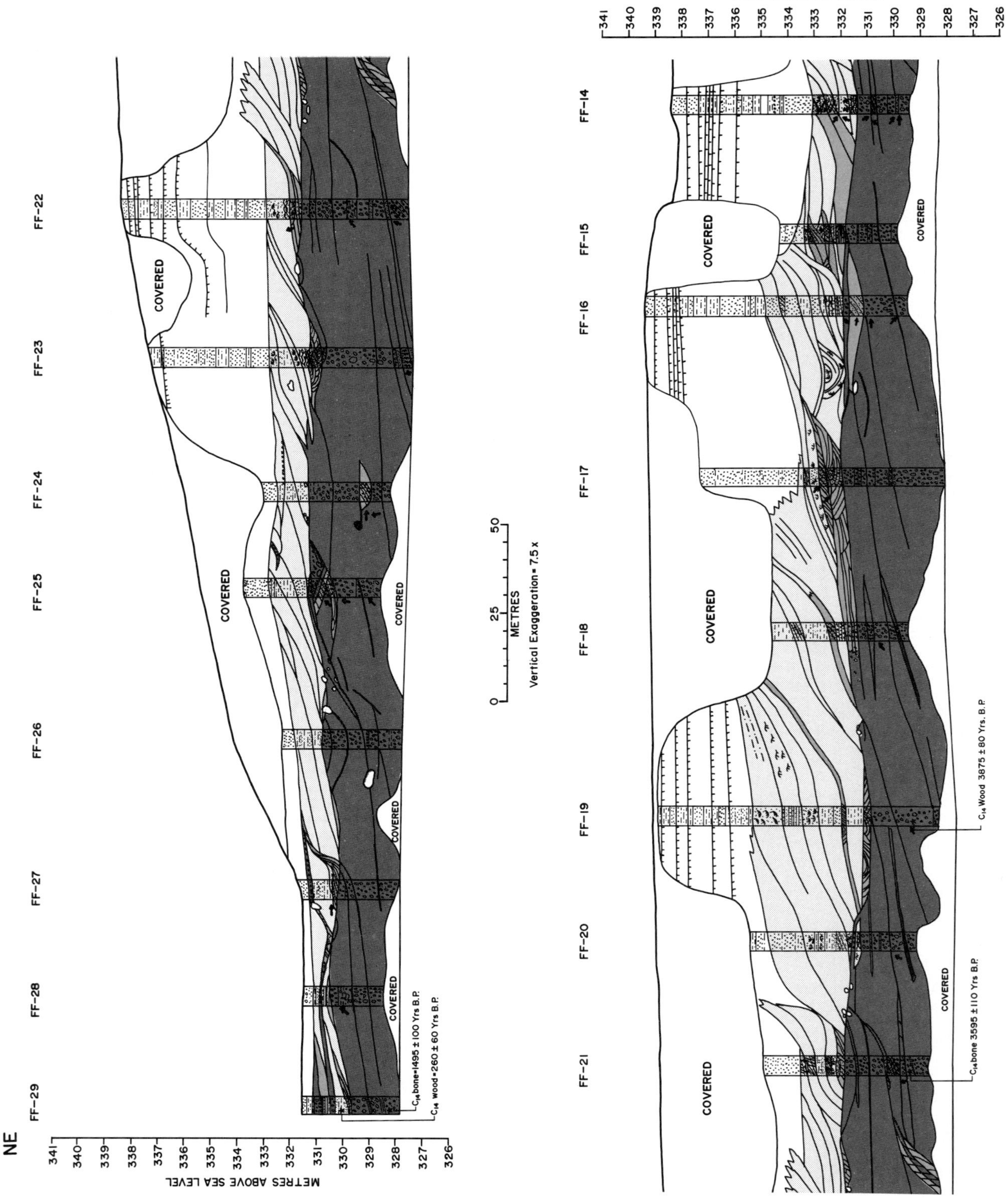
NE
FF-29
FF-28
FF-27
FF-26
FF-25
FF-24
FF-23
FF-22
FF-21
FF-20
FF-19
FF-18
FF-17
FF-16
FF-15
FF-14
COVERED
METRES ABOVE SEA LEVEL
0 25 50
METRES
Vertical Exaggeration = 7.5 x
C14 bone=1495 ± 100 Yrs B.P.
C14 wood=260 ± 60 Yrs B.P.
C14 Wood 3875 ± 80 Yrs. B.P.
C14 bone 3595 ± 110 Yrs B.P.

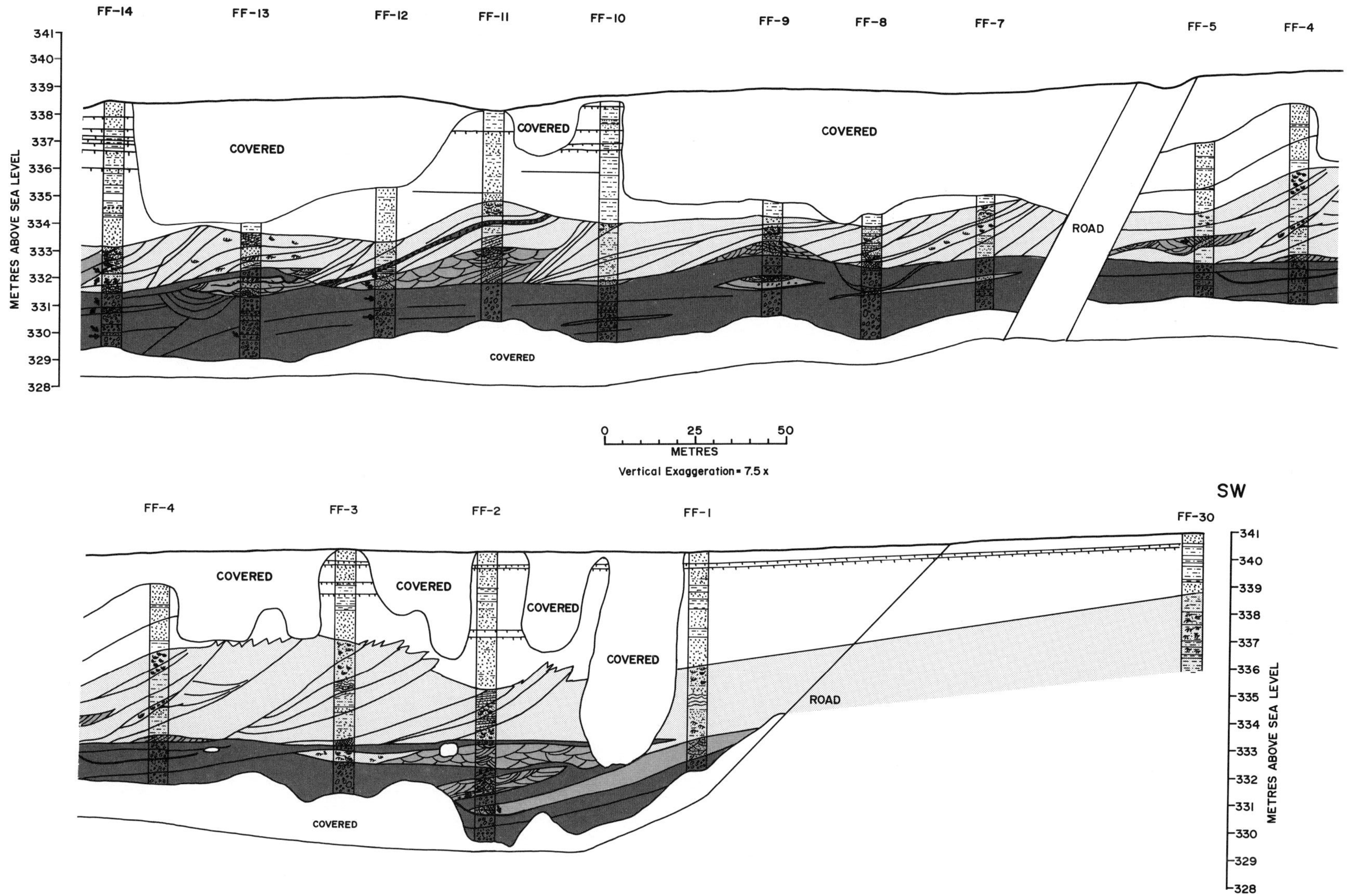

FIG. 4B.—Cross section along the northwest side of Foxe's Flat. The four portions of the section overlap slightly. Dark shading—gravels of facies 1; medium shading—sands and gravelly sands of facies 2; light shading—sands and silts of facies 3; unshaded—sands and silts of facies 4. Note the lateral accretion bedding, dipping consistently to the northeast, in facies 1 and facies 3. Legend is on preceding page.

FIG. 5.—Lateral accretion bedding in the gravels of facies 1, and in the sands and silts of facies 3. Note that the bedding in facies 3 dips more steeply to the northeast than the bedding in facies 1. Numbers refer to facies defined in text.

LAB 1 is visible in outcrop, but because it is inclined at a very low angle (less than 5°) it is difficult to see and is best displayed in the reconstructed cross section with vertical exaggeration (Fig. 4). It is picked out by variations in clast size and matrix composition or proportion in the gravels, by mud drapes on former point bar surfaces, and by sand layers in the gravels. LAB has not been found in modern gravel-bed streams (Jackson, 1978), but it probably is present (Arche, 1983). It is possible that the lack of reports from modern rivers results from the difficulty of exposing sufficiently large and weathered trenches to render the stratification visible.

The preservation of inclined accretion surfaces in the gravels is probably a result of the mode of transport of the gravels through the channel bend. In many sandy or gravelly point bars most of the sediment moves as megaripples or sand bars, and the development of scour surfaces under the crossbed sets prevents the preservation of strata that mark the point bar surface. There is very little high angle cross-stratification in any of the gravel of Foxe's Flat, and the gravel likely was transported as sheets, a mode of transport observed and inferred by many authors (Hein and Walker, 1977; Gustavson, 1978; Forbes, 1983). The sandy cross-stratified layers probably formed at the downstream edges of sheets. The LAB 1 has mainly parallel stratification but the low angle truncation of some layers shows that erosion removed gravel from wide areas of the point bar, and scours as much as 60 cm deep with mud on the base indicate local deep erosion of the gravel surface. The preservation of inclined strata throughout the gravel unit could have occurred only if the sheets of gravel moved significant distances across the point bar during high flows.

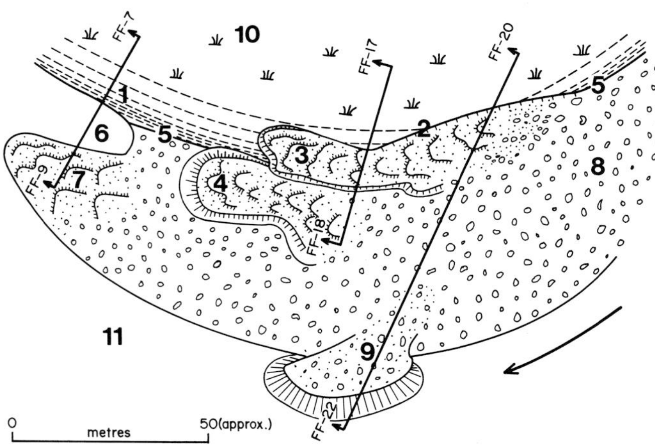

FIG. 6.—Reconstruction of the depositional environments on a meander bend in the Saskatchewan River during formation of Foxe's Flat. 1—Inner Accretionary Bank (IAB); 2—scarp eroded in IAB; 3—megarippled sand sheet lapping onto IAB; 4—megarippled sand sheet on gravel platform, or filling chute channel; 5—gravel platform buried by encroaching IAB; 6—slough with gravel, sand, or mud on bottom. 7—bar-tail sands; 8—bar-head gravels; 9—microdelta produced by late-stage runoff; 10—floodplain; 11—channel. The three cross sections show parts of Figure 4 that were used to construct the model. At other times in its development, the point bar may have had more or less sand on its surface. Arrow indicates flow direction.

The LAB 1 is not generally continuous with the LAB 2, and it is difficult to trace above the gravel unit because of the abrupt change in texture. In most parts of the section the LAB 1 seems to merge with the top surface of the gravel unit which may be horizontal, or which may dip toward the main channel or toward a slough or chute channel. Only in a very few places (between FF-19 and FF-17, for example) is there evidence of continuous layering from facies 1 to facies 3. Throughout most of the time when Foxe's Flat was forming, the point bar appears to have had a relatively flat surface, or minor channels, that broke the slope between the thalweg and the inner bank. Some of the slough channels likely were formed when gravel sheets migrated downstream along the point bar and cut off back-bar sloughs (Fig. 6, No. 6). The slough channels would then be the preferred location for chute development.

Although inclined bedding may be produced on the inner accretionary bank, there are few documented cases and it appears that the dip of the beds on the inner accretionary bank is commonly greater than that on the surface of the adjacent point bar (Sundborg, 1956; Bluck, 1971; Bridge and Jarvis, 1976, 1982; Jackson, 1978, 1980; Nijman and Puigdefabregas, 1978; Bluck and Ferguson, 1981).

The LAB 2 is considerably more steeply dipping than the LAB 1. In many sections, facies 3 deposits containing LAB 2 appear to overlie facies 1 and 2 unconformably, rather than to merge with them, but in several places facies 2 is interfingered with facies 3 ((FF-14, FF-15). The abundant cross-lamination in facies 3, especially ripple drift cross-lamination, suggests deposition on the bank mainly during floods. The early stages of floods likely were the times when the LAB 2 was eroded to produce major scours (Fig. 6, No. 2). There are scour surfaces in the section (Fig. 7); they are as much as 3 m high and are overlain by the deposits of either facies 2 or 3. Jackson (1980) explained the formation of meter-high scarps midway through bends on point bars of the Vermilion River, Indiana, as a result of erosion during rapidly waning flow. Erosional scarps on the inner bank of the Saskatchewan River were much higher than those on the Vermilion, and required removal of very large volumes of cohesive sediment, but intraformational clasts that could have been derived from the inner bank are

FIG. 7.—Erosional scour (arrows) in sands and silts of the inner accretionary bank. Dashed lines mark lateral accretion bedding. Numbers refer to facies defined in text.

absent from the deposits. Such deep scours probably formed during large floods when clasts torn from the bank were transported further downstream and likely rapidly abraded. Deposition during falling stages would have banked sand against the scoured surface and subsequently blanketed the bank with finer grained sediment. Erosion on the inner bank is not characteristic of flows in meander bends, but in the Saskatchewan River near Nipawin large floods have been known to fill the whole valley, so scour would have been greatest where the sediment was most erodible. The locus of significant scour on the inner bank would have been controlled by both the power of the flow, and its pattern within the bend as well as by the topography of the upstream part of the point bar. Generally, flow that was strongly channelized across the upper part of the bar would be expected to cause much erosion.

Inner accretionary banks have been reported from only a relatively small number of modern streams: the River Endrick (Bluck, 1971); the South Esk (Bridge and Jarvis, 1976, 1982); the Wabash River, Vermilion River, and Little Wind River (Jackson, 1976, 1978, 1980); and the River Klaralven (Sundborg, 1956). They have been reported mainly from meandering streams with relatively narrow, confined floodplains. There are even fewer reports from the geological record (Nijman and Puigdefabregas, 1978; Ori, 1979; Stewart, 1981). In ancient deposits it will be impossible to identify the deposits of an inner accretionary bank unless they have a visible dip. The development of a preservable inner bank will be favored if a stream has a high rate of deposition on the flood plain and a relatively slow rate of channel migration; frequent erosion of the inner bank will help maintain steep slopes.

VERTICAL SEQUENCES

The exposure across Foxe's Flat displays an almost complete section through the alluvium deposited by the meandering Saskatchewan River during one sweep of the channel across its valley. The base of the section is not exposed but auger holes and quarrying operations showed that it lies no deeper than 2 m below the level of the roadway, so the maximum thickness of the sequence is about 12 m. Although the section consists of a clearly differentiated basal coarse member (facies 1 and 2) overlain by a fine member (facies 3 and 4), the changes within vertical sections through the deposit are not consistent. The poor development, or lack of development, of fining-upward sequences in the deposits of gravelly meandering streams has been pointed out many times (McGowen and Garner, 1970; Bluck, 1971; Gustavson, 1978; Jackson 1978; Ori, 1979; Arche, 1983). It is due at least in part to the round-bend (downstream) variation in flow conditions and sediment type and to the direction of lateral migration (Bluck, 1971; Bridge and Jarvis, 1976, 1982; Jackson, 1976), but disruptions to an ideal fining-upward sequence can also be produced by long-term temporal changes or fluctuations in flow (Stewart, 1981; Arche, 1983).

A detailed analysis of the vertical sequences displayed in the Foxe's Flat section is beyond the scope of this paper, but a comment can be made on the characteristics of the sequences and their significance. Of all the sections measured, less than half have fining-upward trends. Deviations from a fining-upward trend occur in a number of ways (Fig. 8).

In sections FF-1 and FF-2 (Fig. 4), facies 1 and 2 are interbedded in the lowermost part of the exposure. These sediments were deposited in a position at the downstream end of a meander bend, where good fining-upward sequences might be expected to develop (Jackson, 1976; Bridge and Jarvis, 1982), but the poor development in this case is likely a result of at least two factors. First, the point bar probably had a downcurrent decrease in particle size, so that at its distal end sediment accumulated by the overlap-

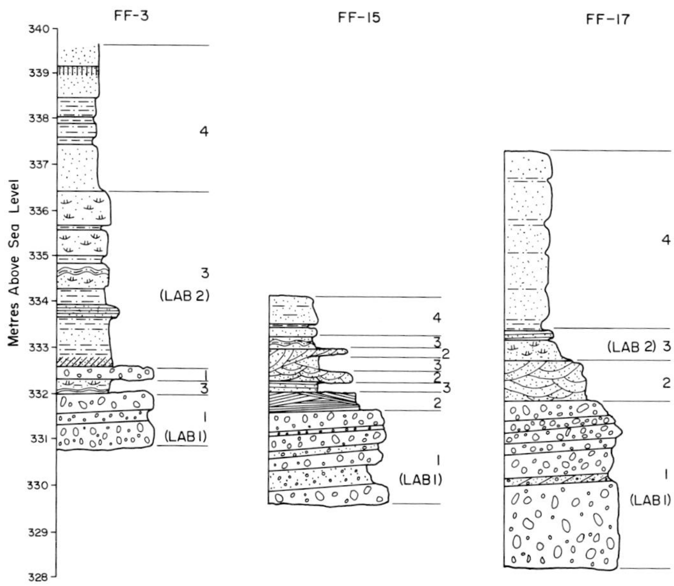

FIG. 8.—Variations in fining-upward sequences because of patchy distribution of sands (facies 2) over the bar. FF-3 was produced in the downstream part of the bar when there were no sand-filled chutes or channels and the inner accretionary bank (IAB) advanced directly over the gravelly bar platform. FF-15 was produced when the advance of the (IAB) alternated with encroachment of bar sands onto the bank. FF-17 is an atypical, but complete, fining-upward sequence with all facies present. The symbols are the same as in Figure 4.

ping of lobate sheets of sand and gravel. Second, wide variations in discharge moved the locus of sedimentation of different grades of sediment higher and lower on the bar with time. Variations in river stage were responsible for deposition of several mud drapes low in the section, and they also account for the presence of sandy lenses produced during late-stage runoff from the bar toward the channel (northeast of FF-21, for example). The discontinuous cover of sand sheets on the gravelly surface of the point bar has produced some sections in which deposits of facies 2 are absent (Fig. 4; Fig. 8, FF-3). Interbedding of facies 2 and 3 on top of the gravelly platform of the point bar and on the lower part of the inner accretionary bank is another source of variation in the vertical grain-size trend (Fig. 8, FF-15). On a point bar such as this, which must have had a well developed gravel bench or platform, variations in the size and duration of floods would expose the zone near the inner bank to a wide range of flow conditions. Deposition of gravels (facies 1) and sands (facies 2) high on the accretionary bank and on the flood plain is a further indication of periodic large floods.

DISCUSSION

Jackson (1978) argued that lithofacies models for meandering streams were inadequate and drew attention to the very few studies of modern gravel-bed meandering streams. Since that time, there have been several studies of modern meandering gravel-bed streams and their deposits (Forbes, 1983; Gustavson, 1978; Arche, 1983; Smith, S. A., this volume), and some ancient gravel deposits have been interpreted as the product of meandering streams (Ori, 1979, 1982).

The deposits of the Foxe's Flat meander lobe in the Saskatchewan River Valley provide a good example of the stratigraphic relationships and sequences that may be produced as a result of lateral migration of a gravel-bed meandering stream. They share many features with other gravel deposits demonstrated to be, or inferred to be, meander belt deposits, namely (a) the wide range of grain sizes and their different modes of transport result in a strong segregation of different size-grades within the deposit (Bluck, 1971; Jackson, 1980); (b) transport and deposition of sand tend to occur either on the higher levels of the point bar, or toward the downstream end of the bar, so that the sequences contain discontinuous lenses of sand resting on a thick gravel unit, or less commonly, interbedded with it (Ori, 1979, 1982; Arche, 1983; Forbes, 1983); (c) deposition of sand during falling stage produces structures with paleocurrent directions at high angle to the trend of the channel; (d) because of the irregular, benchlike platform of gravel on top of the bar, sand commonly fills chute and slough channels (Bluck, 1971, 1976; Ori, 1979, 1982; Arche, 1983); (e) deposits of the inner accretionary bank form a major part of the sequence (Bluck, 1971; Jackson, 1978, 1980; Ori, 1979, 1982); (f) there is abundant evidence of a wide range in discharge of the stream (Arche, 1983); (g) the vertical sequence is generally "fining upward," but there is little regularity in this trend (McGowen and Garner, 1970; Bluck, 1971; Gustavson, 1978; Ori, 1979; Arche, 1983); (h) lateral accretion bedding is present in the gravels, probably because they were transported as sheets rather than relatively high, steep-faced bedforms (Ori, 1979, 1982; Arche, 1983); (i) lateral accretion bedding in the gravels is inclined in one direction over a long distance, about 1000 m on Foxe's Flat (Ori, 1979; Arche, 1983).

A common pattern of features that are associated with deposition in coarse-grained meandering streams seems to be emerging, but it will remain difficult to establish relative channel sinuosity and degree of channel splitting from examination of outcrops. From vertical sequences alone, it will be virtually impossible. Many of the characteristics displayed by the deposits in Foxe's Flat and in the reports cited above are also common to braided streams (Bluck, 1979, 1980; Billi and others, this volume; Calvet and others, 1985; Ramos and Sopeña, 1983), and particularly to braided streams with lateral bars (Bluck, 1976). The relatively small lateral extent of lateral accretion bedding in some conglomerates has been used to argue in favor of deposition in braided streams rather than in meandering streams (Billi and others, this volume; Ramos and Sopeña, 1983), but given the considerable difficulty of recognizing lateral accretion planes in conglomerates, and the possible preservation of only small parts of individual meander lobes, this may not always be a reliable criterion. The thickness of the "fine member" (facies 3 and 4) in the Foxe's Flat section is as much as 50 percent of the thickness of some of the sections. If preserved completely, it would provide direct evidence of the well established flood plain. The most compelling evidence for deposition of a sequence such as that on Foxe's Flat by a meandering stream is found in the lateral relationships between the gravels and the finer grained sediments. The lateral accretion bedding is not continuous from bottom to top of the sequence, and there is no major discontinuity between the coarse and fine members. Both LAB 1 and LAB 2 merge with the bases or tops of lenticular units of facies 2. Vertical sequences show consistent burial of laterally accreted gravel and sand by sand and silt which have accreted laterally in the same direction. Such a sequence indicates long-term stability of one major, laterally migrating channel, with intermittent minor flow adjacent to a relatively stable inner bank bordering a flood plain.

ACKNOWLEDGMENTS

This study was funded by the Saskatchewan Research Council. Drafting of the cross section and photo-mechanical preparation of the figures was done by Carlos Sardinha and Brian McIntosh of Graphic Services, Saskatchewan Research Council. Their time and work are greatly appreciated. The cooperation of the Saskatchewan Power Corporation in allowing access to the gravel pits is gratefully acknowledged. Special thanks are due to Ken Haggard of Nipawin Power Project for his cooperation and assistance throughout this study.

REFERENCES

ALLEN, J. R. L., 1963, The classification of cross-stratified units with notes on their origin: Sedimentology, v. 2, p. 93–114.

———, 1965, The sedimentation and palaeogeography of the Old Red Sandstone of Angelsey, North Wales: Proceedings, Yorkshire Geological Society, v. 35, p. 139–185.

———, 1970a, Studies in fluviatile sedimentation: A comparison of fining-upward cyclothems with special references to coarse-member composition and interpretation: Journal of Sedimentary Petrology, v. 40, p. 298–323.

———, 1970b, A quantitative model of grain size and sedimentary structures in lateral deposits: Geological Journal, v. 7, p. 129–146.

ARCHE, ALFREDO, 1983, Coarse-grained meander lobe deposits in the Jarama River, Madrid, Spain, *in* Collinson, J. D., and Lewin, J., eds., Modern and Ancient Fluvial Systems: International Association of Sedimentologists Special Publication 6, p. 313–321.

BERNARD, H. A., MAJOR, C. F., JR., PARROT, B. S., AND LEBLANC, R. J., SR., 1970, Recent Sediments of Southeast Texas: Bureau of Economic Geology, University of Texas, Guidebook 11, 16 p.

BLUCK, B. J., 1971, Sedimentation in the meandering River Endrick: Scottish Journal of Geology, v. 7, p. 93–137.

———, 1976, Sedimentation in some Scottish rivers of low sinuosity: Transactions, Royal Society of Edinburgh, v. 69, p. 425–456.

———, 1979, Structure of coarse-grained braided alluvium: Transactions, Royal Society of Edinburgh, v. 70, p. 181–121.

———, 1980, Structure, generation and preservation of upward fining, braided stream cycles in the Old Red Sandstone of Scotland: Transactions, Royal Society of Edinburgh, v. 71, p. 29–46.

———, AND FERGUSON, R. I., 1981, Scottish alluvium: Modern and ancient, *in* Elliot, T., ed., Field Guide to Modern and Ancient Fluvial Systems in Britain and Spain, Proceedings, Second International Conference Fluvial Sedimentology, University of Keele, U.K., p. 5.1–5.20.

BRIDGE, J. S., 1985, Paleochannel patterns inferred from alluvial deposits: A critical evaluation: Journal of Sedimentary Petrology, v. 55, p. 0579–0589.

———, AND JARVIS, JACK, 1976, Flow and sedimentary processes in the meandering River South Esk, Glen Clova, Scotland: Earth Surface Processes, v. 1, p. 303–336.

———, AND ———, 1982, The dynamics of a river bend: A study in flow and sedimentary processes: Sedimentology, v. 29, p. 499–541.

BRYANT, I. D., 1983, Facies sequences associated with some braided river deposits of the late Pleistocene age from southern Britain *in* Collinson, J. D., and Lewin, J., eds., Modern and Ancient Fluvial Systems: International Association of Sedimentologists Special Publication 6, p. 267–275.

BURLEY, DAVID, AND MEYER, DAVID, 1982, Nipawin Reservoir Heritage Study, Volume 3, Regional Overview and Research Considerations: Saskatchewan Research Council Publication No. C-805-25-E-82, 318 p.

CALVET, J., MARZO, MARIANO, AND PUIGDEFABREGAS, CAYO, 1985, Lateral bar development during extreme flood in low sinuosity reaches of the Segre River, Southern Pyrenees, Spain (abs.): Third International Fluvial Sedimentology Conference, Fort Collins, Colorado, p. 13.

FISK, H. N., 1944, Geological investigation of the alluvial valley of the Lower Mississippi River: Mississippi River Commission, Vicksburg, Miss., 78 p.

———, 1947, Fine grained alluvial deposits and their effect on the Mississippi River Activity: Mississippi River Commission, Vicksburg, Miss., 82 p.

FORBES, D. L., 1983, Morphology and sedimentology of a sinuous gravel-bed channel system: Lower Babbage River, Yukon coastal plain, Canada, *in* Collinson, J. D., and Lewin, J., eds., Modern and Ancient Fluvial Systems: International Association of Sedimentologists Special Publication 6, p. 195–206.

GUSTAVSON, T. C., 1978, Bedforms and stratification types of modern gravel meander lobes, Nueces River, Texas: Sedimentology, v. 25, p. 401–426.

HEIN, F. J., AND WALKER, R. G., 1977, Bar evolution and development of stratification in the gravelly, braided Kicking Horse River, B.C.: Canadian Journal of Earth Science, v. 14, p. 562–570.

JACKSON, II, R. G., 1976, Depositional model of point bars in the Lower Wabash River: Journal of Sedimentary Petrology, v. 46, p. 579–594.

———, 1978, Preliminary evaluation of lithofacies models for meandering alluvial streams, *in* Miall, A. D., ed., Fluvial Sedimentology: Canadian Society of Petroleum Geologists Memoir 5, Calgary, p. 543–576.

———, 1980, Sediments of the modern meander belt of the Vermilion River near Eugene, Indiana, *in* Fraser, G. S., ed., Field Trip Guidebook, Tenth Annual Meeting, Great Lakes Section, Society of Economic Paleontologists and Mineralogists, p. 17–44.

LEVEY, R. A., 1978, Bedform distribution and internal stratification of coarse-grained point bars, Upper Congaree River, S.C., *in* Miall, A. D., ed., Fluvial Sedimentology: Canadian Society of Petroleum Geologists Memoir 5, Calgary, p. 105–127.

MCGOWEN, J. H., AND GARNER, L. E., 1970, Physiographic features and stratification types of coarse-grained point bars: Modern and ancient examples: Sedimentology, v. 14, p. 77–111.

NANSON, G. C., 1980, Point bar and flood plain formation of the meandering Beatton River, Northeastern British Columbia, Canada: Sedimentology, v. 27, p. 3–29.

NIJMAN, WOUTER, AND PUIGDEFABREGAS, CAYO, 1978, Coarse-grained point bar structure in a molasse-type fluvial system, Eocene, Castisent Sandstone Formation, South Pyrenean Basin, *in* Miall, A. D., ed., Fluvial Sedimentology: Canadian Society of Petroleum Geologists Memoir 5, Calgary, p. 487–510.

ORI, G. G., 1979, Barre di meandro nelle alluvioni ghiaiose del Fiume Reno (Bologna): Bollettino Societa Geologica Italiana, v. 98, p. 35–54.

———, 1982, Braided to meandering channel patterns in humid region alluvial fans deposits, River Reno, Po Plain (Northern Italy): Sedimentary Geology, v. 31, p. 231–248.

———, AND PENNEY, S. R., 1982, The stratigraphy and sedimentology of the Old Red Sandstone sequence at Dunmore East, County Waterford: Journal of Earth Science, Royal Dublin Society, v. 5, p. 43–59.

———, AND RICCI-LUCCHI, FRANCO, 1981, Recognition of ancient point bars in coarse alluvial and deltaic setting (abs.): Second International Fluvial Sedimentology Conference, University of Keele, U.K., p. 92.

RAMOS, AMPARO, AND SOPEÑA, ALFONSO, 1983, Gravel bars in low sinuosity streams (Permian and Triassic, central Spain), *in* Collinson, J. D., and Lewin, J., eds., Modern and Ancient Fluvial Systems: International Association of Sedimentologists Special Publication 6, p. 301–312.

RICCI-LUCCHI, FRANCO, 1975, Miocene palaeogeography and basin analysis in the Periadriatic Apennines, *in* C. H. Squyres, ed., Geology of Italy: Earth Science of the Libyan Arab Republic, Castelfranco Veneto, p. 129–236.

———, 1978, Sedimentologia, Vol. III, Ambienti sedimentari e facies: Cooperative Libraria Univ., Bologna 545 pp.

———, COLELLA, ALBINA, ORI, G. G., OGLIANI, FABIO, AND COLALONGO, M. L., 1981, *in* Ricci-Lucchi, F., ed. Excursion Guidebook with Contributions on Sedimentology of Some Italian Basins: International Association of Sedimentologists, Second European Regional Meeting, Bologna, Italy, p. 79–162.

STEWART, D. J., 1981, A meander belt sandstone of the Lower Cretaceous in Southern England: Sedimentology, v. 28, p. 1–20.

SUNDBORG, AKE, 1956, The River Klaralven, a study of fluvial processes: Geografiska Annaler, v. 38, p. 125–316.

THREE-DIMENSIONAL FACIES GEOMETRY IN PLEISTOCENE OUTWASH SEDIMENTS, WORCESTERSHIRE, U.K.

MARTIN, R. DAWSON[1]
Department of Geography, University of Leicester, Leicester, LE1 7RH England
AND
IAN D. BRYANT[2]
Department of Geology, University of Nottingham, Nottingham, NG7 2RD England

ABSTRACT: Three-dimensional exposures provided by quarrying of aggregate have been used to determine the geometry of genetic units within a sequence of outwash gravels. The genetic units represent lithofacies associations and are of three types. Association A is predominantly composed of gravelly lithofacies interpreted to have been deposited as sheets by laterally migrating bar complexes in the active channel belt of a low sinuosity river. Association B comprises thin, discontinuous sheets of sandy lithofacies interpreted to have been deposited from shallow flows as floodplain or bar top sediments. Association C consists of sands and fine gravels deposited as channel fills. The distribution of units of these various associations indicates that the active channel belt, at any one moment in time, only occupied part of the valley floor and was flanked by areas of overbank deposition. These latter areas were traversed by single thread channels infilled by vertically aggraded sands and gravels.

INTRODUCTION

Glacial outwash environments are well described both in terms of the areal distribution of sediments in present systems and in terms of vertical sequences from older deposits (see review in Smith, 1985). Neither of these perspectives, however, provides a wholly satisfactory model to predict the internal organization of sediments within bodies of glacial outwash, since the sediments most easily observed in modern environments have the lowest preservation potential, whereas the vertical sequence of facies in one section has limited application to the description of lateral variation (Miall, 1985). Prediction of internal variation within such sediment bodies has important implications for their efficient exploitation as reserves of aggregate and as reservoirs for ground water and hydrocarbons. In order to develop a predictive model of facies variation, quantitative data regarding the three-dimensional geometry of the facies are necessary. In this study we present such data from an outwash sequence which underlies the Main Terrace surface of the River Severn in the English Midlands (Fig. 1).

The Main Terrace forms a major geomorphologic[al] feature with a downvalley extent of 120 km and a preserved width which locally exceeds 1 km. The terrace surface is underlain by more than 8 m of fluvial sediments which locally interdigitate with glacial diamict (Shotton, 1977) and proximal fan deposits (Dawson, 1985). This correlation of the Main Terrace with diamict sequences indicates that the Pleistocene sediments beneath the Main Terrace surface were deposited between 25,000 and 18,000 yrs BP during the Devensian (last) Glaciation (Shotton and Coope, 1983).

The data presented in this paper were collected from exposures provided by aggregate extraction at Holt Heath, Worcestershire, some 30 km downstream from the Devensian glacial limit at Eardington (Fig. 1). At this site, much of the original extent of the terrace is preserved as a consequence of diversion of the river to the east subsequent to terrace aggradation (Wills, 1938). During the study period, the quarry exposed as much as 7 m of Pleistocene sediments overlying the Mercia Mudstone (Triassic) bedrock in sections as much as 350 m in length. As aggregate extraction proceeded, successive exposures provided by the retreating quarry faces were recorded as photomosaics on which the distribution of the major lithologic[al] 'units' could be noted. Each of these units represents packages of genetically related sediments separated from adjacent units by 'bounding surfaces' (*sensu* Brookfield, 1977; Allen, 1983). In addition to recording the geometry and composition of these units, 27 detailed vertical sections and 128 paleocurrent directions were measured. In this way information concerning the internal composition of a 105,000 m^3 block of sediment was obtained. This information is summarized below in terms of the constituent lithofacies, their arrangement into genetic units or 'facies associations' and the way in which these elements combine to form the sediment body as a whole.

LITHOFACIES SEQUENCES

The sediments were initially classified into the lithofacies categories proposed by Miall (1977, 1978). Gravelly lithofacies are represented by massive to crudely horizontally stratified units (Gm), trough crossbedded (Gt) and planar cross-bedded (Gp) sandy gravels. Sandy lithofacies occur as horizontally laminated (Sh), trough cross-bedded (St) and planar tabular cross-bedded (Sp) sands and as scour fills (Ss and Se). Fine-grained sediments (muds) occur principally as thin, parallel laminated units (Fl). This classification was used to describe the sediments in the measured vertical sections (Fig. 2).

The relationship between the lithofacies in these vertical sections was further examined by the use of first-order embedded Markov Chain analysis (Krumbein and Dacey, 1969; Harbough and Bonham-Carter, 1970; Davis, 1974; Walker, 1979; Fraser, 1982) in order to identify significant groupings. No attempt was made to take account of the different types of transitions between adjacent units, and the distribution of profiles was such that repeated recording of the same transition between laterally traceable beds was

---

[1]Department of Geography, University College of Wales, Aberystwyth, U.K.

[2]Koninklijke/Shell Exploratie en Productie Laboratorium, Postbus 60, 2280AB Rijswijk (ZH), Nederland.

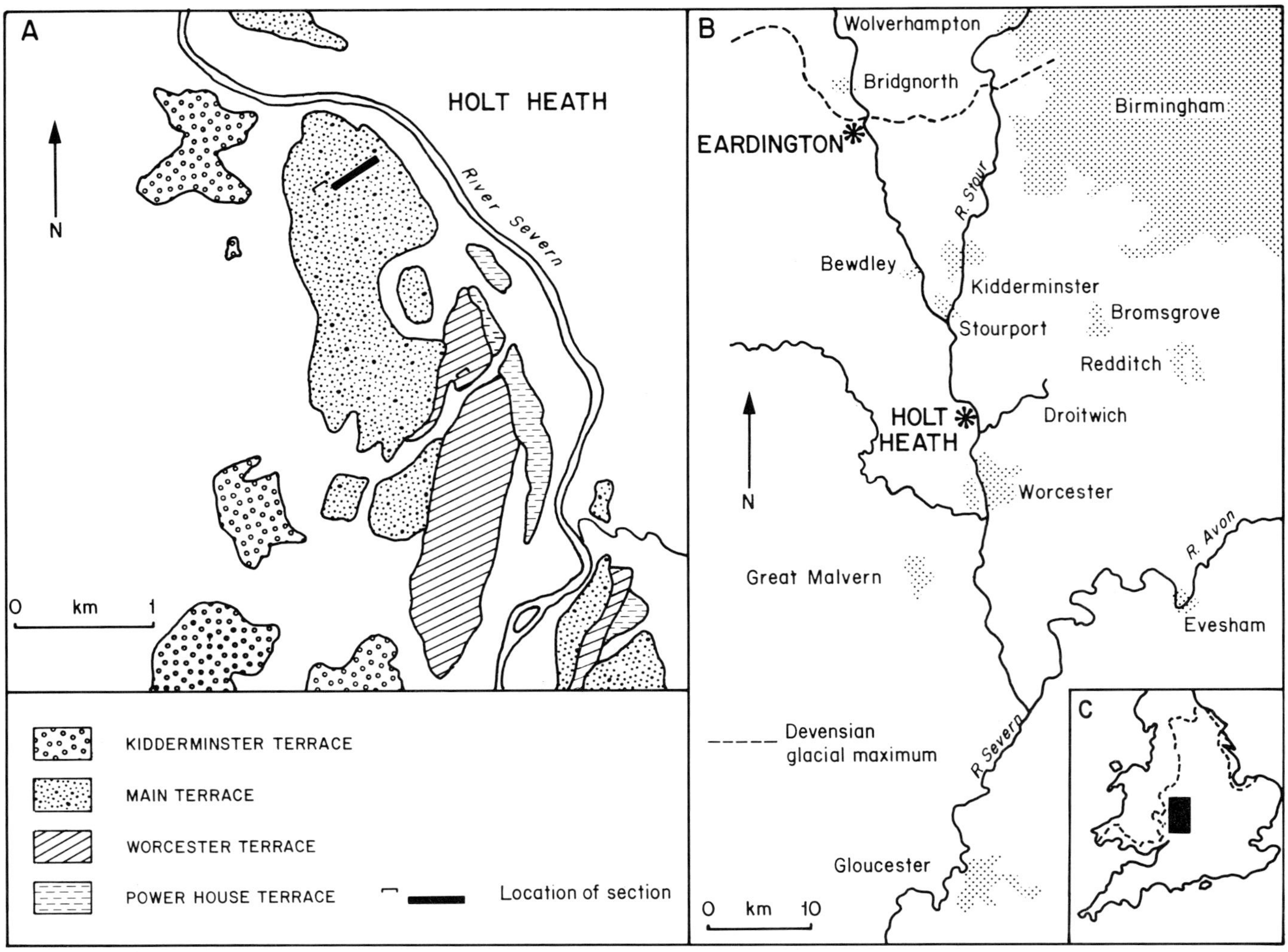

FIG. 1.—A. Location of the study sections with respect to the local terrace sequence. B. Map of the local area to show the relationship of the Severn to the Devensian (last) glacial limit. C. Inset of England and Wales to show the area represented by B. Dashed line represents the Devensian glacial limit.

avoided. A total of 393 transitions between lithofacies states were recorded in the 27 vertical sections (Table 1). The probabilities of transition were calculated and random elements removed (Harbough and Bonham-Carter, 1970; Davis, 1974) to identify significant first-order transitions between states. These transitions are shown in Figure 3, where linkages between lithofacies states are demonstrated relative to a hypothetical facies sequence (Walker, 1979).

In interpreting the sequence at Holt Heath, we prefer to place more emphasis on vertical *and* lateral transitions and less emphasis on Markov analysis of vertical sequences alone than authors of previous studies of Pleistocene outwash (Fraser, 1982; Thomas and others, 1985). The analysis does reveal, however, the relative amount of association between the different lithofacies and is useful in quantifying these relationships.

LITHOFACIES ASSOCIATIONS

The lithofacies associations consist of groups of genetically related facies separated from adjacent groups by 'bounding surfaces.' These surfaces may be classified into a hierarchy such that higher order surfaces truncate lower order surfaces. This concept has attained wide application in the study of aeolian sediments (Brookfield, 1977) and has recently been applied to fluvial sequences (Allen, 1983). The bounding surface concept is used here in a similar way to that described by Allen (1983), such that the unconformity between the Pleistocene sediments and Triassic bedrock is considered to represent a fourth- (high) order surface of composite origin. Groups of different lithofacies associations are bounded by third-order surfaces which represent major erosion surfaces within the Pleistocene sediments. Individual units of a given facies association are defined by second-order surfaces, whereas individual lithofacies units are separated by first-order bounding surfaces. These (lower) first- and second-order surfaces may be represented either by erosional or conformable contacts. The form of these surfaces is discussed below with respect to each of the three lithofacies associations recognized at Holt Heath.

Association A

This association is composed predominantly of gravelly lithofacies (principally Gm and Gp) and is volumetrically

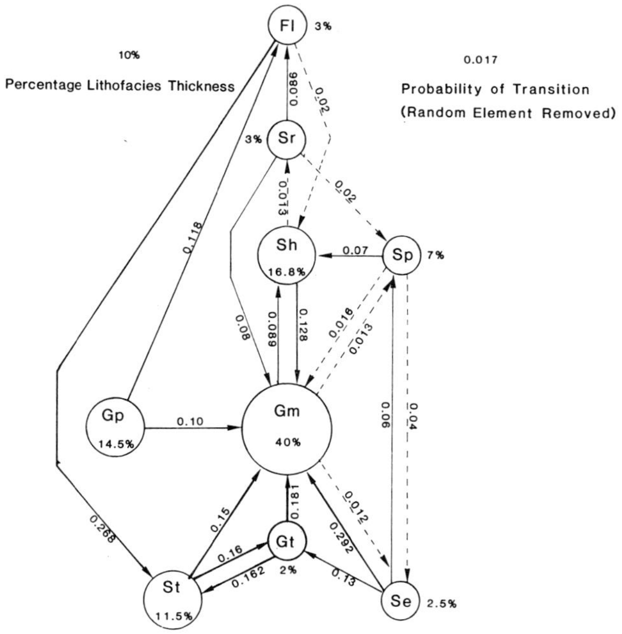

LITHOFACIES TRANSITION DIAGRAM

FIG. 2.—Facies relationship diagram derived by Markov Chain analysis of 393 facies transitions from 27 vertical sections. Gm = massive gravel; Gt = trough crossbedded gravel; Gp = planar crossbedded gravel; St = trough crossbedded sand; Sp = planar crossbedded sand; Sh = horizontally laminated sand; Se = sour-fill sand; Sr = rippled sand; Fl = laminated muds.

the most significant association in the sequence (45.7 percent). The association is poorly defined by the Markov analysis of vertical transitions being represented by the link between Gp and Gm (Fig. 2). The association forms laterally extensive tabular sheets (Fig. 4) as much as 2 m thick which extend as far as 380 m normal to the overall transport direction and in excess of the recorded exposure (110 m) parallel to the transport direction. The lower bounding surface is typically an erosional, planar second- or third-order surface, although conformable contacts were observed where this association overlies association C. Units of association A show little internal grading, consisting predominantly of coarse gravel (mean clast diameter 19–24 mm) with a high matrix content. Their internal structure is predominantly one of poorly defined or lenticular bodies (10–20 cm thick) of both openwork and matrix-rich gravel. Matrix-rich beds may be variously imbricated or disorganized. Locally within this association facies, Gm may show evidence of vertical stacking of thin beds defined by welded horizons (Steel and Thompson, 1983) or discontinuous thin sands. In some sections such units pass laterally into crossbedded sandy gravels as much as 50 m wide and 85 cm thick which show convex-up internal reactivation surfaces. These surfaces separate bundles of concordant, normally graded cross-strata of pseudo-imbricated gravel (Teisseyre, 1976). These crossbeds are, in turn, truncated by channel-like scours infilled by wedge-shaped units of tabular foreset-bedded gravel which are orientated normal to the axes of the scours (Fig. 5).

TABLE 1.—OBSERVED FACIES TRANSITION MATRIX

	Gm	Gp	Gt	Sp	St	Sh	Ss	Sr	Fl	Σ
Gm	–	13	0	18	13	55	8	11	7	125
Gp	14	–	0	2	0	8	1	1	6	32
Gt	4	1	–	0	2	0	0	0	1	8
Sp	13	3	0	–	1	12	3	3	1	36
St	16	2	6	0	–	7	0	1	0	32
Sh	52	9	1	9	7	–	2	10	6	96
Ss	8	1	2	2	0	0	–	0	0	13
Sr	11	1	0	3	2	5	0	–	4	26
Fl	5	2	0	1	9	7	0	1	–	25
Σ	123	32	9	35	34	94	14	27	25	393

Gm = massive gravel; Gt = trough crossbedded gravel; Gp = planar crossbedded gravel; St = trough crossbedded sand; Sp = planar crossbedded sand; Sh = horizontally laminated sand; Ss = scour-fill sand; Sr = rippled sand; Fl = laminated muds.

Many of the units of facies association A are overlain by laterally persistent, thin (10–15 cm) muds showing massive or flat lamination. The thickness of such mud stringers increases into the channel fills where the muds commonly interdigitate with sands. These bands are traceable as discontinuous beds as far as 250 m transverse to the paleocurrent direction.

Interpretation.—

The coarse grade, internal structures and tabular geometry of units of this association suggest that these units formed as part of 'bar cores' (*sensu* Bluck, 1974). Similar units have been described from Quaternary and older wandering or braided, low sinuosity rivers (Costello and Walker, 1972; Eynon and Walker, 1974; Rust, 1978, 1984; Bryant, 1983a, b; Steel and Thompson, 1983; Ramos and Sopeña, 1983). In such rivers relatively thick laterally extensive units are generated as a result of the nucleation and lateral migration of complex bar forms (Bluck, 1974, 1980).

Association B

This association is composed predominantly of sandy lithofacies and forms between 25 and 30 percent of the cross sections (Fig. 4). Most commonly, units of this association are separated from units of association A by second-order bounding surfaces. Occasionally, this association may overlie association C. Typically, units of association B are erosively overlain and form discontinuous, thin sheets composed of thin (5–25 cm) lenticular beds of massive to flat-laminated sand with subordinate amounts of gravel, isolated cross-bedded and ripple-laminated sands and/or thin clay stringers. The association is identified in the Markov analysis by mutual transitions between lithofacies Sp, Sr, and Fl. Individual beds are rarely laterally continuous but are occasionally traceable for 15–25 m. Shallow (<70 cm deep) channels within this association are as wide as 15 m and are frequently filled with isolated cosets of planar tabular crossbedded sand. Channel fills exhibit both vertical and laterally accreted components, such that channel margins are characterized by wedges of tabular crossbedded sands indicating cross-channel flow, whereas the axial zone tends to be infilled by crossbedded sand indicating transport parallel to the channel axis. Paleocurrents within this association consequently show a relatively high dispersion (SD = 26°). Synsedimentary soft sediment deformation is indicated in some places by truncated disharmonic compressional folds of uncertain origin.

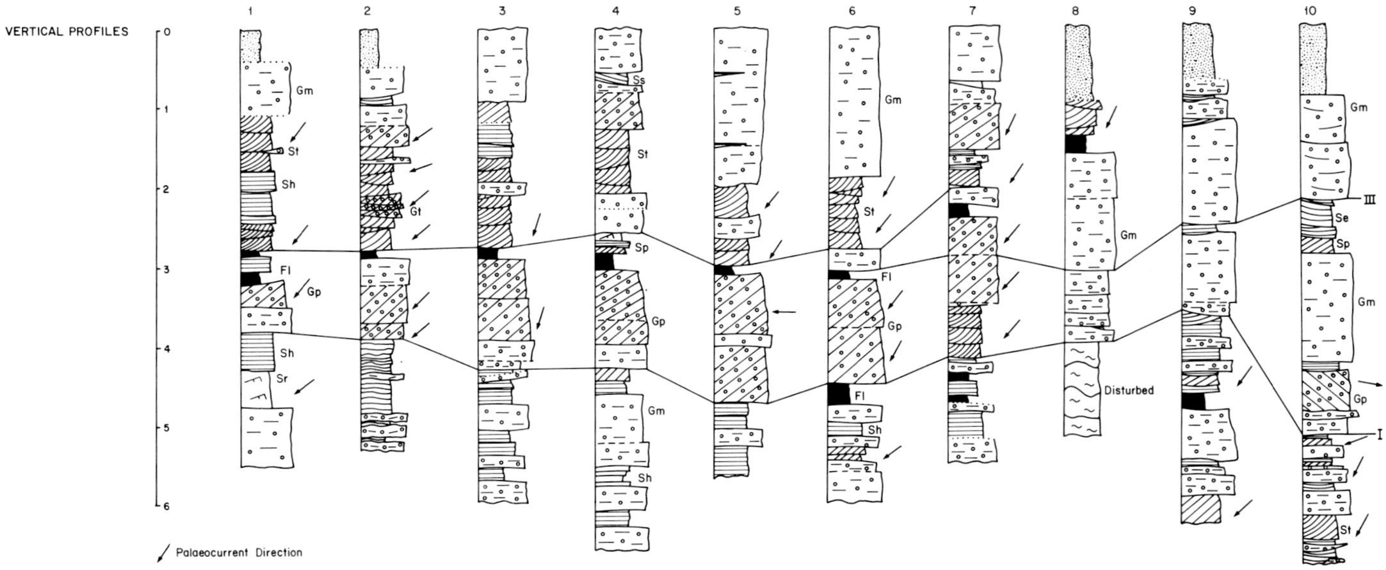

FIG. 3.—Representative vertical sections recorded at Holt Heath. The positions of the sections are indicated in Figure 4. I and III mark the position of third-order bounding surfaces discussed in the text.

Interpretation.—

The thin sheetlike geometry and predominance of flat lamination in units of this association suggest deposition from shallow, low velocity flows which were rarely erosive and spatially discontinuous. The association closely resembles supra-bar sediments described by Bluck (1979) and the medium to fine argillaceous sandstones described from the Sherwood Sandstone by Steel and Thompson (1983). The association at Holt Heath is interpreted as resulting from shallow flows which deposited sandy sediments over preexisting sheetlike gravel bodies left by abandonment of part of the active channel belt or on top of such bodies in the final stages of aggradation.

Association C

This association comprises 20–25 percent of the cross sections and is predominantly composed of vertically aggraded cosets (0.5–1.25 m thick) of cross-stratified pebbly sand. Units of this association typically overlie third-order bounding surfaces (Figs. 4, 5). Typically, units of this as-

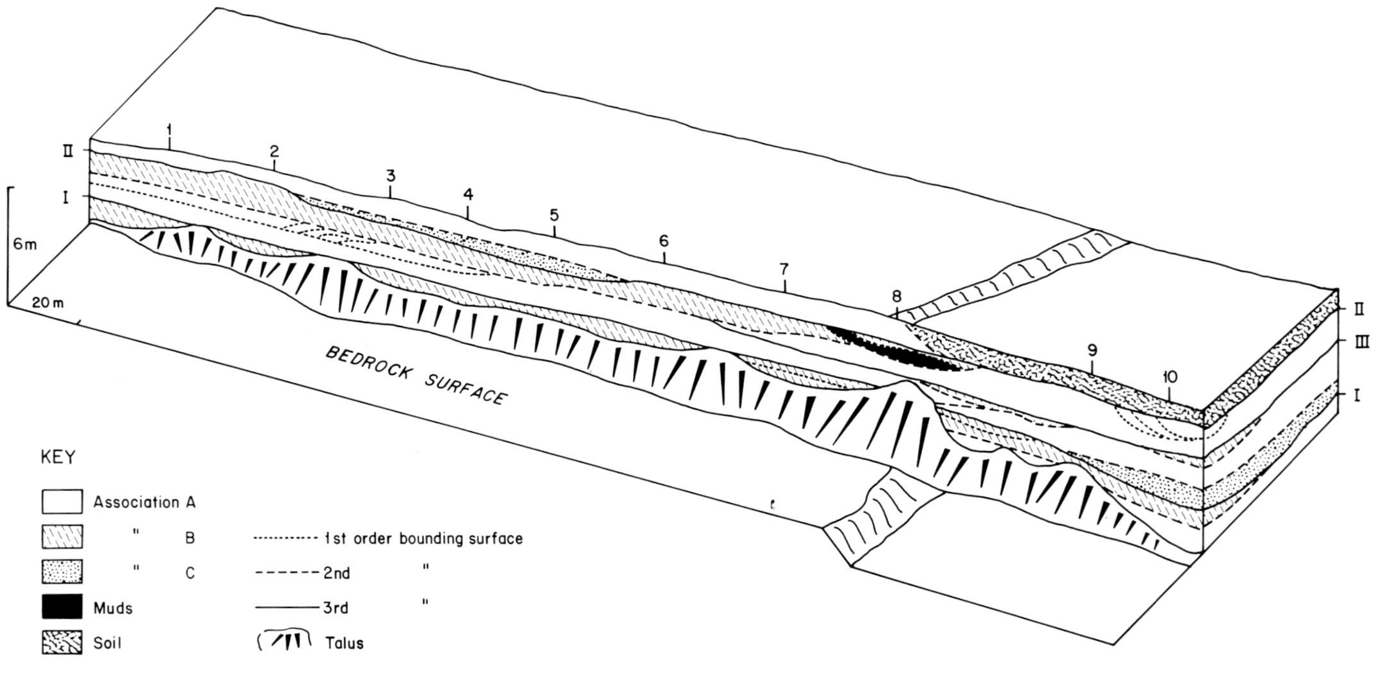

FIG. 4.—Block diagram to illustrate the disposition of the facies associations. I, II and III mark major erosion surfaces (third-order bounding surfaces) discussed in the text. Numbers 1–10 mark the position of the vertical sections shown in Figure 3.

FIG. 5.—Photograph of a quarry face to show characteristics of the three types of facies association (labelled A, B and C). I and III mark the positions of major erosion surfaces (third-order bounding surfaces). Staff graduated at 0.5 m intervals.

sociation are truncated by overlying units of association A, although occasionally they are laterally or vertically conformable upon one another. They form flat-topped lenticular bodies as much as 2 m thick and 40–50 m wide. The association is depicted in the Markov analysis by links between lithofacies Gt, St and Se which show mutual transitions between themselves and a probability of upward transition to facies Gm (Fig. 2). The sediments show both trough and planar cross-stratified sets (10–45 cm high) of fine sand to fine gravel. Gravel lags are common at set and coset boundaries. Paleocurrents show little variance (SD = 12°).

Interpretation.—

The internal characteristics of this facies and its containment within channel-form scours suggest infill of broad, shallow channels by unidirectional currents. The variety of lithofacies present indicates migration of bedforms of varying dimensions and grade which may partly be due to variations in sediment source (Allen, 1983) and partly attributable to fluctuations in discharge. The latter explanation is supported by frequent reactivation surfaces and gravel lags.

ARCHITECTURAL ARRANGEMENT OF LITHOFACIES ASSOCIATIONS

The way in which the units of the various lithofacies associations combine together provides information on the mechanisms whereby the sediments aggraded to form the sediment body as a whole. The units of the different associations are not randomly distributed but show distinct relationships to the different categories of bounding surface and to each other (Fig. 4).

Three third-order bounding surfaces occur within the sections (Fig. 4; I–III). Two of these surfaces (I and II) are subhorizontal erosion surfaces which are traceable throughout the exposures. Surface I has a stepped form and is overlain by two distinct units of association A; it may therefore be of composite origin. All of the third-order surfaces are overlain by laterally extensive sheets of association A, which are interpreted to have been deposited in the lowest levels of the depositional system. These surfaces are therefore interpreted to represent migrations of the major channel belt akin to those recognized by Bluck (1980) and Bryant (1983a). The restricted extent of units of association A above such surfaces indicates that the active channel belt was relatively narrow by comparison with the total width of the valley.

Units of association A occur preferentially in the eastern part of the cross sections, i.e., in the axis of the Severn valley, where they are commonly superimposed upon each other. Deposition of units of association A outside of this area seems to have resulted either from avulsion consequent upon aggradation in the central area of the valley (Bryant, 1983a) or extension of the active zone during exceptional floods. The presence of units of association C comprising unidirectional, vertically aggraded channel fills resting directly on third-order bounding surfaces (Fig. 4) indicates that initial avulsion may have been in the form of single channels cut and filled by upstream breaching of the banks of the main channel belt.

CONCLUSIONS

The outwash sediments comprise three main lithofacies associations attributable to deposition as: a) bar complexes in the lower levels of the active channel belt, b) bar top and floodplain sediments and, c) channel fills. Each of these associations is characterized by different lithofacies types and hence by different mean grain sizes, and porosity and permeability characteristics. In order to develop a predictive model of the distribution of these characteristics within outwash deposits, more data regarding the three-dimensional disposition of lithofacies associations will be necessary.

ACKNOWLEDGMENTS

This study was carried out while MRD was in receipt of a Natural Environmental Research Council postgraduate studentship award. IDB wishes to acknowledge the support of a research grant from Texaco which provided partial support for this work. The figures were skillfully drafted by Ramy de Groot Heupner. C. L. Pillmore and N. Wells are thanked for their helpful reviews of the manuscript.

REFERENCES

ALLEN, J. R. L., 1983, Studies in fluviatile sedimentation, bars, bar complexes and sandstone sheets (low sinuosity braided streams) in the Brownstones (L. Devonian), Welsh Borders: Sedimentary Geology, v. 33, p. 237–293.

BLUCK, B. J., 1974, Structure and directional properties of some sandur deposits in southern Iceland: Sedimentology, v. 21, p. 533–554.

———, 1979, Structure of coarse grained braided stream alluvium: Transactions, Royal Society of Edinburgh, v. 70, p. 181–221.

———, 1980, Structure, generation and preservation of upward fining braided stream cycles in the Old Red Sandstone of Scotland: Transactions, Royal Society of Edinburgh, v. 71, p. 29–46.

BROOKFIELD, M. E., 1977, The origin of bounding surfaces in ancient aeolian sandstones: Sedimentology, v. 24, p. 303–332.

BRYANT, I. D., 1983a, Facies sequences associated with some braided river deposits of late Pleistocene age from southern Britain, *in* Collinson, J. D., and Lewin, J., eds., Modern and Ancient Fluvial Systems: Sedimentology and Processes: International Association of Sedimentologists Special Publication 6, p. 267–275.

———, 1983b, The utilization of Arctic river analogue studies in the interpretation of periglacial river sediments from southern Britain, *in* Gregory, K., ed., Background to Palaeohydrology: A Perspective: Wiley, New York, p. 413–431.

COSTELLO, W. R., AND WALKER, R. G., 1972, Pleistocene sedimentology, Credit River, southern Ontario: A new component of the braided river model: Journal of Sedimentary Petrology, v. 42, p. 389–400.

DAVIS, J. C., 1974, Statistics and Data Analysis in Geology: Wiley, New York, 575 p.

DAWSON, M. R., 1985, Environmental reconstructions of a late Devensian terrace sequence: Some preliminary findings: Earth Surface Processes and Landforms, v. 10, p. 237–246.

EYNON, G., AND WALKER, R. G., 1974, Facies relationships in Pleistocene outwash gravels, southern Ontario: A model for bar growth in braided rivers: Sedimentology, v. 21, p. 43–70.

FRASER, J. Z., 1982, Derivation of a summary facies sequence based on Markov Chain analysis of the Caledon outwash: A Pleistocene braided glacial fluvial deposit, *in* Davidson-Arnott, R., Nickling, W., and Fahey, B. D., eds., Research in Glacial, Glacio-Fluvial and Glacio-Lacustrine Systems: Proceedings, Sixth Guelph Symposium, Norwich, Geobooks, p. 175–202.

HARBOUGH, S. W., AND BONHAM-CARTER, D., 1970, Computer Simulation in Geology: Wiley, New York, 575 p.

KRUMBEIN, W. C., AND DACEY, M. F., 1969, Markov Chain and imbedded Markov Chains in geology: Mathematical Geology, v. 1, p. 79–96.

MIALL, A. D., 1977, The braided river depositional environment: Earth Science Reviews, v. 13, p. 1–62.

———, 1978, Lithofacies types and vertical profile models in braided river deposits: A summary, *in* Miall, A. D., ed., Fluvial Sedimentology: Canadian Society of Petroleum Geologists Memoir 5, p. 597–604.

———, 1985, Architectural—element analysis: A new method of facies analysis applied to fluvial deposits: Earth Science Reviews, v. 22, p. 261–308.

RAMOS, A., AND SOPEÑA, A., 1983, Gravel bars in low sinuosity streams. (Permian and Triassic, central Spain), *in* Collinson, J. D., and Lewin, J., eds., Modern and Ancient Fluvial Systems: Sedimentology and Processes: International Association of Sedimentologists Special Publication 6, p. 301–312.

RUST, B. R., 1978, The interpretation of ancient alluvial successions in the light of modern investigations, *in* Davidson-Arnott, R., and Nickling, W., eds., Research in Fluvial Geomorphology: Proceedings, Fifth Guelph Symposium, Norwich, Geobooks, p. 65–105.

———, 1984, Proximal braidplain deposits in the Middle Devonian, Malbaie Formation of eastern Gaspe, Quebec, Canada: Sedimentology, v. 31, p. 675–696

SHOTTON, F. W., 1977, The English Midlands: [INQUA] International Quaternary Association Guidebook, Xth INQUA Congress, Birmingham, U.K.

———, AND COOPE, G. R., 1983, Exposure in the Power House Terrace of the River Stour, Wilden, Worcestershire, England: Proceedings Geologists Association, v. 94, p. 33–44.

SMITH, N. D., 1985, Proglacial fluvial environment, *in* Ashley, G. M., Shaw, J., and Smith, N. D., eds., Glacial Sedimentary Environments: Society of Economic Paleontologists and Mineralogists Short Course 16, p. 85–134.

STEEL, R. J., AND THOMPSON, D. B., 1983, Structures and Textures in Triassic braided stream conglomerates in the Sherwood Sandstone Group, North Staffordshire, England: Sedimentology, v. 30, p. 341–368.

TEISSEYRE, A. K., 1976, Pebble fabric in braided stream deposits with examples from recent and 'frozen' Carboniferous channels (Intrasudetic Basin, Central Sudetes): Geologica Sudetica, v. 10, p. 7–56.

THOMAS, G. S. P., CONNAUGHTON, M., AND DACKOMBE, R. V., 1985, Facies variation in a Late Pleistocene supraglacial outwash sandur from the Isle of Man: Geological Journal, v. 20, p. 193–213.

WALKER, R. G., 1979, Facies and facies models, *in* Walker, R. G., ed., Facies Models: Geoscience Canada Reprint Series 1, p. 1–7.

WILLS, L. J., 1938, The Pleistocene development of the Severn from Bridgnorth to the sea: Quarterly Journal of the Geological Society, v. 94, p. 161–242.

COARSE-GRAINED LOW-SINUOSITY RIVER DEPOSITS: EXAMPLE FROM PLIO-PLEISTOCENE VALDARNO BASIN, ITALY

PAOLO BILLI
Dipartimento di Ingegneria Civile, Universita' di Firenze
MAURIZIO MAGI
AND
MARIO SAGRI
Dipartimento Scienze della terra, Universitá di Firenze

ABSTRACT: The coarse-grained fluvial deposits of the Plio-Pleistocene Valdarno intermontane basin consist of massive or cross-stratified coarse gravel grading upward into cross-laminated sand overlain by massive overbank silt and clay with widespread pseudogley pedogenesis. The overall characteristics of such deposits (fining upward, abundance of overbank sediments, occurrence of extensive planar cross-stratification) would suggest deposition by a coarse-grained meandering river.

A detailed analysis of the relation between the depositional surface and the fabric of pebbles in planar crossbeds indicates downstream rather than lateral accretion. In fact, the dip of the intermediate axis of the pebbles parallels the foreset dip. Such a feature and the basin-wide divergence between trough and planar cross-stratification suggests deposition by lateral-diagonal bars in a low-sinuosity river.

The Pleistocene fluvial deposits of the Valdarno basin have a facies association and sedimentary structures similar to modern low-sinuosity rivers of the Northern Apennines.

INTRODUCTION

Coarse-grained fluvial deposits were generally interpreted in the past as the product of braided rivers. Recently, several authors have shown modern and ancient examples of coarse-grained sediments deposited by meandering rivers (McGowen and Garner, 1970; Levey, 1978; Jackson, 1978; Bluck, 1979; Ricci Lucchi and others, 1981; Ori, 1982; Arche, 1983) or low-sinuosity rivers (Gustavson, 1978; Bluck, 1979; Ori, 1982; Ferguson and Werritty, 1983; Calvet and others, 1985). The interpretation of ancient coarse-grained deposits is still uncertain, however (Ferguson and Werritty, 1983; Friend, 1983). Models based on studies of present rivers and their facies associations are very useful in recognizing ancient fluvial environments, but a more detailed investigation on the internal arrangement of the sedimentary structures is needed to interpret any facies according to the processes through which they originated.

This paper demonstrates the relation between gravel fabric and cross-stratification as an aid to interpreting the deposits of low-sinuosity rivers. For this purpose, the Pleistocene coarse-grained fluvial deposits (Monticello Group) of the Valdarno fluvio-lacustrine basin (Fig. 1) were studied and compared with modern examples of low-sinuosity rivers of the Northern Apennines. The Valdarno (Fig. 1) is an intermontane basin located southeast of Florence, consisting of fluvio-lacustrine deposits (Castelnuovo Group, middle Pliocene) unconformably overlain by lower Pleistocene lacustrine Montevarchi Group and fan delta deposits (Pratomagno Fans), followed by middle Pleistocene fluvial deposits (Monticello Group) that, together with the ultimate deposition of the fans, mark the basin's infilling (Abbate, 1983). Paleocurrent measurements indicate the main river flowed from the northeast and turned toward the west in its downstream reach (Fig. 2).

FACIES DESCRIPTION

The Monticello Group ranges between 30 to 50 m in thickness and consists of calcareous poorly sorted gravels with fine to coarse sandy matrix, grading upward into crossbedded, medium to coarse sands capped by massive overbank silt and clay with widespread pseudogley pedogenesis (Billi and others, 1983). This sequence clearly fines upward (Fig. 3). The mean size of the largest clasts ranges from 120 mm in the upstream exposures to 80 mm in downstream sections.

Three distinct sedimentary cycles were identified. They are clearly marked in the downstream exposures, where sand largely prevails on gravel, by the basin-wide occurrence of erosional surfaces cut into clay plug deposits. Commonly, in the upstream conglomeratic units such fine sediments are completely eroded and occur as scattered clay boulders which are generally associated with two main erosional surfaces (Fig. 3). Only the third cycle shows the entire sequence, whereas the other two have the upper portion truncated by erosion.

Cycle I.—

The conglomerates of the first cycle are underlain by erosive surfaces cut into the lacustrine deposits of the Montevarchi Group and range from 3.5 to 8 m in thickness (Fig. 3). They either show no internal stratification or are crudely bedded.

In the upstream exposures (Petrognano), massive clast- and subordinately matrix-supported gravels prevail, whereas downstream planar and trough crossbedded conglomerates are more common (Fig. 3). The massive horizons generally show fairly well imbricated pebbles and locally coarsen upward. Lenses of massive sand occur in the conglomerates which may be topped, mostly downstream, by trough crossbedded sand. Upstream the planar crossbeds are limited in their lateral extent and generally dip over 20°. The foresets may be reverse-graded with flat pebbles lying on the accretionary surface. In the downstream exposures, the planar crossbeds are more laterally extensive and dip at a lower angle (about 20°, Fig. 4). They are locally marked by thin drapes of sand whose laminae parallel the dip of the gravel crossbeds.

The dip of the planar crossbeds and that of the flat pebbles inside them is the same, but the latter tends to diverge

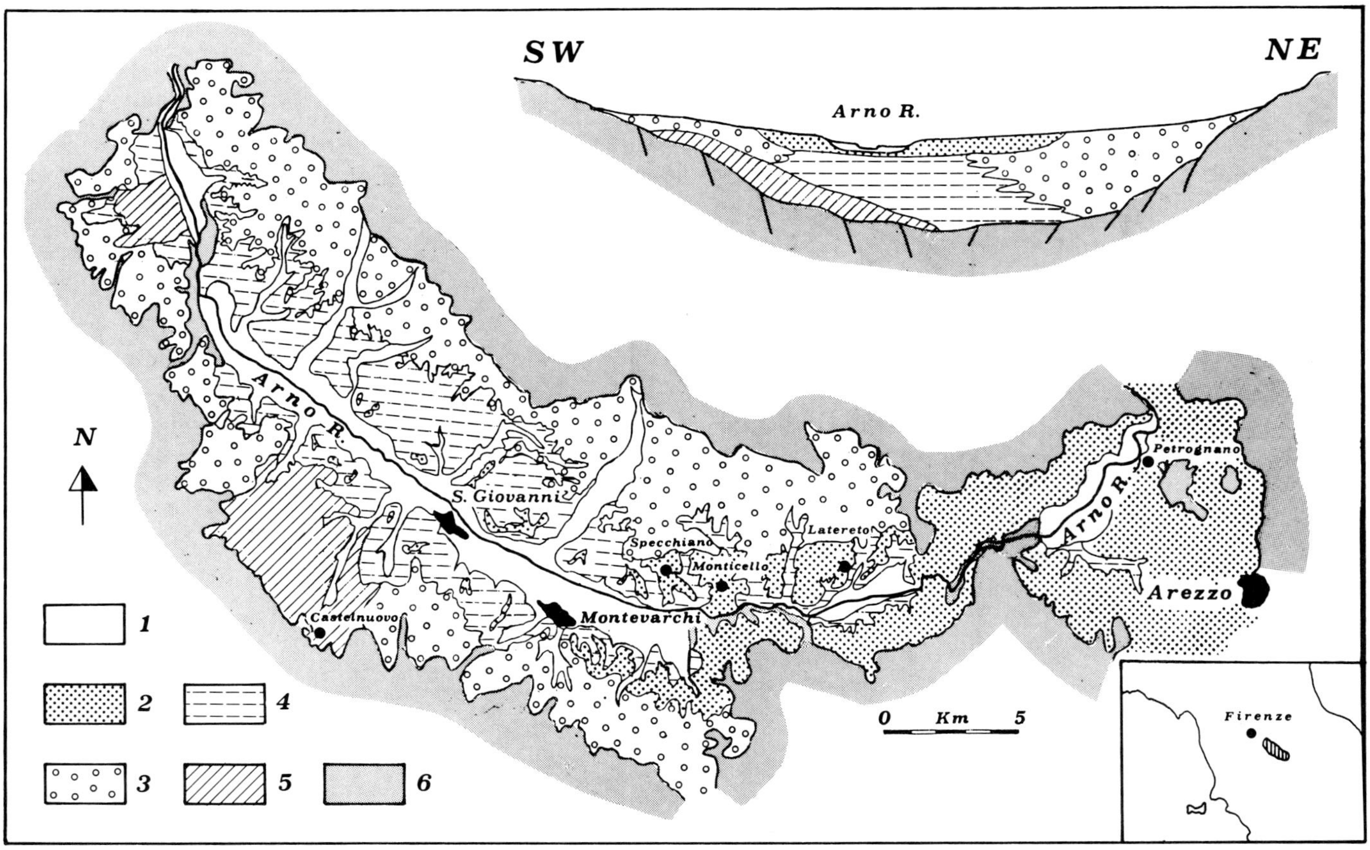

FIG. 1.—Geologic sketch map of the fluvio-lacustrine Valdarno Basin. 1) Recent alluvium (Holocene); 2) Monticello Group fluvial deposits (Middle Pleistocene); 3) Pratomagno Fans (Lower-Middle Pleistocene); 4) Lacustrine Montevarchi Group (Lower Pleistocene); 5) Fluvio-lacustrine Castelnuovo Group (Middle Pliocene); 6) Bedrock: Macigno Formation (Middle Oligocene—Lower Miocene). The relationships among the stratigraphic units are represented in the schematic cross section.

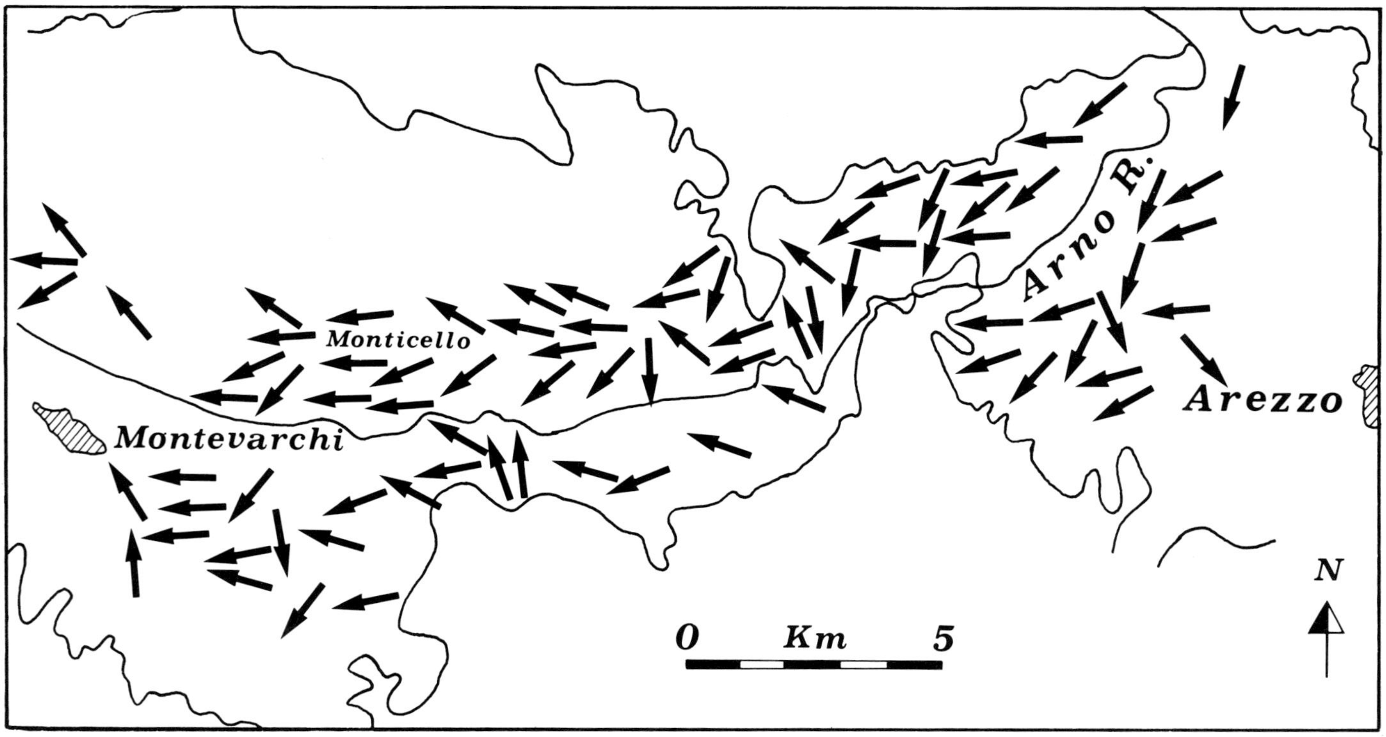

FIG. 2.—Paleocurrent distribution for the Monticello Group. Each arrow represents 100 paleocurrent measurements based on the imbrication of flat pebbles of unstratified or crudely horizontally-bedded gravels.

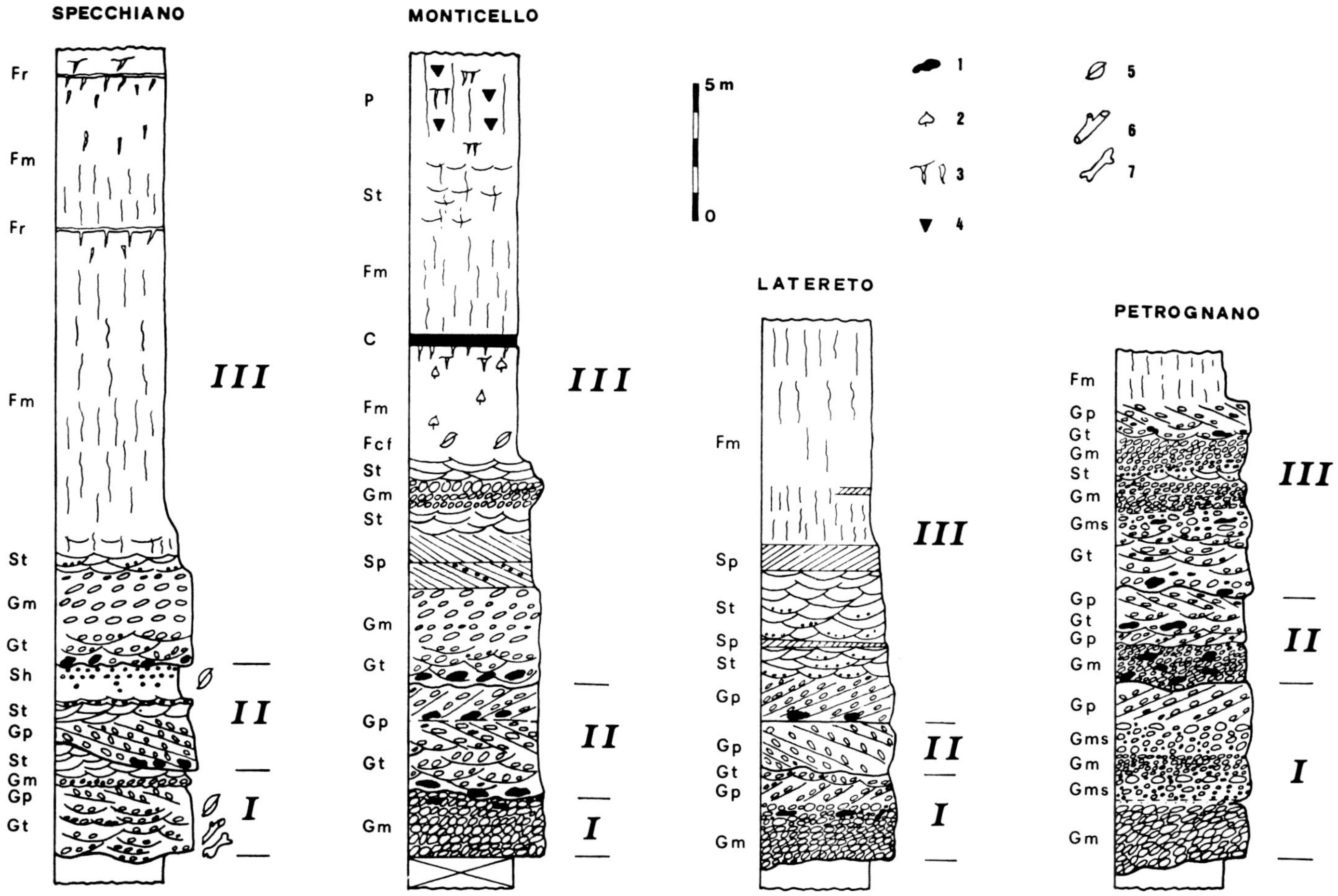

FIG. 3.—Generalized stratigraphic sections of the Monticello Group (for location see Fig. 1). 1) Clay boulders; 2) Plant remains; 3) Root trace fossils; 4) Calcrete nodules; 5) Freshwater clams (*Unio*); 6) Tree trunks; 7) Mammal bones. Lithofacies codes according to Miall's (1977) classification. The sedimentary cycles are indicated by Roman numerals.

(45° on average) near the crossbed's upper surface. These features are so common in the planar crossbed sets as to make them the most distinctive characteristic of the Monticello Group conglomerates.

The planar cross-stratified conglomerates locally have a scoured top filled with trough crossbedded gravel and sand. Moreover, a few large-scale trough crossbeds (2–10 m wide and 0.5–2 m deep) associated with channels occur in the upstream sections, becoming more common and smaller in size in a downstream direction. Orientations of their axes are quite clustered and correspond to the general flow direction of the main river.

Cycle II.—

The second cycle deposits range from 3 to 5 m in thickness and cut those of the first with an erosive surface clearly marked by clay boulders as much as 1 m in size (Fig. 3). Lenses of sand are more common and gravel is a little finer and better sorted with more disc-shaped pebbles than in the first cycle.

Crudely horizontally-bedded and unstratified conglomerates are rather rare and occur only in the upstream exposures. They are clast-supported with an abundant coarse sand matrix.

Second-cycle planar crossbeds are the most common sedimentary structure and are much more extensive than in the first cycle; when well exposed they can be followed for tens of meters (Fig. 5). The bed's dip is generally lower (5°–15°, Fig. 4), and elongated sand drapes commonly occur between the gravel beds. These sands are massive or characterized by plane laminae which again dip approximately in the same direction of the gravel planar crossbeds (Fig. 6).

The planar crossbeds have flat bases, and they are never associated with lag deposits. The latter are poorly distinguishable in the field because they never occur as a distinct coarser gravel unit but rather as an accumulation of clay boulders. Planar cross-stratified sets pass laterally into small channels filled with massive silt and clay resting directly on the gravel of the channel pavement (Fig. 7).

As observed in the foresets of the first cycle, pebble orientation varies depending on the pebble's position in the planar crossbeds. Along the foreset, the dip of the intermediate axis of the pebbles parallels the dip of the foreset, whereas near the bed's upper surface the dips distinctively diverge (Fig. 8). The top of the planar crossbed sets is cut by trough crossbedded sands whose flow direction is approximately at a right angle to the set's dip.

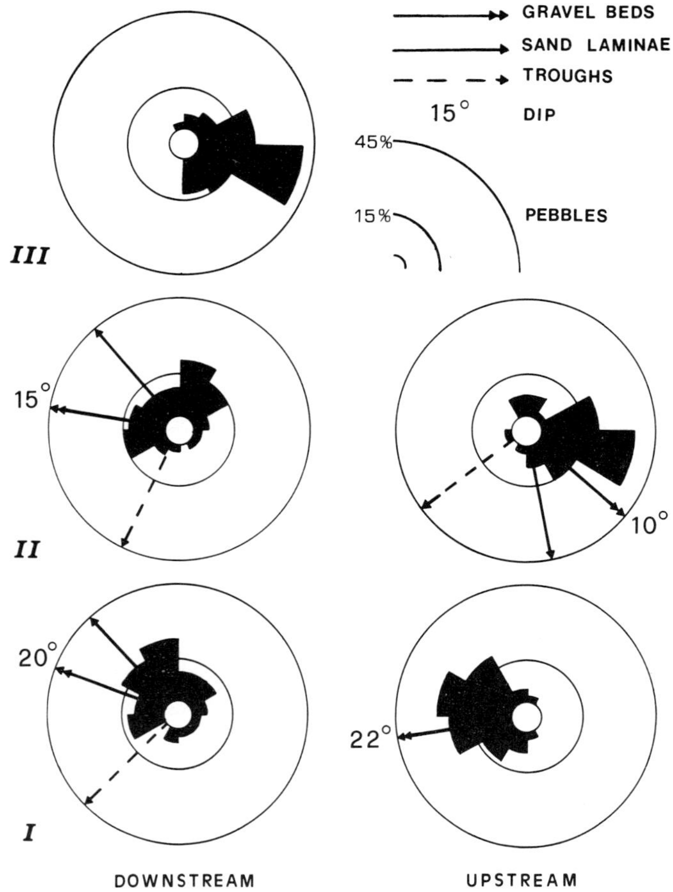

FIG. 4.—Rose diagram showing the dip of the *ab* plane of discoidal pebbles (100 measurements each diagram). The arrows indicate the average dip of the gravel planar crossbeds (double-headed arrow) of the plane laminae in the sand drapes interbedded within the gravels (single-head arrow) and of the troughs axes (dashed arrow). The arrows are based on as many measurements as possible (generally 20–30) for each exposure. The Roman numerals indicate the sedimentary cycles.

In the downstream exposures, the second cycle is topped by trough or horizontally-laminated sand truncated by sediment of the third cycle.

Cycle III.—

The overall thickness of the gravel-and-sand units in the third cycle varies from 5 to 10 m, whereas the largest por-

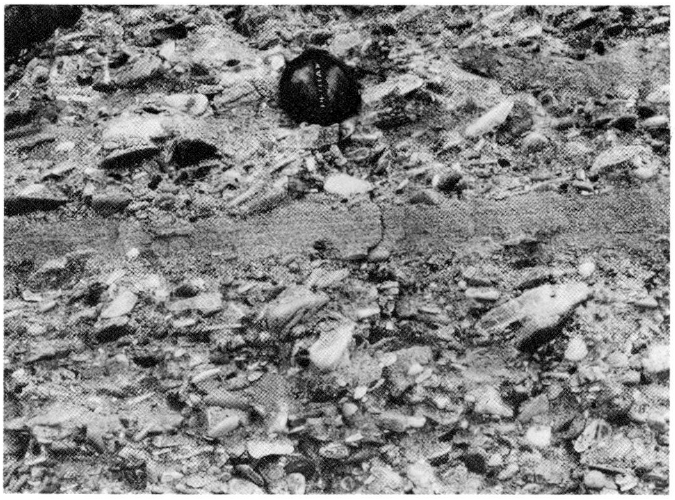

FIG. 6.—Sand drapes interbedded with gravel planar crossbeds of cycle II. The sand laminae dip is the same as the gravel bed's dip (Specchiano Quarry).

FIG. 7.—Lag gravel of the slough channel (see Fig. 5) capped by massive clay. The overlying planar cross-laminated sand represents channel reactivation.

FIG. 5.—Gravel planar crossbeds (*pc*) of cycle II merging into the slough channel (*sc*) (for a closeup view of the slough channel deposits, see Fig. 7). The axes of the large troughs in the sand (*s*) on the top of the sequence are perpendicular to the dip direction of the gravel foresets (Latereto Quarry).

FIG. 8.—Pebble orientations: (*a*) pebbles deposited during the bar accretion as planar crossbeds. Their dip is downcurrent; (*b*) flat pebbles resting on the bar surface during the waning flow. They are imbricated and dip at 60°–90° away from the foreset dip.

tion of the sequence consists of pedogenically altered silt and clay as much as 20 m thick (Fig. 3). In the upstream sections unstratified and trough crossbedded gravels prevail. Downstream conglomerates occur mainly as crudely horizontally-bedded units. The pebbles are smaller in size and better sorted than those of the lower cycles. The matrix is generally a fine, rather scarce sand. The disc-shaped particles are more abundant and, in the horizontal beds, very well imbricated. Trough and planar cross-stratified gravels are less common. They are very similar in scale to those of the second cycle but the planar sets dip at a lower angle (4°–10°).

The sand unit is trough and planar cross-laminated and passes upward into unbedded silt and clay with features typical of overbank deposits. These features include root trace fossils, lignite beds and calcrete nodules. Individual troughs in the sand unit are as much as 2 m wide and 1 m deep. The transition between the underlying conglomerates and the sands is abrupt and in places slightly erosive. In fact, the base of the overbank deposits is commonly marked by fine gravel consistent with overbank scour-and-fill processes during extreme floods. During such floods, the overbank flow has generally a straight downvalley direction.

INTERPRETATION AND DISCUSSION

The unstratified and crudely horizontally stratified gravels, which are most abundant in the upstream exposures, may be interpreted as lag and longitudinal bar deposits, as their characteristics match the description given by several authors (see Miall, 1977, for a complete review). When such deposits are associated with gravels containing trough and high angle, planar tabular crossbeds, they are commonly interpreted as coarse-grained braided stream deposits. The interpretation of the laterally extensive planar crossbeds with foreset dip of 5°–20° is not straightforward. A few authors (McGowen and Garner, 1970; Ricci Lucchi and others, 1981; Ori, 1982; Arche, 1983) indicate similar deposits as typical of point bars in a coarse-grained meandering river.

A detailed analysis of the low angle planar crossbeds of the Monticello Group, however, indicates downcurrent rather than lateral accretion. This conclusion is based on two observations: first, the laminae of the sand drapes between the gravel beds commonly parallel the dip of the planar crossbeds (Figs. 4, 6). Second, along the foreset the dip of the intermediate axis of the flat pebbles parallels the dip of the foreset, whereas near the bed's upper surface their dips diverge (Fig. 8). This results in a bimodal distribution of the dip of discoidal pebbles (Fig. 4). It can be explained as follows: during the downcurrent accretion of the lateral diagonal bar at high flood flow, the pebbles are laid down on the depositional surface and hence they are not imbricated; during the waning flow, diverging currents on the top of the bar orient imbricated pebbles at 60°–90° away from the bed's dip.

All these characteristics may be diagnostic of the planar crossbeds described in this paper, which are typical lateral diagonal bars, and differentiate them from the epsilon crossbeds which are generally associated with point bars (Allen, 1965; Arche, 1983; Jackson, 1978; McGowen and Garner, 1970; Ori, 1982; Ricci Lucchi and others, 1981). On a point bar, bedforms generally migrate along the accretionary surface and indicate a flow oblique to the direction of point bar growth.

In a few good exposures, planar crossbed sets laterally merge into channel deposits characterized by a massive clay plug resting abruptly on a gravel lag (Fig. 7). Such association is consistent with the filling of a slough channel as described by Bluck (1979) and Ferguson and Werritty (1983) for low- to medium-sinuosity gravel rivers. Low angle planar cross-stratified gravels result then from the accretion of lateral diagonal bars (Church and Gilbert, 1979; Bluck, 1979; Ferguson and Werritty, 1983; Forbes, 1983) in a single-channel, low-sinuosity river. This interpretation is also supported by the occurrence of large channels as solitary and rare features even in the wider quarry exposures. Furthermore, the flow directions indicated by the gravel-and-sand troughs associated with the planar crossbeds are quite consistent and run at about a right angle on either side of the foreset dip direction of planar crossbedding in the gravel (Fig. 4). Trough crossbeds are interpreted as either forming in the main channel or in chute and transverse channels and result from the filling of trough scours and megaripple migration.

Thin subhorizontal, matrix-poor gravel beds with very well sorted and imbricated pebbles, occurring in the downstream exposures of the third cycle, suggest a straight river with alternate bars. This river was laterally confined by the late progradation phase of the Pratomagno fans which were connected with a further uplifting of the basin margins.

Most of the time, the river which deposited the Monticello Group generally had a low-sinuosity form, and its facies architecture matches that proposed by Allen (1965, Fig. 35 C) for low-sinuosity rivers. Coarse-grained low-sinuosity rivers are quite common in the Northern Apennines. They are characterized, in their intermediate reaches, by a single channel and by lateral diagonal bars (Fig. 9) that migrate

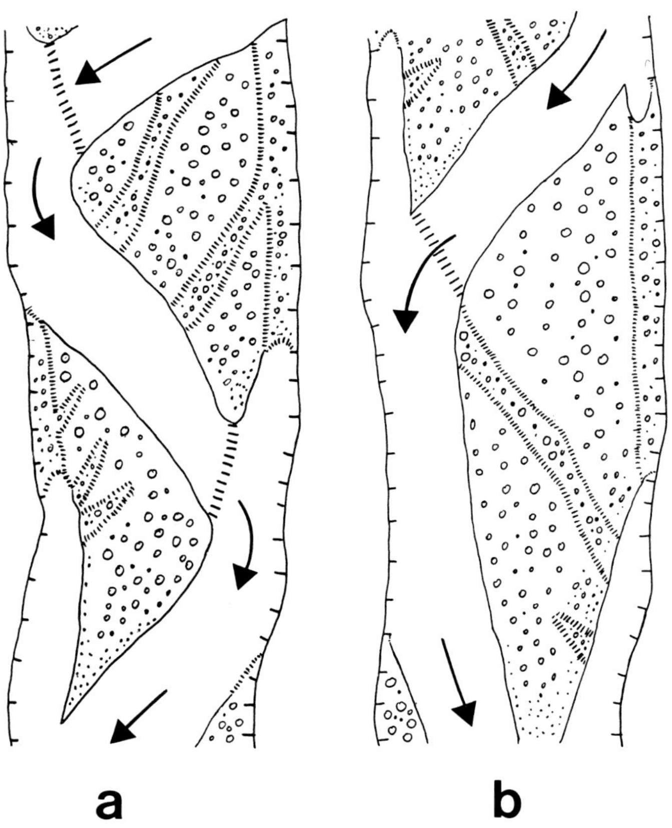

FIG. 9.—Schematic diagrams of the evolution and bar migration of a modern coarse-grained, low-sinuosity river of the Northern Apennines: initial (*a*) and final (*b*) stages. The lateral diagonal bars may migrate, by downcurrent accretion, either into the main channel or into the slough channel. Lateral channel migration is moderate. Avulsion mainly occurs as a consequence of chute channel enlargement.

in the slough channel. Sometimes the slough channel is reactivated during large floods as a chute channel that subsequently may develop into the main channel. The accretion of the lateral diagonal bars produces low angle planar

FIG. 10.—Alluvial deposits of the Albegna River. Low angle planar crossbeds have many characteristics similar to those of the Monticello Group conglomerates (see Fig. 5). Geologic hammer for scale.

crossbedding with characteristics similar to those observed in the Monticello Group conglomerates (Fig. 10). Figure 10 shows the recent deposits of a studied reach of the Albegna River, a modern example of coarse-grained low-sinuosity river of the Northern Apennines. The low-sinuosity path the river had taken in the recent past can be still recognized on the actual alluvial plain. In the outcrop shown in Figure 10, which approximately corresponds to a longitudinal section of a lateral diagonal bar, the dip of the flat pebbles parallels the foreset dip as well as that observed in the planar crossbeds of the Monticello deposits. Moreover, planar crossbedded gravel in modern low-sinuosity rivers of the Northern Apennines is likewise capped by trough crossbedded sand and gravel. Trough axes in this sand and fine gravel are oriented parallel to the axis of the main channel.

CONCLUSIONS

Low-sinuosity or "wandering" river morphology, according to the terminology of Church (1983), is considered a transition between the more stable braided and meandering configurations in a generally stable setting (Church and Jones, 1982; Church, 1983; Ferguson and Werritty, 1983). In the Northern Apennines, wandering rivers are a common and persistent feature, as shown by contemporary and Quaternary examples. Their persistence depends on the moderate, though still active, uplift of the chain that prohibits the rivers from reaching a stable state. The modern wandering rivers of the Northern Apennines are characterized by large annual variations in discharge and by a relatively high sediment supply. Their flood plain is usually small and confined and the channel migrates slowly across it. Avulsion occurs only as a widening of the chute channel that subsequently becomes the main channel.

The most significant features of the deposits of the modern and ancient coarse-grained, low-sinuosity rivers of the Northern Apennines can be summarized as follows:

i) Laterally extensive sets of low angle (5°–20°) crossbedded gravels are characterized by a bimodal distribution in the dip of the intermediate axis of the pebbles. Where the planar crossbeds, the pebbles and the laminae of the associated sand drapes all dip in the same direction, the accretion of lateral diagonal bars was mainly in a downcurrent direction.

ii) Slough channels commonly occur. They have a thin gravel lag abruptly capped by a clay plug. Often, the planar crossbeds merge in the slough channel.

iii) Gravel-and-sand trough crossbeds show a clustered paleocurrent orientation. It is consistent with the valley slope and at about 60°–90° from both sides of the dip direction of the gravel planar crossbeds. This indicates a river that wandered around alternating lateral bars.

iv) The lag deposit generally consists of scattered clay boulders.

v) Gravel, sand, and overbank deposits have a quite constant thickness across the basin.

ACKNOWLEDGMENTS

R. H. Blodgett and J. F. Hubert are greatly acknowledged for the review of the manuscript and for their constructive criticism. P. Bruni and C. Bartolini read an earlier version of the manuscript and made helpful suggestions. The authors are also indebted to C. Marri for the prepration of the figures and to P. Manetti for typing the manuscript. This research was supported by CNR grant 83.00575.05 and 84.00204.05, CNR "Centro studi per la geologia dell'Appennino' contribution n. 180 and MPI grant 09.01.00070 and 09.01.00097.

REFERENCES

ABBATE, E., 1983, Schema stratigrafico della successione neoautoctona del Valdarno Superiore e del Bacino di Arezzo, *in* I Depositi Fluvio-Lacustri del Valdarno Superiore, Guida all' Escursione del Gruppo Informale "Sedimentologia" del C.N.R.: Istituto di Geologia e Palentologia, Universita' di Firenze, p. 1–6.

ALLEN, J. R. L., 1965, Origin and classification of recent alluvial sediments: Sedimentology, v. 5, p. 89–191.

ARCHE, A., 1983, Coarse-grained meander lobe deposits in the Jarama River, Madrid, Spain, *in* Collinson, J. D., and Lewin, J., eds., Modern and Ancient Fluvial Systems: International Association of Sedimentologists Special Publication 6, p. 313–321.

BILLI, P., MAGI, M., AND SAGRI, M., 1983, Dati sedimentologici relativi al Gruppo di Monticello, *in* I Depositi Fluvio-Lacustri del Valdarno Superiore, Guida all' Escursione del Gruppo Informale "Sedimentologia" del C.N.R.: Istituto di Geologia e Palentologia, Universita' di Firenze, p. 13–18.

BLUCK, B. J., 1979, Structure of coarse grained braided stream alluvium: Transactions, Royal Society of Edinburgh, v. 70, p. 181–221.

CALVET, J., MARZO, M., AND PUIGDEFABREGAS, C., 1985, Lateral bar development during extreme flood in low sinuosity reaches of the Segre River, Southern Pyrenees, Spain (abs): Third International Fluvial Sedimentology Conference, Fort Collins, Colorado, p. 13.

CHURCH, M., 1983, Pattern of instability in a wandering gravel bed channel, *in* Collinson, J. D., and Lewin, J., eds., Modern and Ancient Fluvial Systems: International Association of Sedimentologists Special Publication 6, p. 169–180.

———, AND GILBERT, R., 1979, Proglacial fluvial and lacustrine environments, *in* Jopling, A. V., and McDonald, B. C., eds., Glaciofluvial and Glaciolacustrine Sedimentation: Society of Economic Palentologists and Mineralogists Special Publication 23, p. 22–100.

———, AND JONES, D. P., 1982, Channel bars in gravel-bed rivers, *in* Hey, R. E., Bathurst, J. C., and Thorne, C. R., eds., Gravel-Bed Rivers: Fluvial Processes, Engineering and Management: Wiley, New York, p. 291–324.

FERGUSON, R. I., AND WERRITTY, A., 1983, Bar development and channel changes in the gravelly River Feschie, Scotland, *in* Collinson, J. D., and Lewin, J., eds., Modern and Ancient Fluvial Systems: International Association of Sedimentologists Special Publication 6, p. 181–193.

FORBES, D. L., 1983, Morphology and sedimentology of a sinuous gravel-bed channel system: Lower Babbage River, Yukon coastal plain, Canada, *in* Collinson, J. D., and Lewin, J., eds., Modern and Ancient Fluvial Systems: International Association of Sedimentologists Special Publication 6, p. 195–206.

FRIEND, P. F., 1983, Towards the field classifications of alluvial architecture or sequence, *in* Collinson, J. D., and Lewin, J., eds., Modern and Ancient Fluvial Systems: International Association of Sedimentologists Special Publication 6, p. 345–354.

GUSTAVSON, T. C., 1978, Bed forms and stratification types of modern gravel meander lobes, Nueces River, Texas: Sedimentology, v. 25, p. 401–426.

JACKSON, R. G. II, 1978, Preliminary evaluation of lithofacies models for meandering alluvial streams, *in* Miall, A. D., ed., Fluvial Sedimentology: Canadian Society of Petroleum Geologists Memoir 5, p. 543–576.

LEVEY, R. A., 1978, Bed-form distribution and internal stratification of coarse-grained point bars, Upper Congaree River, S. C., *in* Miall, A. D., ed., Fluvial Sedimentology: Canadian Society of Petroleum Geologists Memoir 5, p. 105–127.

MCGOWEN, J. H., AND GARNER, L. E., 1970, Physiographic features and stratification types of coarse-grained point bars: Modern and ancient examples: Sedimentology, v. 14, p. 77–111.

MIALL, A. D., 1977, A review of the braided-river depositional environment: Earth Science Reviews, v. 13, p. 1–62.

ORI, G. G., 1982, Braided to meandering channel patterns in humid-region alluvial fan deposits, River Reno, Po Plain (northern Italy): Sedimentary Geology, v. 31, p. 231–248.

RICCI LUCCHI, F., OGLIANI, F., ORI, G. G., AND COLELLA, A., 1981, Braided versus meandering channel deposits in Pliocene fan delta systems, Bologna intra-Apenninic basin (abs): International Association of Sedimentologists Second European Regional Meeting, p. 161–163.

SEDIMENTOLOGY OF THE WHITE CHANNEL GRAVELS, KLONDIKE AREA, YUKON TERRITORY: FLUVIAL DEPOSITS OF A CONFINED VALLEY

STEPHEN R. MORISON AND FRANCES J. HEIN†

Exploration and Geological Services Division, Department of Indian Affairs and Northern Development, Whitehorse, Yukon, Canada Y1A 3V1; Department of Geology, University of Alberta, Edmonton, Alberta, Canada T6G 2E3

ABSTRACT: The Pliocene-Early Pleistocene White Channel gravels in the Klondike area are divided into nine facies, interpreted as proximal braided to distal braidplain sequences and colluvial sediment-gravity flow deposits within a confined valley setting. Results of Markov chain analysis show that the White Channel deposit shows only three well defined trends. These consist of: (1) disorganized gravel overlain by massive to crudely stratified gravel, which passes back into disorganized gravel. These facies are both poorly sorted and represent high flow, tractional bedload and/or mass flow deposition; (2) scoop-shaped crossbedded cobble-gravel overlain by trough crossbedded pebbly sand and gravel. This sequence is interpreted as infilling of braid channels within a main valley setting; and (3) stratified gravel interbedded with thin, convolute-bedded mud and fine sand. This records overbank deposition during waning flow on braid bar deposits or on diffuse gravel sheets within secondary braid channels.

Compared with other braided river deposits of glaciofluvial settings, the White Channel deposit is more disorganized, very poorly sorted, and generally lacks good sequence development. Thus, the applicability of existing gravelly braided river models, derived mainly from unconfined glaciofluvial braidplains, may not be valid for gravelly sequences deposited during nonglacial periods in confined, upland valley settings.

INTRODUCTION

Yukon placer deposits produced 348 million grams of gold between 1885 and 1982 (Debicki, 1983) and are the most economically significant placers in Canada, contributing as much as 5 percent of Canada's total gold production during 1981 (Morin and Debicki, 1982; Debicki, 1983). The Klondike area, famous for its gold rush of 1896, has consistently been the major producer of Yukon gold and is located between the Klondike and Indian River drainage basins near Dawson City (Fig. 1). Production has mainly been from the White Channel gravel terraces along Bonanza and Hunker Creeks (Fig. 1).

To date, surprisingly little work has been done on the sedimentology of the White Channel sequence. McConnell (1907) interpreted the gravel as being deposited by winding streams with low gradients and slack currents. Milner (1976), Naldrett (1981), and Tempelman-Kluit (1982) described the White Channel sequence as poorly sorted, dominantly massive gravel deposited in a proximal braided environment, similar to the 'Scott-type' system of Miall (1978). Morison (1983, 1985a) described the sedimentology of the placer deposits of the Clear Creek drainage basin in the central Yukon and noted similarities of the Clear Creek deposits to the White Channel gravel in the Klondike.

The purpose of this study is threefold: (1) to describe the sedimentology of the White Channel deposit within the Bonanza Creek and Hunker Creek valleys near Dawson (Fig. 1); (2) to develop a facies scheme and summary which characterizes observed facies relationships; and (3) to construct a paleogeographic model of the depositional environment. The relationships between gold distribution and facies interpretations are discussed by Morison (1985b).

From a sedimentological point of view, the results of this study are important for understanding processes responsible for sedimentation in proximal, confined upland valleys. Until now, most of our understanding of proximal alluvial sedimentation has been based on studies of glacially-influenced settings (Boothroyd and Ashley, 1975; Boothroyd and Nummedal, 1978; Bradley and others, 1972; Casshyap and Tewari, 1982; Church and Jones, 1982; Fahnestock, 1963; Fahnestock and Bradley, 1973; Gustavson, 1974; Hammer and Smith, 1983; Hein, 1984; Hein and Walker, 1977; Rust, 1972; N.D. Smith, 1974; Southard and others, 1984) or on studies of alluvial fans (Bull, 1977; Nilsen, 1982; Steel and others, 1977). Notable exceptions to these studies of deposits of proglacial streams and alluvial fan systems include Steel and Thompson's (1983) work on the Triassic Bunter Pebble Beds in Britain; S. V. Johnson's (1984) study on Eocene alluvial sediments within a rapidly subsiding basin in Washington; and Rust's (1984) study on the Devonian Malbaie Formation of Gaspé, Canada.

GENERAL GEOLOGY

The Klondike placer area is in the Yukon Cataclastic Complex (Tempelman-Kluit, 1981), which is a structurally sheared and metamorphosed terrane. North of the Klondike area is the major Tintina fault, which shows as much as 450 km of dextral strike-slip displacement, most of which occurred during the Tertiary. The Tintina fault forms the Tintina Trench, a narrow graben which was probably active up to the Pliocene (Tempelman-Kluit, 1981). The Klondike Plateau is an uplifted erosional surface of Tertiary age, characterized by dissected, rolling terrain of accordant ridges (Bostock, 1948, 1970; Tempelman-Kluit, 1980). This subdued, mature landscape is the result of extensive subaerial erosion which ended around the Miocene (Tempelman-Kluit, 1980). This plateau was not glaciated during the Pleistocene. The White Channel sequence was emplaced as a valley-fill component of the paleo-Yukon River drainage basin.

In the Dawson City area, the White Channel deposit occurs as terraces on bedrock benches. Despite much attention to finding the 'Mother Lode' since discovery of gold in the Klondike, there appears to be no relationship between bedrock mineral occurrences and the amount of extracted gold from the overlying unconsolidated gravels. Even the most recent exploration programs have failed to locate

†Present address: Center for Marine Geology, Dalhousie University, Halifax, Nova Scotia, Canada B3H3J5

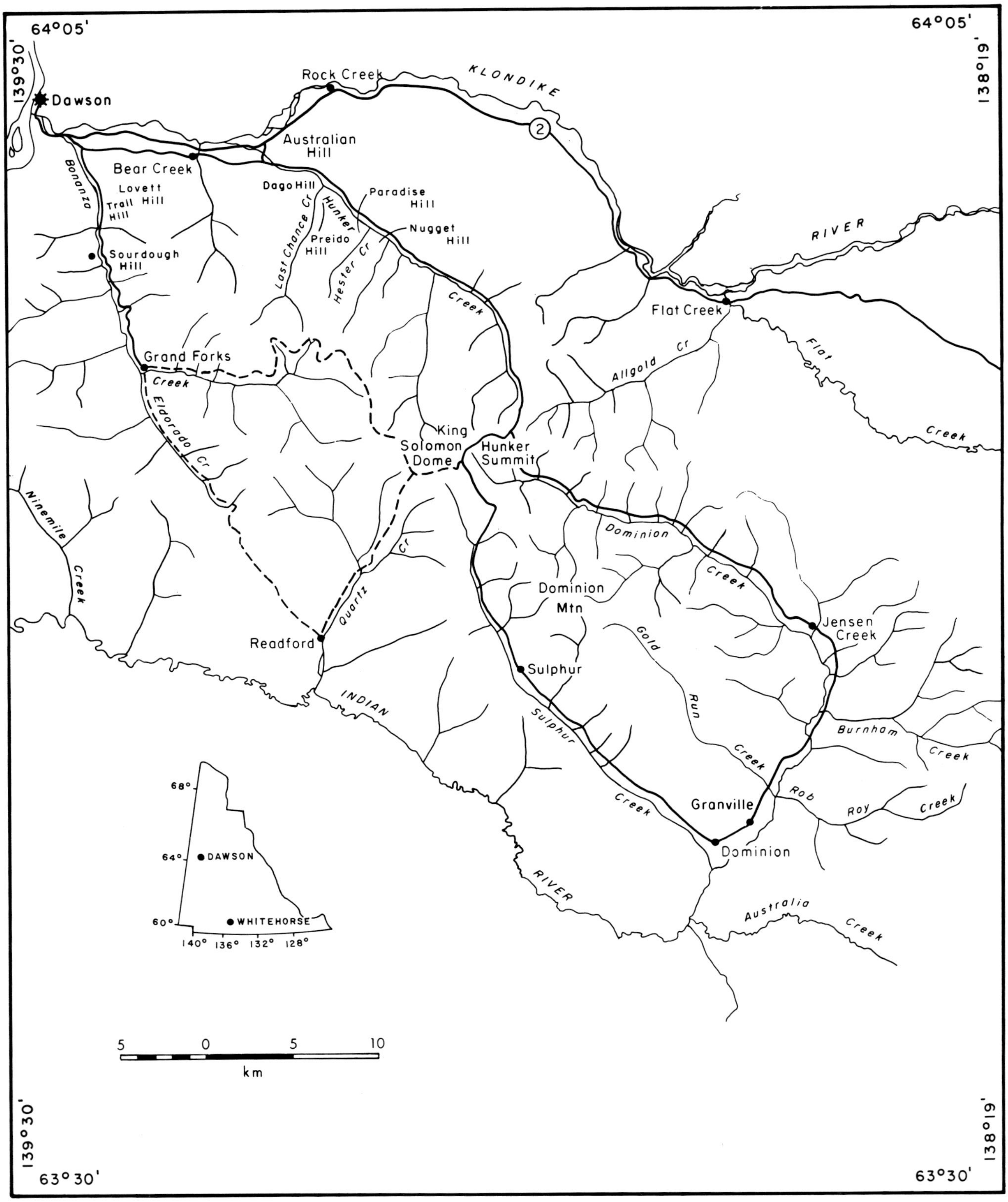

FIG. 1.—Drainage network of the Klondike area showing the White Channel outcrops along Bonanza Creek and Hunker Creek near Dawson City, Yukon Territory.

significant local bedrock sources for the Klondike placer gold (Morison, 1985b).

McConnell (1905, 1907) divided the auriferous gravel deposits in the Klondike area into stream gravel, terrace gravel, White Channel gravel, and high level river gravel or Klondike gravel (Fig. 2). The White Channel placer deposit forms terraces 50 to 100 m above the present stream levels, with a maximum thickness of 35 m for the White

Channel deposit. An additional 10 m of interbedded White Channel and Klondike gravels occur at the mouth of Bonanza Creek (Fig. 2).

McConnell (1907) further subdivided the White Channel deposit into the "white" and "yellow" gravel units (Fig. 2, bottom d and e), suggesting that these gravel units are stratigraphically distinct. As shown by Morison (1985b), however, there is no distinct break in gravelly sedimentation between these units. The yellow gravel unit is interpreted as an altered, "stained" equivalent of the white unit of the White Channel Formation. The yellow staining is thought to be the result of weathering and supergene alteration from surface meteroic fluids (Dusfresne and others, pers. commun.). At Dago Hill (Fig. 1), a Pleistocene, iron-stained fluvial gravel unconformably overlies silty clay overbank deposits of the White Channel Formation (Dusfresne and Morison, 1985; Morison, 1985b).

In southern and central Yukon there have been at least three Cordilleran ice advances: a late Wisconsin McConnell advance, an early to pre-Wisconsin Reid advance and the oldest, 'pre-Reid advance' (Hughes and others, 1972). The Klondike area was affected during pre-Reid time by sedimentation of the glaciofluvial 'Klondike Gravel' derived from the Ogilvie Mountains to the north (Hughes and others, 1972). These glaciofluvial gravels are interbedded with the underlying White Channel sediments, suggesting that deposition of the White Channel continued to early to middle Pleistocene time. Furthermore, cryogenic features such as ice wedge casts and frost cracks occur about half way up the White Channel sections and are indicative of a significant cooling of the climate in this area, perhaps coinciding with the Pliocene-Pleistocene boundary in central Yukon. Downcutting of the Klondike and White Channel deposits was initiated after the pre-Reid glacial interval and continued to at least Reid time. An initial phase of downcutting ended with alluvial terrace sedimentation and a second phase of incision ended with deposition of Holocene-Recent valley-bottom deposits (Fig. 2). The Mosquito Creek tephra (dated at 1.22 Ma, Naeser and others, 1982) is found on a terrace which formed after deposition of the Klondike gravels. Thus, the White Channel sequence in the Klondike area is Pliocene to early Pleistocene in age (Morison, 1985b).

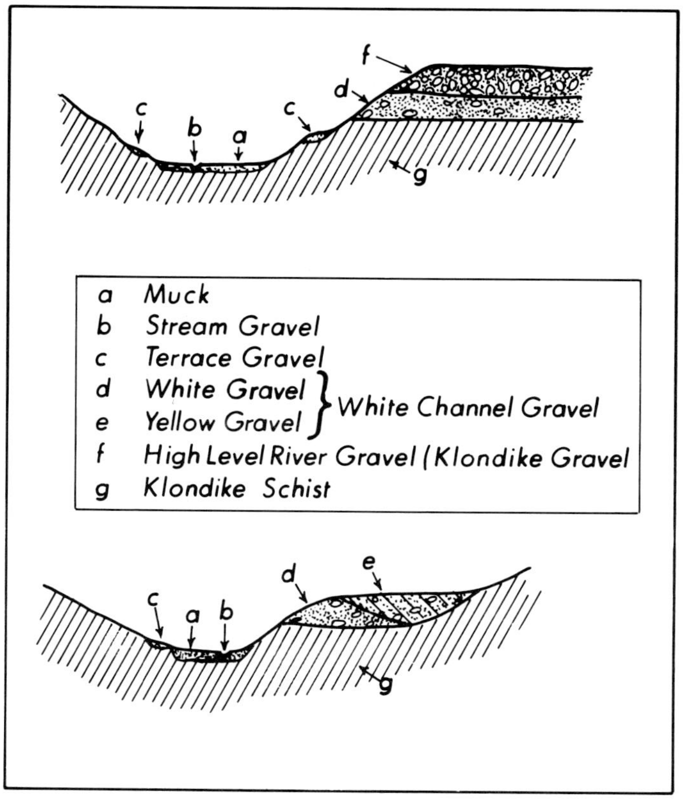

FIG. 2.—Generalized gravel stratigraphy of Pliocene—Recent deposits in the Klondike area, Yukon Territory (from McConnell, 1907).

METHODS

At all of the White Channel outcrops in the Bonanza and Hunker Creek drainage basins (Fig. 1), detailed stratigraphic sections were measured on a bed-by-bed basis. Sourdough, Trail and Lovett Hills showed the best outcrops. Sampling of individual beds included *a*-axis measurement of the 10 largest pebbles in the field; subsampling of representative beds for different facies of 500 gms of material finer than pebble size for grain size analysis; and 1-cu ft bulk samples for heavy mineral analyses and gold content assessment. Orientations of ab-planes of elongate and flattish pebbles were measured to assess gravel fabric patterns.

The main lithologies in the White Channel sequence include clay and silt, fine sand, pebbly sand, and pebble- to cobble-size gravel. Gravelly deposits comprise 80 percent of all beds measured in this study, with sandy and muddy units occurring only as discontinuous interbeds. Nine lithofacies are recognized, with the major features summarized in Table 1. More detailed descriptions and interpretations of individual facies are given in Morison (1985b).

DESCRIPTIONS AND INTERPRETATIONS OF LITHOFACIES

Disorganized gravel (facies 1).—

Facies 1 has a disorganized appearance, distinct cluster bedforms (Fig. 3A), a high matrix content of sand and mud, and scoured bases. Clasts are subrounded to well rounded, with an average clast size of 14 cm (Table 1). These sedimentary features suggest very rapid rates of deposition under highly erosive, sediment-laden flows. Cluster bedforms are interpreted to represent bedload transport of exceptionally large clasts during flood conditions (Brayshaw, 1984). Infilling of the gravel cluster occurs during falling stage from flood levels, when finer suspended particles are trapped in the interstices between the larger clasts. This interpretation of the cluster bedforms in addition to the disorganized texture and scoured bases suggests that this gravel was deposited by channellized, sediment-laden gravelly flows, which contained a large amount of suspended sediment (possibly hyperconcentrated flood flows, G. A. Smith, 1986).

Massive and normally graded gravel (facies 2).—

Beds within facies 2 are very poorly sorted, clast-supported, and matrix-filled gravels (Fig. 3B). Crude normal

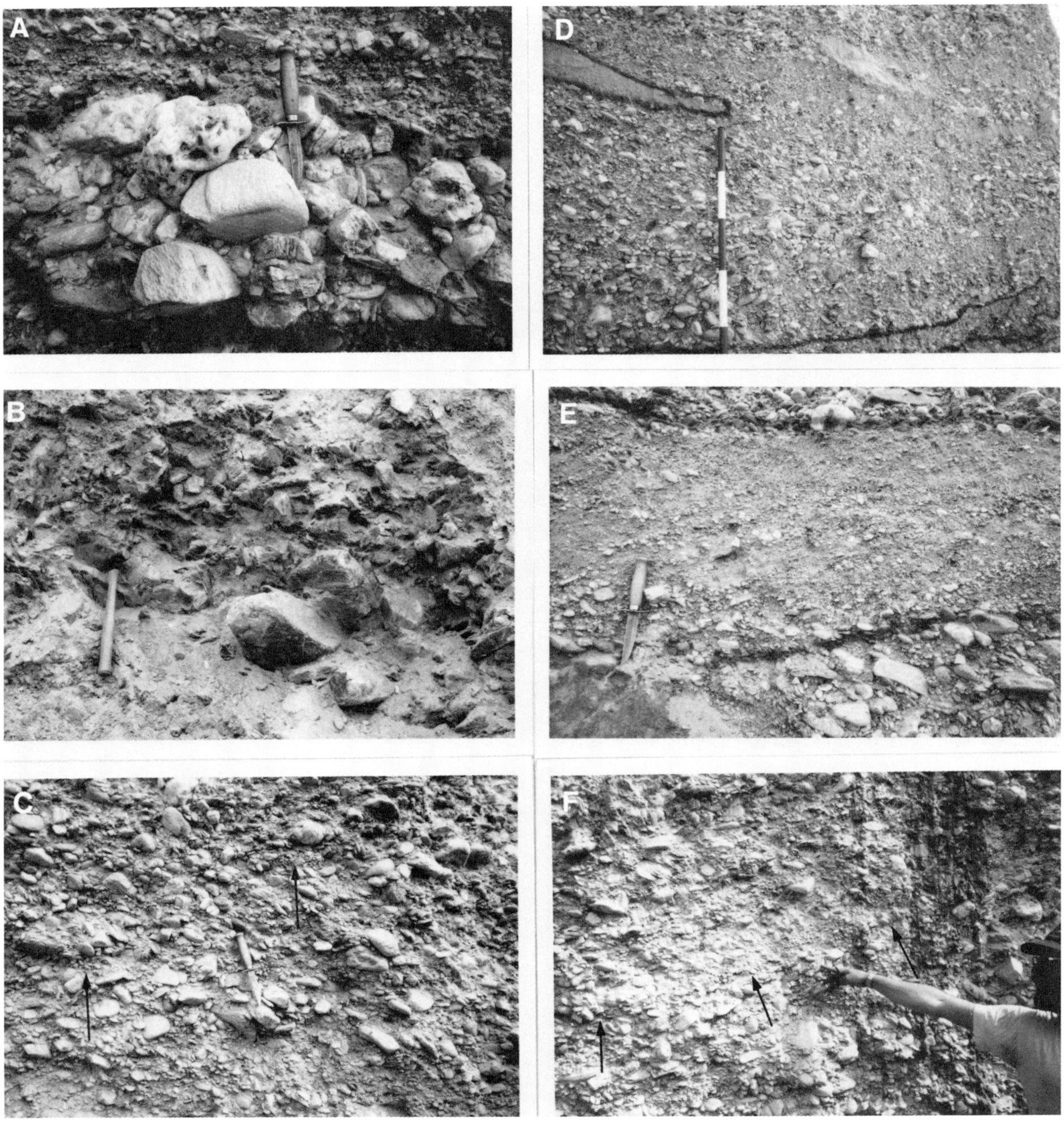

FIG. 3.—Representative photographs of the major gravelly facies within the White Channel deposit. A) Facies 1: cluster bedform of cobble-size quartz clasts at the base of a disorganized gravel bed; B) Facies 2: massive, poorly sorted, matrix-filled gravel with crude normal grading; C) Facies 3: crudely stratified, clast-supported and matrix-filled gravel with lenses of finer gravel (arrows); D) Facies 4: stratified, clast-supported, matrix-filled gravel with improved sorting (compared with facies 3); E) Facies 5: trough crossbedded granule sand- to pebble-size gravel; flow is from right to left; F) Facies 6: scoop-shaped crossbedded cobble- to pebble-size gravel. Note: arrows indicate tangential lower set boundary of the right limb of one crossbed. Scale bars in (D) and knife in (A, E, and C) are 20 cm long; hammer in (B) is 40 cm long.

grading is present. Beds are laterally continuous and bases are flat to slightly scoured. Clasts are very angular, sometimes boulder size (maximum 65 cm *a*-axis length), and lithologies generally reflect local bedrock composition (Morison, 1985b). Some of the massive White Channel gravel beds have clayey to silty zones at their base, which are irregularly laminated, have a "sheared" appearance, and may contain dispersed coarser sandy or granule-size clasts. Similar features have been reported from the base of debris-flow deposits (A. M. Johnson, 1970). The massive gravel facies is always found along steeper slopes of valley walls near confluences with smaller tributaries.

The above facies characteristics and geomorphic location of this facies suggest that the massive gravel was deposited by sediment-gravity flows, which moved downslope in a direction perpendicular to the trend of the main valley axis. Such sediment-gravity flows may be sandy debris flows or hyperconcentrated flood flows (G. A. Smith, 1986). Such flows were initiated by high seasonal flood runoff on unstable, weathered slopes. Fluid properties of these sedi-

TABLE 1.—SUMMARY OF WHITE CHANNEL FACIES

FACIES	LITHOLOGY	% ABUNDANCE	BED THICKNESS AVER.	RANGE (m)	CLAST SIZE MEAN	MAX. (cm)	BASES	GRADING	SEDIMENTARY STRUCTURES	INTERPRETATION REMARKS
1	Clast-supported, matrix-filled gravel	17	0.6	0.1-1.2	14.0	70.0	scoured	none	chaotic, cluster bedforms	abundant rounded quartz clasts and high content of sandy matrix, erosive channel scour during flood intervals
2	Clast-supported, matrix-filled gravel	2.6	0.9	0.2-2.0	15.0	65.0	flat, slightly scoured	crude normal	massive	found along valley walls on steeper slopes, very poorly sorted clayey zones at base of massive units, massive sediment-gravity flows such as sandy debris flows
3	Clast-supported, matrix-filled gravel	23	0.7	0.2-2.0	14.0	42.0	flat, slightly scoured	none	massive to crude planar stratification, discontinuous sandy lenticular beds	main valley settings, a-axis imbrication, massive sediment gravity flow or stream channel sedimentation with high current velocities
4	Clast-supported, matrix-filled gravel	27	0.5	0.1-2.0	9.2	32.0	flat, slightly scoured	none	planar stratification	bedload sedimentation as diffuse gravel sheets or low relief unit bars
5	pebbly sand and gravel	11	0.5	0.2-2.0	5.0	12.0	scoured	none	trough cross-stratification	channel bedform sedmentation as large-scale three-dimensional ripples
6	Clast-supported, matrix-filled gravel	5	1.6	0.4-2.4	10.0	28.0	flat, slightly scoured	none	tangential, scoop-shaped cross-stratification	cut-and-fill of scours and shallow pools
7	Clast-supported, matrix-filled gravel	2	0.8	0.3-1.2	6.0	18.0	flat	none	planar tabular (or wedge) cross-stratification	foreset development on accretion bar surfaces
8	Mud, fine sand	11	0.5	0.02-0.8	-	-	flat, convoluted	none	massive, laminated	overbank deposition associated with local waning discharge conditions, loading from overlying gravelly sedimentation
9	Pebbly sand	2	0.3	0.1-0.5	5.3	11.6	flat	none	massive	rapid channel sedimentation

ment-gravity flows are intermediate between viscous muddy debris flows and more fluid watery flows. Normal grading is developed as denser particles settle directly from the dispersion or suspension during more fluid phases of sedimentation. Similar "intermediate" types of deposits have been reported from sediment-laden volcaniclastic deposits associated with the Mt. St. Helens eruptions (G. A. Smith, 1985, 1986) and are also common in deep-water sediments, in which there appear to be gradations in flow properties from purely viscous to tractional regimes (Lowe, 1979; Nardin and others, 1979; Hein, 1984).

Massive to crudely stratified gravel (facies 3).—

Beds within facies 3 are poorly sorted, clast-supported and matrix-filled gravel, with crude horizontal to low angle stratification (Fig. 3C). Stratification is planar, discontinuous and is discernable through changes in grain size. Discontinuous sandy interbeds also occur. Beds are laterally continuous; bases are flat to slightly scoured. Clasts are subrounded to well rounded, with a mean clast size of 14 cm.

The crudely stratified gravels have scattered strike orientations of the *a*-axes of elongate and flattish pebbles. Both *a*-axis and *b*-axis imbrications occur, with 60–70 percent of the clasts (out of 300 counts) showing an *a*-axis upstream imbrication and 30–40 percent *b*-axis upstream imbrication (Fig. 4A). This pattern is very unusual for a tractional "rolling-type" fabric, where most of the gravel-size clasts have a *b*-axis dipping upstream imbrication (Hein, 1984; Johansson, 1976; Rust, 1975). Similar bimodal imbrication patterns have been reported from deep-water gravels emplaced from sediment-gravity flows (Hein, 1984; Hubert and others, 1977).

The massive to crudely stratified gravel is difficult to interpret clearly because the origin of planar horizontal stratification in gravels is poorly understood. Several workers describe the internal stratification of braid bars and low relief diffuse gravel sheets as internally massive to crudely stratified (Eynon and Walker, 1974; Hein, 1984; Hein and Walker, 1977; N. D. Smith, 1985; Southard and others, 1984). The development of *a*-axis imbrication in crudely stratified gravel deposited as tractional bedload is possible only if the currents can rework and pivot clasts to a more

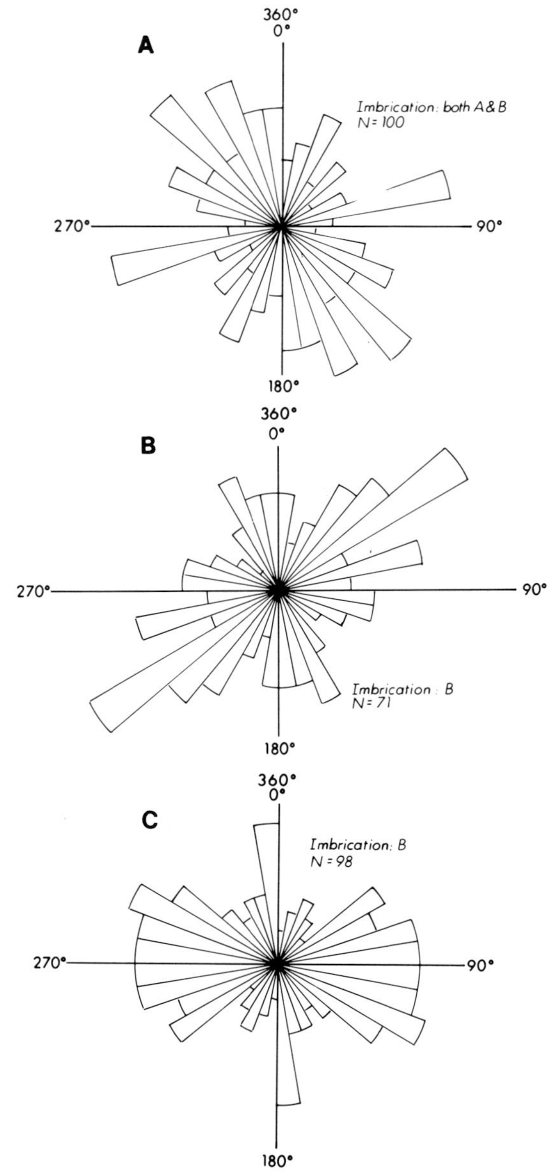

FIG. 4.—Representative rose diagrams of *a*-axis, strike orientations of elongated and flattish pebbles within the massive to crudely stratified facies 3 gravels (A) and the distinctly stratified facies 4 gravels (B and C). Imbrication patterns are not drawn. Facies 3 gravels have both *a*-axis and *b*-axis upstream imbrication patterns; Facies 4 gravels have *b*-axis upstream imbrications.

"streamlined" orientation after initial deposition on the bed. Because the average clast size of gravel within this facies is quite coarse (14 cm), this implies that such current velocities would have to be very large to develop such a fabric.

Alternatively, G. A. Smith (1985, 1986) records crude horizontal stratification from hyperconcentrated flood flows. The characteristics of the facies 3 gravels are more akin to sediment gravity flows, with the poor sorting, dominant *a*-axis imbrications, and scattered strike orientations of *a*-axes. Thus, the massive to crudely stratified gravels are interpreted as due to deposition from either hyperconcentrated flood flows or stream flows with very high sediment concentrations and high current velocities. The general facies characteristics do not provide conclusive evidence for the depositional origin of this facies.

Stratified gravel (facies 4).—

Gravel in this facies is clast-supported and matrix-filled, with distinct stratification defined by clear horizontal banding of gravels of differing grain sizes (Fig. 3D). Stratification is commonly defined by alternations of pebble/cobble sizes with granule/pebble sizes. Less commonly, the stratification is marked by the alternation of clast-supported/matrix-filled layers with clast-supported/openwork layers. Beds are laterally continuous, and bases are flat to slightly scoured. Clasts are subrounded to well rounded, with an average clast size of 9 cm. Fabric analyses of the stratified gravel shows a well organized *a*-axis orientation perpendicular to the local paleoflow, and a well developed *b*-axis upstream imbrication pattern (Figs. 4B, 4C).

The distinct gravelly stratification along with the good *b*-axis upstream imbrication suggests sedimentation under high discharge levels as low relief coarse-grained bedforms, such as diagonal bars or diffuse gravel sheets (Hein and Walker, 1977; Hein, 1984). The lack of planar tabular (or wedge) gravelly crossbeds indicates that the bedforms did not have angle-of-repose foresets. This may be a consequence of deposition of coarse gravels/cobbles under relatively shallow flows. In general, slipface foreset margins develop only if the flow depth/grain size ratio >10 (N. D. Smith, 1985). For the facies 3 gravels, with an average grain size of 9 cm (maximum 32 cm), this implies that flow depths were on average <0.9 m (maximum: <3.2 m) deep.

Alternatively, studies by Hein and Walker (1977) indicate that a period of reworking under lower flow conditions (when the rate of aggradation exceeds the rate of progradation) is also necessary for the development of foreset margins on bars. Thus, rapid deposition of gravels under high flow conditions may not develop bedforms with foresets despite the requisite flow depth/grain size ratio. This would be particularly true where channels are confined by valley walls. Flow velocities in this setting may be higher under comparable discharges than those flows in unconfined braidplain settings.

Cross-bedded gravel (facies 5, 6, and 7).—

The most common types of cross-bedded gravelly facies are trough cross-bedded pebbly sand and gravel (facies 5) (Fig. 3E) and scoop-shaped cross-bedded pebble-to-cobble gravel (facies 6) (Fig. 3F). Planar tabular cross-bedded gravel (facies 4) is rare. Cross-bedded gravel is generally clast-supported and matrix-filled. Trough crossbedding is defined by concave-up crossbeds which have tangentially based lower set contacts. Set thicknesses range from 0.2 to 2 m, and average grain size is 5 cm. In the trough cross-bedded pebbly sand and gravel (facies 5) sorting along strata is well defined (Fig. 3E). The scoop-shaped crossbedding occurs in pebble-to-cobble gravel and is also defined by concave-up crossbeds with tangentially based lower set contacts. Sets range in thickness from 0.4 to 2.4 m. Because of the coarse grain size of this facies (average clast size 10 cm, maximum 28 cm), this is not thought to be true trough crossbedding. The scoop-shaped crossbedded gravels are poorly sorted (Fig. 3F) and may contain some openwork gravelly interbeds or discontinuous sandy interbeds.

Crossbedding in the facies 5, 6, and 7 gravels indicates bedload transport. The trough crossbedded gravel and pebbly sand facies 5 are interpreted as deposits of fields of large-scale three-dimensional ripples or dunes (Harms and others, 1982). The rare planar tabular crossbeds are interpreted as being deposited by the downstream progradation of high relief barforms, with angle-of-repose foresets. The gravelly scoop-shaped crossbeds (facies 6) in the pebble/cobble-size material are too coarse to have been formed by the migration of gravelly dunes. The average grain size in this facies is approximately 10 cm (maximum: 28 cm), which is much coarser than that observed in most gravelly dune fields. Fahnestock and Bradley (1973) reported a D90 (grain size diameter larger than 90 percent of the bed material) of 5.5 cm corresponding to flood flows of 236,000 cfs in the Knik River of Alaska. Under these flood flows, the main channel of the Knik River was floored with gravelly dunes. By analogy for the Klondike deposits, tremendous water depths and very high flood discharges are needed in order for the scoop-shaped crossbedding within the cobble facies to be due to the migration of gravelly dunes.

Alternatively, this crossbedding reflects cut-and-fill of scour holes or pools carved and filled under flood flows within main channels. Such pools are quite common in modern outwash systems, commonly occurring fairly periodically as pool-and-bar couplets (Hein, 1974, 1984; D. G. Smith, 1973); however, the nature of their fill is poorly documented. Scoop-shaped crossbedded gravel (facies 6) is tentatively interpreted as due to "cut-and-fill" deposition in shallow scours or pools within active channels.

Massive to convoluted mud and fine sand (facies 8) and massive pebbly sand (facies 9).—

These facies occur as laterally discontinuous beds which have flat to convoluted bases. Bed thicknesses average 0.5 m for facies 8 and 0.3 m for facies 9 (Table 1). Facies 8 includes massive clays with fine sand laminae and faintly laminated fine sand and silt, with dispersed (or scattered) gravel clasts. Massive pebbly sand (facies 9) rarely occurs. The larger clasts average 5 cm (*a*-axis measurement) in the massive pebbly sand. These massive pebbly sands reflect rapid deposition under high flow conditions. The massive clays with laminated fine sand and silt are interpreted as representing rapid deposition as suspension fallout from possible flood flows. The convolute-bedded silt and sand underlies thin gravel beds. This convolute bedding is therefore interpreted as resulting from pore fluid expulsion generated during loading of overlying gravels onto finer sediment. Comminuted plant debris is also found in association with this facies, suggesting some subaerial exposure of the overbank material prior to burial.

FACIES SEQUENCES AND PALEOGEOGRAPHIC MODEL

The White Channel placer deposit of the Klondike area shows an overall tendency for the gravelly facies to be coarser grained, poorly sorted and more disorganized in the lower part of the section with sorting and stratification (both planar-bedded and crossbedded) improving upsection. To check for the occurrence of possible facies sequences within the main valley setting of the White Channel placer deposit, an embedded semi-Markov chain analysis was done, using the "improved" methods of Harper (1984, 1985). An iterative proportional fitting technique generates the random matrices (see discussions by Harper 1984, 1985), and a binomial expansion fit is used to detect "non-random" transitions (Harper, 1984). These analyses were done on three sets of data and include the vertical facies transitions observed at Trail and Lovett Hills, individually, and a composite pooled data set for all of the main valley White Channel sections measured. The resultant facies relationship diagrams are shown in Figure 5.

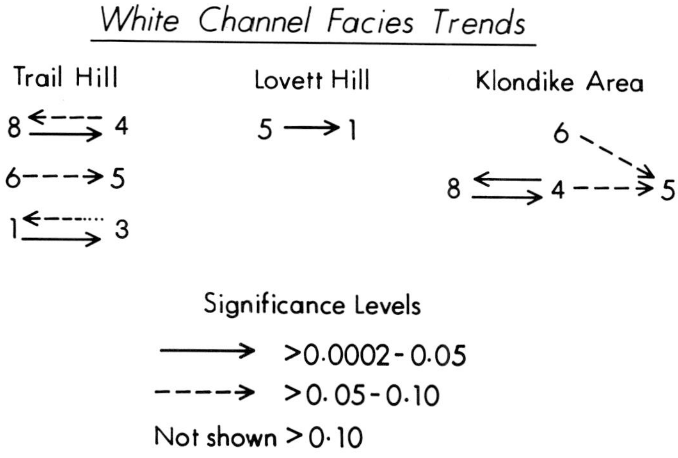

FIG. 5.—Facies relationship diagrams showing results of semi-Markov chain analysis of vertical facies trends from Trail Hill, Lovett Hill and pooled data set (Klondike Area) for the White Channel outcrops. For tally count matrices, see Morison (1985b).

Results from the semi-Markov chain analyses statistically demonstrate that in the main valley setting there are few consistent facies trends (Fig. 5). In the Trail Hill Section there are three facies associations which tend to occur more often than the random occurrences. First, the disorganized gravel (facies 1) tends to be overlain by the massive to crudely stratified gravel (facies 3), which may then pass back into disorganized gravel (facies 1). The scoop-shaped crossbedded gravel (facies 6) tends to be overlain by trough crossbedded pebbly sand and gravel (facies 5). Finally, the convolute-bedded mud, silt and fine sand (facies 8) may be overlain by stratified gravel (facies 4), which in turn passes back into the convoluted finer sediment (Fig. 5).

The transition of disorganized gravel (facies 1) into massive to crudely stratified gravel (facies 3) is common in lower or proximal main valley sequences. This proximal association demonstrates that early White Channel sedimentation was dominated by stream flows with either very high or flood discharge levels which contained large amounts of suspended sediment (Fig. 6). Mid- to upper sequences are dominated by the transition of stratified gravel (facies 4) into convoluted mud and fine sand (facies 8). This shows that medial to distal White Channel sedimentation was dominated by low relief bar deposits capped by overbank-type muddy sediments (Fig. 6).

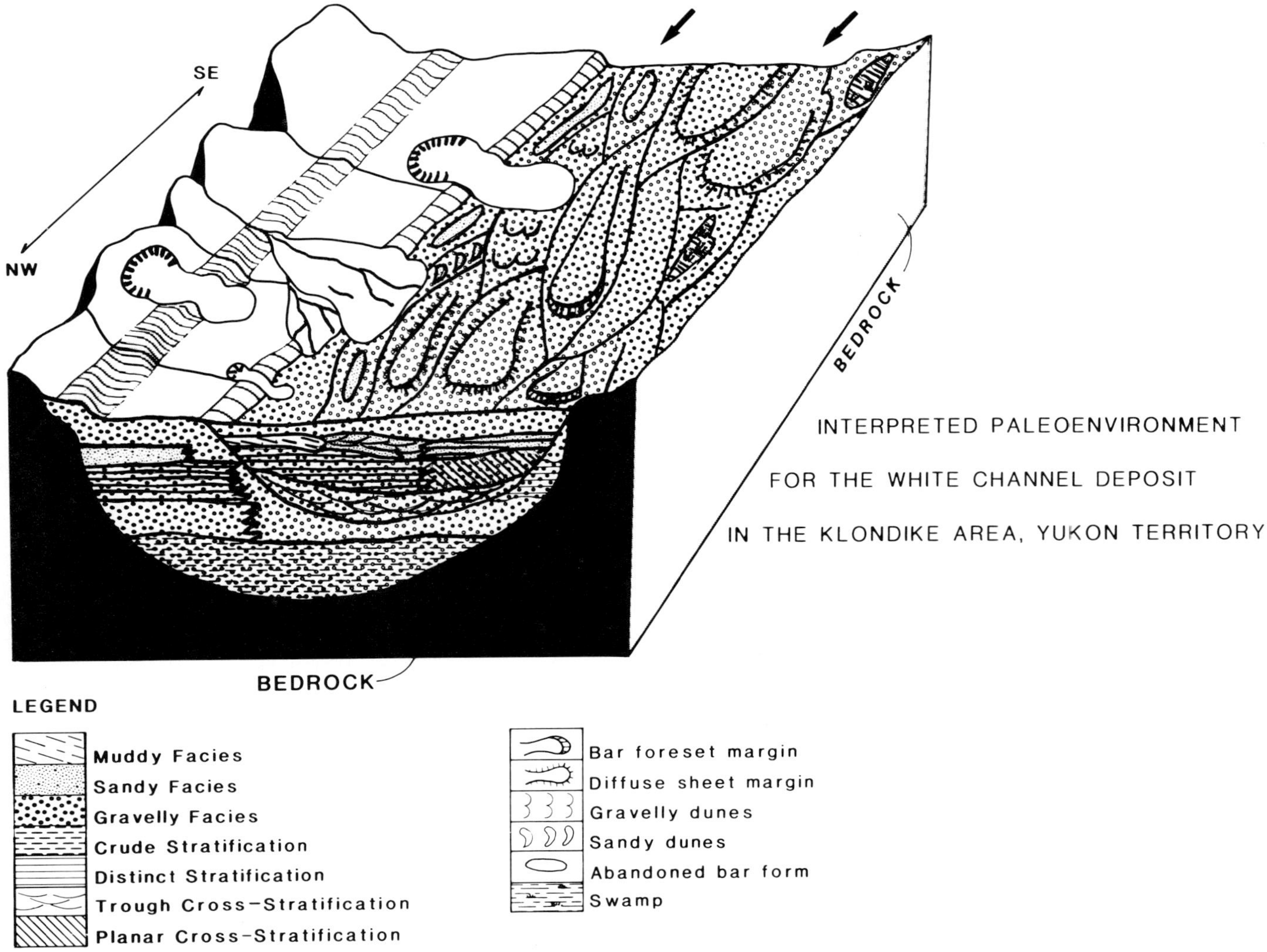

FIG. 6.—Interpreted paleoenvironmental reconstruction for the White Channel deposit, Klondike area, Yukon, showing a gravelly braided river, flanked by alluvial fans and slump or slide deposits from colluvial slopes. Note: lower parts of the valley-fill sequence shown in cross section are crudely stratified and massive or disorganized; gravels further upsection show improved sorting, distinct stratification and crossbedding. This reflects an initial stage of infill by a poorly developed braided channel system which, through time, becomes more fully developed into a mature braidplain with many channels, some floored with gravelly dunes, pools, low relief bars, bars with foresets, and interchannel areas with overbank and swamps.

The transition of the scoop-shaped crossbedded cobble gravel (facies 6) into trough cross-bedded pebbly sand and gravel (Fig. 5) is a fining-upward sequence, suggestive of a decrease in current flow strength. The interpretation of these facies suggests that this transition may record pool deposition within main channels (facies 6) followed by deposition from gravelly dune fields (facies 5). Such a transition may be typical of a fining-upward sequence generated by infilling of a braid channel.

As discussed previously, massive and normally graded gravel (facies 2) is found along valley walls on steeper slopes proximal to tributary drainages of the main valley. These coarse, massive gravelly beds were deposited as sediment-gravity flows which were formed independently of the main valley setting. Sequences of massive gravel (facies 2) beds are usually separated by thinly bedded stratified gravel (facies 4). This suggests that sediment-gravity flow deposition occurred in association with either tributary alluvial fan sedimentation or as a series of valley wall slump-type deposits on the main valley braidplain (Fig. 6).

In summary, White Channel placer deposits were formed in a braided environment (Fig. 6) which had the following major components: 1. A main valley proximal setting which was dominated by high discharge or flood channels in a crudely braided system with few main channels (facies transitions 1 and 3). 2. Main valley medial to distal settings which form well defined braided sequences of channels and low relief bars (facies transitions 4 and 8; 6 and 5). 3. Valley wall sediment-gravity flow deposition in association with tributary alluvial fans and/or slumps within the main valley braidplain.

COMPARISON WITH OTHER PROXIMAL FLUVIAL DEPOSITS

N. D. Smith (1985) gives a good synthesis of the types of sedimentary sequences which can be generated in prox-

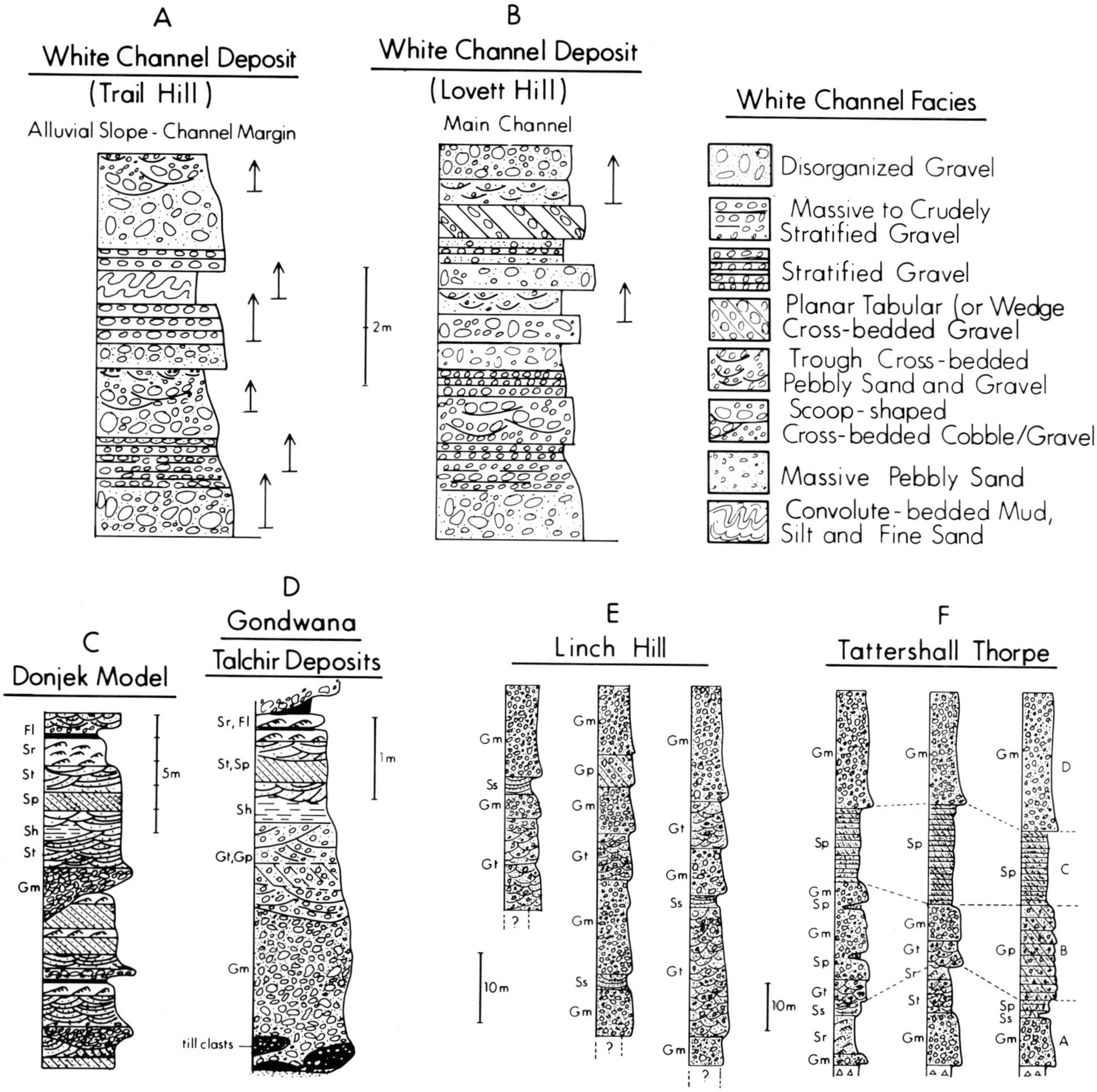

FIG. 7.—Comparison of preferred facies trends of the White Channel deposit at Trail Hill (A) and Lovett Hill (B) with published glaciofluvial sections from the literature. For the White Channel deposit, preferred facies trends are marked with arrows and the facies scheme follows that presented in the text (see Fig. 3). Facies designations for the other published sections follow Miall's (1978) facies code.

imal glaciofluvial settings. Some of these sequences are illustrated in Figure 7 C, D, E, and F along with composite diagrams for the White Channel deposit (Fig. 7 A, B).

The most notable differences between the glaciofluvial sequences and the White Channel is the much better organization and development of sequences within the glaciofluvial gravel sequences. The Donjek model (Fig. 7C) (redrawn from Miall, 1978 by N. D. Smith, 1985) shows scour-based fining-upward sequences of massive gravels (Gm), succeeded by trough crossbedded (St) and horizontally stratified (Sh) sands, with some interbedded planar tabular crossbedded (Sp) sands. The top of the sequences are marked by laminated fine sediment (Fl). This apparent cyclicity is thought by Miall (1978) to record repeated sequences of channel aggradation, possibly by flood events, or for the thicker sequences aggradation of the entire braidplain system.

A much more disorganized type of sequence was recorded by Casshyap and Tewari (1982) in the proximal glaciofluvial Gondwana Talchir Formation of India (Fig. 7D). Here, the lower part of the deposit consists of massive disorganized gravels (Gm), which are succeeded by trough and planar tabular crossbedded gravels (Gt, Gp). The upper part of the sequence resembles that proposed by Miall (1978) for the Donjek model, horizontally stratified sands (Sh), succeeded by planar tabular and trough crossbedded sands (Sp, St). The top of the sequence is marked by finer grained rippled sand (Sr) and laminated fine sediment (Fl). The Gondwana Talchir deposit in India is much more disorganized and lacks the well defined cycles of the Donjek model.

The sequences from Linch Hill and Tattershall Thorpe show similarities to both the Donjek and Gondwana Talchir deposits. The coarser grained Linch Hill section (Fig. 7E) consists of massive gravels (Gm), separated by trough crossbedded (Gt) or rarely planar tabular crossbedded gravel (Gp) or stratified sand (Ss). The Tattershall Thorpe section (Fig. 7F) is finer grained and has a proportionately higher percentage of planar tabular crossbedded gravel (Gp) and sand (Sp). At both sections, the massive gravels were emplaced rapidly under high flow conditions, perhaps recording flood events. Under lower flow conditions within main channels at Linch Hill, sediment was deposited as bedload from gravelly dunes and more rarely bars with angle-of-repose foresets. The Tattershall Thorpe shows more extensive development of planar tabular crossbedding which may be a consequence of the better development of bars with angle-of-repose foresets in the reaches with finer grained material. The sequences at these two sections are not nearly as "cyclic" as those for the Donjek model and seem somewhat intermediate between the Donjek and the Gondwana Talchir types.

In comparison with the Gondwana Talchir strata, the White Channel is even more disorganized, less cyclic and has a higher proportion of the disorganized (facies 1) and massive to crudely stratified gravels (facies 3) (Fig. 7A, B). The proportion of fine-grained mud, silt and fine sand (facies 8) is very minor and is mainly convolute-bedded. Also notable is the occurrence of the scoop-shaped crossbedded gravel and cobble (facies 6) and the near absence of planar tabular crossbedded pebbly sand and gravel (facies 7). Superficially, the sections of the White Channel resemble those facies and sequences developed in fluvial deposits of an interarc sequence in the Cagayan basin, northern Luzon, Philippines (Mathisen and Vondra, 1983). Here, massive matrix-supported pebble-to-boulder conglomerate is interbedded with tuff breccias (ignimbrites) and tuffs, with little evidence of cyclicity (Mathisen and Vondra, 1983). In these generally noncyclic sections, nonmarine volcaniclastic deposits were emplaced rapidly within an interarc basin as braided stream, lahar, and pyroclastic flow and fall deposits.

Within the White Channel deposit, massive and disorganized gravel is more common toward the base of the sections, although thin disorganized gravels (facies 1) are interbedded with the convoluted fine-grained sediments (facies 8) higher upsection (Figs. 6, 7). These observations suggest that for the White Channel deposit, sediment-gravity flows were especially important during deposition of the initial valley fill. During later stages of valley fill (higher upsection), sediment-gravity flows were mainly affecting terrace or valley-wall sites and were not important in the deposition of sediment within the main channels. The types of crossbedding in the White Channel deposits, suggest that bars with well developed foresets were uncommon. Bedload deposition occurred mainly within pools, from gravelly and sandy dunes, or from low relief bars with riffle margins. Such types of geomorphic features are common in proximal reaches of modern glacial outwash systems (Hein, 1984; Hein and Walker, 1977; Boothroyd and Ashley, 1975). By contrast, bars with well developed angle-of-repose foresets are usually common only in the downstream and finer grained reaches (N. D. Smith, 1970, 1985).

In the White Channel deposit, the occurrence of planar tabular crossbedding is rare, suggesting that for this system, bars with true angle-of-repose foresets never developed. This may be a consequence of the coarse grain size of the source material coupled with confined flow conditions within the tributary paleovalleys of Hunker and Bonanza Creek. Under such conditions, flows were probably too confined to develop the bar forms (which occur in sites of flow expansion) most typical of distal braidplain settings.

The lack of good sequences in the White Channel deposit in comparison with proximal glaciofluvial deposits may also be a consequence of the greater longer term cyclicity of discharge fluctuations (monthly? to yearly climatic variations) in glaciofluvial settings. This cyclicity of discharge fluctuations affords an opportunity for greater reworking of bed material into bar or bedforms and also results in common shifting of channels in finer grained reaches (Hein and Walker, 1977). Both the reworking and channel shifting processes would lead to a greater potential for the development of fining-upward sequences of stratified and/or crossbedded sediments in glaciofluvial settings. By contrast during deposition of the White Channel, the climate was a period of cooling, marking the transition from the Pliocene to the Pleistocene, and good cyclicity of discharge patterns would not be expected. Consequently, the sections tend to be very massive, showing very little cyclicity. These comparisons suggest that facies models developed for gravel-bed rivers from glaciofluvial settings may not be directly applicable to nonglacial, confined alluvial valley-fill sequences.

ACKNOWLEDGMENTS

The work in this study was done by S. R. Morison in partial fulfillment for the M.S. degree in Geology at the University of Alberta. Rhys Hughes, Robb Hale and Jesse Duke provided capable field assistance. Most of the work was funded by the Department of Indian Affairs and Northern Development, Exploration and Geological Services Division, Whitehorse. Funds for preparation of this manuscript were from a National Science and Engineering Research Council operating grant to F. J. Hein. The authors thank the placer miners in the Klondike area for their cooperation during the field seasons. Discussions with O. Hughes and B. Nesbitt were useful, and the rocker which G. Gilbert built was valuable during the sampling program. Critical review of the manuscript by colleagues at D.I.A.N.D., by D. J. Cant, S. Johnson and W. Nami is also appreciated.

REFERENCES

BOOTHROYD, J. C., AND ASHLEY, G. M., 1975, Processes, bar morphology and sedimentary structures on braided outwash fans, northeastern Gulf of Alaska, *in* Jopling, A. V., and McDonald, B. C., eds., Glaciofluvial and Glaciolacustrine Sedimentation: Society of Economic Paleontologists and Mineralogists Special Publication 23, p. 193–222.

———, AND NUMMEDAL, D., 1978, Proglacial braided outwash: A model for humid alluvial fan deposits, *in* Miall, A. D., ed., Fluvial Sedimentology: Canadian Society of Petroleum Geologists Memoir 5, p. 641–668.

BOSTOCK, H. S., 1948, Physiography of the Canadian Cordillera, with special reference to the area north of the Fifty-fifth Parallel: Geological Survey of Canada Memoir 247, 101 p.

———, 1970, Physiographic regions of Canada: Geological Survey of Canada Map 1254 A.

BRADLEY, W. C., FAHNESTOCK, R. K., AND ROWEKAMP, G. T., 1972, Coarse sediment transport by flood flows on Knik River, Alaska: Geological Society of America Bulletin, v. 83, p. 1261–1284.

BRAYSHAW, A. C., 1984, Characteristics and origin of cluster bedforms in coarse-grained alluvial channels, *in* Koster, E. H., and Steele, R. J., eds., Sedimentology of Gravels and Conglomerates: Canadian Society of Petroleum Geologists Memoir 10, p. 77–85.

BULL, W. B., 1977, The alluvial fan environment: Progress in Physical Geography, v. 1, p. 222–270.

CASSHYAP, S. M., AND TEWARI, R. C., 1982, Facies analysis and paleogeographic implications of a late Paleozoic glacial outwash deposit, Bihar, India: Journal of Sedimentary Petrology, v. 52, p. 1243–1256.

CHURCH, MICHAEL, AND JONES, D., 1982, Channel bars in gravel-bed rivers, *in* Hey, R. D., Bathurst, J. C., and Thorne, C. R., eds., Gravel-Bed Rivers: Wiley, London, p. 291–338.

DEBICKI, R. L., 1983, An overview of the placer mining industry in Yukon; Yukon placer mining industry 1978 to 1982: Whitehorse, Department of Indian Affairs and Northern Development, Exploration and Geological Services Division, p. 7–14.

DUSFRESNE, M. B., AND MORISON, S. R., 1985, Stratigraphy and alteration of the White Channel Gravel at Dago Hill, A progress report, Klondike area, Yukon: Yukon Exploration and Geology 1983: Whitehorse, Department of Indian Affairs and Northern Development, Exploration and Geological Services Division, p. 55–59.

EYNON, GEORGE, AND WALKER, R. G., 1974, Facies relationships in Pleistocene outwash gravels, southern Ontario: A model for bar growth in braided rivers: Sedimentology, v. 21, p. 43–70.

FAHNESTOCK, R. K., 1963, Morphology and hydrology of a glacial stream—White River, Mount Ranier, Washington: U.S. Geological Survey Professional Paper 422A, p. 1–70.

———, AND BRADLEY, W. C., 1973, Knik and Matanuska Rivers, Alaska: A contrast in braiding, *in* Morisawa, Marie, ed., Fluvial Geomorphology: Proceedings, 4th Geomorphology Symposium, State University of New York at Binghamton, N.Y., p. 220–250.

GUSTAVSON, T. C., 1974, Sedimentation on glacial outwash fans, Malaspina Glacier foreland, Alaska: Journal of Sedimentary Petrology, v. 44, p. 374–389.

HAMMER, K. M., AND SMITH, N. D., 1983, Sediment production and transport in a proglacial stream: Hilda Glacier, Alberta, Canada: Boreas, v. 12, p. 91–106.

HARMS, J. C., SOUTHARD, J. B., AND WALKER, R. G., 1982, Structures and sequences in clastic rocks: Society of Economic Paleontologists and Mineralogists Short Course Notes 9, p. 1-1 to 8-51.

HARPER, C. W., JR., 1984, Improved methods of facies sequence analysis, *in* Walker, R. G., ed., Facies Models: Geological Association of Canada, Geoscience Canada Reprint Series 1, p. 11–13.

———, 1985, Facies models revisited: An examination of quantitative methods: Geoscience Canada, v. 11, p. 203–207.

HEIN, F. J., 1974, Gravel transport and stratification origins, Kicking Horse River, British Columbia: Unpublished M.S. Thesis, McMaster University, 135 p.

———, 1984, Deep-sea and fluvial braided channel conglomerates: A comparison of two case studies, *in* Koster, E. H., and Steel, R. J., eds., Sedimentology of Gravels and Conglomerates: Canadian Society of Petroleum Geologists Memoir 10, p. 33–49.

———, AND WALKER, R. G., 1977, Bar evolution and development of stratification in the gravelly braided Kicking Horse River, B.C.: Canadian Journal of Earth Sciences, v. 14, p. 562–570.

HUBERT, J. K., SUCHEKI, R. K., AND CALLAHAN, R. K. M., 1977, The Cowhead breccia: Sedimentology of the Cambro-Ordovician continental margin, Newfoundland, *in* Cook, H. E., and Enos, Paul, eds., Deep-Water Carbonate Environments: Society of Economic Paleontologists and Mineralogists Special Publication 25, p. 125–154.

HUGHES, O. L., RAMPTON, V. N., AND RUTTER, N. W., 1972, Quaternary geology and geomorphology, southern and central Yukon (Northern Canada): 24th International Geologic Congress, Excursion A11 Guidebook, 59 p.

JOHANSSON, C. E., 1976, Structural studies of frictional sediments: Geographiska Annaler, v. 58, p. 201–301.

JOHNSON, A. M., 1970, Physical processes in Geology: Freeman, San Francisco, 577 p.

JOHNSON, S. Y., 1984, Cyclic fluvial sedimentation in a rapidly subsiding basin, northwest Washington: Sedimentary Geology, v. 38, p. 361–392.

LOWE, D. R., 1979, Sediment gravity flows: Their classification and some problems of application to natural flows and deposits, *in* Doyle, L. J., and Pilkey, O. H., eds., Geology of Continental Slopes: Society of Economic Paleontologists and Mineralogists Special Publication 27, p. 75–82.

MATHISEN, M. E., AND VONDRA, C. F., 1983, The fluvial and pyroclastic deposits of the Cagayan Basin, northern Luzon, Philippines—An example of non-marine volcaniclastic sedimentation in an interarc basin: Sedimentology, v. 30, p. 369–392.

MCCONNELL, R. G., 1905, Report on the Klondike gold fields: Geological Survey of Canada Publication 884, p. 1–71.

———, 1907, Report on gold values in the Klondike high level gravels: Geological Survey of Canada Publication 979, 34 p.

MIALL, A. D., 1978, Lithofacies types and vertical profile models in braided river deposits: A summary, *in* Miall, A. D., ed., Fluvial Sedimentology: Canadian Society of Petroleum Geologists Memoir 5, p. 597–604.

MILNER, N. W., 1976, Geomorphology of the Klondike placer goldfields: Whitehorse, Department of Indian Affairs and Northern Development, Exploration and Geological Services Division, Final Report, Contract OSV 5-0047, 157 p.

MORIN, J. A., AND DEBICKI, R. L., 1982, An overview of Yukon mineral industry 1981: Yukon Exploration and Geology 1981: Whitehorse, Department of Indian Affairs and Northern Development, Exploration and Geological Services Division, p. 3–14.

MORISON, S. R., 1983, A sedimentologic description of Clear Creek fluviatile sediments 115 P, central Yukon: Yukon Exploration and Geology 1982: Whitehorse, Department of Indian Affairs and Northern Development, Exploration and Geological Services Division, p. 50–54.

———, 1985a, Placer deposits of Clear Creek drainage basin 115 P, central Yukon: Yukon Exploration and Geology 1983: Whitehorse, Department of Indian Affairs and Northern Development, Exploration and Geological Services Division, p. 88–93.

———, 1985b, Sedimentology of White Channel placer deposits, Klondike area, West-Central Yukon: Unpublished M.S. Thesis, University of Alberta, 148 p.

NAESER, N. D., WESTGATE, J. A., HUGHES, O. L., AND PEWE, T. L., 1982, Fissiontrack ages of late Cenozoic distal tephra beds in the Yukon Territory and Alaska: Canadian Journal of Earth Sciences, v. 19, p. 2167–2178.

NALDRETT, D. L., 1981, Aspects of the surficial geology and permafrost conditions, Klondike goldfields and Dawson City, Yukon Territory: Unpublished M.S. Thesis, University of Ottawa, 150 p.

NARDIN, T. R., HEIN, F. J., GORSLINE, D. S., AND EDWARDS, B. D., 1979, A review of mass movement processes, sediment and acoustic characteristics, and contrasts in slope and base-of-slope systems versus canyon-fan-basin floor systems, *in* Doyle, L. J., and Pilkey, O. H., eds., Geology of Continental Slopes: Society of Economic Paleontologists and Mineralogists Special Publication 27, p. 61–73.

NILSEN, T. H., 1982, Alluvial fan deposits, *in* Scholle, P. A., and Spearing, D. R., eds., Sandstone Depositional Environments: American Association of Petroleum Geologists Memoir 31, p. 49–86.

RUST, B. R., 1972, Structure and process in a braided river: Sedimentology, v. 18, p. 221–245.

———, 1975, Fabric and structure in glaciofluvial gravels, *in* Jopling, A. V., and McDonald, B. C., eds., Glaciofluvial and Glaciolacustrine Sedimentation: Society of Economic Paleontologists and Mineralogists Special Publication 23, p. 238–248.

———, 1984, Proximal braidplain deposits in the Middle Devonian Malbaie Formation of eastern Gaspe, Quebec, Canada: Sedimentology, v. 31, p. 675–696.

SMITH, D. G., 1973, Aggradation of the Alexandra-North Saskatchewan River, Banff Park, Alberta, *in* Morisawa, Marie, ed., Fluvial Geomorphology: State University of New York, 4th Annual Symposium on Geomorphology, Binghamton, New York, p. 201–220.

SMITH, G. A., 1985, Problems with facies analysis of volcaniclastic sediment in fluvial basins, Pacific Northwest, U.S.A.: Third International Fluvial Sedimentology Conference, Abstracts Volume, Fort Collins, Colorado, p. 35–36.

———, 1986, Coarse-grained nonmarine volcaniclastic sediment: Terminology and depositional process: Geological Society of America Bulletin, v. 97, p. 1–10.

SMITH, N. D., 1970, The braided stream depositional environment: Comparison of the Platte River with some Silurian clastic rocks, north-central Appalachians: Geological Society of American Bulletin, v. 81, p. 2993–3014.

———, 1974, Sedimentology and bar formation in the upper Kicking Horse River, a braided outwash stream: Journal of Geology, v. 81, p. 205–223.

———, 1985, Proglacial fluvial environment, *in* Ashley, G. M., Shaw, J., and Smith, N. D., eds., Glacial Sedimentary Environments: Society of Economic Paleontologists and Mineralogists Short Course Notes 16, p. 85–134.

SOUTHARD, J. B., SMITH, N. D., AND KUHNLE, R. A., 1984, Chutes and lobes: Newly identified elements of braiding in shallow gravelly streams, *in* Koster, E. H., and Steel, R. J., eds., Sedimentology of Gravels and Conglomerates: Canadian Society of Petroleum Geologists Memoir 10, p. 51–59.

STEEL, R. J., MAEHLE, S., NILSEN, H., ROE, S. L., AND SPINNANGR, A., 1977, Coarsening-upward cycles in the alluvium of Hornelen Basin (Devonian), Norway: Sedimentary response to tectonic events: Geological Society of America Bulletin, v. 88, p. 1124–1134.

———, AND THOMPSON, D. B., 1983, Structures and textures in Triassic braided stream conglomerates ('Bunter' Pebble Beds) in the Sherwood Sandstone Group, North Staffordshire, England: Sedimentology, v. 30, p. 341–368.

TEMPELMAN-KLUIT, D. J., 1980, Evolution of physiography and drainage in southern Yukon: Canadian Journal of Earth Sciences, v. 17, p. 1189–1203.

———, 1981, Geology and mineral deposits of southern Yukon: Yukon Exploration and Geology 1979–80: Whitehorse, Department of Indian Affairs and Northern Development, Exploration and Geological Services Division, p. 7–31.

———, 1982, White Channel Gravel of the Klondike: Yukon Exploration and Geology 1981: Whitehorse, Department of Indian Affairs and Northern Development, Exploration and Geological Services Division, p. 74–79.

SEDIMENTOLOGY OF VOLCANISM–INDUCED AGGRADATION IN FLUVIAL BASINS: EXAMPLES FROM THE PACIFIC NORTHWEST, U.S.A.

GARY A. SMITH

Northwest College and University Association for Science Tri-Cities University Center Richland, WA 99352

ABSTRACT: Cenozoic nonmarine sedimentation within and adjacent to the Cascade volcanic arc has produced a varied depositional record. Eocene to early Miocene volcaniclastics accumulated to at least 3 km thick in a longitudinal trough coincident with the arc. Increasing constructional and tectonic relief in the Cascades during the Neogene resulted in development of volcaniclastic aprons adjacent to the arc that grade sourceward into volcanic sections. Quaternary sediments are largely restricted to terraces within valleys near the volcanoes.

Neogene and Quaternary aggradation represents the impact of large eruption-related sediment loads on fluvial systems. There are three notable departures from existing fluvial facies models exhibited by these deposits. First, facies representing processes intermediate between more familiar dilute stream flow and viscous debris flow are important components of these sequences but have not been adequately described in the literature on fluvial processes. These *hyperconcentrated flood-flow* deposits are frequently produced by dilution of debris flows in stream channels, and their recognition is critical to successful evaluation of volcaniclastic sequences. Second, the volcaniclastic aprons and terraces contain considerable thicknesses of debris-flow and flood deposits for which existing models would suggest arid alluvial fan settings, but lateral grain-size variation, paleocurrent data, and paleogeographic considerations fail to support the presence of alluvial fans. Debris flows and floods in volcanic areas are more voluminous and cover much larger areas than alluvial fan counterparts and are not restricted to fan setting or arid climate. Third, aggradation occurs during short periods when eruption-produced sediment loads choke streams with debris and is separated by longer periods of nondeposition and incision. Thus, little or no record of inter-eruption sedimentation occurs and biases evaluation of fluvial style.

INTRODUCTION

Studies of fluvial basins emphasize aggradation of clastic wedges and basin fills related to basin subsidence, high-volume sediment influx from adjacent rising uplifts, or both (Miall, 1981; Nilsen, 1984). These investigations of syntectonic sedimentation, combined with study of Pleistocene-Holocene glaciofluvial sedimentation (Miall, 1977), provide most of the data for distillation of generalized fluvial facies models (Rust and Koster, 1984; Walker and Cant, 1984).

The nature of volcanism-induced sedimentation has been addressed by Davies and others (1978) and Vessell and Davies (1981) with emphasis on modern fluvial systems in Guatemala. This synthesis of studies of late Cenozoic sedimentation in the northwestern United States examines the record of sedimentation in arc-adjacent basins and the applicability of concepts of fluvial sedimentation, developed primarily from study of syntectonic deposits, to synvolcanic sedimentary sequences.

NONMARINE SEDIMENTATION RELATED TO CASCADE VOLCANIC DEVELOPMENT

The Cascade Range extends 1100 km from southern British Columbia to northern California (Fig. 1) and has been an active, continental margin volcanic chain from Eocene to the present. Studies of Cascade development have concentrated on petrologic and tectonic synthesis (McBirney, 1978; Hammond, 1979; Priest and others, 1983). Thus, discussion of sedimentation related to evolution of the range is necessarily generalized.

Early Cascade volcanoes (late Eocene to early Miocene) were constructed on a low-relief coastal plain. Lava flows, ignimbrites, and intercalated sediments accumulated to thicknesses on the order of 3 to 6 km. Volcaniclastic lithologies include fine-grained fluvial and lacustrine facies, products of subaqueous volcanic activity (Fiske, 1963; Wise, 1970; Hammond, 1985), local accumulation of coarse-grained sediments adjacent to some andesitic volcanic centers (Swanson, 1966; Wise, 1970; Buckovic, 1979), and thick ignimbrite sequences. The great thickness of this early Cascade volcaniclastic sequence, in combination with subaqueous environments, indicates subsidence along the axis of the arc to provide a major depositional basin.

Increasing constructional relief and tectonic uplift of the Cascades is indicated by increasing divergence of floras on windward and leeward sides from Oligocene to the present (Wolfe, 1969). Steep gradients of proximal stream valleys along the rising volcanic axis were inappropriate for deposition or preservation of fluvial sediments, and loci of sedimentation generally migrated away from the range.

Aprons of volcaniclastic debris accumulated episodically adjacent to the Cascades during the Neogene. Pliocene uplift and erosion of the Western Cascades in Oregon and Washington have restricted preservation of these aprons to the east flank of the range except in northern Oregon, where middle to late(?) Miocene volcaniclastics are preserved in synclines crossing the volcanic trend (Anderson, 1978; Beeson and Moran, 1979), or crop out along the west flank of the Western Cascades (Thayer, 1939; Hampton, 1972).

Incision of the Columbia River and its tributaries between about 2 and 1 m.y. BP caused dissection of the Neogene volcaniclastic aprons. Erosion of the Western Cascades, which became largely inactive as a volcanic source at about 10 m.y. BP (McBirney, 1978; Hammond, 1979), has produced a rugged, mature topography inappropriate for widespread sedimentation. Therefore, aggradation related to volcanism in the modern High Cascades is recorded by terraces in broad valleys draining the volcanic centers.

GENERAL FEATURES OF THE NEOGENE VOLCANICLASTIC APRONS

The Ellensburg Formation (central Washington), Deschutes Formation (central Oregon) and Tuscan Formation (north-central California) serve as examples of the sedimentation which constructed broad volcaniclastic aprons

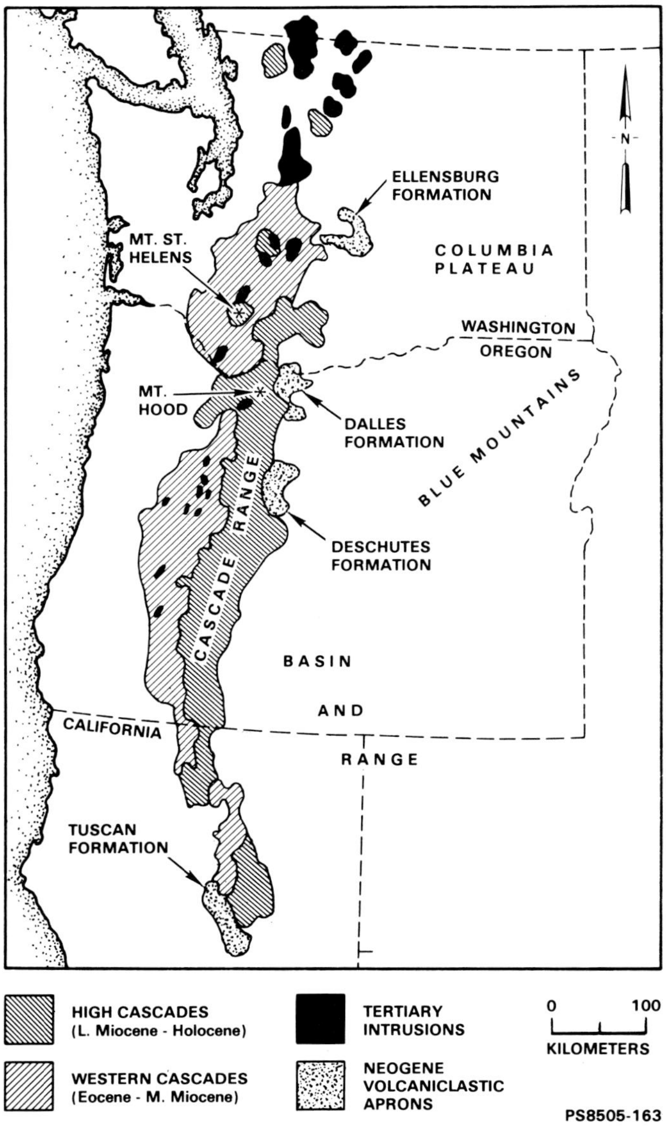

FIG. 1.—Map of the Cascade Range in the Pacific Northwest, U.S.A., showing locations of Neogene volcaniclastic aprons and High Cascade volcanoes discussed in the text.

adjacent to the Miocene and Pliocene Cascades (Fig. 1). These units are dominantly sedimentary in origin but do contain intercalated volcanic units closer to Cascade source areas. A fourth unit, the Dalles Formation in north-central Oregon, has not been the subject of sedimentological studies. If similar aprons developed east of the Cascades in southern Oregon and northern California, they are now buried beneath younger basalts and lacustrine sediments within grabens of the Basin and Range province, which encroach on the east flank of the volcanic chain (Fig. 1).

Sedimentary facies can be divided into two categories: 1) those representing fluvial processes, and 2) those deposited by processes involving sediment loads that are outside the realm of typical stream flow. Normal fluvial facies refer to those which are generally recognized in sedimentologic studies in a variety of settings (e.g., clast-supported conglomerate, crossbedded sandstone) and whose relationship to fluvial processes is generally well understood (Miall, 1977). In general, coarse-grained sandstone and conglomerate dominate over mudstone and probably represent deposition by braided streams.

A common fluvial facies association is 2- to 10-m-thick sequences of plane-bedded and scour-fill crossbedded sandstone and conglomerate, in alternating lenses 5 to 15 cm thick and 0.5 to 2.0 m across (Fig. 2). Planar tabular crossbedding in sets as much as 10 cm high is uncommon but ubiquitous. These sequences can typically be traced transverse to sediment dispersal direction indicated by paleocurrent indicators for over 100 m. The lateral extent, dominance of upper-flow-regime structures and low amplitude of scours and crossbedding, suggest aggradation during shallow super-critical flow over broad areas. These facies resemble sheetflood deposits pictured and described by Bull (1972) and Gloppen and Steel (1981). Although these flows were probably broadly confined, the term sheetflood is applied here to describe this facies association.

In the second facies category, only massive, matrix-supported conglomerate and breccia attributed to mass deposition by debris flows have been generally recognized in studies of nonvolcanic alluvium. Remaining facies are attributed to processes intermediate between normal, dilute streamflow and viscous debris flow and are termed *hyperconcentrated flood-flow deposits* (Smith, 1986a).

Hyperconcentrated flood flows are high-concentration, turbulent dispersions which rapidly deposit sediment to produce units distinct from normal stream-flow or debris-flow facies. Conglomerates (Fig. 3A) are poorly sorted, clast-supported, and typically exhibit normal grading and numerous clasts with a-dimensional axes oriented parallel to flow direction. The conglomerates may grade upward into horizontally stratified sandstones but lack the stratified sandstone lenses typically found within normal fluvial conglomerates (Miall, 1977). Hyperconcentrated flood-flow-

FIG. 2.—Sheetflood deposits in Deschutes Formation. Note poor sorting, thin bedding, and scour-fill crossbedding in foreground (arrow points to hammer) and sheetlike nature of units. "DF" and "HFF" denote debris-flow and hyperconcentrated flood-flow deposits interbedded with the sheetflood facies. "B" and "I" mark a basalt flow and ignimbrite, respectively.

deposited sandstones are poorly sorted, massive to horizontally bedded units, and notably lack cross-stratification (Fig. 3B). Beds are 0.5 to 5 cm thick, have gradational interstratal contacts, and are composed primarily of medium- to coarse-grained sand. Strata thickness, gradational nature of contacts, and coarse grain size distinguish this structure from more familiar plane lamination formed under upper-flow regime conditions (Smith, 1986a). The horizontal bedding may be produced by low-amplitude bedforms favored by rapid deposition from a poorly sorted, high-concentration sediment dispersion (Smith, 1986a). Hyperconcentrated flood-flow deposits often occur as graded-stratified units of massive gravel grading upward into horizontally bedded sand (Fig. 3C).

A genetic relationship between debris flow and hyperconcentrated flood flow is implied by: 1) a proximal-to-distal increase in the abundance and thickness of hyperconcentrated flood-flow facies at the expense of debris-flow deposits; and 2) the common occurrence of debris-flow units that are gradational at the base to hyperconcentrated flood-flow facies (Fig. 3D). These observations suggest that many hyperconcentrated flood flows originate by dilution of debris flows by stream water, a process recently observed at Mount St. Helens (T. Pierson and K. Scott, U.S. Geological Survey, pers. commun.). Because of their close spatial occurrence with debris flows and because hyperconcentrated flood-flow deposits have depositional features atypical of normal stream-flow facies, they have been called *lahars* by some workers. This Indonesian term, however, should only be applied to volcanic debris flows and their deposits. Because of this ambiguity, the term lahar is not used in this paper.

FIG. 3.—Hyperconcentrated flood-flow deposits in the Deschutes Formation. A) Clast-supported pebble-to-boulder conglomerates. Unit with base near figure's head is about 5 m thick and has a finer grained, crudely stratified top. B) Horizontally bedded sandstone; note coarse grain sizes, poorly defined interstratal contacts, and outsized clast. C) Graded-stratified deposit; clast-supported gravel at base grading into horizontally bedded, pebbly, coarse-grained sand. Total thickness of deposit is 8 m and overlies light colored overbank siltstones and darker conglomerates deposited by the ancestral Deschutes River. D) Matrix-supported debris-flow deposit (DF) gradational at the base to horizontally bedded hyperconcentrated flood-flow deposit (HFF).

EXAMPLE 1: ELLENSBURG FORMATION

Description.—

Sedimentary rocks of the Ellensburg Formation are intercalated with, and conformably overly, the middle to late Miocene flood basalts of the Columbia River Basalt Group in central Washington. West of the Yakima River (Fig. 4) the Ellensburg is primarily volcaniclastic material, but micaceous quartzo-feldspathic sandstones, siltstones, and conglomerates (with mostly plutonic and metamorphic clasts) become dominant to the east and represent deposition by the ancestral Columbia River (Schmincke, 1964). Aggradation was the combined response of Columbia Plateau drainage to: 1) slow subsidence of the western plateau and development of east-west trending anticlines and synclines of the Yakima foldbelt; 2) disruption of drainage and temporary establishment of elevated base levels by partly contemporaneous, widespread floodbasalt flows; and 3) episodic introduction of coarse-grained volcaniclastic debris from the Cascade Range.

The Ellensburg Formation near Yakima can be divided into two parts. Lower Ellensburg Formation diatomite, tuff, siltstone, and crossbedded sandstone occurs as thin (<20 m) interbeds between 13 to 16-million-year-old basalts. The upper Ellensburg Formation, interbedded with and overlying basalts younger than 13 m.y., is generally coarser grained and contains extensive debris-flow and hyperconcentrated flood-flow deposits.

The upper Ellensburg west of Yakima (Fig. 4) is principally conglomerate and sandstone, dominated by andesitic and dacitic detritus, occurring in two separate, but probably correlative (Luker, 1985) outcrop areas, west of Nile and in the Naches-Wenas valleys. Scattered outcrops along anticline crests reveal that Ellensburg sediments partly buried contemporary folds and that modern synclinal exposure reflects preferential preservation rather than restricted deposition. In the Naches-Wenas area the Ellensburg is about 400 m thick and the preserved section along Rattlesnake Creek, near Nile, is about 250 m thick (Luker, 1985). Fiske and others (1963) suggested that volcanism associated with subsequently unroofed middle Miocene plutons near Mount Rainier provided the Ellensburg debris (Fig. 4). Columbia River basalt clasts, eroded from the plateau margin and anticlinal ridges within the Columbia basin, systematically decrease in abundance upsection in the upper Ellensburg (Luker, 1985) and the formation is unconformably overlain by Pliocene basaltic gravel (Bentley and Campbell, 1983).

Approximately 40 percent of the section at Rattlesnake Creek (Fig. 5) is composed of debris-flow breccias and conglomerates (Luker, 1985). Debris-flow units are commonly 10 to 12 m thick and contain clasts as much as 2 m across (Fig. 6A). Clast-supported conglomerates comprise 20 percent of the section and the remaining 30 percent is sandstone and conglomeratic sandstone. The sand-dominated facies occurs as 2- to 6-m-thick, amalgamated sheetflood units.

Near Naches the Ellensburg is dominated by crossbedded sheet sandstone (50 percent), implying deposition by sandy braided streams (Fig. 2A; Luker, 1985). Descriptions and

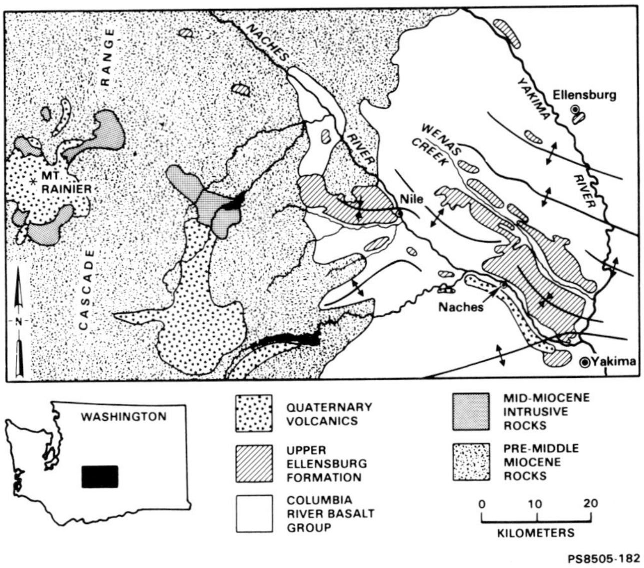

FIG. 4.—Distribution of upper Ellensburg Formation west of the Yakima River, central Washington.

granulometric analyses by Luker (1985) and reexamination of outcrops indicate that many units interpreted as lahars are hyperconcentrated flood-flow deposits comprising approximately 15 percent of the section (Fig. 6B). Debris-flow facies (lahars, *sensu stricto*) also constitute 15 percent of the section but rarely exceed 5 m in thickness and contain mostly pebble-size clasts. Paleosols, as much as 3 m thick, become thicker and more abundant upsection.

Discussion.—

Upper Ellensburg Formation sedimentation west of Yakima records transport of a large volume of fragmental volcanic debris from the Cascades into central Washington. Much of the debris was initially deposited by debris flows and sheetfloods, whose deposits dominate proximal exposures, and was partially reworked by gravel-bedload braided streams. Distal deposition was in a braided-stream environment choked with finer grained sediment derived from erosion of proximal debris. Proximal to distal increase in hyperconcentrated flood-flow deposits is interpreted to reflect dilution of debris flows. The prominence of paleosols in the upper part of the Naches section probably reflects diminished sediment supply as the volcanism which provided the coarse sediment approached quiescence.

Contemporary subsidence of the Nile and Naches synclines may have enhanced deposition and preservation of upper Ellensburg sediment, but tectonism was not the primary control on aggradation. The importance of Cascade volcanism in producing aggradation is indicated by: 1) the widespread debris-flow and hyperconcentrated flood-flow deposits; 2) the paucity of basaltic clasts; 3) deposition along anticline axes; and 4) contrasting style of sedimentation compared to older and younger sediments whose deposition was more closely related to drainage disruption by the flood basalts and intrabasinal tectonism, respectively. Paucity of basalt clasts within upper Ellensburg Formation conglomerates represents temporary dilution of intrabasinal clastics

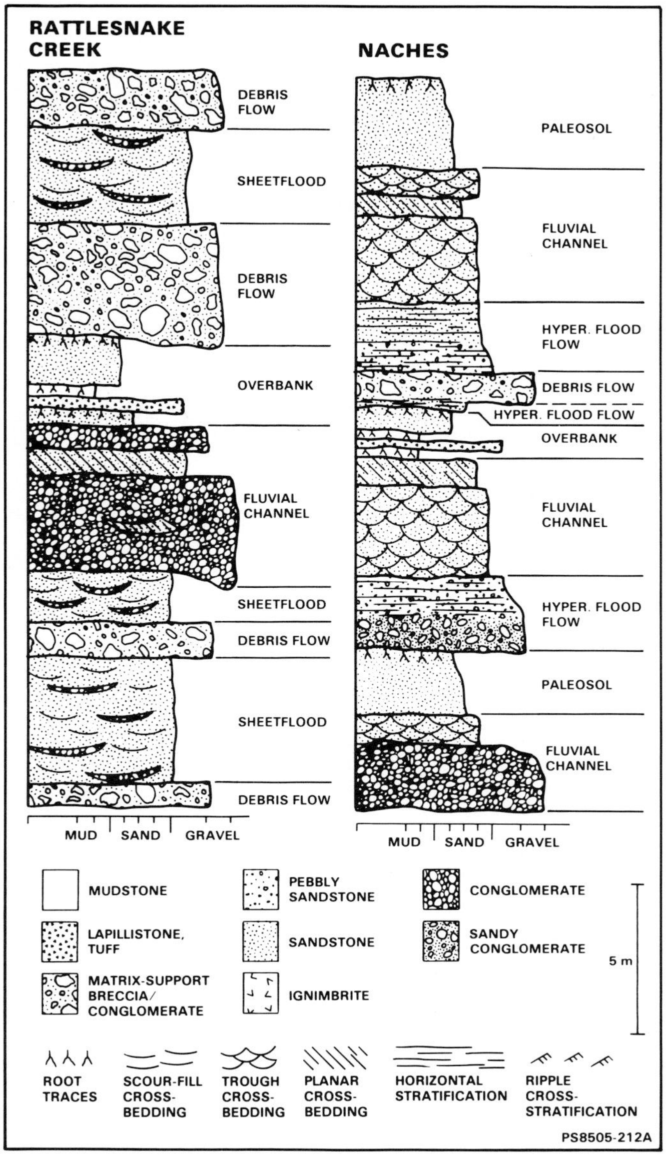

FIG. 5.—Schematic sections of Ellensburg Formation. Relatively proximal Rattlesnake Creek section contains debris-flow and sheetflood deposits with minor stream-flow conglomerate and sandstone. Distal Naches section is mostly fluvial-channel conglomerates and sandstones with numerous paleosols. Hyperconcentrated flood-flow deposits dominate debris-flow facies in Naches section.

by Cascadian sediment, burial of the basalt tectonic highs by the volcaniclastic apron, or both.

EXAMPLE 2: DESCHUTES FORMATION

Description.—

The Deschutes Formation is an assemblage of lava flows, ignimbrites, air-fall tuffs and lapillistones, and coarse-grained volcanogenic sediments deposited on the east flank of the central Oregon High Cascades between about 7.5 and 5.3 m.y. BP (Fig. 7; Taylor, 1980; Smith, 1986b). The formation thins from 700 m along Green Ridge, where the section is dominated by basalt and basaltic andesite lavas, to 250 m along the Deschutes River, where the unit is dominantly sedimentary in origin. The formation also thins northward from the major locus of contemporary volcanism south of modern Mount Jefferson (Smith and Taylor, 1983).

The Deschutes Formation is correlative to the early High Cascade eruptive episode of Priest and others (1983) that culminated in development of an intra-arc graben whose eastern boundary faults form an escarpment at Green Ridge along the west side of the Deschutes basin. Occurrence of over 75 ignimbrites within the Deschutes Formation indicates a major period of pyroclastic volcanism in the adjacent Cascades. Detailed stratigraphy and sedimentology of the Deschutes Formation is considered by Smith (1986b, and in prep.)

Much of the Deschutes Formation was deposited on an eastward tapering alluvial plain dominated by hyperconcentrated flood-flow, debris-flow and sheetflood deposits (Figs. 8, 9). Conglomerate facies typical of braided-river sedimentation (Miall, 1977) occur in some localities but are a minor component of the alluvial plain section. Erosion surfaces, with paleochannels as much as 60 m deep, separate depositional packages (Fig. 9). Debris-flow deposits are more abundant than hyperconcentrated flood-flow facies within 20 km of Green Ridge, but the relationship is reversed at greater distances. Some debris flows travelled at least 50 km from High Cascade source areas however.

Where not removed by later erosion, the uppermost 10 to 50 m of the section consists of superimposed paleosols, tephras, and basalts erupted east of the Cascade crest, and notably lacks ignimbrites and intercalated debris-flow and flood deposits. Stratigraphic study (Smith, 1986b) allows correlation of this transition in the sedimentary and volcanic record to subsidence of the High Cascade intra-arc graben. Abundant, thick tephras in the Deschutes basin indicate that explosive volcanism continued, but ignimbrites and eruption-related sedimentary deposits were ponded in the graben and, thus, not recorded in the basin.

An ancestral Deschutes River flowed northward at the foot of the arc-adjacent alluvial plain and is represented by alternating, equithick sequences of channel-fill conglomerate with coarse-grained sandstone and overbank siltstone and fine-grained sandstone (Fig. 8B). Debris-flow facies are rare within these deposits, but graded-stratified hyperconcentrated flood-flow units as much as 10 m thick (Fig. 3C) are ubiquitous within overbank sequences and/or capping channel-fill facies (Fig. 8).

Adjacent to older Tertiary volcanic highlands along the northern and eastern basin margins, the Deschutes Formation is less than 75 m thick and composed almost entirely of paleosols developed in sand and pyroclastic facies.

Discussion.—

Deschutes Formation aggradation represents the impact of early High Cascade pyroclastic volcanism on a semi-arid fluvial system. Deposition on a broad alluvial plain adjacent to the volcanic chain occurred during flood and debris-flow events. Episodic aggradation during periods of high sediment-load transport, probably induced by explosive volcanic events, was separated by periods of incision when

FIG. 6.—Ellensburg Formation outcrops. A) 4-m-thick debris-flow deposit overlying braided stream cobble conglomerate, in turn overlying thin debris-flow and hyperconcentrated flood-flow deposits. Rattlesnake Creek section. B) Portion of a thick, crudely bedded, hyperconcentrated flood-flow deposit near Naches. Note outsized clasts.

streams attempted to reestablish former, graded elevations as sediment loads diminished.

The record of typical fluvial deposition is scarce in the alluvial plain sequence and is well represented only in the ancestral Deschutes River deposits farther east. The rarity of debris-flow deposits, but common occurrence of thick

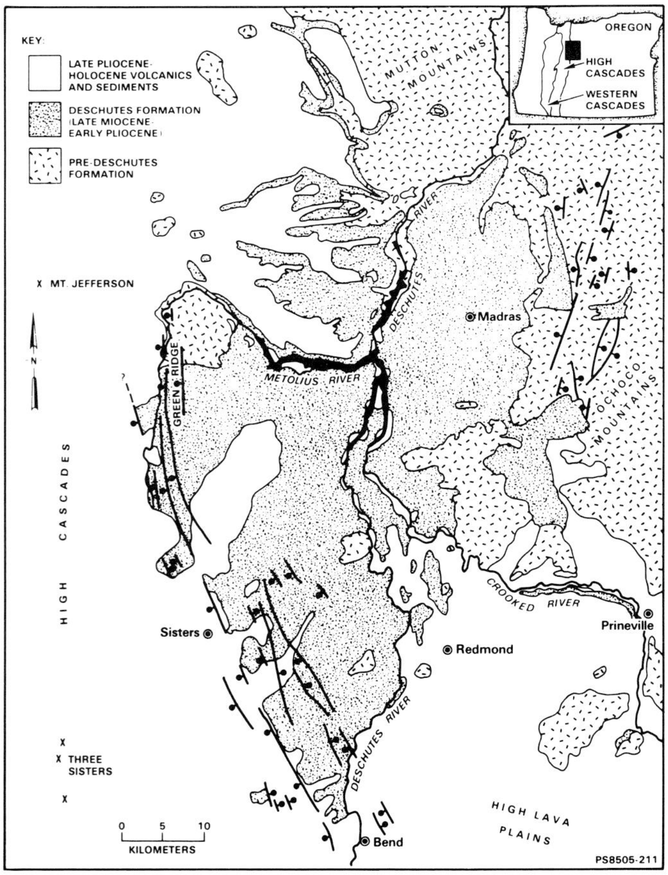

FIG. 7.—Distribution of the Deschutes Formation in central Oregon.

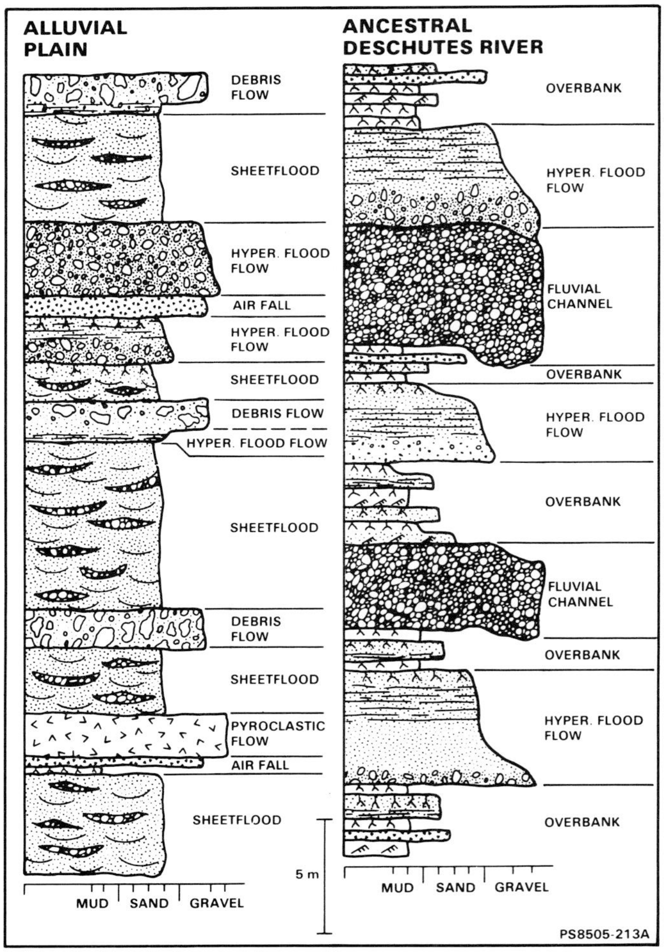

FIG. 8.—Schematic sections of Deschutes Formation; symbols as in Figure 5. Alluvial plain sequence is sheetflood, debris-flow, and hyperconcentrated flood-flow facies. The ancestral Deschutes River sequence is dominated by channel-fill conglomerates and overbank sandstones and siltstones; thick, graded-stratified hyperconcentrated flood-flow deposits occur within the overbank sequences or between channel and overbank facies.

FIG. 9.—Paleochannels in the Deschutes Formation. A) 15-m-deep channel cut through, and filled by, hyperconcentrated flood-flow and sheetflood deposits. Note concentration of large boulders adjacent to channel wall and massive debris-flow deposit at top of exposure. b) Complex relationship of sheet and channel-fill units, numbered in depositional sequence. Units 1 and 4 are sheetflood deposits separated by a discontinuous 1- to 5-cm-thick mudstone not resolvable in this photo. Units 2, 3, 5, 6 are hyperconcentrated flood-flow deposits, 7 is an ignimbrite overlying cogenetic air-fall tuff, and 8 is debris-flow deposits. Note large block of sheetflood sandstone, with bedding in vertical position, in channel-fill, beneath figure's arm.

hyperconcentrated flood-flow facies, in the Deschutes River sediments supports transformation of debris flows to less viscous fluids by the large volume of water in the river channel. The hyperconcentrated flood-flow deposits were reworked in the river channel, thus restricting preservation to overbank sequences. Less common occurrences of these facies between channel-fill and overbank facies associations imply channel diversion by the hyperconcentrated flood-flow event.

Volcanism-induced aggradation is strongly suggested by: 1) increasing formation thickness toward the locus of contemporary pyroclastic volcanism; 2) great lateral extent of debris-flow and hyperconcentrated flood-flow facies; and 3) dominance of paleosols (i.e., slow sedimentation rate) in the formation adjacent to uplifts along the northern and eastern basin margins. Also, the break in sedimentation style on the alluvial plain resulting from structural isolation of the depositional basin from the Cascades documents the importance of volcanism to produce aggradation.

EXAMPLE 3: TUSCAN FORMATION

Description.—

The Tuscan Formation consists of debris-flow breccias, conglomerates, sandstones, and subordinate ignimbrites, lava flows and tuffs that originally covered over 5000 km^2 between the Cascade Range and Sacramento Valley in northern California (Fig. 10) (Lydon, 1968). This unit, at least 500 m thick on the east, thins westward and pinches out in the subsurface near the center of the Sacramento Valley (Lydon, 1968). Eastern exposures are almost entirely debris-flow breccias with a few intercalated basalt and andesite flows. Along the margin of the Sacramento Valley the Tuscan is dominantly sandstone and conglomerate, especially in the lower two-thirds of the formation, interbedded with debris-flow and hyperconcentrated flood-flow deposits (Fig. 11) (Doukas, 1983). Pliocene fossils and radiometric ages suggest that the Tuscan Formation was deposited in a short time period between 3.5 and 2.5 m.y. BP (Harwood and others, 1981). Detailed consideration of the sedimentology of the Tuscan Formation is lacking; however, studies of the debris-flow breccias ("lahars") by Anderson (1933) and Lydon (1968), stratigraphic study by Harwood and others (1981), a detailed measured section near Chico by Doukas (1983), and reconnaissance by the author will suffice for the discussion presented here.

Debris-flow deposits of the Tuscan Formation have a probable volume in excess of 750 km^3 (Lydon, 1968). Along the Sacramento Valley, 50 km away from the Cascades, debris-flow deposits are commonly 5 to 15 m thick and contain clasts several meters across. Associated cobble-to-boulder conglomerate, hyperconcentrated flood-flow facies, and sheetflood-deposited pebbly sandstones become more abundant westward. The section studied by Doukas (1983) exhibits a general coarsening-upward trend with an increasing proportion of fluvial conglomerates and debris-flow facies toward the top. Principal clast lithologies are basaltic andesite and andesite resembling intercalated lava flows farther east. Sedimentary units are laterally discontinuous in the lower part of the formation but become more sheetlike upsection (Anderson, 1933; Lydon, 1968; Doukas, 1983). Paleosols are notably absent in the Tuscan For-

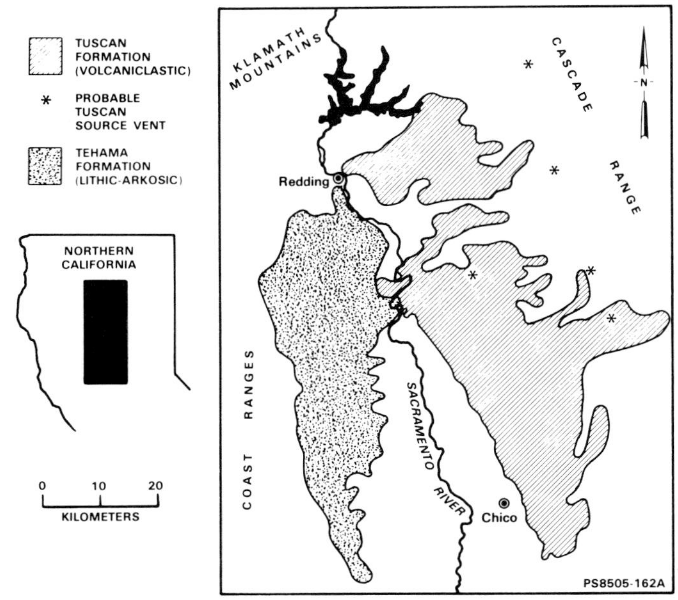

FIG. 10.—Distribution of Pliocene Tuscan and Tehama Formations in the northern Sacramento Valley, California.

mation, but preservation of widespread air-fall tuffs only a few centimeters thick suggests that the paucity of paleosols cannot be explained by erosion.

Tuscan Formation deposition is in contrast with that recorded by the partly contemporaneous Tehama Formation farther west (Fig. 10). Lithic-arkosic fluvial sediments of the Tehama Formation were derived from the Coast Ranges and Klamath Mountains, are at least 800 m thick in the Sacramento Valley, and thin westward to onlap Cretaceous rocks of the Great Valley Sequence (Anderson and Russell, 1939). Correlation of the Tehama and Tuscan formations is based on the occurrence of distinctive pyroclastic units in both formations and an interfingering relationship evident in well logs within the Sacramento Valley and rare surface outcrops (Anderson and Russell, 1939; Lydon, 1968). Pebble conglomerate is common along the western edge of the Tehama outcrop, but the bulk of the formation exposed along the margin of the Sacramento Valley is composed of massive, bioturbated sandstone and mudstone with caliche horizons.

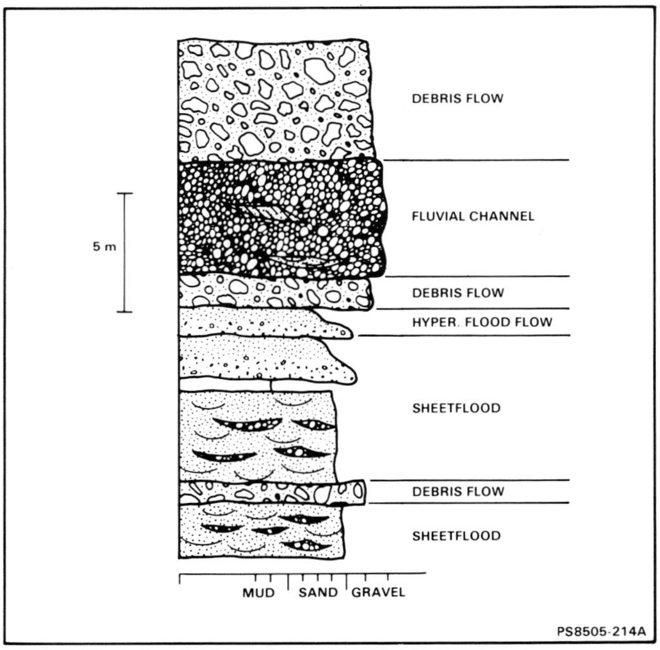

FIG. 11.—Schematic section of a portion of the Tuscan Formation along Big Chico Creek in its most distal exposures; symbols as in Figure 5. The section is dominated by debris-flow, hyperconcentrated flood-flow, and sheetflood facies with minor stream-flow conglomerate.

Discussion.—

The Tuscan Formation represents a rapidly prograding volcaniclastic apron which developed in only about 1 million years. Deposition was primarily by debris flows with other processes recorded only in the most distal exposures. Cascade volcanism produced rapid construction of a sourceward thickening wedge of coarse-grained volcaniclastic sediments, whereas rivers draining tectonic highlands to the west produced a basinward thickening sequence of finer grained lithic-arkosic sediments. Because the Tuscan and Tehama Formations share a common depositional basin, the contrasting sedimentology must reflect source area controls on sedimentation and serves as a dramatic example of the influence of volcanism on fluvial systems.

SEDIMENTATION IN THE MODERN HIGH CASCADES

Lava flows, pyroclastic flows, and sediment derived from Pleistocene and Holocene volcanic centers are largely confined to glacial troughs near the volcanoes and to valleys deeply incised into Tertiary bedrock. Synvolcanic sediments form terraces within these valleys and lowland fills of uncertain lateral extent (Crandell, 1963), in contrast to the broad aprons resulting from Neogene volcanism.

During volcanic-hazard assessment studies, nested terraces dominated by debris-flow facies and poorly sorted, flood-deposited sediment have been recognized at distances in excess of 50 km from volcanic centers (Crandall and Mullineaux, 1978; Crandall, 1980; Beget, 1982). These terraces can often be correlated to proximal ones adjacent to the volcanoes that contain pyroclastic-flow deposits and demonstrate that periods of sedimentation were contemporaneous with explosive volcanic events. Nearly all of the sedimentary facies within proximal terraces record debris-flow, hyperconcentrated flood-flow, and sheetflood events, and there are virtually no typical fluvial facies. The flood and debris-flow deposits become interbedded downvalley with variable proportions of clast-supported gravel and crossbedded sand, suggesting braided stream sedimentation.

Stream channel adjustments following the 1980 eruptions of Mount St. Helens provide insights into the development of terrace sequences related to explosive volcanic events (Lisle and others, 1982). Aggradation in proximal reaches is by eruption-induced debris-flows and floods or occurs during storms that choke streams with fragmental debris eroded from areas affected by earlier eruptions. Gravel-bedload stream channels become incised during intervening periods and show a downstream-increasing tendency to widen. Downstream aggradation is also discontinuous and related to eruption events and storm periods. Although sedimentation episodes are very short-lived in proximal positions, distal valleys may aggrade for weeks or months following eruption and storm events as recently deposited flood deposits upstream are eroded and resedimented.

Observation of recent deposits in the Toutle River valley west of Mount St. Helens reveals several characteristics common to the Neogene aprons. Terraces 8 km from the volcano are dominated by poorly sorted, lenticular, sandy gravel deposits produced by rapid aggradation in broad braided channels whose bar and thalweg morphology is often preserved on the terrace surface (Fig. 12A). Interbedded with these facies are debris-flow and hyperconcentrated flood-flow deposits (Fig. 12B). Farther downstream, the Toutle River is a broad, braided stream with extensive inactive tracts that aggrade only during floods (Fig. 12C). Because of channel widening, recent debris-flow and hyperconcentrated flood-flow deposits have been reworked except where they occupy the top of high-standing, rarely active bars (Fig. 12C), or cap older terraces. The broad, braided channel with sand comprising as much as 85 percent of the bed material contrasts with the pre-eruption, narrow, pool-and-riffle, gravel-bedload Toutle River (Lisle and others, 1982).

FIG. 12.—Sedimentation in North Fork Toutle River valley west of Mount St. Helens. A) View downstream in September 1984, 16 stream-km from the crater, showing incised gravel-bedload channel and terraces composed of debris-flow and flood deposits. Note preservation of braided channel pattern on terrace surfaces formed during floods; river is confined by hummocks on May 18, 1980 debris avalanche surface, about 100 m apart. B) Exposure of terrace sediments near location in A). Dark lines outline lenticular-bedded sheetflood deposits overlying debris-flow and hyperconcentrated flood-flow facies. A thin debris-flow unit with an underlying finer grained hyperconcentrated flood-flow deposit, of equal thickness, caps the terrace. C) View of North Fork Toutle River 30 km from crater in October 1983. Note broad, largely inactive, braided channel. Vegetated bar labeled "1" is the only remaining unreworked deposit of May 18, 1980 debris flows in this reach. Light colored terrace labeled "2" is composed of two ~3000-year-old debris-flow deposits.

DISCUSSION

This paper emphasizes sedimentation resulting from volcanism-induced sediment loads choking fluvial systems. It is important to remember, however, that not all nonmarine volcaniclastic sequences represent volcanism-induced aggradation. Subsidence of forearc or interarc basins adjacent to volcanic chains may result in deposition of thick sequences of nonmarine volcaniclastics. Study of such a setting by Mathisen and Vondra (1983) suggests that tectonically-induced volcaniclastic sedimentation is similar to that in nonvolcanic settings.

Hyperconcentrated flood-flow facies are an important component of volcaniclastic sequences in the Pacific Northwest but have not been broadly recognized in nonvolcanic alluvium. Because of the unparalleled scale of rapid, coarse-sediment transport in volcanic terrains, hyperconcentrated flood-flow deposits are thicker and more extensive than on nonvolcanic alluvial fans. Another factor contributing to the abundance, thickness, and extent of hyperconcentrated flood-flow deposits in volcanic settings is the probable generation of many such flows by dilution of debris flows. Similar transformation is unlikely for debris flows on dry alluvial fan surfaces but is probable for debris flows following channels away from continental margin volcanoes which, as orographic barriers, are rarely located in arid environments.

The presence of debris-flow facies as a criterion for recognizing arid alluvial fan sequences, a common premise of existing facies models (Bull, 1972; Collinson, 1978; Nilsen, 1982; Rust and Koster, 1984), is invalid in volcanic settings. Debris-flow and associated hyperconcentrated flood-flow and sheetflood facies are common to dominant constituents of the Neogene volcaniclastic aprons and Quaternary valley fills which clearly were not alluvial fans. The gradual lateral facies changes over distances of as much as 50 km from the source are atypical of alluvial fan sequences which exhibit abrupt transitions from debris-flow-dominated sections to fine-grained floodbasin or lacustrine facies in lateral distances of 10 km or less (Gloppen and Steel, 1981; Rust and Koster, 1984). Gradual facies changes and gradation to proximal volcanic-dominated sections indicate the lack of abrupt mountain fronts necessary for alluvial fan development. Lateral grain-size variation of fluvial conglomerates in the Deschutes and Ellensburg formations resembles alluvial plain settings (Fig. 13), and paleocurrent data fail to represent a divergent drainage pattern diagnostic of alluvial fan morphology (Smith, 1986b; Luker, 1985).

The Neogene volcaniclastic aprons and Quaternary valley fills contain facies common to alluvial fans but were deposited as thin, widespread sequences lacking the thickness, rapid facies changes, and morphology of alluvial fan deposits. Debris flows are more widespread adjacent to continental margin volcanoes than on arid alluvial fans because explosive volcanism leads to the mobilization of sediment on a scale unparalleled in other settings, and water is readily available from crater lakes, melted snow, perennial streams, and/or heavy rainfall in maritime climates. Whereas debris flows on alluvial fans rarely extend off the fan surface, generally 10 km or less in radius (Nilsen, 1982), volcanic debris flows may travel for over 100 km.

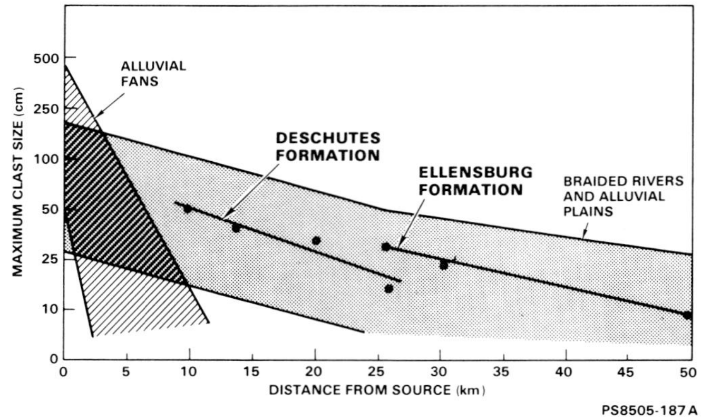

FIG. 13.—Lateral clast size variation in stream-flow conglomerates in the Deschutes and Ellensburg Formations. Fields for alluvial fans and alluvial plains from Rust and Koster (1984).

Evaluation of fluvial style in volcanism-induced volcaniclastic sequences is difficult because of the discontinuity of sedimentation. Most sedimentation occurs during episodes when large sediment loads choke streams and is separated by periods of incision as streams adjust to diminished load. The restricted preservation of Quaternary sedimentation to terraces illustrates the punctuated nature of aggradation adjacent to explosive volcanoes as does the record of the arc-adjacent alluvial plain in the Deschutes basin, where debris-flow and flood deposits occur in sequences separated by erosion surfaces with as much as 60 m of relief. Alternating aggradation and degradation on this scale is atypical of syntectonic alluvium which usually accumulates in a relatively continuous fashion within subsiding basins. Normal fluvial facies are common in distal exposures of the Neogene aprons and Quaternary terraces but may represent short-term (i.e., years to decades) reworking and redeposition of volcanism-induced sediment loads in low gradient reaches of streams which are largely nondepositional during the longer periods between eruptive episodes that cross geomorphic thresholds allowing aggradation (Vessell and Davies, 1981; Lisle and others, 1982).

Volcaniclastic aprons, as represented by the upper Ellensburg, Deschutes, and Tuscan formations, record geologically short time periods when volcanism induces aggradation in adjacent fluvial basins. In the case of the Tuscan Formation, sedimentation rates were very high and produced a rapidly prograding wedge of coarse-grained sediment dominated by debris-flow deposits. Periods of nondeposition were short and thus recognizable paleosols are not evident. Lack of significant intervening periods of erosion led to burial of previous topography and allowed debris flows to spread out as sheets. The increase in occurrence of paleosols near the top of the Deschutes and Ellensburg formations probably reflects the diminishing influence of volcanism on sedimentation. For the Deschutes Formation, explosive volcanism continued, but subsidence of the Cascades isolated the volcanoes from the fluvial basin and removed the volcanic influence on aggradation. In the Ellensburg Formation, decrease in volcanism-induced sedimentation probably represents diminished volcanic activity.

CONCLUSIONS

Subsidence of the Cascade volcanic axis during the early stage of arc development caused thick accumulation of Eocene to early Miocene nonmarine volcaniclastics. Broad, thin volcaniclastic aprons developed next to the higher standing Neogene Cascades and represent aggradation directly induced by periods of volcanic activity which choked streams with fragmental debris. Quaternary sedimentation is largely confined to valleys within the range but involves depositional processes similar to those which produced the more extensive Neogene aprons.

Volcanism-induced aggradation of nonmarine sedimentary sequences includes widespread debris-flow, hyperconcentrated flood-flow and sheetflood deposits. Normal streamflow facies are generally dominant only in distal settings, tens of kilometers from the volcanic chain. Although covering areas on the order of 10^3 km^2, these units are of modest thickness (<1000 m) and form wedge-shaped aprons that taper away from the volcanic axis. Proximal-to-distal facies changes are gradual, and lateral grain size changes are similar to those reported on alluvial plains in nonvolcanic areas. Debris-flow deposits are replaced by hyperconcentrated flood-flow facies with increasing distance from source, and both are subject to erosion by high-gradient proximal streams and reworking and redeposition by lower gradient, distal braided rivers. The sedimentary section may grade sourceward to lava flows and ignimbrites. Incision between episodes of deposition induced by volcanic events may be represented by deep paleochannels.

Successful evaluation of nonmarine volcaniclastic sequences must include consideration of the influence of volcanism-induced sediment loads on fluvial systems, influences which are not adequately explained by existing facies models developed primarily from study of syntectonic fluvial sequences. The two primary weaknesses of these models are: 1) the failure to consider volumetrically important facies representing deposition by processes intermediate between debris flow and normal stream flow, and 2) restriction of subaerial debris flows to arid alluvial fan environments. Explosive volcanism, directly or indirectly, provides large volumes of sediment for transport on a scale unparalleled in nonvolcanic settings, and water is readily available in continental margin arcs. Thus, debris-flow deposits are thicker and more widespread than on arid alluvial fans, where most studies of subaerial debris flows have been conducted, and not restricted in occurrence to fan settings.

ACKNOWLEDGMENTS

Most of this research was conducted during the tenure of a research fellowship awarded by the Northwest College and University Association for Science (University of Washington) under contract DE-AM06-76-RL02225 with the U.S. Department of Energy. Manuscript preparation was supported by Rockwell Hanford Operations, Basalt Waste Isolation Project. Research in the Deschutes basin was funded by grants from the Geological Society of America, Sigma Xi, Oregon State University with support from Sohio Petroleum Corporation and Shell Oil Company, and the National Science Foundation (EAR-8313986). The manuscript has been improved by the comments of P. E. Hammond, D. W. Andersen, J. A. Schmitt, and K. R. Fecht. Drafting was done by L. Lang.

REFERENCES

Anderson, C. A., 1933, The Tuscan Formation in northern California with a discussion concerning the origin of breccias: University of California Publications in the Geological Sciences, v. 23, p. 215–276.

———, and Russell, R. D., 1939, Tertiary formations of the northern Sacramento Valley, California: California Journal of Mines and Geology, v. 35, p. 219–253.

Anderson, J. L., 1978, The stratigraphy and structure of the Columbia River Basalt in the Clackamas River drainage: unpublished M.S. Thesis, Portland State University, 136 p.

Beeson, M. H., and Moran, M. R., 1979, Stratigraphy and structure of the Columbia River Basalt Group in the Cascade Range, Oregon, *in* Hull, D. A., and Riccio, D. F., ed., Geothermal Resource Analysis of Mount Hood: Oregon Department of Geology and Mineral Industries Open-File Rept. 0-79-8, p. 5–77.

Beget, J. E., 1982, Post-glacial volcanic deposits at Glacier Peak, Washington, and potential hazards from future eruptions: U.S. Geological Survey Open-File Report 82-830, 77 p.

Bentley, R. D., and Campbell, N. P., 1983, Geologic map of the Yakima quadrangle, Washington: Washington Division of Mines and Geology, Map GM-29, Scale 1:62,500.

Buckovic, W. A., 1979, The Eocene deltaic system of west-central Washington, *in* Armentrout, J. M., Cole, M. R., and Terbest, H., Jr., eds., Cenozoic Paleogeography of the Western United States: Pacific Section Paleogeography Symposium #3, Society of Economic Paleontologists and Mineralogists Pacific Section, p. 147–164.

Bull, W. B., 1972, Recognition of alluvial fan deposits in the stratigraphic record: Society of Economic Paleontologists and Mineralogists Special Publication 16, p. 63–83.

Collinson, J. D., 1978, Alluvial sediments, *in* Reading, H. G., ed., Sedimentary Environments and Facies: Blackwell Scientific Pub., Oxford, p. 15–79.

Crandell, D. R., 1963, Surficial geology and geomorphology of the Lake Tapps Quadrangle, Washington: U.S. Geological Survey Professional Paper 388-A, A1-A84 p.

———, 1980, Recent eruptive history of Mount Hood, Oregon and potential hazards from future eruptions: U.S. Geological Survey Bulletin 1492, 81 p.

———, and Mullineaux, D. R., 1978, Potential hazards from future eruptions of Mount St. Helens, Washington: U.S. Geological Survey Bulletin 1383-C, 26 p.

Davies D. K., Vessell, R. K., Miles, R. C., Foley, M.G., and Bonis, S. B., 1978, Fluvial transport and downstream sediment modification in an active volcanic region, *in* Miall, A. D., ed., Fluvial Sedimentology: Canadian Society of Petroleum Geologists Memor 5, p. 61–83.

Doukas, M. J., 1983, Volcanic geology of Big Chico Creek area, Butte County, California: Unpublished M.S. Thesis, San Jose State University, 157 p.

Fiske R. S., 1963, Subaqueous pyroclastic flows in the Ohanapecosh Formation, Washington: Geological Society of America Bulletin, v. 74, p. 391–406.

———, Hopson, C. A., and Waters, A. C., 1963, Geology of Mount Rainier National Park, Washington: U.S. Geological Survey Professional Paper 444, 93 p.

Gloppen, T. G., and Steel, R. J., 1981, The deposits, internal structure and geometry in six alluvial fan—fan delta bodies (Devonian-Norway)—A study in the significance of bedding sequences in conglomerates, *in* Ethridge, F. G., and Flores, R. M., eds., Recent and Ancient Nonmarine Depositional Environments: Models for Exploration: Society of Economic Paleontologists and Mineralogists Special Publication 31, p. 49–70.

Hammond, P. E., 1979, A tectonic model for evolution of the Cascade Range, *in* Armentrout, J. M., Cole, M. R., and Terbest, H., Jr., eds., Cenozoic Paleogeography of the Western United States: Pacific Section Paleogeography Symposium #3, Society of Economic Paleontologists and Mineralogists Pacific Section, p. 219–237.

———, 1985, What types of volcanoes erupted the basal volcanic rocks of the proto-Cascade arc? (abs.): EOS, Transactions, American Geophysical Union, v. 66, p. 24.

HAMPTON, E. R., 1972, Geology and groundwater of the Mollala-Salem slope area, northern Wilamette valley, Oregon: U.S. Geological Survey Water-Supply Paper 1997, 79 p.

HARWOOD, D. S., HELLEY, E. J., AND DOUKAS, M. P., 1981, Geologic map of the Chico monocline and northeast part of the Sacramento Valley, California: U.S. Geological Survey Miscellaneous Investigation Map I-1238.

LISLE, T. E., LEHRE, A. K., MARTINSON, H. A., MEYERS, D. F., NOLAN, K. M., AND SMITH, R. D., 1982, Stream channel adjustments after the 1980 Mount St. Helens eruptions: Proceedings, Symposium on Erosion Control in Volcanic Areas: Public Works Research Institute of Japan, Technical Memorandum 1908, p. 33–72.

LUKER, J. A., 1985, Sedimentology of the Ellensburg Formation northwest of Yakima, Washington: unpublished M.S. thesis, Eastern Washington University, 184 p.

LYDON, P. A., 1968, Geology and lahars of the Tuscan Formation, northern California: Geological Society of America Memoir 116, p. 441–475.

MATHISEN, M. E., AND VONDRA, C. F., 1983, The fluvial and pyroclastic deposits of the Cagayan basin, northern Luzon, Philippines—An example of nonmarine volcaniclastic sedimentation in an interarc basin: Sedimentology, v. 30, p. 369–392.

MCBIRNEY, A. R., 1978, Volcanic evolution of the Cascade Range, *in* Donath, F. A., ed., Annual Reviews of Earth and Planetary Science, v. 6, p. 437–456.

MIALL, A. D., 1977, A review of the braided river depositional environment: Earth Science Reviews, v. 13, p. 1–62.

———, 1981, ed., Sedimentation and tectonics in alluvial basins: Geological Association of Canada Special Paper 23, 272 p.

NILSEN, T. H., 1982, Alluvial fan deposits, *in* Scholle, P. A., and Spearing, D., eds., Sandstone depositional environments: American Association of Petroleum Geologists Memoir 31, p. 49–86.

———, 1984, ed., Fluvial sedimentation and tectonics in the western United States: Sedimentary Geology, v. 38.

PRIEST, G. R., WOLLER, N. M., BLACK, G. L., AND EVANS, S. H., 1983, Overview of the geology of the central Oregon Cascade Range, *in* Priest, G. R., and Vogt, B. F., eds., Geology and Geothermal Resources of the Central Oregon Cascade Range: Oregon Department of Geology and Mineral Industries Special Paper 15, p. 3–28.

RUST, B. R., AND KOSTER, E. H., 1984, Coarse alluvial deposits, *in* Walker, R. G., ed., Facies Models: Geological Association of Canada Reprint Series 1, Second edition, p. 53–70.

SCHMINCKE, H.-U., 1964, Petrology, paleocurrents, and stratigraphy of the Ellensburg Formation and interbedded Yakima basalt flows: Unpublished Ph.D. Thesis Johns Hopkins University, 426 p.

SMITH, G. A., 1986a, Coarse-grained nonmarine volcaniclastic sediment: Terminology and depositional process: Geological Society of America Bulletin, v. 90, p. 1–10.

———, 1986b, Stratigraphy, sedimentology and petrology of Neogene rocks in the Deschutes basin, central Oregon: The record of continental-margin volcanism and its influence on fluvial sedimentation in an arc-adjacent basin: Unpublished Ph.D. Dissertation, Oregon State University, 467 p.

———, AND TAYLOR, E. M., 1983, The central Oregon High Cascade graben: What? where? when?: Transactions, Geothermal Resources Council, v. 7, p. 275–279.

SWANSON, D. A., 1966, Tieton Volcano, a Miocene eruptive center in the southern Cascade Mountains, Washington: Geological Society of America Bulletin, v. 77, p. 1293–1314.

TAYLOR, E. M., 1980, Volcanic and volcaniclastic rocks on the east flank of the central Cascade Range to the Deschutes River, Oregon, *in* Oles, K. F., and others, eds., Geologic Field Trips in Western Oregon and Washington: Oregon Department of Geology and Mineral Industries Bulletin 101, p. 1–7.

THAYER, T. P., 1939, Geology of the Salem Hills and the North Santiam River basin, Oregon: Oregon Department of Geology and Mineral Industries Bulletin 15, 40 p.

VESSELL, R. K., AND DAVIES, D. K., 1981, Non-marine sedimentation in an active fore arc basin: Society of Economic Paleontologists and Mineralogists Special Publication 31, p. 31–45.

———, AND CANT, D. J., 1984, Sandy fluvial systems, *in* Walker, R. G. ed., Facies Models: Geological Association of Canada Reprint Series 1, second edition, p. 71–90.

WISE, W. S., 1970, Cenozoic volcanism in the Cascade Mountains of southern Washington: Washington Division of Mines and Geology Bulletin 60, 45 p.

WOLFE, J. A., 1969, Neogene floristic and vegetational history of the Pacific Northwest: Madrono, v. 20, p. 83–110.

CYCLICITY AND FACIES CHARACTERISTICS OF ALLUVIAL SEDIMENTS IN THE UPPER PALEOZOIC MONONGAHELA-DUNKARD GROUPS, CENTRAL WEST VIRGINIA

SANTOSH K. GHOSH*

Department of Geology, State University of New York at Binghamton, Binghamton, N.Y., 13901

ABSTRACT: The fluvial Monongahela-Dunkard sediments in the southern Dunkard Basin consist of two facies: a channel belt, and a much thicker overbank facies assemblage. The channel belt consists of stacked and amalgamated sequences of fining-upward sandstones of point bar origin. The overbank sequence consists of thick floodplain sediments, freshwater limestones, silty levee-splay units, and crevasse channel and splay lobe lithofacies. Proximal crevasses consist of single depositional units, whereas distal ones are composite, thinner and finer grained, with shaly intercalations. Thick red mud-clay rocks with paleosol horizons imply persistent floodplain sedimentation punctuated by periods of very low sedimentation and subsidence.

Statistical analysis of the lithofacies sequence reveals coarsening-upward, fining-upward, and oscillatory sequential patterns. The coarsening-upward motif develops due to crevasse-levee progradation toward interchannel areas. The fining-upward sequence is mainly due to a succession of point bar and top-stratum deposits, and secondarily to silting of crevasse systems. Oscillatory sequences owe their origin to interbedding of distal crevasse and levee-floodplain deposits.

Abundant red floodplain deposits and paleosol horizons in the study area imply sedimentation on a relatively higher (and drier) alluvial plain relative to the low and lacustrine northern Dunkard area, where coal, limestone and gray mudrocks abound. Crevasse progradation is the main cause of cyclicity in the study area. In contrast, channel sandstone is central to the cyclothems in the northern Dunkard area.

INTRODUCTION

Strata of the Monongahela Group (Upper Pennsylvanian) and the basal part of the Dunkard Group (Permian) exposed in excellent roadcuts along Interstate 77, north of Charleston, West Virginia, offer a good opportunity to study alluvial plain cyclicity and relate it to recurring patterns of depositional facies sequence. The study area is located in the southern part of the Dunkard Basin in central West Virginia (Fig. 1). In the road section, 200 m of strata of the alluvial Monongahela-Dunkard Groups are exposed along an 11-km stretch of Interstate 77 (Fig. 1). The Monongahela and Dunkard sediments represent rather similar alluvial depositional regimes, and no marker coal or fossiliferous unit separates one from the other.

The primary focus of this paper is an interpretation of alluvial cyclicity with special reference to the coarsening-upward sequence, and secondly, an interpretation of lithofacies assemblages and depositional complexes in the overbank facies.

Early workers (Donaldson, 1974; Cross and Schemel, 1956; Beerbower, 1961) studying the Appalachian Plateau concentrated on the marine deltaic northern end of the Dunkard Basin. Little published information exists on characteristics of the deposits in the southern Dunkard area with respect to nature of cyclicity, internal facies organization and facies variation (Beerbower, 1961, 1969). Moreover, the cyclic aspects of Dunkard sediments have been, until now, based only upon visual appraisal of outcrops.

GEOLOGIC SETTING

Donaldson (1974) made a detailed study of the paleogeographic and paleoenvironmental evolution of the Pennsylvanian-Permian rocks in the Central Appalachians. The sequence from Pottsville time up to the Dunkard (Fig. 1D) shows a progressive change from extensive seas to relatively small bay (Allegheny-Lower Conemaugh), to essentially river-influenced low salinity bay-lake (Upper Conemaugh-Monongahela), to the nonmarine lacustrine flood-basin environment of Dunkard time (Fig. 2). The Dunkard Basin developed on a relatively stable platform north of a subsiding unstable geosynclinal trough in southern West Virginia (Fig. 1C). Sediments from a southeasterly source area (Old Appalachia or conterminous portions of southwest Europe and Africa, according to Cross, 1972) fed the unstable trough and then, with sediment supply exceeding subsidence rate, filled the Dunkard Basin on the platform. Cross (1972) postulates that with progressive lowering of sea level through time (due to either eustatic sea level fall or uplift in the source area), Dunkard redbeds and freshwater lacustrine facies encroached progressively upon northern and western areas.

The exposed sequence consists of a spectrum of near horizontal depositional units representing a nonmarine, cyclic, fluvial-dominated lithofacies sequence. Diagnostic criteria of fluviatile origin include extensive redbeds with pedogenic calcareous concretionary horizons, freshwater limestone, and absence of brackish or marine fossils coupled with the presence of freshwater amphibian remains and ostracodes.

LITHOFACIES ANALYSIS

Twenty-five closely spaced sections were measured in detail in an area (Fig. 1) where seven different lithofacies have been defined and recognized by a combination of physical properties such as grain size, composition, color, scale and type of sedimentary structures, internal organization, and overall geometry. The seven lithofacies are: conglomerate (A), sandstone (B), sandy siltstone (C), silty shale (D), mud-clay rocks (E), interbedded heterolithic facies of sandstone, siltstone and mudstone (F), and limestone (G). The mudrocks (sandy siltstone, silty shale and mud-clay rocks) were classified using the terminology of Blatt and others (1980). The mud-clay rocks (fissile and nonfissile) possess similar characters and display the same mode of occurrence, whereas the silty shale (D) and sandy

*Present address: INTEVEP S.A., Box 76343, Caracas 1070A, Venezuela.

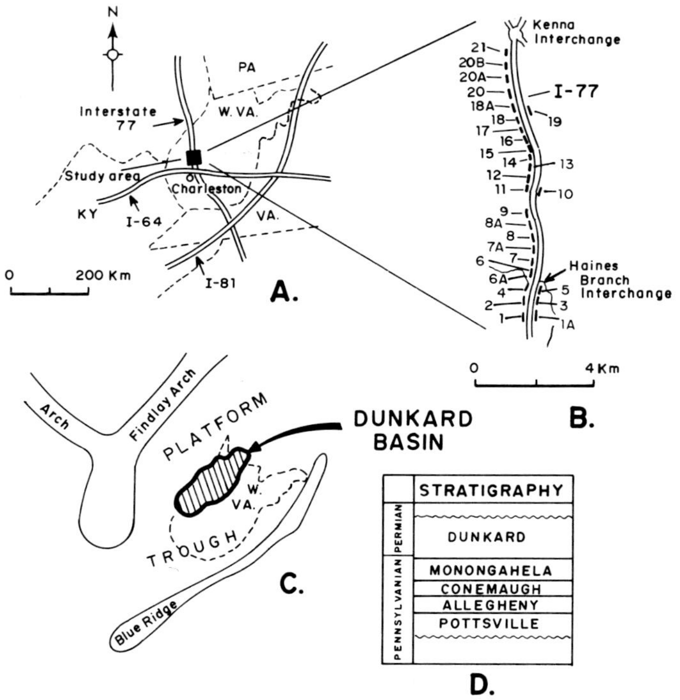

FIG. 1.—A: Location map of the study area. B: Detailed location of the roadcuts in the study area. C: Tectonic framework of the Dunkard Basin (based on Donaldson, 1974). D: Upper Paleozoic stratigraphy in the Dunkard Basin.

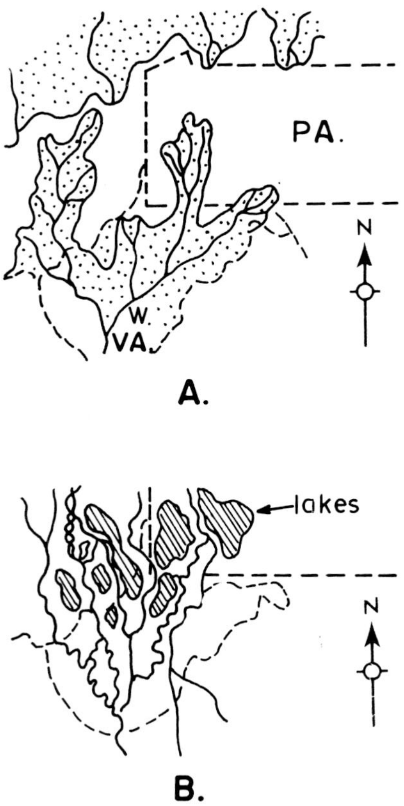

FIG. 2.—Paleogeographic map of West Virginia during Upper Paleozoic time. A: Upper Pennsylvanian and B: Permian (based on Donaldson, 1974).

siltstone (C) differ considerably from each other and also from the mud-clay rocks (E) in terms of physical properties and facies association.

Lithofacies A: Conglomerate

Lithofacies A consists of thin beds of conglomerate and pebbly sandstone overlying a finer substratum and underlying a fining-upward sandy sequence (Fig. 3A). Significant features include a sharp irregular erosional base, lenticular geometry, polymictic pebble composition (mud clasts, quartz and feldspar pebbles), coal and plant fragments, and poor sorting. Large-scale trough cross-strata characterize the pebbly sandstone units. Reactivation surfaces are very common and may reflect abrupt changes in river stage (Hobday and others, 1981).

The presence of a basal sharp erosional surface, lag deposits, abundant trough cross-strata and plant and coal fragments suggests a fluvial origin. Lateral and vertical association with coarse grained, fining-upward sandstones implies deposition as channel-lag facies in a point bar environment (Hobday, 1978).

Lithofacies B: Sandstones

The sandstones, which compose about 25 percent of the Monongahela-Dunkard strata, vary from very fine to coarse-grained and range from quartzose arenite to wacke variably cemented by calcite. Different physical parameters such as nature of basal contact, shape of sand body, vertical grain size trend, sedimentary structures and facies association, were utilized in the recognition of different types of sandstone.

Type 1.—

Type 1 sandstone is a medium- to coarse-grained arenite which overlies the conglomerate and pebbly sandstone facies (Lithofacies A). Type 1 sandstone along with the underlying conglomerate forms less than 30 percent of the total alluvial sandy facies. It shows an upward decrease in grain size as well as in scale of sedimentary structures. Large-scale cross-strata are abundant in the lower part, whereas small-scale cross-strata are characteristic of the upper part. Some low angle parallel lamination and cross-cutting scour surfaces also occur. The lower contact of this sandstone is generally sharp and planar, whereas the upper contact with overlying fine-grained cross-laminated sandstone is transitional. The general paleocurrent trend in this sandstone is shown in Figure 4A, which suggests a general north-northeasterly trend.

The characteristics of Type 1 sandstone, for example, the vertical decrease in grain size and sedimentary structures, and its stratigraphic position between pebbly channel lag facies and overlying finer sandstone and top-stratum deposits, are indicative of a fluvial point bar origin (Jackson, 1976; Levey, 1978; Bridge and Jarvis, 1976). The vertical stacking of several partially preserved fining-upward sequences, as shown in Figure 3A, may be due to superposition of point bar deposits within a meander belt complex (Jackson, 1978).

Figure 4A shows rather low variance in paleocurrent data

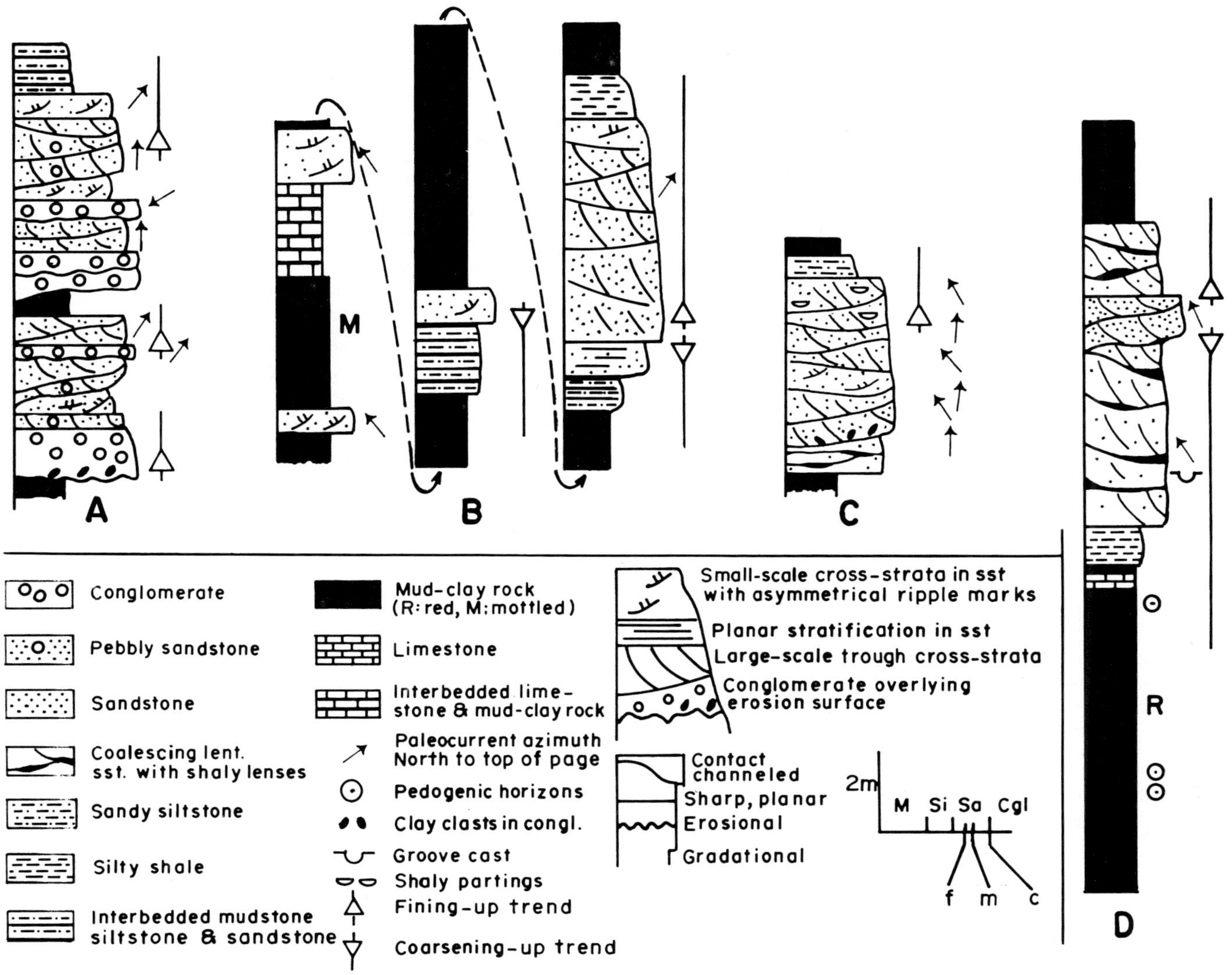

FIG. 3.—Examples of coarsening- and fining-upward sequences in the study area. A: Stacked fining-upward point bar sequence in roadcut #12. B: Coarsening-upward sequences in roadcut #2 due to levee-crevasse progradation on floodplain areas. A fining-upward point bar sequence overlies the upper coarsening-upward sequence. C: A crevasse channel sandstone (Type 3) in roadcut #6 overlying the coalescing lenticular sandstone unit (Type 4 sandstone). D: Development of a coarsening- and fining-upward couplet due to crevasse progradation and abandonment, respectively. Groove cast trend N40° W–S40° E.

for a typical point bar deposit. Such paleocurrent distribution is perhaps more typical of relatively straight braided bedload streams (Miall, 1978). The Monongahela-Dunkard paleochannels do not show structures of typical braided streams, however, nor do they show a high sand-mud ratio characteristic of braided low-sinuosity streams. Galloway (1981) reported similar low sinuosity but nonbraided stream deposits from the Gulf Coast Oligo-Miocene fluvial deposits.

Type 2.—

Type 2 sandstone is a thin, cross-laminated wacke commonly less than a meter thick. It is laterally persistent and contains silty and carbonaceous partings. It underlies mud-clay rocks of floodplain origin and overlies either Type 1 sandstone or more commonly silty levee rocks, interbedded facies and red mud-clay rocks of flood basin origin (Fig. 3B). Ripple-drift lamination with climbing sets of leeside lamina and eroded stoss slopes is abundant. Other sedimentary structures include small-scale slump structures, cosets of small-scale trough and tabular cross-strata (some with overturned foresets), and occasional flutes at the base. Texturally, it is an immature matrix-rich (as much as 20 percent) wacke with appreciable mica content. Burrows and root mottling are common, especially in finer mud drapes.

Type 2 sandstones show features typical of a distal splay or upper point bar deposits. The predominance of ripple lamination in the upper part of point bars is probably a result of gradual abandonment of channel by chute cutoff (Gersib and McCabe, 1981). As flow decreases, it forms a sequence of climbing and nonclimbing ripple lamination (Walker, 1976). Flutes in crevasse-splays at the base of this sandstone originated by the interaction of eddies within turbulent flow and the bottom sediment and scoured flute marks

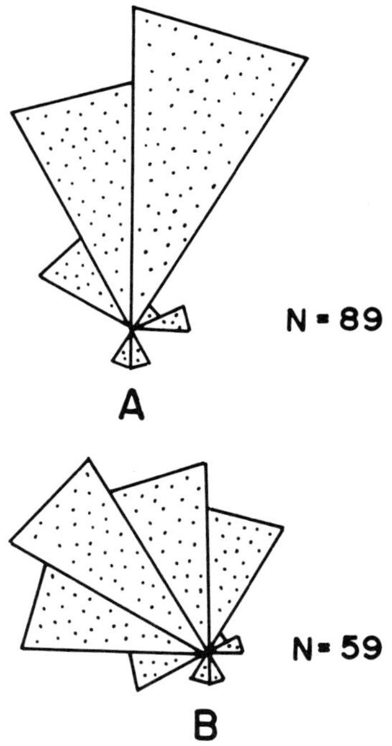

FIG. 4.—Paleocurrent rose diagrams based on cross-strata azimuth. A: Point bar deposits in roadcuts 11–18. B: Crevasse sandstones in roadcuts 18–21. North to the top of page.

on the muddy substratum (Leeder, 1982). This unit occurs in most of the cases as isolated sand bodies forming the distal equivalent of proximal crevasse channel sandstone facies (Type 3 sandstone).

Type 3.—

Type 3 sandstone from 1 to 5 m thick is a fine- to medium-grained unit with a sharp planar or channeled concave-up base and a planar or convex-top upper surface (Figs. 3C, 5A). Channel lag mud clasts are very common at the base (Fig. 3C). No fining-upward grain-size trend has been observed, though shaly intercalations are common toward the top. It overlies interbedded heterolithic levee-splay facies or splay lobe sandstones and underlies finer levee and flood basin facies. Sedimentary structures include small- and large-scale cross-strata, ripple lamination, parallel lamination and occasional root marks. No orderly variation of sedimentary structures has been noted, though burrows and root marks tend to occur near the top. A relatively immature texture, high matrix content (as much as 22 percent) and a low width-depth ratio (5 to 20) mark this sandstone. Measurement of trough cross-strata azimuth suggests a general northwesterly paleoccurent trend (Fig. 4B). Also, the trend of Type 3 sandstones (where exposed on both sides of the road, as in roadcuts 6 and 21) shows an apparent W-E to WNW-ESE orientation of the channels. The northwesterly paleocurrent trend in this sandstone shows relatively low divergence with the north-northeasterly paleocurrent trend in the channel belt facies (Fig. 4A).

Type 3 sandstone is interpreted to be a crevasse channel-fill deposit. Its occurrence above levee and splay lobe facies and below finer levee-floodplain deposits reinforces its crevasse channel origin (Hobday, 1978). Lack of a fining-upward trend implies low sinuosity of the paleostreams. A low width-depth ratio, lenticularity, irregular basal scour, and variability of sedimentary structures and facies thickness are diagnostic features of crevasse channel deposits (Galloway, 1981; Peterson and Tyler, 1985). Mud clasts at the base are of intraformational origin and are derived locally from the levee and floodplain sediments through which the channel was cut. Textural immaturity implies rapid sedimentation from mixed-load streams and minimum winnowing. Upward increase of shaly lenses, burrows and root traces may be due to progressive crevasse channel abandonment and waning of current energy.

Low divergence in paleocurrent trend between crevasse and main channel facies is not common, though it has been reported from the Rhone delta (Oomkens, 1970) and may be ascribed to the relatively low sinuosity of main channel facies (Galloway, 1981).

Type 4.—

Type 4 sandstone is a matrix-rich (as much as 23 percent), highly burrowed unit with multiple mutually erosive lenticular subunits separated by numerous shaly drapes (Figs. 3D, 5B, 5C). Individual subunits are impersistent and are typically less than 5 m wide and less than a meter thick. They either scour the underlying sandy subunits or silty and shaly interbeds and drapes. Internally many show trough cross-strata at the base and ripple lamination at the top. Calcareous concretions are common in the shaly interbeds. This facies normally underlies crevasse channel sandstone or floodplain mud-clay rocks and overlies silty levee and floodplain mud-clay rocks (Fig. 3D).

Type 4 sandstone shows features and facies association characteristic of crevasse-splay lobes (Coleman, 1969; Hobday, 1978; Tyler and Ethridge, 1983). This sandstone may represent the distal part of a splay lobe where large single channels of proximal area fan out into a maze of multiple, anastomosing smaller channels (Farrell, this volume). Frequent overflooding of main channels introduces sediments into the splay system and gives rise to the stacked and overlapping sand bodies. Stacking of subunits is evidently caused by overlapping lateral migration of individual splay channels. Pulsatory sedimentation is indicated by abundant clay drapes and lenses, reactivation and truncation surfaces, and waning flow sequences of primary sedimentary structures (Galloway, 1981; Peterson and Tyler, 1985). The calcareous concretions imply caliche formation during periods of little sedimentation (Leeder, 1975).

Lithofacies C: Sandy Siltstone

Lithofacies C occurs both as thin sheets (average 50 cm thick) and lenticular units. Lower contact with finer mud-clay rocks is generally sharp, planar and erosional. It grades upward into floodplain mud-clay rocks or discordantly underlies crevasse sandstones (Lithofacies B, Type 3 and 4 sandstones). Small-scale cross-strata, ripple- and parallel-lamination, and slump structures are characteristic. Burrows and carbonate concretions are also common.

The sandy siltstone units are probably distal offshoots of large crevasse channel facies. This lithofacies may also be

FIG. 5.—Examples of crevasse sandstones in the study area. a: View of a large crevasse channel in roadcut #6A. A low width-depth ratio and a lenticular external form characterize this sandstone. b: Exposure of coalescing sandstone of splay lobe origin (Type 4 sandstone) overlying mud-clay rocks of floodplain origin (covered by vegetation). c: Closeup view of the sandy subunits of the previous figure. Note small channel-form sandstone (A) scouring the underlying sandy subunit. The shaly interbeds (B) show truncation surfaces due to channel scouring. Thickness of the exposure is 3 m. d: Thin coarsening-upward sequence due to levee progradation. Mud-clay rocks (A) of floodplain origin are overlain successively by thinly interbedded siltstone and mud-clay rocks (B) and very fine sandstone (C) of distal crevasse origin (Type 2 sandstone). The exposure is 2.5 m thick.

of levee origin, especially when it is thinner and interbedded with silty shale. Its common occurrence within finer mud-clay rocks corroborates its distal splay origin (Horne, 1979). The sandy siltstone assumes a sheetlike form when the crevasse sediments approach the adjacent flood basins as unconfined sheet floods (McKee and others, 1967; Suttner and others, 1985). Abundance of ripple lamination may be due to the combined effects of currents and waves in the vicinity of floodplain lakes.

Lithofacies D: Silty Shale

Lithofacies D occurs as blanketlike thick beds or as thinner sheets and is characterized by calcareous concretions, burrows and root structures. It passes gradationally into mud-clay rocks of flood basin origin. Contact with sandy siltstone or interbedded heterolithic facies is gradational or erosional when they succeed and gradational when they precede.

Apparently, the silty shale represents levee deposits. Upward transition into heterolithic facies implies superposition of levee-splay sequence above uniform levee deposits (Cavaroc and Saxena, 1979; Horne, 1979). Lateral facies transition into floodplain mud-clay rocks suggests its origin in a low-energy overbank depositional environment such as that of a distal levee.

Lithofacies E: Mud-Clay Rocks

Lithofacies E, as much as 17 m thick, is persistent laterally and traceable for 5 km along the road section. It is the most abundant of all lithofacies in the area (30 percent). It is mostly red or mottled in color and may contain plant fossils or comminuted plant detritus. The mud-clay rocks are closely associated with sandstone (Lithofacies 2, 3 and 4) and interbedded heterolithic facies of levee-distal splay origin. Interbedding between mud-clay rocks and sandy siltstone lithofacies is also common. In some places the mud-clay rocks are multistoried in which vertical burrows and caliche horizons are abundant toward the top of an individual story (Fig. 3D). Associated variant lithologies include thin and impersistent coal and carbonaceous shaly lenses.

Lateral and vertical association, geometry, predominant red color, extensive pedogenic horizons, as well as scarcity of coal and carbonaceous shales imply a flood basin origin for the mud-clay rocks in a well drained oxygenated environment. The pedogenic caliche horizons developed during periods of little sedimentation and subsidence. Similar calcareous concretions have been found in many ancient and Recent caliche (Friend and Moody-Stuart, 1970; Allen, 1974; Leeder, 1982) within red mudrock units of alluvial origin. Extensive burrowing near the top of individual sedimentation units made the mudrocks sufficiently porous to facilitate formation of concretions.

Thick mud-clay rock units in the study area imply copious sedimentation; at the same time, formation of extensive soil horizons suggests rather slow subsidence. This may indicate strong seasonality of Dunkard climate (Allen, 1974). Vertisols in Monongahela-Dunkard floodplains may be related to low rates of channel migration and relatively high rates of suspended sediment deposition (Blodgett, 1985). The lower proportion of channel belt facies (<10 percent) in the overall sequence is compatible with thick floodplain deposits. Relatively rapid formation of vertisol horizons (in $<10^2$ yrs, according to Blodgett, 1985) and their higher resistence to subsequent erosion helped preserve the thick flood basin sequence in spite of rather slow subsidence.

Lithofacies F: Interbedded Heterolithic Unit

The interbedded mudstone, siltstone and sandstone unit (maximum of 6 m thick) is persistent as a unit, though subunits are variable in thickness laterally. Cross-lamination, asymmetric ripples and microslump structures are characteristic of the sandy and silty layers, whereas parallel lamination, burrows, root marks and calcareous concretions are typical of the finer mudrock subunits. This lithofacies is normally overlain by the sandstones of Type 2, 3 and 4, and underlain by silty shale and mud-clay rocks of levee and flood basin origin, respectively (Fig. 3B).

The interbedded heterolithic unit probably represents levee and minor distal splay deposits, as indicated by the abundant burrow and root structures, alternating laminae of mainly sandy silt and mud-clay rocks, small-scale hydroplastic deformation structures and abundant calcareous concretions (Coleman, 1969; Ethridge and others, 1981; Suttner and others, 1985). Friend and Moody-Stuart (1970) interpreted limestone concretions in a similar setting from the Devonian rocks of Spitsburgen as a characteristic of ancient levee deposits. Levee sediments should be particularly suitable for concretion development because leaching and precipitation would be most intense in well drained areas above mean annual water table. Also, extensive burrows and rootlets acted as pathways for movement of calcareous solutions (Ethridge and others, 1981).

Lithofacies G: Limestone

Lithofacies G, intimately associated with mud-clay rocks (Lithofacies E), is as much as 3 m thick and traceable for nearly 2 km along the road section (Fig. 6). Texturally, it is an intrapelmicrite with intraclasts and pellets forming more than 50 percent of the rock. Ostracodes, fish scales, and vertebrate fossils are common, though their concentrations vary laterally. The limestone is nodular with shaly drapes. Local variants of the nodular type include an aphanitic lime mudstone with an intraclast-rich top part, and a thinly interbedded limestone and mudrock unit. The interbedded limestone and mudrock subfacies is as much as 5 m thick and consists of alternate beds of nodular, blocky-fracturing limestone (as much as 20 cm thick) and mottled reddish-green concretionary mudstone (10–20 cm thick). Tepee-like fold structures are also common (Leeder, 1982).

The limestones owe their origin to chemical precipitation in freshwater ephemeral lakes (Russell, 1967; Flores, 1981). Patchy fossil distribution implies deposition in poorly connected lakes in the deeper parts of the floodplain. Abundance of intraclasts suggests shrinkage due to desiccation. Interbedding of nodular limestone and red or mottled mud-clay rocks is thought to be due to alternate chemical and detrital sedimentation. Oscillation of water level in the lakes, due to compaction, subsidence, climatic variation or fluctuation in river flood stage, could have caused the interbedding of mud-clay rocks and limestone. Vertebrate fossils and carbonaceous detritus are more common at the lateral extremes of limestone beds, implying proximity to paleoshoreline. Figure 6 shows a reconstruction of the extent and geometry of one such floodplain lake, based on closely spaced roadcuts over a distance of nearly 2 km.

LITHOFACIES SEQUENCE

The lithofacies sequence was statistically determined using Markov chain analysis (Miall, 1973). The basic step in the analysis is reconstruction of a transition count matrix which tabulates observed lithic transitions in the sequence (Table 1). Data in the transition count matrix are then processed into a separate transition probability matrix, independent trials matrix, and difference matrix (see Gingerich, 1969 for details). Positive values in the difference matrix identify transitions which have occurred more frequently than by chance associations. Linking of the lithofacies with positive residues permits identification of repetitive sequential patterns (Fig. 7). Also, positive residues define significant asociations that imply close genetic relationship of lithofacies.

Results

Figure 7 shows obvious pattern groupings into fining-upward, coarsening-upward, and two-way transitions between

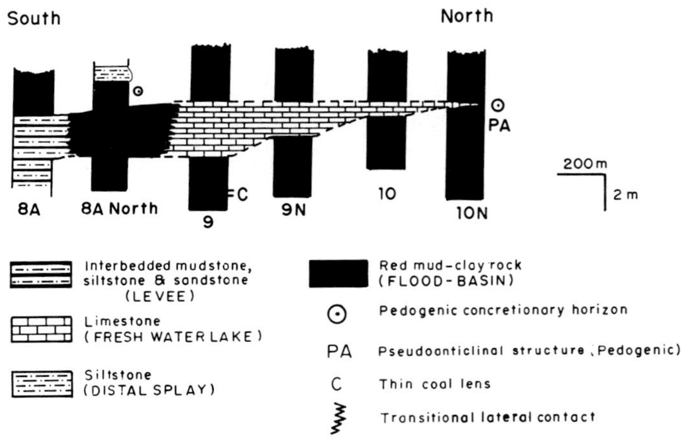

FIG. 6.—Reconstruction of the geometry of a floodplain lake through 6 roadcuts (8A to 10N). Width of section is approximately 2 km.

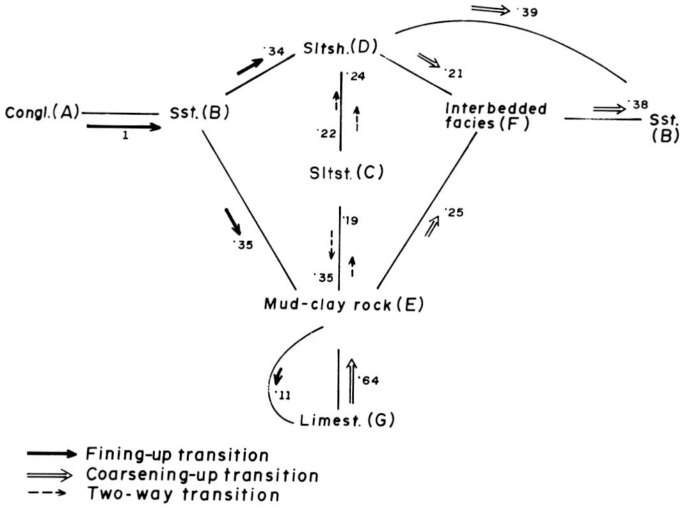

FIG. 7.—Lithic transition diagram for the Monongahela-Dunkard strata. The numbers indicate values in the transition probability matrix corresponding to positive residues in the difference matrix. Arrow length is proportional to transition weight.

several lithofacies. The various fining-upward, coarsening-upward and two-way lithic transitions (cited in order of decreasing transition weight in each class) are as follows:

(1) fining-upward sequence: conglomerate to sandstone (AB), sandstone to mud-clay rocks (BE), sandy siltstone to mud-clay rocks (CE), sandstone to silty shale (BD).
(2) coarsening-upward sequence: limestone to mud-clay rocks (GE), silty shale to sandstone (DB), interbedded facies to sandstone (FB), mud-clay rocks to interbedded facies (EF), silty shale to interbedded facies (DF).
(3) two-way transition pattern: sandy siltstone and silty shale (CD), sandy siltstone and mud-clay rocks (CE).

Fining-upward sequence.—

The fining-upward sequence either involves fining-upward point bar sandstones (Lithofacies A, and Type 1 and 2 sandstones of Lithofacies B) and associated top stratum deposits or alternatively consists of levee-splay silts (Lithofacies C and D) and flood basin muds (E) overlying sands of crevasse origin (Type 2, 3 and 4 sandstones of Lithofacies B).

In the channel belt assemblage ABE type, a fining-upward motif has developed. The conglomerate-to-sandstone (Type 1 of Lithofacies B) transition is the strongest of all fining-upward transitions and is due to transition from channel lag deposits to coarse-grained trough cross-stratified sandstone. Figure 3A shows typical fining-upward sequences related to point bar and associated finer deposits. The multistoried and amalgamated channel belt sequence is laterally traceable over a length of approximately 2 km through nine adjacent roadcuts. Superposition of the cycles (some partly eroded) represents stacking within a meander belt complex of mixed-load fluvial systems (Puigdefabregas and Van Vliet, 1978; Galloway, 1981).

In the Monongahela-Dunkard strata, the proportion of channel belt sandstone is less than 10 percent of the total exposed section. Thus, paucity of point bar sandstones precludes a channel belt origin of most of the fining-upward transitions. These probably originated from deposition of levee silts (Lithofacies C and D) and floodplain muds (E) above crevasse sandstones (channel or splay lobe facies Type 2, 3 and 4 sandstones) due to progressive abandonment of individual splay lobes (Cavaroc and Saxena, 1979). The crevasse-related fining-upward sequences invariably lack the basal channel lag conglomerate. Figure 3C shows an example of a fining-upward sequence of crevasse origin. The crevasse channel facies has a lenticular shape, a low width-depth ratio (Fig. 5A) and resembles the cross-stratified and massive amalgamated channel-form sandstone (Facies B) described by Cross and Smith (1985). The fining-upward sequence in the crevasse facies does not involve an upward decrease in sand size probably because of relatively low sinuosity of the semipermanent crevasse channels (Elliot, 1974). Shaly intercalations, however, have been found to increase in the upper part of the channel facies (Fig. 3C). Accretion in these channels took place by vertical aggradation as well as by bedload transport (Flores, 1981).

In summary, the fining-upward sequences are produced by a) lateral accretion in the channel belt environment, or b) initial crevassing and subsequent abandonment of crevasse channels leading to subsidence and silting by levee and flood basin fines (Gersib and McCabe, 1981; Galloway, 1981). Likewise, abandonment of a splay lobe (Type 4 sandstone of Lithofacies 2) in favor of an adjacent one will also generate a similar sequence (Bridge, 1984).

Coarsening-upward sequence.—

Among the coarsening-upward patterns, the limestone- to mud-clay-rock transition is the strongest, followed by the transitions from silty shale (D) and interbedded heterolithic facies (F) to sandstones (Types 2, 3 and 4). Upward transitions from mud-clay rocks (E) and silty shale (D) to interbedded facies (F) are also significant, though not to the extent as the other transitions.

The freshwater limestones represent deposition in the deeper parts of the flood basin during periods of little detrital influx. When clastic sedimentation renews, mud-clay sediments will be the most likely to reach the distal lake area. The silty shale to sandstone transition (DB) probably represents crevasse deposition on levee tops in areas proximal to the main channel belt. The interbedded facies-to-sandstone (Type 2, 3 and 4) transition (FB), on the other

TABLE 1.—TRANSITION COUNT MATRIX

	A	B	C	D	E	F	G
A	0	19	0	0	0	0	0
B	7	0	9	41	42	20	0
C	1	13	0	13	19	7	2
D	2	35	20	0	12	19	1
E	3	22	19	21	0	26	11
F	6	27	7	11	21	0	0
G	0	3	0	2	9	0	0
Sums	19	119	55	88	103	72	14

A: Conglomerate; B: Sandstone; C: Sandy siltstone; D: Silty shale; E: Mud-clay rocks; F: Interbedded heterolithic facies; G: Limestone.

hand, may represent deposition of crevasse channel and splay lobe sandstones on top of distal splay or levee deposits which, in turn, prograded on the floodplain muddy substratum (ED and EF transitions). Internally, the interbedded facies (F) may show an upward-coarsening trend with progressive increase in silt and sand content (Horne, 1979). Sandstones of splay lobe origin (Type 4 sandstone) may show a gradational contact with the underlying heterolithic unit (F).

Lithic transitions in the study area clearly show that coarsening-upward sequences develop exclusively in overbank sediments. Rarely a fining-upward channel belt sandstone may overlie the coarsening-upward sequence of overbank origin (Fig. 3B), probably due to avulsion along a pre-existing crevasse route. Some coarsening-upward sequences show a complete gradation from lacustrine limestone at the base successively through flood basin mud-clay rocks, levee facies, splay lobe facies to crevasse channel sandstone facies (Fig. 3D). The overlying fining-upward cycle in Figure 3D implies abandonment of crevasse channel and reversion of depositional regime back to the original splay lobe domain. In some places levee progradation toward flood basin areas also gives rise to coarsening-upward cycles topped by a thin sheetlike distal crevasse sandstone (Lithofacies B, Type 2), as shown in Figure 5D.

In summary, the coarsening-upward sequences are produced mainly by crevassing of levees and consequent transfer of trunk channel sand load into flood basin areas. Crevasse facies commonly form a progradational vertical sequence grading from massive flood plain mudstone upward into interbedded heterolithic levee and distal crevasse beds, capped in turn by lenticular or thick-bedded proximal crevasse units (Galloway, 1981). Such gradational coarsening-upward (and also fining-upward) sequences reflect progressive levee and crevasse progradation (and abandonment) while the main channel belt remains unchanged in position (Bridge, 1984). Thickness variation in successive cycles may reflect superposition of proximal and distal sequences due to avulsion and consequently, a shift in the point of sediment supply to the crevasse (i.e., the channel belt).

Two-way transitions.—

A two-way transition pattern or interbedding between several finer facies is noteworthy (Fig. 7). The sandy siltstone facies (C) shows a two-way transition pattern with silty shale (D) and, to a lesser extent, with mud-clay rocks (E). These alternations are interpreted as due to interbedding of distal crevasse-and-levee floodplain deposits. The sandy siltstones may represent suspension load deposition of closing stages of crevasse splays in distal areas (Bridge, 1984). Termination of crevasse sedimentation coincides with renewed flood basin mud-clay deposition, thus giving rise to the facies alternations.

SEDIMENTOLOGIC SYNTHESIS

Table 2 compares and contrasts salient features of the channel belt and crevasse deposits in relation to their geometry, facies and internal organization. These criteria could be used successfully to distinguish between the two facies. Such an analysis is necessary, as some thick coarsening-upward overbank sandstones may superficially resemble channel belt deposits (Collinson, 1978; Bridge, 1984).

The channel belt sandstones are characterized by: a tabular external form, an erosive basal contact, a relatively high width-depth ratio, abundant polymictic channel lag conglomerate, abundant large-scale trough cross-strata with foresets commonly marked by aligned pebbles, notable upward decrease in grain size and scale of sedimentary structures, ubiquitous plant detritus, and coaly stringers. The general paleocurrent trend is north-northeasterly. Longitudinal facies variation is negligible compared to that in the overbank facies. Vertically stacked and amalgamated sequences of fining-upward cycles imply development within a meander belt complex.

Thick, laterally persistent red mud-clay rocks form the framework within which the crevasse deposits occur. The argillaceous, fine-grained, texturally immature crevasse channel sandstones are much more abundant than the channel belt facies. Characteristic features of crevasse sandstones include: symmetrical lenticular external form, absence of an upward decrease in grain size, and a rather low width-depth ratio. Mud clasts of intraformational origin, and flutes, load-and-groove casts characterize the scoured base

TABLE 2.—CHARACTERISTIC FEATURES OF CHANNEL BELT AND CREVASSE SANDSTONE

Feature	Channel-belt	Crevasse
Frequence of occurrence	Less common	Abundant (more than 70% of all sandstones)
External Form	Tabular	Mostly symmetrical, concave-up base; top planar or convex
Thickness	Multistoried, as much as 15 m	Channel facies 5 m or less; splay lobe facies 3 to 4 m thick (average)
Width-depth ratio	Greater than 40	Very low, between 5 and 20
Basal contact	Erosional scoured	Erosional scoured or planar
Channel lag deposit	Well developed; polymictic	Only scattered mud clasts
Grain size	Pebbly coarse to fine sand (Type 1 and 2)	Fine to very fine silty sand (Type 2, 3 and 4 sandstones)
Grain size trend	Sandstone fines upward	Sandstones do not fine upward; lateral fining observed in some beds
Texture	Better sorted	Poorly sorted, micaceous and matrix-rich (20%)
Sedimentary structures	Large trough cross-strata at base & ripple lamination at top	Mostly medium-scale trough cross-strata and ripple lamination. Some flutes, load-and-groove casts at base
Bioturbation	Not common	Vertical traces very common
Nodules	Siderite and pyrite nodules in some beds	Calcareous nodules very common
Paleocurrent azimuth	N to N20° E	N40° W; Paleochannel trend N70° W–S70° E
Longitudinal facies variation	Minimum variation	Proximal to distal variation notable; % of thinner beds, shale and matrix increases distally
Sequence association	Fines upward due to superposition of F.U. channel sand by top-stratum deposits	Both F.U. (due to splay lobe abandonment) and C.U. (due to progradation toward floodplain) sequences

of the crevasse channel facies. No orderly vertical variation of sedimentary structures is observed. Longitudinal facies variation, however, is noteworthy: proximal (with respect to channel belt) sandstones are larger in dimension and relatively coarser (fine- to medium-grained). The distal equivalents of splay lobe facies, on the other hand, consist of: 1) composite depositional units of thinner and lenticular, mutually overlapping very fine sandy subunits separated by shaly lenses and drapes, 2) relatively thin ripple-laminated and parallel-bedded very fine-grained sandstones, and 3) interbedded heterolithic facies. Increase of matrix content, shaly interbeds, and thinner beds including silty sheetflood deposits typify the distal splay facies which may occur totally enclosed in floodplain sediments.

The nature of the lithofacies sequence is strongly controlled by the processes operating within individual alluvial subenvironments. In the channel belt facies, lateral accretion gives rise to fining-upward sequences, whereas in the overbank environment levee and crevasse progradation typically results in coarsening-upward cycles. Likewise, progressive abandonment of crevasse lobes give rise to fining-upward cycles in overbank environments. Thickness of cycles (and therefore of the constituent lithofacies) in overbank sequences appears to diminish progressively with increasing distance from the channel belt. Thus, contiguous but dissimilarly thick cycles may result from superposition of distal and proximal overbank sequences.

REGIONAL IMPLICATIONS

It is worthwhile to compare the style of cyclicity and facies variation between the northern and southern parts of the Dunkard Basin. Cyclothemic development in the southern area is not as complete as in the northern Dunkard area (Cross and Schemel, 1956; Beerbower, 1961). The northern Dunkard area is characterized by thick coals, numerous limestones (some of brackish origin), pervasive gray color of mudrocks, and scarcity of paleosol horizons and oxidized red color (Table 3). All these probably imply that the alluvial southern Dunkard Basin was relatively higher (and drier) than the low, less oxidizing, extensively lacustrine environment of the northern part.

The ideal cyclothem in the northern Dunkard area consists of an asymmetrical pair of coarsening- and fining-upward sequences (Cross and Schemel, 1956; Beerbower, 1961; Fig. 1). The basal coarsening-upward sequence consists of: coal (1), limestone (2), carbonaceous and clay shale (3 & 4), silty shale (5), siltstone (6), and sandstone (7). The overlying fining-upward sequence on top of the channel sandstone (7) includes: siltstone (8), claystone/limestone (9 & 10), and under clay (11). Beerbower (1961) interpreted the basal assemblage of coal, limestone and shales (1–4) as swamp to brackish in origin and the overlying silty facies assemblage (5–6) as delta-lacustrine in origin. The channel belt facies (7) and overlying siltstones (8) represent delta distributary, alluvial channels and associated framework deposits. The overlying claystone and limestone (9 and 10) are of pond and lake origin in a predominantly alluvial setting (Beerbower, 1961). The facies sequence within the northern cyclothem suggests that the environment shifted from brackish water deltaic conditions to a progressively more alluvial environment. In the northern Dunkard area the main channel facies dominates cyclothem development. In contrast, cyclicity in the southern Dunkard area is of alluvial plain origin and developed extensively in overbank facies due to predominant vertical accretion within prograding crevasse systems. Thus, the nature of rhythmic fluctuations in the depositional facies (i.e., cyclothems) in the two areas is different because of regional variations of environments (i.e., facies).

TABLE 3.—DISTINCTIVE FEATURES IN NORTHERN AND SOUTHERN DUNKARD BASIN

Feature	Southern Dunkard	Northern Dunkard*
Channel sandstone	Fluvial mixed-load type	Mainly mixed-load distributary type
Pebble composition	Polymictic	Mostly clay clasts & limestone nodules
Overbank sandstone	Abundant (crevasse type)	Less common
Coal	Rare thin lenses of drifted origin	Thickest and most numerous; both in situ and drifted origin
Limestone	Nodular & impersistent	Thicker and more numerous
Typical fossils	Vertebrates & ostracodes	*Spirorbis*, ostracodes, gastropods and brackish water brachiopods (lingulid)
Color of mudrocks	Mostly red	Mostly gray and green
Paleosol horizons	Abundant	Not common
Paleoenvironment	Alluvial plain	Fluvio-deltaic; lake deposits abundant
Cyclicity	C.U. cycles in overbank (crevasse-levee) facies abundant	Channel facies central to cyclothem (a pair of C.U. and F.U. sequences)

*Data on northern Dunkard mostly summarized from Beerbower (1961).

CONCLUSIONS

The Monongahela-Dunkard sediments in the southern Dunkard area appear to be entirely of alluvial origin, in contrast to the more deltaic northern area. The lithofacies sequence and association suggest that the study area formed part of a northward-flowing fluvial system. In this fluvial environment thick overbank facies of red floodplain and sandy crevasse deposits dominated over the main channel deposits. Dominance of the former probably implies channel entrenchment and vertical stacking, probably related to climate and discharge factors. Seasonal excess supply was accommodated by continuous levee-crevasse progradation from fixed sources in the main channel. Stabilization of channel banks was probably accomplished through early pedogenesis in a strongly seasonal climate.

Statistical analysis of the lithofacies sequence reveals cyclicity (both coarsening- and fining-upward) and preferential two-way transition behavior among certain lithofacies. As each lithofacies in a cycle is the product of a particular depositional environment, the Dunkard sediments require a cyclic depositional mechanism. Autocyclic causes, such as crevassing and stream avulsion, generated the cyclic deposits within a framework of channel belt, levee-splay, floodplain and lake environments.

ACKNOWLEDGMENTS

The author expresses his gratitude to Drs. J. R. Beerbower, Paul Enos, A. C. Donaldson and I. Karcz, who visited the field area and along with Drs. E. Cotter and J. Sorauff provided valuable suggestions on the original thesis manuscript. A. Isea of Intevep, Venezuela made helpful suggestions on an earlier version of this manuscript. Particular mention should be made of Dr. J. Schmitt of Montana State University and an anonymous reviewer whose comments and suggestions have significantly improved the manuscript. The author, however, accepts full responsibility for the opinions expressed in this paper.

REFERENCES

ALLEN, J. R., 1974, Studies in fluviatile sedimentation: Implication of pedogenic carbonate units, Lower Old Red sandstone, Anglo-Welsh Outcrop: Geological Journal, v. 9, p. 181–208.

BEERBOWER, J. R., 1961, Origin of cyclothems of the Dunkard Group (Upper Pennsylvanian-Lower Permian) in Pennsylvania, West Virginia, and Ohio: Geological Society of America Bulletin, v. 72, p. 1029–1050.

———, 1969, Interpretation of cyclic Permo-Carboniferous deposition in alluvial plain sediments in West Virginia: Geological Society of America Bulletin, v. 80, p. 1843–1848.

BLATT, H., MIDDLETON, G. V., AND MURRAY, R., 1980, Origin of Sedimentary Rocks: New Jersey, Prentice-Hall, p. 782.

BLODGETT, R. H., 1985, Paleovertisols—Their utility in reconstructing ancient fluvial floodplain sequences (abs.): Third International Fluvial Sedimentology Conference, Fort Collins, Colorado, p. 10.

BRIDGE, JOHN, 1984, Large scale facies sequences in alluvial overbank deposits: Journal of Sedimentary Petrology, v. 54, p. 583–588.

———, AND JARVIS, J., 1976, Flow and sedimentary processes in the meandering River Esk, Glen Cova, Scotland: Earth Surface Processes, v. 1, p. 303–336.

CAVAROC, V. V., AND SAXENA, R. S., 1979, Crevasse splay deposits in the State Line area of northeast Pennsylvania and northeastern Ohio, *in* Ferm, J. C., and Horne, J. C., eds., Carboniferous Depositional Environments in the Appalachian Region: University of South Carolina, Columbia, S.C., p. 160–172.

COLEMAN, J. M., 1969, Brahmaputra River: Channel processes and sedimentation: Sedimentary Geology, v. 3, p. 129–239.

COLLINSON, J. D., 1978, Vertical sequence and sand body shape, *in* Miall, A. D., ed., Fluvial Sedimentology: Canadian Society of Petroleum Geologists Memoir 5, p. 577–586.

CROSS, A. T., 1972, The Dunkard in perspective: Geology, sedimentation and life, *in* Burlow, J. A., ed., I. C. White Memorial Symposium—The Age of the Dunkard: West Virginia Geological Survey Publications, p. 21–27.

———, AND SCHEMEL, M. P., 1956, Geological and Economic Resources of the Ohio River Valley in West Virginia: West Virginia Geological Survey, v. 22, p. 1–149.

CROSS, T. A., AND SMITH, N. D., 1985, Comparative anatomy of contemporary and ancient crevasse splay and anastomosed crevasse channel complexes (abs.): Third International Fluvial Sedimentology Conference, Fort Collins, Colorado, p. 14.

DONALDSON, A. C., 1974, Pennsylvanian sedimentation of central Appalachians, *in* Geological Society of America Special Paper 148, p. 47–78.

ELLIOT, T., 1974, Interdistributary bay sequences and their genesis: Sedimentology, v. 21, p. 611–622.

ETHRIDGE, F. G., JACKSON, T. J., AND YOUNGBERG, A. D., 1981, Floodbasin sequence of a fine grained meanderbelt subsystem: The coal-bearing Lower Wasatch and Upper Fort Union Formations, Southern Powder River Basin, Wyoming, *in* Ethridge, F. G., and Flores, R. M., eds., Recent and Ancient Nonmarine Depositional Environments: Models for Exploration: Society of Economic Paleontologists and Mineralogists Special Publication 31, p. 191–212.

FLORES, R. M., 1981, Coal deposition in fluvial paleoenvironments of the Paleocene Tongue River Member of the Fort Union Formation, Powder River area, Wyoming and Montana, *in* Ethridge, F. G., and Flores, R. M., Recent and Ancient Nonmarine Depositional Environments: Models for Exploration: Society of Economic Paleontologists and Mineralogists Special Publication 31, p. 169–190.

FRIEND, P. F., AND MOODY-STUART, M., 1970, Carbonate deposition on the river floodplains of the Wood Bay Formation (Devonian) of Spitsburgen: Geological Magazine, v. 107, p. 181–191.

GALLOWAY, W. E., 1981, Depositional architecture of Cenozoic Gulf coastal plain fluvial systems, *in* Ethridge, F. G., and Flores, R. M., eds., Recent and Ancient Nonmarine Depositional Environments: Models for Exploration: Society of Economic Paleontologists and Mineralogists Special Publication 31, p. 127–155.

GERSIB, G. A., AND MCCABE, P. J., 1981, Continental coal-bearing sediments of the Port Hood Formation (Carboniferous), Cape Linzee, Nova Scotia, Canada, *in* Ethridge, F. G., and Flores, R. M., eds., Recent and Ancient Nonmarine Depositional Environments: Models for Exploration: Society of Economic Paleontologists and Mineralogists Special Publication 31, p. 95–108.

GINGERICH, P. D., 1969, Markov analysis of cyclic alluvial sediments: Journal of Geology, v. 39, p. 330–332.

HOBDAY, D. K., 1978, Fluvial deposits of the Ecca and Beaufort Groups in the Eastern Karoo Basin, Southern Africa, *in* Miall, A. D., ed., Fluvial Sedimentology: Canadian Society of Petroleum Geologists Memoir 5, p. 413–430.

———, WOODRUFF, C. M., JR., AND MCBRIDE, M. W., 1981, Paleogeographic and structural constrasts on nonmarine sedimentation of the Lower Cretaceous Antlers Formation and correlatives, North Texas and southeastern Oklahoma, *in* Ethridge, F. G., and Flores, R. M., eds., Recent and Ancient Nonmarine Depositional Environments: Models for Exploration: Society of Economic Paleontologists and Mineralogists Special Publication 31, p. 71–81.

HORNE, J. C., 1979, Criteria for the recognition of depositional environments, *in* Ferm, J. C., and Horne, J. C., eds., Carboniferous Depositional Environments in the Appalachian Region: University of South Carolina, Columbia, South Carolina, p. 295–300.

JACKSON, R. G., II, 1976, Depositional models of point bars in the Lower Wabash River: Journal of Sedimentary Petrology, v. 46, p. 579–594.

———, 1978, Preliminary evaluation of lithofacies models for meandering alluvial streams, *in* Miall, A. D., ed., Fluvial Sedimentology: Canadian Society of Petroleum Geologists Memoir 5, p. 543–576.

LEEDER, M. R., 1975, Pedogenic carbonate and flood-plain sediment accretion rates: A quantitative model for alluvial arid zone lithofacies: Geological Magazine, v. 112, p. 257–270.

———, 1982, Sedimentology: Processes and Product: George Allen and Unwin, Boston, 394 p.

LEVEY, R. A., 1978, Bed form distribution and internal stratification of coarse grained point bars, Upper Congaree River, S. Carolina, *in* Miall, A. D., ed., Fluvial Sedimentology: Canadian Society of Petroleum Geologists Memoir 5, p. 105–127.

MCKEE, E. D., CROSBY, E. B., AND BERRYHILL, H. L., JR., 1967, Flood deposits, Bijou Creek, Colorado, June 1965: Journal of Sedimentary Petrology, v. 37, p. 829–851.

MIALL, A. D., 1973, Markov chain analysis applied to an ancient alluvial plain sedimentation: Sedimentology, v. 20, p. 347–364.

———, 1978, Lithofacies types and vertical profile models in braided river deposits: A summary, *in* Miall, A. D., ed., Fluvial Sedimentology: Canadian Society of Petroleum Geologists Memoir 5, p. 597–604.

OOMKENS, E., 1970, Depositional sequences and sand distribution in the post-glacial Rhone delta complex, *in* Morgan, J., ed., Deltaic Sedimentation: Society of Economic Paleontologists and Mineralogists Special Publication, 15, p. 198–212.

PETERSON, F., AND TYLER, NOEL, 1985, Field Guide to the Upper Salt Wash alluvial complex, *in* Flores, R. M., and Harvey, M., eds., Field Guidebook to Modern and Ancient Fluvial Systems in the United States: Third International Fluvial Sedimentology Conference, Fort Collins, Colorado, p. 45–64.

PUIGDEFABREGAS, C., AND VAN VLIET, A., 1978, Meandering stream deposits from the Tertiary of the Southern Pyrenees, *in* Miall, A. D., Fluvial Sedimentology: Canadian Society of Petroleum Geologists Memoir 5, p. 469–486.

RUSSELL, R. J., 1967, River plains and sea coasts: University of California Press, Los Angeles, 164 p.

SUTTNER, L. J., LANGFORD, R. D., AND MACK, G. H., 1985, Guide to the field study of alluvial fan and fan-delta deposits in the Fountain Formation (Pennsylvanian-Permian), Colorado, *in* Flores, R. M., and Harvey, M., eds., Field Guidebook to Modern and Ancient Fluvial Systems in the United States: Third International Fluvial Sedimentology Conference, Fort Collins, Colorado, p. 65–86.

TYLER, NOEL, AND ETHRIDGE, F. G., 1983, Depositional setting of the Salt Wash Member of the Morrison Formation, southwest Colorado: Journal of Sedimentary Petrology, v. 53, p. 67–82.

WALKER, R. G., 1976, Facies models 3, sandy fluvial systems: Geoscience Canada v. 3, p. 101–109.

PART V
TECTONIC INFLUENCES

ACTIVE TECTONIC CONTROL ON ALLUVIAL ARCHITECTURE

JAN ALEXANDER[1] AND MIKE R. LEEDER
Department of Earth Sciences, Leeds University, LS2 9JT England

ABSTRACT: Tectonic control of alluvial architecture is commonplace in extensional, transcurrent and compressional tectonic terrains. The primary influence of tectonic activity on floodplain behavior is tilting, which varies areally and temporally and is superimposed on the pre-existing channel gradient. These effects are considered theoretically and illustrated by Pleistocene and Holocene examples from southwest Montana and the Mississippi Valley.

A transverse slope will cause the movement of channel belts toward the area of maximum subsidence either 1) by processes of downslope cutoff and preferential erosion, producing an abnormally wide and asymmetric sandstone body, or 2) in steps, by avulsion. Avulsion may be triggered by tilting or may occur subsequently through preferential flooding of the down-tilted side of the floodplain. Although the channel will move toward the position of maximum subsidence, it will often be offset by fans constructed from the footwall or it may flow into a lake occupying the floodplain low. Differential subsidence strongly affects deposition rates, groundwater behavior, pedogenesis and flood distribution, in addition to channel migration.

INTRODUCTION

The existence of a tectonic control on alluvial sedimentation may be observed in many tectonic settings. Miall (1981) observed that rivers tend to run either parallel or perpendicular to the structural strike and he defined nine hypothetical, qualitative basin-fill models based on this idea and the style of coastal environment. These essentially 'static' models are useful in identifying the broad setting of alluvial deposits, but they cannot shed further light upon the kinematic evolution of alluvial sedimentation resulting from tectonic subsidence. As yet, field studies of tectonosedimentary interactions and their implications for alluvial stratigraphy are few in number.

Tectonic control of alluvial architecture was first considered in a quantitative theoretical manner by Bridge and Leeder (1979) who considered the effects of floodplain tilting in graben and half-graben structures. Tilting was assumed to partly control the position of a low-seeking channel after avulsion. In their computed flow-normal sections, Bridge and Leeder found a remarkable spatial control of channel sandbodies in the half-graben with axial drainage. Recently one of us (JA) extended these conclusions in three-dimensional model experiments involving downvalley changes in subsidence magnitude (Fig. 1). The importance of tilting to facies distribution and sandbody interconnectedness was demonstrated by comparing interconnectedness in simulated models with the same sedimentation rate and channel dimensions and by varying the flow-normal tilt rate.

Since Bridge and Leeder's paper, it has been widely recognized that the half-graben tectonic style is widespread in extensional terrains (see Jackson and McKenzie, 1983: Ziegler, 1983). As discussed later, however, asymmetric subsidence is also produced during the formation of both strike-slip and foreland basins.

Tectonic uplift and subsidence control the three-dimensional distribution of fluvial facies in a number of ways: 1) they control the shape, size and orientation of the basin; 2) they control the rate of subsidence and the relative height of local source and sink areas, which in turn are contributing factors to the local sedimentation rate; 3) they affect the detailed surface topography of the alluvial plain, which in turn affects channel pattern. In any tectonic setting facies associations and distributions will also be controlled by the source region geology, eustasy and the climate.

BASIN TECTONICS, MORPHOLOGY AND EVOLUTION

As noted above, a major control upon alluvial architecture is the production of an asymmetric surface slope by tectonic movements. Such asymmetric tilting may be produced in a number of different tectonic settings summarized in Figure 2:

1. Grabens and half-grabens occur most commonly in areas of lithospheric extension (Jackson and McKenzie, 1983), chiefly during the process of lithospheric attenuation (Fig. 2A). In their simplest form, transverse slopes are produced by the cumulative effects of motions along boundary normal faults. Basins may be isolated and scoop-shaped or linear aggradations of scoop-shaped basins bounded by normal faults which commonly extend about 30 km or so along strike. In most cases the hanging-wall slope will be complicated by antithetic and synthetic faults and warps.

2. Asymmetric grabens also occur as pull-apart basins in strike-slip fault terrains (Fig. 2B). Here, the surface tilting occurs because of the downward component of motion induced by deformation of re-entrants or salients during strike-slip movement. The basins are of half-graben type, usually isolated and partly fault-bounded (Mann and others, 1983).

3. The near surface effects of large-scale thrust faulting include asymmetric subsidence brought about by load-induced motion of the footwall crust (Fig. 2C) (Beaumont, 1980). As shortening deformation proceeds, the locus of maximum subsidence migrates and the basin fill is progressively overridden and internally deformed by thrust tectonics. The formation of thrust duplexes will be accompanied by basin migration which is often episodic and usually sudden. Foreland thrust basins are regionally extensive, and the asymmetric subsidence normal to the thrust front will be accompanied by differential subsidence parallel to the thrust front if the magnitude of fault motion changes along the thrust. Axial rivers which collect drainage from alluvial fans coming off the thrust hanging-wall uplands will be strongly influenced by this complex subsidence pattern.

In any of the above settings, if tectonic movement is rapid, fault scarps or sharp monoclines may be formed at the depositional surface forming local topographic features (Fig.

---

[1]Present Address: Department of Geology, University College, Cardiff, Wales, U.K.

6972

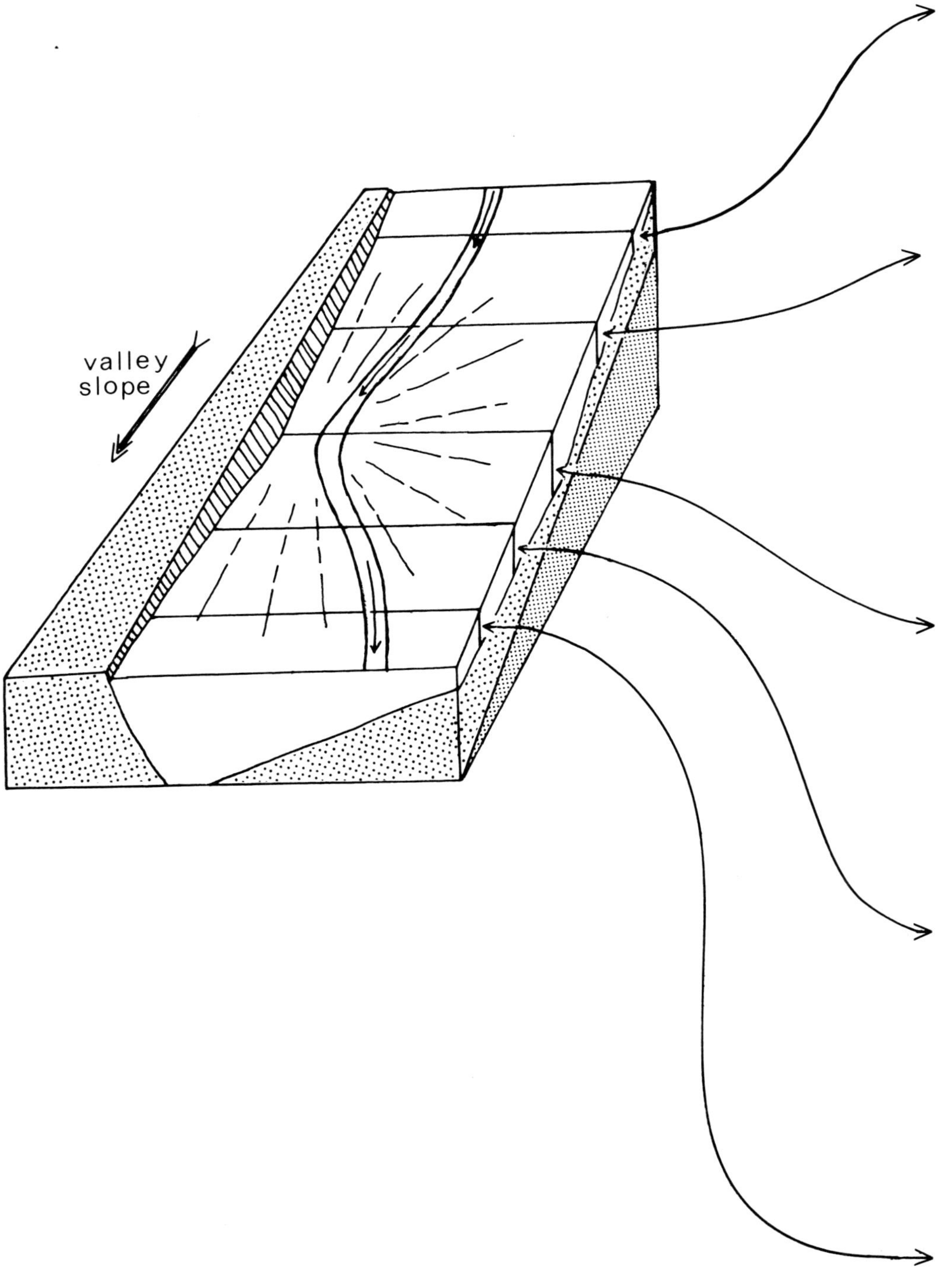
valley
slope

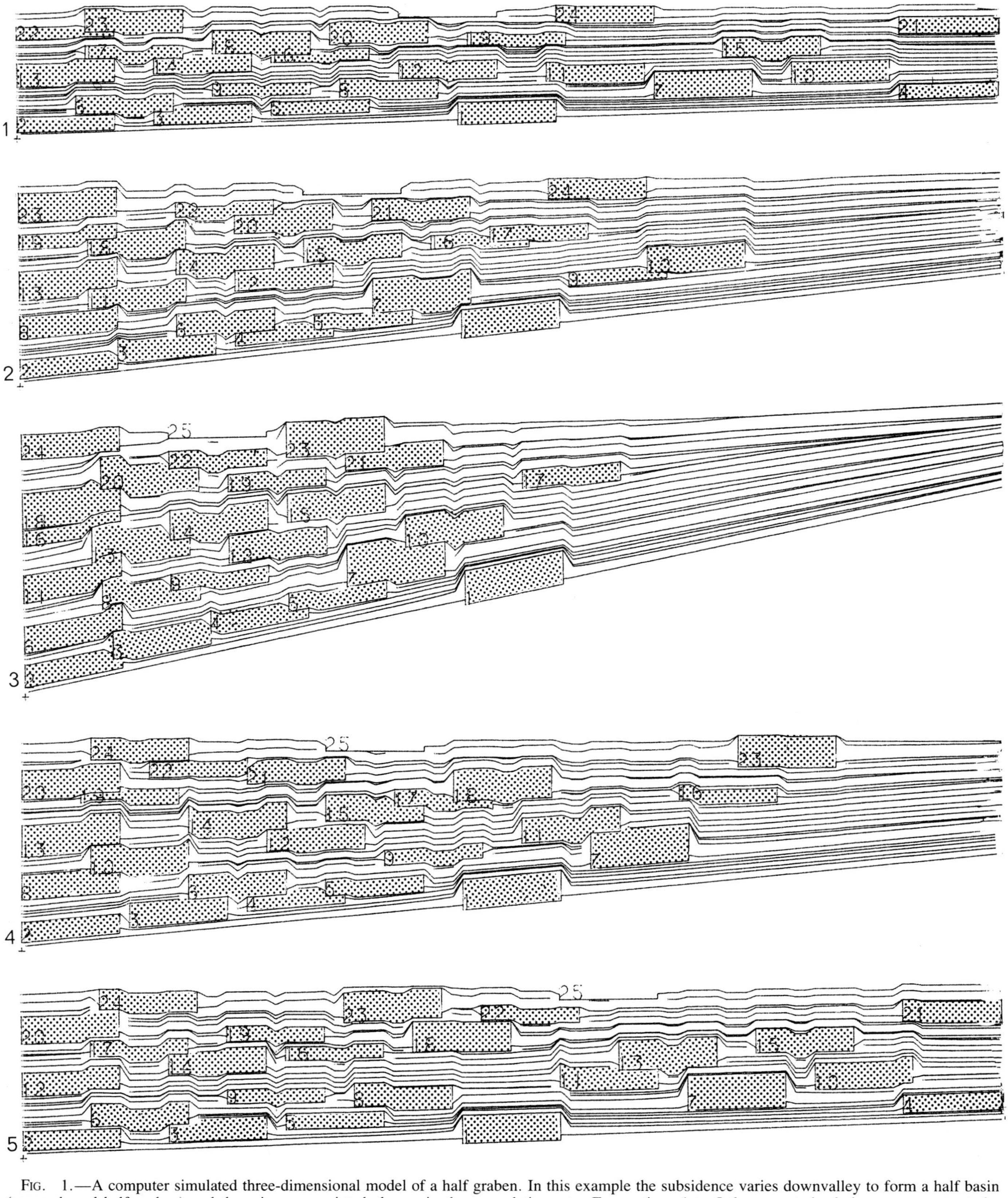

FIG. 1.—A computer simulated three-dimensional model of a half graben. In this example the subsidence varies downvalley to form a half basin (scoop-shaped half graben) and there is an associated change in the aggradation rate. For sections 1 to 5 the expected relative movements at the sides of the floodplain are 0.5, 2.0, 4.0, 2.0, 5.0 m and the mean channel belt aggradation rates are 0.015, 0.020, 0.025, 0.022, 0.020 m, respectively. These sections are generated independently of each other and only represent the general trend that may be expected.

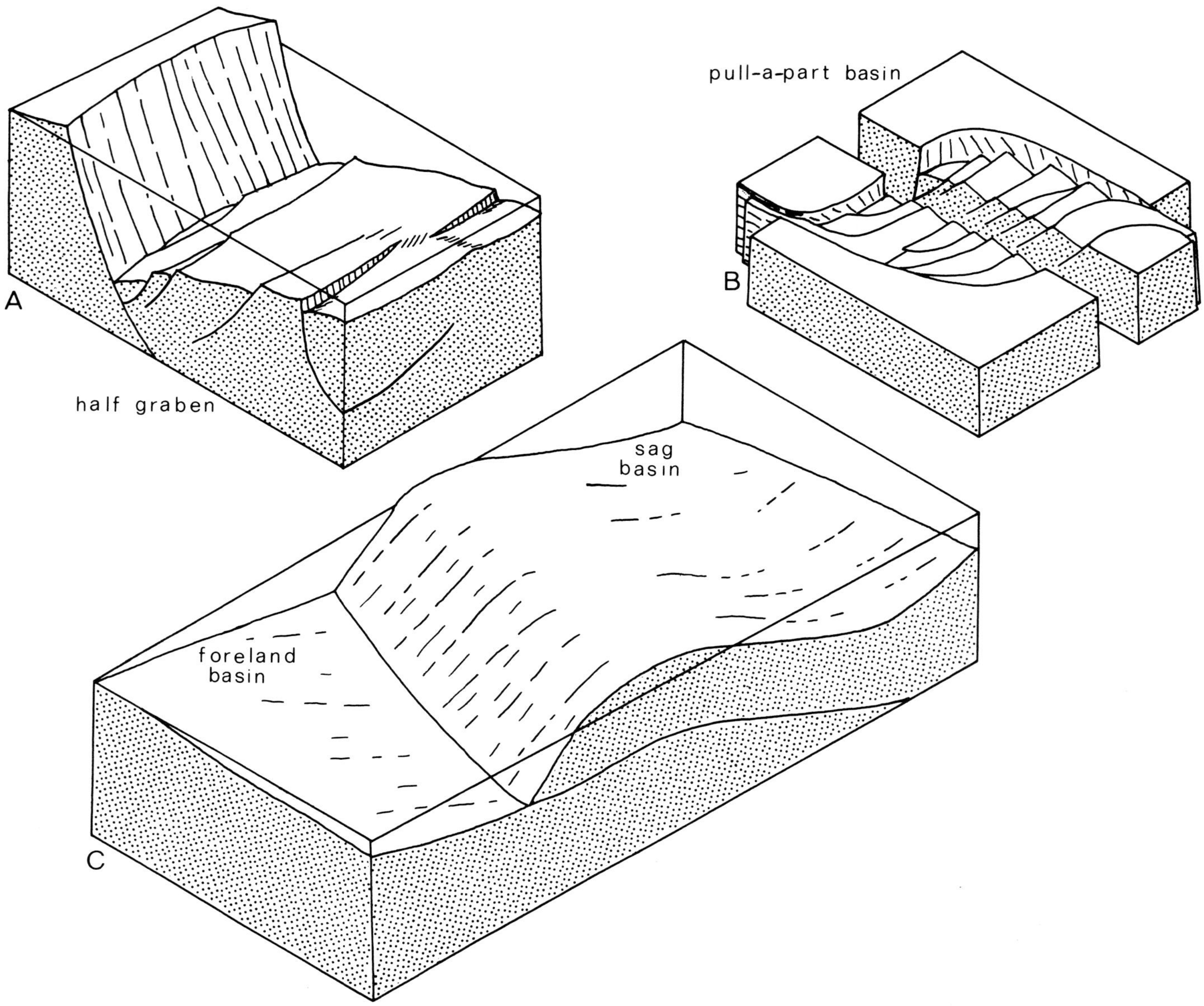

FIG. 2.—Basin morphology. A) Half graben with complex intrabasinal deformation by antithetic and synthetic faults. B) complex pull-apart basin with intrabasinal faults at an angle to the major axis of the basin. C) foreland thrust sheet basin produced by deformation of the footwall and a thrust top sag basin formed over a thrust plane ramp. In each case tectonic motions have induced important additional gravity slopes (see text for discussion).

3). In various areas of the world, extension-induced subsidence is commonly far in excess of local sedimentation rates. Above a critical value of differential movement that is dependent on the sedimentation rate, the basin floor topography is controlled by the developing structure. As displacement decreases through a critical value, the detailed topography becomes progressively less affected by deformation and more by local meso- and microscale geomorphic and depositional topography. In such situations of low displacement rate, the surface expression of tectonic activity will be undramatic, but it may be a significant control of the three-dimensional distribution of facies over a long period and produce considerable thickness variations over the basin.

The evolution of basins with time will be reflected in the facies distribution of the basin fill. In the case of an extensional half-graben system, fault movement will occur at different points along the graben and on different faults with time, whereas at the end of the major faulting phase, thermal subsidence will occur over a larger area (McKenzie, 1978). As a half graben develops, the cumulative effect of the repeated subsidence events causes a migration of the hanging-wall pivot away from the footwall. As sedimentation proceeds, this is accompanied by onlap of the sediments in the half graben up the hanging wall, a feature seen on many seismic sections.

In flexural basins formed on the footwall of thrust faults the locus of subsidence will shift in tandem with the overthrusting hanging wall. Thus, any section through such a developing basin will ideally show upward facies change reflecting 1) increasing subsidence with time, 2) passage from footwall-derived to axial to hanging wall-derived sediments, and 3) the ultimate onset of internal deformation of the sediments. Such a sequence has been observed recently

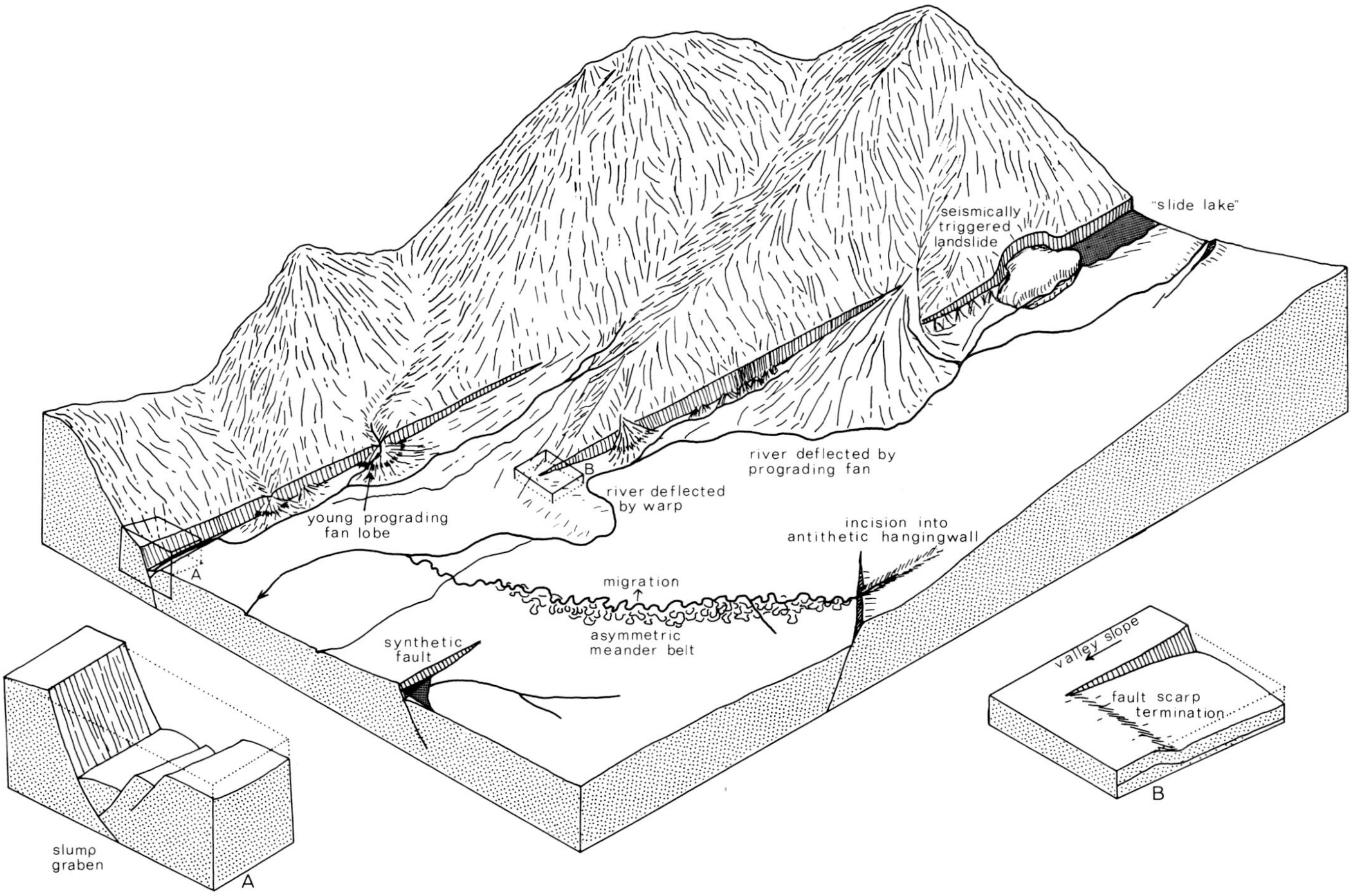

FIG. 3.—A sketch of a half graben to show some common geomorphologic features. Some of these will be inherited into the sedimentary record and control architectural trends. See text for full discussion.

by one of us (MRL) in the Tertiary thrust basin of central Tibet.

In pull-apart basins formed along major strike-slip faults (Mann and others, 1983; Ballance and Reading, 1980), sedimentary evolution is closely controlled by the passage of the basin past a salient re-entrant on the opposite side of the fault. Subsidence is rapidly initiated, deformation occurs and subsidence ceases and may be followed by inversion of the basin as strike-slip movement continues and the area becomes one of transpression.

If the main controlling fault (thrust or normal fault) has a ramp and flat morphology, then sag basins will develop and the position of the depocenter will move with time as the hanging wall moves over the ramp (G. Williams, pers. commun.) (Fig. 2C). A similar effect would be produced where there was a variation in the displacement gradient on the fault (Higgs and Williams, pers. commun.).

EFFECTS OF BASIN MARGIN AND INTRABASINAL STRUCTURES

The detailed geometry of faulted basin boundaries has a strong, if only localized, effect on the architecture produced. At the enechelon intersection of two normal faults, large drainage systems can be established (Fig. 3). Extensive footwall-sourced alluvial fans may prograde obliquely from these sites into one or other of the adjacent scoop-shaped basins formed by the faulting (Leeder and Gawthorpe, in prep.). A feature similar to this was observed by Nichols (this volume) at re-entrants of thrust fronts. The fan apex of his Luna depositional system can be traced back to a point where the controlling thrust changed from surface to subsurface termination. The surface expression of faults may also act as barriers to river flow, either as a constraint on lateral migration or by damming the flowing water to form lakes. Transfer faults in a half-graben system, as in the Locride area of Greece (MRL, personal observations), can be responsible for major changes in channel direction, resulting in some cases in rivers flowing at a high angle to the expected trend.

A river tends toward the position of topographic minimum, which is commonly the site of maximum subsidence. The stream can be deflected from this position if the sedimentation rate more than compensates for the subsidence or if the river is diverted by intrabasinal structures. In a simple half graben, for example, the site of maximum subsidence will be adjacent to the main fault. This area is commonly rapidly occupied by fans prograding from the footwall that may push the axial river path further from the locus of maximum subsidence (Fig. 3). Jackson and McKenzie (1983) observed that small-amplitude gentle folding is often associated with extensional faulting and it is also common in transpressional and compressional terrains. These features will commonly produce local effects large enough to disrupt the general pattern of sedimentation. This appears to have been the case in the Middle Jurassic of the Yorkshire Basin, England, where small displacements on faults affected the position of the channels but where in the intervening areas streams and overbank facies were controlled by gentle surface flexuring.

Intrabasinal faulting can have a strong effect on alluvial architectural pattern since mini-graben or subsidiary propagating thrust tips act to trap the channels away from their optimum gravitational position in the basin. They may also act as floodwater dams and produce adjacent areas of raised topography where more rapid pedogenesis and localized erosion will occur.

INCISION

Where a stream descends from the uplifted area produced by basin margin or intrabasinal structures, tectonic movement will result in incision by nick point retreat (Fig. 3). Such incision locally stabilizes the channel position which may, in turn, encourage nodal avulsion at the fault scarp. As an example, Grayling Creek in southwest Montana descends into the valley of the Madison River from the footwall of the Red Canyon Fault. In the 1959 Hebgen Lake earthquake, a 3-m scarp superimposed on a prehistoric feature cut the stream profile, and incision of the footwall rapidly followed. The Grayling Creek Fan radiates from this point and abandoned channels on its surface demonstrate repeated nodal avulsion at the fault and random avulsion lower on the fan (Alexander and others, in prep.). Similarly, streams flowing down the tilted surface from the opposite side of the basin will undergo incision above the pivot point. Galloway (1981) also observed that the entry point of rivers onto the Texas Coastal Plain is stable for long periods of time because of continuous seaward tilting of the coastal plain and the resulting entrenchment of the rivers above the hinge line.

The abandonment of one crustal-scale normal fault and the development of another (Jackson and McKenzie, 1983) will cause radical changes in the pattern of channel incision and resulting architecture. If, for example, a new fault develops in the hanging wall of the pre-existing fault, then part of the basin fill will be uplifted into the footwall of the new fault and incision of streams into it may occur. Switching between faults can result in alternate periods of incision and rapid deposition with a resulting sandstone percentage higher than that predicted by any model that uses only a pattern of deposition.

ALLUVIAL PLAIN TOPOGRAPHY

Away from active faults and over most depositional areas, tilting strongly affects alluvial processes by superimposing a tectonic slope on the pre-existing floodplain topography. Similar effects result from subsidence due to fluid extraction (Coates, 1983), evaporite solution, magmatic production by partial melting and igneous intrusion. Any change in the gravitational potential at a point will immediately affect the fluid and sediment transport rates which govern erosion and deposition in the area.

The importance of tilting on alluvial channel processes may be shown by a comparison of the tectonic gradients produced by single events and stream gradients. In many cases the tectonic and stream gradients may be of the same order of magnitude, and locally the tectonic slope is much greater. In other areas the tectonic gradient, though less than the stream gradient, is comparable to the sedimentary dip perpendicular to the channel and will affect the channel's lateral stability. For example, in the Hebgen Lake area of southwest Montana, tectonic tilts of as much as about 0.4° were produced in the 1959 earthquake; this compares with the South Fork gradient of 0.05° below the Interstate 191 road bridge (Fig. 6), and the Madison River gradient of 0.4° below Earthquake Lake. In the New Madrid area the tectonic tilts produced in the 1811–1812 earthquake period were as much as about 3.4×10^{-3} degrees and the channel gradient is about 3.8×10^{-3} degrees. The cumulative effect of repeated tectonic movement has a more pronounced effect on the topography than a single seismic event, if the repeat interval is small with respect to the sedimentation rate. If sedimentation can compensate for the subsidence before the next event occurs, then the influence on alluvial architecture will be less pronounced but still of major importance.

One way to examine the effect of a single tilting event (or repeated tilting) is to draw the contours of the area before and after the movement, then to construct gradient lines perpendicular to these. The gradient lines represent the idealized path that flood water would take and may be compared to evaluate the importance of a tectonic event. Figure 4 shows the result of a small sideway tilt on an evenly sloping surface. Figure 5 shows the idealized flood flow lines for the New Madrid area affected by differential tectonic movement during the 1811–1812 earthquakes and during at least two prehistoric events with magnitudes greater than 6.2 on the Richter scale. Sites of rapid lateral erosion are expected where the gradient lines are at a high angle to the channel, and preferred sites of erosion and deposition and the possible course of a river after avulsion will tend to follow gradient lines. The gradient lines are also useful in flood control, drainage and irrigation systems (see below).

The bed shear stress, τ, is directly proportional to the gradient of the river bed, S, in the tractive stress equation, $\tau = \rho g h S$, where ρ is water density and h is water depth. Any tectonically-induced change in the stream gradient will produce a corresponding change in the bed shear stress. This will lead to an increased sediment transport rate with an increased gradient or decreased sediment transport rate with a decrease of slope. Ouchi (1985) and Watson and others (1983) examined in more detail the effects of stream gradient change on meandering and braided streams. It is clear that there is some relationship between channel pattern and gradient and that tectonic activity may well produce a change in both channel pattern and facies. The sinuosity of the Mississippi River appears to be dependent upon the amount of tectonic back tilt, the sinuosity decreasing to maintain a steady gradient (Fisk, 1944).

THE RATE AND STYLE OF CHANNEL LATERAL MOVEMENT

The rate and style of lateral channel movement is dependent, in part, on the amount of tilt perpendicular to the

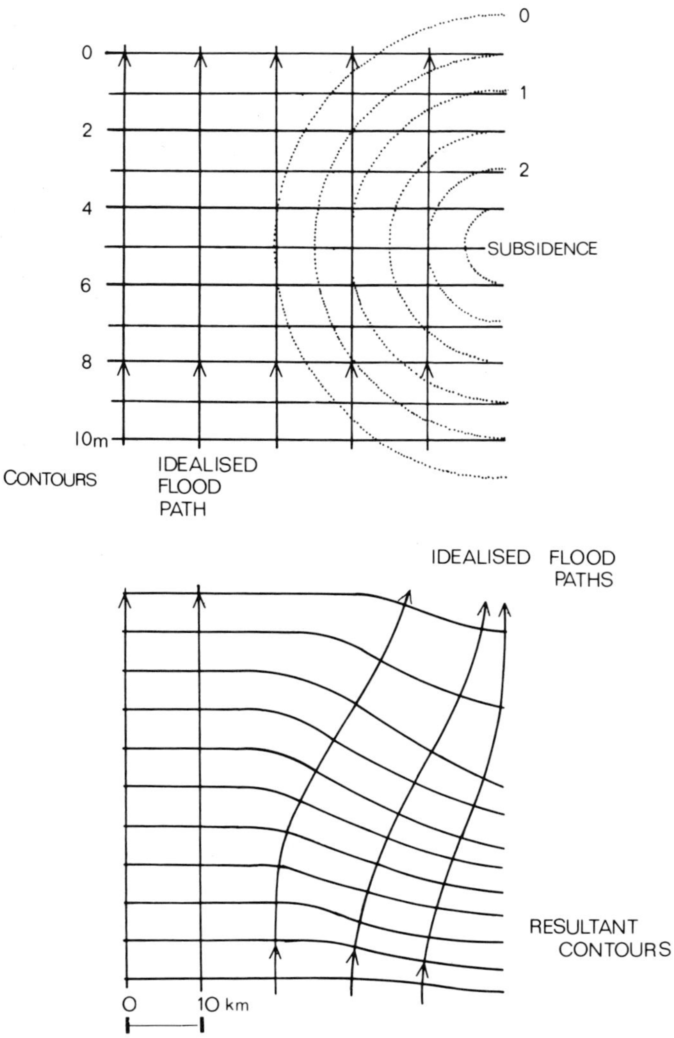

FIG. 4.—Contours and gradient lines before a) and after b) deformation of an initially uniformly tilting surface. Flood water will follow the gradient lines which have become generalized streamlines. Meander migration will be influenced by the resultant contours.

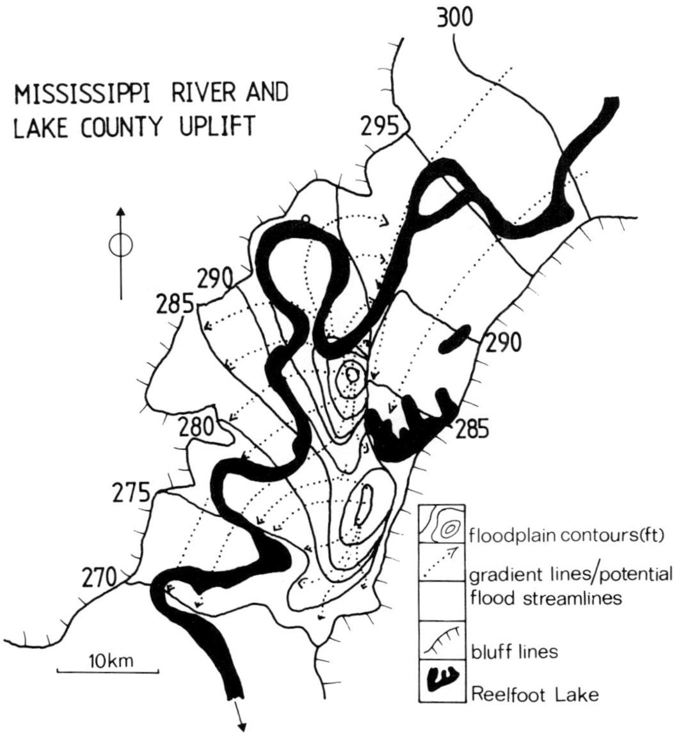

FIG. 5.—Map of the New Madrid area in the Mississippi floodplain to show the effects of surface tilting caused by displacement along the Lake County Uplift. Contours graphically derived from the floodplain and uplift contours given by Russ (1981, Fig. 2). We have drawn gradient lines normal to the contours to indicate potential flood streamlines and to show how Holocene meander migration has progressively adjusted to the tilting provided by the Lake County Uplift. Note especially the abandoned oxbow of Reelfoot Lake and its position on the receiving end of several streamlines.

channel. Three styles of lateral migration may be defined, but the controlling factors that determine which style predominates are not clear.

1) Instantaneous downtilt avulsion triggered by a single tectonic event. This is brought about by a combination of factors including; a sudden hydraulic response to the seismic event; tectonic bank breaching (as, for example, observed in a canal in the New Madrid area by Morse, 1941); shock-triggered bank collapse (see Youd and Wieczorek, 1982) and the gravitational advantage of the new course. If the resulting avulsion interval is short, this will lead to the preservation of immature channel-belt sand bodies and isolated channel sand bodies.

2) Topographically triggered avulsion. With time, a river will build an increased topography until a point is reached when the channel will improve its profile by switching its position down a path determined by the resultant regional tectonic tilt and the downstream gradient. The rate of topographic growth may be speeded by tectonic activity. A trigger is needed to produce bank breaching; this could be an inherent alluvial process such as crevassing. Tectonic motions are suspected to have caused the river Brahmaputra to avulse in this way (Coleman, 1969).

3) Slow migration of a channel belt by processes of preferential downslope erosion and meander cutoff, or anabranch abandonment. This produces an asymmetric channel belt and an abnormally wide channel sand deposit in which there will tend to be a predominance of lateral accretion surfaces facing up the paleoslope. Channel plugs will occur with a relatively small amount of associated laterally accreted sandstone representing the preferentially abandoned meander loops. We have studied the South Fork of the Madison River south of Hebgen Lake (SW Montana), which is slowly migrating northeast down the tectonic tilt in this way (Fig. 6). Abandoned channels can be distinguished clearly on aerial photogrpahs, whereas the tributaries are building gentle fans down the tilted surface. The South Fork is incised into a steep cut bank on its northeast side, and it is possible that this is encouraging the slow migration rather than a stepwise movement. As noted previously, very marked meander-belt asymmetry produced by floodplain tilting is seen in the Mississippi floodplain around New Madrid, Missouri (Fig. 5). A complex history of channel migration was discussed by Mike (1975) from abandoned channel traces

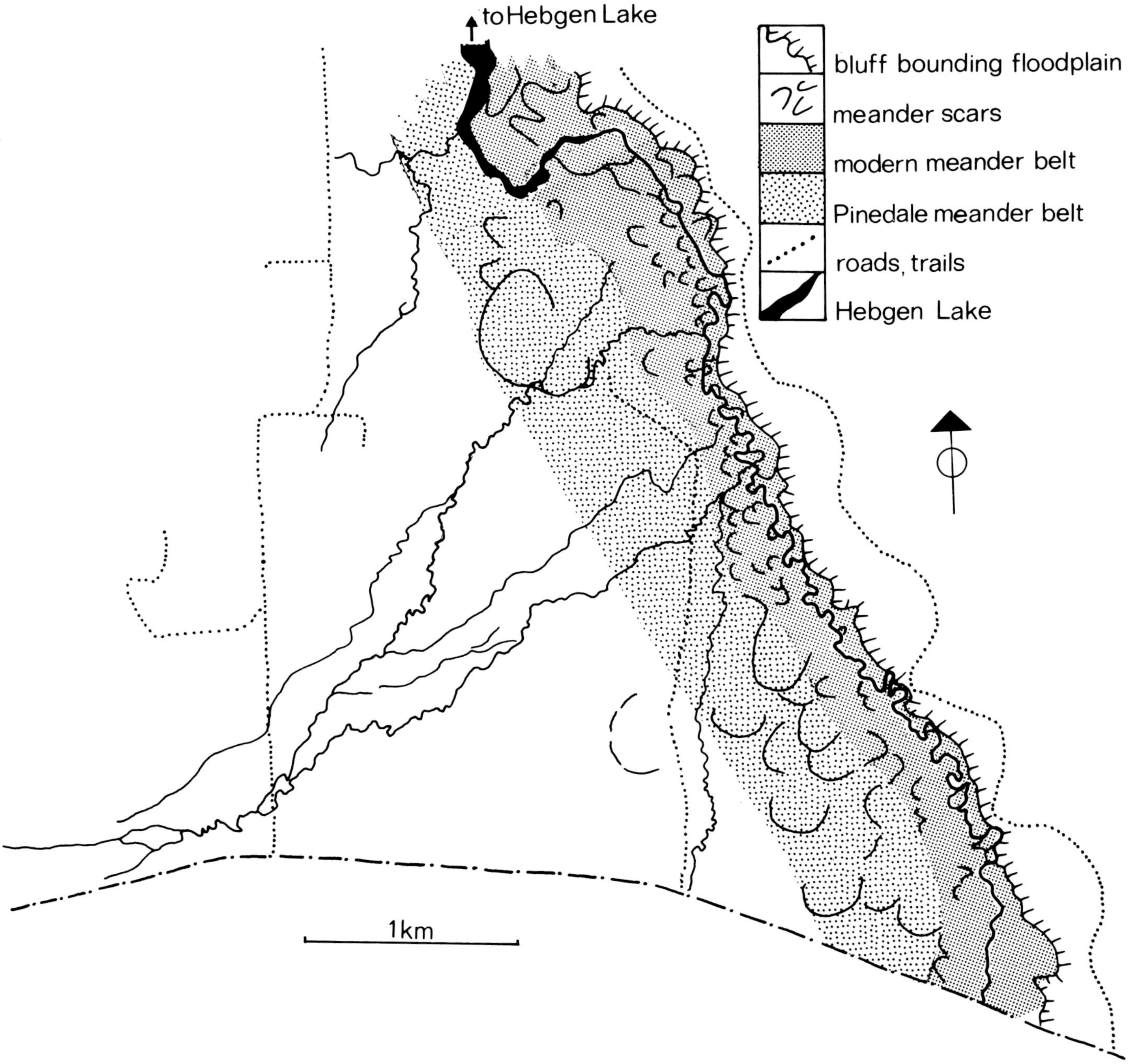

FIG. 6.—Map (traced from aerial photographs) to show the South Fork Madison River floodplain in SW Montana. The Pleistocene and Recent meander belts are clearly asymmetric. Note especially the position of the modern (1981) river close to the NE bluffline and the SW vergence of the convex portions of the abandoned meander scars (see text). The river has gradually migrated NE owing to the tectonic tilting in the Hebgen half graben, as originally proposed by Myers and Hamilton (1964). Several alluvial cones are seen entering the river from the SW. Note the hydrological change implied by the differing scale of the Pinedale and modern meanders (Alexander and others in prep.).

observed on the Great Hungarian Plain. This area is continuously subsiding, and movement on different faults resulted in a complex surface deformation history and a corresponding complexity of channel migration patterns.

FLOODING, FLOOD DEPOSITS AND TILTING

Flood water that escapes the confines of the channel will flow down the gradient lines of the floodplain, as outlined above. In practice the current will be locally influenced by small perturbations in the surface, such as those produced by abandoned bedforms or vegetation. Happ (1948) observed that most overbank flow occurred to the east of the river in the Middle Rio Grande valley. This is the side of the floodplain tilted toward a major valley-bounding fault. Fisk (1944) observed that in the New Madrid earthquake area several crevasses along the Mississippi River occurred at points where regional fracture zones intersected artificial levees. The direction of crevassing was aligned with the trend of the faulting, although no surface deformation was observed around the breach points at the time. Low areas produced by tectonic activity, for example, the Reelfoot depression southeast of New Madrid (Russ, 1979, 1981), may act to trap flood water and form alluvial plain lakes (Fig. 5). The Reelfoot Lake was thus enlarged during the 1811–1812 earthquakes and consists of a number of coalesced oxbow lakes in a structural low (Fig. 5). In these areas distinguishable facies can be deposited which can be used in the rock record to indicate a floodplain low. Overbank deposits will preferentially be deposited in low areas, increasing the rate of aggradation at these sites and producing lateral variations in thickness and facies.

Low rates of overbank deposition on the raised areas, such as on the dip slope of a half graben (above the pivot point) or an intrabasinal horsts or domes, will allow the accumulation of organic material and the development of more mature soils. Areas that are not affected by recurrent overbank deposition undergo more rapid pedogenesis (Bown and Kraus, 1981; Steel, 1974). Above the pivot point, incision will occur and the groundwater table will be lowered, therefore allowing better drained soils to develop. The pivot point is seldom stable, and alternate periods of incision and sedimentation will produce complex stacked soil horizons. Russ (1979, 1981) recorded that better developed soil horizons were found on the Tiptonville Dome in the New Madrid area than in the adjacent floodplains. Read and Dean (1982) observed that in the Upper Carboniferous of the Midland Valley of Scotland the number of rooted horizons is greatest in areas that underwent less subsidence, and the thickness of coals decreases into the areas of higher subsidence.

Tilting will have instantaneous effects on the shore-lines of lakes and the sea. Alteration in the lake floor configuration may also affect the position of the shore. This instantaneous movement was observed around Hebgen Lake in southwest Montana after the 1959 earthquake. Leeder and Gawthorpe (in prep.) observed that important features will result from instantaneous marine transgressions over delta lobes and adjacent shorelines following motion along a normal fault. The effects of the rapid shore movements on sedimentary processes are further exaggerated by the action of seiches that may carry marine-derived sediment and microfauna far inland.

THE EFFECT OF TILTING IN FANS

Fans in grabens react to tilting in different ways, depending on whether they are located and sourced from the uplifted footwall or from the downthrown hanging wall. The only published detailed study of such fan reactions is that of Hooke (1972) in Death Valley, California. He established that surface tilting caused fan segmentation and that characteristic offlap and onlap sequences result if the tilt process occurs with sufficient regularity and magnitude. These effects are predictable in terms of a tilt block model and may be amplified briefly as follows (see Leeder and Gawthorpe, in prep., for a more complete discussion). Fans sourced in the uplifted footwall but depositing sediment on the hanging wall undergo a decrease in gradient. A new fan segment develops closer to the apex. This may prograde with time over the old fan surface, the exact sequence of events being determined by the rate of downfaulting verses the local sedimentation rate (Fig. 7). The reverse process occurs in fans sourced in the uptilted portion of the hanging wall. Here, incision above the pivot line occurs and a new lobe laps onto the previous fan segment.

Atkinson (1983) observed megacyclicity of fan deposits in the Tertiary South Pyrenean Basin of northern Spain. He attributed this to discontinuous thrust loading along the basin margins and sub-basinal uplift above blind thrust culmination structures.

In pull-apart basins fan deposits often form a large proportion of the valley fill and may be used to indicate the development of the basin. As the major fault moves, the source area is displaced relative to the fan apex and a new fan develops in an adjacent position. In this way any locally

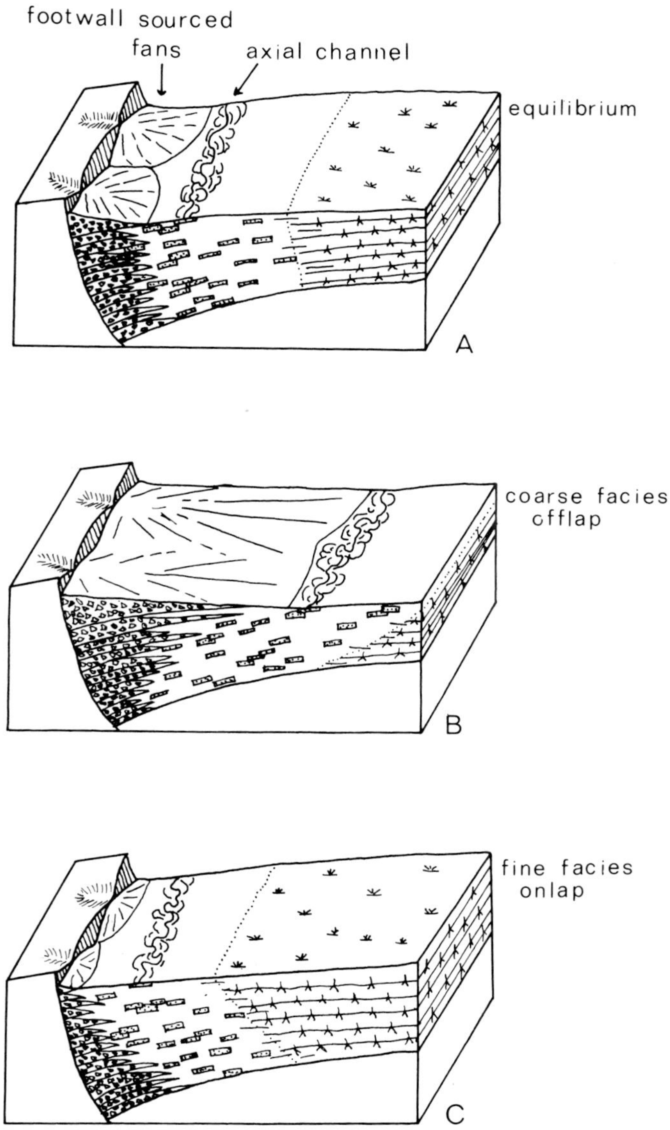

FIG. 7.—Block diagrams to show how the subsurface architectural style of fluvial systems is dependent upon the magnitude and growth of alluvial fans prograding from the graben footwall uplands. Note especially how the position of the axial river channel sands is controlled by the fans.

derived sediment will be deposited in a body that passes diagonally up the section away from the direction of fault movement.

CONCLUSION

Tectonic activity influences the shape, size and orientation of alluvial basins, the rate of subsidence and the relative height of local source and sink areas, the distribution and magnitude of surface displacement vectors, and the detailed surface topography. The surface expression of fault activity has a strong local effect on river orientation and gradient and also controls the distribution of alluvial fan facies.

Tectonic tilting produced by asymmetric subsidence alters the gradient line pattern and will therefore affect the flood distribution, channel migration and post-avulsion channel position. The component of tilt parallel to the channel will act to change its form, whereas the tilt component perpendicular to this will affect the lateral stability of the channels and may result in progressive migration or down-tilt avulsion.

ACKNOWLEDGMENTS

J.A. thanks British Petroleum for a Postgraduate Studentship and travel expenses that enabled her to present this paper at the Colorado Fluvial Conference. M.R.L. thanks Texaco USA and the University of Leeds for financial support. Useful discussions over the years with the following people have sharpened our initial ideas; John Bridge, Rob Gawthorpe, Charlie Bristow, Phil Nell.

REFERENCES

ATKINSON, C. D., 1983, Comparative sequences of ancient fluviatile deposits in the Tertiary, South Pyrenean Basin, Northern Spain: Unpublished Ph.D. Thesis, University of Wales, 350 p.

BALLANCE, P. F., AND READING, H. G., eds. 1980, Sedimentation in oblique-slip mobile zones: International Association of Sedimentologists Special Publication 4, Blackwell Scientific Publications, 265 p.

BEAUMONT, C., 1980, Foreland basins: Geophysical Journal of the Royal Astrological Society, v. 65, p. 291–329.

BOWN, T. M., AND KRAUS, M. J., 1981, Lower Eocene alluvial paleosols (Willwood Formation, Northwest Wyoming, U.S.A.) and their significance for paleoecology, paleoclimatology and basin analysis: Palaeogeography, Palaeoclimatology, Palaeoecology, v. 34, p. 1–30.

BRIDGE, J. S., AND LEEDER, M. R., 1979, A simulation model of alluvial stratigraphy: Sedimentology, v. 26, p. 617–644.

COATES, D. R., 1983, Large-scale land subsidence: *in* Gardner, R., and Scoging, H., eds., Mega-Geomorphology, p. 212–234.

COLEMAN, J. F., 1969, Brahmaputra River; Channel processes and sedimentation: Sedimentary Geology, v. 3, p. 131–239.

FISK, H. K., 1944, Geological investigation of alluvial valley of the Lower Mississippi River: Mississippi River Commission, Vicksburg, Mississippi, 78 p.

GALLOWAY, W. E., 1981. Depositional architecture of Cenozoic Gulf Coastal Plain fluvial systems, *in* Ethridge, F. G., and Flores, R. M., eds., Recent and Ancient Nonmarine Depositional Environments: Models for Exploration: Society of Economic Paleontologists and Mineralogists Special Publication 31, p. 127–155.

HAPP, S. C., 1948, Sedimentation in the middle Rio Grande Valley, New Mexico: Geological Society of America Bulletin, v. 59, p. 1191–1216.

HOOKE, R. LeB., 1972, Geomorphic evidence for Late-Wisconsian and Holocene tectonic deformation, Death Valley, California: Geological Society of America Bulletin, v. 83, p. 2073–2098.

JACKSON, J. A., AND MCKENZIE, D., 1983, The geometrical evolution of normal fault systems: Journal of Structural Geology v. 5, p. 471–482.

MANN, PAUL, HEMPTON, M. R., BRADLEY, D. C., AND BURKE, KEVIN, 1983, Development of pull-apart basins: Journal of Geology, v. 91, p. 529–554.

MCKENZIE, D., 1978, Some remarks on the development of sedimentary basins: Earth and Planetery Science Letters, v. 40, p. 25–32.

MIALL, A. D., 1981, Alluvial sediment basins: Tectonic setting and basin architecture, *in* Miall, A. D., ed. Sedimentation and tectonics in alluvial basins: Special Publication of the Geological Society of Canada, v. 23, p. 1–33.

MIKE, K., 1975, Utilization of the analysis of ancient river beds for the detection of Holocene crustal movements, *in* Pavoni, N., and Green, R., eds., Recent Crustal Movements: Tectonophysics, v. 29, p. 359–368.

MORSE, W. C., 1941, New Madrid earthquake craters: Bulletin of the Seismic Society of America, v. 31, p. 304–319.

MYERS, W. B., AND HAMILTON, W., 1964, Deformation associated with the Hebgen Lake earthquake of August 17, 1959: U.S. Geological Survey Professional Paper 435, p. 55–98.

OUCHI, S., 1985, Response of alluvial rivers to slow active tectonic movement: Geological Society of America Bulletin, v. 96, p. 504–515.

READ, W. A., AND DEAN J. M., 1982, Quantitative relationships between numbers of fluvial cycles, bulk lithological composition and net subsidence in a Scottish Namurian Basin: Sedimentology, v. 29, p. 181–200.

RUSS, D. P., 1979, Late Holocene faulting and earthquake recurrence in the Realfoot Lake area, Northwestern Tennessee: Geological Society of America Bulletin, v. 90, p. 1013–1018.

———, 1981, Style and significance of surface deformation in the vicinity of New Madrid, Missouri: U.S. Geological Survey Professional Paper 1236, p. 95–114.

STEEL, R. J., 1974, Cornstone (fossil caliche)—Its origin, stratigraphic and sedimentological importance in the New Red Sandstone, Western Scotland: Journal of Geology, v. 82, p. 351–369.

WATSON, C. C., SCHUMM, S. A., AND HARVEY, M. D., 1983, Neotectonic effects on river pattern, *in* River Meandering: Proceedings of the Conference on Rivers, American Society of Civil Engineers, p. 55–66.

YOUD, T. L., AND WIECZOREK, G. F., 1982, Liquefaction and secondary ground failure: U.S. Geological Survey Professional Paper 1254, p. 223–246.

ZIEGLER, P. A., 1983, Faulting and graben formation in Western and Central Europe: Philosophical Transactions of the Royal Society of London, v. A305, p. 113–143.

CONTRASTING ARCHITECTURE OF TWO ALLUVIAL SUITES IN DIFFERENT STRUCTURAL SETTINGS

MARY J. KRAUS AND LARRY T. MIDDLETON

Department of Geological Sciences, University of Colorado, Builder, CO 80309; and Department of Geology, Northern Arizona University, Flagstaff, AZ 86011

ABSTRACT: The architecture of alluvial suites is governed by a variety of autocyclic and allocyclic mechanisms that work in concert. Careful evaluation of source lithologies, climatic conditions, and tectonic setting must be combined with sedimentologic analysis of three-dimensional exposures of fluvial rocks to rigorously determine controls on the architecture of individual sequences. Consequently, although study of alluvial architecture is a potentially effective and important means of elucidating the depositional and tectonic history of continental basins, there are few well documented field studies.

The lower Eocene Willwood Formation of Wyoming and the Plio-Pleistocene Glenns Ferry Formation of Idaho offer contrasting examples of alluvial architecture even though both sequences were generated by meandering streams in tectonically active intermontane basins. Independent evidence for the chronostratigraphy, provenance, climate, and tectonic setting is documented for each unit. Evaluation of these parameters demonstrates that differences in tectonic framework, specifically differences in the rate of basin subsidence, were responsible for the contrasting architectures. Decreasing rates in the Bighorn Basin produced a sequence with a thick and strongly multistory sand body plus mature paleosols in the Willwood Formation. In contrast, single or weakly multistory channel deposits, encased by vast quantities of overbank sediments, characterize the Glenns Ferry Formation and reflect more rapid rates of subsidence and sediment accumulation in the Snake River Plain of Idaho.

INTRODUCTION

Alluvial architecture and the degree of interconnection of channel sandstone bodies in fluvial rocks are controlled by a variety of interrelated parameters that may be either allocyclic (extrabasinal) or autocyclic (intrabasinal) in character (Beerbower, 1964). Allocyclic controls include tectonic adjustments in the depositional basin and/or its source areas, source area lithologies, and climatic conditions in both sediment provenance areas and the basin of deposition. These factors work in concert to regulate local variations in stream discharge and caliber, sediment size and load, valley slope, sediment vegetation cover and erodibility and, ultimately, alluvial architecture. Autocyclic mechanisms, on the other hand, are inherent to the river regime and include channel migration, neck and chute cutoff, and avulsive events. In many alluvial suites, it is not possible to identify a single parameter as the dominant control on sandstone geometries and architecture. Information concerning source areas, climate, and position of particular sequences within the geometric framework of the depositional basin is commonly speculative. Nonetheless, knowledge of these factors is critical to a meaningful evaluation of the contributions of the many differing allocyclic and autocyclic constraints to overall architectural motifs.

Because simple models stressing discrete allocyclic or autocyclic controls on sediment accumulation are not satisfactory in explaining sediment relations and geometries seen in the field, there has been a recent surge of interest in modeling alluvial regimes in response to a great number of stimuli of both intra- and extrabasinal origins. For example, Allen (1974) examined the possible effects of climatic change and base level fluctuations on meandering stream sequences. Paradigms incorporating one or another of these allocyclic controls, in conjunction with simulated random avulsive events (autocyclic mechanism), generated hypothetical alluvial sequences that were comparable to actual stratigraphic relations observed in sections of the Old Red Sandstone. A second theoretical study (Leeder, 1978) investigated the effects of independently varying avulsive frequency and alluvial plain width in deposits of meandering streams.

Computer simulations of alluvial stratigraphy (e.g., Allen, 1978, 1979; Bridge and Leeder, 1979) also demonstrate that varying rates of basin subsidence and/or basement uplift can significantly influence the percentage of coarse-grained facies and the degree of interconnection of channel sandstones. Not only does source area uplift affect alluvial architecture, especially through its regulation of stream style and process, but tectonically controlled basin subsidence can also modify fluvial sequences in more distal environments. Under equilibrium conditions, floodplain surfaces maintain a stable level by continuously balancing net sediment accumulation with basin susidence and sediment compaction. Consequently, the rate of net alluvial accretion (and thus preservation potential of floodplain deposits) is directly proportional to the rate of basin subsidence. Computer models (Allen, 1978; Bridge and Leeder, 1979) predict that stream deposits dominated by fine-grained overbank sediments and with widely spaced sand bodies reflect rapid basin subsidence, whereas alluvial suites with high sandstone density and a high degree of interconnection or erosional overlap among channel deposits are produced by decreased rates of basin subsidence. In addition to overall rates of basin subsidence, local variability is also important. Because channel belts tend to localize in areas of greatest subsidence, larger sandstone-to-mudstone ratios and sand bodies with greater interconnection can be produced close to basin axes rather than along basin margins (Bridge and Leeder, 1979; Read and Dean, 1982).

These computer models and excellent summary of fluvial cyclicity by Miall (1980) emphasize that a suite of variables determines the nature of facies models to interpret ancient fluvial sequences and their depositional histories. Simulation studies further demonstrate that alluvial successions that coarsen upward can reflect solely tectonic processes without a concomitant change in stream pattern. Thus, interpretation of fluvial style simply on the basis of sandstone geometry or mudstone/sandstone ratios can be misleading.

Though analysis of fluvial architecture is a potentially

effective and important means of understanding the depositional and tectonic history of continental basins, the study of alluvial architecture, especially in field examples, is still in its infancy. Field analyses are few and include those of Puigdefabregas and Van Vliet (1978), Galloway (1981), Read and Dean (1982), and Blakey and Gubitosa (1984). Unfortunately, these studies were unable to document adequately several important allocyclic controls of architecture, especially paleoclimate, which is commonly difficult to establish in ancient fluvial settings. Field study of three-dimensional exposures of fluvial rocks, in conjunction with knowledge of ancient source areas, climatic and tectonic settings, and detailed stratigraphy are critical to a meaningful evaluation of alluvial architecture; however, many or all of these factors are not commonly examined. Consequently, there are few well documented field studies which rigorously test the validity or usefulness of the computer models.

This study focuses on the geologic histories of two alluvial sequences deposited in intermontane basins with different structural histories. The examples discussed here, the upper Pliocene Glenns Ferry Formation of southwestern Idaho and the lower Eocene Willwood Formation of northwest Wyoming, offer contrasting examples of fluvial architecture. Portions of the principally lacustrine Glenns Ferry Formation are alluvial in origin and consist of widely scattered lenticular channel deposits enveloped by a larger volume of thick overbank mudstones. Conversely, portions of the Willwood Formation have numerous complex sheet sandstones that are generally thick, laterally extensive, and strongly multistory. The depositional histories of these fluvial units are examined and determined through analyses of the external geometry and internal morphology of sand bodies in conjunction with paleocurrent patterns. These sedimentologic studies demonstrate that tectonic controls strongly influenced sediment accumulation rates, and thus alluvial architecture, in both basins. More importantly, the significance of tectonism is evaluated in light of independent evidence for provenance, tectonic history, and paleoclimatic conditions for both the Glenns Ferry and Willwood Formations.

WILLWOOD FORMATION

Geologic setting.—

The Willwood Formation is a lower Eocene alluvial unit exposed throughout the Bighorn Basin of northwest Wyoming (Fig. 1). In the study area in the northern part of the basin, it consists of approximately 850 m of variegated mudstones, lignites, drab sandstones, and local conglomerates (Van Houten, 1944; Neasham and Vondra, 1972). Willwood deposition accompanied structural development of the Bighorn Basin, which is a Laramide intermontane basin approximately 200 km long and 80 km wide in the north. The Beartooth Range to the west, Pryor Mountains to the northeast, and Bighorn Mountains to the east (Fig. 1) rose during Paleocene and early Eocene time (Foose and others, 1961). The basin was open to the south and west until development of the Owl Creek Mountains, probably in early Eocene time, and formation of the volcanic Absaroka Range, which began during the middle early Eocene (Bown, 1980; Wing and Bown, 1985). Gravity anomaly studies demonstrate that the Beartooth Range has been thrust over the northwestern part of the Bighorn Basin by as much as 12 km (Bonini and Kinard, 1983). The markedly asymmetric Tertiary structural axis of the basin passes approximately 10 km southwest and west of the study area (Fig. 1) and, farther north, descends beneath the overthrust margin of the Beartooth Range (Andrews and others, 1947; Osterwald and Dean, 1961).

The Willwood sequence preserved in the northern Bighorn Basin encompasses the middle and late Clarkforkian (earliest Eocene) and earliest Wasatchian (early Eocene) land mammal ages (e.g., Rose, 1981; Gingerich, 1983). Paleomagnetic sampling of the strata (Butler and others, 1981), together with the biostratigraphic zonation and measured stratigraphic sections, provides calculations of sediment accumulation rates and changes in those rates through time. On this basis, Gingerich (1983) determined that sediment accumulation rates in the northern Bighorn Basin increased from early Paleocene through late Paleocene time and then decreased through the early Eocene. Similar trends were reported by Hickey (1980) in his study of lower Tertiary floral zones.

Most sandstones in the study area are lithic arenites with abundant crystalline and sedimentary rock fragments. The mineral fractions indicate that a primary source was the southeastern Beartooth Mountains. The rock fragments in sandstones in the northwest Bighorn Basin and composition of conglomerates flanking the eastern Beartooth Mountains (Flueckinger, 1971) indicate that the range had been stripped to its crystalline core by late Paleocene (Clarkforkian) time. Paleocurrent patterns show that major Willwood streams in the northern Bighorn Basin flowed to the north (Neasham, 1970; Kraus, 1980) and thus paralleled the trend of the Tertiary structural axis. Consequently, Willwood deposits in the study area contain sediment derived from additional sources that probably included the rising Bighorn Mountains.

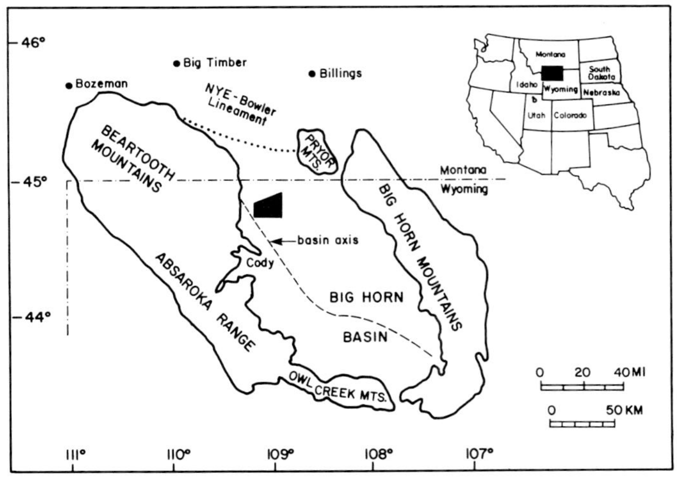

FIG. 1.—Base map of Bighorn Basin showing major structural features. Black trapezoid represents study area. Basin axis following Osterwald and Dean (1961).

Floral and faunal evidence demonstrates that late Paleocene-early Eocene climates in the Bighorn Basin were warm-temperate to subtropical with seasonal precipitation. Latest Paleocene and earliest Eocene time are marked by subtropical taxa, accompanied by an increase in floral diversity, indicating significantly warmer conditions than during most of the Paleocene (Hickey, 1980; Wing, 1981). Hickey inferred a mean annual temperature of 13.5° C during the Clarkforkian and Wing suggested that similar temperatures continued into the early Eocene. The nature of Willwood paleosols corroborates these climatic interpretations and further indicates conditions of relatively abundant, but seasonally spaced, rainfall (Bown and Kraus, 1981).

Depositional setting.—

Willwood deposits in the northern Bighorn Basin are dominated by mudstones and siltstones of overbank origin which have undergone extensive pedogenic modification. The mudrock sequences are interspersed with very fine-grained sandstones interpreted as levee, crevasse channel and crevasse splay deposits. Channel sandstones comprise approximately 20–25 percent of the sequence in the study area and exhibit both sheet and ribbon geometries (sensu Friend and others, 1979; Friend, 1983).

Sheet sandstones, with width/thickness ratios that typically greatly exceed 20, are common in the Willwood sequence (Fig. 2). Most sheet sand bodies are 6 to 7 m thick, although there is variability in thickness becaue of local channel scouring. Sheets are commonly overlain or flanked by levee deposits of interbedded siltstones and very fine-grained sandstones. The levee deposits may exhibit a depositional dip ranging from only several to more than 10°. If not extensively bioturbated, the thin sandstones exhibit small-scale cross lamination. Levee deposits show little pedogenic modification other than bioturbation, indicating very rapid accumulation.

One of the larger sheet sandstones, which occurs 180 m below the Clarkforkian-Wasatchian boundary, illustrates the geometry and internal structure characteristic of Willwood sheets. Although the top of this sandstone body is commonly eroded, it is as much as 12 m thick and has a preserved lateral extent of 1.5 km perpendicular to paleoflow direction, yielding a width/thickness ratio in excess of 1000. The base is highly irregular because it is much more deeply scoured in some places than others.

Extensive scour surfaces, with pebble conglomerates consisting of mudclasts and carbonate nodules eroded from developing soils, subdivide the sand body into as many as four stories. Local relief for these major scours is usually no more than 1 m. Individual stories commonly range from 3 to 4 m in thickness and are subdivided by minor scour surfaces into smaller sedimentary packages or bedsets (Bridge and Deimer, 1983). Whereas local relief on the major scours between stories can be as much as 2 m, it is no more than a few tens of centimeters for the minor scours. Although bedsets show great lateral variability, they consist dominantly of large-scale sets of trough crossbedding and parallel stratification with well developed parting lineation. Some parallel bedding is nearly horizontal; however, most dips at low angles (6° to 12°) and is grouped into units from 20 to 80 cm thick which are stacked discordantly. Contorted bedding is common, especially in the upper story of this and other Willwood sheets. The most poorly represented facies is small-scale cross-stratification which occurs uncommonly at the top of stories. With the exception of pebble conglomerates, most facies are developed in fine or medium sand, and in many stories grain size diminishes upward (Kraus, 1980).

Most ridges in which the sheet sandstone is exposed are oriented perpendicular to the strike of the regional paleoslope, and individual stories appear as relatively continuous bands in these exposures. In such cases, recognition of stories is enhanced by significant differences in paleocurrent directions between adjacent stories. For example, it is not unusual for paleoflow in vertically adjacent stories to vary by as much as 140°. In those exposures oriented perpendicular to paleoflow, lateral accretion surfaces are evident (Fig. 2). Some lateral accretion deposits can be traced laterally into channel fills which usually consist of massive sandstone (Fig. 2) or less commonly of mudrocks.

The presence of lateral accretion surfaces demonstrates that Willwood sheet sandstones were generated by meandering streams. This interpretation is corroborated by the marked differences in paleoflow direction observed in vertically adjacent stories (Puigdefabregas and Van Vliet, 1978) and by sequences of depositional facies in the sheet sandstones (Kraus, 1980). Each story represents an episode of lateral migration and point bar accretion. The proximity of the study area to the Tertiary structural axis of the basin (Osterwald and Dean, 1961; Gingerich, 1983; Fig. 1) and the generally northward paleoflow indicators show that the sheet sandstones were deposited by major axial streams.

In contrast to the sheet sandstones, ribbon sand bodies are no more than 6 m thick and are characterized by width/thickness ratios of 15:1 or less. Ribbon sandstones show no evidence of steady lateral stream migration and are interpreted as the deposits of laterally fixed streams (Kraus, 1985). The absence of evidence for lateral migration and the multistory character of many Willwood ribbons indicate that channels filled vertically and that several episodes of erosion and deposition occurred before the channel switched its position on the alluvial plain. The laterally fixed nature and smaller scale of the ribbon sand bodies, as well as ribbons demonstrating flow at a large angle to the general northerly trend of the sheets, suggest that these were tributaries to the trunk stream.

Significant sedimentologic changes occur at the Clarkforkian-Wasatchian boundary, which is a profound faunal break. Through the middle and late Clarkforkian part of the Willwood sequence, sheet sand bodies are usually less than 10 m thick, although the example described above is 12 m thick. The channel sandstones are enveloped by overbank deposits on which paleosols developed. Comparison with the five-stage developmental or maturation sequence established by Bown (1985) for paleosols in the central Bighorn Basin reveals that these paleosols range from stage 1 (entisols) to stage 3 (alfisols) and thus attain only moderate levels of pedogenic maturity. A major change in sand body size and complexity, accompanied by a significant increase in paleosol maturity, marks the boundary between the

FIG. 2.—Willwood sheet sandstone with well developed lateral accretion surfaces dipping to the east (right). Arrows denote channel fill consisting of massive sandstone.

Clarkforkian and Wasatchian land mammal ages. The major sand body at this faunal boundary, referred to as the boundary sandstone, is at least twice as thick as any other sheet sandstone. It ranges in thickness from 12 to 31 m (averaging 21 m) and can be traced continuously for nearly 12 km in a direction oblique to paleoflow. As many as six stories, each on the order of 3 to 5 m thick and thus similar in scale to those in other sheets, comprise the boundary sandstone (Fig. 3). At one locality, the sheet splits to form two sandstone bodies separated by several meters of pedogenically modified overbank mudrocks. Where the boundary sheet sandstone is thickest, it overlies a sequence of moderately mature (stage 3) paleosols of late Clarkforkian age (Fig. 4A). In marked contrast, where the sheet sandstone is thinner, it is overlain and laterally adjacent to mature soils of at least stage 4 development, which are otherwise rare in the northern Bighorn Basin (Fig. 4B).

FIG. 3.—Sheet sandstone at Clarkforkian-Wasatchian boundary showing its thickness and multistory character. Black lines separate individual stories.

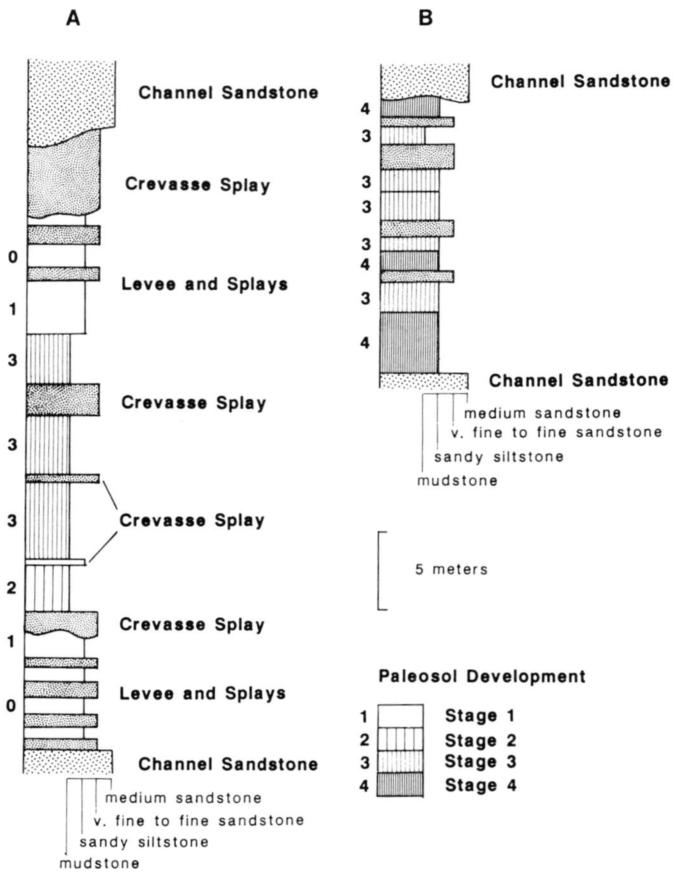

FIG. 4.—Willwood paleosol sequences. A. Latest Clarkforkian section directly below the boundary sandstone. Paleosols range from immature (stage 0) to moderately mature (stage 3). B. Earliest Wasatchian section directly above the boundary sandstone. Stage 3 and mature stage 4 paleosols are characteristic.

GLENNS FERRY FORMATION

Geologic setting.—

Thick sequences of Late Tertiary sediments interbedded with basaltic lavas are exposed throughout the western Snake River Plain in southwestern Idaho and adjacent Oregon (Fig. 5). Sediments of the Glenns Ferry Formation accumulated in a subsiding basin during Pliocene and Pleistocene time (Malde, 1959) and consist dominantly of siliciclastics and volcaniclastics ranging in texture from coarse gravels to clays. Locally, thick carbonates, volcanic ashes, lava flows, and diatomites also occur. The Glenns Ferry Formation is about 600 m thick at the type locality near Glenns Ferry, Idaho (Malde and Powers, 1962) and accumulated in a variety of lacustrine and fluvial settings (Malde, 1972; Kimmel, 1979; Middleton and others, 1985). Complex intertonguing of facies is common over comparatively short distances; Malde (1972) reported over 200 m of uniform silts and fine sands of lacustrine origin only a few kilometers from coarser grained sediments of floodplain origin.

The study area is located on the southern flank of a northwest-trending grabenlike structure (Hill, 1963) and occupies an axial position within the basin. Displacement along high angle faults dropped the western Snake River Plain relative to the highlands to the north. Malde (1959) reported 1.5 km of displacement during the early and middle Pliocene, with an additional 1.2 km of movement continuing into early Pleistocene time. Based on more recent work, Malde (pers. commun.) believes these values are too high. Nevertheless, movement on the bounding faults was episodic and resulted in periods of rapid subsidence alternating with times of relative basin stability. This rift basin received large volumes of material from the surrounding highlands to the north and south.

Petrologic analysis of the sediments suggests a complex and varied source terrain (Middleton, 1976). A volcanic provenance is indicated by felsic volcanic fragments, glass shards, sanidine, pseudohexagonal plates of biotite, and well developed pseudohexagonal dipyramids of quartz. Mixed-layer illite-smectite with a high proportion of expandable layers and nearly pure smectite are the most abundant clays and support a volcanic source. Reworked sedimentary clasts include sandstone and siltstone fragments as well as silty carbonate grains. The sources of these minerals and lithic fragments are Miocene and early Pliocene rocks (Idavada volcanics) south of the study area as well as older Tertiary deposits along the southern margin of the basin. Source interpretations are further substantiated by paleocurrent patterns that indicate north and northwest transport. In addition, the sandstones contain quartzo-feldspathic plutonic fragments derived from the Idaho Batholith.

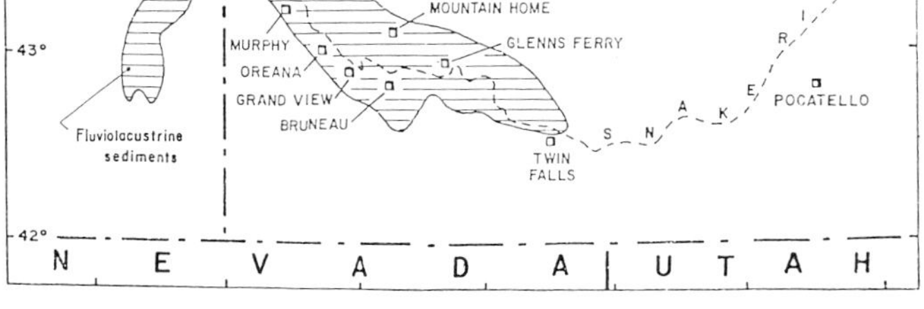

FIG. 5.—Base map of western Snake River Plain showing distribution of Late Tertiary fluvial/lacustrine deposits.

The Glenns Ferry was deposited during the late Pliocene with the formation becoming younger to the west and northwest, as evidenced by a variety of lithostratigraphic (Malde, 1972), paleontologic (Hibbard and Zakrzewski, 1967; Bjork, 1970), paleomagnetic (Neville and others, 1979), and fission-track (Swirydczuk and others, 1982; Kimmel, 1982) data. Magnetostratigraphic information, coupled with detailed studies of the vertebrate fauna, indicate that the formation in the study area near Grand View, Idaho (Fig. 5) was deposited during the upper Gauss and lower Matuyama magnetic epochs, between approximately 2.6 and 2.1 m.y. ago.

Climatic and paleoecologic conditions during Glenns Ferry deposition near Grand View, Idaho are inferred from a variety of paleontologic evidence. Both the mammalian fauna and the floral evidence suggest that the Glenns Ferry paleoclimate was temperate and moderately moist (Conrad, 1980). Ostracodes from the fluvial deposits also indicate moderately moist conditions (R. Forester, pers. commun.).

Depositional setting.—

Although the majority of the Glenns Ferry Formation accumulated in large lakes (Malde, 1972; Swirydczuk and others, 1979, 1980), fluvial deposits are common and locally thick (Malde, 1972). The western outlet of the Glenns Ferry basin was periodically blocked by lava flows (see discussion in Kimmel, 1982); however, regional paleocurrent trends in the fluvial deposits (west-northwest) suggest that streams flowed down the axis of the basin rather than being centripetally oriented as would be expected if the basin were "closed" during fluvial deposition. Of particular interest to this study are a series of ribbon and thin sheet sandstones enclosed by mudstone that are exposed along the Snake River west of Grand View, Idaho. Sandstone constitutes less than 30 percent of any section in the study area with the bulk of the fluvial deposits in the Glenns Ferry Formation comprised of white siltstones and olive claystones of overbank origin. The sandstones were deposited within channels and on levees and crevasse splays (Middleton and others, 1985).

Ribbon sandstones are common in the Glenns Ferry Formation west and north of Grand View, Idaho. Two ribbon sand bodies are described and, although their internal structures differ, each was deposited within a sinuous channel system characterized by active lateral accretion and numerous avulsive events. Most ribbon sandstones overlie an irregular, concave-upward surface which truncates mudstones and rarely other sand bodies. Relief on the ribbon sands is 8 m and width/thickness ratios are approximately 15:1.

A cross section of one of the ribbon sand bodies is shown in Figure 6. The basal scour surface is covered with a thin gravel veneer consisting of volcanic and claystone clasts as much as 3 cm in maximum dimension. Some claystone clasts are armored with a thin rind of medium to coarse sand. Presence of angular to subrounded clay clasts in this basal gravel indicates that scour was contemporaneous with deposition because these grains were clearly derived from channel bank muds.

Internal stratification is dominated by cosets of trough crossbedding as much as 3 m thick with sets ranging from 8 to 30 cm thick (Fig. 7). Set thickness decreases upward as does grain size. The sandstones are medium- to coarse-grained at the base and fine- to medium-grained near the top. The upper parts of most sand bodies consist of ripple cross-stratified fine sand. Climbing ripple laminations are common, as is convolute bedding. Horizontal laminations occur near the top at several localities, and root casts and calcite concretions tend to be more abundant near the top of the sand bodies. Most of these sandstones are gradationally overlain by either siltstone or claystone. In a few instances, however, thin sheetlike very fine-grained sandstones cap the coarser grained ribbon sand bodies.

Lateral facies relationships are obscure in many localities because of cover. Trenching of exposures, however, has documented a downcurrent (northwest) transition from large-scale trough cross-stratification to small-scale trough-and-ripple cross laminations. Similar changes were observed in sections normal to flow.

A second facies association is shown in Figure 8. This sand body is characterized by gently dipping (10° to 15°) lateral accretion deposits with thicknesses of 2 to 4 m. Internally they consist of small- and large-scale trough cross-stratification in sets as much as 35 cm thick. Both scale and grain size of these sets decrease vertically up the lateral accretion surfaces. The gently dipping surfaces typically

FIG. 7.—Coset of trough crossbedded sands and granules at base of ribbon sand body near Grand View, Idaho. Entire sequence is 1.1 m thick. Crossbedding measurements indicate a northwest flow direction. These sets overlie a 0.4-m-thick conglomerate layer and are overlain by 1.5 m of small-scale trough and ripple crossbedded fine-grained sandstones.

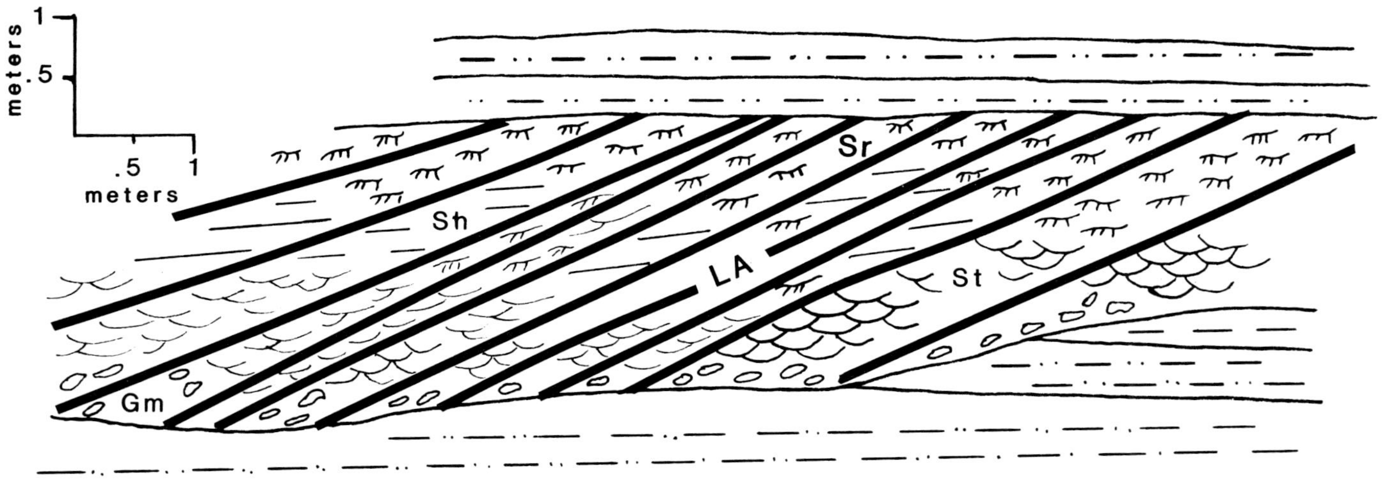

FIG. 6.—Field sketch through a Glenns Ferry ribbon sandstone showing lateral and vertical facies relationships. Gm = structureless conglomerate, St = large- and medium-scale trough crossbedded sandstone, Sh = planebedded sandstone, Sr = ripple crossbedded sandstone. Bold, low-angle dipping surfaces represent lateral accretion surfaces.

FIG. 8.—Lateral accretion (point bar) bundle on Black Butte approximately 12 km west of Grand View, Idaho along the Snake River. The set comprises low-angle (12°) dipping surfaces that internally exhibit large-scale trough crossbedding at the base grading upward along the surfaces into smaller scale and finer grained crossbeds. Entire set is 4 m thick. Overlying horizontally laminated sandstones and siltstones represent proximal overbank deposits.

overlie a gravel bed and are conformably overlain by either ripple cross-laminated sands and silts or plane-bedded sands. The large surfaces are oriented at approximately right angles to the sense of flow indicated by the trough crossbedding.

The large-scale surfaces formed on a point bar in a meandering system (e.g., Allen, 1963; Puigdefabregas and van Vliet, 1978). The internal structures and paleocurrent trends likewise are consistent with such an interpretation. In some instances there are two such sequences vertically juxtaposed indicating that the channel migrated back across a previous point bar deposit. In most cases, however, these lateral accretion bundles are overlain by thin sheet sandstones of crevasse splay origin or by siltstones and claystones deposited during periods of overbank flooding.

Although the first sequence described lacks lateral accretion surfaces, it also reflects point bar deposition as indicated by facies characteristics and association with extrachannel deposits such as levees, crevasse splays and fine-grained floodbasin sediments. Obvious lateral accretion surfaces are not common in most modern and many presumed ancient point bars (Jackson, 1978).

The fluvial deposits of the Glenns Ferry Formation were deposited in a meandering channel system near the depositional center of the Glenns Ferry basin. Vertical accretion deposits, however, constitute the bulk of the alluvial succession attesting to major episodes of overbank flooding. These flooding events were punctuated by periods of nondeposition on the floodplain as evidenced by pedogenically modified mudstones. Although poorly studied, the degree of pedogenic modification appears to have been relatively minor, suggesting that deposition from floodwaters was common across the floodplain and that overall rates of sediment accumulation were relatively rapid.

SYNTHESIS

In general, recent computer simulation studies predict that with rapid basin subsidence the degree of sand body interconnection is diminished, resulting in an alluvial succession dominated by overbank fine-grained clastics with minor volumes of tabular sand bodies. As the rate of basin subsidence decreases, sand body interconnection increases and, under relatively stable conditions, the models predict that the degree of channel sand body connection is very high. Stable conditions produce an alluvial fill comprised largely of sheetlike sand bodies and a small volume of overbank sediments because prolonged periods of basin stability allow time for extensive reworking of finer floodplain deposits by the lateral migration of the channel system.

The degree of channel deposit interconnection, and thus of sand body complexity, changes vertically through the lower Eocene Willwood Formation and also contrasts strongly between the Willwood Formation and the Upper Pliocene Glenns Ferry Formation. Because the controls on floodplain construction are complex and interdependent, it is difficult to recognize and isolate the factor(s) that governed the genesis of individual alluvial suites. Independent evidence for source lithology, climatic conditions, and sediment accumulation rates for these units, however, in conjunction with sedimentologic analyses, allows the contribution of tectonism to be more confidently evaluated both within the Willwood sequence and between the Willwood and Glenns Ferry Formations.

The Willwood Formation in the northern Bighorn Basin consists of sheet and ribbon channel sandstones and large volumes of overbank deposits. Whereas the ribbon sandstones reflect vertical accumulation in laterally fixed channels, the internal structure of sheet sandstones indicates deposition in laterally mobile or meandering streams. The northerly paleocurrent directions observed in most Willwood sheets and the position of the study area directly adjacent to the Tertiary axis of the basin show that the sheet sandstones represent a major axial stream system. Significant sedimentologic changes occur at the Clarkforkian-Wasatchian boundary, including the appearance of an unusually large and strongly multistory sheet sandstone and mature paleosols, which are otherwise rare.

The sedimentologic changes are attributed to tectonic controls, specifically a decrease in the rate of basin subsidence. Influence of other allocyclic mechanisms was minimal. Petrologic studies demonstrate that source rocks produced an unchanging supply of sedimentary and crystalline detritus to Willwood sediments during late Clarkforkian and earliest Wasatchian time. Although it is true that climatic conditions became warmer across the Paleocene-Eocene boundary, changes in the megaflora, which heralded the Eocene warming trend, predated formation of the boundary sheet sandstone by approximately 2 million years (Hickey, 1980). Thus, attributing development of the boundary sandstone to climatic change is inconsistent with this large time gap. Finally, the Willwood study area is geographically restricted and the entire area is proximal to the structural axis of the basin. Although this axis shifted through time as the basin evolved structurally, by Clarkforkian and Wasatchian time it had attained a stable position (Gingerich, 1983). Consequently, changes in alluvial architecture through the Willwood sequence do not reflect local structural controls.

The sedimentologic changes do accompany a decrease in sediment accumulation rate. Hickey's (1980) study indicated that accumulation rates across the northern Bighorn Basin began to decrease in the Clarkforkian and were even more reduced during Wasatchian time. Stratigraphic thicknesses (Gingerich, 1983) in conjunction with revised time estimates for subdivisions of the Paleogene (Berggren and others, 1985) yield sediment accumulation rates of 0.28 mm/year for the Tiffanian, 0.26 mm/year for the Clarkforkian and 0.22 mm/year for the first half of the Wasatchian. Because these figures are calculated for the land mammal ages in their entirety, they provide only gross estimates of accumulation rates for the late Clarkforkian and earliest Wasatchian. Willwood paleosol sequences provide compelling evidence, however, for decreased sediment accumulation rates across the Clarkforkian-Wasatchian boundary.

Kraus and Bown (1985) have demonstrated that the developmental stage attained by an individual Willwood paleosol depends largely on the short-term rate of sediment accretion, which in turn reflects local patterns of sedimentation determined by proximity to the channel and allocyclic factors such as subsidence rate. The degree of pedogenic development should be greater during periods of slow basin subsidence and slow sediment accumulation than during periods of rapid basin subsidence and rapid sediment accumulation. Paleosols in the lower part of the sequence below the boundary sandstone are no more than moderately mature (stages 1 through 3 of Bown, 1985), indicating fairly rapid rates of sediment accumulation. This is true even of paleosols that formed far from the channel, where pedogenic maturity is usually at a maximum. In marked contrast, at the southeast margin of the boundary sandstone is a sequence of unusually mature (stage 4 and some stage 5) paleosols which are temporally equivalent to that sand body (Gingerich, pers. commun.). The maturity of these paleosols indicates considerably slower rates of sediment accumulation as compared to older Willwood paleosols (Kraus, 1985). Because the evidence shows that these soils formed not far from the channel and because paleosols of this maturation stage are rare in the northern Bighorn Basin, their presence strongly suggests that decreased rates of accretion reflect diminished rates of basin subsidence rather than proximity of the developing soil to the channel.

Kraus (1980) suggested that decreased rates of basin subsidence were responsible for development of the sheet sandstone at the Clarkforkian/Wasatchian boundary. The thickness of individual stories and extent of lateral accretion deposits (where exposed) are no greater in the boundary sandstone than in other Willwood sheets. Hence, the three-dimensional magnitude of the boundary sandstone cannot be attributed to a larger channel system; rather, the impressive size of the boundary sandstone is the result of greater vertical stacking and lateral overlap among stories. This reflects lateral migration of the channel and, as indicated by the locality where two large sand bodies coalesce to form a very thick part of the boundary sandstone, probably also reflects channel switching and the superposition of two channel belts.

A second example of contrasting styles of alluvial architecture is offered by the Willwood and Glenns Ferry Formations. Whereas the Willwood is characterized by sheet sandstones, the fluvial deposits in the Glenns Ferry Formation comprise ribbon sandstones engulfed by large volumes of floodplain fines. Comparison between the Willwood and Glenns Ferry sequences reveals significant similarities in certain aspects of their geologic settings, including depositional style and climatic conditions. Deposition of the Glenns Ferry Formation occurred both in lakes and, in the study area, in a meandering stream system that flowed generally north and northwest across the western Snake River Plain. Meandering stream sedimentation also persisted throughout deposition of the Willwood sequence. Channel sizes in the two sequences were similar, as estimated from lateral accretion deposits which are 4 to 5 m thick in the Willwood sequence and 4 m thick in the Glenns Ferry Formation. Although the early Eocene in Wyoming was warmer than Late Pliocene time in Idaho, both formations were deposited in moderately moist climates in which precipitation was seasonal.

Petrologic study indicates that both units had sedimentary and crystalline source areas; however, the Glenns Ferry also had a major volcanic source. Nonetheless, both sequences had plentiful supplies of coarse- and fine-grained detritus and there is no indication that size availability played any role in differing alluvial architectures. Furthermore, the Glenns Ferry channel system was located near the center of the basin a considerable distance from the bounding faults and associated highlands and the Willwood stream system flowed near the structural axis of the basin. As such, the alluvial architectures reflect processes operative across the entire basin rather than simply the influence of nearby source terrains.

A reasonable explanation for the different styles of alluvial architecture therefore appears to be the tectonic settings. Both formations were deposited in tectonically active basins; however, comparison of sediment accumulation rates shows significant differences in the rates of basin subsidence. During fluvial deposition, the Glenns Ferry basin was undergoing periods of rapid subsidence asociated with movement along basin margin fault systems (Malde, 1959; Swirydczuk and others, 1980; Middleton and others, 1985). Although it is difficult to assess sediment accumulation rates for the Glenns Ferry Formation, the paleomagnetic studies of Neville and others (1979) provide a rough calculation of 0.35 mm/year for lacustrine deposits near Hagerman. On the basis of biostratigraphic and paleomagnetic data, Conrad (1980) has suggested even higher rates for fluvial deposits in the Glenns Ferry Formation. In contrast, rates for the Willwood Formation were slower (only 0.22 to 0.28 mm/year), indicating concomitantly slower rates of basin subsidence. The effect of more rapid sediment accumulation was generation of ribbon rather than sheet sand bodies in the Glenns Ferry Formation. As predicted by Allen (1978, 1979) and Bridge and Leeder (1979), the more rapid rates of basin subsidence resulted in Glenns Ferry streams being unable to rework the overbank alluvium to generate sheet-like deposits. With rapid accumulation, channels avulsed more frequently, prior to significant lateral migration or amalgamation of point bar deposits within the meander belt.

Consequently, only simple or weakly multistory sand bodies were generated.

A ribbon geometry resulted because the Glenns Ferry streams remained relatively fixed in their channels. The streams were relatively immobile not only as a result of high rates of basin subsidence, which precluded active lateral migration, but also as a result of the large volumes of silts and clays in overbank areas proximal to the channels. These fine-grained deposits constitute as much as 70 percent of the fluvial associations in some areas and attest to numerous, and most likely frequent, episodes of stream flooding. The general absence of well developed soil profiles and rhythmic intercalations of silt and clay support the concept of a floodplain that experienced numerous periods of suspension deposition by flood waters.

As shown by Bridge (1975), using computer simulations of sedimentation in meandering streams, the percentage of silt and clay in overbank and bank margin deposits profoundly influences the ability of a stream to migrate across its floodplain. The high percentage of silt and clay obviously increases the cohesiveness of the channel margins and thereby retards lateral erosion by the stream. Studies by Smith (1976) have shown that if these areas were vegetated, the lateral migration rates would be reduced even further.

It seems reasonable, therefore, to conclude that different rates of basin subsidence, determined from sediment accumulation rates, led to contrasting styles of alluvial architecture in the Willwood and Glenns Ferry Formations. Rapid basin subsidence and sediment accumulation produced simple or weakly multistory ribbon sand bodies enveloped by overbank deposits in the Glenns Ferry sequence. In the Willwood Formation, slower rates of basin subsidence and accumulation generated multistory sheet sand bodies. Furthermore, the degree of erosional overlap within Willwood sheets increased in earliest Wasatchian time in response to reduced rates of basin subsidence. Independent evidence for climate, position in the basin, and source-rock lithologies demonstrate that these factors were of secondary importance in generating different alluvial stratigraphies; rather, the different styles of alluvial architecture resulted principally from tectonic processes.

ACKNOWLEDGMENTS

We thank W. Bilodeau and G. A. Smith for constructive review of the manuscript. Kraus gratefully acknowledges T. M. Bown, P. D. Gingerich, and K. D. Rose for discussion on biostratigraphic aspects of the Willwood Formation and S. Rose, K. Pereira, and S. Johnson for assistance in the field. Field work in the Bighorn Basin was funded by a Faculty Career Development Award from the University of Colorado and NSF Grant #83-19421.

REFERENCES

ALLEN, J. R. L., 1963, The classification of cross-stratified units, with notes on their origin: Sedimentology, v. 2, p. 93–114.

———, 1974, Studies in fluviatile sedimentation; Implications of pedogenic carbonate units, Lower Old Red Sandstone, Anglo-Welsh outcrop: Geological Journal, v. 9, p. 181–208.

———, 1978, Studies in fluviatile sedimentation; An exploratory quantitative model for the architecture of avulsion-controlled alluvial suites: Sedimentary Geology, v. 21, p. 129–147.

———, 1979, Studies in fluviatile sedimentation: An elementary geometrical model for the connectedness of avulsion-related channel sand bodies: Sedimentary Geology, v. 24, p. 253–267.

ANDREWS, D. A., PIERCE, W. G., AND EARGLE, D. H., 1947, Geologic map of the Bighorn Basin, Wyoming and Montana: Oil and Gas Investigations Preliminary Map 71, U.S. Geological Survey.

BEERBOWER, J. R., 1964, Cyclothems and cyclic depositional mechanisms in alluvial plain sedimentation, *in* Merriam, D. F., ed., Symposium on Cyclic Sedimentation: Geological Survey of Kansas Bulletin, v. 169, p. 31–42.

BERGGREN, W. A., KENT, D. V., FLYNN, J. J., AND VAN COUVERING, J. A., 1985, Cenozoic geochronology: Geological Society of America Bulletin, v. 96, p. 1047–1418.

BJORK, P. R., 1970, The Carnivora of the Hagerman local fauna (late Pliocene) of southwestern Idaho: Transactions of the American Philosophical Society, v. 60, 54 p.

BLAKEY, R. C., AND GUBITOSA, R., 1984, Controls of sandstone body geometry and architecture in the Chinle Formation (Upper Triassic), Colorado Plateau: Sedimentary Geology, v. 38, p. 51–86.

BONINI, W. E., AND KINARD, R. E., 1983, Gravity anomalies along the Beartooth Front, Montana: Wyoming Geological Association, 34th Annual Field Conference Guidebook, p. 89–94.

BOWN, T. M., 1980, Summary of latest Cretaceous and Cenozoic sedimentary, tectonic, and erosional events, Bighorn Basin, Wyoming: University of Michigan Papers on Paleontology, No. 24, p. 25–32.

———, 1985, Maturation sequences in lower Eocene alluvial paleosols, Willwood Formation, *in* Fiores, R. M., and Harvey, M., eds., Field Guidbook to Modern and Ancient Fluvial Systems in the United States: Proceedings, Third International Fluvial Sedimentology Conference, p. 20–26.

———, AND KRAUS, M. J., 1981, Lower Eocene alluvial paleosols (Willwood Formation, northwest Wyoming, U.S.A.) and their significance for paleoecology, paleoclimatology and basin analysis: Palaeogeography, Palaeoclimatology, Palaeoecology, v. 34, p. 1–30.

BRIDGE, J. S., 1975, Computer simulation of sedimentation in meandering streams: Sedimentology, v. 22, p. 3–43.

———, AND DIEMER, J. A., 1983, Quantitative interpretation of an evolving ancient river system: Sedimentology, v. 30, p. 599–623.

———, AND LEEDER, M. R., 1979, A simulation model of alluvial stratigraphy: Sedimentology, v. 26, p. 617–644.

BUTLER, R. F., GINGERICH, P. D., AND LINDSEY, E. H., 1981, Magnetic polarity, stratigraphy and biostratigraphy of Paleocene and lower Eocene continental deposits, Clark's Fork Basin, Wyoming: Journal of Geology, v. 89, p. 299–316

CONRAD, G., 1980, The biostratigraphy and mammalian paleontology of the Glenns Ferry Formation from Hammet to Oreana, Idaho: Unpublished Ph.D. Thesis, Idaho State University, 327 p.

FLUECKINGER, L. A., 1971, Stratigraphy, petrography, and origin of Tertiary sediments off the front of the Beartooth Mountains, Montana-Wyoming: Unpublished Ph.D. Thesis, Pennsylvania State University, 249 p.

FOOSE, R. M., WISE, D. U., AND GARBARINI, G. S., 1961, Structural geology of the Beartooth Mountains, Montana and Wyoming: Geological Society of America Bulletin, v. 72, p. 1143–1172.

FRIEND, P. F., 1983, Towards the field classification of alluvial architecture or sequence, *in* Collinson, J. D., and Lewin, J., eds., Modern and Ancient Fluvial Systems: International Association of Sedimentologists Special Publication 6, p. 345–354.

———, SLATER, M. J., AND WILLIAMS, R. C., 1979, Vertical and lateral building of river sandstone bodies, Ebro Basin, Spain: Journal of the Geological Society of London, v. 136, p. 39–46.

GALLOWAY, W. E., 1981, Depositional architecture of Cenozoic Gulf coastal plain fluvial systems, *in*, Ethridge, F. G., and Flores, R. M., eds., Recent and Ancient Nonmarine Depositional Environments: Models for Exploration: Society of Economic Paleontologists and Mineralogists Special Publication 31, p. 127–156.

GINGERICH, P. D., 1983, Paleocene-Eocene faunal zones and preliminary analysis of Laramide structural deformation in the Clark's Fork Basin, Wyoming: Wyoming Geological Association, 34th Annual Field Conference Guidebook, p. 185–195.

HIBBARD, C. W., AND ZAKRZEWSKI, R. L., 1967, Phyletic trends in the Late Cenozoic microtine *Ophiomys* gen. nov., from Idaho: University of Michigan Paleontology Contributions, v. 21, p. 255–271.

HICKEY, L. J., 1980, Paleocene stratigraphy and flora of the Clark's Fork

Basin: University of Michigan Papers on Paleontology, No. 24, p. 33–50.

HILL, D. P., 1963, Gravity and crustal structure in the western Snake River Plain, Idaho: Journal of Geophysical Research, v. 68, p. 5807–5819.

JACKSON, R. G., 1978, Preliminary evaluation of lithofacies models for meandering alluvial streams, *in* Miall, A. D., ed., Fluvial Sedimentology: Canadian Society of Petroleum Geologists Memoir 5, p. 543–576.

KIMMEL, P. G., 1979, Stratigraphy and paleoenvironments of the Miocene Chalk Hills Formation and Pliocene Glenns Ferry Formation in the western Snake River Plain, Idaho: Unpublished Ph.D. Thesis, University of Michigan, 331 p.

———, 1982, Stratigraphy, age, and tectonic setting of the Miocene-Pliocene lacustrine sediments of the western Snake River Plain: Oregon and Idaho, *in* Bonnichsen, B., and Breckenridge, R. M., eds., Cenozoic Geology of Idaho: Idaho Bureau of Mines and Geology Bulletin 26, p. 559–578.

KRAUS, M. J., 1980, Genesis of a fluvial sheet sandstone, Willwood Formation, northwest Wyoming: University of Michigan Papers on Paleontology, No. 24, p. 87–94.

———, 1985, Sedimentology of Early Tertiary rocks, northern Bighorn Basin, *in* Flores, R. M., and Harvey, M., eds., Field Guidebook to Modern and Ancient Fluvial Systems in the United States: Proceedings, Third International Fluvial Sedimentology Conference, p. 26–33.

———, AND BOWN, T. M., 1985, Relationships of paleosol maturity to autocyclic and allocyclic controls on sediment accumulation rates, Willwood Formation (lower Eocene), Wyoming, U.S.A.: Geological Society of America, Abstracts with Programs, v. 17, p. 634.

LEEDER, M. R., 1978, A quantitative stratigraphic model for alluvium, with special reference to channel deposit density and interconnectedness, *in* Miall, A. D., ed., Fluvial Sedimentology: Canadian Society of Petroleum Geology Memoir 5, p. 587–596.

MALDE, H. E., 1959, Fault zone along northern boundary of western Snake River Plain, Idaho: Science, v. 130, p. 272.

———, 1972, Stratigraphy of the Glenns Ferry Formation from Hammett to Hagerman, Idaho: U.S. Geological Survey Bulletin 1331-D, 19 p.

———, AND POWERS, H. A., 1962, Upper Cenozoic stratigraphy of western Snake River Plain, Idaho: Geological Society of America Bulletin, v. 73, p. 1197–1220.

MIALL, A. D. 1980, Cyclicity and the facies model concept in fluvial deposits: Canadian Petroleum Geologists Bulletin, v. 28, p. 59–80.

MIDDLETON, L. T., 1976, Depositional environments of the Glenns Ferry Formation near Jackass Butte, Idaho: Unpublished M.S. Thesis, Idaho State University, 58 p.

———, PORTER, M. L., AND KIMMEL, P. G., 1985, Depositional settings of the Chalk Hills and Glenns Ferry Formations, west of Bruneau, Idaho, *in* Flores, R. M., and Kaplan, S. S., eds., Cenozoic Paleogeography of West-Central United States: Society of Economic Paleontologists and Mineralogists, Rocky Mountain Section, p. 37–53.

NEASHAM, J. W., 1970, Sedimentology of the Willwood Formation (lower Eocene): an alluvial molasse facies in northwestern Wyoming: Unpublished Ph.D. Thesis, Iowa State University, 98 p.

———, AND VONDRA, C. F., 1972, Stratigraphy and petrology of the lower Eocene Willwood Formation, Bighorn Basin, Wyoming: Geological Society of America Bulletin, v. 83, p. 2167–2180.

NEVILLE, C., OPDYKE, N.D., LINDSEY, E. H., AND JOHNSON, N. M., 1979, Magnetic stratigraphy of Pliocene deposits of the Glenns Ferry Formation, Idaho, and its implications for North American mammalian biostratigraphy: American Journal of Science, v. 279, p. 503–526.

OSTERWALD, F. W., AND DEAN, B. G., 1961, Relation of uranium deposits to tectonic pattern of the central Cordilleran foreland: U.S. Geological Survey Bulletin 1087-I, p. 337–390.

PUIGDEFABREGAS, CAYO, AND VAN VLIET, ARTHUR, 1978, Meandering stream deposits from the Tertiary of the southern Pyrenees, *in* Miall, A. D., ed., Fluvial Sedimentology: Canadian Society of Petroleum Geology Memoir 5, p. 469–485.

READ, W. A., AND DEAN, J. M., 1982, Quantitative relationships between numbers of fluvial cycles, bulk lithological composition and net subsidence in a Scottish Namurian basin: Sedimentology, v. 29, p. 181–200.

ROSE, K. D., 1981, The Clarkforkian Land-Mammal Age and mammalian faunal composition across the Paleocene-Eocene boundary: University of Michigan Papers on Paleontology, no. 26, 196 p.

SMITH, D. G., 1976, Effect of vegetation on lateral migration of anastomosed channels of a glacial meltwater river: Geological Society of America Bulletin, v. 87, p. 857–860.

SWIRYDCZUK, KRYSTYNA, WILKINSON, B. H., AND SMITH, G. R., 1979, The Pliocene Glenns Ferry Oolite: Lake-margin carbonate deposition in the southwestern Snake River Plain: Journal of Sedimentary Petrology, v. 49, p. 995–1004.

———, ———, AND ———, 1980, The Pliocene Glenns Ferry Oolite II: Sedimentology of oolitic lacustrine terrace deposits: Journal of Sedimentary Petrology, v. 50, p. 1237–1248.

———, LARSON, G. P., AND SMITH, G. R., 1982, Volcanic ash beds as stratigraphic markers in the Glenns Ferry and Chalk Hills Formations from Adrian, Oregon, to Bruneau, Idaho, *in* Bonnichsen, B., and Breckenridge, R. M., eds., Cenozoic Geology of Idaho: Idaho Bureau of Mines and Geology Bulletin 26, p. 543–558.

VAN HOUTEN, F. B., 1944, Stratigraphy of the Willwood and Tatman Formations in north-western Wyoming: Geological Society of America Bulletin, v. 55, p. 165–210.

WING, S. L., 1981, A study of paleoecology and paleobotany in the Willwood Formation (early Eocene, Wyoming); Unpublished Ph.D. Thesis, Yale University, 391 p.

———, AND BOWN, T. M., 1985, Fine scale reconstruction of late Paleocene-early Eocene paleogeography in the Bighorn Basin of northwestern Wyoming, *in* Flores, R. M., and Kaplan, S. S., eds., Cenozoic Paleogeography of the West-Central United States: Society of Economic Paleontologists and Mineralogists, Rocky Mountain Section, p. 93–105.

FLUVIAL RESPONSE TO EOCENE TECTONISM, THE BRIDGER FORMATION, SOUTHERN WIND RIVER RANGE, WYOMING

PATRICIA E. GROLL AND JAMES R. STEIDTMANN
Department of Geology and Geophysics, University of Wyoming, Laramie, Wyoming 82071

ABSTRACT: Depositional style of the middle Eocene Bridger Formation on the south flank of the Wind River Range, Wyoming, changed in response to volcanism to the north in the Absaroka volcanic field. Lower Bridger sediments were deposited in a fluvio-deltaic environment during the waning stage of Lake Gosiute. Channel lag sediments, trough crossbedding and well developed epsilon crossbedding indicate a high sinuosity (meandering) stream pattern. In contrast, upper Bridger sandy channels are simple to multistory ribbons composed dominantly of trough cross-strata. Tuffs and tuffaceous siltstones form channel fills and overbank deposits and constitute a major portion of the unit. Upper Bridger deposition occurred under a variety of sediment/water ratios, from debris flow to normal turbulent stream flow. Gravels in the upper Bridger contain well rounded clasts that were derived from the Absaroka volcanic field. Paleocurrents indicate that these volcanic epiclasts were transported across the Wind River Range through a low divide into the basin. This low divide in the range served as a node for avulsion, and rapid deposition caused formation of short-lived channels.

Syndepositional movement along the Continental fault zone in the study area provided intrabasinal control on upper Bridger deposition. On adjacent fault blocks, the South Pass Formation overlies upper Bridger strata and disconformably overlies lower Bridger strata. The disconformity omits 80 m of upper Bridger section. In the upper Bridger on the downthrown block there is a coarse clastic wedge that may have been reworked off the adjacent uplift. Paleocurrents deflect around the upthrown block, suggesting the faulting was contemporaneous with upper Bridger deposition.

INTRODUCTION

The Bridger Formation along the southern flank of the Wind River Range, Wyoming (Fig. 1), displays a fluvial response to both extrabasinal volcanism and intrabasinal faulting. It is a predominantly fine-grained fluvial unit that contains a significant amount of volcanic material. The rapid influx of pyroclastic debris from the Absaroka volcanic field provided controls that radically changed depositional style.

The middle Eocene upper Bridger Formation (Fig. 2) crops out in a series of low sage-covered buttes along the north side of the Continental fault, a southeast-trending fault zone that flanks the southern margin of the range (Nace, 1939; Bradley, 1964) (Fig. 1). At present, the Continental fault is a normal fault system associated with the "collapse" of the southern portion of the range (Zeller and Stephens, 1969; Love, 1970; Steidtmann and others, 1983). Our analysis of sedimentation patterns indicates that the Continental fault provided intrabasinal controls on deposition of upper Bridger sediments. This intrabasinal control documents a multiphase movement history for the Continental fault (Groll and Steidtmann, 1985).

SEDIMENTOLOGY OF THE BRIDGER FORMATION

Matthew (1909) divided the Bridger into five faunal levels, A through E from bottom to top. Bridger A, B and C are informally grouped into the "lower" Bridger, whereas Bridger D and E comprise the "upper" Bridger.

Lower Bridger.—

Deposition of lower Bridger sediments occurred during the waning stage of Lake Gosiute (Bradley, 1964; Wood, 1966; West, 1969; Love, 1970). The lower Bridger conformably overlies the Laney Shale of the Green River Formation (Fig. 2). Predominant rock types include siltstone, sandstone, marlstone, limestone and bentonitic claystone. Ostracodes and algal structures are common in the limestones, which formed during fluctuations in lake level and provide excellent marker beds (Wood, 1966).

Deposition of the lower Bridger in the Bridger Basin occurred primarily in a fluvio-deltaic environment (Wood, 1966). Initially, the fluvial contribution was relatively small. Clastics brought in by drainage from the west prograded eastward into Lake Gosiute (Wood, 1966). As sediment supply fluctuated or subsidence rates changed, there were minor transgressions of the lake but in general lacustrine deposition decreased upward. The delta plain developed better channel systems that drained from the northwest (113° vector mean, Gustav, 1974). These channels exhibit well developed epsilon crossbedding in association with lag deposits and trough crossbedding, indicating a meandering (high sinuosity) stream system.

Upper Bridger.—

The upper Bridger contains a significantly higher volume of volcanogenic material than the lower Bridger. Upper Bridger deposits of the study area (Bridger D of Matthew, 1909) are entirely fluvial, and sedimentation patterns are distinctly different from those in lower Bridger sediments. Sandstone bodies are simple to multistory lenticular ribbons (average width to depth ratio of 11:1) which consist primarily of trough crossbedding. Apparently, dune migration was the dominant transport mechanism (Harms and others, 1982). Channel sands are medium to coarse, have a bimodal size distribution, and display layer differentiation between sand-rich laminae and coarser, probably hydrodynamically equivalent pumiceous material (Fig. 3). Paleocurrent data from large-scale trough cross-strata have a low variance and are unimodal (Fig. 4). A vector mean of 182° indicates a paleoslope from the north into the basin.

Lenses of pebbly sandstone and gravel containing both volcanic and plutonic clasts are also present. Volcanic epiclasts range from 1 to 3 cm in long dimension, are fairly well rounded and moderately to highly spherical. They have variable composition but intermediate types, such as dacite and andesite predominate. Plutonic epiclasts are slightly smaller (0.5–2.0 cm maximum dimension) and much more angular. The difference in roundness suggests that the volcanic clasts were transported farther. Although the smaller size of the plutonic clasts seems inconsistent with this, it

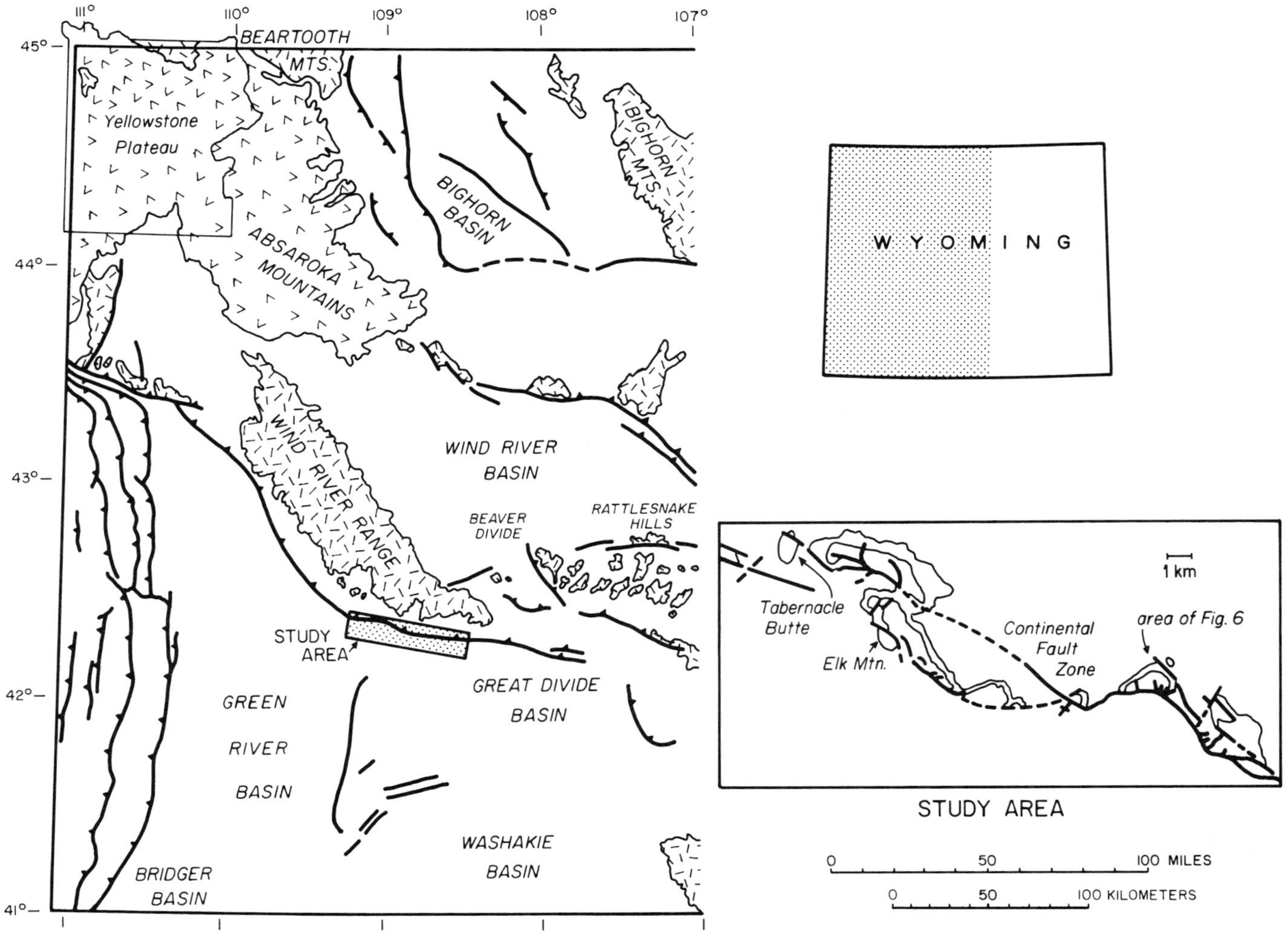

FIG. 1.—Location map of western Wyoming showing major basins, uplifts, and structural trends. Inset is a more detailed map of study area showing locations discussed in text.

may simply be a function of what was available in the source, or of reduced competency in a stream already overloaded by the large volume of volcanics.

The dominant sediment supplied to the upper Bridger deposystem was pyroclastic material, primarily ash and pumice. Tuffs and tuffaceous siltstones form a major portion of the unit, both as overbank deposits and as channel fills. Most of the ash deposits are massive.

In the crossbedded sandy channels, it is clear that deposition took place under fully turbulent, dilute stream flow. Sedimentation patterns elsewhere suggest there were a variety of sediment/water ratios, ranging from viscous debris flows to fully turbulent normal stream flow. In volcanic areas, debris flow deposits have been shown to extend in excess of 100 km from their source (Janda and others, 1981). They are characterized by poor sorting, lack of internal stratification, and inverse grading. In the distal setting of the study area, debris flow deposits occur as inverse graded or massive layers of tuff which, in places, contain large pumice fragments dispersed in the silt to fine sand-size matrix. These deposits are often stacked, indicating recurrent surges of debris were rapidly emplaced. Beverage and Culbertson (1964) described flows with sediment content of 40 to 80 weight percent and used the term "hyperconcentrated stream flows" for those intermediate between debris flow and normal turbulent flow. A related term, "hyperconcentrated flood flow," is used to describe a depositional process involving high discharge and rapid deposition which, unlike debris flow, does not occur *en masse* (Smith, 1986). Unfortunately, there are no well documented fine hyperconcentrated flood flow deposits for comparison. Those we interpret in this study are massive tuffs with very lenticular shapes, as if the ash had filled channels progressively. These ash deposits locally show indistinct horizontal laminations.

We envision that immediately following an eruption, the sediment/water ratio in the streams increased abruptly, and high sedimentation rates resulted. As time passed, debris flows became dilute and made the transition to hyperconcentrated flood flows. The variety of deposits in the area, from crossbedded sandy channel deposits to inverse graded debris flow deposits, suggests this range of conditions existed during deposition of upper Bridger sediments.

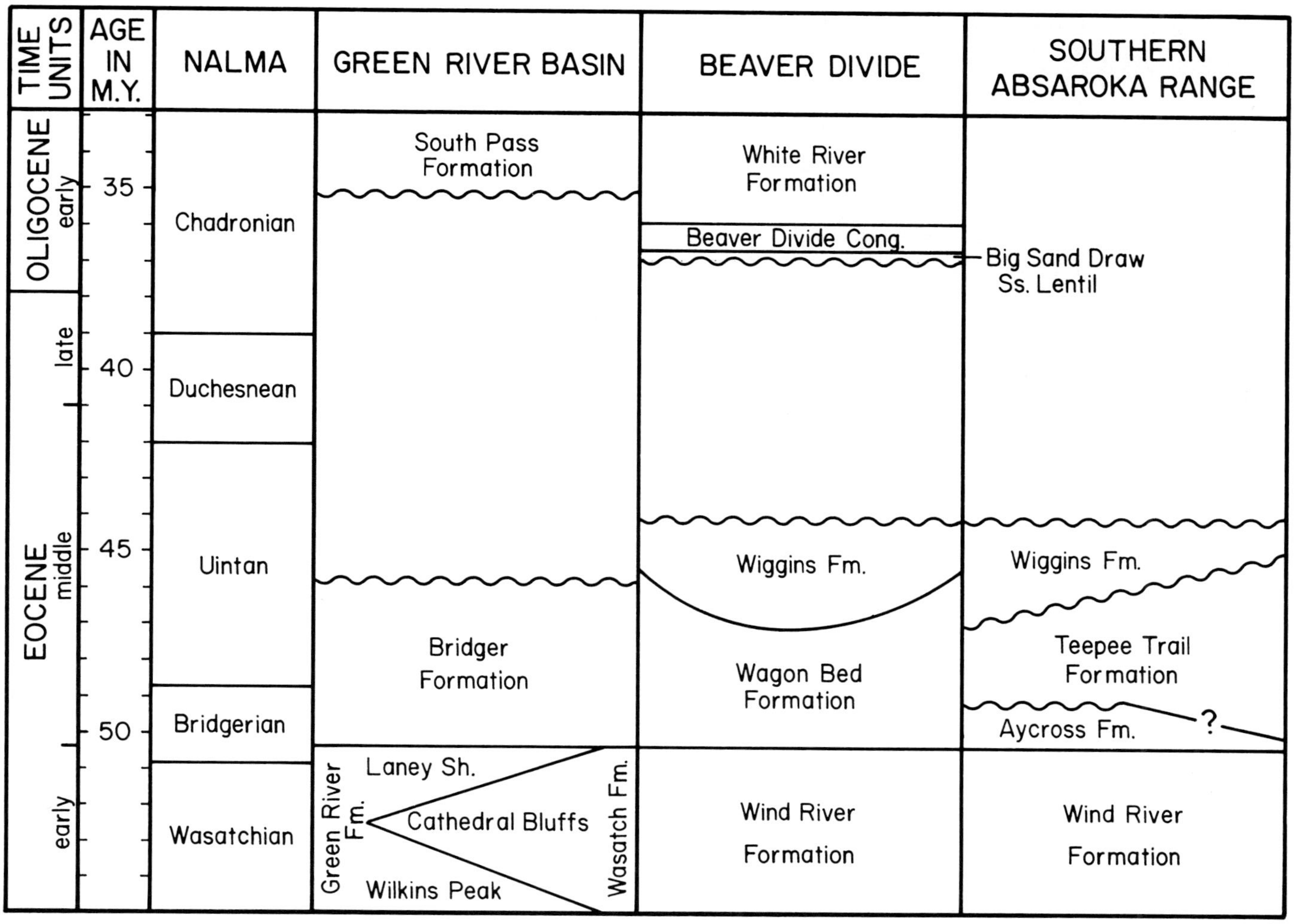

FIG. 2.—Stratigraphic chart for part of western Wyoming during the Eocene and early Oligocene. NALMA = North American Land Mammal Age. Compiled from Zeller and Stephens, 1969; Love, 1939; Smedes and Prostka, 1972; Van Houten, 1964; Evernden and others, 1964; Houston, 1964; Mauger, 1977; Bottjer, unpublished material, 1982.

EXTRABASINAL CONTROL

Throughout deposition of Bridger sediments, an increasing amount of volcanism was occurring to the north in the Absaroka volcanic field (Fig. 1). Major volcanic activity began in middle Eocene time and persisted into the Miocene (Wilson, 1971). An older to younger age trend from northwest to southeast, identified by Love and others (1974), is accompanied by a trend from acidic to more basic composition (Love, 1970; Wilson, 1971) with predominant rock types being dacite and pyroxene- and/or hornblende-andesite. Early eruptions in the north were primarily responsible for the pyroclastic material in the region. Structural relief caused by folding, faulting and subsequent erosion may have contributed to the influx of epiclastics in many Eocene sediments in the region (Wilson, 1971). In the southern half of the Absaroka volcanic field, andesite flows and breccias grade laterally into volcanic sandstones and conglomerates. Correlative deposits farther south in the Beaver Divide area (Fig. 2) may be the distal portion of the same depositional system of volcaniclastics (Emry, 1975).

Surdam and Stanley (1980), working to the southeast in the Great Divide, Green River and Washakie basins, documented volcanic clasts of andesitic composition in Eocene deposits that interfinger with the Green River Formation (late early to early middle Eocene). Their paleocurrent measurements indicate drainage from the north, and they postulated a drainage from the Absaroka volcanic field along the east margin of the Wind River Range and into the Great Divide and Washakie basins during Eocene time (Fig. 1). This drainage may be the same one described by Emry (1975) for the Beaver Divide area.

The composition of volcanic epiclasts in the Bridger gravels of this study are also compatible with an Absaroka provenance, and our paleocurrents also suggest drainage from the north. The clast size is intermediate between the more proximal boulder deposits of the Wind River Basin and the medium sands of the Piceance Basin to the south. To derive these clasts from the Absarokas would require drainage *across* the Wind River Range. If the clasts had been transported along the drainage postulated by Surdam and Stanley, the channel would have to have been deflected to the

FIG. 3.—Bridger channel deposit showing bimodal components: sand-rich laminae and coarser, hydrodynamically equivalent pumiceous material. Trough cross-stratification typifies channel deposits.

west, around the southern tip of the range. This is not substantiated by our paleocurrent data (Fig. 4). Likewise, drainage along the west side of the range would require a drainage deflection toward the east, also not indicated by our data. Bridger strata lap onto the hanging wall of the Wind River fault, suggesting that the range had low relief at the time. We infer that low relief on the Wind River Range during the middle Eocene allowed direct southward transport of volcanic epiclastic material from the Absaroka volcanic field across a low divide in the Wind River Range and into the northeast Green River and northwest Great Divide basins. This may have been accomplished by headward erosion which ultimately caused stream capture of the drainage system proposed by Surdam and Stanley (1980), or there may have been more than one breach of the range. These drainages were established because the Wind River Basin had been completely filled with pyroclastics by this time (Love, 1970; Surdam and Stanley, 1980).

This localization of flow across a divide may have served as a node for the avulsion pattern of upper Bridger fluvial systems (Fig. 5). As flow reached the Bridger alluvial plain, it was no longer constrained in discrete valleys by crystalline basement. When surges of pyroclastic debris were transported through the divide, channels filled and flow avulsed. This resulted in the single story lenticular ribbon deposits characteristic of upper Bridger channels. By filling the Wind River Basin with pyroclastics, Absaroka volcanism allowed formation of a drainage across the Wind River

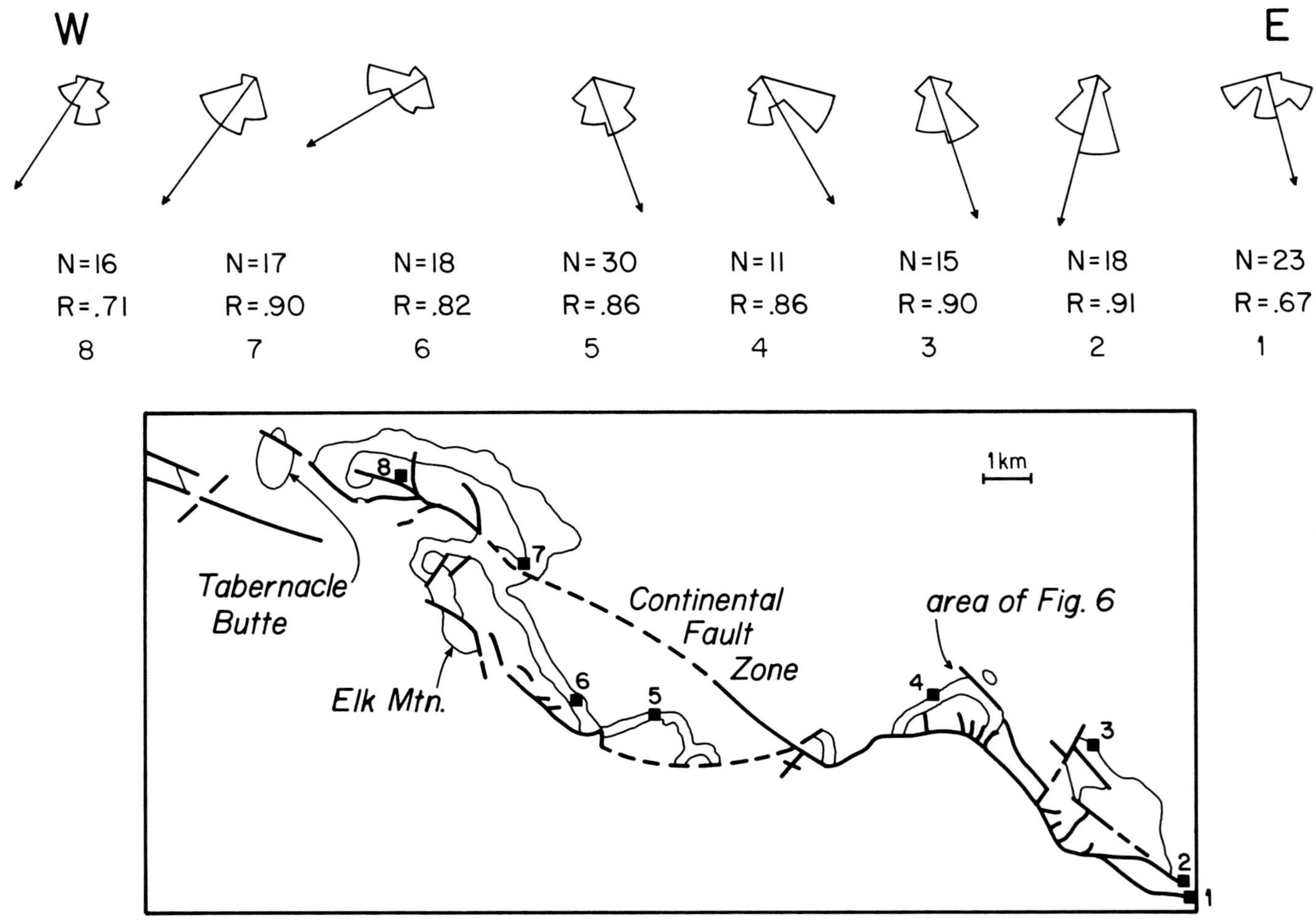

FIG. 4.—Upper Bridger paleocurrent rose diagrams from eight localities across the study area. N = number of measurements. R = dispersion value, 1.0 = 0 percent dispersion. Measurements obtained on trough cross-strata.

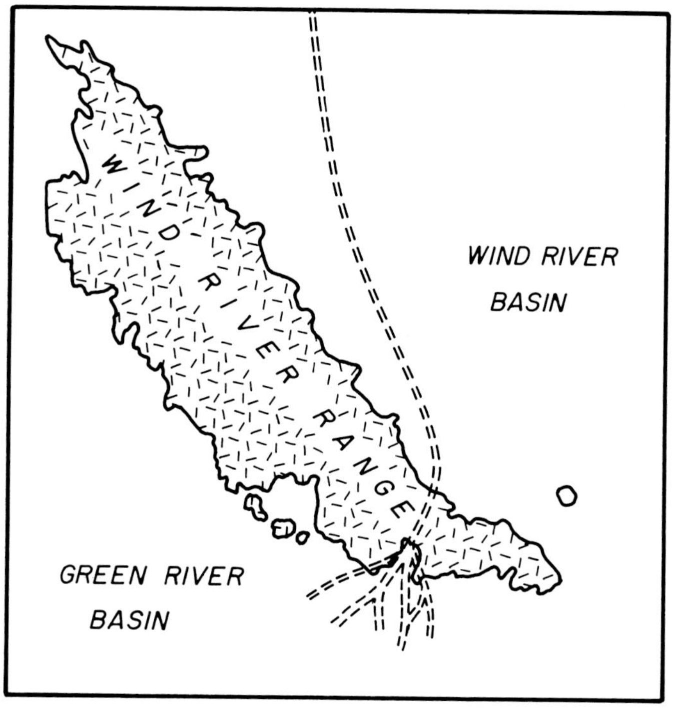

FIG. 5.—Simplified general model for upper Bridger deposition. Drainage from Absarokas was constricted by divide in Wind River Range. This constriction acted as node for avulsion pattern in the depocenter. Flow direction was generally to the south in immature channels. Avulsion occurred downstream from the node.

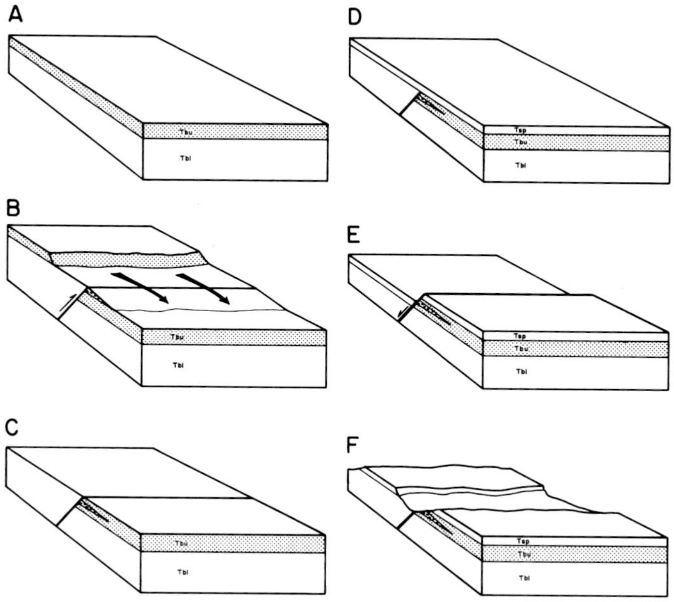

FIG. 6.—Sequence of events in development of disconformity. A) Bridger section was deposited conformably. B) Reverse movement on Continental fault system, formation of clastic wedge on south block. C) Upper Bridger completely removed from north block. D) South Pass deposition. E) Reactivation of fault in normal sense. F) Erosion of fault valley shows present configuration.

Range. This drainage brought a large amount of extrabasinal detritus into the Bridger deposystem. The result was alteration of both depositional style and lithologic composition.

INTRABASINAL CONTROL

Structural modification of the Bridger alluvial plain occurred along many segments of the Continental fault. A long and complex history of movement for this fault zone generated both disconformities and angular unconformities within the Bridger and adjacent units. A close look at these stratigraphic relations and related sedimentation patterns reveals the sequence of causitive tectonic events. There is a disconformity between the lower Bridger and overlying late Oligocene or early Miocene South Pass Formation in the eastern part of the study area (Fig. 1). On a downthrown fault block to the south, the upper Bridger rests conformably on the lower Bridger, and is overlain by the South Pass. On the adjacent upthrown block to the north the South Pass is directly on lower Bridger rocks. Nearly 80 m of upper Bridger strata are absent on this block, only 0.25 km from the complete section on the south block. Apparently, upper Bridger sediments were removed from the north block prior to South Pass deposition. The reworking of these sediments is evidenced by a pebble conglomerate wedge approximately 30 m above the base of the upper Bridger on the south block. This wedge fines upward into sand, silt and clay and is wider and thinner than other clastic bodies in the upper Bridger. It contains clasts of volcanic, plutonic and sedimentary rocks, as well as what appear to be intraformational clasts reworked from the Bridger on the upthrown block.

A scenario illustrating the development of the stratigraphic and structural setting for the Bridger in this area is shown in Figure 6. Bridger beds were deposited conformably throughout the area. Reverse motion along the Continental fault uplifted the north side, causing flow to deflect around this block and upper Bridger sediments to be removed and redeposited as a clastic wedge on the south block. Eventually all the upper Bridger was removed. After South Pass sediments were deposited over the entire area, the Continental fault was reactivated in a normal sense with the north side dropping as the Wind River Range began its post-compressional collapse.

CONCLUSION

The great volume of volcanic debris extruded in northwestern Wyoming during the Eocene strongly controlled both style and mechanism of syntectonic deposition in southwest Wyoming. Bridger Formation deposits changed from fluvio-deltaic bentonite, marl and epsilon crossbedded sand in the lower sequence to coarse pumiceous trough crossbedded sand, tuffaceous channel fill and flood plain tuff in the upper sequence. These upper Bridger rocks represent deposition by normal turbulent flow, debris flow and probably hyperconcentrated flood flow. The presence of a bimodal suite of clasts in gravels (large round volcanics and smaller, more angular plutonics), coupled with the paleocurrent data suggest that drainage carried volcanic material from the Absarokas through a divide in the Wind River Range, where the stream picked up plutonic clasts and carried both into the Green River and Great Divide basins. Other workers (Van Houten, 1964; Love, 1970; Emry, 1975; Surdam and Stanley, 1980) have also documented Absaroka-type clasts of decreasing size southward from the volcanic center.

This divide across the Wind River Range also functioned as an avulsion node for upper Bridger fluvial systems. Flow conditions alternated between debris flow and relatively clear flow as sediment content fluctuated. Sporadic volcanism caused rapid deposition in poorly formed channels. Because sedimentation was so rapid, flow frequently avulsed to lower areas producing a system of fairly straight short-lived channels. Sand facies are dominated by trough crossbedding, indicating that dune migration was the common transport mechanism in these streams.

The presence of both disconformities and angular unconformities between upper Bridger and adjacent strata, as well as within Bridger strata suggest that the Continental fault was an active intrabasinal control on deposition. Contemporaneous faulting and deposition are indicated by paleocurrents deflected around the uplifted block and a coarse intraformational wedge on the lower block. These fluvial sedimentation patterns document a multiphase movement history for the Continental fault zone.

ACKNOWLEDGMENTS

This study was supported by National Science Foundation Grant No. EAR-8108939 to Steidtmann and by a grant to Groll from Amoco Production Company, Denver, Colorado.

REFERENCES

BEVERAGE, J. P., AND CULBERTSON, J. K., 1964, Hyperconcentrations of suspended sediment: Journal of Hydraulics Division, Proceedings, American Society of Civil Engineering, v. 90, p. 117–128.

BRADLEY, W. H., 1964, Geology of the Green River Formation and associated Eocene rocks in southwestern Wyoming and adjacent parts of Colorado and Utah: U.S. Geological Survey Professional Paper 185-I, p. 163–199.

EMRY, R. J., 1975, Revised Tertiary stratigraphy and paleontology of the western Beaver Divide, Fremont County, Wyoming: Smithsonian Contributions to Paleobiology, no. 25, 20 p.

EVERNDEN, J. F., SAVAGE, D. E., CURTIS, G. H., AND JAMES, G. T., 1964, Potassium-argon dates and the Cenozoic mammalian chronology of North America: American Journal of Science, v. 262, p. 145–198.

GROLL, P. E., AND STEIDTMANN, J. R., 1985, Tertiary deformation history of the Continental fault zone, southern Wind River Range, Wyoming: Geological Society of America, Rocky Mountain Section Meeting, Abstracts with Programs, v. 17, no. 4, p. 221.

GUSTAV, S. H., 1974, The sedimentology and paleogeography of the Bridger Formation (Eocene) of southwest Wyoming Unpublished M.S. Thesis, University of Massachusetts, Amherst, 82 p.

HARMS, J. C., SOUTHARD, J. B., AND WALKER, R. G., 1982, Structures and sequences in clastic rocks: Society of Economic Paleontologists and Mineralogists Short Course No. 9, 249 p.

HOUSTON, R. S., 1964, The petrographic calendar, Part 3 of Parker, R. B., ed., Non-paleontological methods of correlation of rocks of Tertiary age in Wyoming: Contributions to Geology, v. 3, p. 15–26.

JANDA, R. D., SCOTT, K. M., NOLAN, K. M., AND MARTINSON, H. A., 1981, Lahar movement, effects and deposits, *in* Lipman, P. W., and Mullineaux, D. R., eds., The 1980 Eruptions of Mt. Saint Helens, Washington: U.S. Geological Survey Professional Paper 1250, p. 461–478.

LOVE, J. D., 1939, Geology along the southern margin of the Absaroka Range, Wyoming: Geological Society of America Special Paper 20, 134 p.

———, 1970, Cenozoic geology of the Granite Mountains area, central Wyoming: U.S. Geological Survey Professional Paper 495-C, 154 p.

LOVE, L. L., KUDO, A. M., AND LOVE, D. W., 1974, Petrochemical and age trends in the Absaroka-Gallatin Volcanic Province: EOS, American Geophysical Union Transactions, v. 55, p. 487.

MATTHEW, W. D., 1909, The Carnivora and Insectivora of the Bridger Basin, middle Eocene: Memoirs of the American Museum of Natural History, v. 9, Art. 9, p. 311–328.

MAUGER, R. L., 1977, K-Ar ages of biotites from tuffs in Eocene rocks of the Green River, Washakie and Uinta basins, Utah, Wyoming and Colorado: Contributions to Geology, v. 15, p. 135–139.

NACE, R. L., 1939, Geology of the northwest part of the Red Desert, Sweetwater and Fremont Counties, Wyoming: Wyoming Geological Survey Bulletin No. 27, 48 p.

SMEDES, H. W., AND PROTSKA, H. J., 1972, Stratigraphic framework of the Absaroka Volcanic Supergroup in the Yellowstone National Park region: U.S. Geological Survey Professional Paper 729-C, 33 p.

SMITH, G. A., 1986, Coarse-grained nonmarine volcaniclastic sediment: Terminology and depositional process: Geological Society of America Bulletin, v. 97, p. 1–10.

STEIDTMANN, J. R., MIDDLETON, L. C., AND MCGEE, L. T., 1983, Laramide sedimentation, folding and faulting in the southern Wind River Range, Wyoming, *in* Lowell, J. D., and others, eds., Rocky Mountain Foreland Basins and Uplifts, Rocky Mountain Association of Geologists, p. 161–167.

SURDAM, R. C., AND STANLEY, K. O., 1980, Effects of changes in drainage basin boundaries on sedimentation in Eocene Lakes Gosiute and Uinta of Wyoming, Utah and Colorado: Geology, v. 8, p. 135–139.

VAN HOUTEN, F. B., 1964, Tertiary geology of the Beaver Rim area, Fremont and Natrona Counties, Wyoming: U.S. Geological Survey Bulletin 1164, 99 p.

WEST, R. M., 1969, Geology and vertebrate paleontology of the northeastern Green River Basin, Wyoming: 21st Annual Field Conference, Wyoming Geological Association Guidebook, p. 77–92.

WILSON, W. H., 1971, Volcanic geology and mineralization, Absaroka Mountains, northwest Wyoming: 23rd Annual Field Conference, Wyoming Geological Association Guidebook, p. 151–155.

WOOD, C. B., 1966, Stratigraphy and paleontology of the Bridger Formation northeast of Opal, Lincoln County, Wyoming: Unpublished M.S. Thesis, University of Wyoming, Laramie, 112 p.

ZELLER, H. D., AND STEPHENS, E. V., 1969, Geology of the Oregon Buttes area, Sweetwater, Sublette and Fremont Counties, southwestern Wyoming: U.S. Geological Survey Bulletin 1256, 60 p.

STRUCTURAL CONTROLS ON FLUVIAL DISTRIBUTARY SYSTEMS— THE LUNA SYSTEM, NORTHERN SPAIN

GARY J. NICHOLS

Department of Geological Sciences, University of Liverpool, P.O. Box 147, Liverpool, L69 3BX, U.K.

ABSTRACT: Flat-lying Lower Miocene deposits in the northwestern part of the Ebro Basin, Spain, are considered to be the deposits of a single, radial distributary system, the Luna System, which formed in this subsiding foreland basin. Facies distribution patterns and vertical sequences were controlled by the structure of the adjacent south Pyrenean thrust belt, the rate of basin subsidence and the rates of evaporation and infiltration. The apex of the Luna System is well defined and is adjacent to a structural low in the southern Pyrenean front. Proximal pebbly braided stream deposits show two major pulses of sedimentation which are related to two tectonic episodes and rejuvenation of erosion in the drainage basin within the thrust belt. The medial part of the system is dominated by sandy channel-fill and overbank facies. The channel bodies show a low divergence of paleocurrent directions. They contain little evidence for lateral migration and have a low interconnection. This suggests that the channels were relatively straight and stable and that the rates of subsidence and sediment accumulation were high compared to the frequency of avulsion. Evaporation and infiltration in the basin may have resulted in a radial decrease in discharge. Poorly channelized sheets are more abundant in the distal parts of the system (40 km from the apex) where fluvial sediments interdigitate with lacustrine muds and carbonates. It is suggested that the vertical sequences and lateral facies patterns in this exceptionally well preserved system may be characteristic of subsiding foreland basins with internal drainage.

INTRODUCTION

In this study the deposits of an early Miocene fluvial distributary system in northeast Spain are considered. Both facies distribution patterns and vertical sequences can be related to tectonic controls. The deposits are generally flat-lying, and it is possible to trace variations in the style of fluvial deposition from the most proximal parts of the system at the basin margin to a zone of interdigitation with lacustrine deposits in the basin center. Variations in the extent of these facies belts through time reflect temporal changes in the system. Some of these changes affecting only a sector of the system are regarded as autogenic. Other changes which influenced sedimentation across the whole system are attributed to tectonic (allogenic) controls.

The fluvial distributary system considered here forms part of the molasse of the Ebro Basin, an area of continental sedimentation in the Oligocene and Miocene (Riba and others, 1973; Puigdefabregas, 1975). The deposits of this fluvial system have been mapped as the Uncastillo Formation by previous workers (Riba and others, 1973; Puigdefabregas, 1975). The Uncastillo Formation is largely unfossiliferous, but occasional vertebrate remains suggest an Aquitanian (Early Miocene) age (Crusafont and Pons, 1969). The fluvial distributary system is here termed 'The Luna System' after one of the larger towns in the area (Fig. 1).

The sediments deposited by the Luna System are dominantly clastic with chemically precipitated carbonates and evaporites occurring in the more distal parts. The clastic component ranges in size from boulders to clay material, the coarser material being restricted to the proximal parts of the system. The sandstone component occurs as ribbon or sheet bodies (width: thickness ratios of <15:1 and >15:1, respectively; Friend and others, 1979). The sandstone bodies often have an erosional basal contact with the enclosing mudstones. The bulk of the coarse fraction occurs within fluvial channels, whereas the fine fraction occurs mainly as overbank material.

The area of these deposits forms an approximately semicircular region with a radius of as much as 40 km bounded by the basin margin to the north. The thickness of the Luna System deposits is at least 1190 m near the basin margin. The total thickness of Oligocene and Miocene deposits in the center of the Ebro Basin is about half that found at the northern margin (Riba, 1971). This may be accounted for by differential subsidence in the basin. If this subsidence pattern operated during the Early Miocene, it may be expected that the thickness of the Luna System deposits would be around 600 m near the basin center at the distal margins of the system.

Tectonic and palaeogeographic setting.—

The Ebro Basin is the southern foredeep of the Pyrenees, which is a late Cretaceous to late Tertiary mountain belt. The southern Pyrenean zone was a zone of southward-directed thrusting from the early Eocene to early Miocene (Camara and Klimowitz, 1985). During this time, the Ebro Basin was a subsiding foredeep receiving detritus from the Pyrenees. Subsidence in the Ebro Basin was asymmetric, occurring at a much higher rate adjacent to the southern Pyrenean front than elsewhere; this may be attributed to tectonic loading of the northern margin by the stacking of thrust sheets in the southern Pyrenean zone and also sedimentary loading of the basin by detritus eroded from the Pyrenees.

To the southwest and southeast, the Iberian and Catalan chains flanked the Ebro Basin such that through most of the Tertiary there was no breach into either the Bay of Biscay or the Mediterranean. Drainage from the Pyrenees and, on a smaller scale, the Catalan and Iberian chains flowed into an enclosed basin. A balance between the rate of evaporation in the basin and the input of water (by rivers from the surrounding mountains and by direct precipitation) must have been maintained during this period. Evaporite deposits contemporaneous with the Luna System deposits are not well developed. Extensive gypsum and other evaporite deposits in other units, however, indicate that at times the climate may have been warmer and drier.

Paleosol development is indicated by color banding and occasional mottling in some Luna System deposits. These paleosols resemble the spodosols and entisols recognized

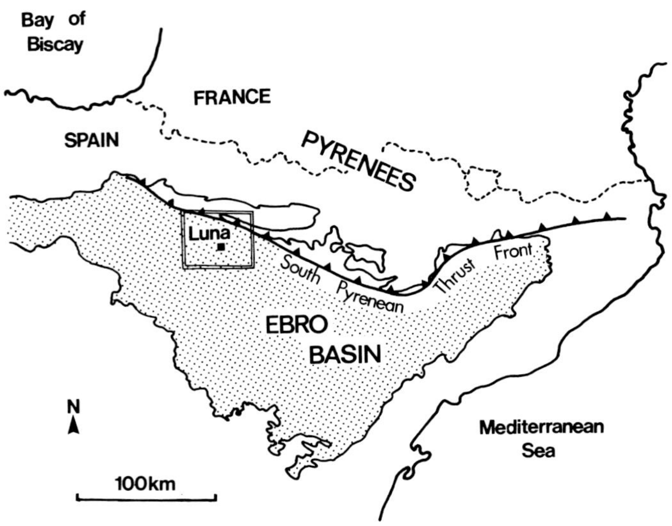

FIG. 1.—Location map of the study area in the northwestern part of the Ebro Basin, Spain.

by Bown and Kraus (1981), who suggested a relatively humid environment for their development. Humid conditions are also indicated by the absence of calcretes; calcrete development would be expected under arid or semi-arid conditions because of the abundant supply of calcium carbonate in Luna System alluvium. It is suggested that the paleoclimate in the early Miocene in this part of the Ebro Basin was humid (to allow the humid paleosols to develop) but with seasonal rainfall, allowing the rivers to dry up in the dry seasons.

Correlation.—

Deformation of the Luna System deposits is limited to a narrow belt along the basin margin and can be attributed to late thrust movements. Over the remainder of the area, the maximum dip is 1° to 2° to the southwest. By making corrections to take into account this slight dip, stratal correlation is possible between isolated outcrops. This makes it possible to examine both vertical and lateral trends in the Luna System deposits.

THE DISTRIBUTARY PATTERN

To establish the pattern of fluvial sedimentation, paleocurrent data were collected throughout the area. These data are summarized on Figure 2. Each reading of paleocurrent direction presented on the rose diagrams is an assessment of the paleoflow recorded within a sandstone body using a series of paleocurrent indicators. For each sandstone body the axial orientation was determined using one or both channel margins and the direction of flow determined using trough and planar crossbedding. This represents a certain rank of paleoflow (Allen, 1966; Miall, 1974), equivalent to the orientation of a short reach of a channel. In places plan exposures allowed the assessment of the trend of a channel over hundreds of meters. In these cases each exposure was divided into 100-m stretches, a direction recorded for each stretch and each exposure treated as equivalent to a single, limited exposure of a sandbody. It is apparent from the distribution of paleocurrent directions in the rose diagrams (Fig. 2) that the overall form of the Luna System is an approximately radial, fluvial distributary system.

Back-plotting of all the data establishes the apex of the system as a point at the basin margin, as marked on Figure 2. This calculated apex lies in the area where the Luna System deposits are both at their thickest and show their most proximal characteristics (see next section). Hence, it is considered to be a reliable estimate of the position of the actual apex of the fluvial system in the early Miocene.

THE LUNA SYSTEM FACIES ZONES

The fluvial system is here divided into three facies zones on the basis of lithology and sandbody type (Fig. 3): a proximal zone in which pebbly beds are dominant, a medial zone of channel fill ribbon and sheet sandstone bodies enclosed by overbank mudstones, and a distal zone in which the sheet sandstones are more abundant and ribbon bodies are smaller. The distal zone interdigitates with a lacustrine zone dominated by muds, marls and limestones in which there are no channelized sandstone bodies.

The proximal zone.—

The deposits in the proximal zone can be divided into three units (Fig. 4). Two conglomerate units make up the bulk of the deposits in this zone; they are separated by a generally finer interval approximately 200 m thick. The conglomerates are clast-supported and mainly massive or crudely bedded. They show imbrication of oblate pebbles in well sorted beds and some scouring/minor channel fill occurs on a decimeter scale. Medium to very coarse sandstone lenses occasionally show planar cross-stratification. Within the finer interval, conglomerates make up only 10 percent of the sequence; they occur as discrete bodies with erosional bases and limited lateral extent. In some cases distinct cut banks can be identified, defining pebbly units 50 to 100 m across and 4 to 5 m thick enclosed by sandstones and mudstones.

The proximal zone deposits are considered to be the deposits of braided streams. Sedimentation occurred mainly on gravelly bars within broad river channels. Both the bar forms and the channel forms have become amalgamated. Coarse input into the system was restricted for a period, resulting in an interval of sandy braided stream deposits and overbank sandstones and mudstones. The two distinct phases of conglomerate deposition in the proximal zone reflect two episodes of rejuvenation of erosion in the drainage basin.

The medial zone.—

The deposits of the medial zone are mainly fine to very coarse sandstone bodies enclosed by mudstones and fine sandstones. The sandstone bodies display a variety of geometries and may be ribbons, sheets or amalgamated complexes (Friend and others, 1986). Some of the ribbons show a single sandstone fill with very few internal structures. Others show decimeter-scale trough and planar cross-stratification representing a buildup of sand by migrating mega-

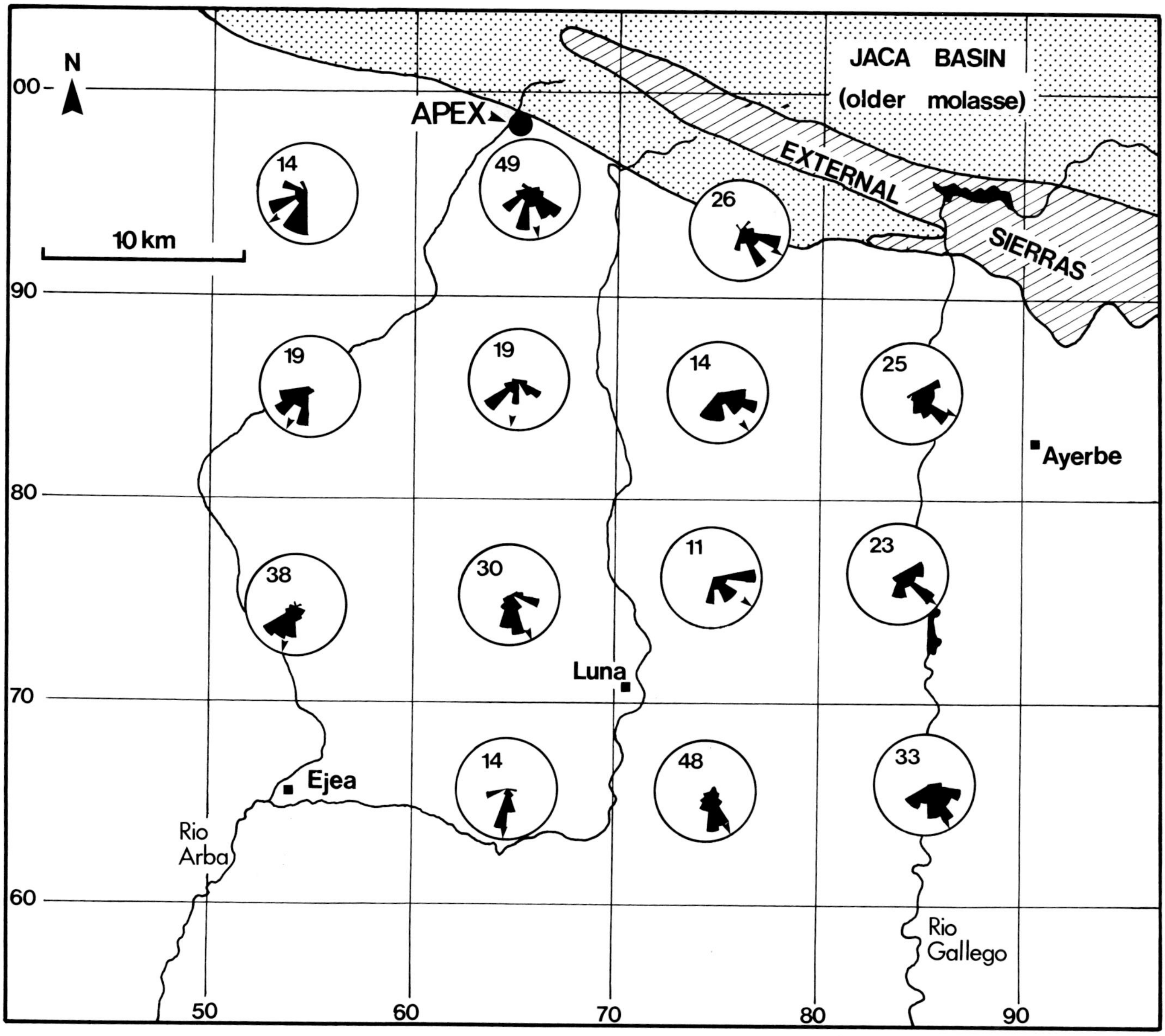

FIG. 2.—Rose diagrams for the paleoflow of short reaches of paleochannels in the Luna System. The data are grouped geographically into sets within 10 km by 10 km grid squares. The number of paleochannels measured in each group is indicated on the rose diagram. Arrowheads represent the vector mean paleoflow direction for each group. The position of the apex has been calculated by statistical analysis of the directions of all 363 paleochannel orientations.

ripple bedforms. Some of the larger channels show bar forms in cross section. The sheet sandstone bodies are of several types. The larger sheet sandstone bodies are amalgamated complexes of channels and bars with a multistory or multilateral internal structure. Sheet sandstones with large-scale epsilon cross-stratification (Allen, 1963) are rare in the Luna System deposits. Many of the thinner sheets do not exhibit internal structures and are likely to be the products of unchannelized or poorly channelized flow (Friend and others, 1986).

Sandstone bodies in the inner parts of the medial zone are mainly complex forms showing evidence of accumulation on sandy bars in braided streams. These may be seen as the more distal equivalents of the large pebbly braided stream deposits in the proximal zone. Multilateral sandstone bodies showing several stages of scour and fill are also characteristic of this inner zone. Throughout the rest of the medial zone there is a mixture of ribbons, sheets and amalgamated sandstone bodies. In some channels the fill was episodic. Desiccated mud layers occur above rippled sandstones at several stages before the channel was filled by medium sandstone (Fig. 5).

The channels cut into beds of mudstone and silty mudstone with centimeter-thick beds of well cemented siltstone, very fine and fine sandstone. Minor scouring sometimes occurs at the beds of the sandstone horizons, although most

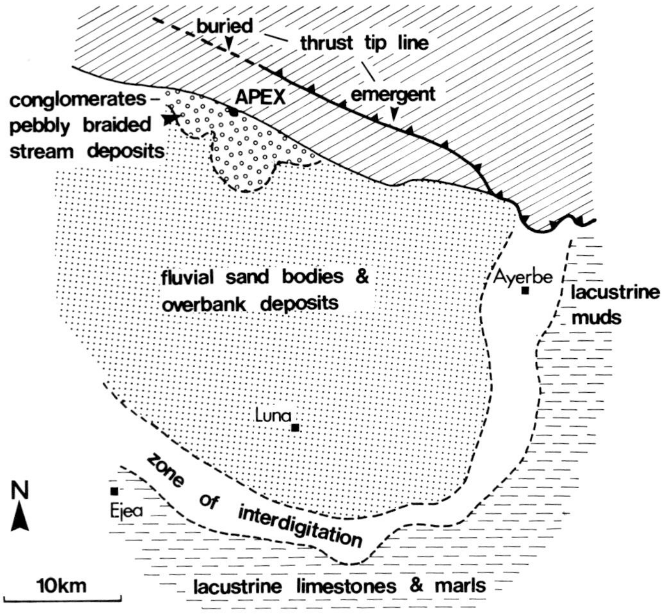

FIG. 3.—Facies belts in the Luna System.

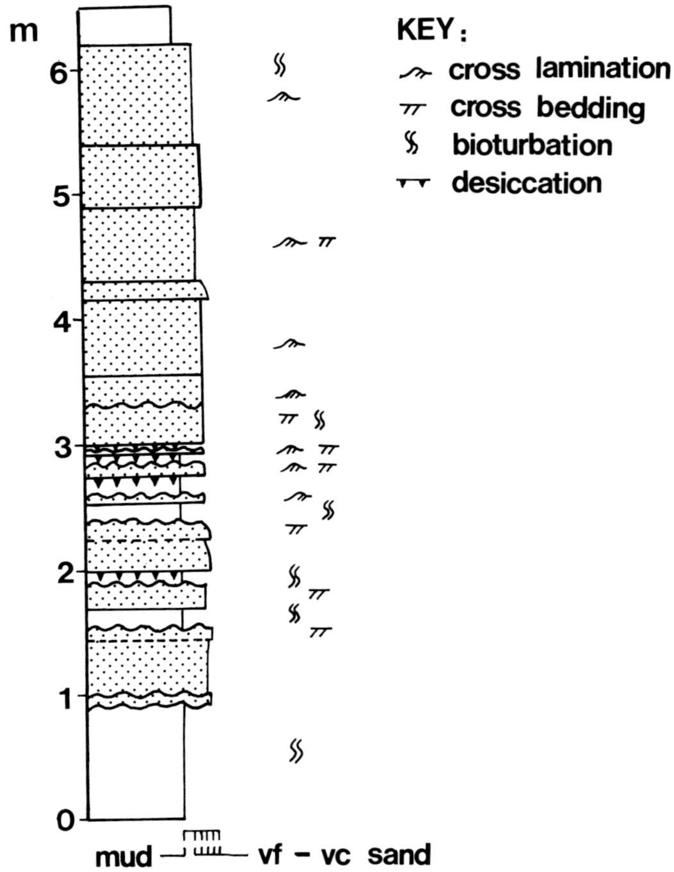

FIG. 5.—A vertical section through a channel-fill sandbody in the medial part of the system. Ephemerality of flow in the channel is indicated by the presence of desiccation surfaces in the lower half of the body.

beds have nonerosional contacts with the underlying material. Ripple laminations occasionally occur and bioturbation is fairly common. These deposits are the products of overbank sedimentation during overbank flow from the river channels. Color mottling frequently occurs, similar to that attributed to pedogenic processes by Bown and Kraus (1981).

Channel-fill sandstone bodies in the medial zone range from a few meters wide and less than half a meter thick to over a hundred meters wide and as much as seven meters thick. The ratio of width to thickness of these sandbodies ranges from under 3:1 to over 100:1. There is a general tendency for the dimensions of the sandstone bodies to decrease distally (e.g., in thickness; Fig. 6A). This trend reflects a decrease in channel size across the system due to repeated bifurcation of channels and possibly a decrease in discharge through evaporation and infiltration. There is also a distal increase of the width:thickness ratio of the sandstone bodies (Fig. 6B). This probably reflects a tendency for the channels to become less confined distally; with decreasing gradient and flow velocity, deeply incised channels formed less frequently and the streams spread out into poorly channelized sheets.

The principal types of sandbodies recorded in the medial zone are those produced by the fill of a simple channel resulting in a ribbon sand body (Fig. 6C). Ribbon sand bodies in the proximal parts of the system tend to be larger and show evidence of deposition on mid-channel bars. Sandstone bodies containing lateral accretion surfaces occur in the medial zone but are not common. This suggests that most channels were laterally stable, although they may have been sinuous (as plan exposures indicate in places). The degree of sinuousity can be estimated from the divergence of paleocurrents from the vector mean in any given area. Few paleocurrent readings diverge more than 90° from the vector mean for the area, and the circular standard deviation of the groups of data is relatively low (between 31 and 46, with the exception of readings from the apical area, which show a higher standard deviation of 52). The bulk of the channels in the Luna System were moderately sinuous and were only occasionally laterally migrating.

The distal zone.—

This zone is characterized by fewer and smaller sandbodies than occur in the medial zone, a high proportion of thin (less than 50 cm) sandstone sheets and the occurrence

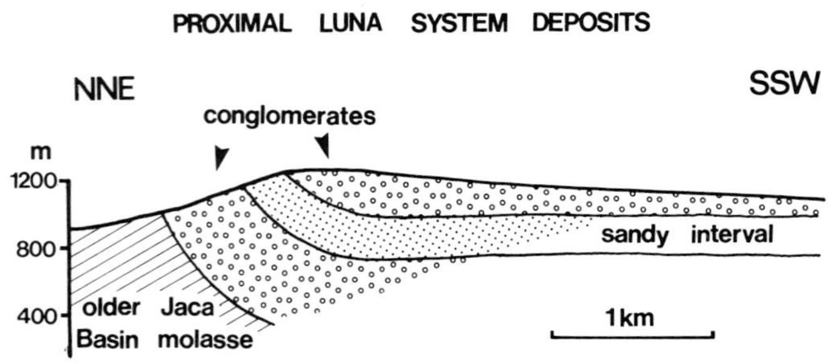

FIG. 4.—A cross section through the most proximal part of the system. The upper conglomerate unit is considered to be a result of a rejuvenation of the system. The tilting of the beds is due to late thrust movements which only affected a narrow fringe along the basin margin.

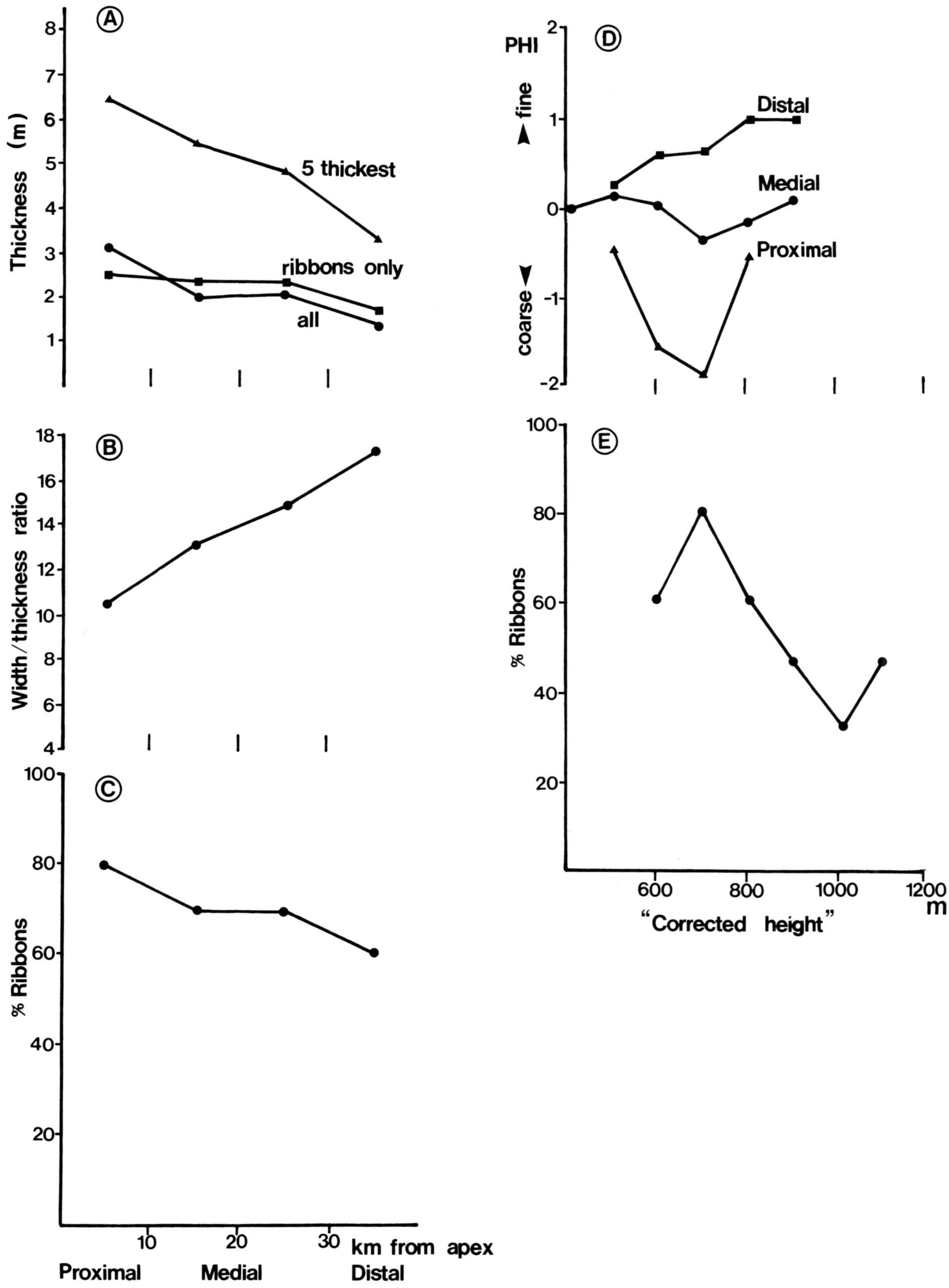

FIG. 6.—(A–C) Graphs constructed from calculations of the average thickness, width/thickness ratio and percentage ribbons (as a proportion of all sandbodies) from four 10-km-wide annular bands concentric with the apex of the system. (D) A plot of the average maximum clast size in 100-m vertical units in the proximal (0–10 km), medial (10–30 km) and distal (30 km+) zones of the Luna System. (E) The percentage of ribbon sandbodies (as a proportion of the total number of sandbodies) in 100-m vertical units in the Luna System. The value of the present height above sea level of an exposed sandbody has been altered to take into account the slight tectonic dip across the area, such that the 'corrected heights' are equivalent to depositional time-lines. The data base for these five graphs is 363 sand bodies throughout the Luna System.

of lacustrine fine sandstones and limestones interbedded with the fluvial deposits. This zone represents a transition between wholly fluvial and wholly lacustrine facies. Bioturbation is very common, and it is suggested that the environment supported a reasonably abundant fauna, more abundant than in the fluvial facies of the medial and proximal zones. A vegetated area and less oxidizing conditions are indicated by the presence of carbonaceous material and some well developed pedogenic horizons. Sandstone units with erosional bases and containing trough cross-stratification are interpreted as channelized incursions into a background sedimentation of overbank and/or lacustrine muds and carbonates.

Lacustrine zone.—

The Miocene deposits of the central Ebro Basin are principally lacustrine limestones, marls, muds and gypsiferous muds (Quirantes, 1969; Riba and others, 1973). Up to the late Miocene, the Ebro Basin was a wholly enclosed basin with internal drainage. Evaporation from the lakes in the basin center in a periodically arid climate resulted in the deposition of gypsum and enhanced the chemical precipitation of carbonates (Quirantes, 1969). Evaporation, infiltration and differential subsidence in the basin combined to limit the extent of the fluvial facies. To the south of the Luna System, the lacustrine deposits are largely very pale grey micrites, gastropodal limestones and calcareous mudstones. Fine to medium sandstone sheets occur but without erosional bases and with wave-ripples preserved. Along the eastern margin of the fluvial system, the lacustrine facies are not as well developed. The bulk of the deposits are red and buff mudstones and siltstones. This higher clastic input may be attributed to the proximity of the basin margin. Alluvial fans along the basin margin provided some detrital input in addition to that supplied through the Luna System (Hirst and Nichols, 1986).

TEMPORAL VARIATIONS WITHIN THE LUNA SYSTEM

Precise correlation across the whole system is difficult, but deposits in the paleogeographic zones described above are approximately coeval (Fig. 6D). (The data presented in Figure 6 are from beds equivalent to the upper phase of conglomerate sedimentation in Figure 4.) Channel-fill bodies in the medial and distal zones tend to be filled with finer sandstone in the higher parts of the sequence of sediments (Fig. 6D). This appears to be accompanied by a decrease in the proportion of ribbon (as opposed to sheet) sandbodies upward (Fig. 6E). These trends suggest that the medial parts of the system took on more distal characteristics higher in the sequence, possibly reflecting a decrease in the energy of the fluvial system.

These large-scale variations are considered to be tectonically induced. The onset of sedimentation in the fluvial system in this part of the Ebro Basin can be attributed to the emplacement of the Guarga thrust sheet, which occurred in the Late Oligocene (Camara and Klimowitz, 1985). The frontal ramp of this thrust sheet formed the uplift boundary separating the Ebro Basin from the erosional realm of the southern Pyrenean Zone, similar to the situation recognized by Eisbacher and others, (1974) in the Canadian Cordillera. After the first major pulse of sedimentation in the Luna System, degradation of the erosional realm resulted in a quiet phase before a rejuvenation of the system, represented by the later phase of conglomeratic sedimentation. This rejuvenation could not have been in response to further southward movement of the Guarga thrust sheet because of the lack of significant deformation of the earlier Luna System deposits along the basin margin; there is no unconformity between the lower and upper parts of the conglomerate sequence. The cause of the rejuvenation is most probably out-of-sequence thrusting (Butler, 1982) in the southern Pyrenean belt which uplifted parts of the Luna System drainage basin. The general upward trends (Figs. 6D, 6E) in the system recognized in the medial and distal deposits may be attributed to a decrease in discharge. This would have been caused by a reduction in the source area topography due to degradation subsequent to the rejuvenation episode.

Vertical sequence changes on a smaller scale may not be tectonically induced. In the southeastern part of the Luna System, there is a marked change from lacustrine to fluvial facies upward through 50 m. Gastropodal limestones and rootlet horizons indicating a lacustrine and/or marsh environment pass upward into a coarsening-upward sequence of sandstones and mudstones. The sandstone bodies are larger and more incised into the overbank deposits at higher levels. The apparent absence of similar fluvial sandbodies in the study area at the same level suggests that these were part of an isolated lobe of fluvial deposits impinging upon the lacustrine zone. Lobes of this kind seem to have been a common feature of the outer part of the fluvial system. The geomorphic form of the distributary system at any one time would have been similar to a river-dominated delta. The relatively uniform radial form of the Luna System (Figs. 2, 3) is a function of the time-averaging of these lobes across the whole of the distal part of the system. The shifting of deposition from one lobe to another is likely to have been autogenic resulting from major avulsion of the main rivers in the medial or proximal part of the system. Avulsion of channels must have been an important process in the Luna System, resulting in channel infill sandstone bodies building up the pattern seen today.

SANDSTONE BODY DISTRIBUTION

Although the average sandstone body dimensions show a decrease distally (Figs. 6A, 6B), there are some notable exceptions to this pattern. Small ribbon bodies occur in the inner part of the medial zone, and the largest channel-fill body recorded (over 100 m wide and 8 m thick) occurs some 27 km from the apex of the system. Two explanations for this are considered. First, the discharge across the system may have varied through time. During periods of high discharge, large river channels would have reached distal parts of the system, but when discharge was low, only small channels formed and were later plugged to form small sandstone bodies in the more proximal parts. Second, bifurcation of channels probably occurred. The splitting of the channels into major and minor distributaries at various points along their courses would have produced a pattern of varying channel dimensions across the Luna System. This could result in the preservation of small channels near the apex and relatively large 'trunk' channels in distal parts of the system.

The interconnection of sandbodies has been shown by Leeder (1978) to depend upon the rates of sediment accumulation on the floodplain and the rate of channel avulsion. The observed interconnection of the Luna System sandbodies in the field is very low in the distal and medial parts of the system and increases in the proximal zone. Such a pattern is to be expected in a radial distributary system. The discharge through river channels would remain constant or decrease distally, resulting in a constant or decreasing in-channel component. The area over which the channels occur would increase distally, resulting in a decrease in the concentration of channel-fill bodies.

As stated above, sandstone bodies in the outer part of the system tend to be concentrated into discrete lobes. It is suggested that these lobes are the result of periods when sedimentation in channels and on the floodplain was restricted to certain areas. Major avulsion events would have shifted activity to another sector of the system, creating a new lobe of sandstone bodies. This pattern probably reflects a relatively high rate of floodplain accretion compared to the rate of major channel avulsion. The degree of interconnection of sandstone bodies is much higher in these lobes than between them, and hence, the interconnection varies considerably both vertically and laterally in the Luna System.

CONTROLS ON SEDIMENTATION

Basin subsidence.—

The facies distribution patterns show some variation with time, but the overall relationship between the positions of the proximal, medial and distal facies remains constant. Paleocurrent data from all levels within the exposed parts of the Luna System suggest that the location of the apex of the system stayed relatively fixed throughout the early Miocene. In order to account for the accumulation of at least 1200 m of sediments near the basin margin, there must have been a net balance between the subsidence and sedimentation. The boundaries between the facies belts (e.g., between dominantly pebbly braided streams and mainly sinuous streams and between fluvial and lacustrine deposits) are very sensitive to changes in slope. These boundaries varied little during long periods of sedimentation in the Luna System. This suggests that the basin slope remained fairly constant. This situation could only have been achieved by the attainment of a fine balance between sedimentation and subsidence rates. Both sedimentation and subsidence rates were about twice as high at the basin margin compared to those in the basin center.

Subsidence in a foreland basin is primarily controlled by tectonic loading of the basin margin by the stacking of thrust sheets and by the sedimentary loading caused by detrital material filling the basin. The rate of subsidence appears not to have changed dramatically while the Luna System was active. It may be inferred that no major emplacement of thrust sheets occurred and that the basin margin was tectonically quiescent during that time. The absence of synsedimentary deformation in the Luna System deposits adjacent to the basin margin is consistent with this.

Structural controls.—

In the study area the northern margin of the Ebro Basin is formed by the frontal thrust complex of one of the south Pyrenean thrust sheets. The tip line of the frontal thrust is emergent in the eastern part of the area. Limestone units in the hanging wall of the thrust form a prominent ridge along the basin margin. To the west, the thrust tip line becomes buried and the limestone ridge is not developed (Figs. 2, 3). Reconstructions of the paleogeography and paleogeology show that the general structural pattern of the frontal thrust complex observed today existed in the early Miocene (Puigdefabregas and Soler, 1973; Nichols, 1984; Hirst and Nichols, 1986). The apex of the Luna System is adjacent to this structural low (Figs. 2, 3). This structural low is the re-entrant point, where rivers from the drainage basin cross the uplift boundary and pass into the aggradational basin (Eisbacher and others, 1974). The position of the apex was therefore controlled by the geometry of the thrust front.

DISCUSSION

Comparison with other distributary systems.—

The term 'fluvial distributary system' is used here in preference to the more general term 'alluvial fan' which may be applied to any fan-shaped cone of alluvial material. These distributary systems are distinct from arid alluvial fans because the dominant processes are the incision and later infill of fluvial channels, plus associated overbank deposition. Alluvial fans may be composed of a combination of debris flow, sheetflood, sieve and channel deposits (Bull, 1972). A distinction may also be made between these distributary systems and large humid fans, such as the Kosi River fan, which are formed by the lateral migration of a single river channel (Gole and Chitale, 1966).

A variety of ancient fluvial sequences have been interpreted as the deposits of fluvial distributary systems (Mullen and Freeman, 1957; Friend, 1978). The principal features of these systems are a distributary pattern of channels, a lobate topographic form and a downcurrent decrease in both grain size and channel size. Terminal fans of material which show some of the characteristics of distributary systems have been described by Mukherji (1976) and Parkash and others (1983), but these are relatively small, 64 km^2 in the case of the Markanda River terminal fan (Parkash and others, 1983). They are, therefore, only partly comparable with larger distributary systems envisaged as responsible for depositing many ancient fluvial sequences. For example, the Devonian Wood Bay Formation is considered to have been deposited over an alluvial area of 5400 km^2 (Friend and Moody-Stuart, 1972) and the Saltwash Member of the Morrison Formation over an area of at least 250,000 km^2 (Mullen and Freeman, 1957).

SUMMARY OF THE LUNA SYSTEM

The principal features of the Luna System are as follows:

1) The fluvial system deposited sediments over an area of 2500 km^2.
2) Paleocurrent data show that this fluvial system was radial, forming a semicircular region of fluvial channels and overbank deposits adjacent to the northern margin of the Ebro Basin.
3) The size of the channels shows a consistent downslope decrease, away from the apex; this may be due to bifurcation of channels and/or a distal decrease in the volume of water within the channels as a result of evaporation and infiltration.
4) The grain size of material deposited within channels decreases distally, indicating a downslope decrease in energy within individual channels as the slope approaches horizontal in the most distal parts.
5) Pebbly braided streams dominated the proximal parts of the system within 5 km of the apex. Channels in the medial part of the system (approximately 5 to 35 km from the apex) were mostly only moderately sinuous and laterally stable, although there were some laterally migrating stream courses in places. Sandy braided rivers also occurred in the inner part of this zone, and broader, less confined streams depositing sheets of material became increasingly common distally. At the distal margins of the system, the fluvial deposits interdigitate with lacustrine deposits. These are mainly marls and limestones, except close to the basin margin where there was a higher clastic input.
6) The degree of interconnection decreases downslope as a result of a distal decrease in the concentration of channel-fill material.
7) The concentration of sandbodies in discrete lobes in the distal parts of the system is attributed to periods of activity in particular sectors of the system. Major avulsion events occurring at or near the apex would have caused shifts in the activity from one sector to another.
8) Paleosols indicate a humid environment, whereas the occurrence of desiccated river channels suggests that the rainfall must have been seasonal.

These patterns of channels are characteristic of terminal distributary systems where there is no external drainage from the basin. Areas of internal drainage, such as the Ebro Basin in the early Miocene, form where the basin is bounded on all sides by mountain chains, and evaporation is high enough to balance the input from rivers and direct precipitation. For the distributary system to be maintained and deposit hundreds of meters of fluvial sediments, there must have been continued subsidence in the basin. Subsidence in the basin was largely controlled by tectonic loading of the northern margin of the basin by thrust sheets and by sedimentary loading by material derived from the Pyrenean chain to the north.

Major pulses of sedimentation, identifiable in the proximal parts of the system, can be attributed to uplift in the source area which was probably caused by out-of-sequence thrusting in the southern Pyrenean zone. The thrust front formed an orographic barrier, and the drainage system entered the Ebro Basin at a structural low in this barrier; hence, the position of the apex of the system is controlled by the geometry of the thrust front.

ACKNOWLEDGMENTS

The calculation of the position of the apex from the paleocurrent data was undertaken by Peter Jupp and Barrie Spurr at the University of St. Andrews using a directional statistics package currently under development. The author is very grateful for the helpful comments of Peter Friend and Philip Hirst. The manuscript benefited considerably from the constructive comments of Timothy Cross. This work was conducted during the tenure of a U.K. Natural Environment Research Council studentship at the University of Cambridge.

REFERENCES

ALLEN, J. R. L., 1963, The classification of cross-stratified units, with notes on their origin: Sedimentology, v. 2, p. 93–114.

———, 1966, On bedforms and palaeocurrents: Sedimentology v. 6, p. 153–190.

BOWN, T. M., AND KRAUS, M. J., 1981, Lower Eocene alluvial palaeosols (Willwood Formation, Northwest Wyoming, U.S.A.) and their significance for palaeoecology, palaeoclimatology and basin analysis: Palaeogeography, Palaeoclimatology, Palaeoecology, v. 34, p. 1–30.

BULL, W. B., 1972, Recognition of alluvial fan deposits in the stratigraphic record, *in* Rigby, J. K., and Hamblin, W. K., eds., Recognition of Ancient Sedimentary Environments: Society of Economic Paleontologists and Mineralogists Special Publication 16, p. 63–83.

BUTLER, R. W. H., 1982, The terminology of structures in thrust belts: Journal of Structural Geology, v. 4, p. 239–245.

CAMARA, P., AND KLIMOWITZ, J., 1985, Interpretacion geodinamica de la vertiente Centro-occidental Surpirenaica: Estudios Geologicos, v. 41, p. 56–69.

CRUSAFONT, M., AND PONS, J. M., 1969, Nuevos datos sobre el Aquitainenses del norte de la provincia de Huesca: Acta Geologica Hispanica, 4, p. 124–125.

EISBACHER, G. H., CARRIGY, M. A., AND CAMPBELL, R. B., 1974, Palaeodrainage pattern and late-orogenic basins of the Canadian Cordillera, *in* Dickinson, W. R., ed., Tectonics and Sedimentation: Society of Economic Paleontologists and Mineralogists Special Publication 22, p. 143–166.

FRIEND, P. F., 1978, Distinctive patterns of some ancient river systems, *in* Miall, A.D., ed., Fluvial Sedimentology: Canadian Society of Petroleum Geologists Memoir 5, p. 531–542.

——, HIRST, J. P. P., AND NICHOLS, G. J., 1986, Sandstone-body structure and river processes in the Ebro Basin of Aragon, Spain: Fluvial Sediments in Spain, Cuadernos de Geologia, Madrid, p. 121–142.

——, AND MOODY-STUART, M., 1972, Sedimentation of the Wood Bay Formation (Devonian) of Spitsbergen: Regional analysis of a late orogenic basin: Norsk Polarinstitut Skrifter, v. 157, 77 p.

——, SLATER, M. J., AND WILLIAMS, R. C., 1979, Vertical and lateral building of river sandstone bodies, Ebro Basin, Spain: Journal of the Geological Society, v. 136, p. 39–46.

GOLE, C. V., AND CHITALE, S. V., 1966, Inland delta building of the Kosi River: Journal of the Hydraulics Division, American Society of Civil Engineers, HY-2, p. 111–126.

HIRST, J. P. P., AND NICHOLS, G. J., 1986, Thrust tectonic controls on Miocene alluvial distribution patterns, southern Pyrenees, *in* Allen, P. A., and Homewood, P., eds., Foreland Basins: International Association of Sedimentologists Special Publication No. 8, p. 153–164.

LEEDER, M. R., 1978, A quantitative stratigraphic model for alluvium, with special reference to channel deposit density and interconnectedness *in* Miall, A. D., ed., Fluvial Sedimentology: Canadian Society of Petroleum Geologists Memoir 5, p. 587–596.

MIALL, A. D., 1974, Palaeocurrent analysis of alluvial sediments: a discussion of directional variance and vector magnitude: Journal of Sedimentary Petrology, v. 44, p. 1174–1185.

MUKHERJI, A. B., 1976, Terminal fans of inland streams in Sutlej-Yamuna Plain, India: Zeitschrift fur Geomorphologie, Neue Folge, v. 20, p. 190–204.

MULLEN, T. E., AND FREEMAN, V. L., 1957, Lithofacies of the Saltwash Member of the Morrison Formation, Colorado Plateau: Geological Society of America Bulletin, v. 68, p. 505–526.

NICHOLS, G. J., 1984, Thrust tectonics and alluvial sedimentation, Aragon, Spain: Unpublished Ph.D. Thesis, Cambridge, 247 p.

PARKASH, B., AWASTHI, A. K., AND GOHAIN, K., 1983, Lithofacies of the Markanda terminal fan, Kurukshetra district, Haryana, India: *in* Collinson, J. D., and Lewin, J., eds., Modern and Ancient Fluvial Systems: International Association of Sedimentologists Special Publication 6, p. 337–344.

PUIGDEFABREGAS, C., 1975, La sedimentation molasica en la cuenca de Jaca: Monographias del Instituto de Estudios Pirenaicos Numero 104, 188 p.

——, AND SOLER, M., 1973, Estructura de las Sierras Exteriores Pirenaicas en el corte del rio Gallego (Prov. Huesca): Pirineos, v. 109, p. 5–15.

QUIRANTES PUERTAS, J., 1969, Estudio sedimentologico y estratigrafico del terciaro continental de los Monegros: Tesis (C.S.I.C.), Zaragoza, 101 p.

RIBA, O., 1971, Mapa Geologico de Espana Escala 1:200000: sintesis de la cartografia existente, Lerida (primera edition): Departamiento de publicaciones del Instituto de Geologico y Mineralogico de Espana, Madrid, 22 p.

——, PUIGDEFABREGAS, C., SOLER, M., QUIRANTES, J., AND MARTI, C., 1973, Mapa geologico de Espana. E.1:200000, sintesis de la cartografia existente, Tudela. Departamiento de publicaciones del Instituto de Geologico y Mineralogico de Espana, Madrid. 16 p.

FLUVIAL–SANDSTONE ARCHITECTURE AND THRUST–INDUCED SUBSIDENCE, NORTHERN GREEN RIVER BASIN, WYOMING

MARK W. SHUSTER AND JAMES R. STEIDTMANN
Department of Geology and Geophysics University of Wyoming Laramie, Wyoming 82071

ABSTRACT: Facies analysis of the Lenticular Sandstone and Shale Sequence in the northern Green River basin, Wyoming provides the basis for testing the concept that subsidence controls fluvial-sandstone architecture. This sequence consists of three genetic facies including (1) low sinuosity fluvial channel sandstones of both ribbon and multistory geometries, (2) levee and splay complexes, and (3) interfluve swamp, bog, and lake deposits. The fluvial channels are thick (>5 m), lenticular bodies consisting of fine- to medium-grained sandstone with basal scour lags. The absence of point bar lateral accretion surfaces and the low variance of paleocurrent directions suggest low sinuosity. Ubiquitous levee and splay sandstones are thin (0.5 m) and contain small-scale trough crossbedding and microtrough cross-lamination. The interfluve rocks consist of dark gray mudstone, carbonaceous shale, thin coal beds, and thin, nonmarine micritic limestone.

We interpret these deposits to be the products of a fixed-channel fluvial system that developed in a rapidly subsiding foreland basin. High subsidence rates are indicated by the large proportion of overbank mudstone, ribbon to multistory channel geometries and lack of paleosol horizons. Subsidence analysis (backstripping) also shows that basin subsidence was rapid during the Campanian and therefore that the architecture of the Lenticular Sandstone and Shale Sequence may have been tectonically controlled. We attribute this subsidence to regional flexure induced by tectonic loading during major thrust emplacement to the west.

INTRODUCTION

Controls on the geometric distribution of lithofacies in alluvial sequences are both autocyclic and allocyclic (Beerbower, 1964). Autocyclic controls include variations in flow strength and flow direction in the alluvial environment, whereas allocyclic include major changes in base level, gradient, discharge and sediment load type directly related to external factors such as climate, tectonism and eustacy. The spatial distribution of sandstone channels within overbank deposits is a function of the scale and mode of behavior of the river system, the rate of avulsion, and the subsidence rate of the alluvial plain (Allen, 1978). Hence, given a constancy of river type and rate of avulsion for an alluvial sequence, sandstone geometry is dependent on subsidence rates. In this paper we construct a working model for the depositional environment of Campanian strata in the northern Green River basin (Fig. 1) and test the validity of interpreting subsidence rates from alluvial sandstone architecture by using an independent subsidence analysis.

The northern Green River basin lies in the zone of impingement between the thrust belt to the west and the Gros Ventre Range of the foreland to the east (Fig. 1). Campanian rocks of fluvial origin are exposed along the trace of the Game Hill fault, an eastward dipping reverse fault mechanically related to latest thrusting events (Dorr and others, 1977). Until pollen dating by Guennel and others (1973), these strata were thought to be Paleocene in age and constituted a major part of the type section of the Paleocene Hoback Formation, as defined by Dorr (1952). The redefinition of stratigraphic nomenclature by Guennel and others (1973) restricts the Hoback Formation to the Paleocene deposits and assigns the lower units, including the Coaly Sequence, the Lenticular Sandstone and Shale Sequence, and the White Sandstone to the Cretaceous (Fig. 2). The Lenticular Sandstone and Shale Sequence is the focus of our study. It consists of approximately 900 m of interbedded mudstone, siltstone, carbonaceous shale, thin coal beds, micritic limestone, and medium-grained sandstone in large lenticular bodies (Fig. 3). Spearing (1969a, b) interpreted these rocks as the depositional product of meandering rivers. This interpretation was incorporated into a synthesis of deformation and depositional patterns in the Hoback Basin (Dorr and others, 1977).

FACIES IDENTIFICATION

Three major associations of lithofacies based on grain size, assemblages of sedimentary structures, and geometric relationships were distinguished in the Lenticular Sandstone and Shale Sequence.

Facies Association A.—

Facies Association A consists of thick (>5 m) lenticular sandstone bodies of fine- to medium-grained sand (Fig. 4). The basal contacts are erosive and are commonly directly overlain by lag deposits which consist of very poorly sorted clasts of mudstone, siltstone, and micritic limestone ranging in diameter from 1 to 15 cm. Typically these isolated clasts are set in medium-grained sand, but clast-supported layers are not uncommon. Log casts as much as 1 m in length and 15 cm in diameter are present in these lag layers. Most of these casts are aligned subparallel with other flow indicators in a given lag horizon, but evidence for preferential parallel alignment is not conclusive. Clasts are common throughout Facies Association A, and there is a decrease in maximum clast diameter upsection.

Other sedimentary structures in Facies Association A include planar-horizontal to sub-horizontal lamination, planar-tabular to wedge-planar cross-stratification, trough cross-stratification, contorted bedding associated with massive sandstone and some microcross-stratification. Sediments containing planar-horizontal bedding and trough forms comprise more than 50 percent of Facies Association A, and planar-tabular cross-stratification is locally abundant.

Sediments with microcross-stratification of the planar-tabular and trough types comprise less than 5 percent of Facies Association A. Of these, the microtrough sets are most abundant. No preserved ripples or biogenic structures were found in this facies.

The internal arrangement of sedimentary structures is shown in outcrop sketches of major sandbodies (Fig. 4).

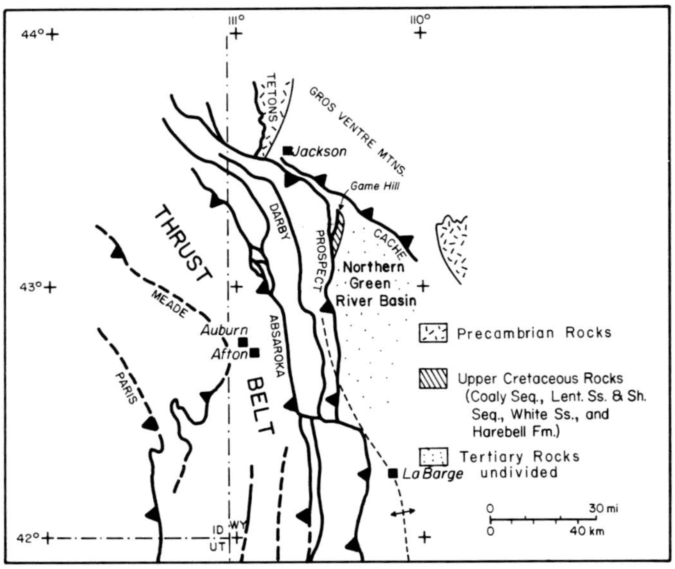

FIG. 1.—Location of study area showing major structural elements of the Idaho-Wyoming thrustbelt.

FIG. 3.—Photograph of Game Hill showing the geometry of channel sandstones in the Lenticuler Sandstone and Shale Sequence.

Generally, massive sandstone, scour horizons and planar bedding are localized in the lower parts of these sandbodies while planar-tabular, trough- and microcross-stratification are more abundant in the upper parts. Due to the variability of structure succession both vertically and laterally, however, it is difficult to identify "typical" transitions. An idealized succession of sedimentary structures is shown in Figure 5.

Facies Association B.—

Facies Association B consists of thin (<5 m) sandstone bodies with trough cross-stratification and microcross-stratification, including climbing ripple sets (Fig. 6). Sediments of Facies Association B range in grain size from silt to fine-grained sandstone, and no pervasive fining- or coarsening-upward successions are present. Facies Association B sandstone bodies can be traced laterally into Facies Association A. In all cases Facies Association B thickens and grain sizes increase toward the larger Facies Association A sandstone bodies. Occasionally, there are scour surfaces at the base of Facies Association B sandstone bodies. Biogenic structures are present at the tops of some of the Facies Association B bodies where "rooting" traces are recognized by the presence of coaly stringers or casts. The textural consistancy of the deposits did not permit unequivocal recognition of burrows, but bioturbation by animals could not have been extensive because layering is preserved in the deposit.

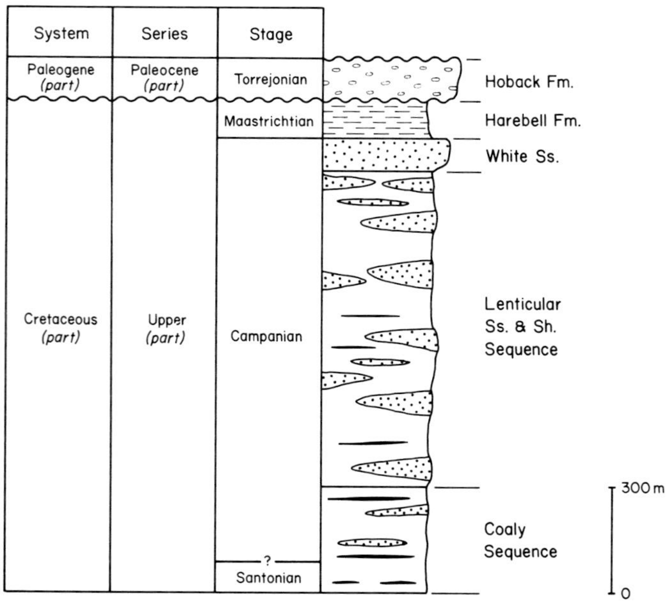

FIG. 2.—Upper Cretaceous stratigraphy, Hoback Basin, Wyoming (see text for references).

Facies Association C.—

Facies Association C consists of interbedded siltstone, mudstone, shale, organic shale, thin coal beds, and thin (0.3 m) massive, dark gray micritic limestone beds that comprise from 60–75 percent of the total section (Fig. 6). The limestone is in isolated bodies (0.3 m thick and as much as 70 m wide) within the shale and siltstone. It consists of very pure micrite with little (<1 percent) clastic material except where it intertongues with shale and siltstone. The limestones are non-modular, resembling interfluve lacustrine micrites documented by Flores (1981) in the Fort Union Formation. Coal beds are thin and discontinuous, and the number and size decrease upsection.

INTERPRETATIONS

Facies.—

In most general terms these three facies associations are indicative of a floodplain environment, as suggested by Spearing (1969a); however, we disagree with his interpre-

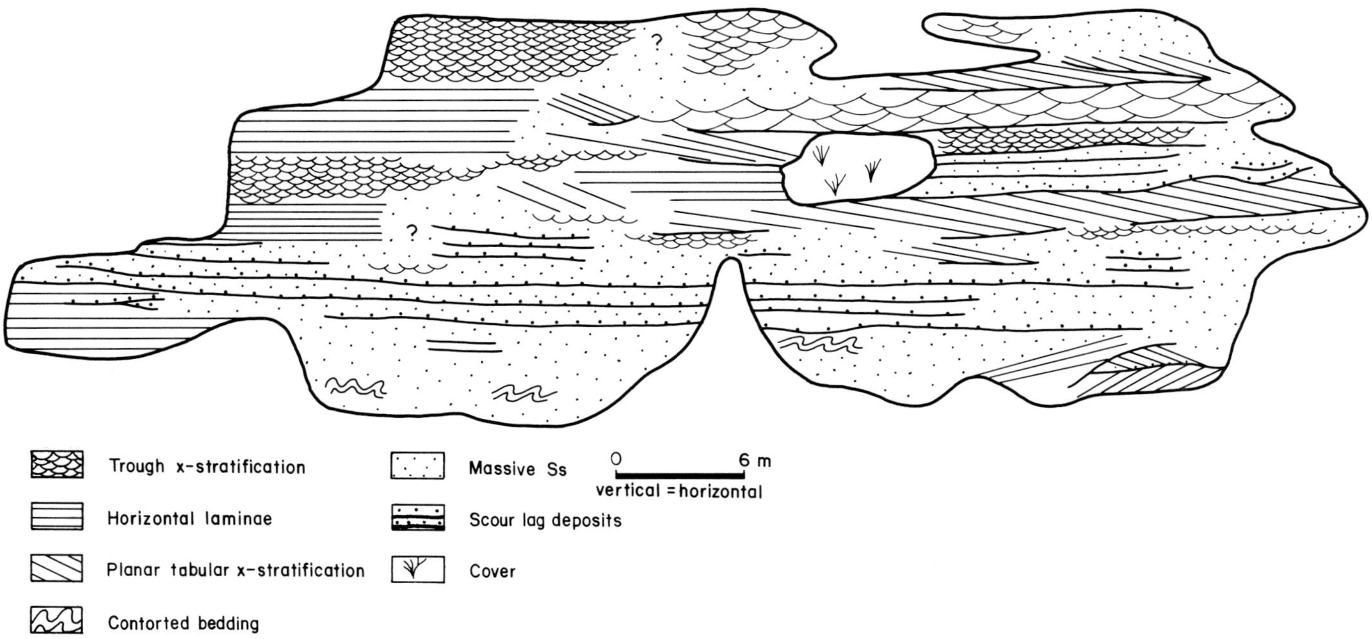

FIG. 4.—Outcrop sketches of select channel sandstones, Lenticular Sandstone and Shale Sequence.

tation that the sandstone bodies represent high sinuosity, meandering channels. We propose that the Lenticular Sandstone and Shale Sequence was deposited by a series of low sinuosity, fixed-channel rivers with associated levee, splay, and interfluve deposition (Fig. 7).

Modern fixed channels have been described in the anastomosed river systems of western Canada by Smith (1983); Smith and Putnam (1980); Smith and Smith (1980); and Smith (1976) and in Australia by Rust (1981). The Canadian fixed channels are stable, have low sinuosity, are interwoven, and have well developed levee and crevasse-splay subsystems (Smith, 1983). Their channel deposits are typically thick (5–15 m) "shoestring" sands isolated in fine-grained levee and interfluve deposits. The internal stratification is typified by planar-tabular set geometries and numerous fining-upward flood cycle deposits. Splay deposits consist of lobate, laterally extensive, coarsening-upward sequences of silt- to medium-grained sand with unidirectional ripple stratification and high angle, large-scale crossbed sets in the upper parts (Smith, 1983). Levee deposits consist of "laminated fine sand and silt" with occasional organic lenses. Fine-grained lacustrine, marsh and swamp deposits (silt, shale, and peat) volumetrically comprise the largest part of these anastomosed systems (Smith, 1983; Smith and Smith, 1980).

By similarity to these characteristics, we interpret Facies Association A as channel deposits, Facies Association B as levee/splay complex deposits, and Facies Association C as interfluve lacustrine deposits of a fixed-channel system (Fig. 7). Specifically, facies association deposits contain no pervasive fining-upward trends or lateral accretion sets, and paleocurrent data show little variation between sandstone units (Fig. 8). It is unlikely that the trends in mean flow directions would be this stable for a system of stacked high sinuosity channel deposits, and we propose that internal geometries of Facies Association A sandstone are largely a result of vertical aggradation rather than the lateral migration of point bar sequences. The internal geometry of these sandstones can be explained by periodic changes in flow strength due to periodic flood events. Periodic flooding is also indicated by scour lag surfaces throughout the channel bodies. The ubiquitous presence of splay sandstones and well developed levee complexes also supports the interpretation of a fixed-channel system (Moody-Stuart, 1966). Although levees and splays are also associated with meandering river deposits, they are frequently not well preserved because of the migratory, and hence destructive, nature of these systems.

Alluvial sandstone architecture.—

Both single-story ribbon and multistory ribbon channel sandstone geometries (terminology of Friend and others, 1979; Blakey and Gubitosa, 1984) are present in the Lenticular Sandstone and Shale Sequence (Fig. 3). Width-to-height ratios range from 10 to 15. The ribbon sandstones are separated by sediments of lithofacies associations B and C, and there is little interconnection between these sandstone bodies.

Fluvial sand body interconnection is a function of both the frequency of avulsion and the rate of subsidence (Allen, 1978; Bridge and Leeder, 1979). Given a constant avulsion rate (frequency) and constant stream characteristics (i.e., depth, channel width, and meander-belt width), the spacing and geometry of fluvial sandstones should be dependent on subsidence rate. During periods of relatively rapid subsidence, fluvial sandstone bodies are more or less isolated in large volumes of overbank fines. Conversely, during pe-

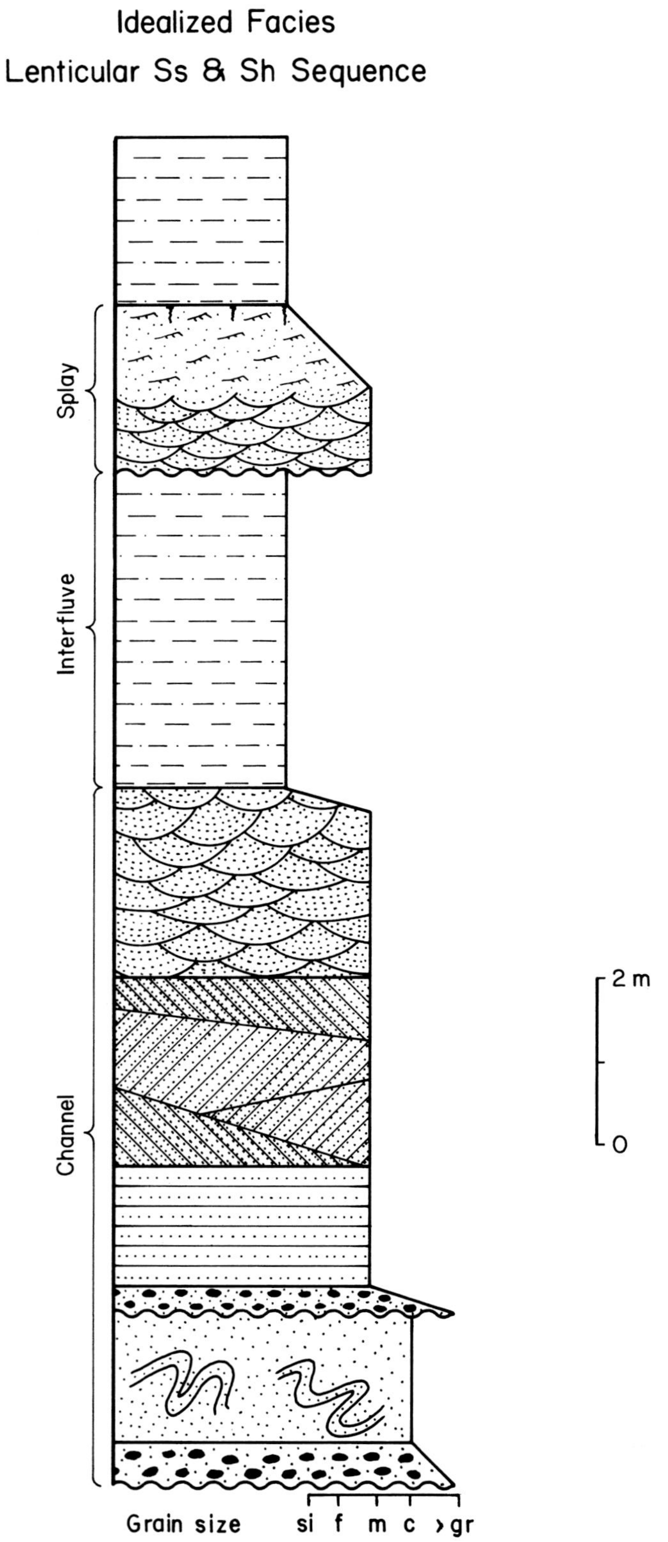

FIG. 5.—Idealized stratigraphic section through channel (Facies Association A), interfluve (Facies Association C), and splay facies (Facies Association B), Lenticular Sandstone and Shale Sequence.

riods of low subsidence rates, sandstone bodies with high degrees of interconnection are expected.

Although we do not propose a one-to-one comparison between the modeling results of Allen (1978) and Bridge and Leeder (1979) and the sandstone architecture of the Lenticular Sandstone and Shale Sequence, we do expect similar patterns. Assuming that the avulsion rate was constant throughout deposition, the interpreted fixed-channel sandstone architecture should reflect subsidence rate. This assumption is probably justified inasmuch as there is no petrologic or fossil evidence of major climatic or vegetational changes during this time. Thus, based on low degree of interconnection, the single-story ribbon and multistory ribbon sandbodies which characterize the Lenticular Sandstone and Shale Sequence are herein interpreted as the products of alluvial sedimentation on a floodplain where subsidence rate exceeded that of avulsion. This interpretation is further supported by the absence of well developed paleosol horizons. Presumably, well developed paleosol horizons, either calcareous or noncalcareous, would indicate slower subsidence (Allen, 1974; Bown and Kraus, 1981).

SUBSIDENCE ANALYSIS

In order to test our hypothesis that the fluvial sandstone architecture of the Lenticular Sandstone and Shale Sequence is indicative of relatively high subsidence rate, we constructed subsidence curves (Fig. 9) for the Mesozoic section of the northern Green River basin following the method of Van Hinte (1978). Locally compensated components due to the overlying sedimentary units were removed following an Airy-type isostatic model (i.e., backstripping) as discussed by Watts and Ryan (1976) and Steckler and Watts (1978). For backstripping, the densities used were 3.3 for mantle and 1.0 for water. Porosity-depth corrections used in backstripping were modified from exponential functions presented by Rieke and Chillingarian (1974) and Sclater and Christie (1980). Moreover, these functions were used to calculate changes in sediment density with increasing burial depth. The porosity corrections were done iteratively using a BASIC computer program. The time scale is that of Palmer (1983). Water depths were estimated by faunal assemblages or interpretations of depositional environments as follows: fully marine facies = 60 to 300 m, shallow marine facies = 0 to 60 m, and nonmarine facies = 0 to 300 m above sea level. The results (Fig. 9) are presented as total subsidence and tectonic subsidence curves.

Because of the lack of detailed age control in the studied strata, the curves probably represent minimum subsidence rates. Most likely, the actual duration of deposition was significantly less. Assuming the maximum (9.5 m.y.) duration (see Palmer, 1983), the subsidence rates are still relatively high (172 m/m.y.) during the Campanian (84-74 m.y.; see stippled area on Fig. 9). Results of this analysis corroborate our conclusion of relatively high subsidence rates as indicated by the facies and alluvial sandstone architecture.

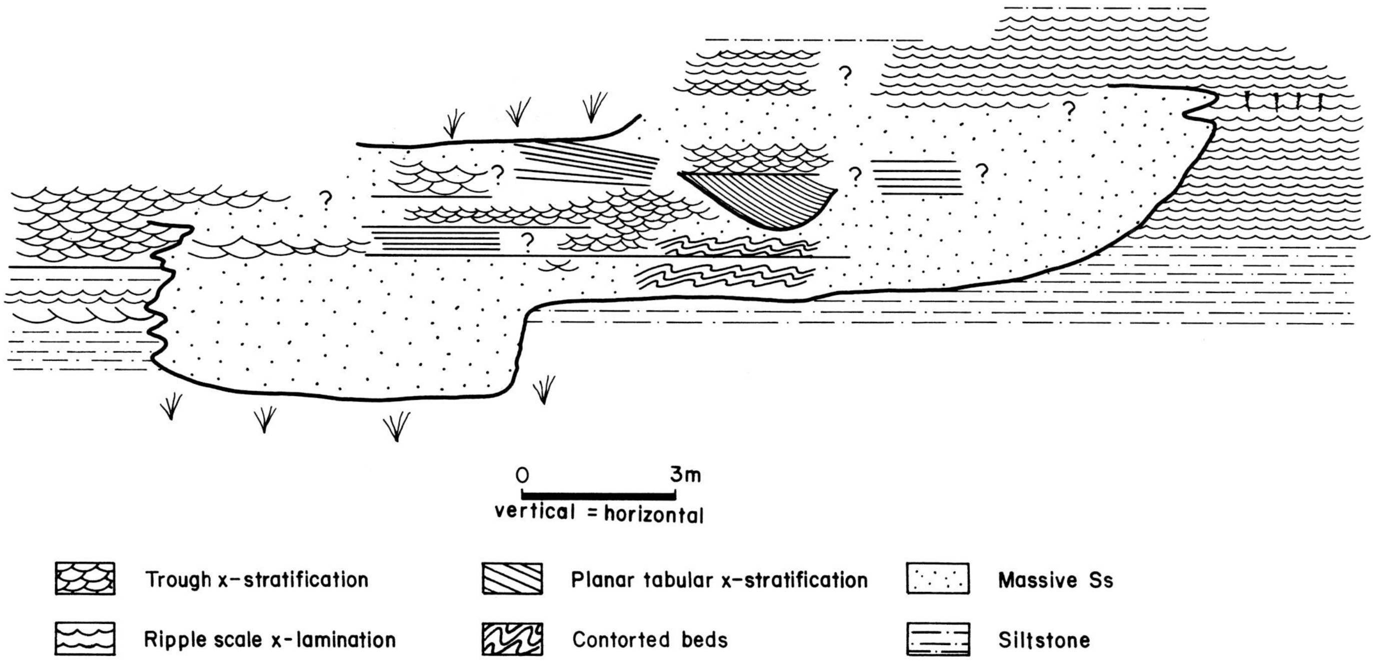

FIG. 6.—Outcrop sketch showing internal geometry of splay, levee, and interfluve deposits (Facies Associations B and C).

RELATIONSHIP TO THRUSTING

The Campanian strata described herein represent part of the foreland molasse deposits derived from the thrust belt to the west. Thrusting in the Idaho-Wyoming salient of the thrust belt began in Cretaceous time and continued at least through the early Eocene (Dorr and others, 1977).

Jordan (1981) and Beaumont (1981) modeled subsidence

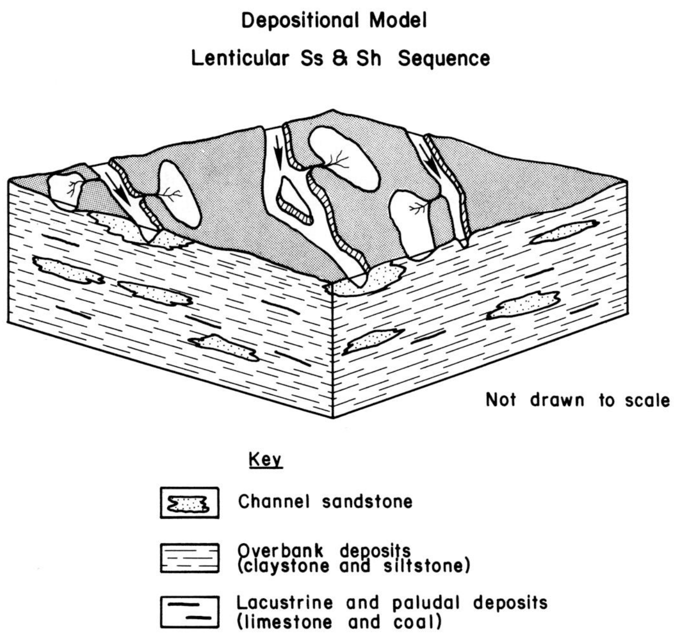

FIG. 7.—Postulated depositional model for the Lenticular Sandstone and Shale Sequence. Note low sinuosity channels and isolated sandbodies.

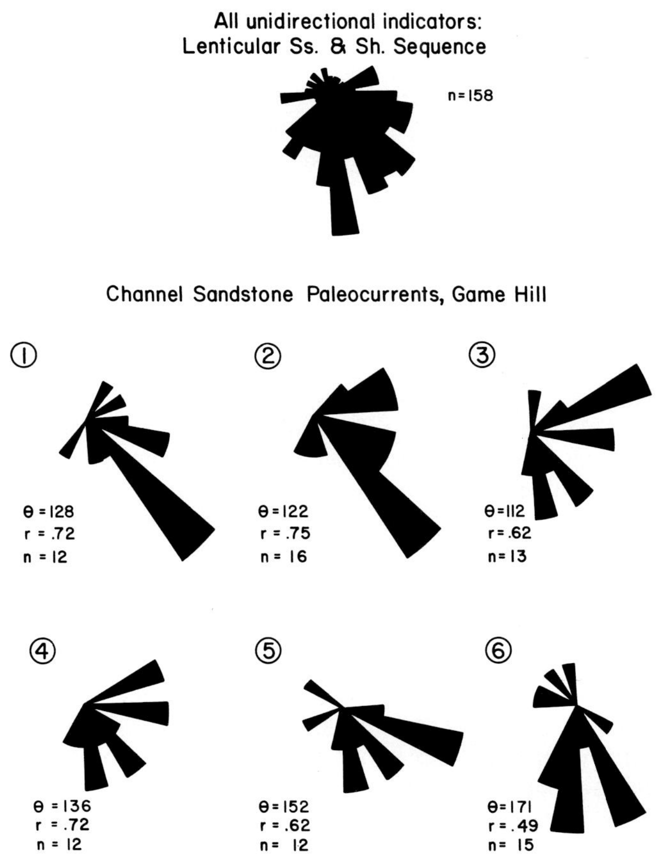

FIG. 8.—Paleocurrent rose diagrams from the Lenticular Sandstone and Shale Sequence. a) all unidirectional indicators; b) data from six vertically stacked channel sandstones. r = mean azimuth, θ = vector length (dispersion), n = number of observations. Note range of r's is less than 60°.

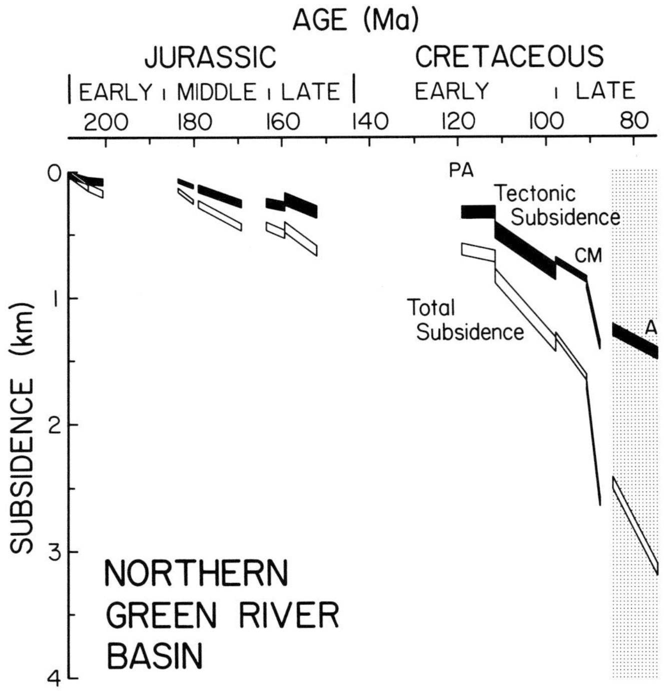

FIG. 9.—Subsidence curves, northern Green River basin, constructed by plotting water depth and cumulative sediment thickness (data not included) through time. Stippled area equals duration of deposition of Coaly Sequence, Lenticular Sandstone and Shale Sequence and White Sandstone. PA, CM, and A represent the timing of initial movements of the Paris, Crawford-Meade, and Absaroka thrusts (Wiltschko and Dorr, 1983).

of the foreland basin as the flexural product of thrust and sediment loading. Although the assumptions of lithospheric behavior differed between the two studies, results indicate that subsidence of the Western Interior Foreland basin is a consequence of tectonic loading and subsequent sediment loading due to redistribution of thrust-derived sediments in the basin. No evidence for intra- or subcrustal loads (Karner and Watts, 1983) has been documented for the Western Interior Foreland basin. Similar work by Shuster and Steidtmann (1983), Hagen and others (1985), and Quinlan and Beaumont (1984) substantiates the importance of tectonic loading in the development of thrust-bounded basins.

This genetic relationship between thrusting and subsidence suggests that periods of rapid subsidence are synchronous or slightly postdate major episodes of thrusting. Because of the problems involved in dating thrust events and in constructing detailed subsidence curves, however, a direct correspondence between the timing of individual thrusts and rapid subsidence is difficult to make. Both the Absaroka and Crawford-Mead thrusts (Fig. 1) were apparently active during the Campanian (Wiltschko and Dorr, 1983) and thus, either or both may have induced the rapid subsidence observed.

CONCLUSIONS

Analysis of the Campanian Lenticular Sandstone and Shale Sequence in the northern Green River Basin, Wyoming provides insight into the response of a fine-grained fluvial system to tectonics in a foreland basin. Facies analysis indicates that the Lenticular Sandstone and Shale Sequence was deposited by a series of low sinuosity/low gradient channels with associated levee, splay, and interfluve sedimentations. These facies represent deposition in a fixed-channel fluvial system in response to rapid subsidence of the alluvial plain. Results of subsidence analysis (backstripping) substantiate that the Campanian was a time of rapid subsidence. Comparison of times of rapid subsidence with published timings of initial thrust movements in the thrust belt to the west suggests a temporal relationship between thrusting and subsidence and possibly a causal relationship with fluvial-sandstone architecture.

These conclusions suggest that combined analyses of facies, architecture and subsidence in fine-grained alluvial systems can be extremely useful in understanding the interaction of thrusting, subsidence, and consequent sedimentation. Although coarser (proximal) tectogenic sediments often provide useful information regarding timing and nature of uplifts, they provide little information concerning basin subsidence. Analysis of fine-grained distal deposits, in the manner described herein, may eventually prove to be more useful in understanding the interaction of thrusting, subsidence, and consequent sedimentation.

ACKNOWLEDGMENTS

This study was supported by National Science Foundation Grant No. EAR-8108939 to Steidtmann and by grants to Shuster from Amoco Production Company, Marathon Oil company and ARCO Exploration Company. We thank Dave Marshall for his help in the field.

REFERENCES

ALLEN, J. R. L., 1974, Studies in fluviatile sedimentation: Implications of pedogenic carbonate units, Lower Old Red Sandstone, Anglo-Welsh outcrop: Geological Journal, v. 9, p. 181–208.

———, 1978, Studies in fluviatile sedimentations: An exploratory quantitative model for the architecture of avulsion-controlled alluvial suites: Sedimentary Geology, v. 21, p. 129–147.

BEAUMONT, C., 1981, Foreland basins: Royal Astronomical Society Geophysical Journal, v. 65, p. 291–329.

BEERBOWER, J. R., 1964, cyclothems and cyclic depositional mechanisms in alluvial plain sedimentation: Kansas Geological Survey Bulletin, v. 169, p. 31–42.

BLAKEY, R. C., AND GUBITOSA, R., 1984, Controls of sandstone body geometry and architecture in the Chinle Formation (upper Triassic), Colorado Plateau: Sedimentary Geology, v. 38, p. 51–86.

BOWN, T. M., AND KRAUS, M. J., 1981, Lower Eocene alluvial paleosols (Willwood Formation, northwest Wyoming, U.S.A.) and their significance for paleoecology, paleoclimatology, and basin analysis: Paleogeography, Paleoclimatology, Paleoecology, v. 34, p. 1–30.

BRIDGE, J. S., AND LEEDER, M. R., 1979, A simulation model of alluvial stratigraphy: Sedimentology, v. 26, p. 617–644.

DORR, J. A., Jr., 1952, Early Cenozoic stratigraphy and vertebrate paleontology of the Hoback Basin, Wyoming: Geological Society of America Bulletin, v. 63, p. 59–94.

——, SPEARING, D. R., AND STEIDTMANN, J. R., 1977, Deformation and deposition between a foreland uplift and an impinging thrust belt: Hoback Basin, Wyoming: Geological Society of America Special Paper No. 177, 82 pp.

FLORES, R. M., 1981, Coal deposition in fluvial paleoenvironments of the Paleocene Tongue River Member of the Fort Union Formation, Powder River Area, Powder River Basin, Wyoming: Society of Economic Paleontologists and Mineralogists Special Publication 31, p. 169–190.

FRIEND, P. F., SLATER, M. J., AND WILLIAMS, R. C., 1979, Vertical and lateral building of river sandstone bodies, Ebro Basin, Spain: Journal of the Geological Society, v. 136, p. 39–46.

GUENNEL, G. K., SPEARING, D. R., AND DORR, J. A., Jr., 1973, Palynology of the Hoback Basin: Wyoming Geological Association Guidebook, 25th Annual Field Conference, p. 173–185.

HAGEN, E. S., SHUSTER, M. W., AND FURLONG, K. P., 1985, Tectonic loading and subsidence of intermontane basins: Wyoming foreland province: Geology, v. 13, p. 585–588.

JORDAN, T. E., 1981, Thrust loads and foreland basin evolution, Cretaceous, western United States: American Association of Petroleum Geologists Bulletin, v. 15, p. 2500–2520.

KARNER, G. D., AND WATTS, A. B., 1983, Gravity anomalous and flexure of the lithosphere at Montana ranges: Journal of Geophysical Research, v. 88, p. 10449–10477.

MOODY-STUART, M., 1966, High- and low-sinuosity stream deposits, with examples from the Devonian of Spitzbergen: Journal of Sedimentary Petrology, v. 36, p. 1102–1117.

PALMER, A. R., 1983, The Decade of North American Geology 1983 Geologic Time Scale: Geology, v. 11, p. 503–504.

QUINLAN, G. M., AND BEAUMONT, C., 1984, Appalachian thrusting, lithospheric flexure, and the Paleozoic stratigraphy of the Eastern Interior of North America: Canadian Journal of Earth Sciences, v. 21, p. 973–996.

RIEKE, H. H., III, AND CHILINGERIAN, G. V., 1974, Compaction of argillaceous sediments: Developments in Sedimentology, v. 16, 424 p.

RUST, B. R., 1981, Sedimentation in arid-zone anastomosing fluvial system: Cooper's Creek, Central Australia: Journal of Sedimentary Petrology, v. 51, p. 745–755.

SCLATER, J. G., AND CHRISTIE, P. A. F., 1980, Continental stretching: An explanation of the post-mid-Cretaceous subsidence of the central North Sea Basin: Journal of Geophysical Research, v. 85, p. 3711–3739.

SHUSTER, M. W., AND STEIDTMANN, J. R., 1983, The origin and development of the northern Green River basin: A stratigraphic and flexural study (abs.): American Association of Petroleum Geologists Bulletin, v. 67, p. 1356.

SMITH, D. G., 1976, Effect of vegetation on lateral migration of anastomosed channels of a glacier meltwater river: Geological Society of America Bulletin, v. 87, p. 857–860.

———, 1983, Anastomosed fluvial deposits: Modern examples from western Canada, *in* Collinson, J. D., and Lewis, J., eds., Modern and Ancient Fluvial Systems: International Association of Sedimentologists Special Publication 6, p. 155–168.

SMITH, D. G., AND PUTNAM, P. E., 1980, Anastomosed river deposits: Modern and ancient examples in Alberta, Canada: Canadian Journal of Earth Science, v. 17, p. 1396–1406.

——, AND SMITH, N. D., 1980, Sedimentation in anastomosed river systems: Examples from alluvial valleys near Banff, Alberta: Journal of Sedimentary Petrology, v. 50, p. 157–164.

SPEARING, D. R., 1969a, Stratigraphy, sedimentation, and tectonic history of the Paleocene-Eocene Hoback Formation of western Wyoming: Unpublished Ph.D. Dissertation, University of Michigan, 179 p.

———, 1969b, Stratigraphy and sedimentation of the Paleocene-Eocene Hoback Formation, western Wyoming: Wyoming Geological Association, 21st Annual Field Conference Guidebook, p. 55–63.

STECKLER, M. S., AND WATTS, A. B., 1978, Subsidence of the Atlantic-type continental margin off New York: Earth and Planetary Science Letters, v. 41, p. 1–13.

VAN HINTE, J. E., 1978, Geohistory analysis—Application of micropaleontology in exploration geology: American Association of Petroleum Geologists Bulletin, v. 62, p. 201–222.

WATTS, A. B., AND RYAN, W. B. F., 1976, Flexure of the lithosphere and continental margin basins: Tectonophysics, v. 36, p. 25–44.

WILTSHKO, D. V., AND DORR, J. A., 1983, Timing of deformation in overthrust belt and foreland of Idaho, Wyoming, and Utah: American Association of Petroleum Geologists Bulletin, v. 67, p. 1304–1322.

EVOLUTION OF DRAINAGE SYSTEMS IN RESPONSE TO ACADIAN DEFORMATION: THE DEVONIAN BATTERY POINT FORMATION, EASTERN CANADA

DAVID A. LAWRENCE* AND BRIAN P. J. WILLIAMS
Department of Geology, University of Bristol, Bristol BS8 1RJ, U.K.

ABSTRACT: The 2.3-km-thick Lower Devonian Battery Point Formation conformably overlies shallow marine sandstones of the York River Formation and coarsens upward into proximal braidplain deposits of the Malbaie Formation. Whereas the lower part consists largely of west- to northwesterly-draining distal sandy braidplain and meandering fluvial systems, the uppermost Fort Prével Member (760 m thick) contains a variety of fluvial deposits formed in direct (tectonic) and indirect (climatic) response to early Acadian deformation.

Four lithofacies associations are recognized in the Fort Prével Member. The lowest (Association 1, 80 m thick) consists of isolated and amalgamated sheet sandstones with parallel lamination and trough crossbedding deposited in low sinuosity, north- to northeastward-flowing, sand-bed fluvial channels.

Association 2 (235 m thick) largely comprises laterally and vertically variable thin channel-fill and sheetflood sandstones deposited on extensively desiccated ephemeral mudflats. In the lower part, these sediments interbed with mainly parallel-laminated, amalgamated sheet sandstones, with cross-cutting channel fills flowing toward the north and northeast. At the top of the association, axial ESE-flowing, more perennial fluvial systems traversed ephemeral mudflats.

Association 3 (165 m) consists of stacked planar and trough cross-bedded sandstones deposited in major ESE-flowing axial drainage systems, overstepped by more proximal, though still axially flowing, sandy braidplain facies of Association 4 (260 m thick).

A tectonic model is proposed for the reversal in dispersion direction of axial drainage systems in the Battery Point Formation, incorporating recent views on the Acadian Orogeny as involving oblique continental collision of several microcontinents. Collision with the St. Lawrence promontory initiated subsidence in the Gaspé Basin and supplied longitudinal fluvial systems flowing westward into a coastal embayment during lower Battery Point Formation time. Later collision of the New York promotory to the southeast reversed basin axis slope during deposition of the Fort Prével Member. This supplied sediment longitudinally eastward and southeastward into an uplifted, probably enclosed semi-arid basin. Marginal basin uplift to the south later generated the Malbaie Formation conglomerates.

INTRODUCTION

The direct correlation of nonmarine depositional episodes with tectonic events has only recently progressed beyond a general level, with the application of plate tectonic theory (e.g., Miall, 1981, 1983). Alluvial successions give probably the best indications available of those tectonic events giving rise to terrestrial relief (Rust, 1981a). Hence, it should be possible to relate changes in fluvial style and dispersal direction to changing basin configuration as a result of tectonic activity.

The 2.3-km-thick Battery Point Formation (McGerrigle, 1950) of the Gaspé Peninsula, E. Québec forms part of a coarsening-upward clastic wedge shed in response to early Acadian deformation (Rust, 1981a). Conformably overlying marginal marine sandstones of the York River Formation, it passes abruptly upward into proximal gravelly and sandy braidplain facies of the Malbaie Formation (Rust, 1981a, 1984). The facies associations recognized in this study clearly, though indirectly, document changes in basin shape, basin axis slope, and orientation as Acadian deformation progressed. The Battery Point Formation itself is made up of a spectrum of mainly distal alluvial deposits. The lower 1 to 2 km consist of distal sandy braidplain facies associations at the base (Cant and Walker, 1976). These are overlain by a thick (as much as 640 m) succession of moderate to high sinuosity channel deposits which accumulated in low-lying, well vegetated coastal plain settings (Lawrence and Williams, 1984).

On the south shore of Gaspé Bay (Fig. 1), the upper 760 m of the Formation comprises mainly redbed facies assigned to the Fort Prével Member (Sikander, 1978). It is the variations in fluvial style and paleocurrent patterns within this Member which best record the changing basin configuration and climatic response resulting from Acadian events.

DESCRIPTION AND INTERPRETATION OF FACIES ASSOCIATIONS

Introduction

The Fort Prével Member has been divided into four distinct lithofacies associations (Fig. 2), varying in thickness from 80 to 260 m. In the following sections the facies, geometry and paleocurrent characteristics of these associations are described and interpreted.

Association 1

Description.—

This 80-m-thick, sandstone-dominated association (Fig. 2) abruptly overlies the lower part of the Battery Point Formation. The lower 35 m contains well defined fining-upward sequences averaging 6 m thick (Fig. 3A) in mainly medium to coarse sand. Sequences are generally sheetlike to broadly lenticular in geometry within outcrop with undulating to gently scoured basal erosion surfaces. Trough crossbedding dominates (sets vary from 20 cm to 1 m) coarser sets containing abundant mudclasts (Fig. 4A). These facies often occur above multiple erosion surfaces as much as 1 m in relief (Fig. 4B), particularly toward the top of sequences. Paleocurrents show a marked unidirectional dispersion toward the NNE (Fig. 5) with low directional variance between sequences. Associated with trough crossbedding are broad finer grained trough fills, often with a current lineation paralleling trough axes. Parallel to low angle lamination and small-scale (20–25cm) planar crosssets occur at the bases of some sequences (Fig. 3A). Dip directions of low angle sets diverge greatly or may even

*(Present Address B.P. Petroleum Development Ltd, Farburn Industrial Estate, Dyce, Aberdeen, AB2 0PB, U.K.)

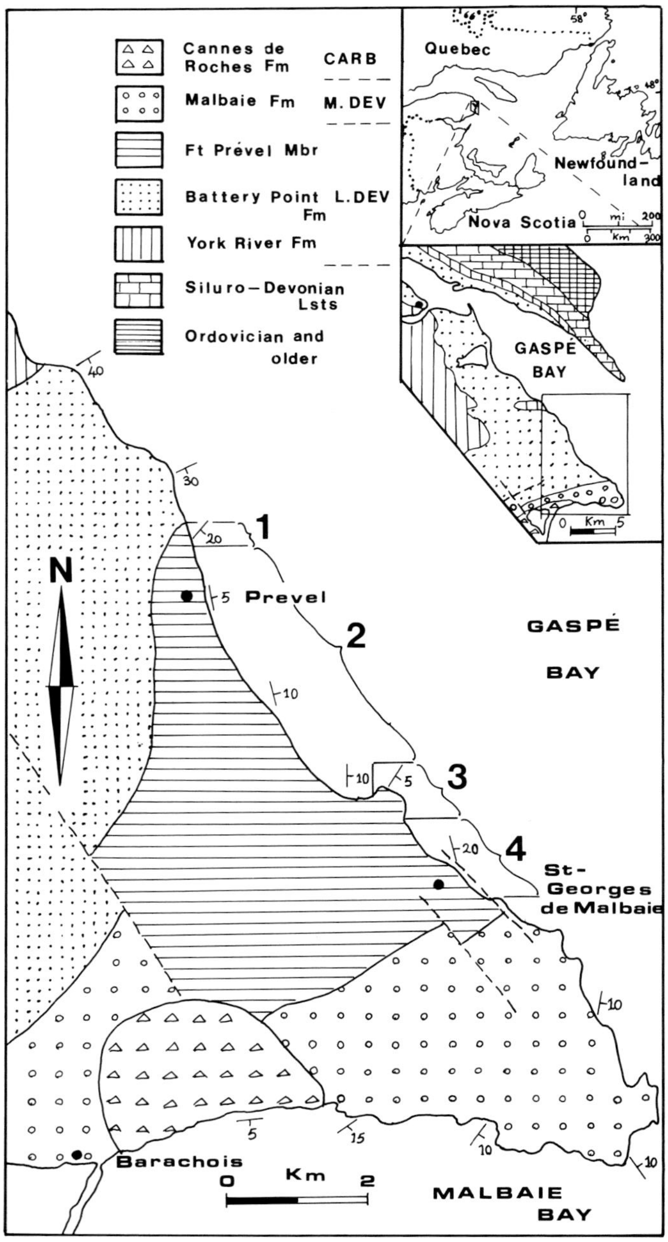

FIG. 1.—Location map of eastern Canada, and below, a geological location map of the Fort Prével Member (Gaspé Bay inset). Numbers on geological map refer to locations of Associations 1-4.

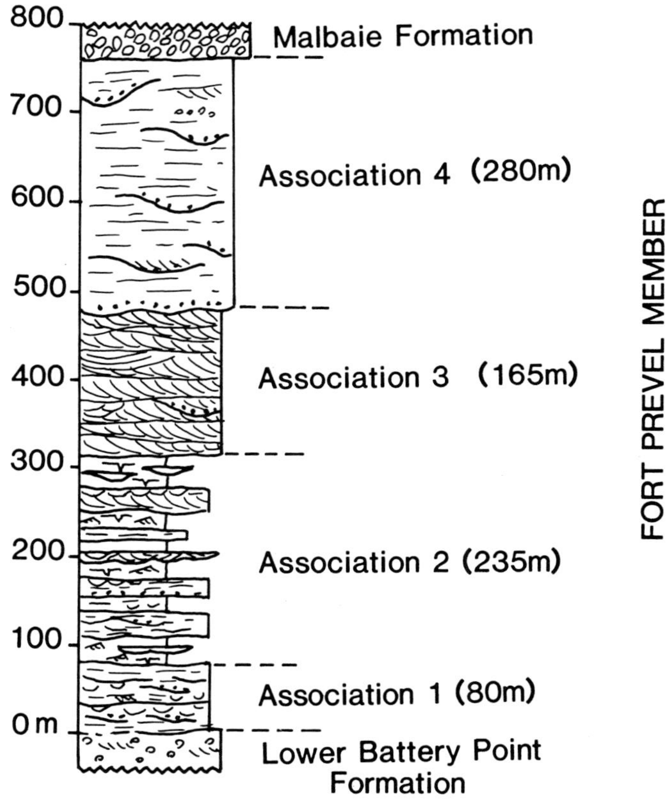

FIG. 2.—Schematic stratigraphic section through associations in the Fort Prével Member.

dip upstream from local paleocurrent trends. Sequences are capped by thin (as much as 1 m) associations of small-scale planar and trough cross-bedded and parallel laminated sandstones with frequent desiccated mud drapes. These are overlain by homogeneous to current ripple-laminated mudstone and very fine sandstone.

The upper 45 m of Association 1 lack fining-upward sequences, depositional units ranging from isolated sheet and channel sandstones (averaging 0.8 m thick) to multistory sheet sandstones between 3.5 and 4.7 m thick (Fig. 3B). Sheet sandstones have flat (Fig. 4C) to steeply channelized and intraclast-lined erosion surfaces (Fig. 4D). Current-lineated parallel to low angle lamination dominates in mainly medium sand with rare concentrically filled scours (Fig. 3B). Current lineation measurements from these sandstones, and thinly bedded facies at the top of sequences in the lower part of the association, indicate an ENE to E current direction (Figs. 3A, 3B). Trough crossbedding with common intraclasts occurs at the base and cuts the tops of sheet sandstones (Figs. 3B, 4C) and fills small-scale channels. Thinly bedded (3 to 15 cm), ripple drift laminated finer sandstones with frequent desiccated mud partings (Fig. 4C) terminate sequences.

Interpretation.—

Sequences in the lower 35 m of the association were probably generated by multiple scouring and filling of low sinuosity stream systems, with characteristic internal geometries of concave-upward erosion surfaces (Leeder, 1973; Campbell, 1976; Friend and others, 1979; Tunbridge, 1981a; Allen, 1983). Trough crossbedding is recorded from the fills of many modern ephemeral stream channels (Williams, 1971) and described from ancient inferred low sinuosity stream systems associated with parallel lamination (Bryhni, 1978; Hiller and Stavrakis, 1984). Its occurrence here suggests rapid channel cutting and filling events, mainly by vertical accretion of straight crested dune bedforms (Tunbridge, 1981a, 1984). Abundant intraclast material indicates rapid channel shifting and reworking of mud from abandoned channel areas (Rust, 1984).

Parallel to low angle lamination (with current lineation) in the lower parts of some sequences suggests initial de-

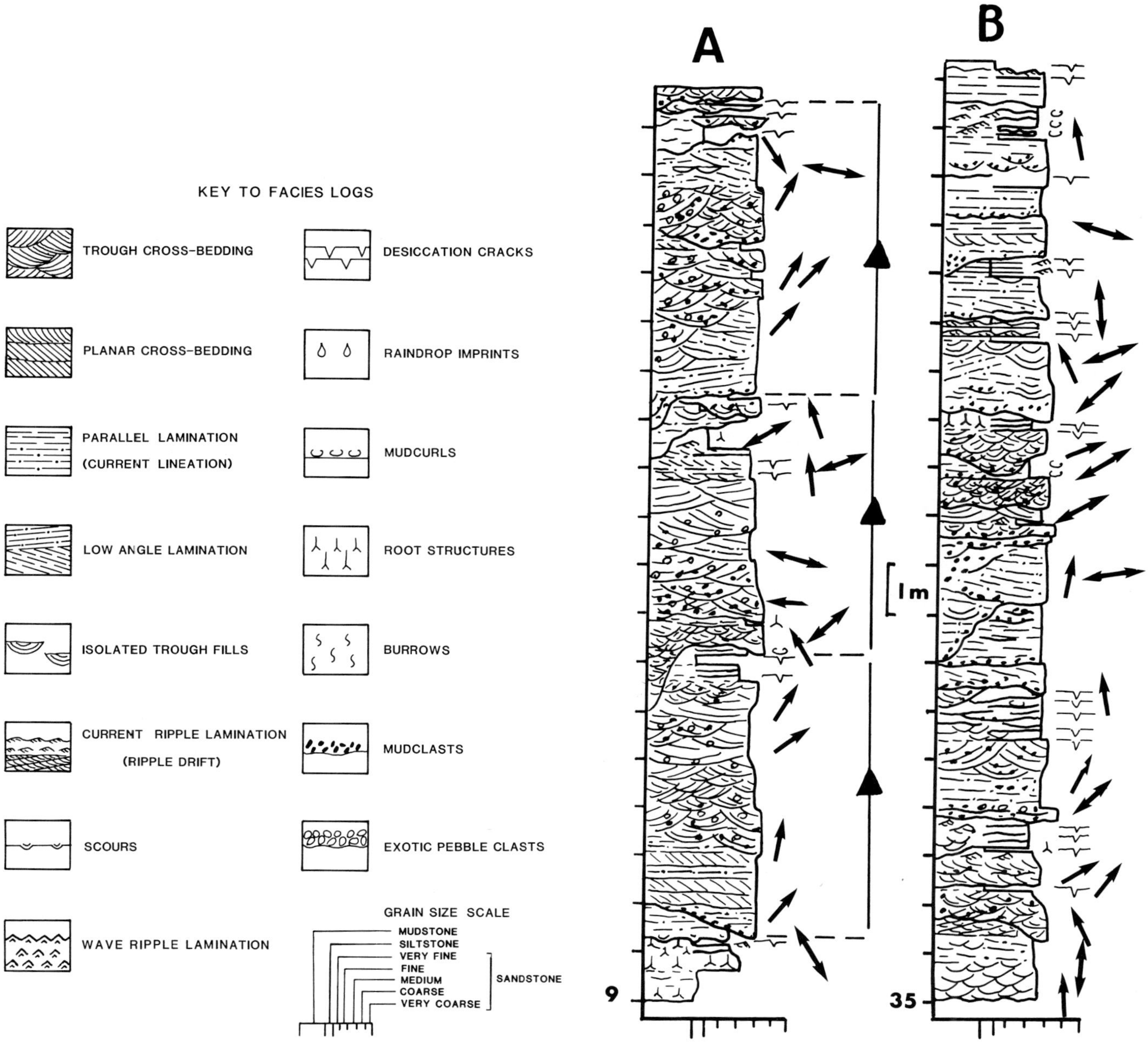

FIG. 3.—Association 1. A) Fining-upward sequences from the lower part. B) Sheet sandstones from the upper part. Numbers refer to meters above the base of the association.

position from broad, relatively unconfined flows under upper flow regime conditions (Harms and others, 1982). Some low angle sets, continuous with parallel lamination, also suggest an upper flow regime origin (Rust, 1978, 1984). Planar-tabular cross-sets were accreted on shallow channel or point bars (Picard and High, 1973).

In the upper 55 m of the association, flat basal erosion surfaces with sparse intraclasts imply deposition of sandstones from laterally unconfined to broadly channelized sheetfloods (McKee and others, 1967; Tunbridge, 1981b). Low angle lamination infills shallow scours (Rust, 1984) or occurs at steep channel margins (Picard and High, 1973). Trough crossbedding fills small channels (<1 m deep) often formed by waning flood dissection. Mud draping and thin ripple drift lamination capping sandstones suggests rapidly falling flood stage (Picard and High, 1973; Tunbridge, 1981b). Alternating mudstone/sandstone lithologies are compared to terminal floodplain sequences of modern ephemeral streams (Sneh, 1983).

Association 2

Description.—

Association 2, as much as 235 m thick, is characterized by interbedding of heterolithic facies and discrete, usually sheetlike sandstones. In the basal 85 m, sheet sandstones alternate with equal to greater amounts of heterolithic facies. The former are often multistory (Fig. 6A, 7), varying

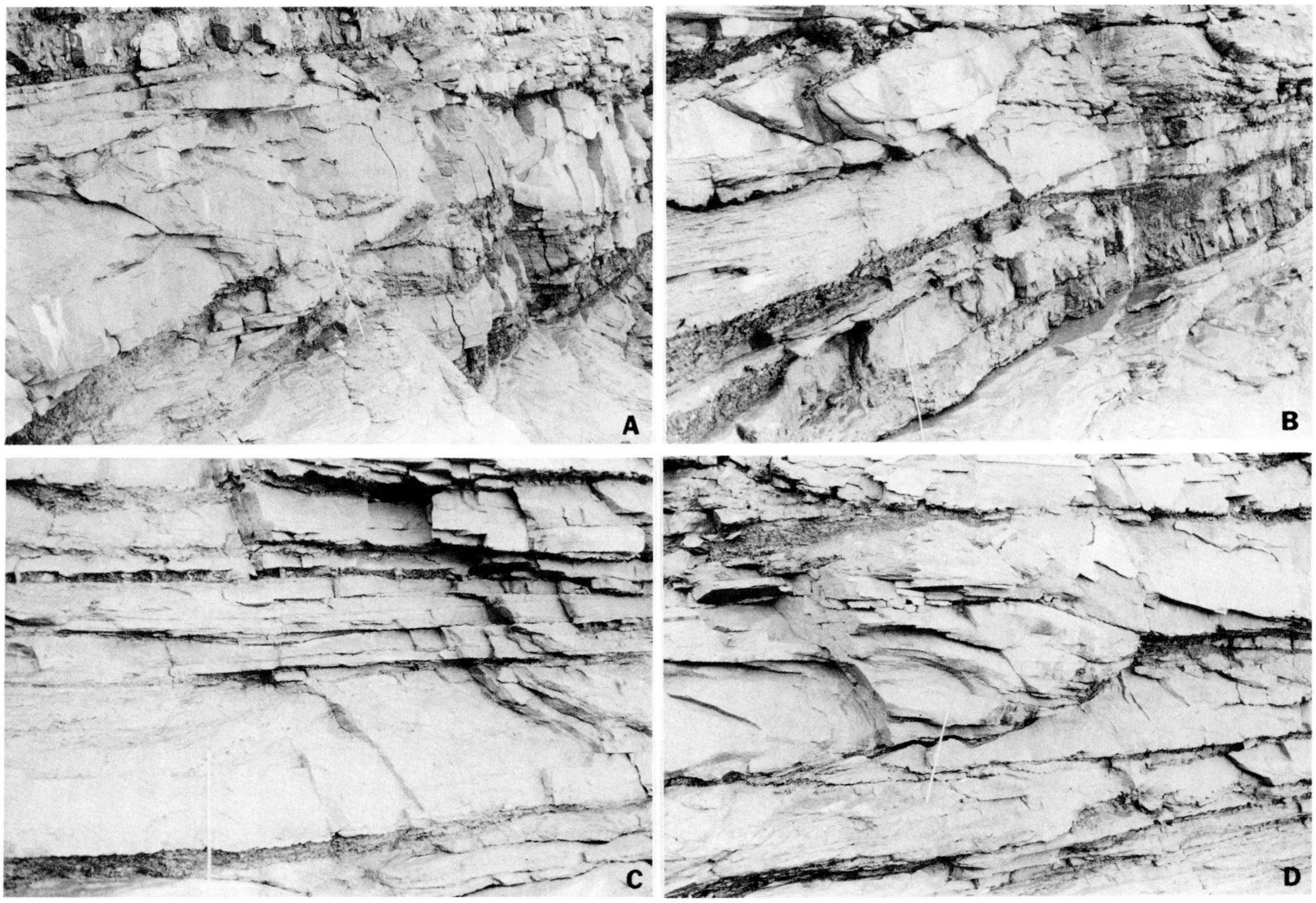

FIG. 4.—Association 1. A) Intraclast-rich large-scale (as much as 1 m) trough cross-sets. B) Isolated channel fill cutting low angle laminated sandstones. C) Erosive-based parallel laminated sheet sandstone, overlain by thin sheet sandstones and desiccated mudstones. D) Channel margin (stepped) in parallel to low angle laminated sandstones. Rule, in all photos, is 1 m long.

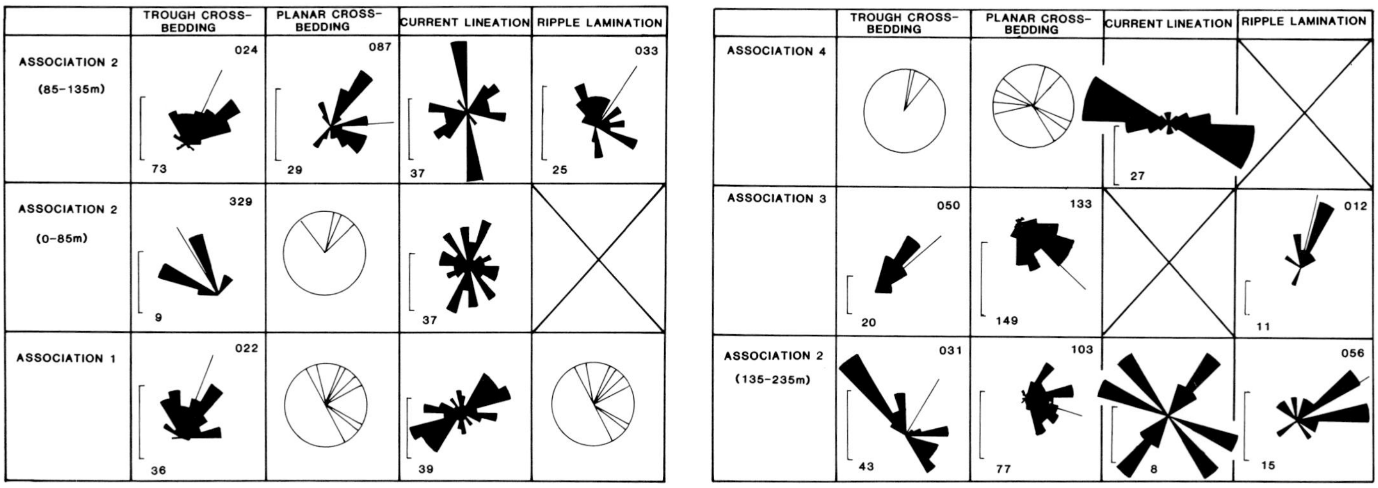

FIG. 5.—Combined paleocurrent plots for different sedimentary structures from the four associations. Note that broad trough sets have been grouped with planar cross-sets in Association 3, and the rose diagram for trough crossbedding comprises readings from the upper, erosive-based channel fills in this association. Number in bottom left of each rose = number of readings. Number in top right = vector mean. Bar scale for rose diagrams represents 20 percent of readings.

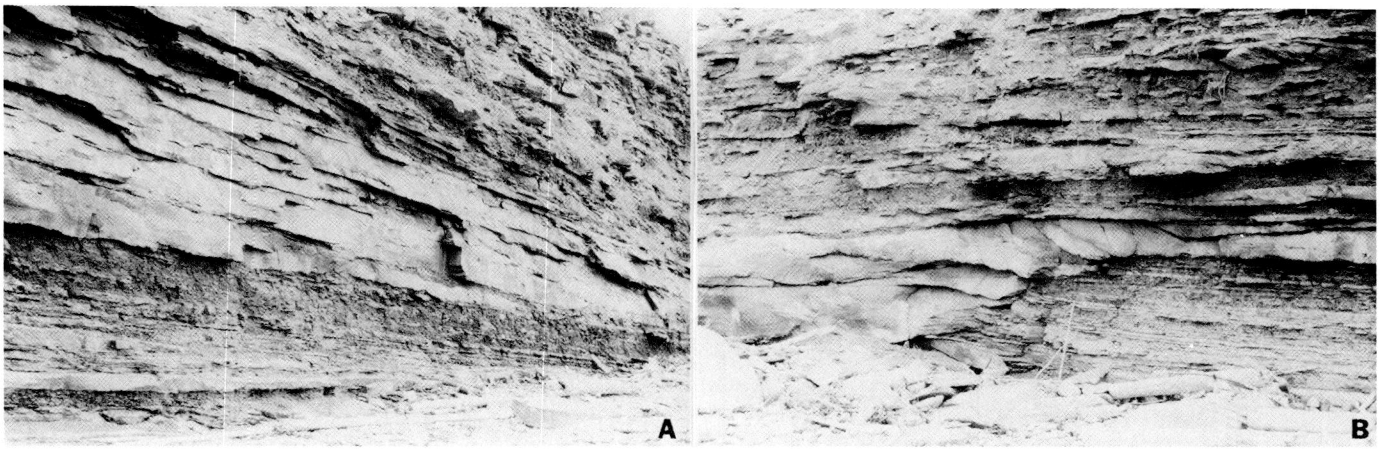

FIG. 6.—Association 2. A) Sheet sandstone (3 m thick) with internal erosion surfaces, encased in mainly mudstones. B) Small-scale channel fill within mudstones and thinly bedded sandstones (rule = 1 m long).

from 1 to 6 m thick, and are traced for as much as 740 m perpendicular to paleoflow. Internal channeled erosion surfaces with as much as 3 m relief have parallel laminated fills, though some have draping lamination with slumped channel margins (Fig. 7B). Trough crossbedding fills irregular relief at the base of some erosion surfaces near channel margins or rarely truncates the tops of sheet sandstones (Fig. 7a). Thin desiccated mud drapes between stories are reworked laterally into mudflake or mudclast horizons. Paleocurrents from all structures within this interval compare with those of Association 1, showing strong northerly modes (Fig. 5).

Heterolithic facies, similar to those found in the lower part of the association, where they contain small (<1 m thick) channel fills (Fig. 6B), dominate the upper 150 m of Association 2 (Figs. 8A and 9A). Mudstones are interbedded with thin sandstones (usually <80 cm) containing three basic internal structures. First, small-scale (<1 m thick) trough cross-bedded channel fills have scoured basal erosion surfaces, abundant intraclast lags and planar to mud draped tops. Second, isolated planar cross-sets (as much as 70–80 cm thick) have flat basal erosion surfaces (often incorporating mudflakes) and tabular to asymptotic foresets (Fig. 9B). Planar sets may be truncated by isolated scour-fills or trough cross-bedding, whereas muds drape common reactivation surfaces and sets. The third sandstone type, disappearing in the upper half of Association 2, consists of parallel to low angle laminated, mud draped sheet sandstones with rare trough crossbedding filling local, intraclast-lined scours. Paleocurrents from trough crossbedding still show mainly northerly trends, whereas planar cross-sets and current lineation (below 127 m) display subsidiary but important E to ESE modes.

Thin sandstones are interbedded with common laminated to massive mudstones and current ripple-laminated fine to very fine sandstones. Desiccated mud films are often reworked into mudcurls (Fig. 10A). Other indications of exposure include raindrop impressions, adhesion ripples, and wave-rippled surfaces. Rarely within heterolithic facies are laterally accreted channel units preserved (Fig. 11A). Desiccated mud drapes separate as much as 1.5-m-thick bundles of low angle to trough cross-bedded sandstones. In the

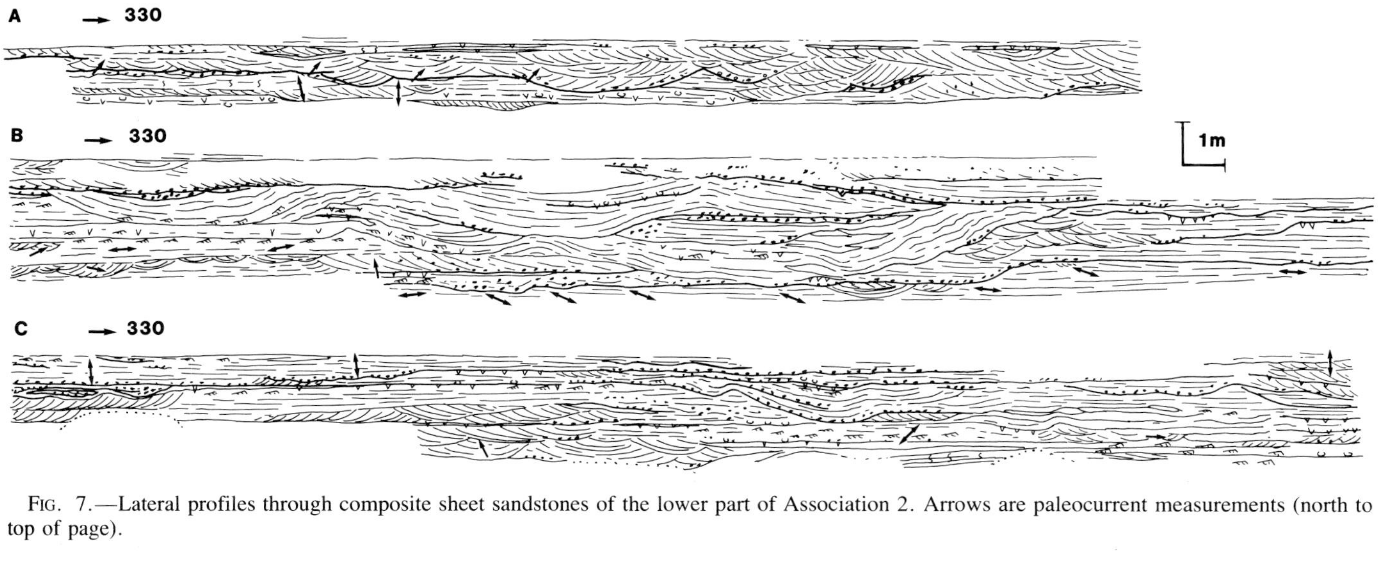

FIG. 7.—Lateral profiles through composite sheet sandstones of the lower part of Association 2. Arrows are paleocurrent measurements (north to top of page).

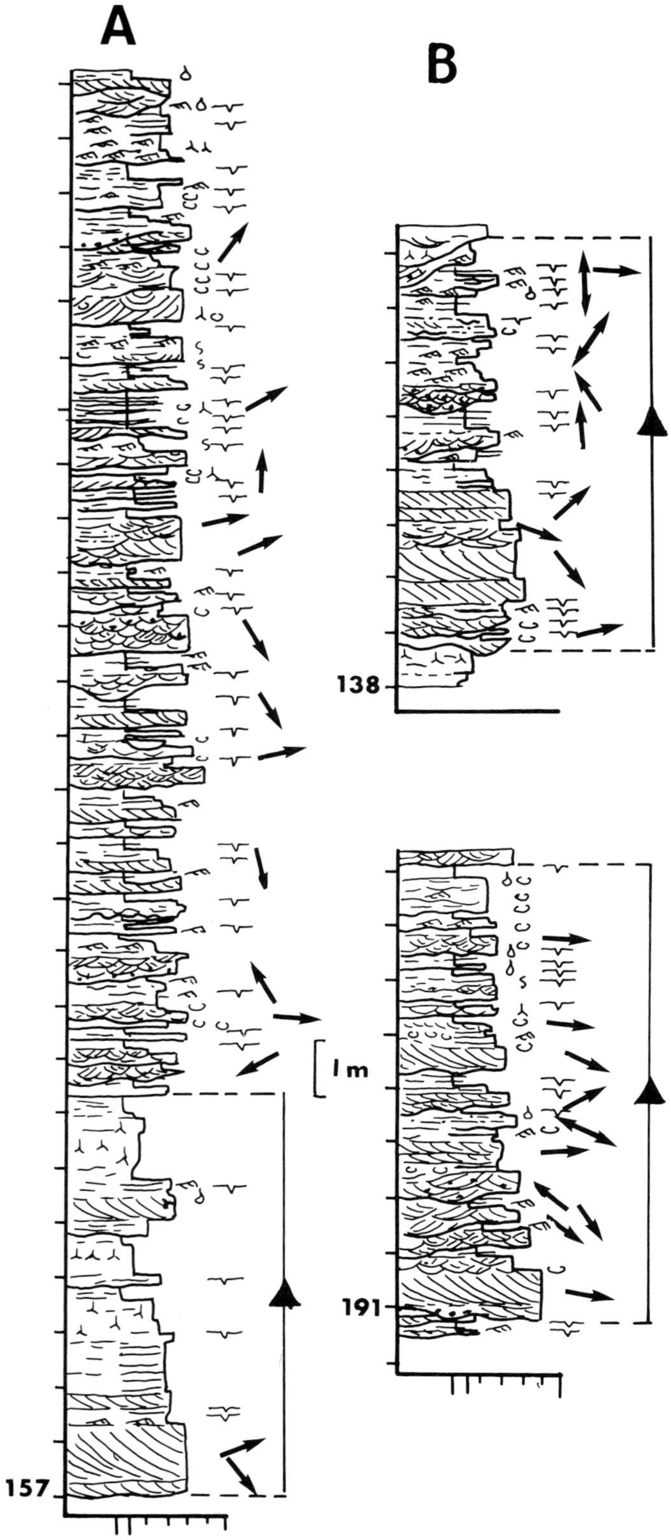

FIG. 8.—Association 2. A) Characteristic log of heterolithic facies underlain by a channel unit. B) Two examples of fining-upward and thinning sequences from the upper part of the association.

example in Figure 11A, the cutbank is preserved, the channel being mud-filled.

Above 132 m in Association 2, thicker (average 6.8 m) channel sequences appear, with thinning and fining-upward trends (Fig. 8B). Large-scale (as much as 80 cm thick) usually planar cross-sets associated with trough crossbedding and parallel lamination overly flat erosion surfaces. Paleocurrents for both trough and planar crossbedding show strong ESE and SE directions (Fig. 5), completing a clockwise rotation of dominant trends from northerly in Association 1 to ESE at the top of Association 2. Cross-bedded intervals pass upward into heterolithic facies. These consist of small-scale trough cross-bedded minor channel fills, isolated planar-tabular cross-sets, and parallel to ripple laminated sandstones, all with common desiccated mud drapes. Fining-upward sequences, which increase in abundance toward the top of the association, have relatively thick (as much as 3 m) caps of massive very fine sand to siltstone and thin current-rippled sandstones with frequent desiccated mud drapes.

Interpretation.—

Sheet sandstones in the lower part of Association 2 probably formed in low sinuosity channel networks incised at high stage, and then either abandoned (mud draped) and later filled, or filled during the same event by vertical accretion of plane bed sand (Tunbridge, 1981a, 1984; Rust, 1984). High pore fluid pressure during falling stage, in sands draping steep channel margins, may have led to their slumping in places. Common thick (as much as 0.5 m) intraclast lags testify to extensive reworking of suspension-deposited muds during subsequent flooding (Rust, 1984) and also attest to their truly ephemeral nature (Picard and High, 1973). Isolated cross-sets occur where weak separation eddies formed, for example, above irregular scours in the channel bed (Tunbridge, 1981a, 1984) or at channel margins. Cross-bedded channel fills cutting the tops of some sheet sandstones may indicate waning flow degeneration of flood waters into channel networks.

Interbedded fine-grained facies are compared to distal sheetflood episodes on unconfined floodplains (Hubert and Hyde, 1982; Sneh, 1983; Tunbridge, 1984) with mainly suspension fallout of muds and current ripple laminated sands. The floodplain was probably traversed by small-scale channel networks (Sneh, 1983).

In the upper 150 m of the association, parallel laminated sheet sandstones are uncommon but likened to high-energy sheetflood deposits (McKee and others, 1967; Tunbridge, 1981b, 1984), with rapidly waning flow velocities evidenced by mud draping. The origin of isolated planar tabular cross-sets is, however, less clear. Similar bedding types have been reported from small bars in modern ephemeral stream channels (Williams, 1971; Picard and High, 1973). In Association 2, however, sets extend for tens of meters and may have formed as more extensive slipface-bounded sand sheets, with margins reworked and mud draped during falling stage. Small-scale trough cross-bedded channel fills are compared to larger scale examples of Rust (1981b), probably forming channel networks whose stability was influenced by mud-rich floodplains.

Rare laterally accreted channel fills (Fig. 11A) are com-

FIG. 9.—Association 2. A) View of typical heterolithic facies (height of cliff approximately 10 m). B) Isolated planar-tabular cross-set encased in desiccated mudstones (scale = 1 m).

FIG. 10.—Association 2. A) Curled mudflakes preserved at the base of a cross-bedded sandstone (FIG. 9B). Scale = 50 cm. B) Desiccation cracks controlled by wave ripple crests in a thin-bedded sandstone (numbered divisions on scale are centimeters).

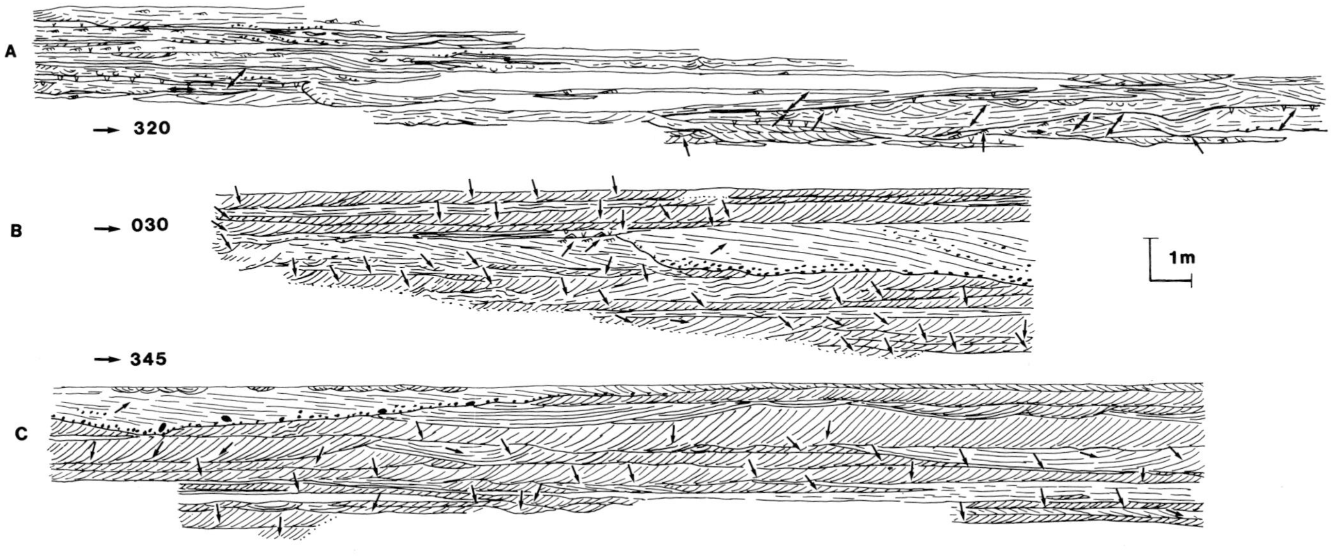

FIG. 11.—A) Profile through laterally accreted channel unit in Association 2. B) and C) Profiles through cross-bedded channel sandstones of Association 3.

parable to those of the modern Rio Puerco (Shepherd, this volume), though with less eolian influence, and similar ancient examples cited by Stear (1983) and Tunbridge (1984). Channel filling was by pulses of upper flow regime flooding, with falling stage mud draping and desiccation of accretion surfaces. Channel depths were on the order of 2 m and widths of 35 m.

Exposure phenomena are ubiquitous within finer grained facies. Adhesion ripples and desiccation crack infills of probable eolian sand grains suggest some wind activity, whereas raindrop impressions and desiccation cracks are abundant. Massive mudstones may have been disrupted by shrinking and swelling associated with desiccation (Taylor and Woodyer, 1978; Rust, 1981b). Mudcurls and flakes, derived from desiccation of mud films, seem to have undergone some subaqueous transport and are even preserved in upper flow regime flood deposits. It may be that calcareous ground waters led to early cementation of mudcurls on drying, possibly enhanced by algal binding (Fagerstrom, 1967).

Fining-upward and thinning sequences toward the top of Association 2 seem to represent the filling of more permanent stream courses, though still with highly ephemeral discharge. Basal large-scale planar cross-sets may have originated as avalanche sets on side or longitudinal bars (Williams, 1971; Karcz, 1972; Picard and High, 1973). This was followed by punctuated aggradation by small channel fills, isolated planar cross-sets and low stage, mud draped ripple lamination. Finer sediment accumulated in abandoned channel reaches, with common plant rootlet traces suggesting concentration of sparse vegetation along stream courses, as in modern anastomosing ephemeral channels of Cooper's Creek, Central Australia (Rust, 1981b).

Association 3

Description.—

This 16-m-thick association is sand-dominated, with thin mudstones occurring only toward the base and top of the association. It consists of 7- to 9-m-thick sequences of planar and broad trough crossbedding (Fig. 12A) interbedded with 2- to 3-m-thick more heterolithic facies.

Cross-bedded sequences average medium to coarse in grain size, and planar cross-sets are commonly graded with coarser foreset laminae thinning down-foreset dip. Trough sets average 0.5 to 1 m thick and are recognized in flow parallel

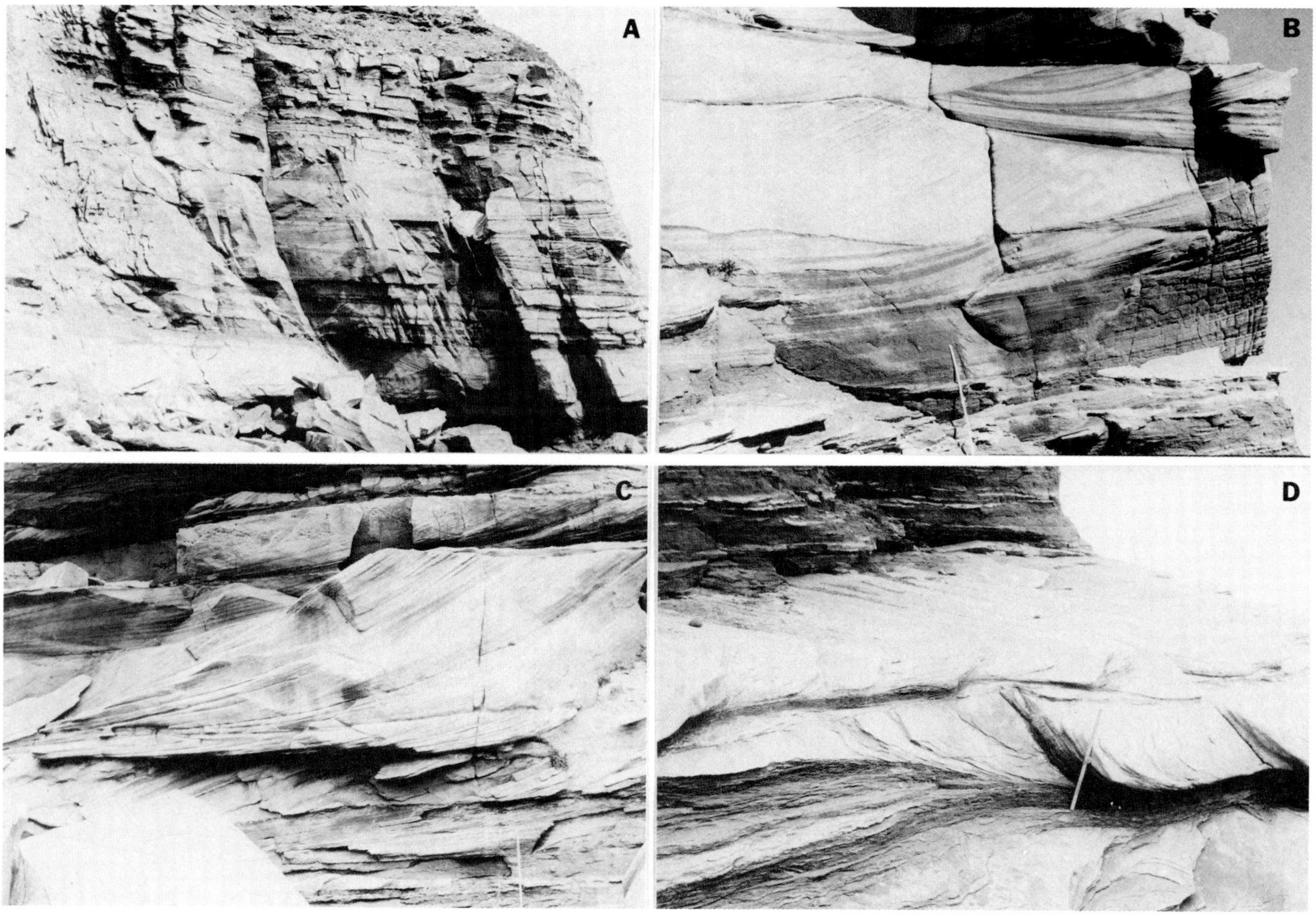

FIG. 12.—Association 3. A) View of cliff face (8 m high) showing cross-bedded sandstone complexes and horizons with interbedded mudstone. B) Planar cross-set truncated above by trough cross-sets (rule = 1 m). C) Finer grained foreset drapes (darker) to large-scale planar-tabular cross-set (rule = 1 m). D) High paleocurrent divergence (90°) shown by planar-tabular cross-sets at the top of a sandstone complex (rule = 50 cm).

sections by downcutting, scoop-shaped lower contacts and tangential toesets and can be traced for 15 to 25 m parallel to paleoflow. Low within-set paleocurrent variance and rare cross sections suggest a large radius of curvature for such sets (widths as much as 10 to 15 m).

Planar cross-sets often exceed 1 m in thickness and are characterized by flat lower bounding surfaces, tabular foresets, and low within-set paleocurrent variation with straight crested foreset plan views. Individual sets can be traced for as much as 55 m parallel to paleoflow. Variance between planar cross-sets within sandstone complexes is low, although smaller scale sets vary in orientation by as much as 90 at the top of some complexes (Fig. 12D). Larger planar cross-sets are truncated by trough cross-sets above, (Figs. 11C, 12B), whereas others have periodically spaced foreset drapes thickening toward toesets (Fig. 12C). Planar foresets are occasionally oversteepened or slumped (Fig. 11B), with deformation extending for as much as 8 to 9 m parallel to paleoflow. Combined paleocurrent plots for planar and trough cross-sets (Fig. 5) show a marked southeasterly trend, similar to channel sandstones in the upper part of Association 2.

Erosive contacts and intraformational debris are rare in Association 3, though localized channel fills (as much as 2 m deep and 15–17 m wide) contain low angle to massive sandstone (Figs. 11B, C). Interbedded heterolithic facies consist mainly of small-scale (as much as 40 cm) form sets of cross-bedding interbedded with parallel to current ripple laminated fine sandstones and occasional desiccated mud drapes, the latter particularly abundant toward the top of the association.

Interpretation.—

Dominantly unidirectional paleocurrents, sedimentary structures and low between-sequence paleocurrent variance suggest deposition of this association in low sinuosity sandbed rivers. Broad trough sets probably originated from filling of leeside scours of large dune bedforms (Cant, 1978; Harms and others, 1982). The form and distribution of planar cross-sets resemble those of loosely periodic or non-repetitive bars described from modern braided streams as transverse (Smith, 1970, 1971, 1974; Williams, 1970) or cross-channel (Cant, 1978; Cant and Walker, 1978). Rare descending cross-sets are analogous to structures formed by downstream accretion on the lee slopes of large, slower moving bars (Banks, 1973; Haszeldine, 1983). Similarity in orientation of trough and planar cross-sets (compare with the Hawkesbury Sandstone, Jones and Rust, 1983) suggests formation of the latter by downstream accretion of probable mid-channel bars. High paleocurrent spreads between smaller scale sets at the tops of some sequences suggest shoaling over bars and lateral migration of small-scale straight crested bedforms (Cant and Walker, 1978). Finer foreset drapes on some sets testify to stage fluctuations, though reactivation surfaces are rarely found. Channel scouring events are rare, the one prominent example (Fig. 11B) containing low angle lamination dipping perpendicular to channel orientation, suggesting filling by lateral accretion (compare with Allen, 1983, fig. 14).

Deformation of foresets mostly involved down-foreset slumping (the second type of Allen and Banks, 1972), lowering foreset slopes which are later overstepped by subsequent bedform migration. Fluid escape structures in the crests of more extensively deformed foresets suggest high pore fluid pressure and rapid dewatering (Lowe, 1976). Stage-induced fluctuations in water level provide the most likely trigger mechanism for foreset failure in this case (Doe and Dott, 1980).

Toward the top of the association, cross-cutting intraclast-lined erosion surfaces and common desiccated mud drapes suggest reversion to a more ephemeral discharge regime. Some of these channel fills cut sequences at high angles (as much as 90°), reminiscent of anastomosed channel systems superimposed on relict braids of Cooper's Creek, Central Australia (Rust, 1981B).

Association 4

Description.—

Although the thickest of the four associations at 260 m, it will only be briefly mentioned here due to its uniformity in facies and paleocurrent trends and marked similarity to the Malbaie Formation sandstones. The association consists of multistory channel sandstones (mainly medium to coarse sands) overlying abundant intraclast-lined, concave-upward erosion surfaces with as much as 4 m relief. Parallel to low angle lamination dominates with abundant current lineation and less common planar and trough cross-sets. Sandstones occasionally have sheetlike geometries. Rare in situ mudstones (as much as 1.1 m thick) show abundant evidence of desiccation (mudcracks as much as 70 cm deep) associated with rare raindrop imprints and wave-rippled surfaces. Paleocurrents show a dominantly WNW to ESE trend (Fig. 5), similar to channel orientations in Association 3, although planar and trough cross-sets show high dispersions and are mainly directed to the N to NE.

Interpretation.—

Association 4 indicates a return to deposition in low sinuosity ephemeral channels on an extensive sandy braidplain and is likened to Rust's (1978) Si facies assemblage. Multistory sandstones of this nature suggest rapid cutting and filling of ephemeral streams (Tunbridge, 1981a). Flood events with a high concentration of suspended sediment alternated with low stage mud accumulation in abandoned channels that subsequently became desiccated and reworked into intraclast lags (Rust, 1984). Horizontal to low angle lamination with common current lineation, again suggests high-stage vertical accretion of sand, possibly on shallow point bars (Picard and High, 1973) with cross-bedded infilling of local scour pools (Tunbridge, 1981a). Exposure after channel abandonment is evidenced by abundant desiccation phenomena.

ENVIRONMENTAL SYNTHESIS

Facies associations in the lower part of the Battery Point Formation record deposition in distal sandy braided and high sinuosity fluvial systems on the southern flank and in the axial portions of an ESE- to WNW-trending basin which

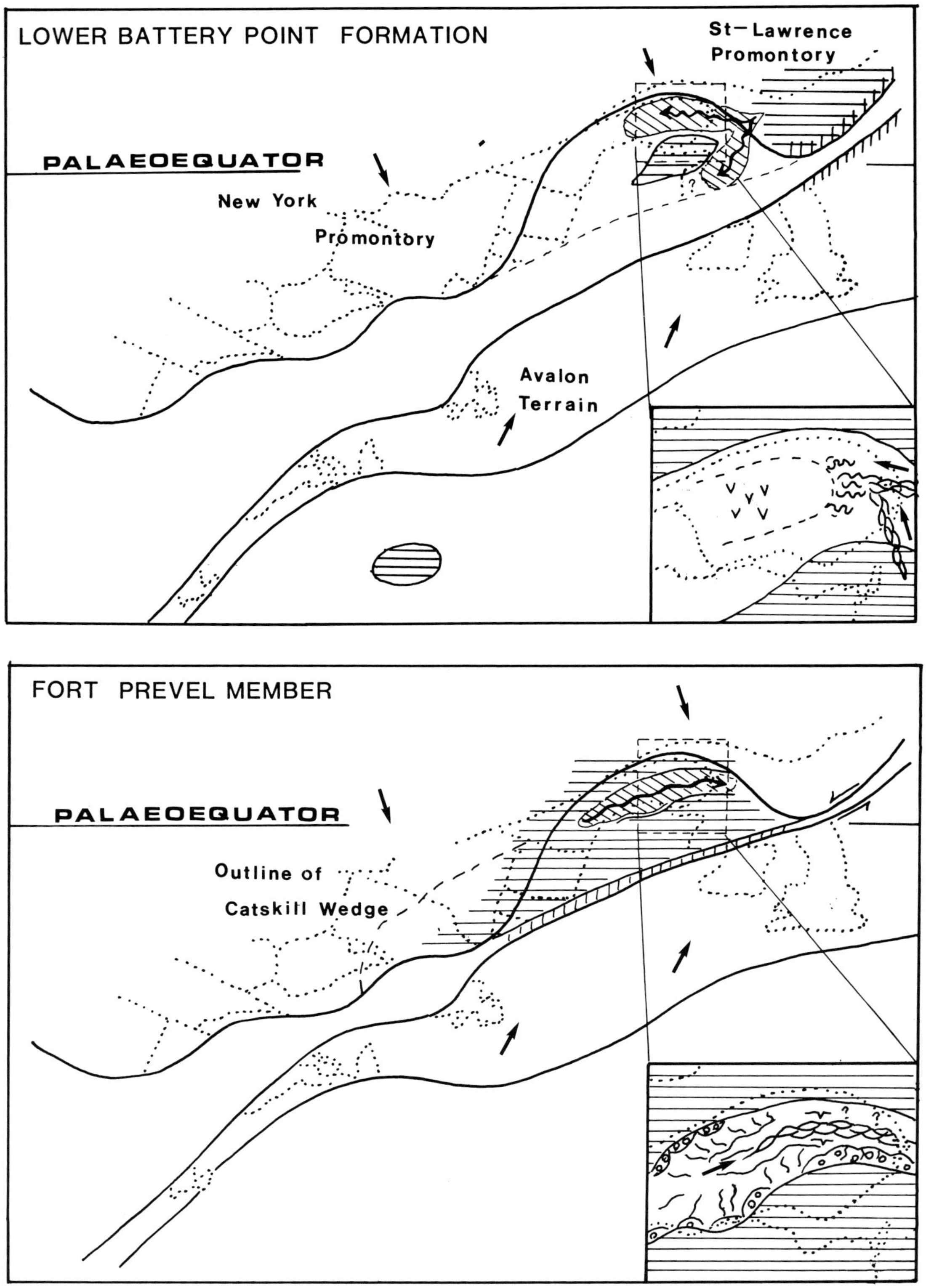

FIG. 13.—Summary diagram of the tectonic evolution of the Gaspé Basin during deposition of the Battery Point Formation, with inset schematic diagram of evolving drainage systems. Arrows refer to directions of convergence of opposing plates.

had subsided rapidly during the late Silurian and early Devonian. A marine connection to the west (Fig. 13) was eliminated prior to deposition of the Fort Prével Member.

Within the Fort Prével Member, Association 1 records a transition from northward-draining low sinuosity channel networks in the lower part to higher energy, more ephemeral and more easterly-flowing sheetflood and small-scale channel fills in the upper part. This transition was probably a response to an increasingly ephemeral discharge regime as a result of severing of the marine connection to the west and development of an enclosed basin. Laterally extensive sheet sandstones of the lower part of Association 2 represent northerly progradation of broadly convex-upward, lobate 'alluvial aprons' (terminal fans of Friend, 1978; Parkash and others, 1983; Tunbridge, 1984). Channels freely migrated across these broad alluvial plains (Friend, 1978; Bluck, 1980), the abandonment of which resulted in interbedded finer grained facies.

The transition from Association 1 to 2, therefore, continued the pattern of increasing ephemerality with the development of extensive mudflats (e.g., modern examples of Sneh, 1983; Williams, 1970 and inferred ancient examples of Hardie and others, 1978; Hubert and Hyde, 1982; and Tunbridge, 1984). The absence of playa lake sediments, evaporites and pedogenic features suggests that long periods of aridity did not occur. Although at this time the basin was probably under the direct influence of wind-blown sand, frequent floods probably removed and recycled any eolian accumulations (Sneh, 1983).

Tectonic controls are, however, sought to explain the increasingly more eastward dispersion of high-energy sheetflood sandstones, beginning in Association 1 and reflecting distal effects of westward-shifting basin margin uplift. Increased uplift to the west was also responsible for initiation of ESE-flowing ephemeral drainage courses (top of Association 2), and the eventual ESE progradation of distal (Association 3) and more proximal (Association 4) sandy braidplain facies across ephemeral mudflats. Renewed uplift to the south, initiated during deposition of Association 4, shed the proximal gravelly braidplain facies of the Malbaie Formation conglomerates during the Middle Devonian (Rust, 1984).

EVOLUTION OF DRAINAGE SYSTEMS: TECTONIC SETTING

The coarsening-upward Devonian clastic wedge in eastern Québec formed in response to uplift resulting from the Acadian orogeny, which culminated in the Middle Devonian in Eastern Canada (Poole and others, 1970). It is compared to broadly similar clastic wedges of the Catskill Mountains, New York, and Old Red Sandstone deposits of Spitzbergen and Britain (Allen and Friend, 1968).

Recent interpretations of the Acadian orogeny favor continental collision and accretion of several microplates, derived probably from Armorica, against the North American continental (Laurentian) margin (Dewey and Kidd, 1974; Van der Voo, 1983; Williams and Hatcher, 1983; Ettonsohn, 1985; Kent, 1985). This collision involved accretion of the Acadia and Avalon terrains (Kent and Opdyke, 1980; Kent, 1985), the northern margins of which underwent substantial sinistral strike-slip displacement during the Carboniferous (Kent and Opdyke, 1978; Van der Voo and others, 1979; Kent, 1982). Nonorthogonal collision is argued for, due to post-collisional strike-slip movement in the Maritime Provinces and the time transgressive nature of Acadian deformation peaking during the mid-Devonian in the Quebec Appalachians, but including late Devonian to early Mississippian events in the central and southern Appalachians (Rodgers, 1967; Johnson, 1971; Ettensohn, 1985). Collision against an irregular Laurentian continental margin, possibly created by late Precambrian rifting (Dewey and Burke, 1974; Rankin, 1976; Thomas, 1977, 1983) is also favored, such that collision of promontories would create areas of more intense deformation and uplift and collision of embayments less intense deformation (Dewey and Burke, 1974; Dewey and Kidd, 1974; Thomas, 1977). Initial transtension, due to oblique collisional convergence, opened peripheral basins, such as the Gaspé Trough, where early subsidence would dominate. The Devonian clastic wedge in eastern Québec, therefore, formed in an intrasuture embayment basin (Miall, 1984), with early subsidence giving way to uplift and mild deformation as collision progressed.

Thomas (1977) noted that the maximum thickness of post-Acadian clastic wedges in the Appalachian belt corresponds to structural re-entrants in the Laurentian continental margin. Moreover, uplifts, which resulted from promontory collisions, are successively younger to the southwest due to non-orthogonal convergence. Longitudinal transport systems, which flow into structural re-entrants, also become younger to the southwest, e.g., the Catskill wedge (Mid- to Late Devonian) and Pocono clastic wedge (Frasnian to Tournaisian) (Ettonsohn, 1985).

The Devonian clastic sequences of the Gaspé Trough, particularly the Battery Point Formation, seem to record changes in basin configuration, and basin axis slope in a convergent plate margin setting. Initially, the collision of the St. Lawrence promontory with the Avalon terrain caused intense deformation in New England and the Canadian Maritime provinces during Siegenian to Emsian times (Ettensohn, 1985) and probably supplied longitudinal, westerly-directed transport systems into the subsiding Gaspé Trough (Fig. 13). As collision migrated southwestward, the sea regressed from the west, and easterly source areas were reduced to low-lying hinterlands. Although Acadian compression impinged little on deposition in the Gaspé Trough (Thomas, 1977; Faill, 1985), local uplift to the south provided source areas for fluvial systems, largely overriding axial drainage in the lower part of the Fort Prével Member. Eventually, with the collision of the New York promontory to the west and southwest against the Avalon terrain (culminating in the Eifelian-Givetian), east- and southeasterly-flowing axial drainage systems were initiated in the Gaspé Trough (e.g., in the upper part of the Fort Prével Member, Fig. 13). Basin axis slope was therefore reversed along what was by that time an intermontane basin, and similar axial drainage trends are recorded from the Middle and Upper Devonian of the Gaspé region (Poole and others, 1970; Rust, 1981a). Stability, uplift and mainly E to ESE basin axis paleoslopes persisted into the Carboniferous, though strike-slip tectonics dominated basin margins (Rust, 1981a).

No modern examples of such drainage reversals are known (Miall, 1981), but major trunk rivers flowing longitudinally through foreland basins are commonly known to reverse in response to differential tectonic movement (e.g., thrust emplacement) along a fold front. Examples are known from the Cenozoic Swiss molasse basin (Van Houten, 1981) and the Late Mesozoic-Early Tertiary molasse basins of the

Cordilleran region of western Canada (McLean, 1977, Eisbacher, 1981). The Battery Point Formation example discussed here differs in the larger scale and lack of internal disconformities within the molasse sequence.

CONCLUSIONS

Gross changes in paleodrainage (e.g., the reversal in axial fluvial systems and northward progradation of the Malbaie conglomerates), along with more subtle trends such as the westerly shift in source areas for sheetflood sandstones in Associations 1 and 2 of the Fort Prével Member, can be directly linked to Acadian tectonic events. Variations in channel type, morphology, and inferred discharge patterns between the lower Battery Point Formation and the Fort Prével Member were, however, direct responses to local climatic change resulting from uplift and isolation of the depositional basin. Although the Gaspé region was equatorial (5° north) at this time (Scotese and others, 1979; Van der Voo, 1983; Kent, 1985), with inferred highly seasonal rainfall (i.e., tropical wet and dry, Rust, 1984; Woodrow, 1985), the ephemeral nature of many fluvial systems in the Fort Prével Member was probably enhanced by rain shadow effects created by the rising Acadian mountains. This probably severely limited plant cover on these upland slopes (Schumm, 1968). The rain shadow effect may not have been marked enough to preclude a relatively high, strongly seasonal rainfall (Rust, 1984), since evidence for prolonged exposure is lacking in Fort Prével sediments.

In conclusion, changes in fluvial style, both between the lower Battery Point Formation and Fort Prével Member, and within the Fort Prével Member itself, are direct responses to allocyclic controls, the most important of which is tectonics, coupled with climatic responses to changing basin configuration. Superficially, the Devonian clastic succession of eastern Gaspé resembles contemporaneous terrestrial basin fills in the Appalachian-Caledonian fold belt, e.g., the Catskill wedge (Allen and Friend, 1968, Friend, 1969), and broadly coarsening-upward successions in SW Ireland, Southern England and Spitzbergen. In evolution, however, it is quite different, developing in a relatively narrow intra-suture embayment basin.

ACKNOWLEDGMENTS

The research on which this paper is based was carried out during tenure of a studentship from the Natural Environment Research Council, which also provided financial support, together with Colorado State University, to attend the Third International Fluvial Sedimentology Conference (I.F.S.C.) (DAL). Generous support was given by Brian Rust in the field, with much logistical support from the Ottawa-Carleton Centre for Geoscience Studies. Financial assistance for field visits, and attendance at the I.F.S.C., provided by the Departments of Geology at Bristol and Colorado State University, and from the Royal Society, is gratefully acknowledged (BPJW). We also thank Joyce Rowland for typing the manuscript.

REFERENCES

ALLEN, J. R. L., 1983, Studies in fluviatile sedimentation: Bars, bar-complexes and sandstone sheets (low-sinuosity braided streams) in the Brownstones (L. Devonian), Welsh Borders: Sedimentary Geology, v. 33, p. 237–293.

———, AND BANKS, N. L., 1972, An interpretation and analysis of recumbent-folded deformed cross-bedding: Sedimentology, v. 19, p. 257–283.

———, AND FRIEND, P. F., 1968, Deposition of the Catskill facies, Appalachian Region: With notes on some other Old Red Sandstone basins, *in* Klein, G. de V., ed., Late Palaeozoic and Mesozoic Continental Sedimentation, North-eastern North America: Geological Society of America Special Paper 106, p. 21–74.

BANKS, N. L., 1973, The origin and significance of some downcurrent-dipping cross-stratified sets: Journal of Sedimentary Petrology, v. 43, p. 423–427.

BLUCK, B. J., 1980, Structure, generation and preservation of upward fining braided stream cycles in the Old Red Sandstone of Scotland: Transactions of the Royal Society of Edinburgh, v. 71, p. 29–46.

BRYHNI, INGE, 1978, Flood deposits in the Hornelen Basin, West Norway (Old Red Sandstone): Norsk Geologisk Tidsskrift, v. 58, p. 273–300.

CAMPBELL, C. V., 1976, Reservoir geometry of a fluvial sheet sandstone: American Association of Petroleum Geologists Bulletin, v. 60, p. 1009–1020.

CANT, D. J., 1978, Beforms and bar types in the South Saskatchewan River: Journal of Sedimentary Petrology, v. 48, p. 1321–1330.

———, AND WALKER, R. G., 1976, Development of a braided-fluvial facies model for the Devonian Battery Point Sandstone, Québec: Canadian Journal of Earth Science, v. 13, p. 102–119.

———, AND ———, 1978, Fluvial processes and facies sequences in the sandy, braided South Saskatchewan River, Canada: Sedimentology, v. 25, p. 625–648.

DEWEY, J. F., AND BURKE, K. C. A., 1974, Hot spots and continental break up: Implications for collisional orogeny: Geology, v. 2, p. 57–60.

———, AND KIDD, W. S. F., 1974, Continental collisions in the Appalachian/Caledonian orogenic belt: Variations related to complete and incomplete suturing: Geology, v. 2, p. 543–546.

DOE, T. W., AND DOTT, R. H., 1980, Genetic significance of deformed cross bedding—With examples from the Navajo and Weber Sandstones of Utah: Journal of Sedimentary Petrology, v. 50, p. 793–812.

EISBACHER, G. H., 1981, Late Mesozoic-Palaeogene Bowser Basin molasse and Cordilleran tectonics, western Canada, *in* Miall, A. D., ed., Sedimentation and Tectonics in Alluvial Basins: Geological Association of Canada Special Paper 23, p. 611–622.

ETTENSOHN, F. R., 1985, The Catskill Delta complex and the Acadian Orogeny: A model, *in* Woodrow, D. L., and Sevon, W. D., eds., The Catskill Delta: Geological Society of America Special Paper 201, p. 15–37.

FAGERSTROM, J. A., 1967, Development, flotation, and transportation of mud crusts—Neglected factors in sedimentology: Journal of Sedimentary Petrology, v. 37, p. 73–79.

FAILL, F. T., 1985, The Acadian orogeny and the Catskill Delta, *in* Woodrow, D. L. and Sevon, W. D., eds., The Catskill Delta: Geological Society of America Special Paper 201, p. 15–37.

FRIEND, P. F., 1969, Tectonic features of Old Red Sandstone sedimentation in North Atlantic borders, *in* Kay, M., ed., North Atlantic Geology and Continental Drift: American Association of Petroleum Geologists Memoir 12, p. 703–710.

———, 1978, Distinctive features of some ancient rivers, *in* Miall, A. D., ed., Fluvial Sedimentology: Canadian Society of Petroleum Geologists Memoir 5, p. 531–542.

———, SLATER, M. J., AND WILLIAMS, R. C., 1979, Vertical and lateral building of river sandstone bodies, Ebro Basin, Spain: Journal of the Geological Society of London, v. 136, p. 39–46.

HARDIE, L. A., SMOOT, J. R., AND EUGSTER, H. P., 1978, Saline lakes and their deposits, *in* Matter, A., and Tucker, M. E., eds., Modern and Ancient Lake Sediments: International Association of Sedimentologists Special Publication 2, p. 7–42.

HARMS, J. C., SOUTHARD, J. B., AND WALKER, R. G., 1982, Structures and sequences in clastic rocks: Society of Economic Paleontologists and Mineralogists Short Course No. 9, 249 p.

HASZELDINE, R. S., 1983, Descending tabular cross-bedded sets and bounding surfaces from a fluvial channel in the Upper Carboniferous of North-east England, *in* Collinson, J. D., and Lewin, J., eds., Modern and Ancient Fluvial Systems: International Association of Sedimentologists Special Publication 6, p. 449–456.

HILLER, NORTON, AND STAVRAKIS, NICHOLAS, 1984, Permo-Triassic fluvial systems in the south-eastern Karoo Basin, South Africa: Paleogeography, Palaeoclimatology, Palaeoecology, v. 45, p. 1–21.

HUBERT, J. F., AND HYDE, M. G., 1982, Sheet-flow deposits of graded beds and sandstone on an alluvial sandflat-playa system: Upper Triassic Blomidon red beds, St. Mary's Bay, Nova Scotia: Sedimentology, v. 29, p. 457–474.

JOHNSON, J. G., 1971, Timing and co-ordination of orogenic, epeirogenic and eustatic events: Geological Society of America Bulletin, v. 82, p. 3263–3298.

JONES, B. G., AND RUST, B. R., 1983, Massive sandstone facies in the Hawkesbury Sandstone, a Triassic fluvial deposit near Sydney, Australia: Journal of Sedimentary Petrology, v. 53, p. 1249–1259.

KARCZ, IAAKOV, 1972, Sedimentary structures formed by flash floods in southern Israel: Sedimentary Geology, v. 7, p. 161–182.

KENT, D. V., 1982, Paleomagnetic evidence for post-Devonian displacement of the Avalon Platform (Newfoundland): Journal of Geophysical Research, v. 87, p. 8709–8716.

———, 1985, Paleocontinental setting for the Catskill Delta, *in* Woodrow, D. L., and Sevon, W. D., eds., The Catskill Delta: Geological Society of America Special Paper 201, p. 9–13.

———, AND OPDYKE, N. D., 1978, Palaeomagnetism of the Devonian Catskill red beds: Evidence for motion of the coastal New England-Canadian Maritime region relative to cratonic North America: Journal of Geophysical Research, v. 83, p. 4441–4450.

———, AND ———, 1980, Palaeomagnetism of Siluro-Devonian rocks from eastern Maine: Canadian Journal of Earth Science, v. 17, p. 1653–1665.

LAWRENCE, D. A., AND WILLIAMS, B. P. J., 1984, Evolution of fluvial style—Lower Devonian Battery Point Formation, Gaspé Peninsula, Québec, Canada (abs.): American Association of Petroleum Geologists Bulletin, v. 68, p. 539.

LEEDER, M. R., 1973, Sedimentology and palaeogeography of the Upper Old Red Sandstone in the Scottish Border Basin: Scottish Journal of Geology, v. 9, p. 117–144.

LOWE, D. R., 1976, Subaqueous liquefied and fluidised sediment flows and their deposits: Sedimentology, v. 23, p. 285–308.

MCGERRIGLE, H. W., 1950, The geology of eastern Gaspé: Québec Department of Mines, Geological Report 35, 168 p.

MCKEE, E. D., CROSBY, E. J., AND BERRYHILL, H. L., 1967, Flood deposits, Bijou Creek, Colorado, June 1965: Journal of Sedimentary Petrology, v. 37, p. 829–851.

MCLEAN, J. R., 1977, The Cadomin Formation: Stratigraphy, sedimentology and tectonic implications: Bulletin of Canadian Petroleum Geology, v. 25, p. 792–827.

MIALL, A. D., 1981, Alluvial sedimentary basins: Tectonic setting and basin architecture, *in* Miall, A. D., ed., Sedimentation and Tectonics in Alluvial Basins: Geological Association of Canada Special Paper 23, p. 1–33.

———, 1983, Basin analysis of fluvial sediments, *in* Collinson, J. D., and Lewin, J., eds., Modern and Ancient Fluvial Systems: International Association of Sedimentologists Special Publication 6, p. 279–286.

———, 1984, Principles of Sedimentary Basin Analysis: Springer-Verlag, New York, 490 p.

PARKASH, B., AWASTHI, A. K., AND GOHAIN, K., 1983, Lithofacies of the Markanda terminal fan, Kurukshetra District, Haryana, India, *in* Collinson, J. D., and Lewin, J., eds., Modern and Ancient Fluvial Systems: International Association of Sedimentologists Special Publication 6, p. 337–344.

PICARD, M. D., AND HIGH, L. R., 1973, Sedimentary Structures of Ephemeral Streams: Developments in Sedimentology 17: Elsevier, Amsterdam, 223 p.

POOLE, W. H., SANFORD, B. V., WILLIAMS, HAROLD, AND KELLY, D. G., 1970, Geology of southeastern Canada, *in* Douglas, R. J. W., ed., Geology and Economic Minerals in Canada: Geological Survey of Canada Economic Geology Report 1, p. 227–304.

RANKIN, D. W., 1976, Appalachian salients and recesses: Late Precambrian continental breakup and the opening of the Iapetus Ocean: Journal of Geophysical Research, v. 81, p. 5605–5619.

RODGERS, JOHN, 1967, Chronology of tectonic movements in the Appalachian region of eastern North America: American Journal of Science, v. 265, p. 408–427.

RUST, B. R., 1978, Depositional models for braided alluvium, *in* Miall, A. D., ed., Fluvial Sedimentology: Canadian Society of Petroleum Geologists Memoir 5, p. 605–625.

———, 1981a, Alluvial deposits and tectonic style: Devonian and Carboniferous successions in Eastern Gaspé, *in* Miall, A. D., ed., Sedimentation and Tectonics in Alluvial Basins: Geological Association of Canada Special Paper 23, p. 49–75.

———, 1981b, Sedimentation in arid-zone anastomosing fluvial systems: Cooper's Creek, Central Australia: Journal of Sedimentary Petrology, v. 51, p. 745–755.

———, 1984, Proximal braidplain deposits in the Middle Devonian Malbaie Formation of Eastern Gaspé, Québec, Canada: Sedimentology, v. 31, p. 675–695.

SCHUMM, S. A., 1968, Speculations concerning paleohydraulic controls of terrestrial sedimentation: Geological Society of America Bulletin, v. 79, p. 1573–1588.

SCOTESE, C. R., BAMBACH, R. K., BARTON, C., VAN DER VOO, R., AND ZIEGLER, A. M., 1979, Paleozoic base maps: Journal of Geology, v. 87, p. 217–277.

SIKANDER, A. H., 1978, Lithological observation on Grande Grève, York River and Battery Point Formations, and hydrocarbon possibilities in Eastern Gaspé, Québec: Ministère des Richesses Naturelles, Gouvenment du Québec, Report GM33690, 59 p.

SMITH, N. D., 1970, The braided stream depositional environment: Comparison of the Platte River with some Silurian clastic rocks, north-Central Appalachians: Geological Society of America Bulletin, v. 94, p. 437–450.

———, 1971, Transverse bars and braiding in the Lower Platte River, Nebraska: Geological Society of America Bulletin, v. 82, p. 3407–3420.

———, 1974, Sedimentology and bar formation in the Upper Kicking Horse River, a braided outward stream: Journal of Geology, v. 82, p. 205–224.

SNEH, AMIHAI, 1983, Desert stream sequences in the Sinai Peninsula: Journal of Sedimentary Petrology, v. 53, p. 1271–1279.

STEAR, W. M., 1983, Morphological characteristics of ephemeral stream channel and overbank splay sandstone bodies in the Permian Lower Beaufort Group, Karoo Basin, South Africa, *in* Collinson, J. D., and Lewin, J., eds., Modern and Ancient Fluvial Systems: International Association of Sedimentologists Special Publication 6, p. 405–420.

TAYLOR, G., AND WOODYER, K. D., 1978, Bank deposition in suspended-load streams, *in* Miall, A. D., ed., Fluvial Sedimentology: Canadian Society of Petroleum Geologists Memoir 5, p. 257–275.

THOMAS, W. A., 1977, Evolution of Appalachian-Ouachita salients and recesses from re-entrants and promontories in the continental margin: American Journal of Science, v. 277, p. 1233–1278.

———, 1983, Continental margins, orogenic belts and intracratonic structures: Geology, v. 11, p. 270–272.

TUNBRIDGE, I. P., 1981a, Old Red Sandstone sedimentation—An example from the Brownstones (highest Lower Old Red Sandstone) of south central Wales: Geological Journal, v. 16, p. 111–124.

———, 1981b, Sandy high-energy flood sedimentation—Some criteria for recognition with an example from the Devonian of S.W. England: Sedimentary Geology, v. 28, p. 79–85.

———, 1984, Facies model for a sandy ephemeral stress and clay playa complex: The Middle Devonian Trentishoe Formation of North Devon, U.K.: Sedimentology, v. 31, p. 679–715.

VAN DER VOO, R., 1983, Paleomagnetic constraints on the assembly of the Old Red continent: Tectonophysics, v. 91, p. 271–283.

———, FRENCH, A. N., AND FRENCH, R. B., 1979, A paleomagnetic pole position from the folded Upper Devonian Catskill red beds, and its tectonic implications: Geology, v. 7, p. 345–348.

VAN HOUTEN, F. B., 1981, The odssey of molasse, *in* Miall, A. D., ed., Sedimentation and Tectonics in Alluvial Basins: Geological Association of Canada Special Paper 23, p. 35–48.

WILLIAMS, HAROLD, AND HATCHER, R. D., 1983, Suspect terranes and accretionary history of the Appalachian orogen: Geology, v. 10, p. 530–536.

WILLIAMS, G. E., 1970, The Central Australian stream floods of Febru-

ary-March 1967: Journal of Hydrology, v. 11, p. 185–200.
———, 1971, Flood deposits of the sand-bed ephemeral streams of Central Australia: Sedimentology, v. 17, p. 1–40.
WOODROW, D. L., 1985, Paleogeography, paleoclimate and sedimentary processes of the Late Devonian Catskill Delta, *in* Woodrow, D. L., and Sevon, W. D., eds., The Catskill Delta: Geological Society of America Special Paper 201, p. 51–63.

PART VI
ECONOMIC APPLICATIONS

EVOLUTION OF FLUVIAL STYLES IN THE EOCENE WASATCH FORMATION, POWDER RIVER BASIN, WYOMING

PETER D. WARWICK
U.S. Geological Survey, Reston, Virginia 22092
AND
ROMEO M. FLORES
U.S. Geological Survey, Denver, Colorado 80225

ABSTRACT: Vertical and lateral facies changes in the lower part of the Eocene Wasatch Formation in the Powder River Basin, Wyoming represent an evolution of fluvial systems that varied from meandering to anastomosing. The meandering facies in the lower part of the study interval formed in a series of broad meanderbelts in a north-northwest-flowing system. Upon abandonment this meanderbelt facies served as a topographic high on which a raised or ombrotrophic Felix peat swamp developed. Peat accumulated until compaction permitted encroachment of crevasse splays from an adjoining transitional facies which consists of deposits of a slightly sinuous fluvial system. Crevasse splays eventually prograded over the peat swamp that was partly covered by lakes. Bifurcation, reunification, and transformation of crevasse channels into major conduits produced an anastomosing system that was characterized by diverging and converging channels separated by floodbasins drowned by lakes and partly covered swamps.

INTRODUCTION

Fluvial environments for Tertiary coal-bearing strata in the Rocky Mountain region have been proposed by a number of workers in recent years (Jacob, 1972; Beaumont, 1979; Galloway, 1979; Ethridge and others, 1981). These studies described facies for meandering trunk-tributary systems that perennially drained the coal basins. Although coal deposits are commonly associated with facies of a meandering fluvial system, Flores (1981), and Flores and Hanley (1984) suggested that they occur in both meandering and anastomosed systems in the Paleocene Fort Union Formation of the Powder River Basin. The purpose of this paper is to document the evolution of these fluvial systems and to relate their facies to the occurrence of economic coal deposits.

The study interval, which averages 250 ft (76 m) thick, includes the Felix coal bed and associated rocks in the lower part of the Eocene Wasatch Formation exposed in a 400 mi^2 (1,036 km^2) area in the Powder River Basin, Wyoming (Fig. 1). Facies analysis was determined from 146 measured sections, 56 geophysical logs, and 118 published coal thickness data. Vertical profiles and stratigraphic cross sections were utilized to establish successions and relationships between facies types. Paleocurrent direction of sandstone channels was determined from trough crossbeds measured at 84 localities.

A generalized description and origin of the rock types of the Wasatch Formation were provided by Love (1952), Elliott (1976), and Toomey (1977) in conjunction with uranium exploration. A more specific analysis of the fluvial facies of the lowermost part of the Wasatch Formation was conducted by Ethridge and others (1981, 1983) in relation to coal exploration. These commodity related studies of the Wasatch Formation are concentrated in small areas in the basin. Seeland (1976) proposed a basinwide depositional setting of the Wasatch Formation as a trunk-tributary system. Flores and Warwick (1984) suggested the tributaries were fed by alluvial fans along the western basin margin. These changes in the fluvial style of the trunk system are stressed in this paper.

LITHOFACIES VARIATIONS

In decreasing order of abundance rocks in the study interval comprise: 1) interbedded claystone, siltstone, and limestone; 2) sandstone; 3) carbonaceous shale; and 4) limestone and ironstone lithofacies. The interval consists mainly of sandstone lithofacies in the lower part with interbedded claystone, siltstone, and sandstone lithofacies in the upper part.

Sandstone lithofacies.—

Three types of sandstone lithofacies can be recognized on the basis of their morphology and internal sedimentary structures. Type 1 lithofacies consists of en echelon lenticular sandstone bodies that range from 40 to 90 ft (12 to 27 m) thick and from 2 to 9.5 mi (3.2 to 15.2 km) in lateral extent (thickness to width ratio from 1:260 to 1:557). These sandstones have an erosional base and contain multiple scoured surfaces floored by granule-size lag conglomerate which often includes fragments of mudballs, and coal spars and bands. The lower part of each unit includes a lag deposit that is overlain by coarse-grained, planar-bedded sandstone which merges upward into large trough and tabular (sets greater than 3 ft or 0.9 m wide) cross-bedded sandstone. Epsilon cross-stratification or lateral accretion units (average dip angle of 15°) and slumped blocks are common in the lower portion of the sandstone bodies. The upper part of the sandstone contains abundant small trough crossbeds (sets less than 1 ft or 0.3 m wide), ripple laminations, and convolute laminations and grades upward into claystone, siltstone, carbonaceous shale, and coal lithofacies.

Type 2 sandstone lithofacies occurs as lenticular bodies that range from 10 to 80 ft (3 to 24 m) thick and 0.2 to 2.5 mi (0.3 to 4.0 km) in lateral extent (ratio from 1:105 to 1:165). This lithotype exhibits sedimentologic features that are similar to type 1 sandstones but differs in that individual sequences are thinner with less epsilon cross-stratification, and more abundant tabular and trough cross-stratification.

Type 3 sandstone lithofacies occurs as tabular bodies generally less than 20 ft (6 m) thick and about 1 mi (1.6

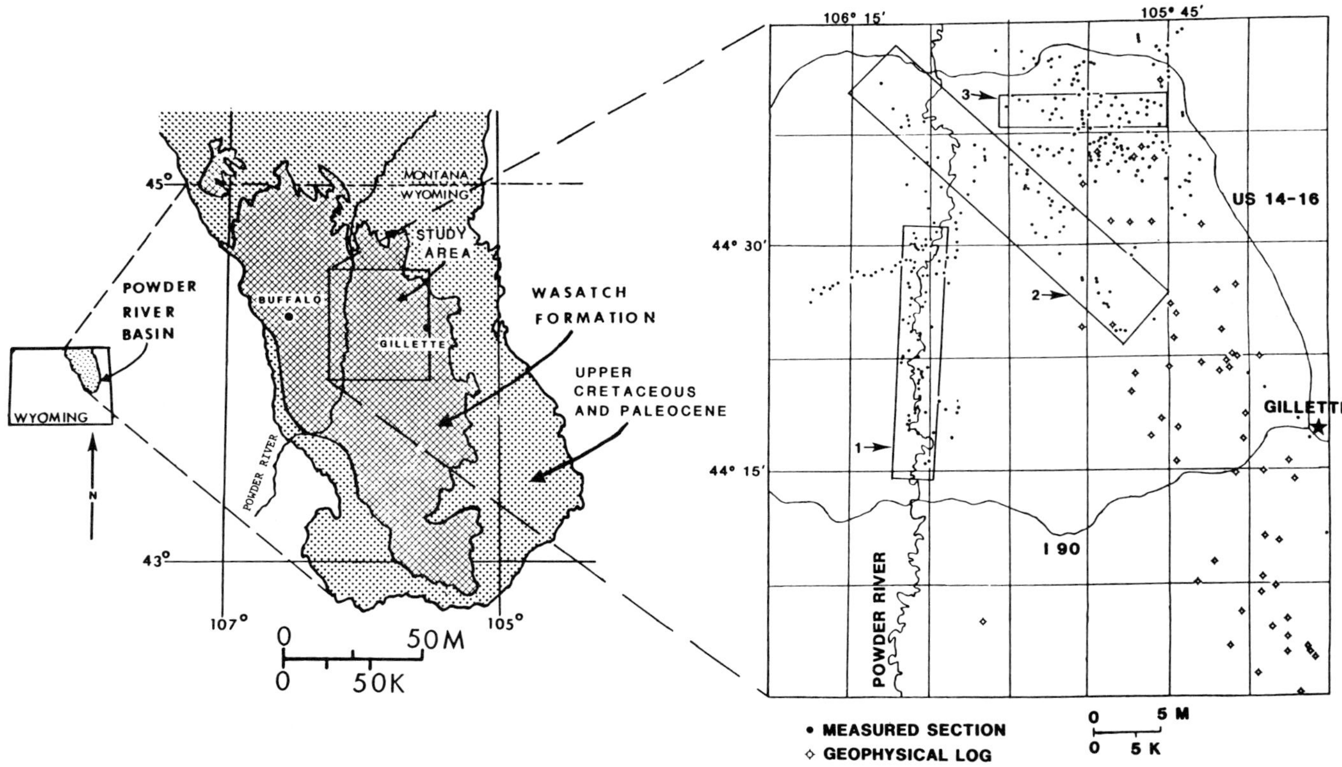

FIG. 1.—Maps showing the localities of the Powder River Basin, Wyoming and study area in the basin. Rectangles 1, 2, and 3 represent areas of cross section shown in Figures 2, 3, and 4.

km) in lateral extent. This sandstone type, which varies from fine- to medium-grained, commonly caps a coarsening-upward sequence and is usually underlain by interbedded siltstone and shale. The sandstone consists of abundant ripple laminations, although small-scale trough and tabular crossbeds are locally common. Laterally, the sandstone type grades into interbedded siltstone, claystone, and limestone containing fossils of freshwater mollusks.

Siltstone lithofacies.—

The siltstone lithofacies is commonly dark to light gray and ranges from a few inches to 30 ft (9 m) in thickness. This lithofacies displays gradational to sharp contacts with underlying and overlying deposits. The siltstone exhibits some flat bedding and usually contains burrows and ripple laminations. This lithofacies contains common whole and fragmented freshwater gastropod and bivalve fossils. It is rooted where overlain by carbonaceous shale and coal lithofacies.

Claystone lithofacies.—

The claystone lithofacies ranges from black to light gray and comprises units which rarely exceed 30 ft (9 m) in thickness. In some localities the claystone contains gradational or sharp upward transitions into silstone and sandstone, and in other locations it underlies coal beds and carbonaceous shale as rooted seatrock. Claystones commonly overlie carbonaceous shale and coal beds. These units commonly contain siderite nodules and freshwater mollusk fossils as well as thin laminations marked by plant fragments.

Limestone and ironstone lithofacies.—

The limestone lithofacies is a silty to mud-rich carbonate rock that usually is light gray. These rocks occur in tabular bodies that are generally less than 3 ft (0.9 m) thick and can be traced laterally for hundreds of feet. The beds are commonly massive, but some display ripple laminations and burrows. The ironstone lithofacies occurs in claystone as reddish brown discontinuous lenses a few inches thick. The facies contains plant fragments and occasionally grades into or replaces limestone.

Carbonaceous shale and coal lithofacies.—

Carbonaceous shale lithofacies is a dark gray to black laminated claystone which contains abundant macerated plant fragments and ranges from a few inches to as much as 3 ft (0.9 m) in thickness. This lithofacies is interbedded with claystone, siltstone, sandstone and coal lithofacies. Carbonaceous shale, however, is commonly laterally transitional into coal beds.

The coal lithofacies is commonly less than 3 ft (0.9 m) thick and can be traced for only a few miles. The thickest coal lithofacies in the study interval is the Felix bed, which is as much as 40 ft (12 m) thick. It is uniformly thick (>10 ft or 3.0 m) over a 2,100 mi^2 (3379.5 km^2) area. The coals range from lignite to sub-bituminous with the Felix bed representing the highest grade. Petrographic analysis of macerals in the Felix bed indicate a direct relationship between huminite and inertinite maceral contents of the lower and upper parts of the bed. Huminite is abundant (92–99 percent) in the lower part of the bed and decreases in the upper part where inertinite is more common (as much as 35 percent). The huminite at the base of the coal bed represents woody plant remains, and the increased inertinite near the top of the bed reflects oxidized plant material that was preserved in a huminite groundmass with other macerals.

FACIES RELATIONSHIPS

The vertical and lateral variations of the lithofacies in the study interval are shown in Figures 2, 3, and 4, and represent cross sections across the length and width of the study area. The study interval in the cross sections shows a general fining-upward sequence. The lower part of the interval, below the Felix bed consists of abundant (>50 percent) type 1 sandstone lithofacies with subordinate amounts of claystone, siltstone, limestone, and ironstone lithofacies (Fig. 2). The type 1 sandstone lithofacies are channel deposits of a meandering stream as indicated by the en echelon arrangement, lateral accretion units, and slumped block deposits. The extensive vertical and lateral distribution of the sandstone lithofacies also indicates deposition in a broad meanderbelt system. Multiple scour surfaces within the sandstone bodies suggest recurrence of cut and fill in the meanderbelt resulting from reoccupation by fluvial channels. The claystone, siltstone, limestone, and ironstone lithofacies that grade vertically and laterally into the channel sandstones represent overbank floodplain deposits adjacent to the streams. These sediments include overbank levee deposits of interbedded rooted claystone, siltstone, and sandstone (type 3) lithofacies that immediately flanked the meanderbelt facies. The levee deposits grade into coarsening-upward crevasse splay deposits of claystone, siltstone, and sandstone (type 3) lithofacies toward the floodplain. These floodplain deposits are commonly interbedded with carbonaceous shale and coal lithofacies that formed in adjacent backswamps. Thus, the lower part of the study interval is dominantly characterized by meanderbelt channel facies with subordinate floodplain-backswamp facies.

The middle part of the interval is dominated by type 2 sandstone and coal lithofacies (Figs. 1, 3, 4). The type 2 sandstone lithofacies is laterally juxtaposed against the Felix bed, which overlies the meanderbelt facies. The presence of some lateral accretion units in this sandstone type indicate that it represents a channel deposit of a locally sinuous stream. This fluvial system was coeval with the Felix peat swamp which formed on abandoned deposits of the meandering stream systems. The Felix peat swamp was probably an ombrotrophic or raised deposit, as is indicated by the decrease in huminite toward the top of the coal bed

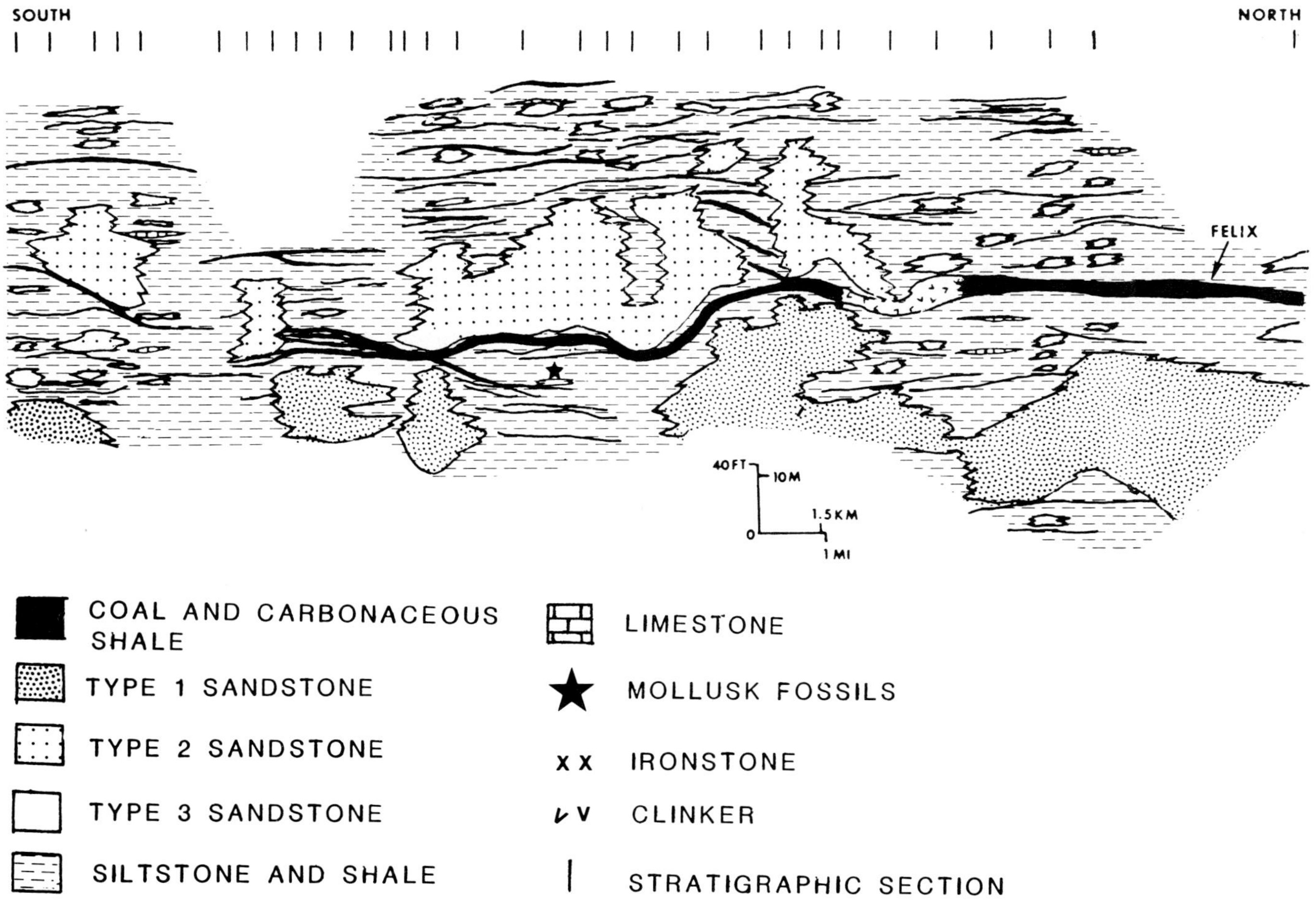

FIG. 2.—A north-south cross section showing environmental-stratigraphic framework of the study interval along the Powder River. Note that the Felix bed splits into a number of beds in the southern area. Vertical exaggeration = 150×.

(Warwick, 1985). Thus, the middle part of the study interval is dominated by channel-backswamp facies that represent a transition between the underlying meanderbelt-dominated facies and the overlying anastomosed fluvial facies.

The upper part of the study interval, above the Felix bed, consists of abundant floodplain deposits of claystone, siltstone, sandstone (type 3), limestone, and ironstone lithofacies (Figs. 2, 3, 4). The floodplain deposits are mainly coarsening-upward sequences of crevasse splays, which contain abundant freshwater mollusk fossils. The fossils suggest debouching of crevasse splays into ponded waters or lakes on the floodplains. Widespread type 3 sandstone directly above the Felix bed, as seen on Figure 4, indicates lakes formed in former sites of peat deposition probably due to peat compaction. These lakes filled with crevasse sediments. Thin coal and carbonaceous shales representing backswamp deposits are interbedded with the floodplain deposits. The floodplain-backswamp facies pass laterally into type 2 sandstones bounded by rooted, fine-grained levee sediments. Type 2 sandstones are both thick and narrow and represent channel deposits of small, locally sinuous streams, an interpretation also indicated by the general absence of lateral accretion units. The presence of lateral accretion units is only a function of preservation, however, and is not a good indicator of channel sinuosity. In addition, this sandstone type as shown in Figures 2, 3, and 4 seems to comprise contemporaneous bodies that are arranged in patterns very similar to those of modern anastomosed streams, described by Smith (1983). Thus, the areal distribution of coeval, small, vertically aggrading, slightly sinuous channels separated by small floodplains drowned

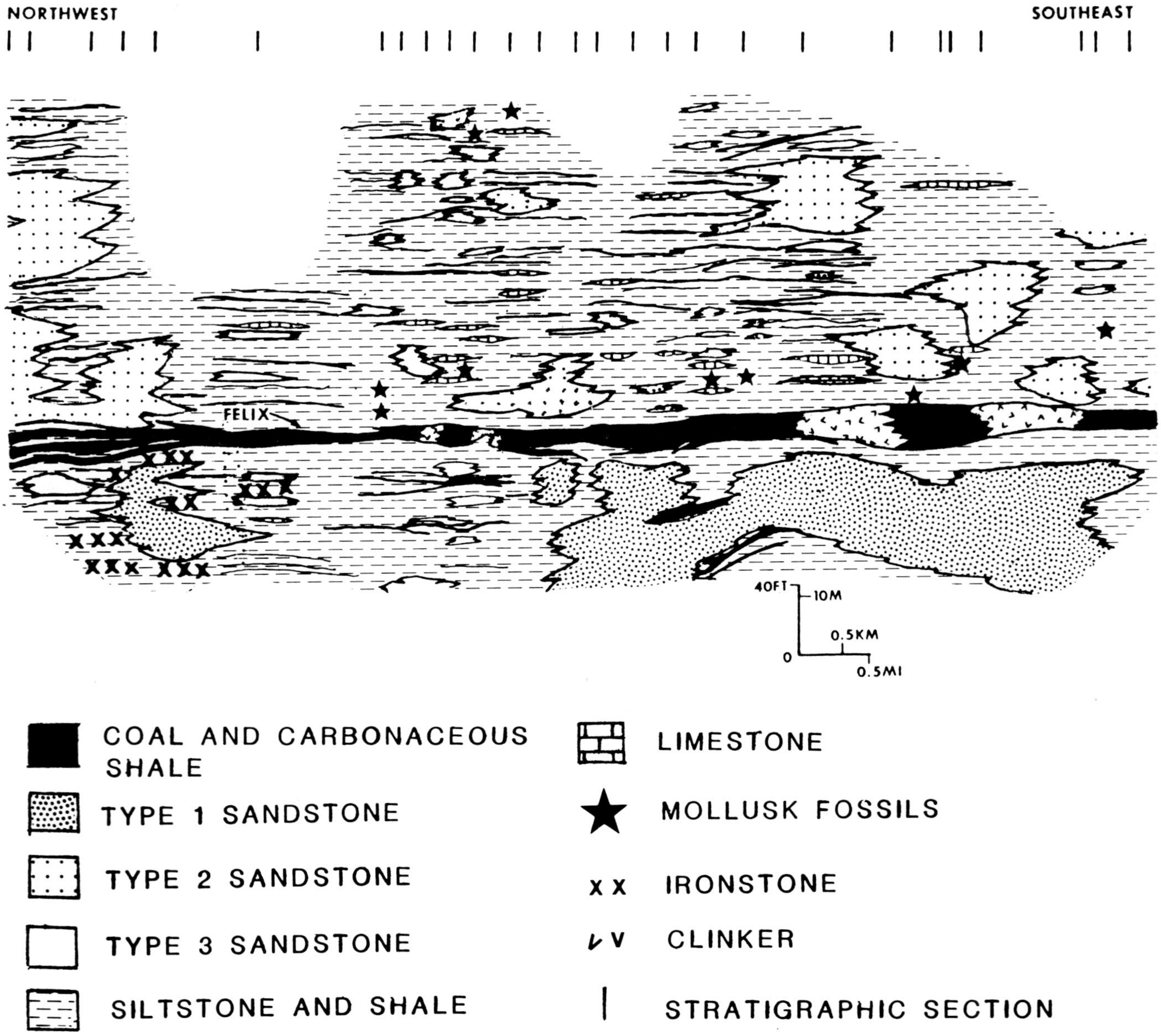

FIG. 3.—A northwest-southeast cross section showing environmental-stratigraphic framework of the study interval along the Wild Horse Creek. Note that the thickest part of the Felix bed is underlain by type 1 sandstone. Vertical exaggeration = 75×.

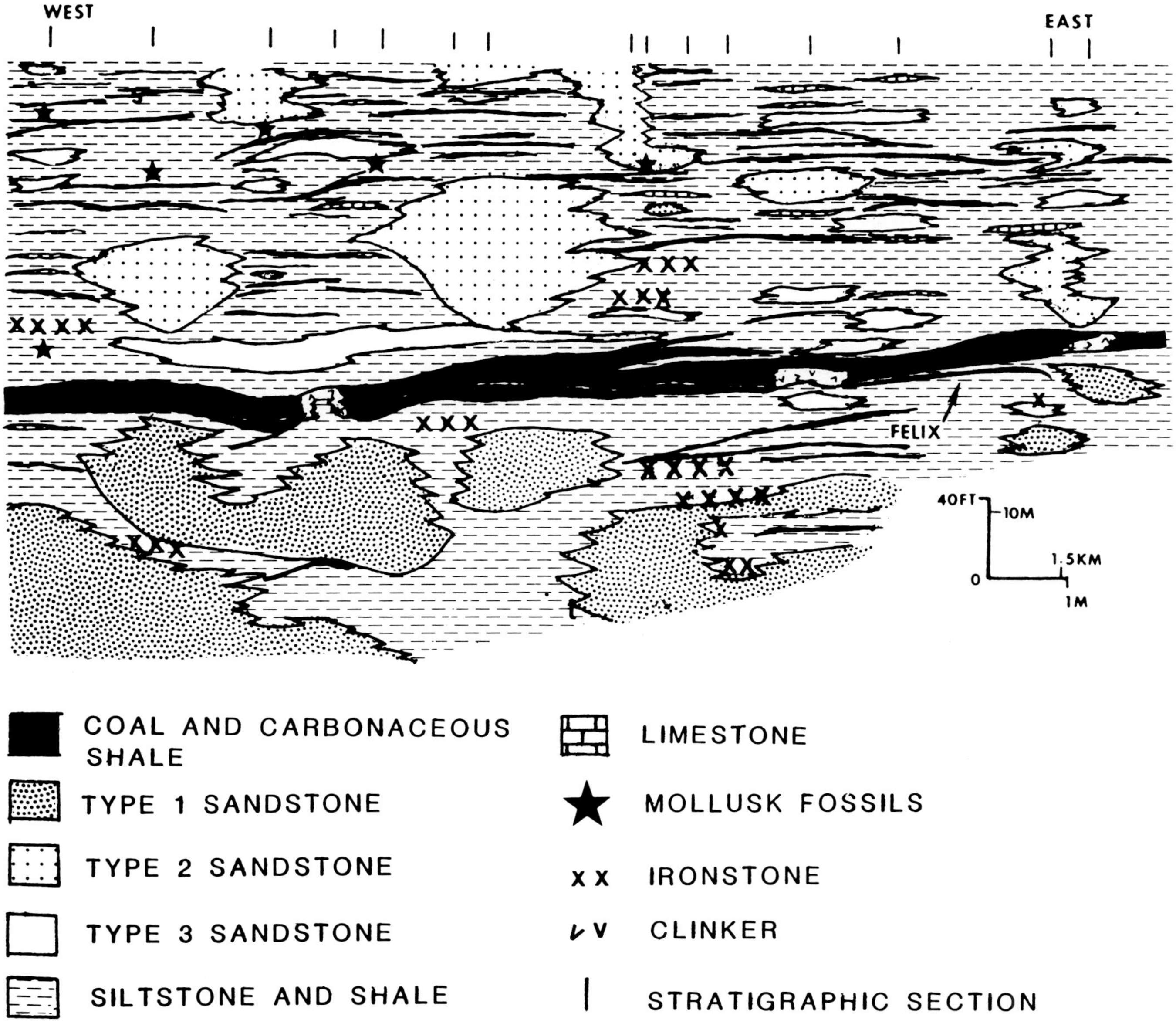

FIG. 4.—A west-east cross section showing environmental-stratigraphic framework of the study interval along the North Prong Wild Horse Creek. Note the widespread type 3 sandstone above the Felix bed. Vertical exaggeration = 150×.

by lakes suggests deposition in an anastomosed belt of the alluvial plain.

SUMMARY OF EVOLUTION OF FLUVIAL STYLES

The vertical succession and lateral relationship of facies documented in this study suggest deposition in a continuum of fluvial environments. The vertical sequence of meandering and anastomosed stream deposits from the lower to upper parts of the study interval, respectively, can be interpreted according to Walther's *Law of Facies* (1894) as products of laterally juxtaposed environments. The presence of local erosional breaks below channel sandstones in the vertical succession indicates removal of parts of the fluvial record and represents short periods of time. These surfaces indicate erosion that occurred during the migration of environments.

Figure 5 is an isolith map of sandstones in the meander-belt facies, probably representing a north-flowing trunk in a meandering stream system. The map shows a broad belt in which areas of sandstone concentration are found to the east and west of the study area. Based on the 25 ft (7.5 m) and greater isolines, the sandstones are elongate bodies oriented in north-south and northwest-southeast directions. These bodies are interpreted as channel deposits of a meandering stream which may or may not be contemporaneous as indicated by the en echelon arrangement of the type 1 channel sandstones. Thus, it could be presumed that these bodies are offset in which one body was being formed while the other was being abandoned. The en echelon depositional pattern as seen resulted from lateral shifts of stream channels into adjoining low-lying areas in the alluvial plain, which resulted from differential compaction of fluvial sed-

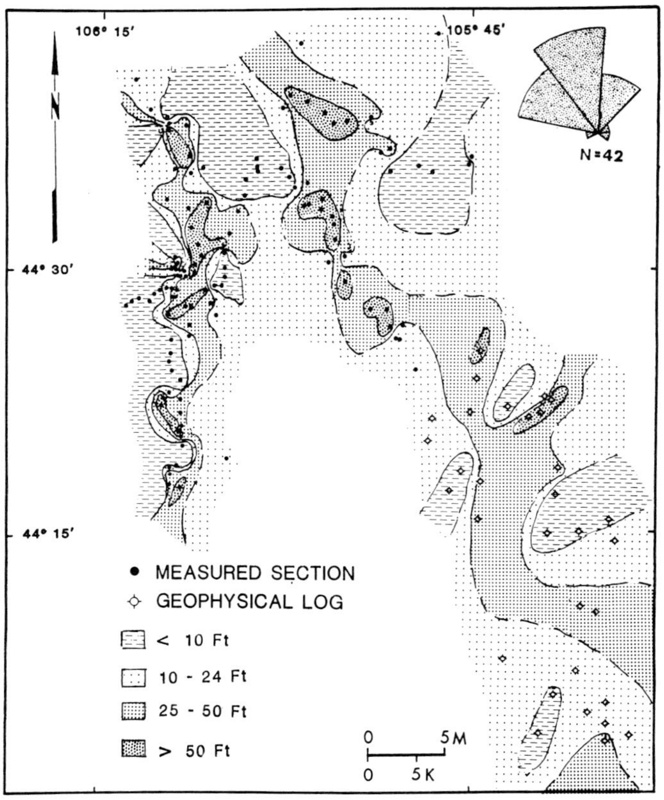

FIG. 5.—Isolith map of sandstone lithofacies below the Felix bed representing the meanderbelt-dominated facies. Rose diagram indicates paleocurrent directions. N = number of observations.

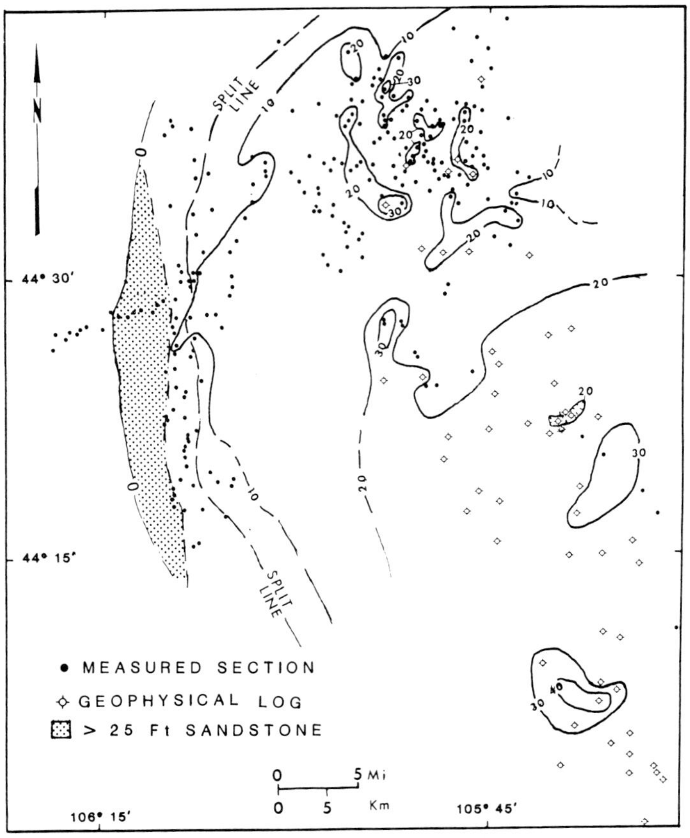

FIG. 6.—Isolith map of the Felix bed and laterally equivalent channel sandstone of the transition facies between the meanderbelt and anastomosed-belt facies. Stippled pattern indicates area where the coal bed is split by sandstone.

iments in a process similar to that proposed by Ferm and Cavaroc (1969). Walker and Cant (1979) suggested that the en echelon pattern of channel sandstones is common for deposits of meandering streams. Crossbed measurements from channel sandstones indicate the direction of flow was toward the north. The trunk meandering stream consisted of a meanderbelt of both active and abandoned channels. The meanderbelt shifted its position on the alluvial plain through time by floods breaching the levee and diverting flow to create an avulsion route. The avulsion is followed by development of a new meanderbelt and abandonment of the old meanderbelt.

The abandoned meanderbelts eventually became a topographic high on which a poorly drained backswamp formed. This is indicated by the accumulation of the Felix bed, which is uniformly thick above the meanderbelt facies (Fig. 6). As noted by Ferm and Staub (1984) in Appalachian coal beds, areas where the Felix bed substrate is sandstone and relatively uncompactible, the swamp surface was topographically high free of detrital influx. This condition, in addition to the development of a raised or domed (ombrotrophic) Felix peat swamp, provided an ideal setting for organic accumulation. The thicker parts of the Felix bed represent peat deposition on a raised platform in compari-

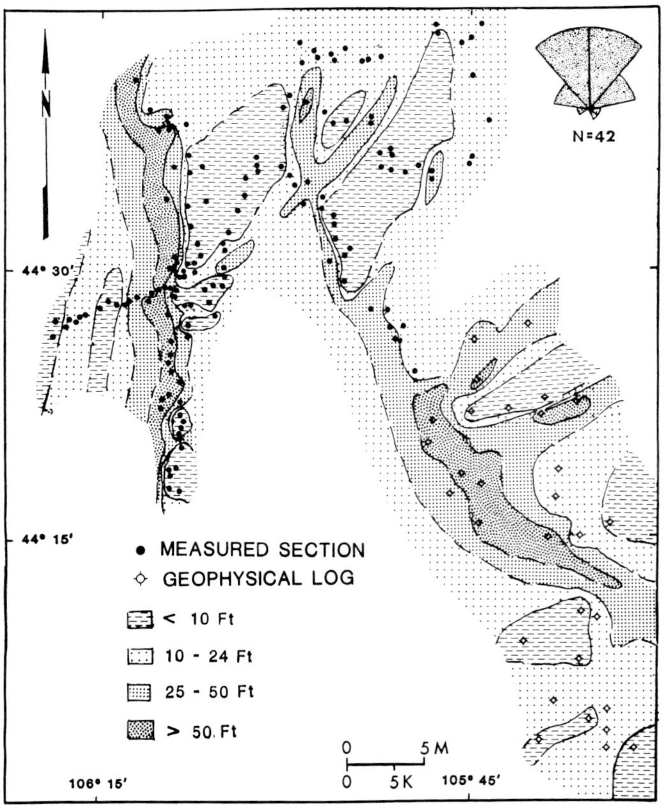

FIG. 7.—Isolith map of the sandstone lithofacies above the Felix bed representing the anastomosed-belt facies. Rose diagram indicates paleocurrent directions. N = number of observations.

son to adjacent areas where the coal bed is thin and split (Fig. 6). Where the detritus underlying the Felix bed is fine-grained and readily compactible, the resulting low-lying areas received water-borne clastic sediments that interrupted peat accumulation. The low-lying swamps to the west of the raised Felix peat dome seem to have controlled aggradation of a north-flowing, slightly sinuous trunk stream, as indicated by the sandstone body lateral to the Felix bed (Fig. 6). The north-trending trunk stream flowed along the western edge of the swamp and was responsible for the influx of sediments that produce splits in the Felix bed.

A combination of vertical stream aggradation and subsidence as well as compaction of the peat allowed the channel to flow into the adjoining swamp. In areas where the peat swamp subsided below the water table, lakes formed and the accumulation of peat in the Felix swamp was terminated by drowning. The lakes were gradually filled with sediments of crevasse splays, and the associated feeder channels served as avulsion routes and major conduits. These channels bifurcated and rejoined as they prograded to form as an anastomosed pattern. Thus, the trunk stream during this period was transformed into a multichannel system of an anastomosed stream, as indicated by an isolith map of the sandstones above the Felix bed (Fig. 7). The areal distribution of the sandstones greater than 25 ft (7.6 m) thick shows merging and diverging bodies similar to channel deposits of modern anastomosed streams described by Smith (1983) for the Columbia River in western Canada, by Flores and Hanley (1984) for Paleocene coal-bearing deposits in the Powder River Basin in Wyoming, and by Rust and others (1984) for Carboniferous coal-bearing deposits in Nova Scotia, Canada.

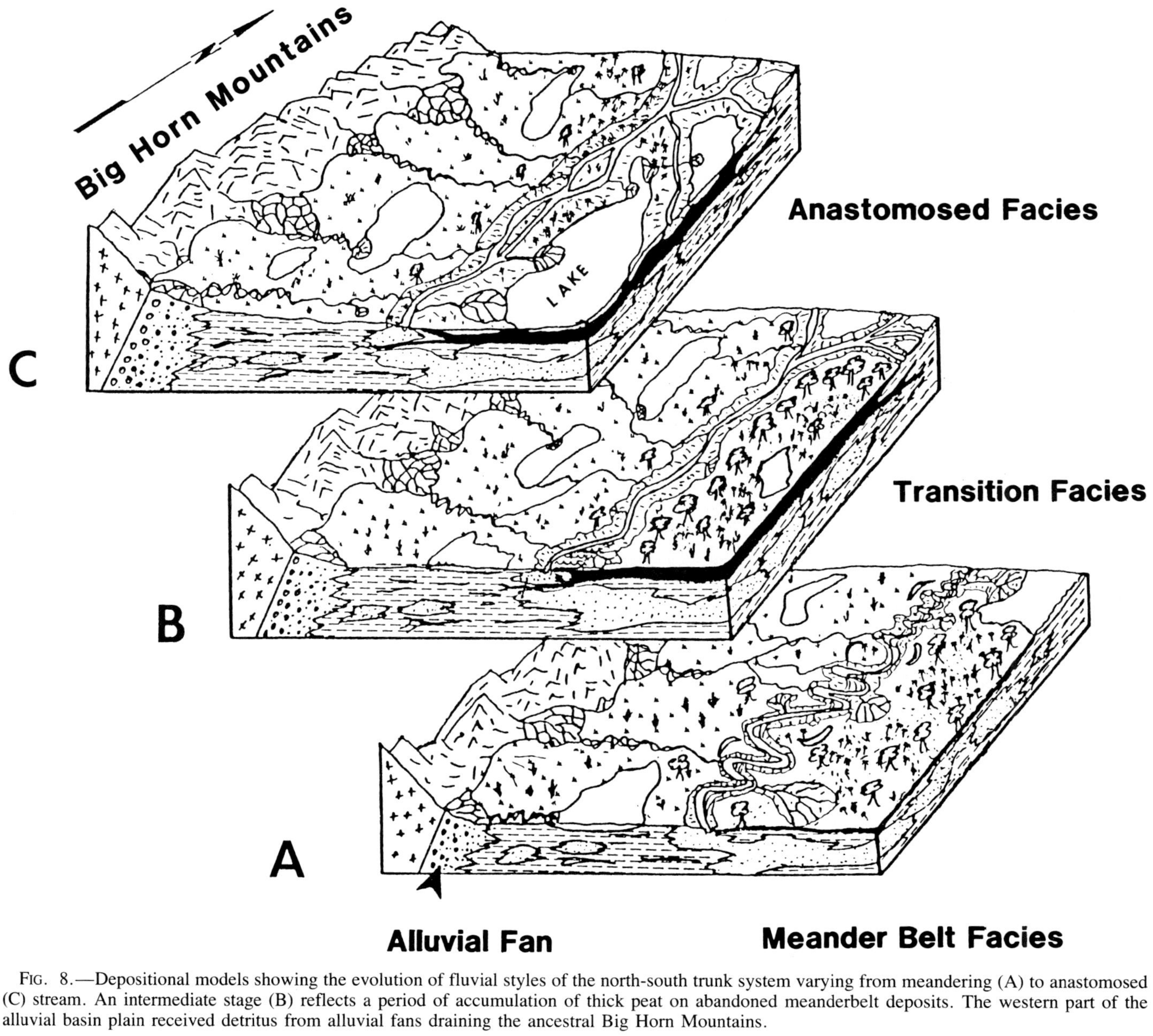

FIG. 8.—Depositional models showing the evolution of fluvial styles of the north-south trunk system varying from meandering (A) to anastomosed (C) stream. An intermediate stage (B) reflects a period of accumulation of thick peat on abandoned meanderbelt deposits. The western part of the alluvial basin plain received detritus from alluvial fans draining the ancestral Big Horn Mountains.

The paleogeographic succession and evolution of fluvial systems of the study interval are summarized in Figure 8. The transition (Fig. 8B) of the trunk system from the meandering stream (Fig. 8A) to the anastomosed stream (Fig. 8C) is similar to the avulsion-controlled process described by Smith and Cross (1985) for the lower Saskatchewan River. The transition period, which followed the abandonment of meanderbelts, provided a stage for accumulation of thick peat. The slightly sinuous stream along the west edge of the Felix peat swamp probably represents deposition during a transition period between development of meandering and anastomosed streams. Anastomosis may have been controlled by lowering of base level by compaction of sediments and/or basin subsidence.

ACKNOWLEDGMENTS

Portions of this work were derived from a Ph.D. dissertation completed by P. D. Warwick at the University of Kentucky under direction of J. C. Ferm. Financial support was provided in part by the U.S. Geological Survey, Wyoming Geological Survey, Geological Society of America, and the University of Kentucky Department of Geology.

REFERENCES

Beaumont, E. A., 1979, Depositional environments of Fort Union sediments (Tertiary, northwest Colorado) and their relation to coal: American Association of Petroleum Geologists Bulletin, v. 63, p. 194–217.

Elliot, E. S., 1976, Environment of deposition of the Wasatch Formation, Powder River Basin, Wyoming: Unpublished M.S. Thesis, Wright State University, Dayton, Ohio, 94 p.

Ethridge, F. G., Burns, L. K., Alexander, W. G., Craig, G. N. II, and Youngberg, A. D., 1983, Overburden characterization and postburn study of the Hoecreek, Wyoming underground coal gasification site and comparison with the Hanna, Wyoming site: U.S. Department of Energy Contract Report DE-AS20-81LC10705, 185 p.

———, Jackson, T. J., and Youngberg, A. D., 1981, Flood basin sequence of a fine-grained meanderbelt subsystem: The coal-bearing lower Wasatch and upper Fort Union Formations, Powder River Basin, Wyoming, *in* Ethridge, F. G., and Flores, R. M., eds., Recent and Ancient Nonmarine Depositional Environments: Models for Exploration: Society of Economic Paleontologists Special Publication 31, p. 191–289.

Ferm, J. C., and Cavaroc, V. V., Jr., 1969, A field guide to Allegheny deltaic deposits in the upper Ohio Valley: Ohio Geological Society and Pittsburgh Geological Society, 27 p.

———, and Staub, J. R., 1984, Depositional controls of minable coal bodies, *in* Rahmani, R. A., and Flores, R. M., eds., Sedimentology of Coal and Coal-bearing Sequences: International Association of Sedimentologists Special Publication 7, p. 257–289.

Flores, R. M., 1981, Coal deposition in fluvial paleoenvironments of the Paleocene Tongue River area, Powder River Basin, Wyoming and Montana, *in* Ethridge, F. G., and Flores, R. M., eds., Recent and Ancient Nonmarine Depositional Environments: Models for Explorations: Society of Economic Paleontologists and Mineralogists Special Publication 31, p. 169–190.

———, and Hanley, J. H., 1984, Anastomosed fluvial deposits: Upper Tongue River Member, Paleocene Fort Union Formation, northern Powder River Basin, Wyoming, *in* Rahmani, R. A., and Flores, R. M., eds., Sedimentology of Coal and Coal-bearing Sequences: International Association of Sedimentologists Special Publication 7, p. 85–104.

———, and Warwick, P. D., 1984, Dynamics of coal deposition in intermontane alluvial paleoenvironments, Eocene Wasatch Formation, Powder River Basin, Wyoming, *in* Houghton, R. L., and Clausen, E. N., eds., 1984 Proceedings of the Symposium on the Geology of Rocky Mountain Coal: North Dakota Geological Society Special Publication 84-1, p. 184–199.

Galloway, W. E., 1979, Early Tertiary-Wyoming intermontane basins, *in* Galloway, W. E., Krietler, C. W., and McGowen, J. H., eds., Depositional and Ground-water Flow Systems in the Exploration for Uranium: Bureau of Economic Geology, The University of Texas at Austin, A Research Colloquium, p. 197–213.

Jacob, A. F., 1972, Depositional environments of parts of the Tongue River Formation, western North Dakota, *in* Ting, F. T. C., ed., Depositional Environments of the Lignite-Bearing Strata in Western North Dakota: North Dakota Geological Survey Miscellaneous Series 50, p. 43–62.

Love, J. P., 1952, Preliminary report on the uranium deposits in the Pumpkin Buttes area, Powder River Basin, Wyoming: U.S. Geological Survey Circular 176, p. 37.

Rust, B. R., Gibling, M. R., and Legun, A. S., 1984, Coal deposition in an anastomosing-fluvial system: The Pennsylvanian Cumberland Group south of Joggins, Nova Scotia, Canada, *in* Rahmani, R. A., and Flores, R. M., eds., Sedimentology of Coal and Coal-bearing Sequences: International Association of Sedimentologists Special Publication 7, p. 105–120.

Seeland, D. A., 1976, Relationships between early Tertiary sedimentation patterns and uranium mineralization in the Powder River Basin: Wyoming Geological Association 28th Field Conference Guidebook, p. 221–230.

Smith, D. G., 1983, Anastomosed fluvial deposits: Modern examples from western Canada, *in* Collinson, J. D., and Lewin, J., eds., Modern and Ancient Fluvial Systems: International Association of Sedimentologists Special Publication 6, p. 155–168.

Smith, N. D., and Cross, T. A., 1985, Avulsion-controlled fluvial evolution: The Cumberland Marshes, east-central Saskatchewan (abs.): Third International Fluvial Sedimentology Conference Abstracts volume, p. 36.

Toomey, W. J., 1977, A stratigraphic analysis of the Wasatch Formation, western Powder River Basin: Unpublished M.S. Thesis, Wright State University, Dayton, Ohio, 96 p.

Walker, R. G., and Cant, D. J., 1979, Facies models 3, sandy fluvial systems, *in* Walker, R. G., ed., Facies Models: Geological Association of Canada Reprint Series 1, p. 23–31.

Walther, J., 1894, Lithogenesis der Gegenwart, Beobachtungun uber die Bildung der Gesteine ander heutigen Erdoberflache: Dritter Teil einer Einleitung in die Geologie als historische Wissenchaft, Jea: Verlag Gustav Fischer, p. 535–1055.

Warwick, P. D., 1985, Depositional environments and petrology of the Felix coal interval (Eocene), Wasatch Formation, Powder River Basin, Wyoming: Unpublished Ph.D. Dissertation, University of Kentucky, 333 p.

TECTONIC CONTROL ON ALLUVIAL PALEOARCHITECTURE OF THE CRETACEOUS AND TERTIARY RATON BASIN, COLORADO AND NEW MEXICO

ROMEO M. FLORES AND CHARLES L. PILLMORE
U.S. Geological Survey Denver, Colorado 80225

ABSTRACT: The Raton basin is bounded on the west by the ancestral San Luis highlands. The highlands were probably uplifted along a high-angle reverse fault zone marginal to the basin. Coarsening-upward megacycles and types of facies sequences and facies associations within the Upper Cretaceous Vermejo Formation and Upper Cretaceous and Tertiary Raton and Poison Canyon Formations indicate that both allocyclic and autocyclic mechanisms influenced alluvial sedimentation in the Raton basin. The megacycles comprise a coal-rich, fine detritus-dominated facies and an overlying coal-poor, sand-dominated facies. The coal-rich, fine detritus-dominated facies was deposited by braided, meandering, and anastomosed fluvial systems during periods of increased basin subsidence associated with a stable source area. The sand-dominated facies was deposited by alluvial fans and by braided and high-bedload meandering fluvial systems during periods of tectonic uplift in the source area and reduced basin subsidence.

The basin alluvial paleoarchitecture (spatial arrangement, patterns, interconnection, distribution, and evolution of the fluvial systems) was mainly controlled by allocyclic mechanisms of uplift in the source area and subsidence in the basin. Climatic change and the presence of vegetation played only minor roles in controlling the alluvial paleoarchitecture of the Raton basin. In this alluvial environment, basin subsidence (an allocyclic mechanism) and fluvial avulsion and abandonment (autocyclic mechanisms) served as the major controlling factors in the accumulation of thick coal beds.

INTRODUCTION

One of the most important controls on siliciclastic sedimentation is tectonism, either the uplift of source areas or basin subsidence, or both. The response of sedimentary processes to tectonism is commonly represented by downslope changes in the patterns of alluvial systems (McLean and Jerzykiewicz, 1978; Steel and Aasheim, 1978). An example of this response is the arrangement of vertical sedimentary successions expressed as coarsening-upward and fining-upward sequences (Steel and others, 1977; Heward, 1978). Three coarsening-upward megacycles occur in the Upper Cretaceous Vermejo and Upper Cretaceous and Paleocene Raton and Poison Canyon Formations in the Raton basin of Colorado and New Mexico (Fig. 1). Each megacycle is made up of a lower, coal-rich, fine detritus-dominated (FDD) facies and an upper, coal-poor, sand-dominated (SD) facies. The megacycles are as much as 760 ft (230 m) thick and generally thicken westward toward the ancestral San Luis highlands (Pillmore and Flores, 1984). Westward-thickening clastic wedges of the SD facies are traceable across the width of the basin.

The purpose of this paper is to analyze the facies sequences and associations of the Vermejo, Raton, and Poison Canyon megacycles to determine the controls of sedimentation. Repetition of the coarsening-upward megacycles may reflect cyclic sedimentation common to an alluvial plain, as recognized by Beerbower (1964). Analysis of a facies within the megacycles will help establish the mechanism of cyclicity. Cyclic sedimentation resulting from changes in energy supply or seidment input into a sedimentary system, which may be caused by tectonic, sea level, or climatic changes, is called allocyclic (Beerbower, 1964). Cyclic sedimentation that results from redistribution of the total energy and sediment input into a sedimentary system, caused by channel avulsion, bar migration, and crevassing, is called autocyclic (Beerbower, 1964).

STRATIGRAPHIC-DEPOSITIONAL FACIES VARIATIONS

Stratigraphic facies variations are best displayed by generalized cross sections across the basin and by detailed outcrop diagrams. The cross sections show the lateral continuity of major clastic wedges of the SD facies and their vertical and lateral relationships with the FDD facies. The outcrop diagrams provide information on stratigraphic-environmental facies variations at individual outcrop locations. These diagrams also record the successions of environmental changes, as well as variation of temporally equivalent depositional settings, for both the SD and FDD facies.

Basinal Facies Variations

Facies variations from west to east across the basin are shown in the cross sections in Figures 1 and 2. Both cross sections show three clastic wedges of the SD facies, which thin eastward and coarsen upward. Figure 1 displays the lateral variation of three coarsening-upward megacycles in the western part of the basin. The first coarsening-upward megacycle, as much as 300 ft (90 m) thick, includes the FDD facies of the Vermejo Formation and the overlying SD facies of the basal Raton Formation (Lee, 1917). In the western part of the basin, the basal Raton Formation is a conglomerate bed resting unconformably on the FDD facies of the Vermejo Formation. Where present, the basal Raton conglomerate consists of quartzose sandstone with chert pebbles derived in large part from the Upper Cretaceous Dakota Sandstone (Flores, 1984). The conglomerate reflects the initial uplift of the source area, an allocyclic mechanism. The same relationship between the Vermejo Formation and the basal Raton is observed in other parts of the basin; however, in some areas, especially near Raton, New Mexico, and Trinidad, Colorado, the conglomerate is not present. In these areas, the lower or first megacycle consists of the Vermejo FDD facies passing into and interfingering with channel and interchannel deposits of the Raton SD facies. The second coarsening-upward megacycle overlies the SD facies of the basal Raton Formation, is as much as 350 ft (100 m) thick, and consists of the FDD facies overlain by an SD facies. The SD facies of this megacycle, described by Pillmore and Flores (1984), was identified by Lee (1917) as a barren zone consisting mostly of sandstones and a few thin minor coal beds. The Creta-

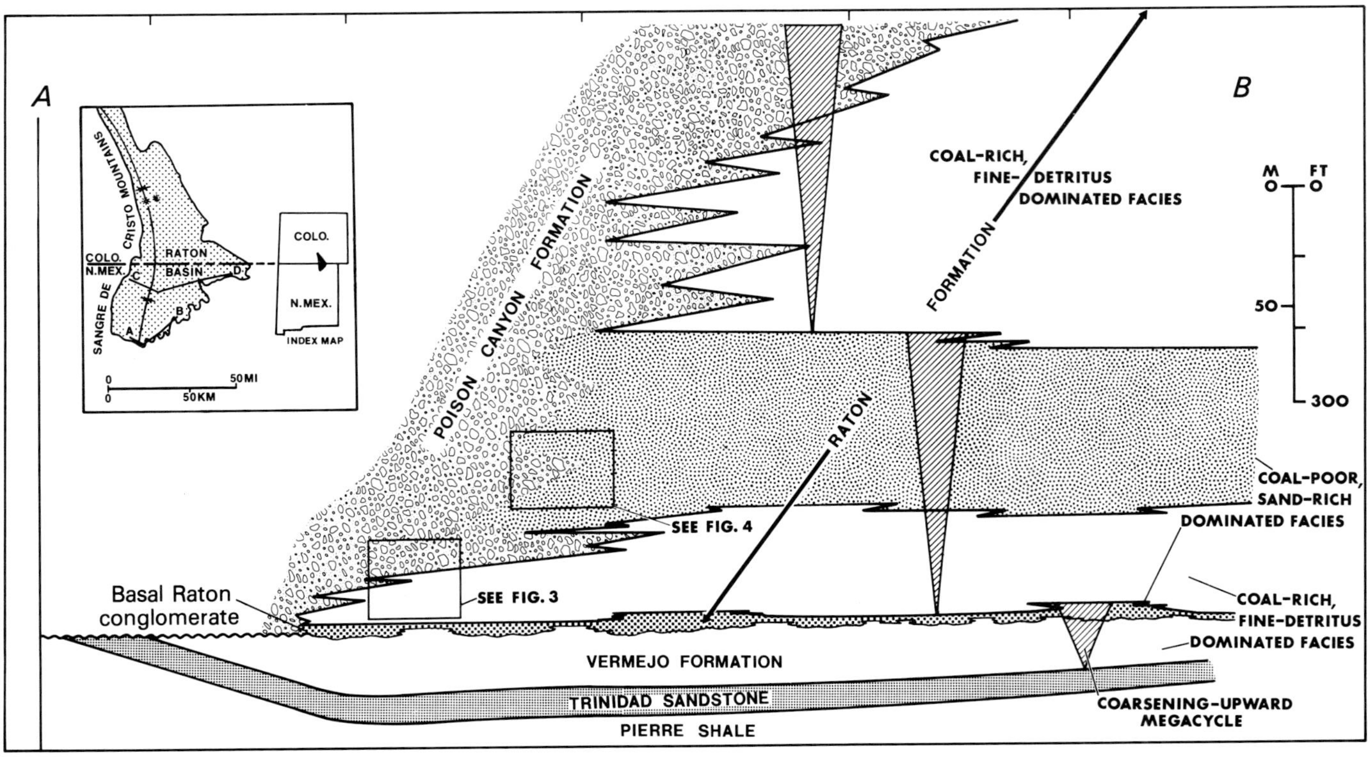

FIG. 1.—Diagrammatic cross section of the western part of the basin, displaying coarsening-upward megacycles (1–3) that contain clastic wedges (coal-poor, sand-dominated facies) thickening to the west. Location shown on the inset map.

ceous-Tertiary boundary clay, a chronostratigraphic marker (Pillmore and others, 1984; Pillmore and Flores, 1987), is located either at or just below the base of the barren zone. The third coarsening-upward megacycle, as much as 1,000 ft (300 m) thick, comprises the FDD facies above the barren zone of the Raton Formation upward through the SD facies in the overlying Poison Canyon Formation. This megacycle differs from the lower two megacycles in that it contains a coal-rich FDD facies three times as thick as the other megacycles.

The sand-dominated facies of the three megacycles probably represents deposits formed during recurring allocyclic episodes (tectonic) with reduced basin subsidence. These episodes accompanied uplift, unroofing, and erosion of the granitic core rocks in the source area. Marked thickening of the sand-dominated facies to the west suggests a source in that direction. Recurrent uplifts of the source area were probably caused by repeated movement along an inferred, ancient, high-angle, reverse fault zone at the western margin of the basin. A younger reverse fault zone to the east has been recognized by Ouellette and Boucher (1983) along the northwestern margin of the basin. Uplift along the fault zone in the source area is reflected in the change from fine to coarse detritus.

The FDD facies of the megacycles probably represents deposition during subsidence induced by differential compaction and related isostatic adjustment that resulted from sediment loading. Basin subsidence is probably indicated by thickening of the FDD facies upward and along the basin axis (Fig. 2). The FDD facies of the third megacycle contains coal beds as much as 12 ft (4 m) thick (Pillmore and Flores, 1984) and is three times as thick as the first and second megacycles. This may suggest more protracted basin subsidence and filling than during deposition of the FDD facies of the two older megacycles. The basin-axis thickening of the FDD facies in all three megacycles may be a response to crustal downwarping.

Facies Sequences and Associations

The facies sequences (vertical successions of facies) and associations (assemblages of facies occurring together as genetically related units) of the Vermejo, Raton, and Poison Canyon Formations are best analyzed from outcrops along the western basin margin, as well as within the basin proper. The western basin margin includes lithofacies proximal to the source area, and the basin proper consists of lithofacies distal to the source area. Photographs and diagrams of outcrops are used to illustrate the detailed facies sequences and associations and the influence of the cyclic mechanisms. The stratigraphic position and basinal distribution of these facies sequences are shown in Figures 1 and 2.

Basin-margin lithofacies.—

The basin-margin facies sequences and facies associations of the second coarsening-upward megacycle in the Raton and Poison Canyon Formations are shown in Figures 3 and 4. Figure 3 shows a sequence of lenticular sandstones interbedded with fine-grained sediments of the Raton Formation overlain by sheetlike sandstones of the Poison Canyon Formation. The lenticular sandstones have erosional

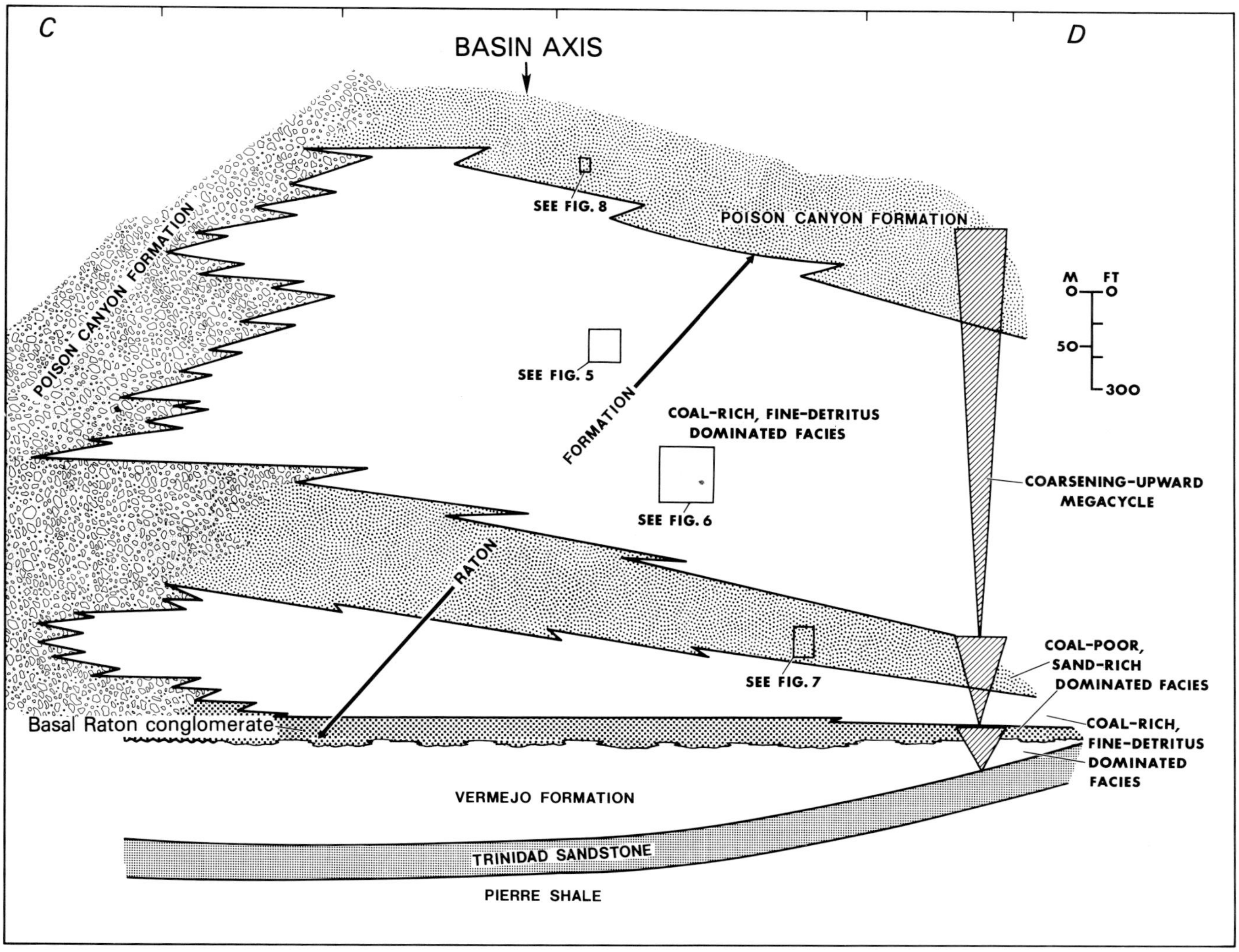

FIG. 2.—Diagrammatic cross section showing the extent of the coarsening-upward megacycles (1–3) across the width of the basin. Location shown on Figure 1.

bases, fine upward, contain mud-draped lateral accretion surfaces, and are locally channeled with mudstone and siltstone infills. The sandstones contain trough crossbeds and convolute laminations in the lower part and ripple laminations in the upper part. In addition, the lenticular sandstones are arranged en echelon. Facies associations (Fig. 3) are displayed by the lenticular sandstones passing laterally and vertically into the sheetlike sandstones. The sheetlike sandstones are multiscoured, floored by granule- to gravel-size particles, and infilled by coarse sandstones that include random successions of trough and planar crossbeds. Both lenticular and sheetlike sandstones laterally merge with fine-grained sediments and coaly carbonaceous shale.

The lenticular sandstones are interpreted as deposits of a meandering fluvial system, whereas the sheetlike sandstones were probably deposited by a braided fluvial system. The vertical succession of facies sequences suggests that the meandering streams gave way to braided streams through time. Yet, the lateral facies associations suggest that the meandering and braided streams may have coexisted within a 2- to 3-mi-wide (3.2- to 4.8-km) belt. The contemporaneity of these tributary systems at the basin margin may be a function of availability of sediment load from upstream. For instance, the presence of alluvial fans upstream may provide high bedload sediments locally to the braided streams during flash floods. The close proximity of the two fluvial systems indicates that neither basin subsidence nor tectonic uplift (allocyclic mechanisms) greatly influenced the overall proportion of coarse-to-fine sediments that accumulated. Tectonic uplift (allocyclic mechanism), however, may have controlled widespread development of braided streams along the basin margin. The coarse detrital influxes (sand-dominated) pervasive along the basin margin may have had their point sources in mountain-front alluvial fans generated by uplift along reverse faults.

Figure 4 shows the basin-margin lithofacies of the Poison Canyon and Raton Formations, representing a stratigraphically higher part of the second coarsening-upward megacycle. This lithofacies consists of sheetlike, coarse-grained, arkosic sandstones interbedded with conglomerates and siltstones. The sheetlike sandstones contain trough and planar crossbeds, are multiscoured, and are floored by lag gravels.

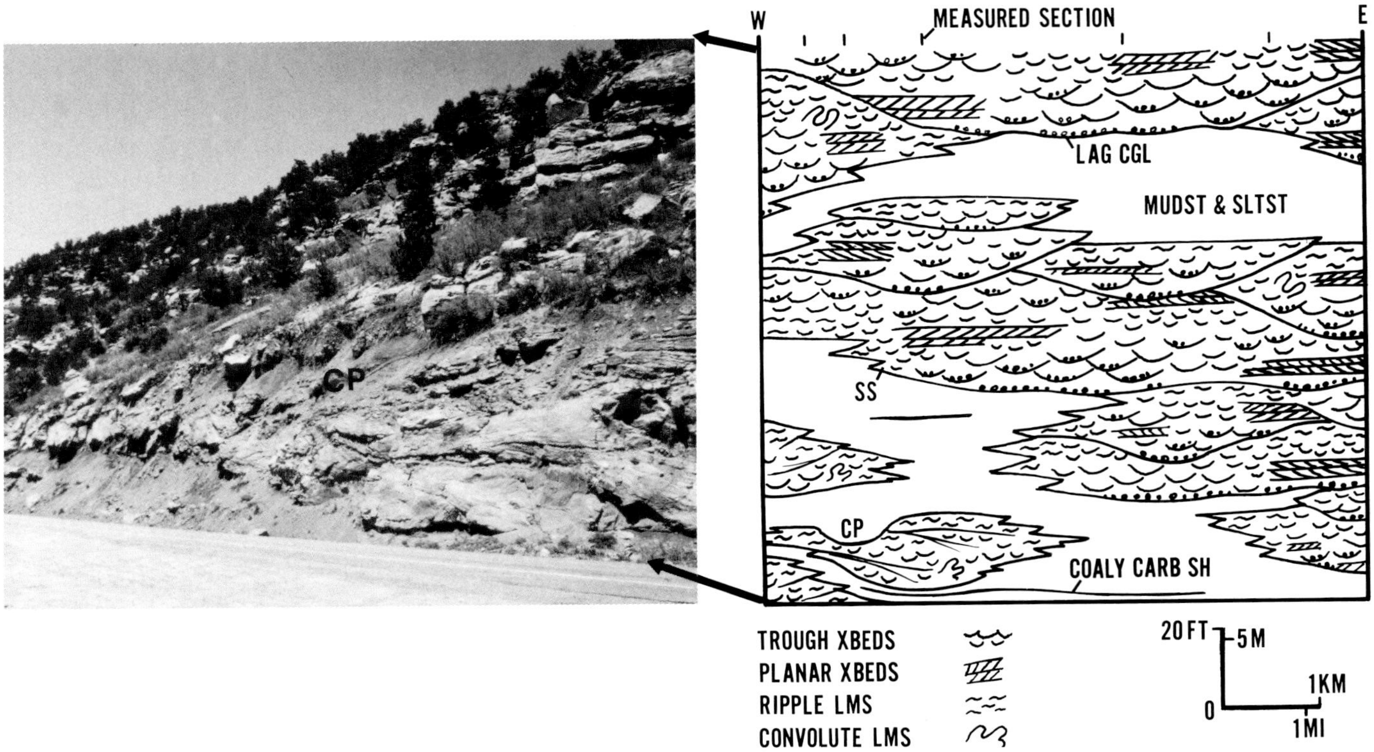

FIG. 3.—Lithofacies of the Raton and Poison Canyon Formations in the basin margin, consisting of channel, channel plug (CP), and overbank deposits of meandering and braided streams. The diagram of the outcrop was constructed from the measured section shown.

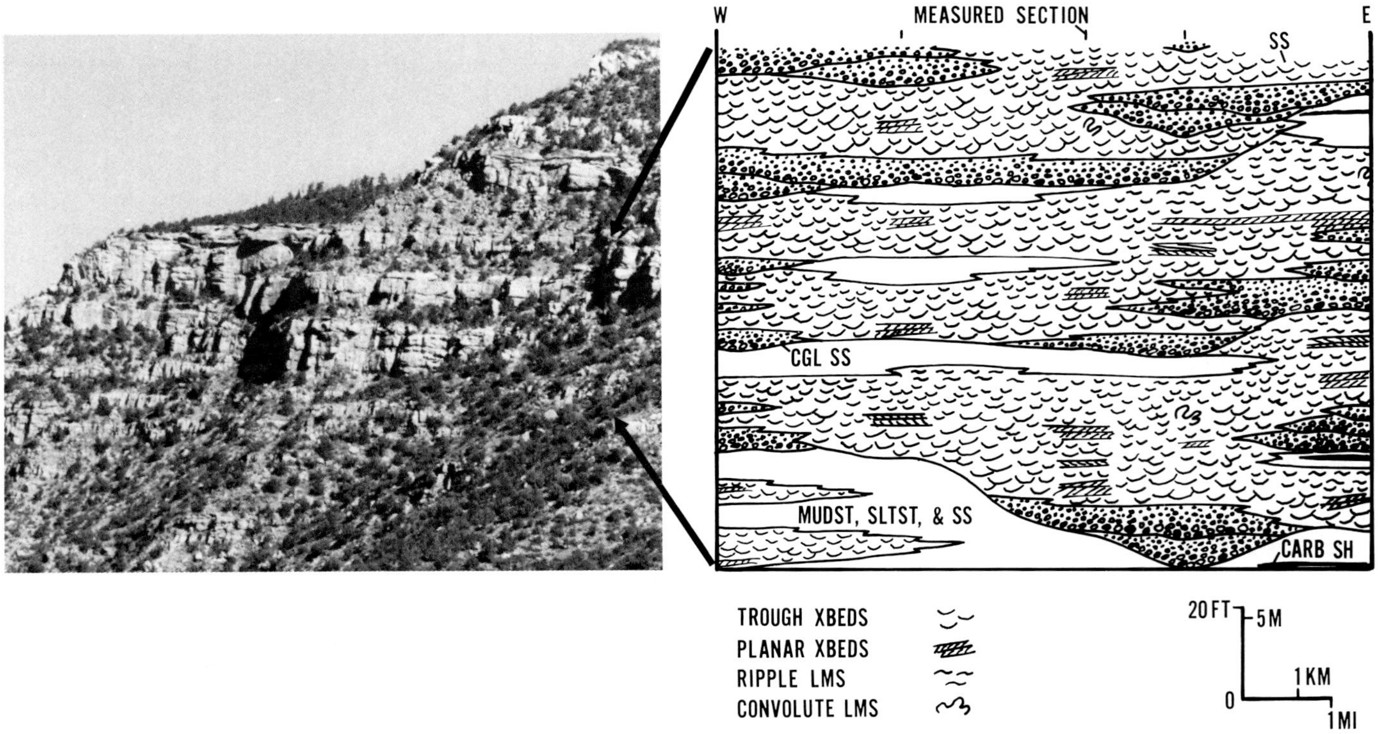

FIG. 4.—Photograph and outcrop diagram of the Raton and Poison Canyon Formations along the basin margin, comprising braided stream deposits. The diagram was constructed from the measured section shown.

The conglomerates have basal erosional surfaces, are crudely horizontally bedded, are imbricated, and fine upward. The conglomerates consist of framework-supported gravel, and gravel-size particles may be present in associated trough crossbedded sandstones. The conglomerates make up 20–50 percent of the succession and occur more frequently in the upper part of the succession. The sandstones and conglomerates grade upward into siltstones that contain very rare carbonaceous detritus.

The facies succession of sandstones and conglomerates is interpreted as deposits of proximal braided streams. The crudely horizontally bedded, framework-supported, imbricated conglomerates represent longitudinal bar deposits. Such bars are deposited as coarse bedload in response to lowering of flood stages (Leopold and Wolman, 1957). Gravel imbrication indicates deposition during high flow, and associated silty sandstone was probably deposited during waning stages of flow. Reworking of the longitudinal bars during lower flood stages is indicated by associated trough-crossbedded sandstones. Paucity of carbonaceous detritus in associated fine-grained sediments suggests that the proximal braid belt was composed mainly of active braid tracts with a few vegetated passive tracts. This proximal braided-stream model resembles the gravelly braid belt of the Donjek River in Canada (Williams and rust, 1969).

Basin-proper lithofacies.—

The basin proper lithofacies consists of coal-bearing and coal-barren deposits, the former deposit being the most common. The sequences and associations of the FDD facies of the Raton Formation (lower part of the third coarsening-upward megacycle) are shown in Figures 5 and 6.

Figure 5 exhibits facies sequences that include a multiscoured, erosional-based, lenticular sandstone laterally juxtaposed with a tilted sequence of sandstone, siltstone, and mudstone that is marked by a sharp, curvilinear basal contact. These facies sequences are underlain and overlain by coal beds as much as 7 ft (2 m) thick. The lenticular sandstones display lateral accretion surfaces delineated by mudstone and siltstone drapes. Internally, these lenticular sandstones contain large-scale trough crossbeds with convolute laminations in the lower part and small-scale trough crossbeds with ripple laminations in the upper part. The lenticular sandstones are associated with erosional-based deposits of mudstone, siltstone, and silty sandstone.

The facies sequences and associations suggest deposition in a meanderbelt of a high-sinuosity meandering stream. The lenticular sandstones indicate that this meandering stream was formed by slow, lateral accretion of point bars. Avulsion and abandonment (autocyclic mechanism) of the meandering stream are indicated by coal beds that overlie the lenticular sandstones. Oversteepening of meander cutbanks produced slumped and rotated overbank deposits of mudstone, siltstone, and silty sandstone. The margins of the meanderbelt are often defined by channel plugs of fine-grained detritus. The abandoned meanderbelt deposit formed by an autocyclic mechanism may have served as a platform on which swamps formed and peat accumulated, undisturbed by detrital influx and vegetation sustained by ground water discharged through underlying permeable and transmissive channel sands.

Figure 6 shows facies sequences and association in the

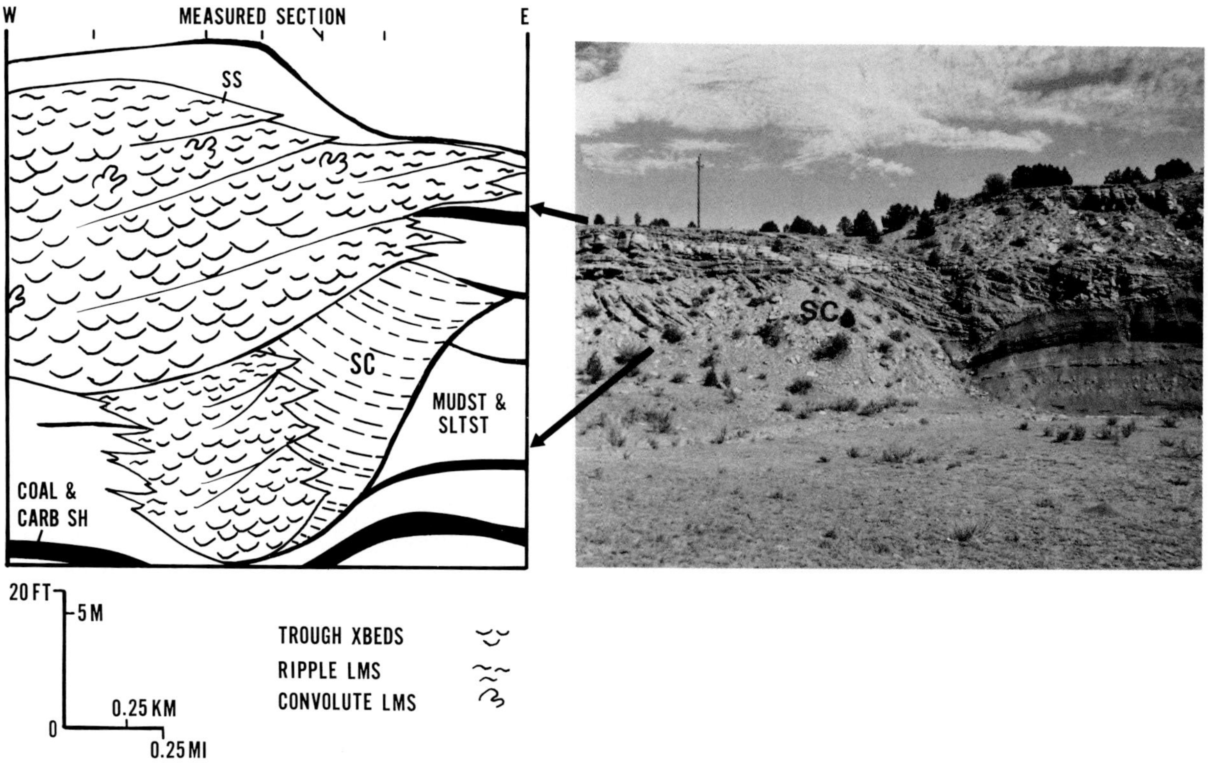

FIG. 5.—Photograph and outcrop diagram of the Raton Formation in the basin proper, containing channel and slumped cutbank (SC) deposits of meandering streams. The diagram was constructed from the measured section shown.

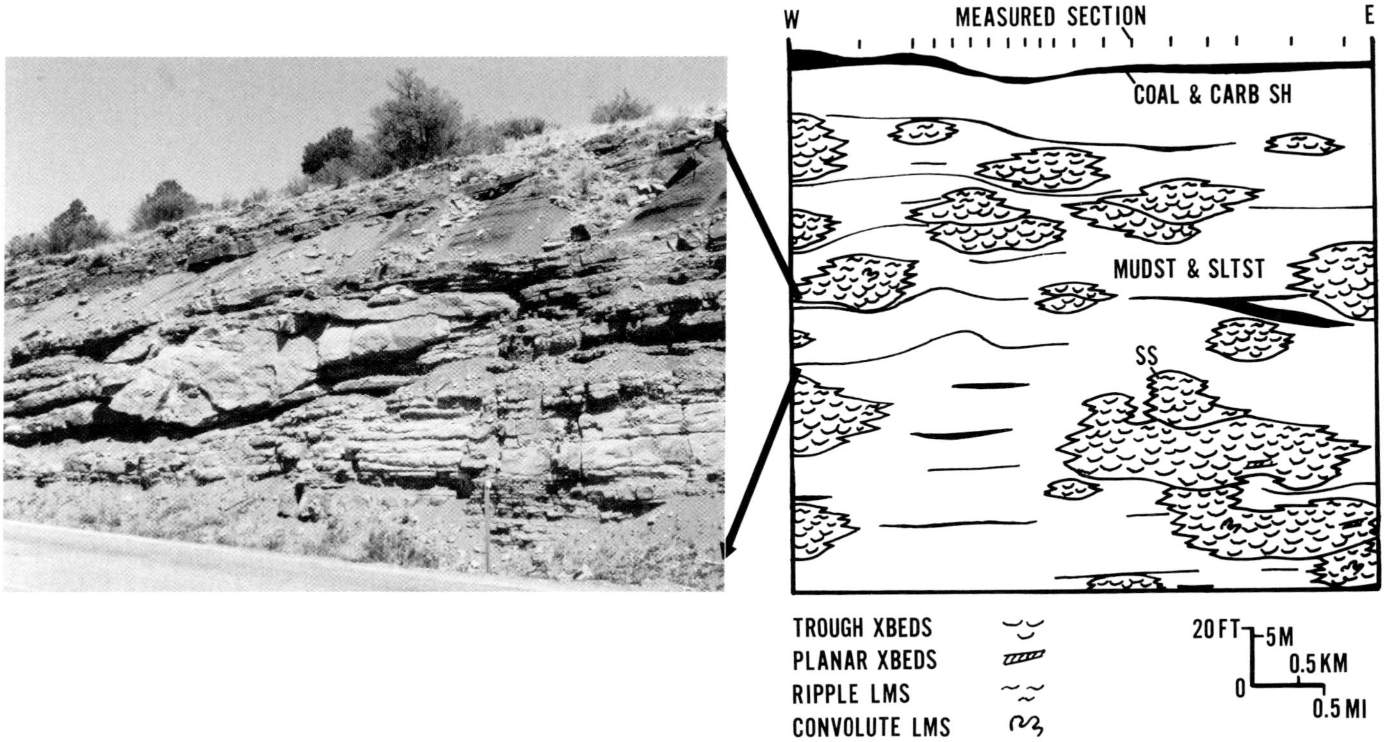

FIG. 6.—Photograph and outcrop diagram of the Raton Formation in the basin proper, consisting of a succession of channel and crevasse-splay deposits of anastomosed streams. The diagram was constructed from the measured section shown.

coal-bearing Ration Formation which are different from those of the meanderbelt. Here, the facies sequence consists of interbedded, fining-upward, lenticular sandstones; coarsening-upward mudstones, siltstones, and sandstones; and coals and carbonaceous shales. The lenticular sandstones have erosional bases and a thickness-to-width ratio ranging from 1:2 to 1:120; internally, they are trough crossbedded. Several of the lenticular sandstones occur as isolated, contemporaneous bodies within a 3- to 4-mi-wide (4.8- to 6.4-km) belt and are encased in thick successions of coarsening-upward sequences. These sequences consist of rippled mudstones and siltstones that are bioturbated in the lower part and rippled (ripple drift and asymmetrical) and convolute laminated in the upper part. The coarsening-upward sequences are overlain by thin, discontinuous coals and carbonaceous shales.

These facies relationships suggest deposition in a crevasse-splay-dominated alluvial environment as indicated by vertically accreted successions of thick coarsening-upward sequences. The crevasse splays probably debouched into subaqueous floodbasins, as evidenced by the bioturbation of the deposits. The lenticular sandstones associated with the crevasse-splay sediments indicate deposition in crevasse channels that enlarged into major conduits with increased discharge. These channels were separated by small floodbasins that were, in turn, infilled by crevasse-splay, lake, and backswamp deposits. One of these channels was a pathway of avulsion (autocyclic mechanism) and was likely transformed into an active, laterally aggrading, meandering stream. This alluvial setting is very similar to the anastomosed streams described by Clough (1983) and Smith and Cross (1985).

The facies sequences and associations of the barren zone of the Raton Formation (SD facies of the second megacycle) are shown in Figure 7. The facies sequence comprises abundant (as much as 75 percent), multistory, lenticular sandstones interstratified with subordinate mudstones, siltstones, and minor coaly carbonaceous shales. The lenticular sandstones have erosional bases, fine upward, are multiscoured, and are trough crossbedded with convolute lamination in the lower part and ripple lamination in the upper part. Although the lenticular sandstones lack lateral accretion surfaces, they show an en echelon arrangement.

The facies associations of the barren zone represent deposition in meandering streams that rapidly incised, filled, and avulsed (autocyclic), as indicated by the en echelon deposition of lenticular sandstones. Multiple scouring and subsequent infilling by channel sandstones indicate reworking of old meanderbelt deposits by younger streams. The meandering fluvial system was a high-bedload stream, as evidenced by the abundance of sand and the scarcity of fine-grained detritus. The input of sand in the fluvial system was probably the result of tectonic uplift (allocyclic mechanism) of the source area. Lateral migration of the meandering streams may account for the occurrence of minor fine-grained detritus. Most of the fine-grained detritus was probably reworked and subsequently retransported and redeposited downstream.

The basin proper lithofacies sequences and associations of the noncoal-bearing Poison Canyon Formation (SD fa-

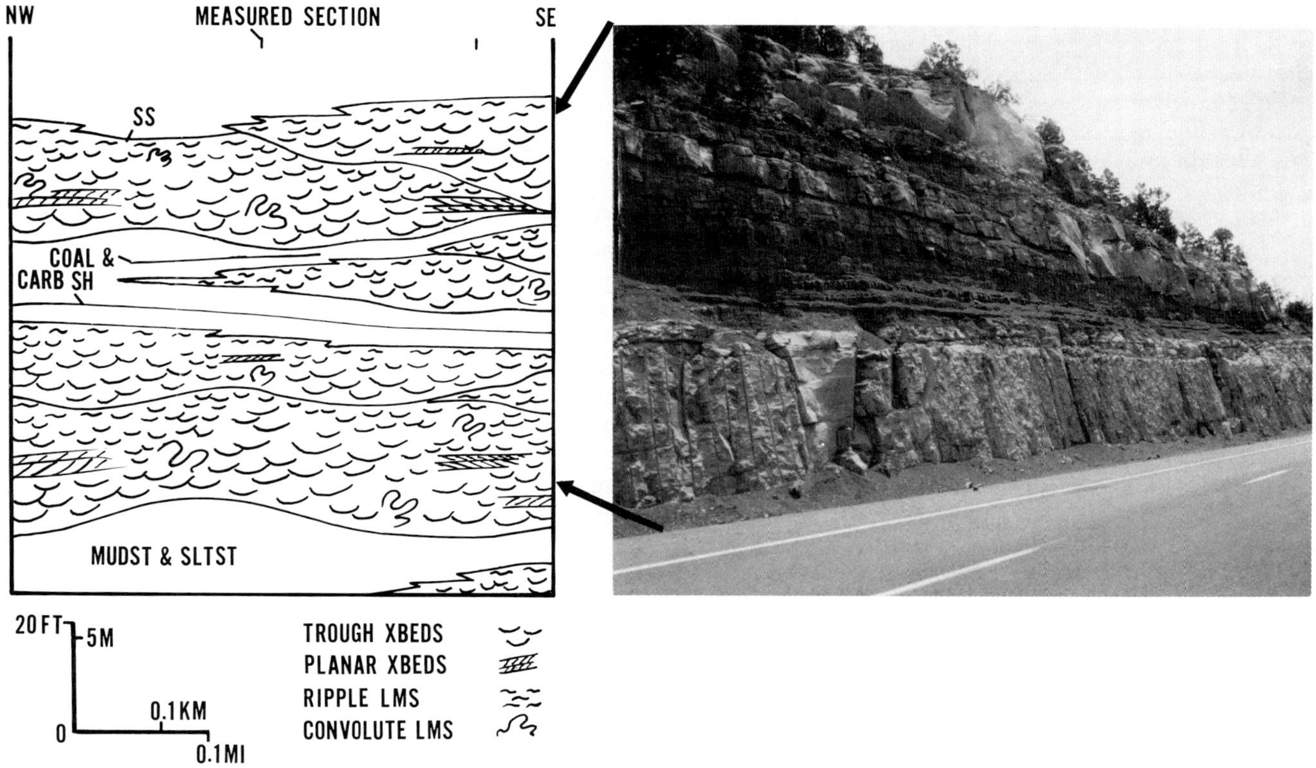

FIG. 7.—Photograph and outcrop diagram of the barren zone of the Raton Formation in the basin proper, comprising multistory channel and overbank deposits of high-bedload meandering streams. The diagram was constructed from the measured section shown.

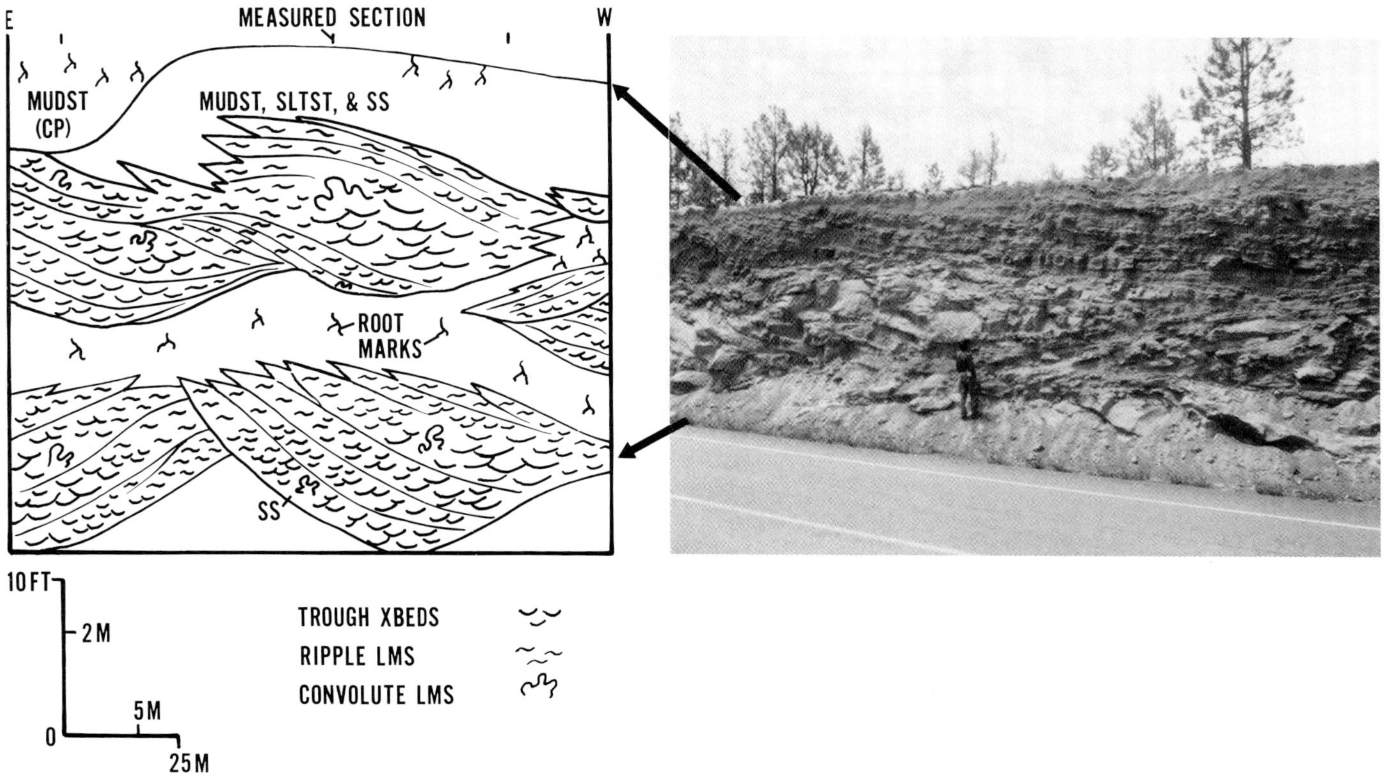

FIG. 8.—Photograph and outcrop diagram of the Poison Canyon Formation in the basin proper, containing channel, overbank, and channel plug (CP) deposits of meandering streams. The diagram was constructed from the measured section shown.

cies of the second megacycle) are illustrated in Figure 8. The facies sequence is composed of lenticular arkosic sandstones interstratified with siltstones and silty sandstones. The lenticular sandstones have erosional contacts at their bases. Texturally, the sandstones are coarse-grained to conglomeratic. Within these sand bodies, lateral accretion surfaces are delineated by siltstones. The sandstones exhibit trough and planar crossbeds in the lower part and ripple laminations in the upper part. The lenticular sandstones, which are stacked en echelon, are juxtaposed with scour-based, muddy siltstones and rooted silty sandstones.

The facies associations of the Poison Canyon Formation in the basin proper suggest deposition in a meandering fluvial system. Lateral accretion of deposits is indicated by point bar sandstones formed in meanderbelts. Abandoned meanders infilled by fine-grained detritus (channel plugs) are associated with the meanderbelt deposits. Scarcity of organic remains in related floodplain deposits may indicate deposition in a less humid climate. This observation is supported by the abundance of feldspar in the arkosic sandstones, since feldspar is less prone to chemical decomposition and is better preserved in a cold or arid climate (Fuchtbauer, 1974).

ALLUVIAL PALEOARCHITECTURE AND SUMMARY

Facies sequences and associations described for the basin margin and basin proper can be integrated into a model that shows temporal deposition of a facies in the coarsening-upward megacycles. The paleoarchitectural model displays the patterns of a paleodrainage network, including the amount of interconnection and areal distribution in a downslope direction. The alluvial paleoarchitecture and the basinal evolution within the Raton basin are discussed with the aid of the depositional models in Figures 9A and 9B.

The depositional model for the FDD facies (Fig. 9A) shows the relationships of coeval-tributary braided and meandering streams along the basin margin. The braided streams were probably controlled by increased influx of bedload sediments due to either floods from alluvial fans, proximity

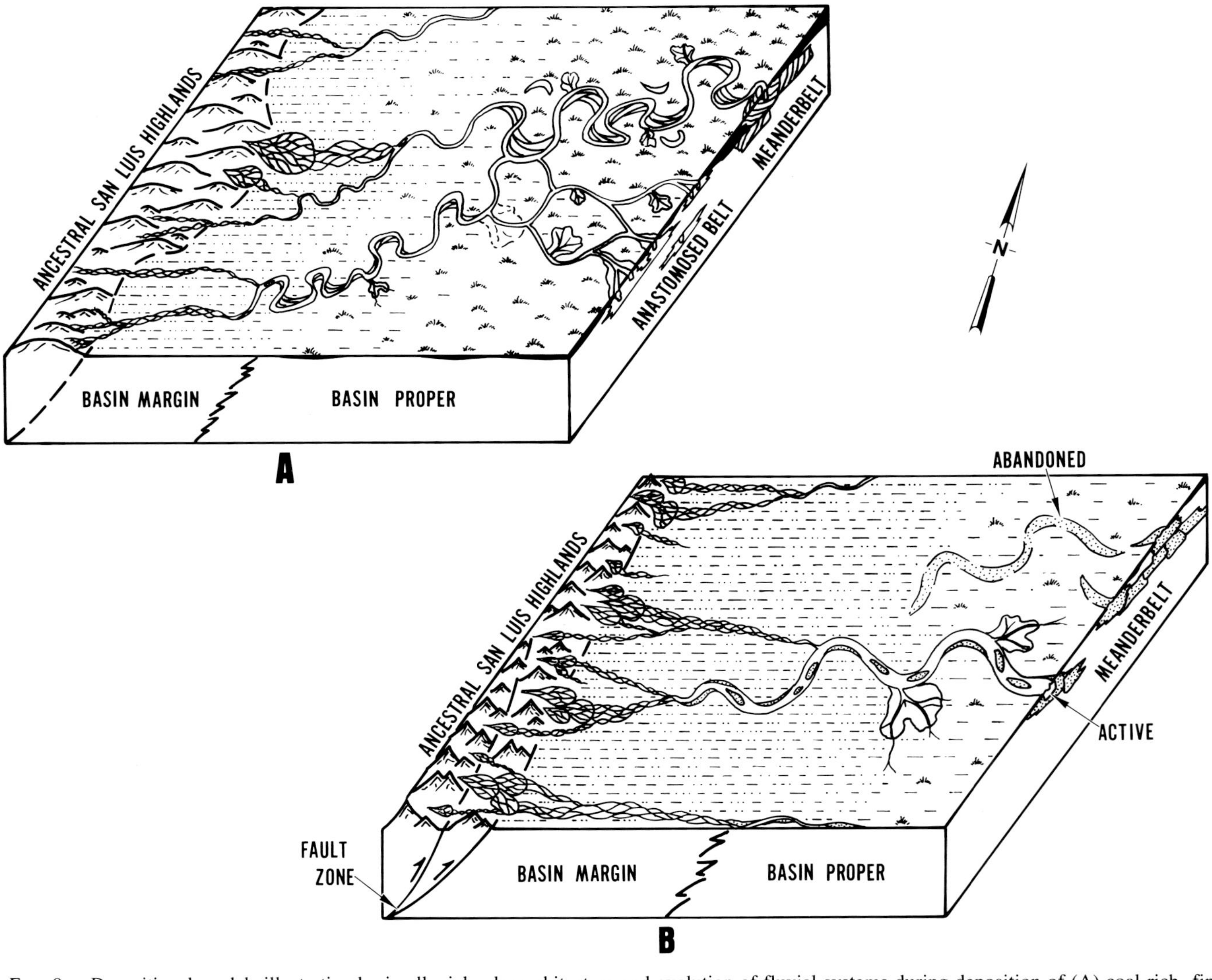

FIG. 9.—Depositional models illustrating basin alluvial paleoarchitecture and evolution of fluvial systems during deposition of (A) coal-rich, fine detritus-dominated facies and (B) coal-poor, sand-dominated facies.

to the source area, or high gradient imposed by local tectonism. In contrast, the meandering streams probably resulted from the increase of mixed-load sediments due to either distance from source area or low gradient. Sparse vegetation, as indicated by organic remains in some carbonaceous shales of floodplain deposits, may have retarded widespread braiding of bedload streams but may not have influenced bank stabilization and formation of mixed-load streams.

The tributary system in the basin margin merged downslope into the basin proper fluvial systems and peat-forming swamps and floodplains. The fluvial systems in the basin proper consisted of meandering and anastomosed streams that were choked with mixed-load and suspended-load sediments. These fluvial systems, shown in Figure 9A, may have formed contemporaneously, with either a meandering stream or an anastomosed stream being dominant. The dominance of one type of fluvial system over the other type is a function of the stage of evolution and the nature of autocyclic mechanism. That is, vertical aggradation in the alluvial plain was primarily developed by anastomosed streams and lateral aggradation by meandering streams. Lateral aggradation was dominant when the alluvial plain was near base level. Rapid vertical aggradation occurred during periods when differential compaction of sediments or basin subsidence had lowered the alluvial plain significantly below base level. The latter allocyclic mechanism was significant during deposition of the thick sequences of the FDD facies. The vertical aggradation in the alluvial plain was initiated by crevassing, which occurs when the levee of a meandering stream is breached during floods. A crevasse splay formed on the low-lying floodplain and prograded as feeder (crevasse) channels bifurcated. Progradation, bifurcation, and reunification of crevasse channels produced anastomosis. The crevasse channels served as avulsion routes (autocyclic mechanism) as progradation ensued. Vertical aggradation continued in channel and overbank-floodplain environments. Peat-forming swamps were, at this stage, restricted by the size of floodbasins and by rapid detrital influx from crevasse splays. The aggradation to near base level would transform branches of the anastomosed channels into major conduits and lead to channel migration and lateral accretion of deposits by meandering streams. Thus, the deposits of a meandering fluvial system were superimposed on older deposits of anastomosed streams (Fig. 9B). Sedimentation on the alluvial plain by a meandering stream proceeded until the next stage of crevassing and avulsion (autocyclic mechanism), together with anastomosis, was repeated.

Although peat-forming swamps occurred in both meandering and anastomosed fluvial systems, poorly drained swamps were better developed on abandoned meanderbelts. In addition, portions of the anastomosed belts, which were abandoned while an anastomosed channel was being transformed into a meandering channel, also served as platforms on which poorly drained swamps formed. These abandoned areas were ideal settings for peat-forming swamps because they were far removed from detrital influx.

Figure 9B is a depositional model of the SD facies that succeeded the FDD facies. The model suggests a downstream continuum from braided to meandering streams. The extensive braiding of bedload streams at the basin margin reflects a response to tectonic uplift (allocyclic mechanism) along a fault zone, which promoted formation of mountain-front alluvial fans that were point sources of bedload sediments. The conglomerates of the proximal braided system deposits represent distal alluvial fan deposits.

The downslope continuum of the braided streams consisted of high-bedload meandering streams choked with side channel and mid-channel bars. The abundance of bedload and lack of overbank floodplain sediments indicate development of slightly to moderately sinuous channels that were rapidly filled and abandoned. The fluvial channels developed on older channel fills and on low-lying floodplains, the positions of which were determined by the positions of previously deposited channel sands. This process of channelization is responsible for reworking meanderbelt and overbank floodplain sediments. The reworking and lateral accretion of deposits by the meandering fluvial system may have been caused by reduced basin subsidence.

The depositional models of braided and meandering fluvial systems (Figs. 9A and 9B) mainly apply to the second and third coarsening-upward megacycles. In these megacycles, the relationship of the alluvial paleoarchitecture to the presence or absence of coal or the remains of land vegetation is not consistent with that observed by Schumm (1968) and Cotter (1978). Schumm proposed that vegetation controls stream patterns by: (1) retarding erosion, altering sediment yields, and changing total runoff, discharge, and flood peaks; (2) decreasing bedload grain size and intensifying production of fine sediments; and (3) stabilizing banks. Thus, the occurrence of vegetation tends to favor development of meandering instead of braided streams. Cotter tested this hypothesis in Paleozoic rocks of the central Appalachian basin and observed changes in stream patterns consistent with the speculation of Schumm. In addition, Cotter found that, in older Paleozoic units in which land vegetation was not an important factor, a sheet-braided fluvial style dominated the entire downslope continuum.

The test of the hypothesis of Schumm and observations of Cotter on the relationship of vegetation to stream patterns yielded mixed results in our study. The development of meandering and anastomosed streams in the FDD facies in the basin proper is consistent with the observations of Schumm and Cotter. The coexistence of tributary braided and meandering streams within a short distance along the Raton basin margin, however, is an enigma. Here, the vegetation, which may be indicated by the presence of some coal beds, was insignificant in controlling the pattern of meandering streams. The downslope changes in alluvial paleoarchitecture associated with the SD facies is especially problematic. In this facies, the development of meandering streams in the basin proper in relation to the presence of some vegetation, as indicated by a few thin coal beds, is inconsistent with the observations of Schumm and Cotter. The meandering pattern and downslope change from braided to meandering streams formed despite some presence of vegetation. The sparse vegetation in the SD facies of the second megacycle may be a function of either sedimentation controlled by tectonism rather than by climatic change or controlled by an oxidized environment, because the facies is enclosed by coal-bearing intervals that contain thick coal beds. The sparse vegetation of the same facies in the third megacycle, however, may be due in part to climatic change.

Finally, basin subsidence (allocyclic mechanism) and fluvial avulsion and abandonment (autocyclic mechanism) are the major factors in controlling the accumulation of thick coal beds in an alluvial environment. Although coal beds formed in swamps of meandering and anastomosed fluvial systems, the meandering system offered the ideal setting, where abandoned meanderbelts served as swamp platforms free of detrital influx. Extensive detrital influx caused by tectonic uplift of the source area (allocyclic mechanism) chocked peat-forming swamps.

ACKNOWLEDGMENTS

We thank R. C. Johnson of the U.S. Geological Survey, H. G. Churnet of the University of Tennessee, and R. G. Goodwin of the University of Nebraska for their technical reviews. We also thank V. H. Sable of the U.S. Geological Survey for her assistance in editing and preparation of the manuscript.

REFERENCES

BEERBOWER, J. R., 1964, Cyclothems and cyclic depositional mechanisms in alluvial plain sedimentation, *in* Merriam, D. F., ed., Symposium on Cyclic Sedimentation: Kansas State Geological Survey Bulletin 169, p. 31–42.

CLOUGH, S. R., 1983, Facies development and evolution of fluvial channels in the Cumberland marshes of Saskatchewan, Canada: Unpublished M.S. Thesis, University of Illinois at Chicago, 80 p.

COTTER, EDWARD, 1978, The evolution of fluvial style, with special reference to the central Appalachian Paleozoic, *in* Miall, A. D., ed., Fluvial Sedimentology: Canadian Society of Petroleum Geologists Memoir 5, p. 361–383.

FLORES, R. M., 1984, Comparative analysis of coal accumulation in Cretaceous alluvial deposits, southern United States Rocky Mountain basins, *in* Stott, D. F., and Glass, D. J., eds., The Mesozoic of Middle North America: Canadian Society of Petroleum Geologists Memoir 9, p. 373–385.

FUCHTBAUER, HANS, 1974, Sediments and sedimentary rocks I: John Wiley and Sons, New York, 464 p.

HEWARD, A. P., 1978, Alluvial fan sequence and megasequence models—With examples from Westphalian D-Stephanian B coalfields, northern Spain, *in* Miall, A. D., ed., Fluvial Sedimentology: Canadian Society of Petroleum Geologists Memoir 5, p. 669–702.

LEE, W. T., 1917, Geology of the Raton Mesa and other regions in Colorado and New Mexico, *in* Geology and Paleontology of Raton Mesa and other regions in Colorado and New Mexico: U.S. Geological Survey Professional Paper 101, p. 2-221.

LEOPOLD, L. B., AND WOLMAN, M. G., 1957, River channel patterns—Braided, meandering, and straight: U.S. Geological Survey Professional Paper 282-B, p. 39–85.

MCLEAN, J. R., AND JERZYKIEWICZ, T., 1978, Cyclicity, tectonics, and coal—Some aspects of fluvial sedimentology in the Brazeau-Paskapoo Formations, Coal Valley area, Alberta, Canada, *in* Miall, A. D., ed., Fluvial Sedimentology: Canadian Society of Petroleum Geologists Memoir 5, p. 441–468.

OUELLETTE, R. G., AND BOUCHER, D. A., 1983, Frontal thrust structure, south-central Colorado, *in* Bally, A. W., ed., Seismic Expression of Structural Styles: American Association of Petroleum Geologists Studies in Geology Series No. 15, p. 3.2.2-25–3.2.2-28.

PILLMORE, C. L., AND FLORES, R. M., 1984, Field guide and discussions of coal deposits, depositional environments, and the Cretaceous-Tertiary boundary, southern Raton Basin, *in* Lintz, J., Jr., ed., Western Geological Excursions: Geological Society of America Annual Meeting, v. 3, p. 1–51.

———, TSCHUDY, R. H., ORTH, C. J., GILMORE, J. S., AND KNIGHT, J. D., 1984, Geologic framework of nonmarine Cretaceous-Tertiary boundary sites, Raton basin, New Mexico and Colorado: Science, v. 223, no. 4641, p. 1180–1182.

———, AND FLORES, R. M., 1987, Stratigraphy and depositional environments of the Cretaceous-Tertiary boundary interval and associated rocks, Raton basin, New Mexico and Colorado, *in* Fassett, J. E. and Rigby, K. B., Jr., eds., Cretaceous-Tertiary Boundary of the San Juan and Raton Basins, Northern New Mexico and Southern Colorado: Geological Society of America Special Paper 209, p. 111–130.

SCHUMM, S. A., 1968, Speculations concerning paleohydrologic controls of terrestrial sedimentation: Geological Society of America Bulletin, v. 79, p. 1573–1588.

SMITH, N. D., AND CROSS, T. A., 1985, Avulsion-controlled fluvial evolution—The Cumberland marshes, east-central Saskatchewan (abs.): International Fluvial Sedimentology Conference, Colorado State University, Fort Collins, p. 36.

STEEL, R. J., AND AASHEIM, S. J., 1978, Alluvial sand deposition in a rapidly subsiding basin (Devonian, Norway), *in* Miall, A. D., ed., Fluvial Sedimentology: Canadian Society of Petroleum Geologists Memoir 5, p. 385–412.

———, MAEHLE, S., NILSEN, H., ROE, S. L., AND SINNANGR, A., 1977, Coarsening-upward cycles in the alluvium of Hornelen Basin (Devonian) Norway—Sedimentary response to tectonic events: Geological Society of America Bulletin, v. 88, p. 1124–1134.

WILLIAMS, P. F., AND RUST, B. R., 1969, The sedimentology of a braided river: Journal of Sedimentary Petrology, v. 39, p. 649–679.

AN APPLICATION OF STATISTICAL MODELLING TO THE PREDICTION OF HYDROCARBON RECOVERY FACTORS IN FLUVIAL RESERVOIR SEQUENCES

CHRISTOPHER R. FIELDING[1] AND RICHARD C. CRANE

Sedimentology Branch, BP Petroleum Development Ltd., Britannic House, Moor Lane, London EC2Y 9BU, UNITED KINGDOM

ABSTRACT: Fluvial channel sandstones commonly form hydrocarbon reservoirs in oil- and gas-bearing basins. Where such sands were the product of deposition in broad, extensive channel systems, their lateral extent is often greater than the area of the producing oil or gas field. In such situations, lateral pinchout of reservoir sandstone bodies is unlikely to reduce the gross reservoir volume of the field. Where fluvial sands were deposited in more laterally restricted, perhaps meandering channel belts, however, lateral pinchouts are common. In the case of laterally restricted fluvial reservoirs, the spacing and distribution of producing wells becomes critical, particularly where numbers of laterally offset, "shoestring" sandstone bodies within a vertical sequence are isolated from each other by floodplain/floodbasin fines.

A method has been developed for predicting the proportion of laterally discontinuous fluvial reservoir sandstones that can be accessed by specified development well patterns. The method is based on (1) relationships between channel belt width, channel depth and deposit thickness for various fluvial regimes; and (2) a statistical description of reservoir sandstone beds encountered in previously drilled wells, and is designed to be generally applicable. A Fortran IV computer program has been written to allow rapid and easy usage of the method.

INTRODUCTION

Fluvial channel sandstones have been found to constitute oil and gas reservoirs in a number of hydrocarbon provinces, particularly in North America (e.g., Brown, 1979; Putnam & Oliver, 1980; Mossop & Flach, 1983). Most fluvial channel deposits occur in laterally confined, elongate belts, recording the limited capability of the depositing channel to expand in a direction perpendicular to its paleoflow. There is, however, considerable variation in modern fluvial channel morphology (Allen, 1965; Fig. 1) and consequently in the geometry of ancient channel belt deposits. In the development of oil and gas fields with fluvial reservoirs, particularly those with multiple pay intervals, questions arise as to the continuity of those beds across the area of the field and the likely effectiveness of planned development well patterns (Fig. 2).

Modern fluvial channels appear to obey certain "natural laws" which, while complex, allow an understanding of the processes and parameters which control channel morphology and ultimately, deposit geometry. Based on observations of channel dimensions and geometry, Carlston, Schumm and others were able to define equations which linked those parameters for meandering and other channels to their hydrological characteristics (reviewed by Collinson, 1978). Extending this concept, Collinson (1978) suggested that, using such equations, the width of ancient channel belt deposits could be predicted from a knowledge of channel bankfull depth or sandstone body thickness.

This paper aims to further expand the concept of channel belt width prediction from basic sandstone thickness data, by examining available dimensional information on modern and ancient fluvial channel deposits. The results of this study, together with simple statistical techniques, have been applied to distributions of sandstone bed thickness within exploration wells to develop a model for predicting the efficacy of a planned pattern of development wells in an oil/gas field.

[1]Present address: Department of Geology and Mineralogy, University of Queensland, St. Lucia, Queensland, Australia 4067

RELATIONSHIP BETWEEN SAND BODY GEOMETRY AND FLUVIAL CHANNEL TYPE

The first stage in the construction of the predictive model is the establishment of sandstone thickness versus deposit width relationships for various types of fluvial channel. If the thickness of a channel sandstone could be said to be equivalent to the depth of its formative channel, that thickness could be used directly to predict channel belt width from relationships described below. For most ancient fluvial deposits this is not the case, however, since channel deposits accrete in the vertical sense with time, as well as spreading laterally. Collinson's (1978) published figure 2, which relates ancient channel sandstone thickness to channel belt width using a 1:1 channel depth: preserved sandstone thickness ratio, is therefore rather oversimplified.

In some cases, it may be possible to accurately determine original channel depths in subsurface wells (e.g., from continuous core, or nearby outcrops). A recent predictive study by Lorenz and others (1985) employed such a technique. For instances where such detailed data are not available, however, a thickness: depth relationship has been distilled from published information (Fig. 3). Sandstone thicknesses determined from wireline logs may then be assigned a formative channel depth.

Published data relating modern and ancient fluvial channel belt widths to channel depths have been compiled for various channel morphological types (Figs. 1, 4) ranging from braided systems through meandering to anastomosed, straight and incised channels. Broad positive relationships between channel depth and channel belt width are identifiable, which are to some extent governed by the different channel types. Five such relationships have been distilled from the data set to represent the spectrum of fluvial channels. These have been named Cases 1a, 1b, 2a, 2b, 3 (Fig. 4).

Cases 1a and 3 are, respectively, the least and most optimistic situations with regard to hydrocarbon recovery, as they represent the lowest and highest width/depth ratios of all the collected data.

From a sandstone body thickness as estimated from wireline logs, it is thus possible to estimate channel belt (i.e.,

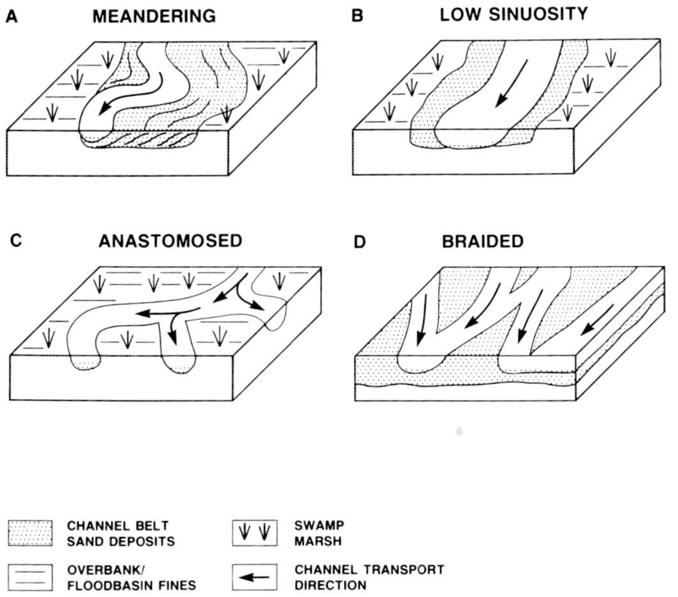

FIG. 1.—Three-dimensional block models showing possible fluvial systems and the morphology of sand bodies which might be deposited by them. A) meandering stream, B) low sinuosity stream, C) anastomosed stream, and D) braided stream. Each has different implications for reservoir geometry.

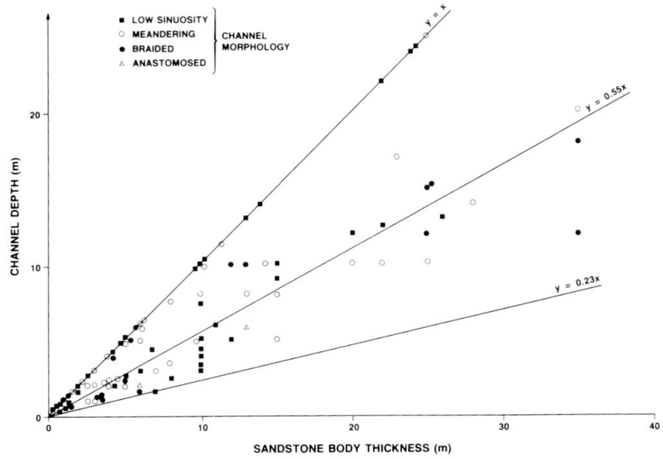

FIG. 3.—Channel depth vs. sandstone bed thickness relationships for deposits of various fluvial channel types. The two outer lines enclose the limited available data compiled from published studies. A median line (channel depth = 0.55 sandstone bed thickness) is taken as a reasonable representation of the data. Where more reliable data are available on the local relationship between the two parameters in any study, these should be used to generate the required figures in preference to the crossplot.

sandstone body) width for various potential channel morphological cases (Figs. 3, 4). To some extent, the range of cases considered should reflect the degree of confidence in the sedimentological interpretation of the considered reservoir sequence. Any available data on the interval of interest should be used in channel belt width computations.

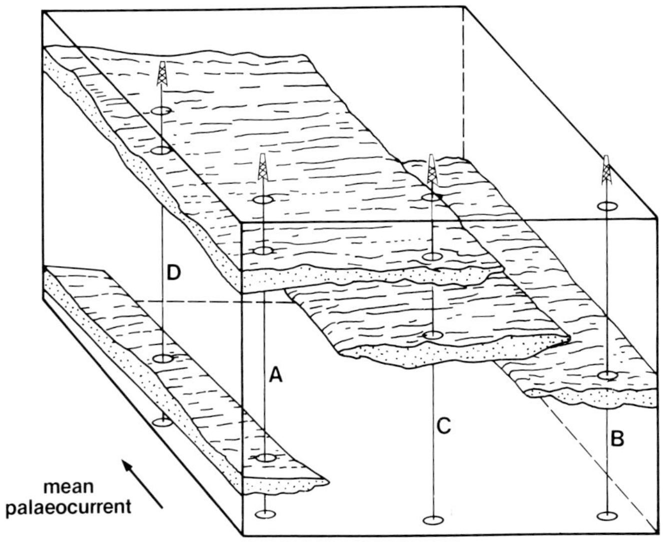

FIG. 2.—Three-dimensional block diagram illustrating the relative effectiveness of different well spacings in penetrating isolated, laterally restricted, fluvial reservoir sandstones. The channel belt sandstone bodies are of the order of 100's of meters wide and several meters thick. Wells A and B collectively penetrate only three of the four sandstones illustrated. A closer well spacing entails a third well, C, which penetrates the fourth sandstone. Well spacing must be measured perpendicular to paleocurrent; wells A and D penetrate the same sand bodies because they overlie one another with respect to paleocurrent.

The graphs of Figure 3 and 4 relating preserved sandstone bed thickness to formative channel belt width via channel depth have been combined in Figure 5 for a generalized case. This is Case 2a of Figure 4 and the median line of Figure 3.

THE STATISTICAL MODEL

The next stage in the development of the model is the assembly of a statistical description of sandstone beds in the intervals of concern, in terms of bed thickness and width. Assuming that continuous core is not available for the interval of interest, wireline logs may be employed as the basis for geological interpretation, calibrated by any cores which may be available.

For many sequences, a gamma ray log, which measures natural gamma ray emissions from rocks downhole, represents the closest imitation of sediment grain size. In some instances, however, this will not be the case, perhaps due to high concentrations of (radioactive) *K*-feldspar in sandstones; in such cases an alternative wireline log such as neutron porosity, spontaneous potential, or resistivity should be used.

Well results must first by simplified to block logs describing only two facies: (1) sandstones of 2 m or greater thickness, and (2) all other sediments, predominantly siltstones and claystones with thin sandstones (Fig. 6). The 2-m cutoff is employed so as to eliminate thin, non-channel sandstone (crevasse-splay deposits). Thicker non-channel sandstones of interdistributary minor mouth bar or levee origin are common in coastal plain sequences (Fielding, 1984). These may be distinguished from channel deposits by their coarsening- or "cleaning" upward character on well logs and eliminated from the model (channel deposits typically show "blocky," "spiky," or fining-upward log responses; Fig. 6). Other sandstones, which for whatever reason (low

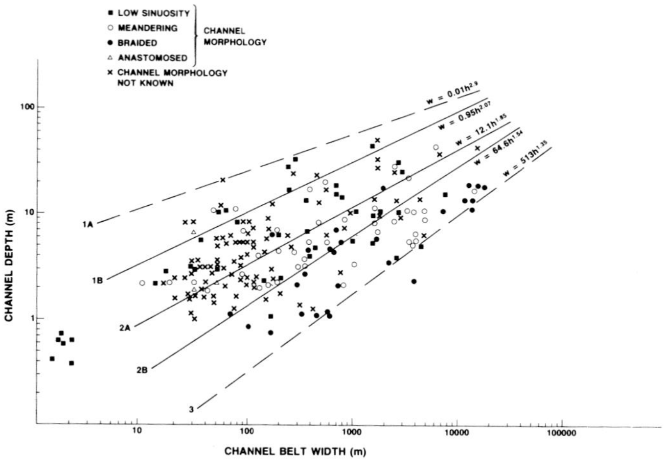

FIG. 4.—Channel depth vs. channel belt width relationships for various types of modern and ancient fluvial channel deposits (data from 45 published sources).

Case 1a: the upper limit of all data collected (i.e., lowest width/depth ratio)—channel belt width (m) = (0.01 channel depth)$^{2.9}$. This relationship describes deposits of incised, straight and non-migrating channels and is an extreme case.

Case 1b: the upper bounding line for meandering channel deposits (i.e., lowest width/depth ratio)—channel belt width (m) = 0.95 (channel depth)$^{2.07}$. This relationship also describes deposits of channels which have not been allowed to migrate laterally, but excludes highly incised and narrow, vertically accreted types.

Case 2a: the best fit line for all the collected data—channel belt width (m) = 12.1 (channel depth)$^{1.85}$. This relationship is a geometric mean of all fluvial channel types, and describes a situation where a variety of channel types are said to exist or alternatively may be used when the formative channel type is not known.

Case 2b: the published empirical relationship for modern, truly meandering streams (Collinson, 1978)—channel belt width (m) = 64.6 (channel depth)$^{1.54}$. This relationship describes deposits of channels which have been allowed to develop fully meandering profiles.

Case 3: the lower bounding line of all data collected (i.e., highest width/depth ratio)—channel belt width (m) = 513 (channel depth)$^{1.35}$. This relationship describes deposits of laterally unrestricted, (i.e., braided) fluvial systems.

Cases 1a and 3 are, respectively, the least and most optimistic situations with regard to hydrocarbon recovery, as they represent the lowest and highest width/depth ratios of all the collected data.

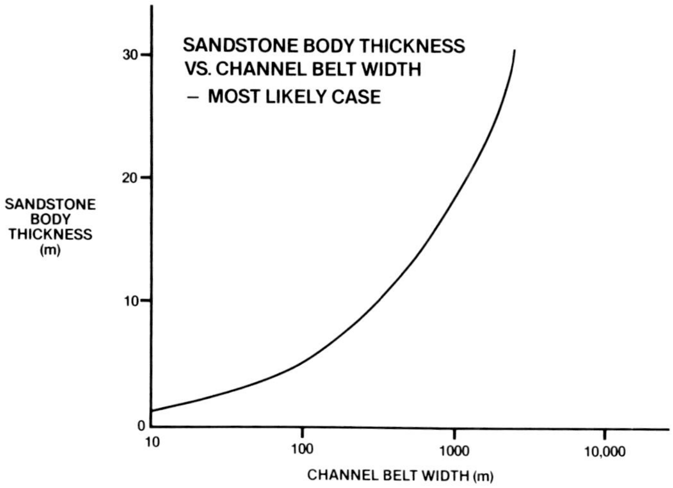

FIG. 5.—Sandstone bed thickness vs. width perpendicular to paleocurrent (generalized case, constructed by using Case 2a from Fig. 4 and the median line of Fig. 3).

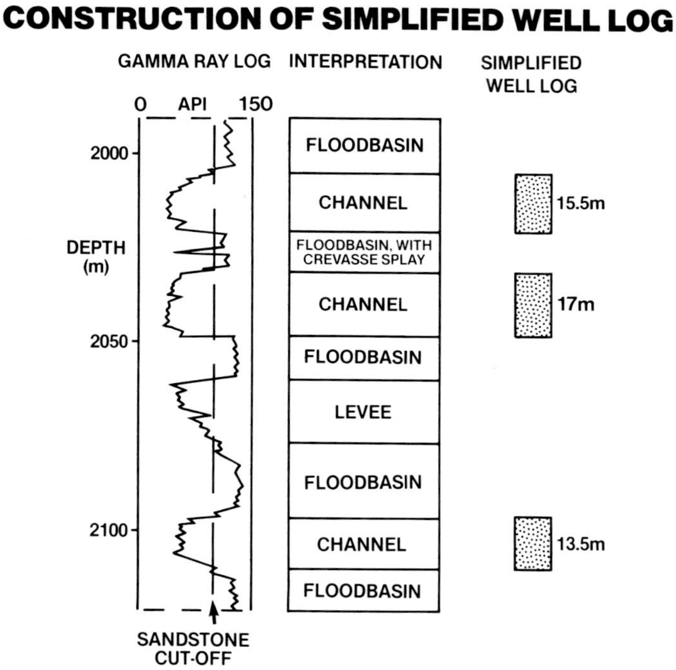

FIG. 6.—Example of a hypothetical fluvial sequence, showing the method of construction of simplified block logs. In this case sandstones are first identified by using an independently determined cutoff value on a Gamma Ray log. Sandstones of non-channel origin are then eliminated, leaving the three channel sandstone bodies displayed along with their thicknesses in the right-hand column.

porosity, high clay content) are considered non-reservoir, can also be removed from the model at this stage.

Multiplying the thickness of an identified reservoir bed by its predicted channel belt width gives the cross-sectional area of that sandstone perpendicular to paleocurrent. Cross-sectional areas of all the sandstones identified in control wells may then be calculated and totalled (Fig. 7a). If, in a given sequence, there are n sandstone beds $s_1, s_2, \ldots\ldots s_{n-1}, s_n$ with thicknesses $t_1, t_2, \ldots\ldots t_{n-1}, t_n$ meters and cross-sectional widths $w_1 \ldots\ldots w_n$ meters, then the total sandstone cross-sectional area in communication with the test wells is $\Sigma_{i=1}^{n} t_i w_i$ (Fig. 7a).

The total cross-sectional area of sandstone existing in the succession is dependent on their statistical distribution. Given that wells previously drilled in an area have penetrated a representative sequence, then any hypothetical well drilled through the interval of interest will penetrate a set of sandstone beds equivalent to those found in the test wells (Fig. 7b). Thus, between two proposed wells of spacing W meters, there exists a total sandstone cross-sectional area of $\Sigma_{i=1}^{n} t_i W$.

The "success rate" for penetrating sandstones in the interval of interest between two proposed wells a certain dis-

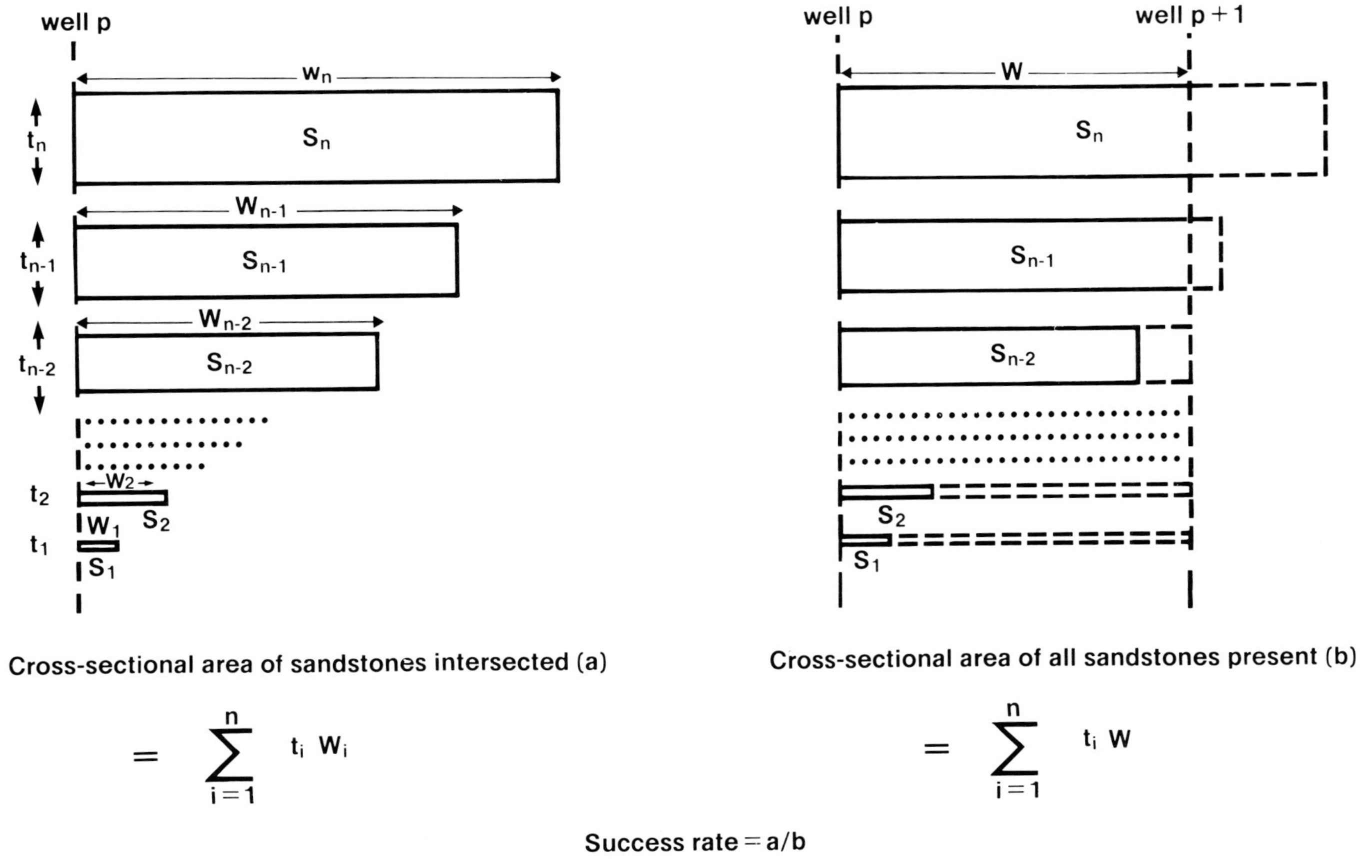

FIG. 7.—The statistical description of sandstone beds in the average well, based on log data from available wells.

(a) Cross-sectional area of sandstones intersected. The average well p intersects n channel belt sand bodies. Their individual widths (w) are computed from the observed thickness (t) using graphs of Figures 3 and 4. The total cross-sectional area of channel belt sandstone bodies perpendicular to paleocurrent is then equal to the sum of all individual cross sections. For this computation the order of occurrence of the sandstones in the well is of no significance.

(b) Cross-sectional area of all sandstones present. The average spacing between proposed wells is W, and across this distance the distribution of sandstones fits the average statistics of the field. The total cross-sectional area is thus equal to the thickness of each sandstone times W, since sand bodies of each thickness are inferred to occur everywhere between the wells.

The success rate or proportion of the total reservoir sandstone volume in a given interval which will be penetrated by a planned pattern of production wells of a certain spacing may then be calculated by dividing the total (a) by the total (b).

tance apart is then calculated by dividing the cross-sectional area of sandstone hit by the total cross-sectional area of sandstones in existence. Since cross-sectional area is proportional to volume, the "success rate" may also be regarded as representing reservoir volumes. It is possible that some of the channel sandstone bodies in a given reservoir interval will be sufficiently wide that they cover more than the distance between two adjacent wells. This must be compensated for in the model by limiting the width of sandstones considered to the distance between adjacent wells (in a direction perpendicular to that of regional paleoflow).

$$\text{Then success rate} = \frac{\sum_{i=1}^{n} t_i \text{ minimum } (w_i, W)}{\sum_{i=1}^{n} t_i W}$$

where "success rate," expressed as a percentage, is defined as *the proportion of the total channel belt sandstone reservoir volume in a given interval which will be penetrated by planned development wells a certain distance apart* (cf. Fig. 2).

It is important to note, though, that further sandstones could be in hydraulic communication with wells through erosional interconnections between channel belt deposits. Theoretical simulation studies indicate that meandering channel belt sandstone bodies are essentially isolated if they comprise less than about 50 percent of the succession (Allen, 1978; Crane, 1982; Fig. 8). Studies of Eocene fluvial deposits in the Ebro basin by Atkinson (pers. commun.), however, suggest that sequences with far lower channel sand fractions may possess considerable interconnection between channel sandstones. In the present study, consideration of interconnection is based on theoretical studies only, and it is acknowledged that the current data base is poor.

For field development cases involving secondary and/or tertiary hydrocarbon recovery, the equation for "success rate" may be rewritten as

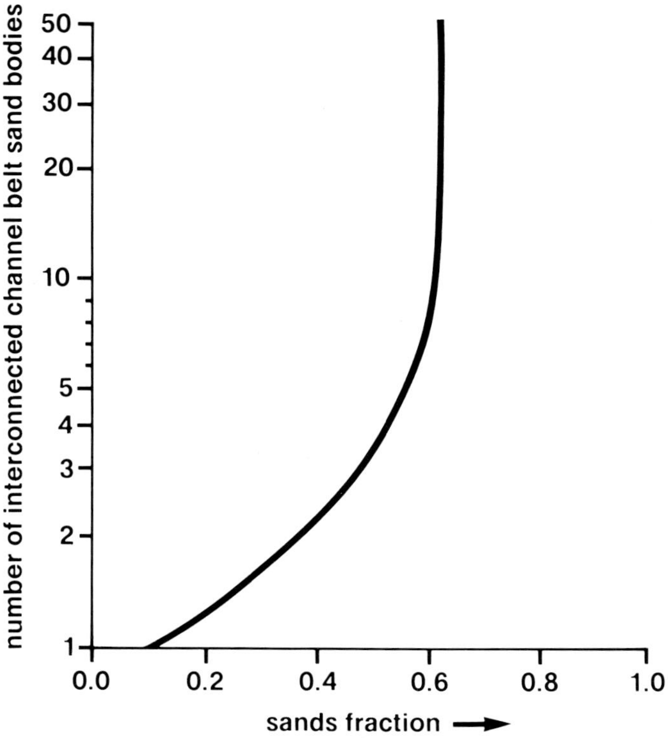

FIG. 8.—Sandstone body interconnection as a function of the net sands fraction of the sequence (after Crane, 1982). The ordinate denotes the average number of channel belts in one interconnected domain. A domain is a set of channel belt sandstones which, during deposition, have eroded into each other. The abscissa records the sands fraction, which is the quantity of channel belt sandstones as a fraction of the total sequence thickness.

$$\frac{\sum_{i} (w_i \geqslant W)\, t_i W}{\sum_{i=1}^{n} t_i\, W}$$

since for effective recovery, reservoir sand bodies must be simultaneously penetrated by at least two development wells.

A VAX computer has been programmed to execute the above computations enabling rapid delivery of results and allowing a number of sensitivities to be easily tested, including:

1. variable well spacing,
2. variable channel belt orientation (and thus effective spacing in a planned well pattern),
3. variable channel morphology (and hence sand body width/thickness ratio),
4. the presence of multistory sandstone bodies. Channel interconnection has also been considered, and lateral thickness variations and the possible presence of fine-grained channel "plugs" are accounted for by the statistical description of test wells.

The model does have a number of limitations, however, which are worth pointing out:

1. Channel belt "wings" or overbank/minor delta deposits, the presence of which could increase the lateral extent and connectivity of channel belt sandstones, have not been incorporated into the model since they are very difficult to quantify.
2. It is assumed that sandstone bodies of a predicted width are homogeneous units with respect to fluid movement and this need not always be the case.
3. Because channel belts are themselves often sinuous, individual reservoir sandstone bodies may intersect with each other at some point up or down paleocurrent, thereby increasing connectivity.
4. The effect of faulting, which, depending on the throws and angles of throw of faults and sands fraction of the sequence, could either increase or decrease the success rate, has not been modelled in this study.
5. The statistical distribution of sandstone bodies is specified by the method used but not the actual location in space of any particular sandstone body.
6. The method is presently applicable only to a purely gravity-driven fluvial system where there is no disruptive tectonic control on deposition across the area of interest (were such an influence demonstrable, the model framework would require adjustment).

The accuracy and resolution of the method are dependent on a number of factors, including: the number of test wells available; the validity of the chosen depth vs. width, and thickness vs. width relationships; and the user's choice of most appropriate depth vs. width relationship. Here, the sedimentological interpretation of a sequence, in terms of environmental setting and channel type, whether based on core, log or even regional geological data, is of the utmost importance in defining the operating parameters for the model. It is essential that any application of this model to fluvial reservoir sequences should first take into account their sedimentologic context.

A HYPOTHETICAL CASE STUDY

To demonstrate the application of the model, a hypothetical example has been devised and examined. The hydrocarbon accumulation of interest is contained within multiple reservoir beds of fluvial channel origin, which are interbedded with fine-grained lithologies and occasional sandstones of non-channel origin. Figure 9 shows simplified well logs for the five exploration and appraisal wells drilled in the field; on these logs the producing formation has been reduced to reservoir and non-reservoir facies, and sandstones believed to be of non-channel origin have been deleted.

The individual sandstone beds vary in thickness from 2 to 65 m, suggesting a variety of channel types and channel belt dimensions. Some of the thicker units may be composite bodies but are treated as the product of deposition in a single channel belt for the purposes of this example. Figure 9 also shows the net "sands fraction" of the reservoir sequence in each well, which varies from 0.23 to 0.33. Individual reservoir sandstones may therefore be considered to be isolated within non-reservoir facies.

Figure 10 illustrates the planned pattern of development wells for the hypothetical field. Dipmeter and orientated

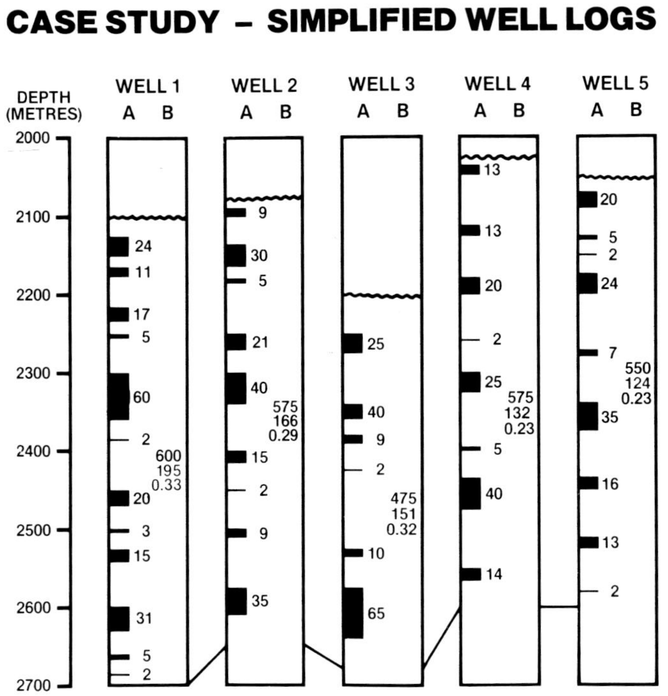

FIG. 9.—Simplified block logs for the five exploration and appraisal wells in a hypothetical oil/gas field. Only inferred channel belt sandstones are shown in column A with their thickness in meters. Column B gives (from top to bottom) (1) total reservoir formation thickness, (2) total reservoir sandstone thickness, and (3) net sands fraction (2 divided by 1).

core data have suggested a northerly inclined paleoslope, and in a north-south direction the maximum distance between wells is 875 m. Allowing for uncertainty in the interpretation of regional paleoflow, and for local variability, the maximum distance between wells in any direction is 950 m. These spacings are used as input parameters to the

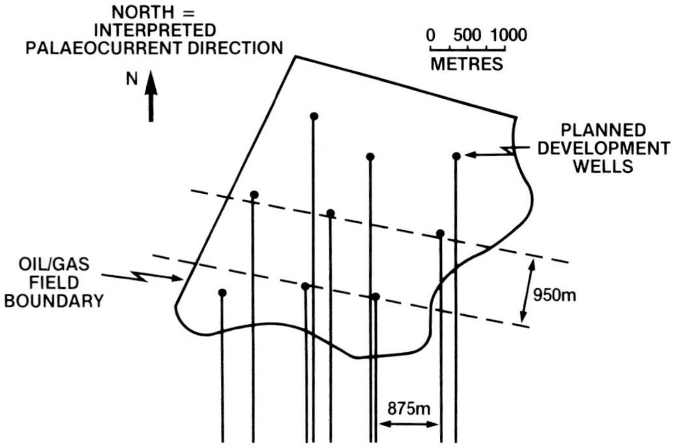

FIG. 10.—Map of the hypothetical oil/gas field, showing the locations of planned development wells (black circles). By projecting lines in a down-paleocurrent (i.e., north-south) direction, the effective well spacings may be determined. The maximum spacing in this orientation is 875 m. Varying the orientation of the projected lines to account for the possibility of a variable paleocurrent direction gives a maximum spacing between wells of 950 m.

TABLE 1.—COMPUTED SUCCESS RATES FOR THE CASE STUDY, SHOWING THE EFFECT OF VARIABLE WELL SPACING. WELL SPACING IN METERS. ALL CASE VALUES IN PERCENT.

Well Spacing	Case 1A	Case 1B	Case 2A	Case 2B	Case 3
500	16	60	94	99	100
875	9	43	88	97	100
950	8	41	87	97	100
1500	5	32	80	94	100

model, together with 500 m and 1500 m for comparison.

The model has been run for all five width-depth cases (Fig. 4). The sequence in question is known to have been deposited on a coastal plain, fed by distributary channels of variable sinuosity, and so the "best fit" Case 2a may be considered a realistic "most likely" situation.

The resultant success rates (defined as "the proportion of the total reservoir volume in a given interval which will be penetrated by a planned pattern of production wells of a certain spacing") suggest that the planned well pattern will allow a large proportion of the recoverable hydrocarbon reserves to be accessed (Table 1). Increasing the density of wells would increase the success rate by 7 percent, while decreasing the number of producing wells to a spacing of 1500 m would reduce the penetrated hydrocarbon volume by 7 percent.

CONCLUSIONS

This paper describes a technique for predicting the proportion of total reservoir volume in an oil or gas field which will be penetrated by development wells a certain distance apart, where the reservoirs are laterally restricted, fluvial channel sandstones.

A series of relationships has been defined, from published data relating fluvial channel deposit width, thickness and formative channel depth for various channel morphological types. These relationships, together with measured sandstone bed thickness distributions and channel sand fractions from exploration and appraisal wells, form the basis of a statistical description of the interval of interest. From this description, the predicted total cross-sectional area of sandstone intersected by wells a certain distance apart can be calculated. Dividing this figure by the total cross-sectional area present in the sequence (also from the statistical description) gives a "success rate" for a given well spacing. Since cross-sectional area is proportional to volume, the "success rates" refer also to reservoir volume.

The model is in the form of a Fortran computer program for VAX machines, designed for rapid and easy usage, and is generally applicable to fluvial reservoir sequences.

ACKNOWLEDGMENTS

We acknowledge the permission of the British Petroleum Company to publish this paper. Valerie Muscutt, Carol McGraill, Jenny Huggett and Selina Britton of BP London are thanked for their help, encouragement and technical support. Cathy Fielding typed the manuscript. S. R. Britton, J. W. Buckee, B. G. Els, Yinan Qiu, P. J. Smith and C. A. Wright reviewed earlier versions of the manuscript.

REFERENCES

ALLEN, J. R. L., 1965. A review of the origin and characteristics of recent alluvial sediments: Sedimentology, v. 5, p. 89–192.

———, 1978, Studies of fluviatile sedimentation—An exploratory quantitative model for the architecture of avulsion-controlled suites: Sedimentary Geology, v. 21, p. 129–147.

BROWN, L. F., 1979, Deltaic sandstone facies of the Mid Continent, *in* Hyne, N.J., ed., Pennsylvanian sandstones of the Mid Continent: Tulsa Geological Society Special Publication 1, p. 35–63.

COLLINSON, J. D., 1978, Vertical sequence and sand body shape in alluvial sequences, *in* Miall, A.D., ed., Fluvial Sedimentology: Canadian Society of Petroleum Geologists Memoir 5, p. 577–588.

CRANE, R. C., 1982, A computer model for the architecture of avulsion-controlled alluvial suites: Unpublished Ph.D. Dissertation, University of Reading, 543 p.

FIELDING, C. R., 1984, Upper delta plain lacustrine and fluviolacustrine facies from the Westphalian of the Durham coalfield, NE England: Sedimentology, v. 31, p. 547–567.

LORENZ, J. C., HEINZE, D. M., CLARK, J. A., AND SEARLS, C. A., 1985, Determination of widths of meander-belt sandstone reservoirs from vertical downhole data, Mesaverde Group, Piceance Creek Basin, Colorado: American Association of Petroleum Geologists Bulletin, v. 69, p. 710–721.

MOSSOP, G. D., AND FLACH, P. D., 1983, Deep channel sedimentation in the Lower Cretaceous McMurray Formation, Athabasca Oil Sands, Alberta: Sedimentology, v. 30, p. 493–509.

PUTNAM, P. E., AND OLIVER, T. A., 1980, Stratigraphic traps in channel sandstones in the Upper Mannville (Albian) of east-central Alberta: Bulletin of Canadian Petroleum Geology, v. 28, p. 489–508.

FLUVIAL SANDSTONE BODIES AS HYDROCARBON RESERVOIRS IN LAKE BASINS

YINAN QIU, PEIHUA XUE, JINGSIU XIAO

Scientific Research Institute of Petroleum Exploration and Development, P.O. Box 910, Beijing, China

ABSTRACT: Fluvial sandstone bodies play an important role in forming hydrocarbon reservoirs, especially in lake basins. They possess more than one-half of the oil in place in known oil fields in Chinese Mesozoic and Cenozoic lake basins. The strong reservoir heterogeneity and restricted lateral continuity of the fluvial sandstone bodies, however, result in some difficulties in secondary and tertiary oil recovery.

Almost all types of fluvial deposits documented from studies of modern river patterns have been identified in ancient Chinese lake basins. The identification of depositional facies models that relate to fluvial sandstone body geometry in the basins has been possible because of close well spacing and complete coring through pay zones. Six depositional-facies models have been identified: (1) short-coursed braided river, (2) long-coursed braided river, (3) highly sinuous meandering river, (4) low sinuosity meandering river, (5) straight distributary channel river, and (6) confined-valley channel. Each depositional group is identified by its depositional character and respective heterogeneities. The depositional composition of the sandstone unit is a result of the depositional processes that took place during its formation in the stream. These processes and heterogeneities thus formed vary from one facies to another and depend on such depositional processes as lateral accretion in point bars, vertical accretion in braided bars, aggradation in confined-valley channels, and channel fill in straight distributary channels.

Although there are variations in lateral continuity of sandstone bodies among fluvial types, all are rather discontinuous in lateral dimension. Only those sandstone bodies deposited during the transition from a period of tectonic uplift to one of subsidence, and during episodes of change from lacustrine regression to transgression, result in a wide areal distribution of the size of coalesced sandstone bodies. The vertical density of channel sandstone bodies of 50 percent and 30 percent may be empirically used to predict laterally extensive and isolated fluvial sandstone bodies, respectively, but much care should be exercised in the prediction of lateral continuity of sandstone units when the density of channel sandstone bodies ranges between 30 and 50 percent.

INTRODUCTION

Fluvial sandstone bodies are important hydrocarbon reservoirs, particularly in lake basins. Clastic rock reservoirs contain more than 90 percent of the oil in place in Mesozoic-Cenozoic lake basins in China; nearly half of the reservoirs are developed in fluvial sandstone bodies. The largest oil accumulations in each lake basin, such as Daqing and Shengtuo fields, generally occur in the basin's largest ancient fluvial-deltaic deposit complex, in which units of a fluvial and deltaic origin are nearly equal in volume. Fluvial reservoirs, however, are the most common reservoir type for "secondary" oil accumulations in those parts of the eastern China faulted lake basins that have never been deeply buried. These "shallow" clastic deposits originated in streams during periods in which the lake shrank or ceased to exist. Most ancient lake basins in western China existed in an arid to semi-arid climate and oil fields developed in rocks formed in these depositional basins are almost entirely in sandstone reservoirs that originated as fluvial and alluvial fan deposits. More fluvial reservoir sandstones undoubtedly remain to be detected in the ancient lake basins of both western and eastern China.

Although to date ancient counterparts of virtually all river patterns documented from modern rivers have been discovered in ancient lake basins, they do vary in a manner specific to environmental factors and to individual basins. First, braided rivers are most extensively developed in smaller intermontane and foremontane lake basins and in major components of crosscutting "transverse depositional systems" in those larger lake basins with a steep depositional slope and short transport distance. Many of the rivers in these basins are braided along their entire course to the point of their intersection with the lake's shoreline. Second, large-scale meandering rivers are rather rare in China's ancient lake-basin deposits except in those cases where a meandering river system formed to flow longitudinally for long distances along the depression's depositional axis. Generally, sandstone bodies formed in highly sinuous streams are not volumetrically significant as petroleum reservoir rocks. Third, straight channels characterize distributary channels on deltaic plains and on terminal fans, the latter frequently formed during the "dying" stage of the lake basins. Fourth, some confined channel-fill sandstone bodies, believed by some to be similar to those of the anastomosing pattern proposed by Smith (1983), have been identified in the longitudinal system that formed during periods of active tectonism in the region of the basin. A generalized model of deposition for a lake basin is shown in Figure 1.

In many Chinese oil fields, it has been proven that the heterogeneity of fluvial reservoirs has caused some difficulties for oil recovery. In secondary or tertiary oil recovery processes, the reservoir heterogeneity can be briefly described from two prospectives: (1) in-layer heterogeneity, and (2) areal heterogeneity. They determine the vertical and lateral conformance factor of injected fluids and the effectiveness of the recovery attempt. The in-layer heterogeneity of a sandstone reservoir concerns (1) vertical variations in permeability, (2) the position of the rock unit in the sandstone body that possesses maximum values of permeability, (3) the ratio of vertical to horizontal permeability in a reservoir body, and (4) the frequency, location, and dimension of discontinuous shaly barriers to flow. Areal heterogeneity mainly concerns a sandstone reservoir's lateral continuity and directional permeability. Both types of heterogeneity are the result of the lithologic and sedimentologic architecture of the rock unit. Therefore, it is important to recognize the origin of fluvial sandstones by depositional processes so that they may be compared to characteristics of reservoir heterogeneity to permit construction of a comprehensive reservoir model that links depositional processes to the characteristics of reservoir performance. These models can be used to improve responses to production tests in fluvial sandstone reservoirs. On the basis of our experiences with

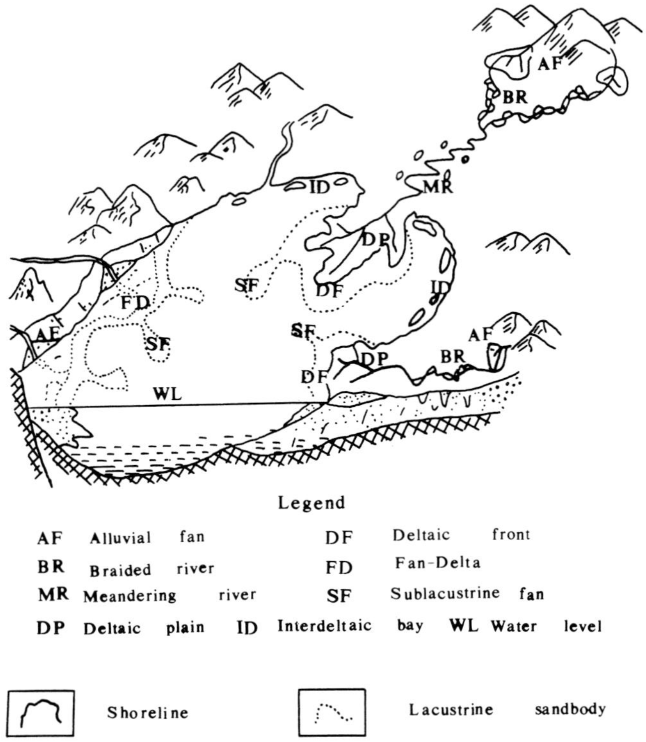

FIG. 1.—A generalized model of depositional systems in a lake basin.

examples from ancient Chinese lake basins, this paper suggests a six-fold division of depositional process-reservoir heterogeneity models for fluvial sandstone bodies. The models are: (1) meandering stream-high sinuosity, (2) meandering stream-low sinuosity, (3) braided stream-long course, (4) braided stream-short course, (5) distributary-straight, and (6) confined channel filled.

MEANDERING MODELS

The main sand units formed in meandering rivers are parts of point bars, of which the characteristic bed pattern is lateral accretion. The gradual growth of a series of lateral accretion units combines to form a point bar. In general, lateral accretion results in a vertical bar sequence in which the topographically lowest components at the unit base are transitional upward to the topographically highest components. Lag deposits at the bar base contain the bar's coarsest grains and the grain size generally becomes finer upward to the bar's overbank deposits. Consequently, the vertical variation of permeability values within a lithified point bar declines upward and the values result in a striking permeability contrast between the maximum values near the point bar base and minimum values near its top. This contrast in permeability within a point bar is very common in rock reservoirs formed in sandstone bodies of the meandering type, regardless of their sinuosity, and it usually results in waterflood performance in which injected water preferentially and rapidly advances along the reservoir's lowermost components and within a thin interval.

Mud drapes formed in the upper part of the point bar sequence as a result of deposition between two flood events are among the unit's most important lithic constituents that influence fluid flow in a sandstone body. Because mud drapes commonly occur at an oblique intersection to the regional bedding and can only be preserved in the upper part of the point bar, the geometry and number of drapes can strongly influence the lateral flow of both hydrocarbons and injected fluid.

In almost all the examples of point bar reservoirs we have studied, mud drapes have been identified but with much variation in the amount of their preservation. Under conditions of strong intermission and unexpectedness of flood events with relatively rapid rates of sediment deposition, as in low sinuousity meandering rivers, mud drapes are frequently broken into intraclasts by subsequent floods. On the other hand, in high sinuousity meandering rivers characterized by relatively infrequent floods and relatively slow rates of deposition, drapes are commonly preserved. Although mud drapes frequently hinder the effectiveness of waterfloods, they may increase the volume of reservoir swept in its upper units during gas injection tests.

The lateral continuity of sandstone units in a genetic unit is essentially determined by the width of the river's meander belt. Obviously, the width/thickness ratio of a highly sinuous sandstone body is greater than that of a relatively straight unit.

Case 1. The highly sinuous sandstone bodies—Subzone PI2 in the eastern part of Sardu field, Daqing oil province.—

A typical cored point-bar sequence is illustrated in Figure 2. A sharp base with an erosional contact with the underlying rock is common. Overlying the erosion surface are lag deposits; the medium-grained sandstone contains scattered pebbles derived from the sediment of the source area. The unit contains a few intraformational pebbles of mudstone. The subrounded shape and scarcity of mud pebbles imply that the pebbles have been transported an uncertain distance and that rate of deposition of the sediments was rather slow. Thereafter, fine- to medium-grained sandstone with large-scale trough or tabular crossbedding, fine-grained sandstone with small-scale crossbedding, and siltstone to fine-grained sandstones with ripple crossbedding are developed in ascending order. Uppermost units contain structureless mudstone, pedogenetic carbonate concretions and rootlets that collectively indicate formation on a floodplain.

Internal values of permeability occur in a sequence of gradually decreasing-upward numbers. Maximum values are located along the unit base within the coarsest lag deposits. The maximum permeability (in lowermost interval)/minimum permeability (commonly in topmost interval) ratio, which is usually called "permeability contrast" in reservoir engineering, ranges from 8 to 50, and is 10–20 in general. The "variation coefficient of permeability" (kv)* is about 0.6–1.3 and mostly 0.8–1.0

A complete point-bar sequence is usually 5 to 7 m thick, and it can commonly be divided into 3 to 5 subcycles. A

---

*note: $Kv = bk/\bar{K}$, bk is mean square deviation and $\bar{K}$ is average value of permeability.

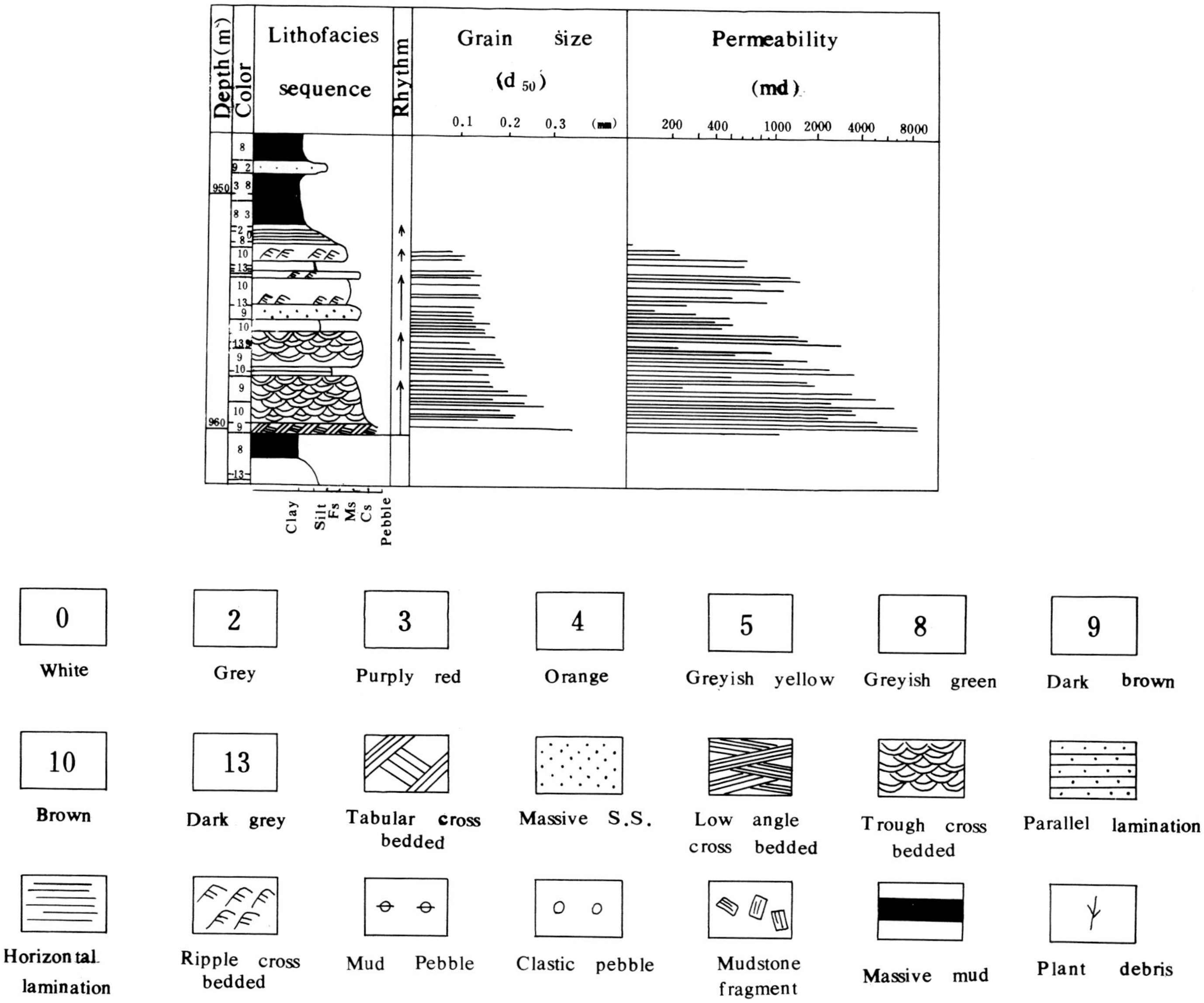

FIG. 2.—Lithofacies sequence of high sinuosity meandering sandstone body (Subzone PI_{1+2}, Well ZJ3-23, Daqing).

trend of thinning in bed thickness (3.0–0.1 m), an upward fining in grain size for each subsequent subcycle, and the presence of silty and clayey streaks between the subcycles in the upper half of the bar sequence collectively indicate that the cycles are as a series of continuously deposited lateral accretion bodies and that the mud streaks are the preserved parts of intercalated inclined mud drapes (Fig. 2).

We assume an "equal elevation of the top bed surface underlying the nearest marker unit" and an analogous response of logging to utilize drill hole logs from closely spaced development wells in the identification of sandstone units formed in individual meander belts. With this method of identification of sandstone units on drill hole geophysical logs, we suggest that the subzone PI2, as shown in Figure 3, is characterized by a rather simple sandstone geometry and that it has good continuity. Our interpretation is that a genetic unit of meander belt is about 800–1000 m wide with a width/thickness ratio 130:170. We assume that the thickness of a complete point bar sequence is 5–7 m and that it is a measure of the bank full depth of the ancient PI2 river. We further assume 5°–10° as the dip angle of the lateral accretion surface, which indicates that the river's bank full width ranged from 90 to 120 m. Thus, the width of the meander belt is about 10–15 times as wide as the channel width. Our inference is that the course of the ancient PI2 river was highly sinuous.

Case 2. Meandering sandstone bodies of low sinuosity—Subzone PI2–3 in the western part of Sardu field, Daqing oil province.—

The vertical rock sequence illustrated in Figure 4 is interpreted to be that of a typical point bar similar to that of Subzone PI2. Important differences can be distinguished between the two cases, however. In the bar's lag deposits, there are many intraformational pebbles associated with the pebbles derived from the sediment of the source area. Most

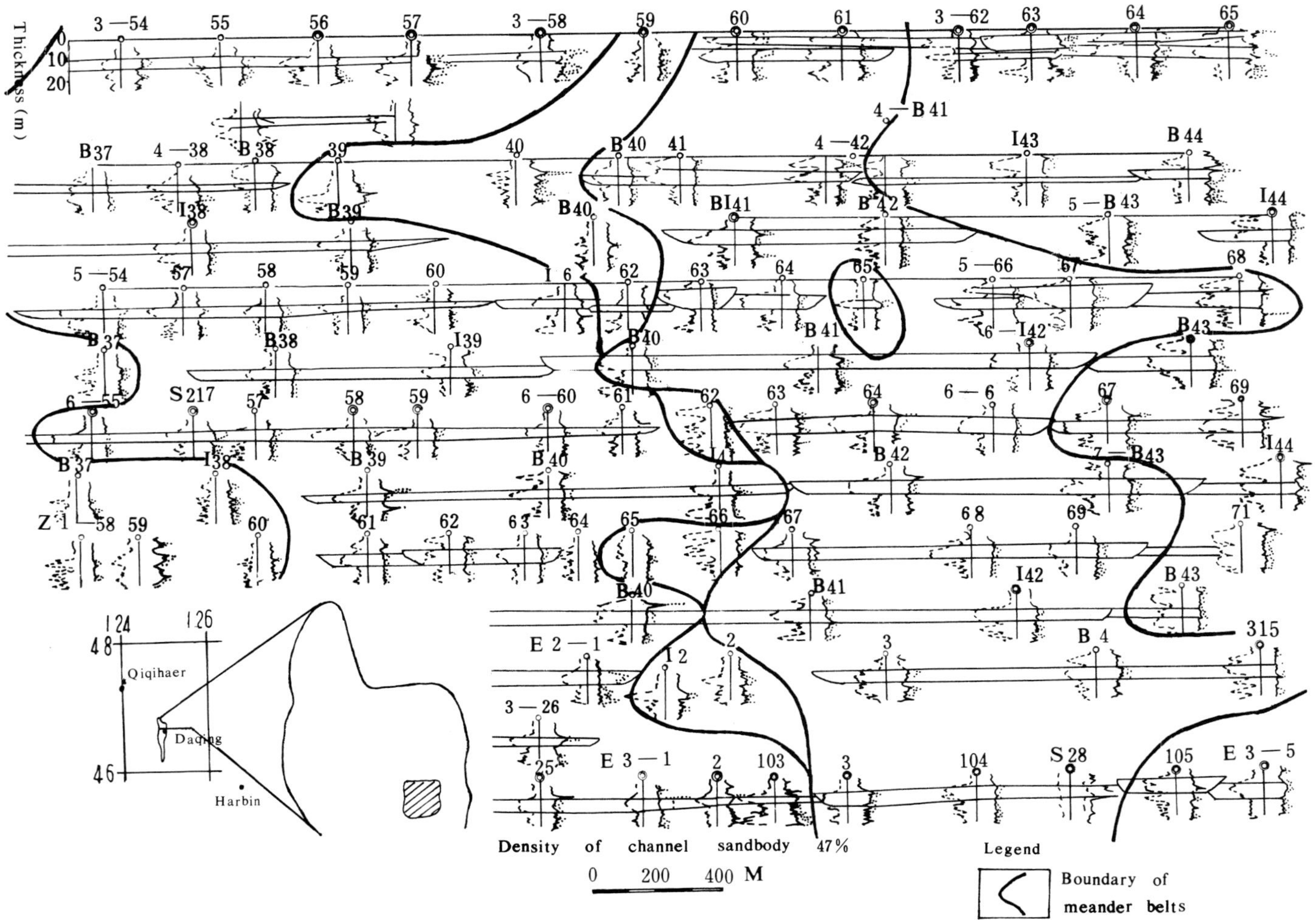

FIG. 3.—Geometry of high sinuousity meandering sandstone body, (Subzone PI_2, Daqing).

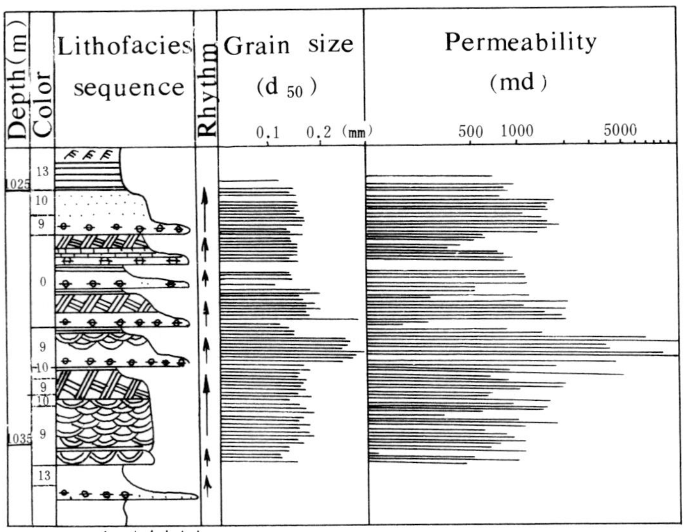

FIG. 4.—Lithofacies sequence of low sinuousity meandering sandstone body (Subzone PI_{2+3}, Well L7-28, Daqing).

intraclasts are composed of mudstone breccia, and their diameters range to as much as 7–8 cm; no regular grain fabric is evident. Intraclasts are abundant at the base of the point bar and through the lower half of the bar sequence. The clasts clearly mark the boundaries between lateral accretion bodies (subcycles). The close association of rock type and sedimentary structures between most intraclasts and mud streaks preserved between some lateral accretion bodies probably indicate that the bar intraclasts were derived from previously deposited mud drapes. The composition of some mudstone fragments is identical to that of the structureless mudstone of flood plain deposits and is most likely derived from bank caving. In addition to brecciated mud clasts, an additional distinctive feature of low sinuosity meandering sandstone bodies is the presence of an irregular and high rugged erosional basal scour surface, as interpreted from the correlation of drill hole logs from closely spaced wells (Fig. 5). Collectively, these facts imply that, as compared with PI2, the PI2-3 river was subject to sudden floods, strong channel-base erosion, high rates of sediment deposition, and frequent channel migration.

As a result, two features of in-layer heterogeneity in low sinuosity meandering units are evident which contrast with

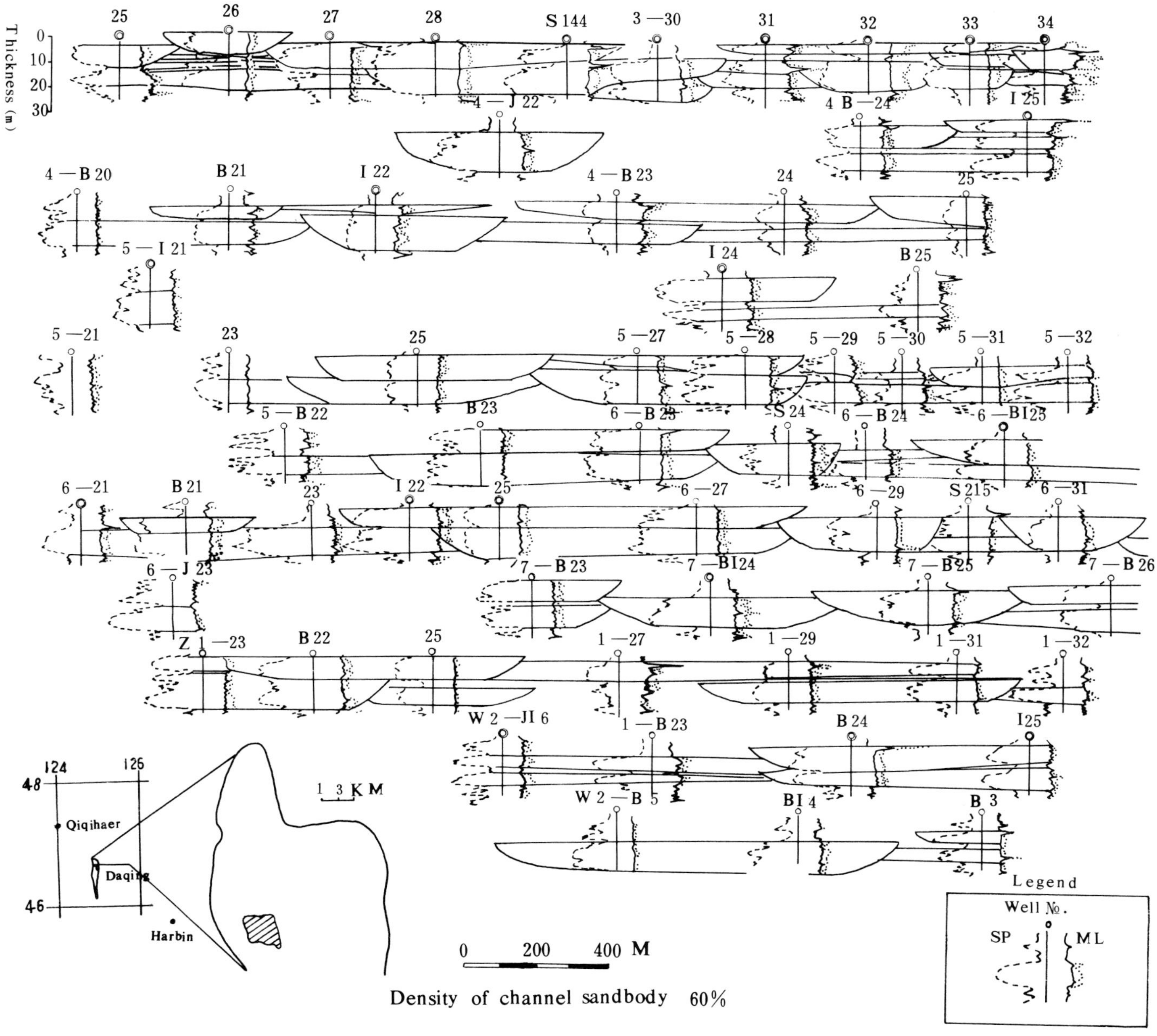

FIG. 5.—Geometry of low sinuousity meandering sandstone body, (Subzone PI_{2+3}, Daqing).

those of bars developed in highly sinuous sandstone bodies. In Case 2 the uppermost meander-belt sandstone body cuts into and was developed upon the underlying bar unit to form a thicker composite sandstone body. The internal heterogeneity of the composite unit was reduced because of the apparent juxtaposition and connection of the upper and lower sandstone units. Although the permeability contrast in the composite unit is 10–20 and as much as 40, approximately equal to that of case 1, the variation coefficient of permeability is apparently declined to 0.6-0.7 and is never over 1. The second contrasting feature is poorer preservation of mud drapes in low sinuosity meandering bodies than in high sinuosity units. The flood plain mudstone between the stacked units, however, may act as a relatively continuous barrier to fluid flow.

As shown in Figure 5, it is likely that composite sandstones with good lateral continuity are composed of numerous genetic units of the meander belt. These individual units are very narrow in width (about 150–400 m, according to log correlation) and they vary in thickness and grain size corresponding to the varied hydraulic conditions of the ancient PI2-3 river phases. A single, complete point-bar sequence ranges from 5 to 11 m thick, thus the width/thickness ratio of a meander belt sandstone body is only 30–60. Additionally, we infer a 150–200 m of bankfull width. From these data, a conclusion could be logically made that the ancient PI2-3 river was a low sinuousity meandering stream with frequent avulsion and migration phases and that it frequently changed hydraulic conditions.

BRAIDED MODELS

The main sandstone bodies deposited in a braided river are mid-channel bars developed dominantly by vertical accretion. Vertical accretion is an accumulation process in which topographic elements of the surface do not shift laterally but instead build upward. Mid-channel bars that accumulate vertically during a series of flood events are characterized by a unique in-layer heterogeneity that contrasts with that of bars formed in meandering streams. Each flood event is characterized by its hydraulic force and capacity to transport grains of varied sizes. Mid-channel bars are deposited through multiflood events such that internal bedform and grain-size stratification exist but do not grade regularly between unit boundaries. In turn, internal variations of permeability are controlled by characteristics of the resultant stratification.

An important characteristic of internal heterogeneity of the braided bars is that few mud streaks (drapes) are deposited and preserved due to the stream's small suspended load/bedload ratio. Mud streaks, which may form in braided bars and constitute barriers to fluid flow, are formed in abandoned channels. The lateral continuity of abandoned channel fills however, seems impossible to extend more than a channel width, which when compared with the entire dimension of a braided bar, is of little significance to fluid flow. Few significant barriers to fluid flow exist in braided sandstone bodies. Thus, the vertical permeability is approximately equal to the horizontal permeability. This permeability character is very important for the cross flow of fluid when secondary or tertiary oil recoveries are attempted. In waterfloods, the injected water always advances rapidly along the lowermost interval with little thickness to the zone being swept, even though the maximum permeability pathway is not generally at the unit's base. These data indicate a great effect of gravity on fluid flow due to the density differentiation between oil and water.

Although a braided river is commonly considered to be able to deposit continuous sandstone bodies due to less cohesive stream banks and frequent channel migration, a number of braided sandstone bodies has been identified in Chinese lake basins in which these sandstone characteristics were not apparently dominant. Both a multilateral sandstone body with a wide areal distribution and an isolated sandstone body with little continuity can be formed in braided rivers depending on the river's tectonic setting. For example, short-coursed braided rivers, extensively developed at the strongly faulted side (transverse to the strike of the tectonic framework) of faulted lake basins in China, commonly produced isolated sandstone bodies.

In faulted basins, several key factors influence the channel sandstone bodies deposited. Those factors are the basin's steep slope (maybe as much as 10 percent), the stream's short course (± 20 km from source area to depocenter), rather small-scale rivers (depth < 10 m), and very active basin margin faults with high subsidence rates in the graben. As a result, the channel sandstone bodies of individual streams generally do not coalesce laterally to form widespread rock units. As a result, in this scenario, many isolated sandstone bodies are embedded in structureless and thick mudstone of the flood plain.

Long-coursed rivers, such as those developed along a basin's longitudinal axis and during the waning stage of a lake basin, may develop widespread, multilateral sandstone bodies as a result of sufficient supply of sediments and rather slow rates of basin subsidence.

Case 3. Short-coursed braided sandstone bodies—SaII2 and SaII3 in Shentuo field, Shengli oil province.—

Subzones SaII2 and SaII3 contain the main rock reservoirs with high productivity (from tens to more than 100 t/d oil/well) in Shentuo field, Shengli oil province. They consist of a series of interbedded pebbly medium- to coarse-grained sandstone units. The reservoir units are easily identified as being of a fluvial facies that extends between the northern alluvial fans (marked by abundant debris flow deposits) and southern lacustrine deltas (with typical river-mouth bars). Tectonically, the reservoirs are located along the steep depositional slope and their formation was controlled by a series of active major basin-boundary faults.

The median grain size of the fluvial sandstone units ranges from 0.3 to 0.6 mm with very poor sorting. The fine gravels, almost all derived from source rocks of granite, are developed throughout the section. Few intraformational clasts can be seen. Vertically, the sequence of grain size is disordered and some abrupt changes are evident. Coarsest grains are not confined to unit bases (Fig. 6). Although crossbeds are well developed in the sandstone, parallel lamination is not uncommon. Most crossbeds seem to be tabular with low (± 5°) foreset dips; some tabular and trough crossbeds with high foreset dips are evident. We did not identify ripple or horizontally laminated siltstone or very fine-grained sandstone as the top stratum of the point bar. Some sandstone sequences, however, are capped by a thin-bedded gravel which in turn is overlain by the flood-derived mudstone. Few mud streaks are identified within the sandstone bodies. We infer from these sedimentary features that the fluvial sandstone bodies were deposited by a braided river system near its source, that it was subject to strong intermittent floods and had a low suspended load/bedload ratio, and that they resulted from a deposition pattern of vertical accretion. We believe that this pattern is similar to that of the Donjek type proposed by Miall (1978).

The geometry of Subzones SaII2 and SaII3, as delineated by production wells (closest well spacing is only 10 m), is obviously narrow and elongated with a width of 200–600 m and a width/thickness ratio of 40–70 (Figs. 7, 8). A few multilateral sandstone bodies may be as wide as 2000 m.

The internal vertical permeability heterogeneity varies greatly and seems to conform to the changes in grain size. Maximum permeability intervals are commonly located in the middle to lower part of the reservoir and are not along unit bases. Five to 20 of permeability contrast and 0.88 of average variation coefficient of permeability indicate that the braided sandstone bodies have slightly less heteroge-

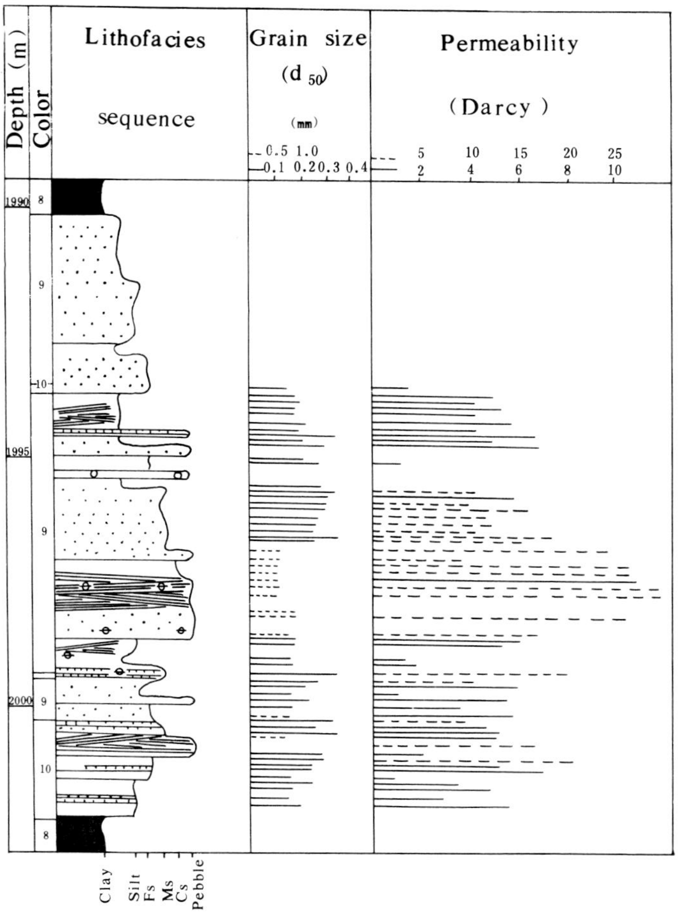

FIG. 6.—Lithofacies sequence of short-coursed braided sandstone body (Subzone SaII$_3^{3+4}$ Well 2-3-J1502, Shengtuo field).

neity than meandering ones. In addition, the vertical/horizontal permeability ratio measured on a core plug is particularly large, ranging from 0.8 to 0.9 due to the lack of clay minerals in the rock's matrices.

Case 4. The long-coursed braided sandstone bodies—Subzone NgI of Gongdong field, Dagong oil province.—

The Lower Neogene Guantao Formation (Ng) was widely deposited in Bohai Bay Basin overlying a regional unconformity developed upon Eocene rocks at a time when the basin's floor was rather flat. There was sufficient erosion of the surrounding mountains, however, to supply a large volume of sediments to this basin. Consequently, fluvial deposits, dominantly of braided and low sinuosity meandering streams, were extensively developed and cover a large area of the basin.

Subzone NgI, has been identified as consisting of braided sandstone units. A main secondary oil reservoir in Gongdong field is at the upstream reach of the northern depositional system, and it is one of the basin's largest braided systems with a course length of 120 km developed along the basin's depositional axis.

NgI is composed of grey and brown conglomeratic sandstone, and coarse- to fine-grained sandstone intercalated with light green and purply-red mudstone. The sandstone units can be classified as arkose and graywacke. They have a median grain size of 0.3–0.4 mm and contain pebbles in amounts that vary with degrees of poor to moderate sorting.

The depositional sequence of NgI sandstone units has an erosional base that is overlain by lag deposits of conglomerate or conglomeratic sandstone (Fig. 9). Most pebbles were originally derived from the source area of the sediments and a few are of intraformational mudstone fragments. Crude imbrication of pebbles can sometimes be identified. Coarse- to medium-grained sandstone that contains pebbles constitutes most of the sequence. Although no clear depositional cycles can be recognized, a crude fining-upward sequence can be identified in some units. Large-scale tabular and trough crossbeds with 20°–30° foreset dips are developed throughout the section; parallel bedding is locally evident. The massive flood mudstone marked by abundant plant debris, rootlets and carbonate concretions, directly overlies the crossbedded rocks. Some thin-bedded (10–20 cm) siltstone occurs with other rock types. Mud streaks have not been found, a phenomenon that may suggest that lack of suspended-load sediments characterized the streams.

Thickness of individual sandstone bodies is generally 5–8 m but ranges upward to 13 m. Some multistory composite units may be more than 20 m thick. Laterally, channel sandstone bodies change rapidly in thickness or elevation. Coalescence of clastic units is extensive, as interpreted from drill hole log signatures (Fig. 10). The very good continuity of reservoirs is evident in production performance by the large volumetric influx of very active edge water. This phenomenon indicates the existence of a large connected water body beyond the zone of oil saturation of the reservoir. It is this good reservoir continuity that marks an important distinction between long-coursed and short-coursed braided sandstone reservoir bodies.

The internal heterogeneity of NgI sandstone bodies is similar to that of Case 3 discussed above in the rather disordered variation of permeability, permeability contrast of 4–14 and as much as 30, and average variation coefficient of permeability 0.74. Silt streaks are believed to be present, and they undoubtedly act as barriers to fluid flow, for the plant debris commonly associated with silt laminae is abundant in most sandstone units.

STRAIGHT DISTRIBUTARY CHANNEL MODELS

The distributary channels of lacustrine deltas commonly are straight because of their small size and rather low energy where they intersect the lake. Meandering distributary channels may be able to develop, especially in the larger delta complex of longitudinal systems. In the latter case, the in-layer heterogeneity is similar to that of the fluvial meandering model described above.

Straight distributary channels consist dominantly of active channel fill. Straight channel-fill sediments appear to be crudely segregated by gravitational forces during their accumulation, so that the process of this deposition results in crude sediment cycles in which grain size decreases upward from a unit's base; permeability values decrease upward as grain size decreases. The degree of lithologic and

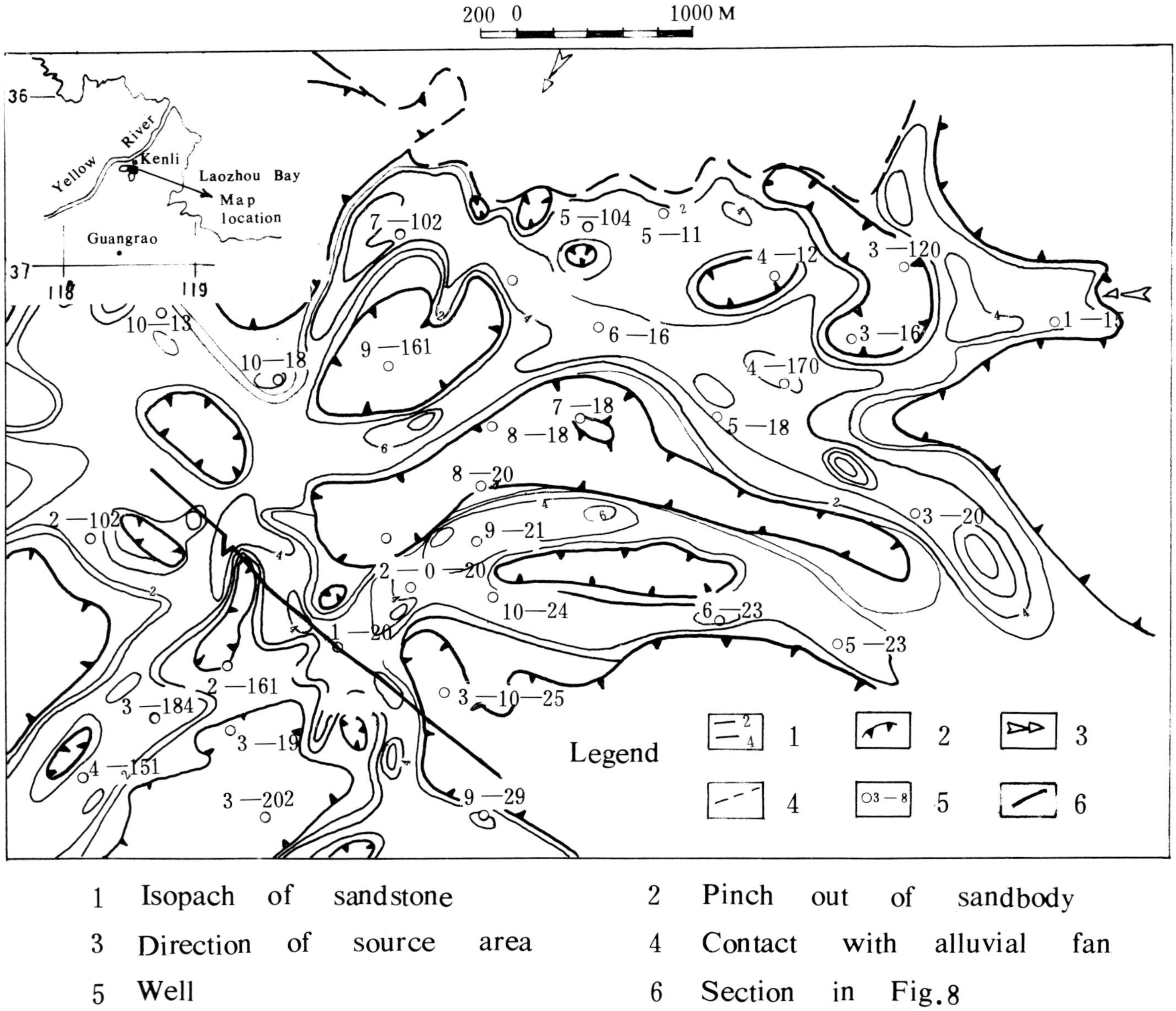

FIG. 7.—Geometry of short-coursed braided sandstone bodies, (Subzone $SaII_2^4$, Shengtuo field).

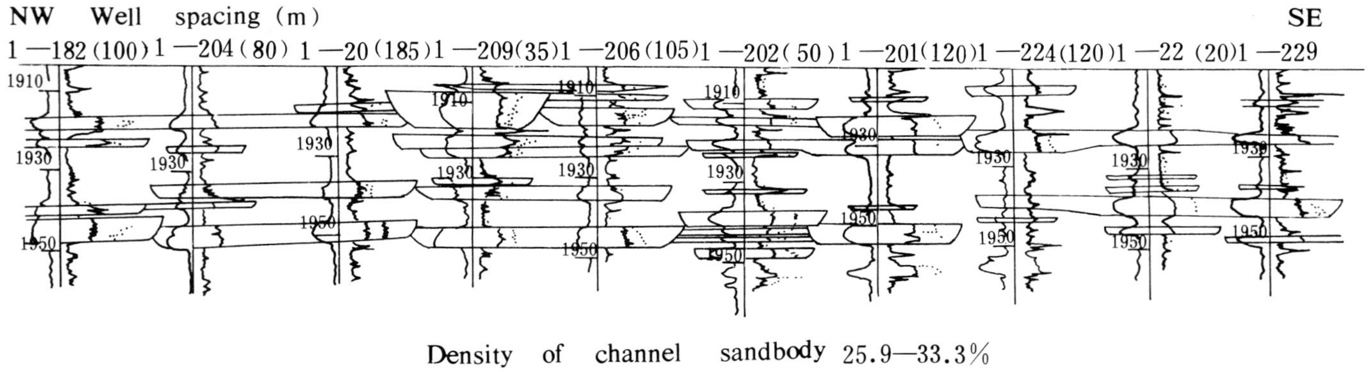

FIG. 8.—Correlation section of subzones $SaII_2$, $SaII_3$, the short-coursed braided sandstone bodies, Shengtuo field.

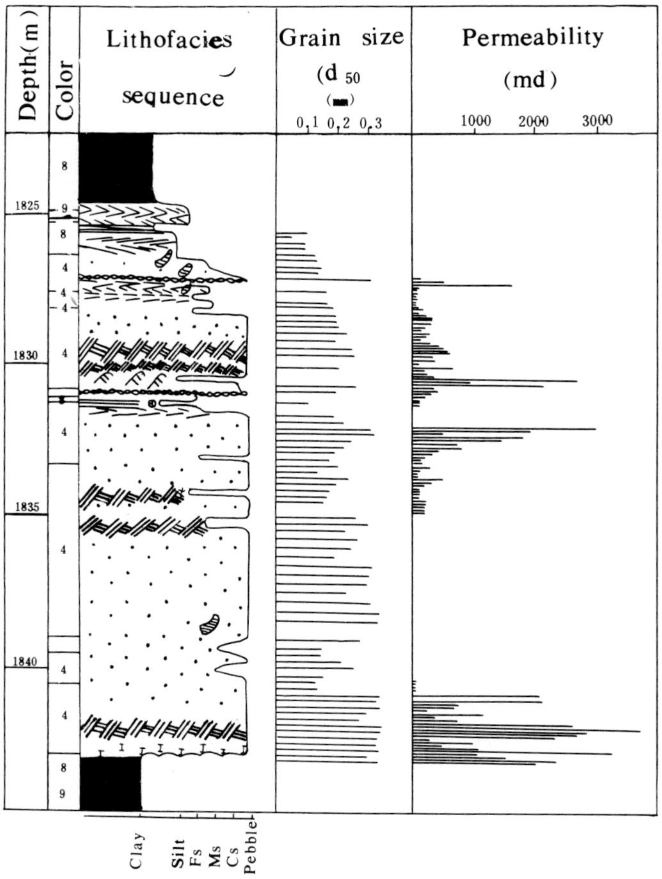

FIG. 9.—Lithofacies sequence of long-coursed braided sandstone body (Subzone NgI, Well G212, Gongdong field).

therefore permeability variation, however, is by far less than those of point bars formed by lateral accretion.

An additional important characteristic of straight distributary channel sandstone bodies is the low width/thickness ratio, which results in a very discontinuous lateral geometry of the rock unit. The low ratio seems to be much more important than the internal heterogeneity of permeability in determining the success of secondary and tertiary recovery attempts. The most difficult problem for recovery attempts is that it is impossible to drain oil effectively from these units in the absence of a very dense well spacing, more dense than that normally used in recovery attempts from most sandstone reservoirs.

Case 5. The sandstone body reservoirs in a set of straight distributary channels—the Subzone SII_{13-16} in Sardu field, Daqing oil province.—

SII_{13-16} is a deltaic unit of the basin's largest depositional system of Early Cretaceous age in Songliao basin. This depositional system contains the majority of reservoirs in Daqing. As shown in Figure 13, most of the deltaic-plain channel sandstone bodies of SII_{13-16} are very narrow, are elongate with a N-S trend, intersect the lake shore, and conform to the regional direction of stream sediment transport. This geometry of zone sandstone bodies illustrates the straight pattern of the ancient SII_{13-16} distributary channels. Interpretation of their lithofacies composition indicates that they originated as channel fills.

In addition to common sedimentary characteristics of these sandstone bodies, such as a basal scour marked by mudstone pebbles, fine-grained and small-scale crossbedded to ripple crossbedded sandstone, two subtypes of lithofacies sequence can be identified. The first lithofacies subtype is a thick sandstone interval overlain by a thinner layer of rock

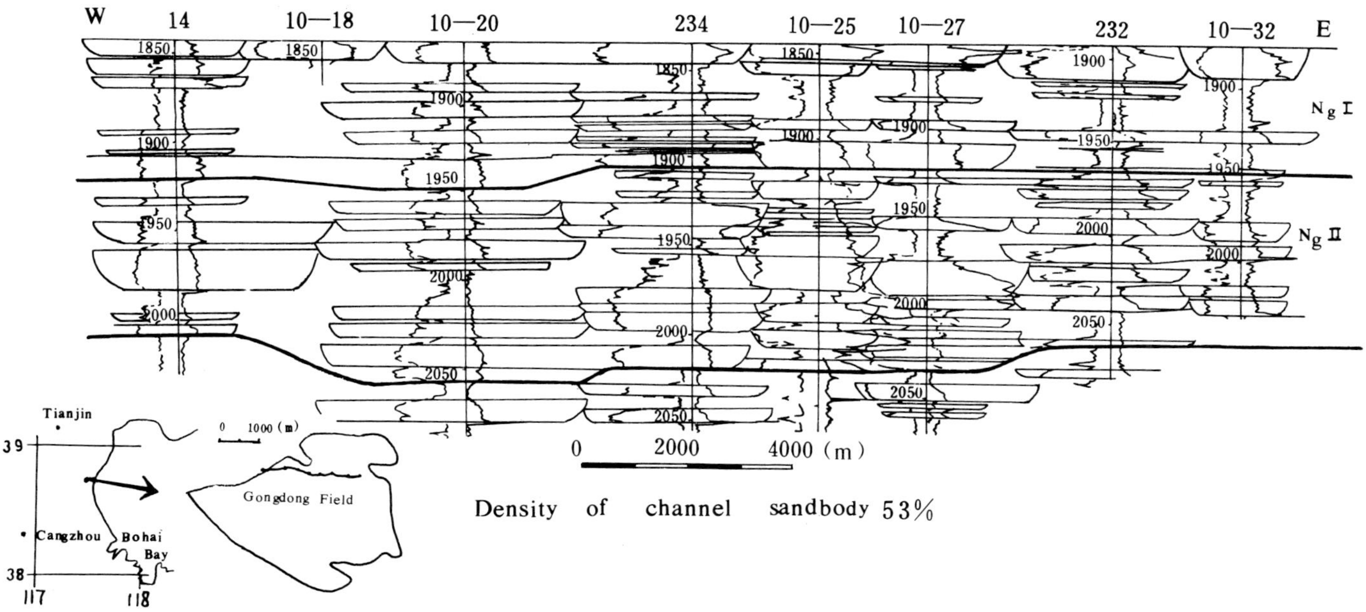

FIG. 10.—Correlation section of Ng, the long-coursed braided sandstone bodies, Gongdong field.

derived from suspended sediments (Fig. 11). Lack of significant vertical variation of grain size in the sandstone unit implies rapid sedimentation in channels; the fact that the top stratum is composed mainly of mud-size grains suggests rapid abandonment of the distributary channel. Moreover, we suggest this sequence indicates that the main distributary channels carried a large volume of sand.

The second lithofacies subtype is composed of a thin sandstone interval overlain by a thicker unit composed of interbedded siltstone to very fine-grained sandstone and mudstone in centimeter-scale thick beds (Fig. 12). The mudstone is developed with well defined flat laminae. This lithologic sequence is formed in subordinate distributary channels that carried sandy sediments which did not fill the channel. The process of channel filling was therefore prolonged and influenced repeatedly by the main distributary channels from which some sandy suspended sediments periodically overflowed into the secondary channels. Straight distributary channel sandstone bodies always feature an asymmetric fining-upward sequence as opposed to the uniformly fining-upward sequence characteristic of most point bars in sinuous systems.

In summary, the internal heterogeneity of Subzone SII_{13-16} is characterized by: (1) normal rhythmic (cyclic), grain size variations and resultant permeability values, (2) a basal interval of maximum permeability corresponding to a coarse basal rock, (3) a great permeability contrast in the unit of as much as 30–50, and (4) a smaller variation coefficient of permeability about $\pm$ 0.5.

Some multistory composite sandstone bodies are developed as a result of vertical stacking. Since downcutting is not great in distributary channels, the complete removal of the mud deposits at the top of the underlying channel sandstone bodies is rare. Thus, these multistory sandstone bodies are commonly separated by thin mud barriers and each genetic unit is an isolated reservoir unit for fluid flow.

Correlation of drill hole logs has shown (Fig. 13) that most sandstone bodies in Subzone SII_{13-16} are 100–150 m wide and change abruptly to mudstone of the interdistributary facies. The thickness of an individual sandstone genetic unit ranges from 3 to 5 m, and we assume a width/thickness ratio of 20–40 by comparison to their modern equivalents. Thus, the fluvial architectural system of this zone is composed of many isolated channel sandstone bodies of various dimensions imbedded in massive mudstone.

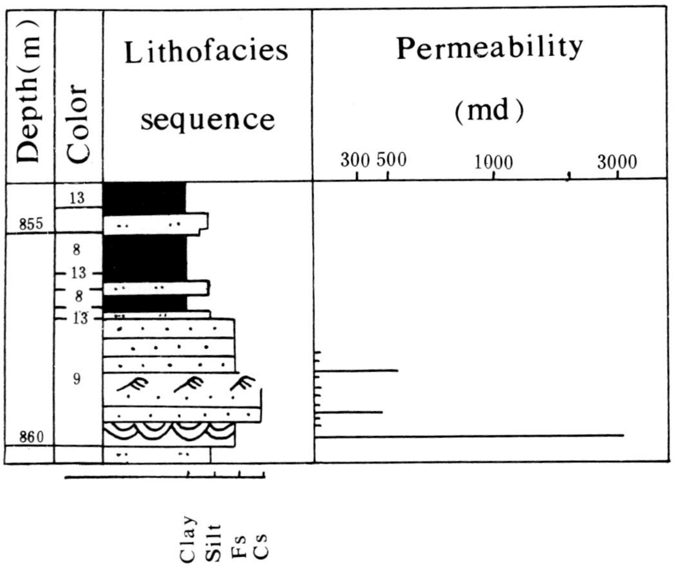

FIG. 11.—Lithofacies sequence of straight distributary channel sandstone body (rapidly abandoned, Subzone SII_{12}, Well Z51-J203, Daqing).

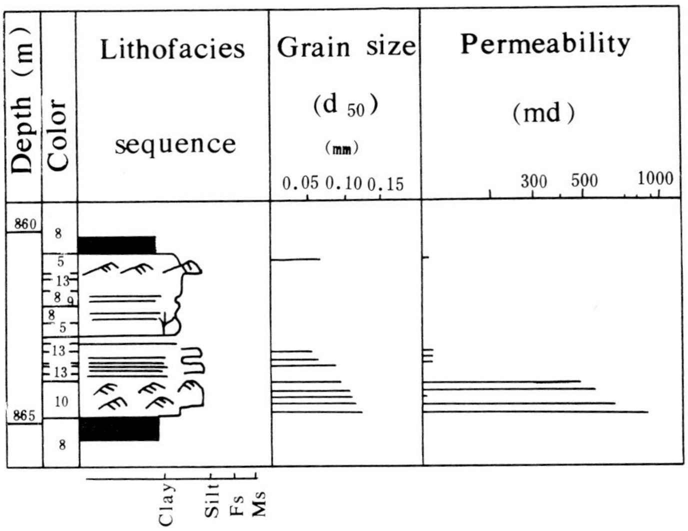

FIG. 12.—Lithofacies sequence of straight distributary channel sandstone body (slowly abandoned, Subzone SII_{14}, Well Z51-J203, Daqing).

THE CONFINED CHANNEL-FILLING MODELS

The anastomosed river pattern as defined by Smith (1983) has not been definitely identified yet in Chinese ancient lake basins. Although some fluvial sandstone bodies have been reported to be of that pattern (Chen, 1984), the relation is not accepted by all. We suggest, however, that many of the "anastomosed" channels reported in the Chinese literature should be cautiously called confined channel-filling deposits, as described here. Examples are found in some oil fields, such as the Zone Y10, the Lower Jurassic of Maline field in Shanganning basin, northwestern China.

The Lower Jurassic sequence in Shanganning basin was formed on a post-Triassic erosional surface, which had geomorphologically formed as a stream valley system flowing to the coast. During the earliest depositional period, this valley system was continuous because of the relatively stable equilibrium between valley downcutting and tectonic uplift. As a result, a set of confined channel-fill sandstone units were deposited to form strata about 300 m thick and, the main reservoir in Maline field is the upper part of this sandstone suite. An interesting phenomenon of these reservoir sandstone bodies, which may be indirectly regarded as the evidence by which they are identified as confined channel-filling deposits, is that the mineralogical and petrophysical properties of sandstone bodies deposited in different paleotributary valleys contrast greatly due to their different source terranes, even though in some cases the distance between two tributary valleys is less than 10 km.

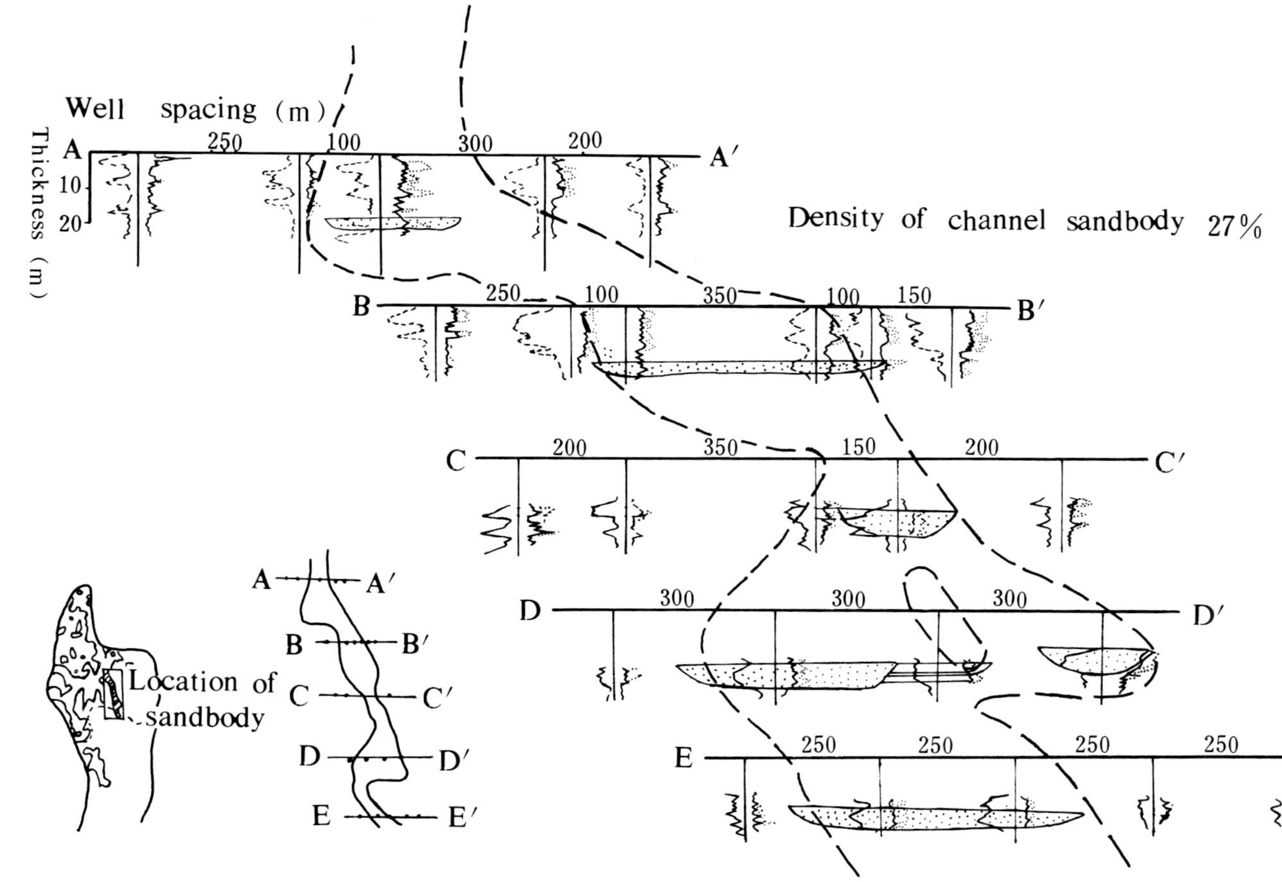

FIG. 13.—Geometry of straight distributary channel sandstone body (Subzone SII_{15+16}, northern Daqing).

The lithology, sedimentology, and reservoir variability of this type of channel sandstone body can be characterized by using Zone Y10 as an example.

Case 6. Confined channel-fill sandstone bodies—Zone Y10 in Maline field, Shanganning basin.—

The Y10 reservoir zone is chiefly composed of sandstone and conglomerate, which collectively constitute more than 70 percent of the total thickness of the section. Both are mineralogically immature and they are rich in clay minerals. In places, the reservoir contains clayey conglomerates with the pebbles floating within the clay matrices. The conglomerates are probably debris flow deposits.

The lithofacies sequence contains a number of thick fining-upward cycles; a number of subcycles can be recognized (Fig. 14). Each subcycle contains a fining-upward sequence that is capped by centimeter, thick flat or ripple laminated siltstone or muddy siltstone. Most cycles are approximately 1.0 m thick but some range to 4.5 m.

All pebbles are derived from one source area, and the units contain little evidence of grain fabric. Carbonaceous plant stems are very common near the bottom of the coarser subcycles. The fact that it is very rare to detect two sets of crossbeds in the sandstone units through a core tens of centimeters long means that the crossbed sets must be of a large scale. Parallel lamination has been identified in some sandstone.

Sedimentary features described herein could be reasonably interpreted to mean that the Y10 channel sandstone bodies are the result of aggradation in confined valleys during flash flood events. Each thin subcycle is the result of a flood event of great magnitude and short duration. Moreover, flat laminated muddy siltstones sharply overlying the top of subcycles illustrate the existence of rapid and intermittent channel abandonment. Their excellent preservation could be considered the result of the relative balance between channel aggradation and tectonic basin-floor subsidence.

The geometry of confined channel sandstone units is typically that of a "shoestring." The thickness of the units can be several times that of the river's depth due to multiple episodes of aggradation and depositional stacking. Consequently, no empirical value of width/thickness ratio could be established for purposes of prediction.

As shown in Figure 15, the Subzone Y10-2 apparently consists of two main channel sandstone bodies which had a width of 400–800 m in its coarser parts.

Sandstones of the Y10 zone have been subjected to mineral diagenesis. Permeability variation in the zone cannot be used as a guide to general depositional characteristics of reservoir heterogeneity. The in-layer heterogeneity of the sandstone bodies of the confined channel-fills, however, suggests they should be similar to that of multistory normal cycles. The smaller variation in grain size in zone units

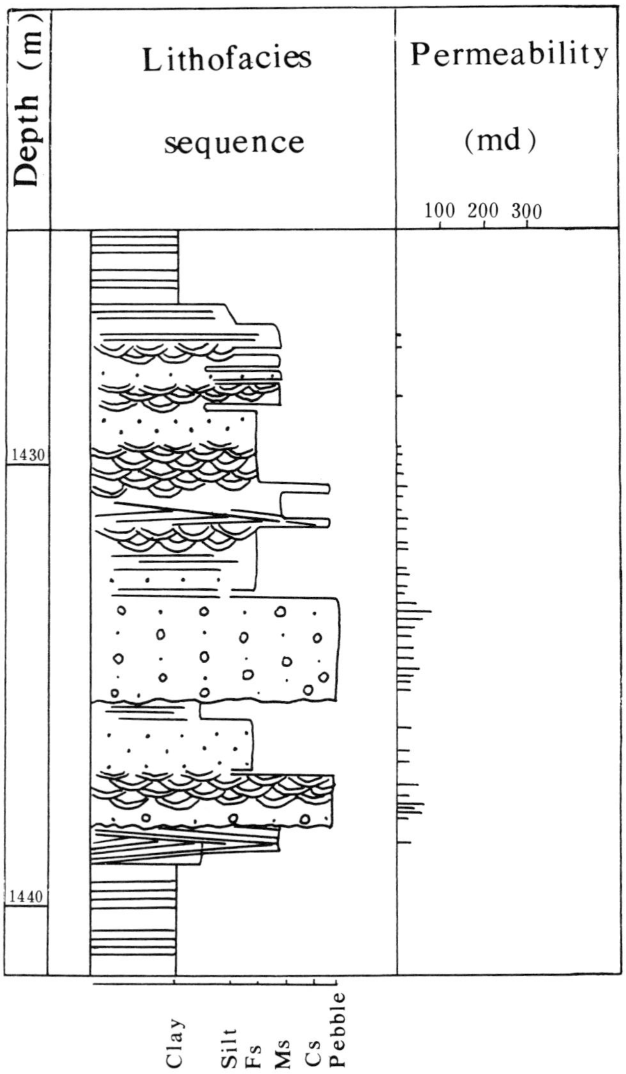

FIG. 14.—Lithofacies sequence of confined channel-filling sandstone body (Zone Y10, Well L68, Maline field).

seems to produce a rather small permeability contrast, and this relation is similar to that of straight distributary channel sandstone bodies. In those beds where debris flow deposits exist at the unit bases, the interval maximum permeable values may be developed in the middle-lower part of the reservoir. Internal thin-bedded muddy siltstone units are barriers to fluid flow where preserved, and they are generally a really extensive relative to the lateral continuity of the overall sandstone body.

LATERAL CONTINUITY

The geometry of fluvial sandstone bodies always occurs in elongated shapes that are substantially determined by the geomorphology of the originating rivers. In general terms, a good lithologic continuity can be expected along the channel's trend. Therefore, the lateral continuity—the width of sandstone bodies—is the key to determining the effectiveness of the fluvial petroleum reservoir.

The width of a genetic unit in a river belt is controlled by the river's paleohydraulic character, and it may therefore differ from one to another depending on the river's size. Generally, the width/thickness ratio of fluvial sandstone bodies can be arranged in a decreasing order: (1) high sinuousity meandering, (2) braided, (3) low sinuousity meandering, and (4) straight channel sandstone bodies. The examples described above are given in this order. The fact, however, that lake-basin rivers are generally small in scale with a depth of no more than several meters and with widths of tens to hundreds of meters indicates that the width of a genetic unit composed of fluvial sandstone bodies cannot exceed several hundred meters. From a practical standpoint, the variability in width of a genetic unit produced by a river pattern in a lake basin will have an insignificant effect on oil production. The degree to which multiple genetic units are connected to form a composite multilateral sandstone body has proven, however, to be the most significant factor in determining the effectiveness and size of rock reservoirs in Chinese lake basins.

Allen (1978) proposed a theoretical density of channel sandstone bodies based on his simulation research, which relates degree of connection of multiple fluvial units. He concluded that 50 percent is a critical value. In our examples, the densities of channel sandstone bodies are estimated using data derived from a large number of production wells. We have found that the density of channel sand bodies in one dimension can be used to predict the connection in three dimensions. A value of over 50 percent, such as in Cases 2 and 4, confirm Allen's hypothesis. We expect these units to be well connected. Values of less than 30 percent, such as in Cases 3 and 5, suggest to us that only isolated genetic units will be expected. Values between 30 to 50 percent, however, should be used carefully because of their uncertainty. In Case 1, the density of channel sandstone bodies is 47 percent, assuming a rock interval consisting of four layers of sandstone bodies, and that one layer will be effectively connected. This tells us that it is important to select a proper stratigraphic interval for estimation. Ideally, the thinner the interval, the more accurate the results should be. It is impossible in practice, however, to correlate in detail thin intervals in the subsurface, because only a few evaluation wells can be used for prediction during the early stages of development drilling.

It should be emphasized that many fluvial units in lake basins are poorly connected. The small scale of rivers and rapid subsidence of the lake basin floor result in restricted units. If we assume 1000–2000 yrs as the avulsion period and 0.005-0.0025 m/yr as the basin's rate of subsidence, fluvial sandstone bodies produced in the same genetic channel belt at a 5 m-thick scale cannot be effectively connected. Such a rate of subsidence is realistic during the main period of deposition in the Mesozoic-Cenozoic lake basins of China. We believe it is these conditions that produced most of the fluvial units in China that are poorly connected and laterally restricted sandstone reservoirs. Some exceptions have been found, such as in Cases 2 and 4. In those cases, however, units were deposited at a time either when the basin changed from being uplifted to a stage of subsidence or from a stage of lake regression to transgression.

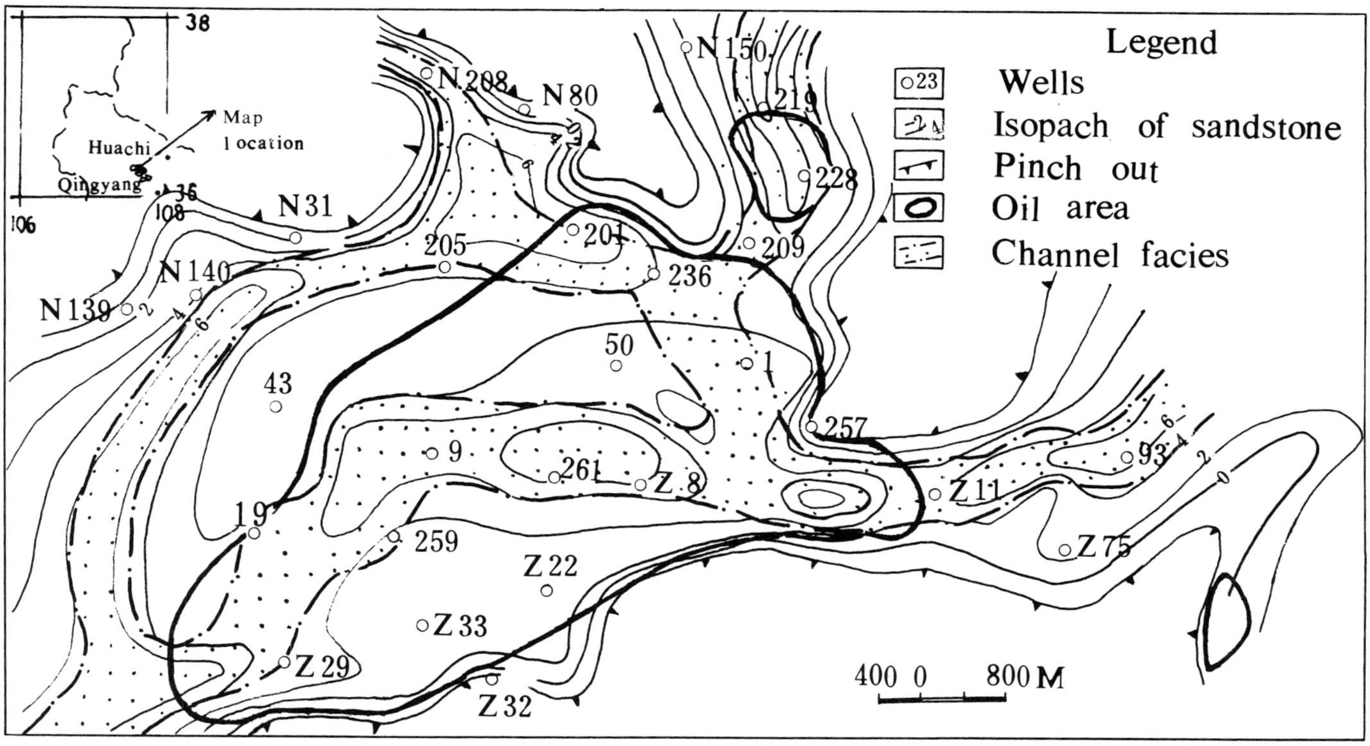

FIG. 15.—Geometry of Subzone $Y10^2$, the confined channel-filling sandstone body.

TABLE 1.—THE DIFFERENCES IN RESERVOIR HETEROGENEITY OF SAND BODIES DEPOSITED IN VARIOUS RIVER PATTERNS

	The High Sinuosity Meandering	The Low Sinuosity Meandering	The Short-coursed Braided	The Long-coursed Braided	The Straight Distributary	The Confined Channel
Sequence in grain size	uniformly fining upward	single or multi-fining upward	disordered	disordered or roughly fining upward	asymmetric fining upward	multiple, asymmetric fining upward
The location of maximum permeability	bottom	bottom	uncertain	bottom or middle-lower	bottom	botom or middle to lower
Variation coefficient of permeability	0.8–1.0 (1.3)*	0.6–0.7 (1.0)	0.4–1.0	0.5–1.0	± 0.5 (0.8)	similar to the straight distributary
Permeability contrast	10–20 (50)	10–20 (40)	5–20 (40)	4–14 (30)	30–50	
Internal muddy barriers	mud drapes of lateral accretion, rich	mud drapes, overbank, rich	channel filling, rare	channel filling, rare	overbank, channel filling, rather continuous	channel filling, rather continuous
Width/thickness ratio of a genetic unit	130–170	30–60	40–80	±100	20–40	Very small

*The value within brackets is the maximum.

During these events, subsidence of the basin started at a very slow rate, thus a widely distributed fluvial sandstone body with multiple connections was easily formed where sediment supply was sufficient.

CONCLUSIONS

From the standpoint of the in-layer heterogeneity and continuity for oil recovery, fluvial sandstone reservoirs can be subdivided into six models based on "depositional processes-reservoir heterogeneity" and the depositional and tectonic conditions in a lake basin. The comparison between them is listed in Table 1 as follows:

The small scale of rivers in a lake basin indicates that the width of fluvial sandstone bodies is in the magnitude of hundreds of meters or less. The connection of multiple genetic units becomes the key link for forming a continuous sandstone body.

The density of channel sandstone bodies could be empirically used to predict their degree of connection. Values of over 50 percent indicate a good connection; only isolated sandstone bodies can be expected when the value is less than 30 percent. Care should be underscored when the value ranges between 30 and 50 percent.

ACKNOWLEDGMENTS

The authors acknowledge the encouragement and discussions of Professor Qin Tonglou and of their colleagues at Daqing, Shengli, Changqing and Dagang oil fields. Sincere thanks are due to Dr. T. Fouch who revised, and put into an understandable form, the English text.

REFERENCES

ALLEN, J. R. L., 1978, Studies in fluviatile sedimentation: An exploratory quantitative model for the architecture of avulsion-controlled alluvial suites: Sedimentary Geology v. 21, p. 129–147.

CHEN, Y. L., 1984, A preliminary analysis on the types of river system depositional characteristics and development results of the Guantao Reservoir in Gudao Oilfield: in Chinese, Petroleum Exploration and Development, v. 11, p. 66–72.

MIALL, A. D., 1978, Lithofacies types and vertical profile models in braided river deposits: A summary *in* Miall, A. D., eds., Fluvial Sedimentology, Canadian Society of Petroleum Geologists Memoir 5, p. 597–604.

SMITH, D. G., 1983, Anastomosed fluvial deposits: Modern examples from Western Canada: *in* Collinson J. D., and Lewin J., eds., Modern and Ancient Fluvial Systems, International Association of Sedimentologists Special Publication 6, p. 155–168.

FLUVIO-PALUDAL DEPOSITS IN THE LOWER KEKIKTUK FORMATION (MISSISSIPPIAN), ENDICOTT FIELD, NORTHEAST ALASKA

JOHN MELVIN
Standard Oil Production Company, 5400 LBJ Freeway, Suite 1200, Dallas, Texas 75240

ABSTRACT: The hydrocarbon reservoir at Endicott Field, located offshore from the Sagavanirktok Delta in northeast Alaska, is the Kekiktuk Formation which is late Mississippian in age. It rests directly on weathered basement of weakly metamorphosed argillite. The lowermost Kekiktuk, known as Zone 1 in Endicott Field, is not recognized at the type section in the northeastern Brooks Range. It consists of coarsely interbedded mudstones, siltstones, coals and very fine- to fine-grained sandstones. Facies relationships suggest that deposition took place upon an undulating peneplaned basement surface in a number of environments. Those included: well and poorly drained swamps; deeper, swamp-plain lakes; partially and wholly abandoned fluvial channels, and a number of overbank environments including levees and possibly crevasse splays/lacustrine deltas. The overall setting was a low-lying swamp plain with numerous lakes, traversed by sluggish, mixed- to suspended-load streams. In the Endicott Field area, the streams appear to have been more dominant in middle Zone 1 time; however, the distinction between streams and overbank environments was not marked, and the area was characterized by common overbank flooding and channel abandonment. By end-Zone 1 time, the entire area appears to have been blanketed by swamps and swamp-plain lakes.

INTRODUCTION

The Endicott Field is located in the Beaufort Sea 1–4 mi (1–6 km) offshore from the Sagavanirktok delta in northeast Alaska (Fig. 1). The field was discovered in January 1978 with the drilling of Sohio Sag Delta 4 well and contains about one billion reservoir barrels of oil. Production from the field is scheduled to start in early 1988. The reservoir is made up of Mississippian fluvial sandstones of the Kekiktuk Formation, which is the lowermost of three formations comprising the Endicott Group (Fig. 2). In the discovery well, the Kekiktuk was encountered at 9779 ft (2981 m) subsea. The trap comprises a tilted fault block within a southwesterly plunging antiform that is truncated updip by a major unconformity, known as the Lower Cretaceous Unconformity, and bounded to the north by the Niakuk Fault Zone.

GEOLOGIC SETTING

The Endicott Group crops out in the Brooks Range of northern Alaska where it has been studied extensively by U.S. Geological Survey workers. In the central and western Brooks Range, it forms an allochthonous (overthrust) terrain, whereas in the northeastern Brooks Range and the North Slope subsurface the Endicott Group is considered to be autochthonous (Nilsen, 1981; Nilsen and Moore, 1982; Moore and Nilsen, 1984). The allochthonous Endicott sediments represent a regressive-transgressive sequence, the lower part of which is composed of marine deposits of the Hunt Fork Shale and the Noatak Sandstone. These rocks are overlain by fluvial deposits of the Kanayut Conglomerate (Moore and Nilsen, 1984). The Kanayut is Devonian in age, and thus is older than the autochthonous lower Mississippian Kekiktuk Formation (lower Endicott Group) at its type section in the northeastern Brooks Range. The Kekiktuk of the subsurface at Endicott Field is still younger, since palynological studies suggest a Visean (Late Mississippian) age for those rocks (J. E. Williams, pers. commun.). In all areas the fluvial coarse clastics of the lower Endicott Group are overlain by shales of the Kayak Formation (Fig. 2) which represent a marine transgression.

The type section of the Kekiktuk Formation was first described by Brosge and others (1962) from near Lake Peters in the northeastern Brooks Range. Nilsen (1981) described how in that location the Kekiktuk may be divided into three members. The basal member, 1 to 3 m thick, represents a thin interval of mixing of weathered basement rocks and other transported clasts. Overlying this basal member is an interval of massively cropping out fining-upward couplets of pebble/cobble conglomerate and sandstone 58 m thick. The highest member of the Kekiktuk at the type section is 20 m thick and is composed of several fining-upward cycles of conglomerate, sandstone and shale.

The Kekiktuk Formation in the subsurface at Endicott Field is somewhat different from the type section. It also rests directly on basement rocks, and where that contact has been cored, about 1 m of altered and weathered basement material can be seen, similar to that described at the type section by Nilsen (1981). Above that, the Kekiktuk is subdivided into three distinct zones that can be easily identified on wireline logs and correlated with core material (Fig. 2). They are numbered from the base upward. Zone 1 rests directly on the weathered basement and is composed of mudstones, siltstones, coals and fine-grained sandstones. The base of Zone 1 is defined by a downward increase in bulk density from about 2.2 to 2.4 g/cm^3 to greater than 2.7 g/cm^3 (K. Woidneck, pers. commun.). Its top is defined by the base of the "barrel"-shaped gamma curve profile that typifies Zone 2, (Fig. 2) along with the highest occurrence of the uppermost Zone 1 coal or shale section. Zone 2 consists primarily of medium- to coarse-grained sandstones with subordinate mudstone and granule conglomerate and is characterized by a blocky gamma curve profile. Zone 3 is composed of varying amounts of sandstone and shale with subordinate coal and conglomerate. Its lower contact with Zone 2 is transitional and difficult to determine on logs; the top is defined by an upward increase in bulk density from about 2.5 g/cm^3 in the Kekiktuk to 2.8 g/cm^3 in the overlying calcareous Kayak or Itkilyariak Formations (K. Woidneck, pers. commun.).

The Kekiktuk in the subsurface differs from that in the type section in that (i) it is generally finer-grained and (ii) it contains the distinctive lowermost interval of mudstone, siltstone, coal and sandstone defined as Zone 1, which is

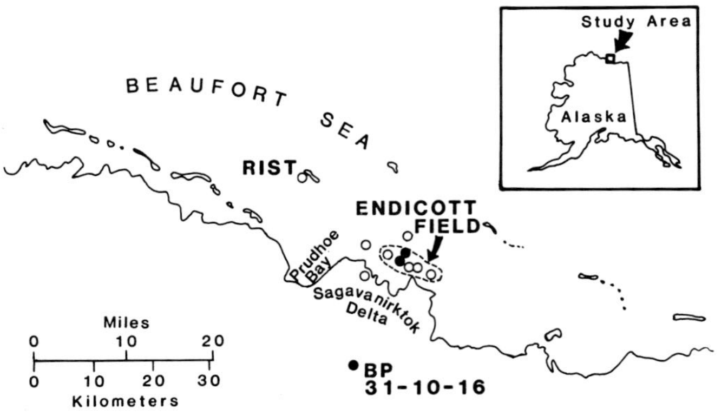

FIG. 1.—Location map of Sagavanirktok Delta area, showing Endicott Field and wells that have penetrated Kekiktuk Formation (open circles). Where Zone 1 has been cored is indicated by black circles.

not present in the type section. Zone 1 ranges in thickness from about 30 m (100 ft) to over 61 m (200 ft) across the Endicott Field area and is the subject of this paper. The sedimentological development of the entire Kekiktuk Formation at Endicott Field in relation to these three zones will be presented by the author elsewhere.

DEPOSITIONAL ENVIRONMENTS

General.—

Since the discovery of Endicott Field in 1978, a number of wells have penetrated the Kekiktuk Formation, as have other wells in the area surrounding the field (Fig. 1). Downhole logging has enabled Zone 1 to be identified in those wells, but only limited coring has taken place throughout its entire thickness. Palynological studies of ditch cuttings and core material in this zone do not reveal the presence of any marine palynomorphs (J. E. Williams, R. L. Ravn, pers. commun.) and it is considered to have been deposited entirely within a continental setting.

Those cores that have complete recovery from Zone 1 have yielded important information regarding lithofacies, vertical grain size distribution and sedimentary structures, allowing considerable appreciation of depositional processes and environment to be gained. Figure 3 illustrates the sedimentary sequence in Zone 1 where it has been completely cored in two wells at Endicott Field (Fig. 1). Those wells are approximately 1.2 km (0.72 mi) apart.

This lower Kekiktuk sequence is dominated by fine-grained sediments (mudstone, siltstone, coal and fine-grained sandstone) arranged in sequences exhibiting a variety of vertical grain-size trends (Fig. 3). The general fine-grained and coaly

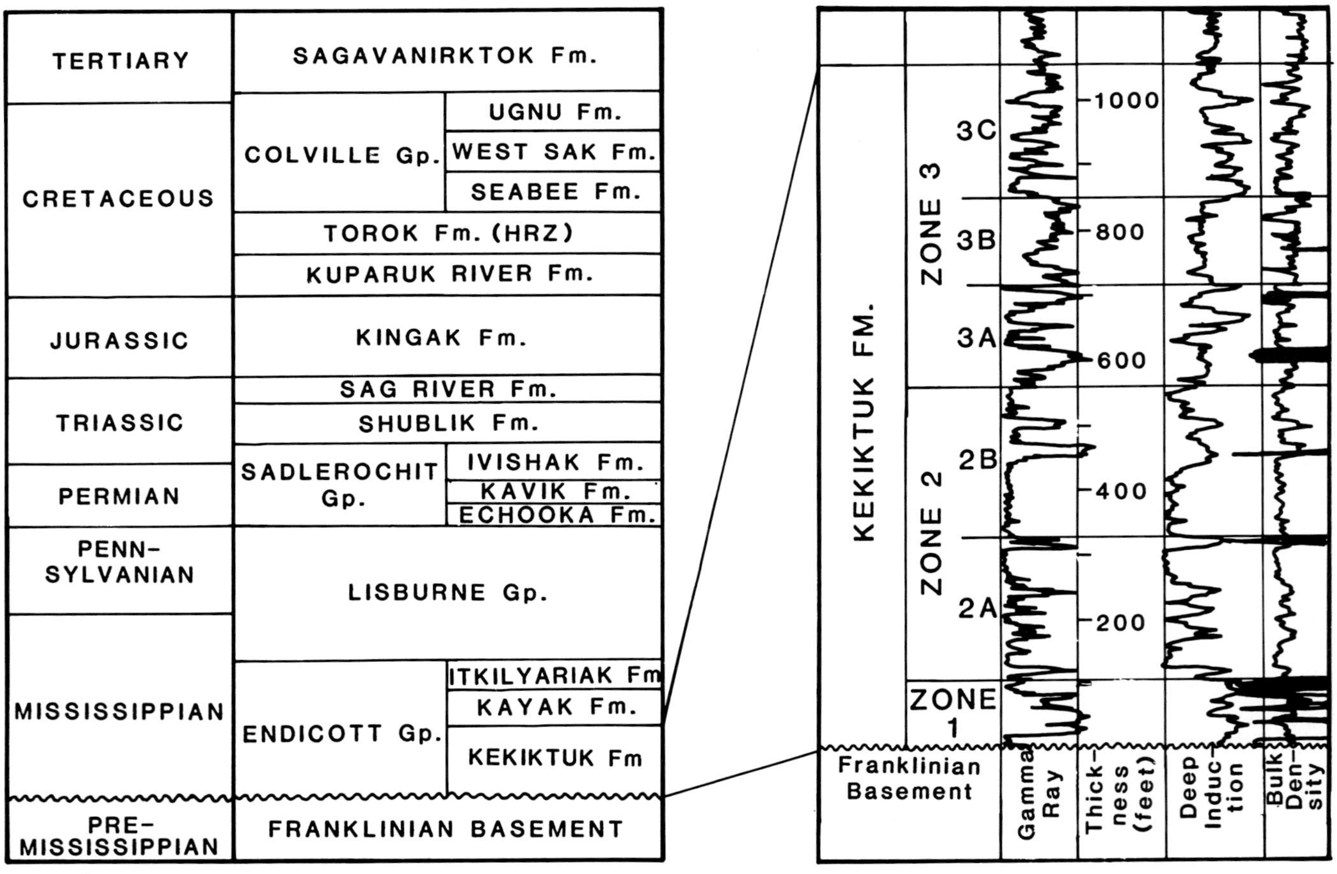

FIG. 2.—Stratigraphic column showing position of Kekiktuk Formation at base of Endicott Group, overlying Franklinian Basement, and subsurface reference section from Endicott Field showing tripartite zonation of Kekiktuk Formation.

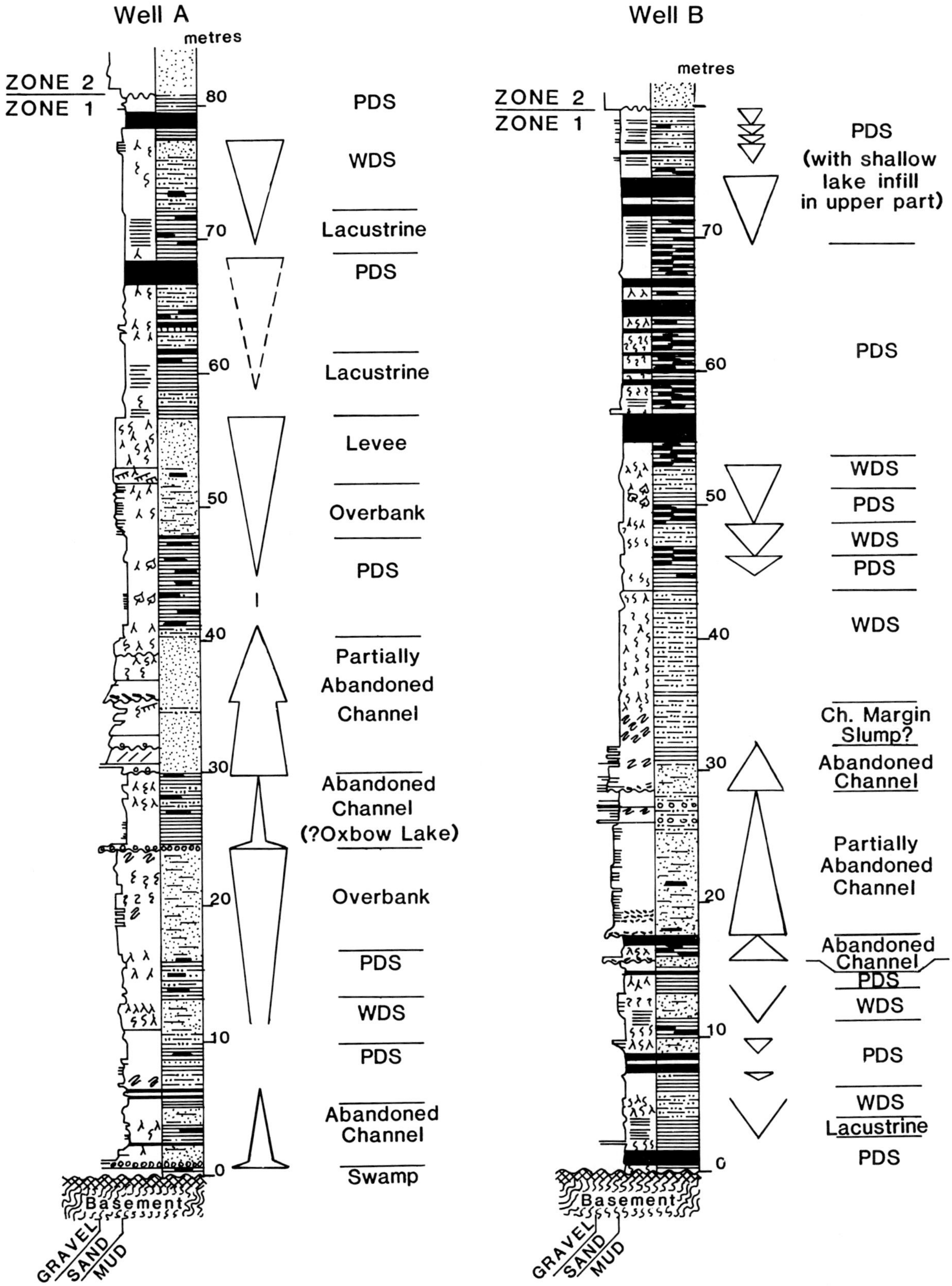

FIG. 3.—Lithostratigraphic sections through lower Kekiktuk (Zone 1) based on complete core coverage in two wells at Endicott Field. Note predominance of lacustrine/paludal sequences in lowest and highest parts and abandoned channels (fining-upward) in middle part of the section. (WDS: well drained swamp; PDS: poorly drained swamp). For legend see Figure 4.

character of these rocks suggests deposition upon a low-lying plain, within which a number of subenvironments are represented. These subenvironments are described below.

Partially abandoned channel deposits.—

Zone 1 is characterized by fine-grained lithologies and a distinct lack of sandstone-dominated fluvial channel deposits. There is evidence for limited channel activity, however. In Well A (Fig. 3), the interval between 30 m and 39 m is reproduced in detail in Figure 4. The sequence commences with a medium- to coarse-grained crossbedded sandstone overlying a sharp erosional base, suggesting active migration of dune forms on the base of a channel. The crossbedded sandstone is only about 1.7 m thick, and fines upward into featureless, medium-grained sandstones that are also underlain by sharp, locally scoured bases. Those sandstones are overlain by a thin interval of siltstone with interbeds of fine- to very fine-grained sandstone (Fig. 4). This entire fining-upward interval is interpreted as point-bar deposits within a fluvial channel (Fig. 4). That sequence is truncated by another erosionally based medium- to coarse-grained sandstone about 1.8 m thick which is abruptly overlain by about 3 m of fine-grained sandstones that are intensely mottled, penetrated by roots and possibly bioturbated (Fig. 4).

The composite sequence described above is about 10 m thick (Fig. 4) and is thought to represent a partially abandoned channel. Following active point-bar deposition (represented in the lower part), the bar top was scoured, possibly by a chute channel which was actively infilled and then abandoned, resulting in gradual accumulation of bioturbated, root mottled fine-grained sand. This indicates that relatively long periods of time passed between depositional events, allowing complete destruction of any primary sedimentary structures by plants and possibly also burrowing organisms.

The interval between 17 m and 27 m in Well B (Fig. 3) is similarly thought to represent partial channel abandonment. There also the sequence commences with a sharp, scoured basal contact that is overlain by about 1 m of coarse sandstone containing abundant mudchips and coalified plant debris. That deposit, however, is abruptly overlain by a thick sequence (about 9 m) of thin (10 to 50 cm), fine- to very fine-grained sandstones separated by very thin (1–3 cm) silty partings. The lowermost of those thin beds contains dense concentrations of mudchips (18–19 m level, Well B, Fig. 3). Following channel cutting and deposition of the basal coarse sand, partial abandonment of the channel confined deposition within the channel to predominantly suspension fallout of fine sand introduced during flood events in the nearby main channel. Complete abandonment is discounted due to the lack of very fine-grained, argillaceous sediment. Similar interpretations have been made on outcrop sections in the Carboniferous of Nova Scotia by Gersib and McCabe (1981).

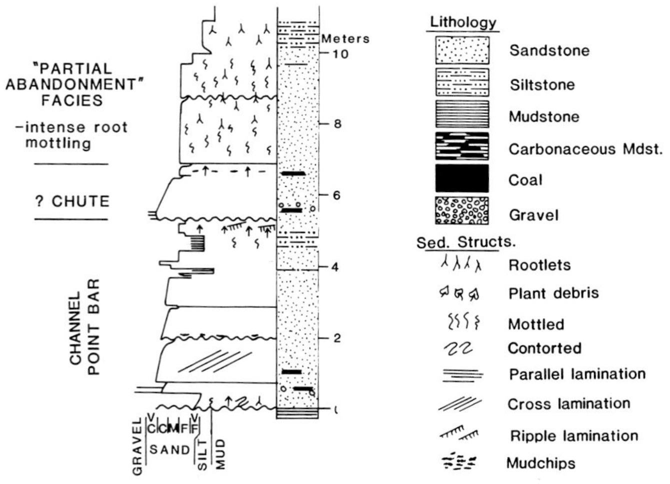

FIG. 4.—Detail of section between 30 m and 39 m in Well A (Fig. 3) showing partially abandoned channel sequence. Note that the only preserved evidence of primary current activity is confined to the very base of the section.

Wholly abandoned channel deposits.—

Ethridge and others (1981) discussed how abandoned channel deposits are difficult to recognize in cores, because they consist of fine-grained, parallel laminated, organic-rich claystone and siltstone with very fine-grained sandstones that are not unique to that environment of deposition. Nonetheless, the recognition of such subenvironments is important to an understanding of the three-dimensional geometry of the interval and distribution of permeability barriers within hydrocarbon reservoirs. A few wholly abandoned channel deposits (?oxbow lakes) have been recognized within the lower Kekiktuk cores examined in this study.

In Well A (Fig. 3), the sequence of deposits between 23 m and 29 m is thought to represent a wholly abandoned channel or oxbow lake fill. It is reproduced in detail in Figure 5. The sequence commences with a granule conglomerate resting upon a sharp scoured base. This is critical evidence for channel incision into underlying sediment. The conglomerate is abruptly overlain by about 2 m of dark grey mudstone within which are a number of very thin (1–2 mm) silty laminae. The silty laminae are undisturbed apart from local water escape structures. The laminated mudstone passes upward into a black, root mottled (unlaminated), carbonaceous mudstone containing siderite nodules and interbedded thin coals. The passage from laminated to mottled carbonaceous mudstone indicates infilling of the oxbow lake to a level where vegetation could maintain a hold in the sediment, and poorly drained swamp conditions (q.v.) became established. Infilling of the abandoned channel continued, and the swamp deposits were overlain by two small (about 1 m) coarsening-upward sequences composed of grey argillaceous siltstone that is intensely mottled as a result of water escape and rootlets (Fig. 5). These coarsening-upward units probably represent the progradation of sediment deposited rapidly by flood events associated with a nearby channel (crevasse-splay or overbank deposits). Such incursions represent only an interruption of the poorly drained swamp conditions, which resumed until the oxbow lake was completely infilled (or invaded by the next cycle of channel formation and infilling).

Other abandoned channel deposits have been identified in the lower Kekiktuk using similar criteria as above, for

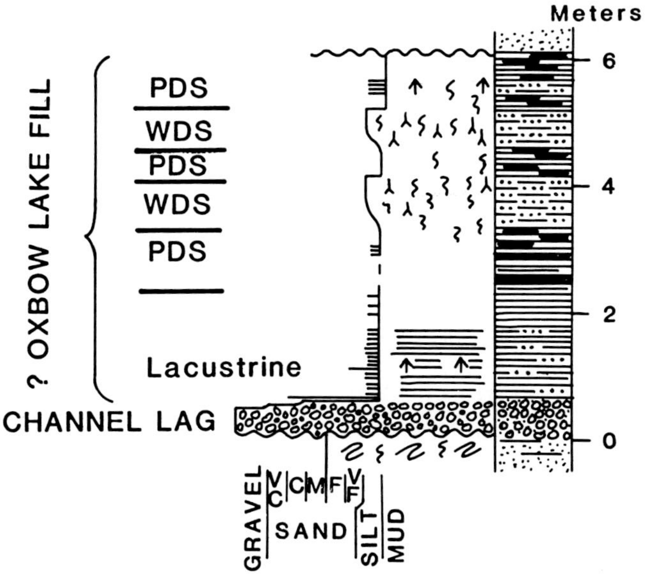

FIG. 5.—Detail of section between 23 m and 29 m in Well A (Fig. 3) showing wholly abandoned channel (?oxbow lake) sequence. Note abrupt shift in grain size from base-of-channel lag conglomerate to overlying mudstones representing progressive lacustrine/swamp infill. (WDS: well drained swamp; PDS: poorly drained swamp). For legend see Figure 4.

example at the very base of Zone 1 in Well A (Fig. 3), and in Well B between 15 m and 17 m (Fig. 3).

Natural levee and crevasse-splay/lacustrine delta deposits.—

These subenvironments of the flood basin are grouped together, because of their depositional similarities and close association in the natural environment. The added difficulties inherent in differentiating such subenvironments in cored sequences were highlighted by Ethridge and others (1981).

Figure 6 is a detailed illustration of the sequence between 46 m and 55 m in Well A (Fig. 3). It is about 10 m thick, commencing with a coarsening-upward transition from siltstones with thin (1 cm) interbeds of very fine-grained sandstone to thin (1–3 cm) very fine- to fine-grained sandstones with silty partings. The thin-bedded sandstones appear to be arranged in minor thickening-upward sequences with some beds 15 cm thick (Fig. 6). They also display an overall upward increase in degree of root mottling. This lower interbedded interval is overlain by a sharp-based, fine-grained sandstone 1.3 m thick characterized by small-scale, ripple cross-lamination. That sandstone is overlain by about 4 m of fine-grained, argillaceous, intensely mottled sandstone. Primary bedding is severely interrupted in this upper interval by abundant root mottling. There is sufficient evidence, however, to show that the sediment was originally deposited as thin beds of fine sandstone and siltstone (Fig. 7).

Elliott (1974) has described the characteristics of a number of minor coarsening-upward sequences in the fluviodeltaic environment, and he attributes them to a variety of flood-generated processes. Such processes include *overbank flooding,* where sediment-laden waters flood over the channel banks as sheet flow with no breaches or crevasse channels; and *crevassing,* by which flood waters are transferred to the flood basin via crevasse channels cut in the levee crest (Coleman, 1969). Fielding (1984) has stated how the relative importance of crevasse, as opposed to overbank, processes is still poorly understood. He has identified a number of depositional facies from the Upper Carboniferous of the Durham Coalfield, England, some of which he attributes to (i) coarse-grained overbank deposits, (ii) medial crevasse splay/minor delta and (iii) distal crevasse splay/minor delta. All can occur in sequences as much as 10 m thick, and all are composed of interbedded claystone, siltstone and fine-grained sandstone (Fielding, 1984). The criteria for distinguishing between these facies and the subenvironments they represent are related in part to lateral facies associations as well as to internal characteristics of the facies themselves, which creates problems in vertical facies and sequence analysis of cores.

The sequence described above from the Kekiktuk Formation (Fig. 6) is interpreted as follows. The lower 5 m of thinly interbedded sandstones and silty mudstones represents pulsatory sedimentation associated with floods. The sandstone layers may have formed as overbank deposits by sheet flow of sediment-laden flood-waters. Alternatively, the minor thickening-upward sequences may represent stacked, distal crevasse-splay deposits (Elliott, 1974; Fielding, 1984). The upward increase in root mottling in these thin beds is indicative of shallowing waters. The sharp-based, ripple cross-laminated fine-grained sandstone is the thickest

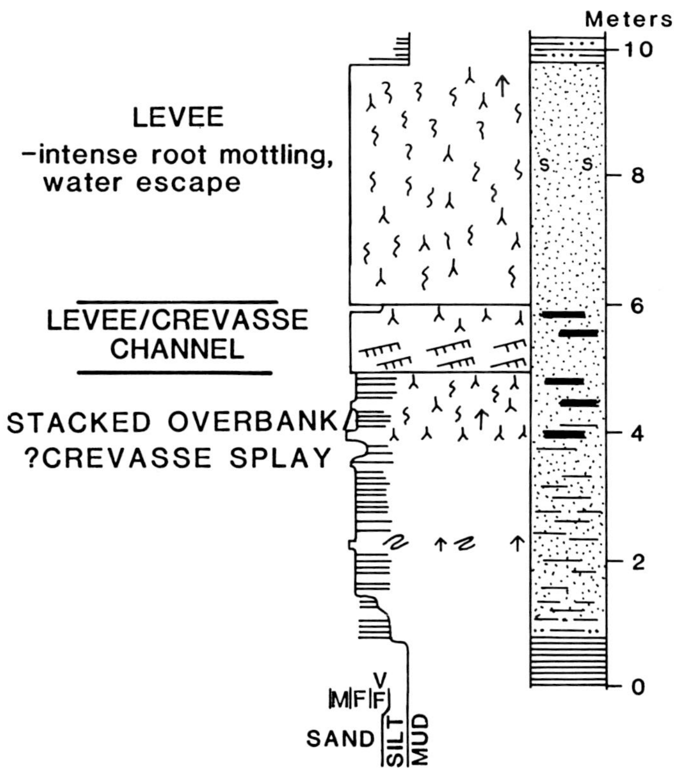

FIG. 6.—Detail of section between 46 m and 55 m in Well A (Fig. 3) showing proximal overbank sequence. Note upward increase in (i) grain size, (ii) bed thickness, (iii) intensity of mottling. For legend see Figure 4.

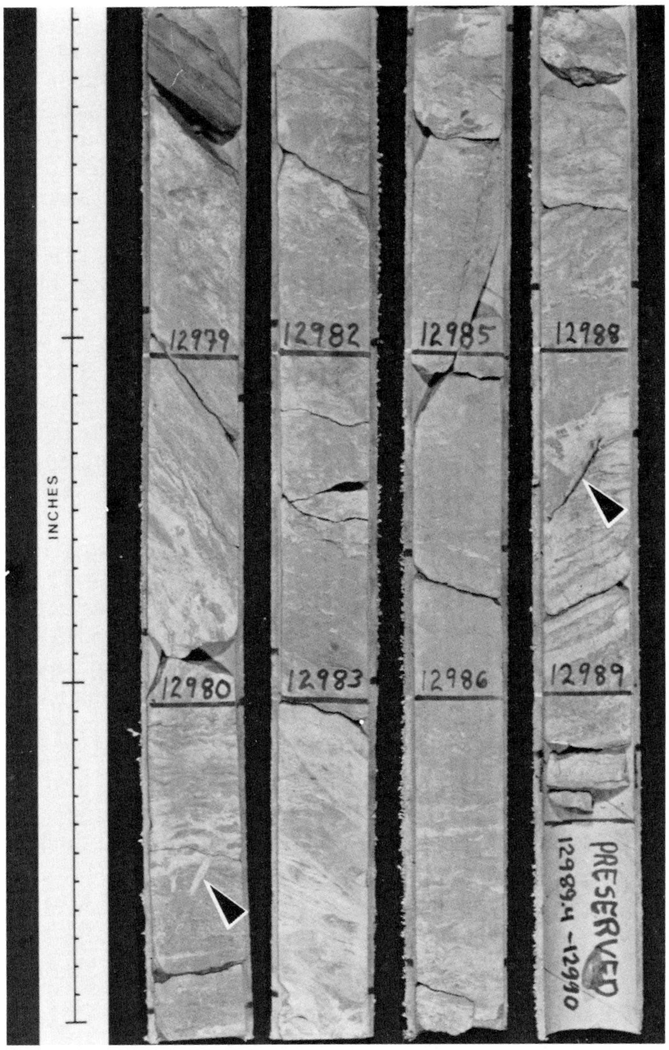

FIG. 7.—Core photograph of levee deposits that occur at top of Figure 6. Note highly mottled appearance and locally well preserved roots (arrowed).

and the coarsest bed of the sequence. As such it may represent traction-dominated deposits of a minor crevasse channel. The highest interval of this sequence, namely the 4-m-thick mottled silty sandstone with rootlets (Fig. 6) was probably originally deposited by pulsatory overbank flow, but deposition diminished allowing colonization by plant life to dominate. This mottled sandstone thus represents the paleotopographically highest part of the sequence, and the entire 10-m sequence (Fig. 6) is interpreted as the progradation of vegetated, shallow (?subaerial) levee deposits over a minor crevasse channel overlying deeper water, subaqueous distal levee or crevasse-splay deposits.

The sequence between 15 m and 23 m in Well A (Fig. 3) shows characteristics similar to the one described above. It is similarly interpreted as a stacked crevasse-splay, or prograding levee, deposit.

Lacustrine deposits.—

Lake floor sediments in the lower Kekiktuk Formation are composed of dark grey, finely laminated, very carbonaceous mudstones and siltstones (Fig. 8a). No current-generated sedimentary structures have been observed, and burrows and other faunal remains are rare to absent, suggesting restricted circulation and anoxic bottom conditions. Although these rocks contain abundant black organic matter, the preservation of delicate lamination implies that the water was too deep to support a permanent rooted plant community such as might be expected in a swamp, and the fine sediment grain size suggests low depositional energy levels. Lacustrine deposits in the lower Kekiktuk commonly occur abruptly above coals and black carbonaceous mudstones of the poorly drained swamp environments (see below). They also very commonly pass upward into such deposits (Fig. 3).

Well drained swamp deposits.—

Two types of swamps have been reported in the Atchafalaya Basin of Louisiana by Coleman (1966), namely well drained and poorly drained swamps. The type of swamp is determined by the level of the groundwater table relative to the depositional interface. When the river floods and the groundwater table rises in the channel margin area, the depositional interface is inundated across the entire backswamp area. During dry periods, river and groundwater table are lower and water drains from the topographically higher well drained swamps. Water continuously covers the topographically low, poorly drained swamps (see Coleman, 1966; Weimer, 1977, p. 13).

In the lower Kekiktuk Formation (Zone 1) at Endicott Field, both types of swamp deposits are recognized. The well drained swamp deposits are characterized by pale grey to light olive brown, silty mudstones and muddy siltstones (Fig. 8b). The rock is characteristically crumbly and breaks apart along curved surfaces that are highly polished and grooved ("slickensided"). Such breaks may be caused by expansion of roots that were later removed by oxidation (Weimer, 1977). A common explanation of such features is that the slickensides form around roots and rootlets during compaction, these being areas of low strength within the sediment pile (Fielding, pers. commun., 1986). In many places, these Kekiktuk mudstones are traversed with a fine network of hairline fractures filled with kaolinite. Siderite is also very common in these deposits. Although these silty mudstones do show abundant evidence of oxidation, in places root mottling is preserved (Fig. 8c). Such occurrences may indicate "intermittently drained" swamp conditions. These sediments of the well drained swamp environment are much less common in Zone 1 of the Kekiktuk Formation at Endicott Field than poorly drained swamp deposits, described below.

Poorly drained swamp deposits.—

The poorly drained swamp deposits of the lower Kekiktuk are characterized by an intimate association of coal, and black, highly carbonaceous mudstone (Fig. 3). The coal is variably argillaceous and generally dull in luster. In places it displays poorly developed banding and a somewhat more shiny luster. The mudstones are extremely carbonaceous, with abundant plant remains including stems and rootlets. In Well B (Fig. 3), a large imprint of *Stigmaria* was identified on a mudstone bedding plane in Zone 1. Pyrite is

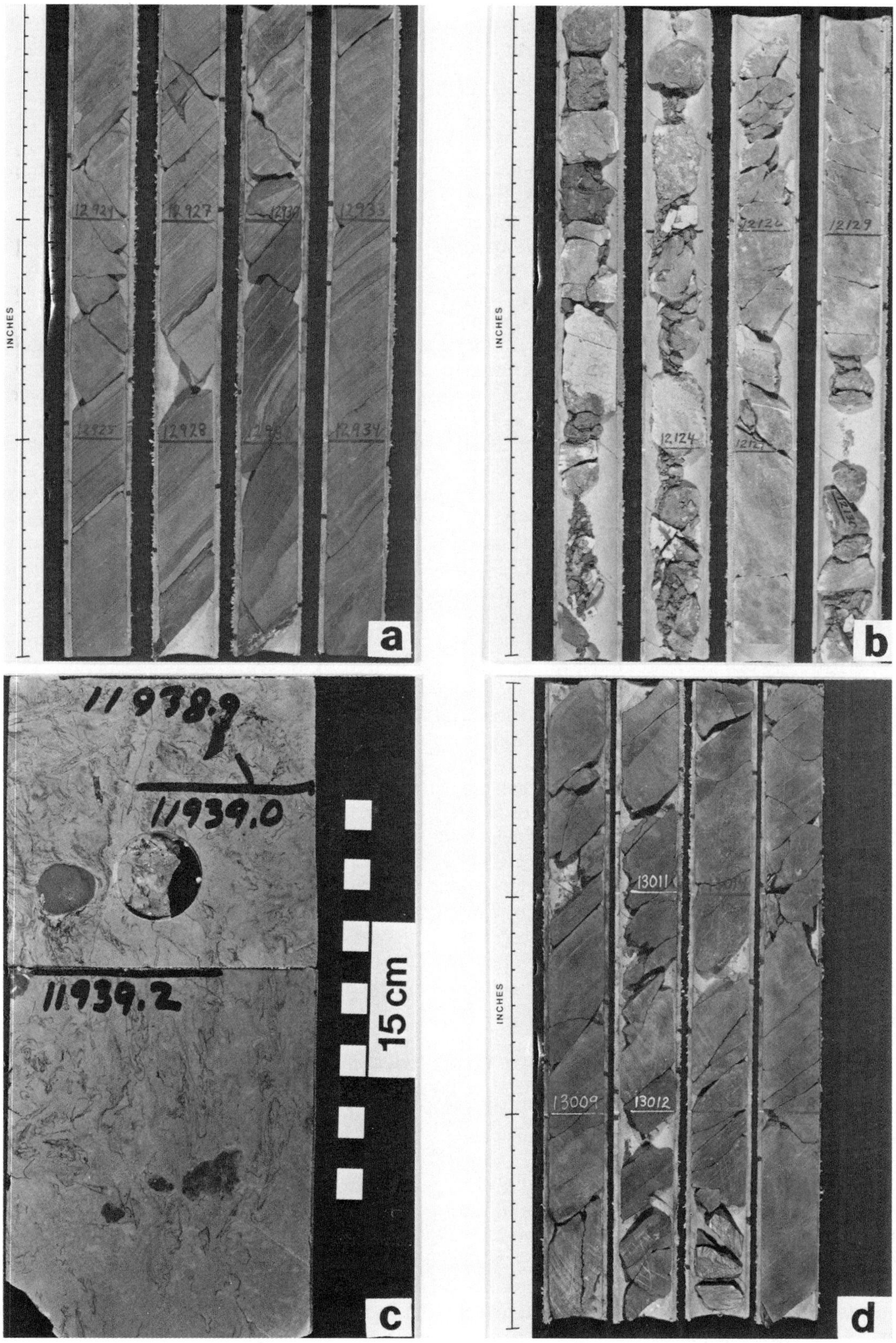

FIG. 8.—Core photographs illustrating various aspects of lacustrine/swamp deposits in lower Kekiktuk at Endicott Field: (a) very well laminated, dark grey to black, carbonaceous mudstones of lacustrine environment. Note: core cut in a deviated well; (b) unlaminated, rubbly, light grey silty mudstone of well drained swamp environment; (c) detail of another well drained swamp siltstone showing abundance of root traces; (d) poorly to non-laminated, dark grey to black carbonaceous mudstones of the poorly drained swamp environment. Note: core cut in deviated well.

quite common, both as nodules and disseminated throughout the rock. These black mudstones are similar to the lacustrine mudstone described above, with the important distinction that they exhibit no lamination but instead are intensely mottled throughout (Fig. 8d).

Lack of drainage in the subenvironment characterized by these coals and carbonaceous mottled mudstones is inferred from the preservation of the abundant organic material. This suggests stagnant, reducing conditions, with highly restricted circulation of the ponded water in what must have been some of the lowest elevations of the overall lower Kekiktuk depositional environment. Similar rocks have been described from the lower Tertiary Wasatch Formation of the Powder River Basin in Wyoming by Ethridge and others (1981), and are analogous to today's organic-rich freshwater clays described from the lowest-lying areas of the Atchafalaya Swamp in Louisiana (Coleman, 1966; Krinitzsky and Smith 1969). Poorly drained swamp deposits are abundant throughout Zone 1 in the lower Kekiktuk at Endicott Field, but it is important to note that they are most common in the lowest, and very highest parts of this stratigraphic interval (Wells A, B. Fig. 3).

The poorly drained swamp deposits are commonly interbedded with the rarer, well drained swamp mudstones. In places, particularly in the upper part of Zone 1 in both Wells A and B (Fig. 3), these mixed swamp deposits overlie laminated, dark grey lacustrine mudstones, suggesting that the lakes slowly filled to a level where plants could take root and flourish in a swamp environment (e.g., Well A, 56 m–58 m, 68 m–77 m; Well B, 67 m–73 m: Fig. 3). In almost all places, the coals and carbonaceous mudstones of the poorly drained swamp environment are also abruptly overlain by well laminated, lacustrine deposits (Well A, 68 m; Well B, 67 m, 73 m: Fig. 3). This suggests that the accumulation of swamp peats commonly ceased by drowning as a result of markedly increased rates of subsidence, particularly toward the end of Zone 1 time. McCabe (1984) has noted how a majority of coals are apparently derived from swamps which were eventually drowned. Lacustrine sediments having relationships similar to those described above in the lower Kekiktuk Formation between swamp deposits and deeper water have been observed by Fielding (1984) and Haszeldine (1984) in the Upper Carboniferous of northern England.

DISCUSSION

Consideration of lithology, lithofacies and lithofacies associations, as well as palynology and paleontology, suggests that the entire Zone 1 succession in the lower Kekiktuk Formation at Endicott Field was deposited within a flood basin environment. A number of distinct paleoenvironments can be recognized, namely: flood basin lakes, well drained swamps and poorly drained swamps, as well as partially and wholly abandoned fluvial channels. Natural levee and crevasse-splay/lacustrine delta deposits are also represented, although specific distinction between them is necessarily difficult considering the nature of the evidence available. The most completely cored sections of Zone 1 occur at Endicott Field, but other cores have been recovered from farther afield at BP Sag Delta 31-10-16 well, and ditch cuttings combined with palynological data and wireline logs have revealed its presence to the northwest at the Reindeer Island Stratigraphic Test (RIST) well (Fig. 1). The lithologies in Zone 1 in those wells are similar to those that have been discussed in this article; hence, the flood basin environment extended over a considerable area.

The evolution of Zone 1 depositional environment may be elucidated by examining its internal stratigraphy and facies organization. The basement of pre-Mississippian deformed argillite that directly underlies the Zone 1 deposits of Endicott Field must have been an extremely low-lying, undulating peneplane. Figure 3 shows how a large proportion of the Zone 1 depositional sequences is characterized by either coarsening-upward trends or by the superposition of shallow-water swamp deposits over deeper water lake floor deposits. In either case, the sequences are interpreted as indicating progressively shallowing water. This suggests that, following drowning of the basement surface, deposition occurred by the infilling of topographic lows upon the lower Kekiktuk swamp plain. Those lows were reflected in an abundance of lakes and poorly drained swamps, and infill took place by either overbank flooding or crevassing of streams which traversed the swamp plain.

Those streams are interpreted from the sharply fining trends observed in the vertical sequence (Fig. 3). The paucity of primary sedimentary structures and the fine grain size typical of those trends suggests that the streams were characterized by mixed to suspended sediment load and prone to widespread flooding. The greatest channel activity occurred around middle Zone 1 time in the Endicott Field area. The sedimentologic evidence and generally low proportion of recognizable channel deposits suggest that even then there was only limited fluvial channel activity. The existence of an environment that was prone to widespread flooding is substantiated by consideration of the coals of the poorly drained swamp environment.

Flores (1983) has observed that frequent interruptions of backswamps in a flood basin by crevasse splay and other overbank sediments, as well as frequent avulsions of meander belts, appear to produce unfavorable conditions for very thick peat (i.e., coal) accumulations. Examination of the cored sequence in Zone 1 (Fig. 3) shows that it can be divided informally into an upper interval characterized by lacustrine and swamp deposits and a lower interval characterized by a greater diversity of environmental settings (channel, overbank, lacustrine, and swamp plain deposits). The coals in that lower Zone 1 interval are generally thin (as much as 25 cm), argillaceous and everywhere associated with an abundance of extremely carbonaceous mudstone. This association suggests that the poorly drained swamps of lower Zone 1 were exposed to regular inundation by mud- and silt-laden floodwaters and that there was little distinction between the fluvial channels and the flood basin. By implication, the streams of the flood basin did not have very high or well developed restraining levees. The relatively diverse number of paleoenvironments interpreted from lower Zone 1 cores included sluggish streams containing a mixed to suspended sediment load, prone to widespread overbank (and possible crevasse) flooding.

Floodwaters invaded numerous floodbasin lakes, which consequently silted up to become well and poorly drained swamps. This setting is illustrated in Figure 9.

The localized infilling of the irregular lower Kekiktuk depositional surface is reflected in the nature of the vertical sequences as discussed above. Infill of the lower Kekiktuk basin can also be considered at a more regional level. Lower Zone 1 deposition has been discussed above. The marked increase in coals and carbonaceous mudstone in the upper parts of this zone (Fig. 3) suggests that in later Zone 1 time a more restricted depositional environment became established that was composed of shallow lakes and poorly drained swamps. Palynological evidence strongly supports the idea of the onset of a widespread swamp environment in latest Zone 1 time.

Two major palynomorph assemblages can be recognized within Zone 1 in a number of wells at Endicott Field (R. L. Ravn, pers. commun.), and they appear to have stratigraphic significance within the zone (Fig. 10). Following initial drowning of the basement surface, the irregular topography became levelled in early Zone 1 time by depositional infill of the hollows as a result of widespread flooding. The variety of depositional subenvironments at that time supported a relatively diverse spore assemblage. Subsequently, the lower Kekiktuk swamp plain became incapable of supporting any active sedimentation other than lacustrine fill and the development of poorly drained swamps. These upper Zone 1 deposits are characterized by the higher of the two palynomorph assemblages, which is more restricted in its character but more widespread in its occurrence (Fig. 10). Coals in the upper Zone 1 interval are more abundant and tend to be thicker than in the lower parts of the zone (Fig. 3), possibly reflecting a marked reduction in argillaceous clastic input in latest Zone 1 time. These better developed coals may have formed in clear water as floating or raised bogs (see McCabe, 1984, for discussion). Certainly, the onset of such widespread swamplands reflects diminished gradient in the lower Kekiktuk depositional basin, and that in turn limited the effectiveness of any rivers that may have existed.

Lower Kekiktuk (Zone 1) sedimentation came to an end with widespread subsidence. Increasingly larger areas of the low-lying basin became inundated by swamps as indicated by the onlapping nature of the higher, but more environmentally restricted, of the two palynomorph assemblages (Fig. 10). Locally, subsidence rates may have fluctuated quite dramatically, as suggested by the repeated drowning of the peat swamps by lakes, discussed previously (Fig. 3). Those fluctuations in subsidence rates may have been related to compactional effects, or they may reflect initial responses to tectonic instability in the Kekiktuk basin in the region of Endicott Field. That instability manifested itself ultimately in the termination of Zone 1 depositional patterns and the abrupt introduction of coarse-grained clastics of Zone 2 (Melvin, 1985).

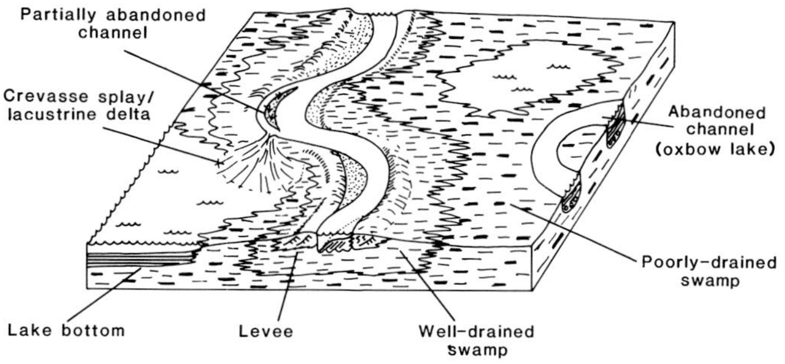

FIG. 9.—Schematic representation of sedimentation in Endicott Field area during lower Kekiktuk (Zone 1) time. The area was dominated by a very low-lying swamp plain with abundant lakes, abandoned meanders and sluggish silt-laden streams.

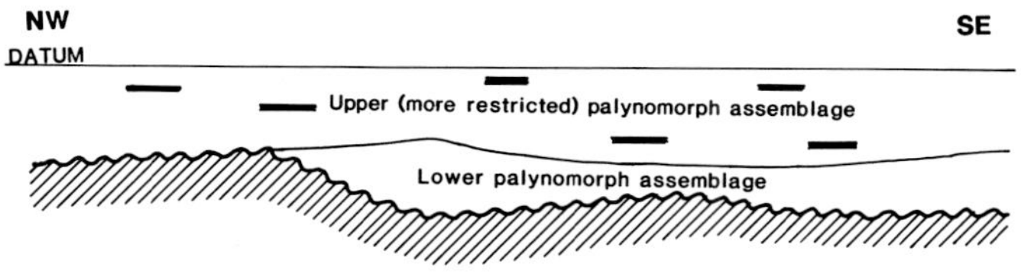

FIG. 10.—Schematic NW-SE biostratigraphic section across Endicott Field area, showing distribution of palynologic assemblages in Zone 1. Note onlapping effect of upper restricted assemblage. Datum is top Zone 1. Black bar ornament indicates dominance of swamp deposits.

ACKNOWLEDGMENTS

This paper is published with the permission of Standard Oil Production Company. The data were originally presented as part of a broader paper on the sedimentological evolution of the Kekiktuk Formation at Endicott Field during the 1985 Meeting in Anchorage, Alaska of the Pacific Sections of the AAPG/SEPM. My thoughts on the Kekiktuk Formation have benefitted from discussions with A. S. Knight, J. E. Williams, K. Woidneck and R. L. Ravn to whom I owe my thanks. I also appreciate the constructive comments of C. R. Fielding and B. P. J. Williams who reviewed the paper. I thank Leta Smith (drafting) and Robin Gardner (typing) for their patient assistance in the production of this paper.

REFERENCES

BROSGE, W. P., DUTRO, J. T., MANGUS, M. D., AND REISER, H. N., 1962, Paleozoic sequence in the eastern Brooks Range, Alaska: American Association of Petroleum Geologists Bulletin, v. 46, p. 2174–2198.

COLEMAN, J. M., 1966, Ecologic changes in a massive fresh-water clay sequence: Transactions of the Gulf Coast Association of Geological Societies, v. XVI, p. 159–174.

———, 1969, Brahmaputra River: Channel processes and sedimentation: Sedimentary Geology, v. 3, p. 129–239.

ELLIOTT, T., 1974, Interdistributary bay sequences and their genesis: Sedimentology, v. 21, p. 611–622.

ETHRIDGE, F. G., JACKSON, T. J., AND YOUNGBERG, A. D., 1981, Floodbasin sequence of a fine-grained meander belt subsystem: The coal-bearing lower Wasatch and upper Fort Union Formations, southern Powder River Basin, Wyoming, *in* Ethridge, F. G., and Flores, R. M., eds., Recent and Ancient Nonmarine Depositional Environments: Models for Exploration: Society of Economic Paleontologists and Mineralogists Special Publication 31, p. 191–209.

FIELDING, C. R., 1984, Upper delta plain lacustrine and fluviolacustrine facies from the Westphalian of the Durham coalfield, NE England: Sedimentology, v. 31, p. 547–567.

FLORES, R. M., 1983, Basin facies analysis of coal-rich Tertiary fluvial deposits, northern Powder River Basin, Montana and Wyoming, *in*

Collinson, J. D., and Lewin, G., eds., Modern and Ancient Fluvial Systems: International Association of Sedimentologists Special Publication 6, p. 501–515.

Gersib, G. A., and McCabe, P. J., 1981, Continental coal-bearing sediments of the Port Hood Formation (Carboniferous), Cape Linzee, Nova Scotia, Canada, *in* Ethridge, F. G., and Flores, R. M., eds., Recent and Ancient Nonmarine Depositional Environments: Models for Exploration: Society of Economic Paleontologists and Mineralogists Special Publication 31, p. 95–108.

Haszeldine, R. S., 1984, Muddy deltas in freshwater lakes, and tectonism in the Upper Carboniferous coalfield of NE England: Sedimentology, v. 31, p. 811–822.

Krinitzsky, E. L., and Smith, F. L., 1969, Geology of backswamp deposits in the Atchafalaya Basin, Louisiana: U.S. Army Waterways Experiment Station, Corps of Engineers, Vicksburg, Mississippi, Technical Report S-68-8, 60 p.

McCabe, P. J., 1984, Depositional environments of coal and coal-bearing strata, *in* Rahmani, R. A., and Flores, R. M., eds., Sedimentology of Coal and Coal-Bearing Sequences: International Association of Sedimentologists Special Publication 7, p. 13–42.

Melvin, J., 1985, Sedimentological evolution of Mississippian Kekiktuk Formation, Sagavanirktok delta area, North Slope, Alaska (abs.): American Association of Petroleum Geologists Bulletin, v. 69, p. 669.

Moore, T. E., and Nilsen, T. H., 1984, Regional variations in the fluvial Upper Devonian and Lower Mississippian(?) Kanayut Conglomerate, Brooks Range, Alaska: Sedimentary Geology, v. 38, p. 465–497.

Nilsen, T. H., 1981, Upper Devonian and Lower Mississippian redbeds, Brooks Range, Alaska, *in* Miall, A. D., ed., Sedimentation and Tectonics in Alluvial Basins: Geological Association of Canada Special Paper 23, p. 187–219.

———, and Moore, T. E., 1982, Fluvial facies model for the Upper Devonian and Lower Mississippian(?) Kanayut Conglomerate, Brooks Range, Alaska, *in* Embry, A. F., and Balkwill, H. R., eds., Arctic Geology and Geophysics: Canadian Society of Petroleum Geologists Memoir 8, p. 1–12.

Weimer, R. J., 1977, Stratigraphy and tectonics of western coals, *in* Murray, D. K., ed., Geology of Rocky Mountain Coal: Colorado Geological Survey Resource Series 1, p. 9–27.

NUMERICAL SIMULATION OF GOLD DISTRIBUTION IN THE WITWATERSRAND PLACERS

M. NAMI
Gold Exploitation Laboratory, Chamber of Mines Research Organization, Johannesburg
AND
C. S. JAMES
Department of Civil Engineering, University of Witwatersrand, Johannesburg

ABSTRACT: The distribution pattern of gold deposits in a fluvial environment can be studied by either a descriptive or a deterministic approach. The deterministic approach is based on physical and mathematical modelling of the processes involved in the formation of the placer and can be applied to the system, channel reach and channel cross section scales. Two numerical models are presented which address the channel reach to cross section scales. These describe the fine particles in suspension in a compound channel reach and the associated patterns of deposition of these particles. The models are used to reproduce observed cross section distributions of gold concentration in the Carbon Leader and Composite Reefs. The results show that the observed distributions can be explained deterministically. The improved understanding of gold distribution gained by this approach will help to improve techniques for valuation and ore reserve estimation and thus lead to more economical exploitation of gold reserves.

INTRODUCTION

An understanding of the distribution and concentration of any economically viable mineral in placer deposits is of fundamental importance to the mining of that deposit. As a result of advance knowledge concerning distribution patterns of these minerals, considerable improvements can be made to valuation of the ore body which leads to the development of more economic methods of mining.

One of the most important, yet difficult, problems is the prediction of the distribution pattern of tracer minerals, such as gold, in a water-laid placer deposit. In order to predict distribution patterns of gold in a fluvial environment, two approaches can be employed: (i) the descriptive approach, and (ii) the deterministic approach. The descriptive approach involves the identification of the sedimentary environment and the relationship between sedimentological characteristics and gold distribution. Using this approach, it is possible to classify the placer into various sedimentary units (facies) with specific characteristics, where each facies represents a unique process of sedimentation. Cognizance of sedimentary facies and incorporation of these facies in the present valuation techniques (e.g., kriging) results in considerable improvements in valuation and ore reserve estimation of the gold-bearing placer deposits.

The deterministic approach is based on physical and mathematical modelling of the prevailing hydraulic conditions and processes which were responsible for deposition and distribution of gold. In this case not only can sedimentary units be classified, but relative distribution patterns can also be estimated. Furthermore, the direction and magnitude of the trends in the placer can be determined with a high degree of confidence.

While the descriptive sedimentological approach is receiving considerable attention in the Witwatersrand gold mining industry, little emphasis has been placed on the deterministic approach. In an effort to investigate the deterministic approach, a major program has been embarked upon by the Chamber of Mines Research Organization. Two numerical models have been developed to simulate the hydraulic transport and deposition processes responsible for the distribution and concentration of gold in Witwatersrand placers. This paper describes the fundamental concept of these models and illustrates case studies from various underground exposures of the placer.

SPATIAL SCALES

The following three-scale levels (Fig. 1) have been identified on which deterministic analysis of gold distribution can be conducted.

(i) The system scale considers whole fluvial systems extending over distances on the order of kilometers.
(ii) The channel reach scale focuses on individual reaches within the total system and would consider longitudinal distributions over hundreds of meters and transverse distributions over distances of less than tens of meters.
(iii) The cross section scale considers detailed transverse distributions at individual sections along a reach.

The distribution of gold at all scales is interrelated and the reach and cross section scale distributions, in particular, cannot be considered independently. It is important to make the scale distinctions, however, because the dominant transport processes are different, and different mathematical models would be required.

Simulation of distribution of gold deposits is probably easiest at the system scale. The system characteristics can be described by average values of parameters, such as gradient, channel size, channel shape, etc. Local deviations from average conditions would not affect the overall distribution of gold concentration through the system. On the smaller scales, it is the local characteristics which determine the patterns of concentration distribution which are of interest. These local characteristics are not easy to predict. From a practical point of view, however, it is the smaller scale distributions which are important in planning mining operations, and for this reason simulation on the reach to cross section scale has been attempted first.

TRANSPORTATION OF GOLD

It has been established (James, 1984) that the fine-grained gold now present in the placer was originally transported in suspension. This was done by comparing the hydraulic

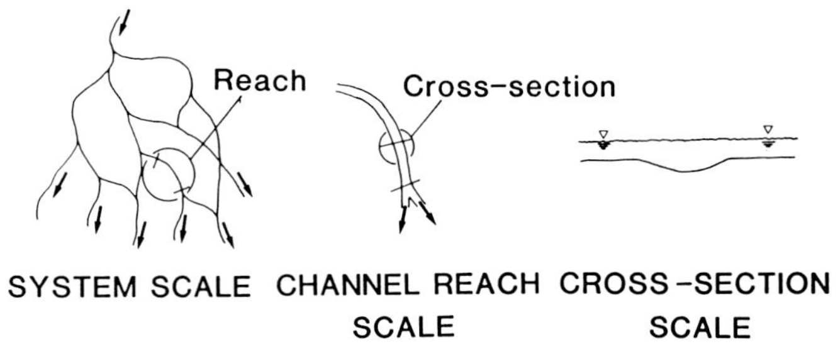

FIG. 1.—Classification into system, reach and cross section scales.

TABLE 1.—MATHEMATICAL NOTATION

c	Local sediment concentration (arbitrary units)
d	Particle size (mm)
Dp	Flow depth on plain (m)
ks	Representative bed roughness (mm)
S	Hydraulic gradient
p	Probability of deposition of suspended particle
u	Transverse velocity component (m/s)
u^*	Shear velocity (m/s)
	Longitudinal velocity (m/s)
w	Particle fall velocity (m/s)
x	Longitudinal direction
y	Vertical direction
z	Transverse direction
ε	Particle diffusivity (m^2/s)

conditions necessary to mobilize the largest quartz-density particles occurring in the placers with the conditions necessary for suspension of gold particles. Figure 2 shows the combinations of flow depth and hydraulic gradient necessary to transport 70-μ gold particles in suspension, according to Engelund's (1973) criterion:

$$w < 0.8\mu* \tag{1}$$

in which w is the particle fall velocity and $u*$ is the shear velocity associated with grain roughness (Table 1). Also shown are the combinations necessary to mobilize 17-mm quartz-density particles according to the criteria of Shields and Meyer-Peter and Muller (see Graf, 1971). For these fairly typical particle sizes, it can be seen that the prevailing hydraulic conditions were probably well in excess of those necessary for suspension of gold.

Because suspension is believed to have been the dominant mode of transport of gold, the distribution of deposits must be closely related to the distribution of suspended gold during transport. The mathematical simulation models describe the distribution of suspended particles under the influence of convective and diffusive transfer components. Interaction between suspended particles and the bed material is described by Einstein's (1950) probability of erosion.

NUMERICAL MODELS

The models developed so far simulate the distribution of suspended and deposited material in a channel-overbank system as illustrated in Figure 3. During periods of overbank flow, a strong interaction takes place between the fast, deep, main channel flow and the relatively slow, shallow flow over the plain sections. This interaction transfers momentum and any suspended particles from the high concentration channel region to the lower concentration plain regions, where suspended particles tend to settle because of the reduced transport capacity of the flow. The presence of a transverse flow velocity component, such as would occur if the channel were inclined to the principal flow direction over the plain, would enhance the transfer process on one side of the channel and inhibit it on the other, resulting in an asymmetrical distribution of deposition.

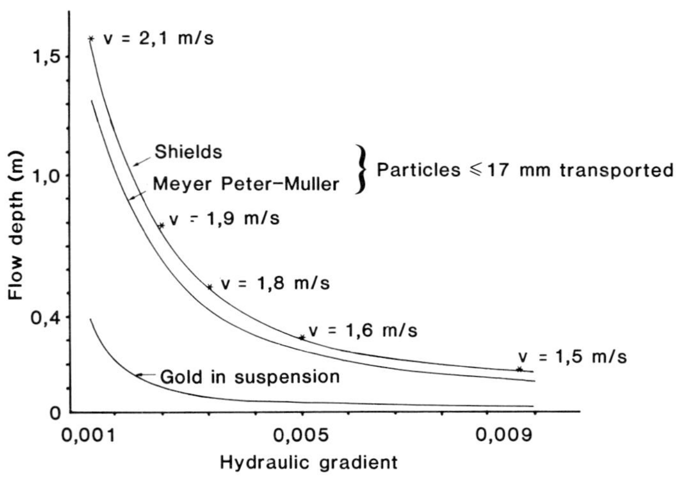

FIG. 2.—Hydraulic conditions for gold suspension.

The transfer of material to the overbank section depletes material in the channel and, therefore, must also affect the distribution of deposits along the length of the channel. The three-dimensional problem of describing the distribution of suspended material in a system as shown in Figure 3 has been separated into two two-dimensional problems. The transverse distribution across the plain and the longitudinal distribution down the channel are described by two separate models.

The transverse distribution is described by the equation:

$$0 = \frac{\partial}{\partial y}\left(\varepsilon y \frac{\partial C}{\partial y}\right) + \frac{\partial}{\partial z}\left(\varepsilon z \frac{\partial C}{\partial z}\right) + w\frac{\partial C}{\partial y} - u\frac{\partial C}{\partial z} \tag{2}$$

in which y and z are the vertical and transverse directions, respectively, εy and εz are the diffusivities for sediment in these directions, C is sediment concentration, w is the particle fall velocity (positive downward) and u is the transverse velocity component. The solution to this elliptic equation is obtained numerically by the method of Successive

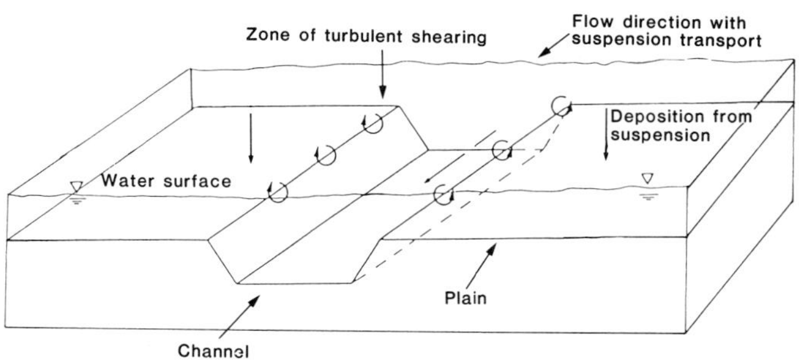

FIG. 3.—Illustration of a compound channel system.

Over-Relaxation (Smith, 1978) and gives the distributions of sediment concentration throughout the flow region over the plain. This numerical solution procedure assumes initial concentration values at grid points throughout the solution domain. These values are adjusted iteratively until equation (2) is satisfied simultaneously for all points within a specified tolerance. Convergence is accelerated by applying an optimization procedure proposed by Apelt and Isaacs (1977). The distribution of deposits can be obtained from the concentration distribution according to the specified bed boundary condition (equation 5). The values of εy and εz are obtained from empirical results presented by other researchers. (Graf, 1971; Rajartnam and Ahmadi, 1981).

The longitudinal distribution is described by the equation:

$$0 = v\frac{\partial C}{\partial x} + w\frac{\partial C}{\partial y} + \frac{\partial}{\partial y}\left(\varepsilon y\frac{\partial C}{\partial z}\right) - u\frac{\partial C}{\partial z} + \frac{\partial}{\partial z}\left(\varepsilon z\frac{\partial C}{\partial z}\right) \quad (3)$$

in which x is the longitudinal direction and v is the velocity in that direction. Longitudinal diffusion is considered negligible compared with longitudinal convection. The last two terms in this equation can be evaluated by the transverse distribution model and treated as constants in the solution of the longitudinal model, making it a two-dimensional rather than a three-dimensional problem. Transverse transfer within the channel is ignored. The solution is obtained by an explicit numerical method in which concentration values at grid points on a vertical section are computed progressively downstream from known values at the preceding section. This yields the vertical and longitudinal distribution of concentration along the channel. The distribution deposits can again be determined from the concentration distribution and the specified boundary condition.

The initial condition for the longitudinal model and the channel-plain interface boundary for the transverse model are specified in terms of concentration values over the flow depth. In both models the boundary condition at the water surface states that no transfer can take place across this surface, i.e.,

$$\varepsilon_y\frac{\partial C}{\partial y} + wC = 0 \quad (4)$$

At the channel bed or plain surface, the rate of transfer across the boundary is defined by the probability (p) that a particle reaching the surface will deposit (entrainment from the bed is not accounted for), i.e.,

$$\varepsilon_y\frac{\partial C}{\partial y} + (1-p)wC = 0 \quad (5)$$

The probability (p) is estimated as the complement of Einstein's (1950) probability of erosion.

The mathematical models and their numerical solutions are described in detail by James (1984, 1985).

Simulation of Gold Distribution

The two models described above have been used together to reproduce some observed distributions in several sections of the Carbon Leader Reef placer at Blyvooruitzicht Gold Mine and the Composite Reef placer at Randfontein Estates Gold Mine. Examples of these simulations follow.

Carbon Leader Reef—east of No. 2 shaft pillar.—

On the cross section shown in Figure 4, the channel shape and size can be clearly seen. The measured gold concentration values plotted on the cross section show a distinct correlation with the channel geometry, with high values on the overbank sections close to the channel edge. The distribution pattern is consistent with the transfer mechanisms described earlier. The distribution is asymmetric, suggesting the presence of transverse convection. This is supported by the increasing trend of concentration within the channel toward the bank with the greater concentration.

The overbank distributions were reproduced using the transverse distribution model by assuming gold particle characteristics and different combinations of uniform plain flow depth (Dp), gradient(s) and transverse convection (u) until a reasonable fit was obtained (as shown). There are, in fact, many different combinations of these parameters which would give just as good a fit. Only one combination, however, will give a satisfactory fit to the overbank distributions and also give the correct proportional relationship between overbank deposits and average channel deposition relative to the plain deposition. The overall distribution thus found and the assumed parameter values are shown on the cross section in Figure 4.

Carbon Leader Reef—western part of 5-22 raise, Blyvooruitzicht.—

The same procedure was used to reproduce another distribution measured in the Carbon Leader Reef. This cross section, shown in Figure 5, has some interesting features. The channel is not symmetrical, possibly because the section is at a bend, in which case significant transverse convection would be expected, as is indicated by the very ex-

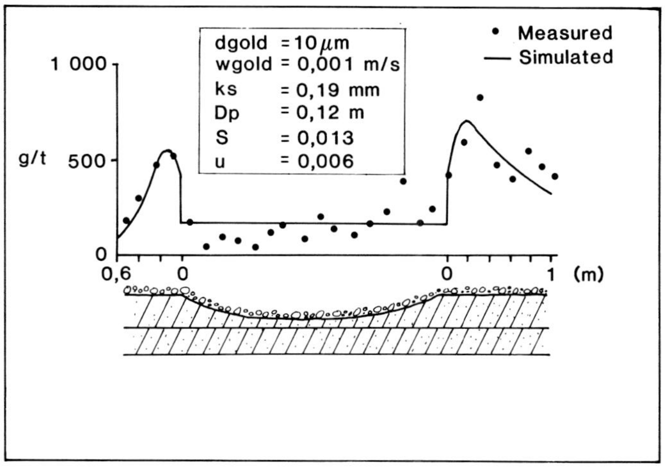

FIG. 4.—Diagram of a section through the Carbon Leader east of the No. 2 shaft pillar showing the measured and simulated gold distributions.

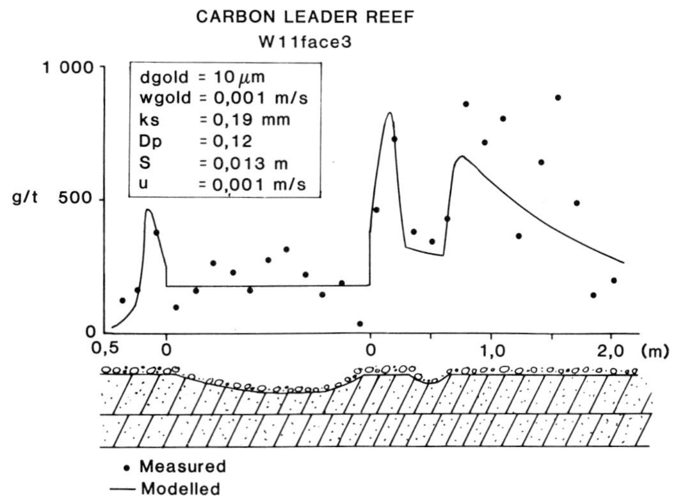

FIG. 5.—Diagram of a section through the Carbon Leader at 5-22 raise, showing the measured and simulated gold distribution patterns.

tensive gold deposition on the one side of the channel. Also, there appears to be a secondary channel a short distance from the main one on one side. Although there is considerable scatter of data, the gold concentration appears to be lower at the location of this secondary channel. The secondary channel was accounted for in the simulation by precalculating the deposition probability appropriate to its depth, and overriding the calculated values with this. The overall simulated distribution is as shown on Figure 5.

Composite Reef—85 west 4.—

This simulation is not as satisfactory as the others because the channel geometry is not well defined. Only one side of the channel is present (Fig. 6) due to faulting, and on this side the channel bed rises gradually to the overbank level, making it difficult to define a distinct intersection between channel and plain. On the other hand, more data were available and the gold concentration distribution represents an average obtained from a number of adjacent cross sections, and the scatter is, therefore, less. The absence of one overbank distribution removes a constraint from the

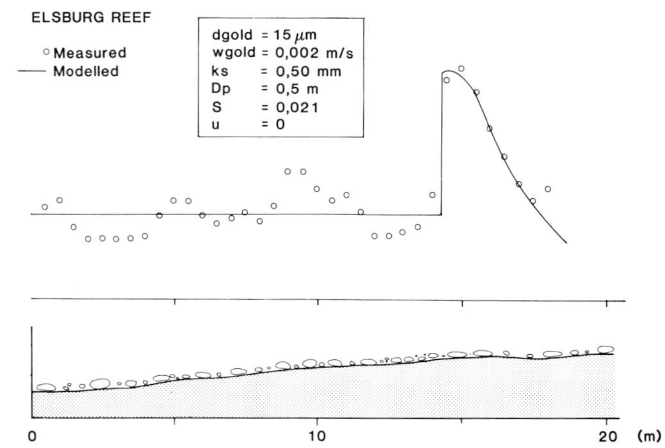

FIG. 6.—Diagram of a section through the composite reef at 85W4 showing the measured and simulated gold distribution patterns.

simulation, and the assumed set of hydraulic parameters is not as nearly unique as for the other simulations. The simulated distribution is shown on the cross section in Figure 6.

DISCUSSION AND CONCLUSIONS

The models presented in this paper provide a deterministic explanation of some gold distribution patterns observed in the Witwatersrand placers. The improved understanding of gold distribution gained by developing and applying such models will lead to better economic exploitation of the gold mine reserves by improving valuation techniques and advance planning of mining operations. Examples of such improvements include selective mining, demarcation of pay and nonpay areas in mines, appropriate blending of high- and low-grade ore for increasing mine life, and accurate siting of stabilizing pillars in deep-level mining.

Although many advances have been made in recent years in geostatistical techniques for valuation and ore reserve estimation, these use sampling data without giving cognizance to the distribution of gold. Many errors in geostatistical applications result from the omission of geological and especially sedimentological information regarding the placer deposit. As our knowledge of the hydraulic conditions prevailing during formation of the placer advances, the use of deterministic models could provide quantitative techniques for predicting gold distribution patterns, thereby improving present valuation techniques.

It must be emphasized that the simulation results presented here are reproductions of observed distributions, obtained by selection of suitable hydraulic parameters (flow depth, gradient, transverse convection, bed roughness), rather than predictions. The deterministic approach will be useful for predictions only when more knowledge is available concerning prevailing hydraulic conditions and the characteristics of gold particles at the time of deposition. Certain aspects of the existing models also still require experimental validation. It is also not entirely realistic to attempt simulations at isolated sections of channels; sediment concentrations can vary along a reach even if flow is relatively uniform. The results do show, however, that observed distributions can be explained deterministically, and the parameter values used to obtain agreement are realistic. It is therefore believed that the deterministic approach has potential for improving our understanding of gold distribution, thereby leading to more economical exploitation of gold reserves.

ACKNOWLEDGMENT

The work described in this paper forms part of a research program of the Research Organization of The Chamber of Mines of South Africa. Their permission for its publication is gratefully acknowledged.

REFERENCES

APELT, C. J., AND ISAACA, C. T., 1977, On the estimation of the optimum accelerator for SOR applied to finite element methods: Computer Methods

in Applied Mechanics and Engineering, v. 12, p. 383–391.
EINSTEIN, H. A., 1950, The bed-load function for sediment transportation in open channel flows: U.S. Department of Agriculture, Soil Conservation Service Technical Bulletin No. 1026, 70 p.
ENGELUND, F., 1973, Steady transport of moderately graded sediment, Lyngby: Technical University of Denmark, Institute of Hydrodynamics and Hydraulic Engineering, Progress Report No. 29, p. 3–12.
GRAF, W. H., 1971, Hydraulics of Sediment Transport: McGraw-Hill Book Company, N.Y., 573 p.
JAMES, C. S., 1984, Numerical modelling of gold transport and deposition: Unpublished Ph.D. Dissertation, University of the Witwatersrand, Johannesburg, South Africa, 257 p.
———, 1985, Sediment transfer to overbank sections: Journal of Hydraulic Research, v. 23, p. 435–452.
RAJARATNAM, N. AND AHMADI, R. M., 1981, Hydraulics of channels with flood plains: Journal of Hydraulic Research, v. 19, p. 43–60.
SMITH, G. D., 1978, Numerical Solution of Partial Differential Equations: Second edition, Oxford University Press, 304 p.

FACIES CHANGES FROM FLUVIAL CONGLOMERATE TO BRAIDED SANDSTONE OF THE EARLY PROTEROZOIC ELDORADO FORMATION, WELKOM GOLDFIELD, SOUTH AFRICA

CHARLES S. KINGSLEY
Department of Geology, Anglo American Corporation, Welkom, South Africa

ABSTRACT: The Eldorado Formation is a wedge-shaped sequence of fanglomerates and arenites which thickens from 200 m in the west to 600 m in the east. The formation rests unconformably on the Aandenk Formation and is overlain by the Ventersdorp lava. The rapid change in thickness is due to intraformational unconformities. Four distinct alluvial fan cycles with radii of between 5 km and 25 km prograded from west to east; these cycles define four members, from the bottom upward: the Rosedale Member, the Van den Heeversrust Member, the EA zone and Uitkyk Member. The lower two members are crude fining-upward megasequences. The topmost two members of the formation in the west consist almost entirely of rudities which, 15 km to the east, grade into arenites.

In the southern part of the area the entire Eldorado Formation is composed of clast-supported, poorly sorted polymictic conglomerates, which show crude stratification, and trough cross-bedded litharenites and subgreywackes. This is called the Welkom facies. In the northern part of the area oligomictic conglomerates and quartzarenites are commonly developed in the lower three members, especially in the EA zone; this is called the Loraine facies. These two facies interfinger. Sedimentation units commonly between 60 cm and 120 cm thick fine upward into matrix-supported conglomerates capped in places by trough cross-bedded arenites. Ephemeral sheet floods produced the gravels, whereas sands were deposited during waning flow conditions. Detrital pyrite, gold and uraninite occur sporadically and in low concentrations in the polymictic Rosedale Placer at the base of the formation. These heavy minerals are also concentrated at various horizons in oligomictic conglomerates of the Loraine facies. Pay zones as much as 400 m long are oriented from west to east perpendicular to the depoaxis of the basin. The amount of gold shows a sympathetic relation with the size and amount of pyrite grains which decreases sharply east of this axis.

Each of the lower two members is interpreted as a geomorphological cycle which was deposited during uplift of the source area. In each case tectonism was followed by recession of the fan and retreat of the uplifted mountains. The upper two members were deposited during more frequent uplifts of the source area. Marginal unconformities and the wedging of units near the basin edge in the west suggest syndepositional folding and thrusting in the source area and accumulation of fanglomerates in a foreland molasse basin. Two entirely different sources, one oligomictic and the other polymictic, are postulated. Gravels and sands were distributed from west to east as alluvial fans from both sources. The sediments were redistributed longitudinally as gravel and sand bars on a broad braid plain down a southeastward paleoslope.

INTRODUCTION

Vertical sequence analysis has only recently been applied in studies of the Witwatersrand Basin and a few authors have attempted to investigate parts of the sequence toward model building. In a broad sense Winter (1984) used the vertical sequence to draw some similarities between the structural styles of the Great Eastern Basin and those of the Witwatersrand Basin. He approached his study from a tectonostratigraphic viewpoint using genetic sequences of strata (GSS), as defined by Busch (1971), in correlating stratigraphic units. Eriksson and others (1981) studied the lower part of the Witwatersrand succession making use of sequence analysis to derive its model of sedimentation. In a detailed sequence analysis of the Dagbreek Formation, Kingsley (1984) constructed the environmental model of this fining-upward megacycle.

In the present study of the Eldorado Formation the author has combined detailed sequence analysis with the concept of tectonostratigraphy. Using this concept, GSS have been studied and reliable interrelationships among facies defined (Kingsley, 1979). Finally, the model of sedimentation and tectonic control has been synthesized based on the observed features and geometry of the sedimentary bodies. By using a genetic approach to depositional cycles combined with tectonics, the sedimentation of the Eldorado sequence and more particularly the Eldorado placers is explained. This synthesis was done from information accumulated from four underground sections and core from approximately 150 drill holes.

The Eldorado Formation is about 500 m thick, covers the entire Welkom Goldfield and extends much farther into the basin. The Welkom Goldfield lies at the southern tip of the Witwatersrand Basin (Fig. 1). This paper describes proximal-distal facies relationships over a 1000 km^2 area in the Eldorado Formation and relates its depositional characteristics to periodic uplift and syntectonic alluvial fan sedimentation. The proximal facies is dominated by fanglomerates and the distal facies by braid plain sands. The author points out first the remarkable facies changes that take place in the Eldorado Formation; second the thinning of this sequence due to intraformational basin edge unconformities; and, third that syntectonic sedimentation took place during folding and thrusting episodes in a foreland molasse basin.

The stratigraphic classification is shown in Figure 2. Only a few reliable previous studies are notable. Winter (1964a) found rapid facies changes in the Loraine area from conglomeratic to arenitic facies, both within the EA conglomerates and the Uitkyk Member. Sims (1969) showed that the percentage conglomerate in the Van den Heeversrust Member decreases in a fanlike fashion eastward from 60 percent to 5 percent within 3.5 km. A study on clast size distribution by Kleynhans (1970) showed that the main source of the Eldorado sediments lay west of the goldfield and that the coarse material was distributed in a fanlike pattern, thus confirming Sims's conclusion.

In a more general sense a conceptual model of sedimentation of Witwatersrand placer deposits was proposed by Pretorius (1976) in which each placer was considered to be the result of cyclic sedimentation, first deposited during progradation, followed by recession of an alluvial fan. In another description of the megasequences and facies changes of the Witwatersrand Supergroup, Vos (1975) concluded that the main depositional environment contained wet al-

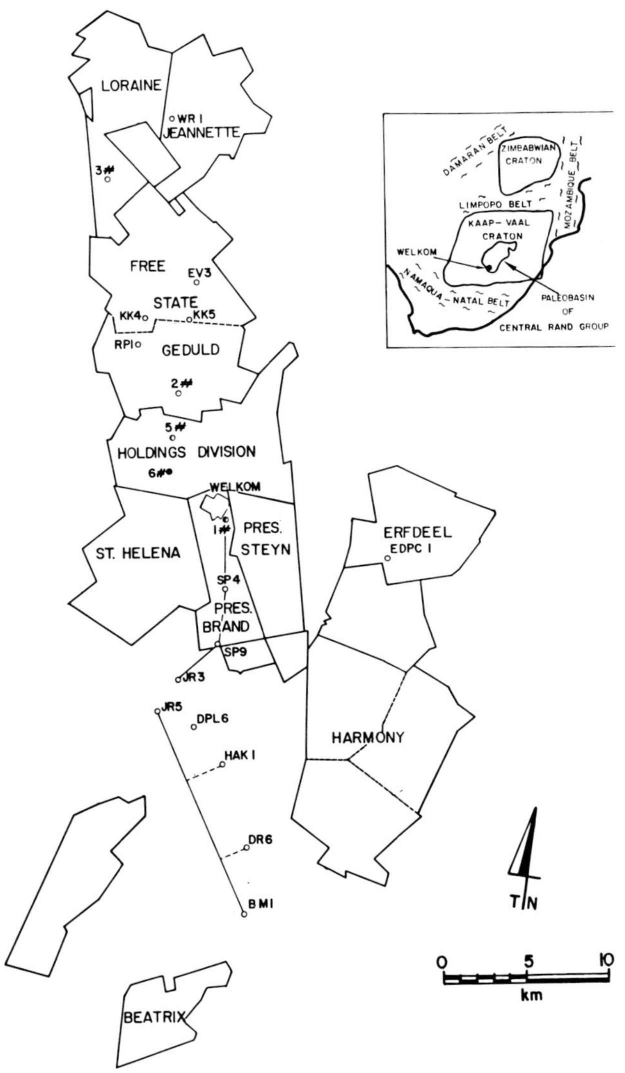

FIG. 1.—Index map of the Welkom Goldfield showing location of drill holes, shafts and two section lines (Figs. 4 and 9).

luvial fans. He postulated a braided alluvial plan with coarse gravels grading downstream into braided channel sands and finally into lacustrine silts and muds.

STRATIGRAPHY

The Eldorado Formation is a wedge-shaped coarsening-upward sequence of fanglomerates and arenites located at the top of the Central Rand Group. It thickens from west to east from 200 m to 600 m over a distance of 5 km. The rapid decrease in thickness westward is mainly due to basin edge unconformities at the bases of the major conglomerate units. Four different sedimentary cycles are recognized in the Welkom Goldfield, namely the ED zone (Rosedale Member), the EC-EB zone (Van den Heeversrust Member), the 'EA' zone, and the Uitkyk Member, each varying between 100 m and 150 m in thickness (Fig. 2).

The Rosedale Member consists of a polymictic conglomerate, the Rosedale Placer, at the base, succeeded by an alternation of quartzarenites and litharenites. The Van den Heeversrust Member is typically a very gritty litharenite and subgreywacke in the lower part with minor quartzarenite beds that increase upward. Subgreywacke and polymictic conglomerates alternating with minor quartzarenites and oligomictic conglomerates characterize the overlying EA zone. Locally, the latter conglomerates are economically mineable deposits on Loraine Gold Mine. Overlying this unit is the Uitkyk Member consisting of polymictic boulder and cobble conglomerates.

The lithofacies of the various units of the Eldorado Formation vary in the different areas of the goldfield because dramatic facies changes take place down the paleoslope, a phenomenon discussed in more detail later. Furthermore, in the Loraine area in the north, quartzarenites predominate over litharenites and subgreywackes, whereas in the south, (the Welkom area) litharenites and subgreywackes are in abundance. Similarly, oligomictic conglomerates are a feature of the Loraine area and polymictic conglomerates that of the Welkom area. These two contrasting facies interfinger.

At the western edge of the basin the base of the Eldorado Formation is an angular unconformity which truncates all the underlying formations of the Central Rand Group. Unconformities are also evident at the bases of the different members, the Uitkyk Member showing the most marked unconformity as it overlaps all the underlying members as well as the older formations (Fig. 3). Both McKinney (1964) and Winter (1964a) discuss the basin edge unconformities and show how the increase in conglomerate westward may cause misinterpretations of the stratigraphy. These unconformities become disconformities eastward deeper into the basin. The unconformable relationships within the EA zone are discussed later. Within the basin south of Welkom the unconformity at the base is less angular, and the Eldorado Formation overlaps the underlying formations of the Central Rand Group over a distance of some 30 km.

DESCRIPTION AND FACIES CHANGES

In the area of investigation two distinct facies are recognized, one of which is predominantly quartzarenite in the north (the Loraine facies) and the other which constitutes mainly litharenite and subgreywacke facies in the south (the Welkom facies). These two contrasting facies are developed in the lower three members, especially in the EA zone. These facies are best illustrated where quartzarenites and litharenites of the EA zone interfinger to form the Rainbow Reefs on Loraine Mine No. 3 Shaft area (Winter, 1964a). The different lithofacies of the Eldorado Formation are repeated vertically through the sequence. A common feature of the Welkom facies is the presence of polymictic conglomerates with a broadly similar pebble assemblage. In contrast, in the Loraine area subordinate oligomictic placer conglomerates occur at various horizons but particularly within the EA zone. Another feature of the Welkom facies is the chloritic matrix of the immature arenites and conglomerates giving it a dark grey color, which is vastly different from the mainly pyrophillitic component present in the underlying Aandenk and Spes Bona Formations.

Marked facies changes from conglomerates to arenites

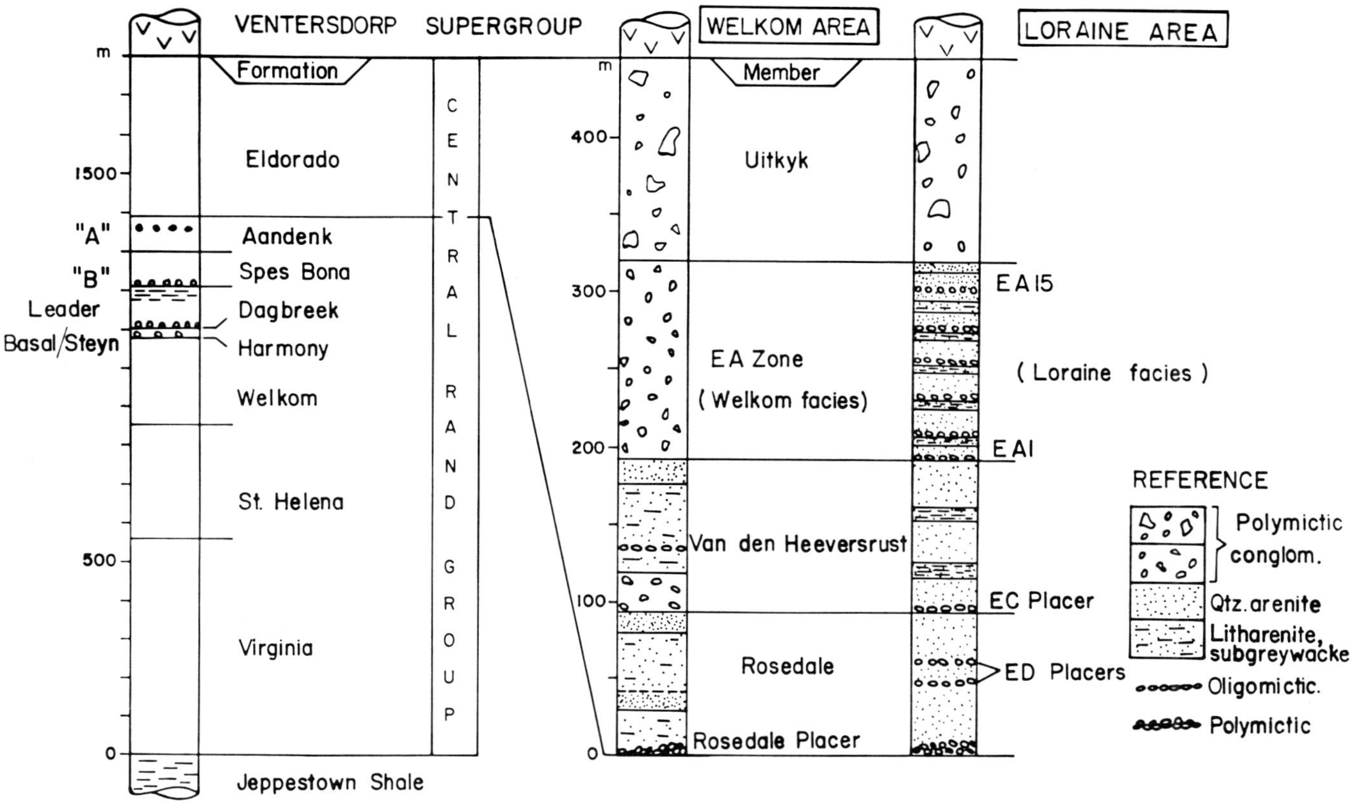

FIG. 2.—Stratigraphic subdivision of the Central Rand Group and the Eldorado Formation showing its main lithologies.

can be demonstrated at many localities and in all directions down the fan paleoslope. Facies changes affecting all the members regionally are illustrated by the section line which goes from the proximal JR 5 drill hole to the distal BM 1 drill hole over 18 km (Fig. 4). Note how the conglomerates of the EA zone reach much farther into the basin than the others. This phenomenon will be described in more detail later in this paper.

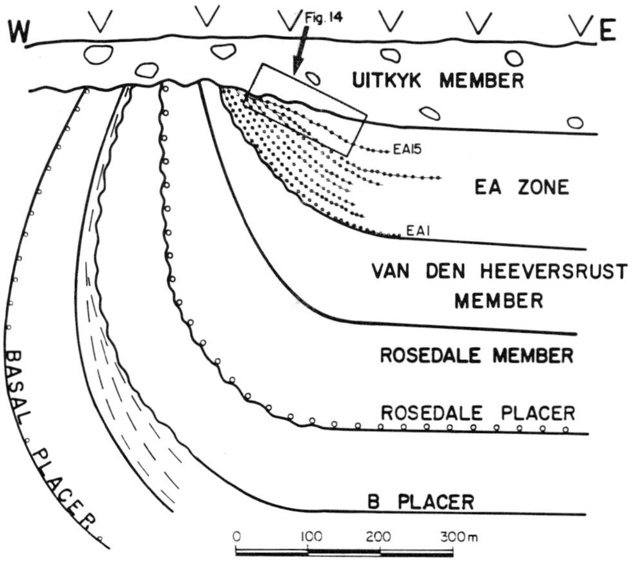

FIG. 3.—Unconformable relationship of the Eldorado Formation upon older formations on Loraine Mine. Note the angular unconformity underneath the Uitkyk Member and the various intraformational unconformities within this formation (modified after Winter, 1964a).

The Rosedale Member

Geometry.—

The isopach map of the combined Rosedale and Van den Heeversrust Members (Fig. 5) shows a distinct southeast-

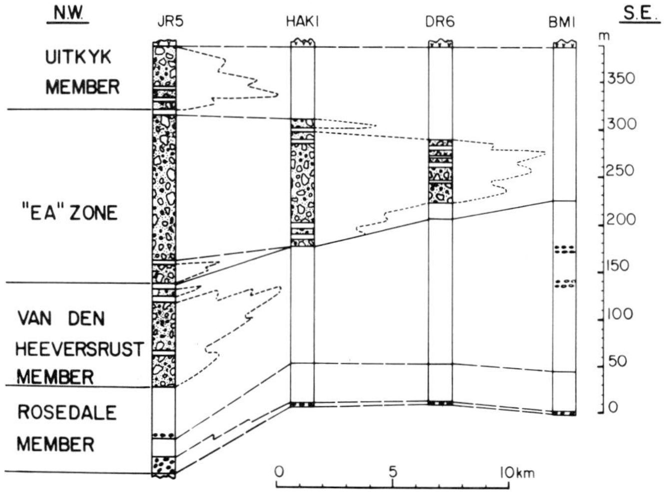

FIG. 4.—Major facies changes in the Eldorado Formation from a proximal conglomeratic facies to a distal arenitic facies over a distance of 18 km. See Figure 1 for location of section line.

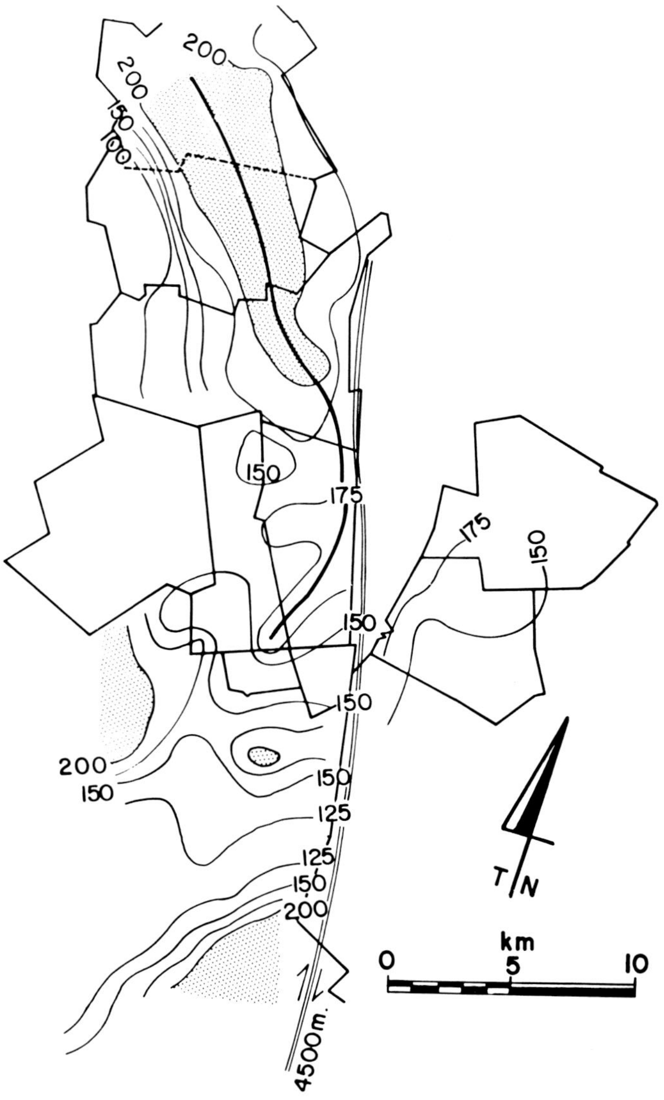

FIG. 5.—Isopach map in meters of the combined Rosedale and Van den Heeversrust Members. Note how the depoaxis curves from a north-south orientation on President Steyn Mine to a northwest-southeast direction on Free State Geduld Mine.

northwest trending depoaxis, indicating that the maximum deposition took place in a strip stretching between FSG (North) and the northern tip of President Steyn Gold Mine. This axis then swings southward through the southeastern corner of President Brand Gold Mine, where the package of sediment thins and the axis becomes indistinct.

The position of this axis parallels the distinct fan shape defined by the cumulative thickness of conglomerate and the maximum quartz clast size (MPS) depicted in Figure 6a. The distribution pattern of the Rosedale Placer as shown by the MPS is the same as that of the fanlike pattern of the cumulative conglomerate thickness within this member. The Rosedale Placer is much more extensive, however, and covers almost the entire goldfield.

The Rosedale conglomerates are intimately associated with facies changes. They interfinger northward, eastward and southward with the overlying arenites and grade into grit and arenite. For example, the Rosedale Placer grades into a poorly developed grit some 20 km to the east of the western edge (Winter, 1964b).

The Rosedale/Beatrix Placer.—

In the southern part of the area an oligomictic placer, the Beatrix Placer, is sporadically developed on the old paleosurface or pediment. This gold placer is older and does not form part of the Eldorado sequence proper and for this reason it is not discussed here.

The actual base, the Rosedale Placer, is a polymictic cobble conglomerate or a pebbly quartzite as much as 5 m thick in the proximal area grading into a medium and small pebble conglomerate 1 m thick or less 15 km down the paleoslope. The conglomerate is absent in some distal areas due to the irregular paleosurface. Clasts are poorly sorted and commonly consist of the following types: quartz = 25 percent, quartzite (various varieties) = 25 percent, chert = 30 percent, yellow silicified siltstone/shale = 15 percent and minor felsic volcanics set in a dark grey, gritty, and argillaceous matrix. Quartz rarely exceeds 25 percent of the pebble population but increases relative to the other components downslope because the nondurable clasts were destroyed during abrasion. The matrix is typically dark grey litharenite and subgreywacke with a coarse-grained to gritty texture, similar to the lenticular arenitic intercalations within the conglomerate body itself. At some localities the lower 1 m of the conglomerate was sorted texturally, thereby upgrading it to a litharenite and in some cases a quartzarenite. As a result of this sorting, concentration of some heavy minerals such as pyrite, gold and uraninite took place. The detrital pyrite grains are as much as 10 mm in the proximal area in the west (Fig. 7a) but decrease to a maximum of 3 mm only 5 km downstream and to 1 mm in the distal area some 15 km downcurrent.

Lithology and sedimentology.—

In the Welkom area the Rosedale Member may be subdivided into two units, the lower one being dark grey and siliceous with siltstone drapes on top of thin sedimentation units. Crossbedding is conspicuous (Fig. 7b). Coset units are between 30 cm and 100 cm thick and in many cases are capped by dark grey silty beds. Pyrite foresets are common in places. The thickness of this lower unit varies from 30 m in the south to about 60 m in the north where the top becomes poorly defined. Some fine-grained to medium-grained arenite intercalations occur near the top. Arenites in this lower unit are typically medium- to very coarse-grained quartzarenites with some litharenites and subgreywackes.

The color changes in the overlying unit to slightly yellowish-grey as a result of an increase in the pyrophyllite matrix but toward the top, grey quartarenites dominate again. Gritty material is very common in the upper part of the unit. In the Loraine facies several thin, mostly oligomictic conglomerates occur at various stratigraphic horizons, largely in the lower part of the member where they are called the ED placers (Fig. 2). Although no clast size data are available, they seem to have been distributed radially on Loraine Mine. These conglomerates grade rapidly northeastward and

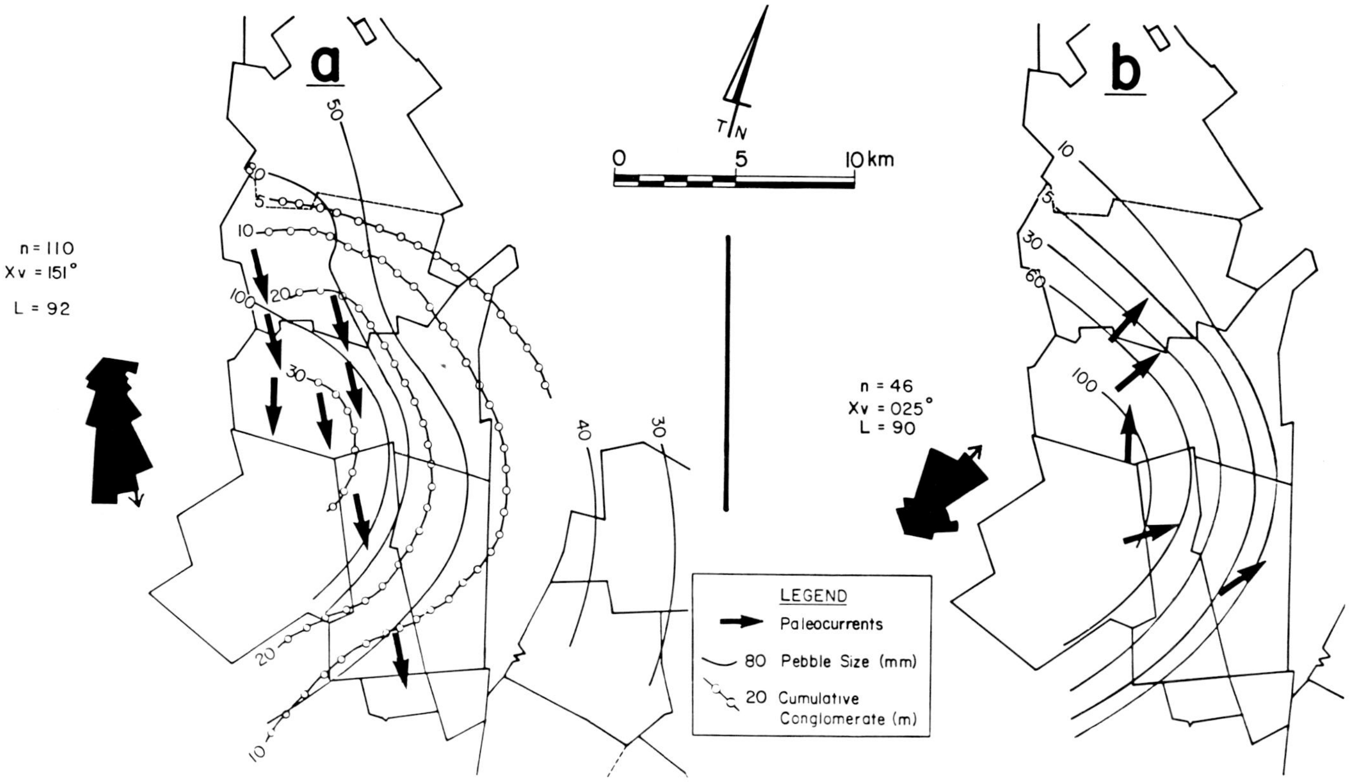

FIG. 6.—a. Cumulative conglomerate thickness of the Rosedale Member and maximum quartz clast size of the Rosedale Placer both show a distinct fanlike distribution for the gravels to the east. Paleocurrents in intercalated and overlying arenities show a completely different direction of transport. b. Maximum quartz clast size as well as paleocurrents in the Van den Heeversrust Member indicate a dispersal pattern to the east.

eastward into grits and sands. Southeastward, however, down the regional paleoslope, as defined by paleocurrents in the arenites (Fig. 6a), this facies change is more gradual. Figure 8 shows some details of typical sequences in the Welkom facies of this member.

The Van der Heeversrust Member

The basal conglomerate of the Van der Heeversrust Member rests disconformably on the Rosedale Member. This relationship is especially noticeable on the western part of

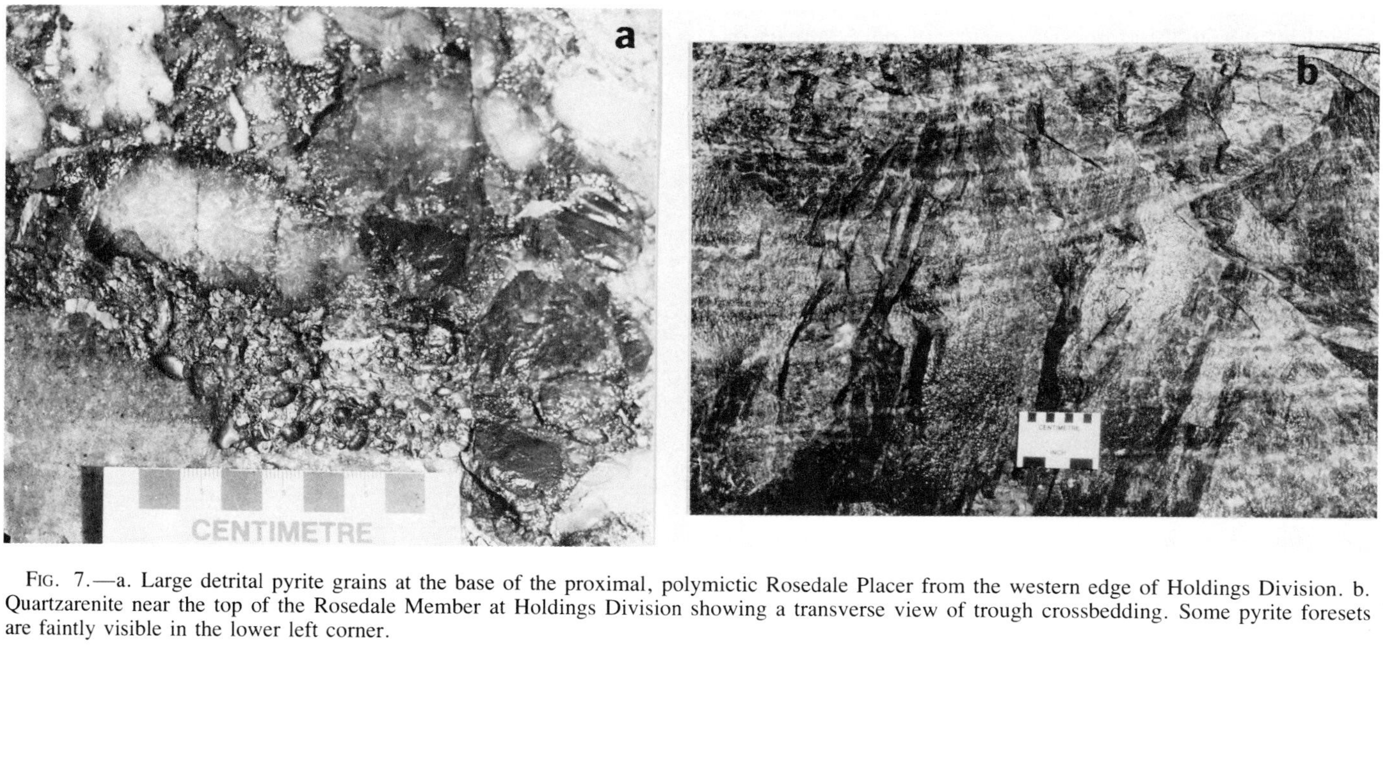

FIG. 7.—a. Large detrital pyrite grains at the base of the proximal, polymictic Rosedale Placer from the western edge of Holdings Division. b. Quartzarenite near the top of the Rosedale Member at Holdings Division showing a transverse view of trough crossbedding. Some pyrite foresets are faintly visible in the lower left corner.

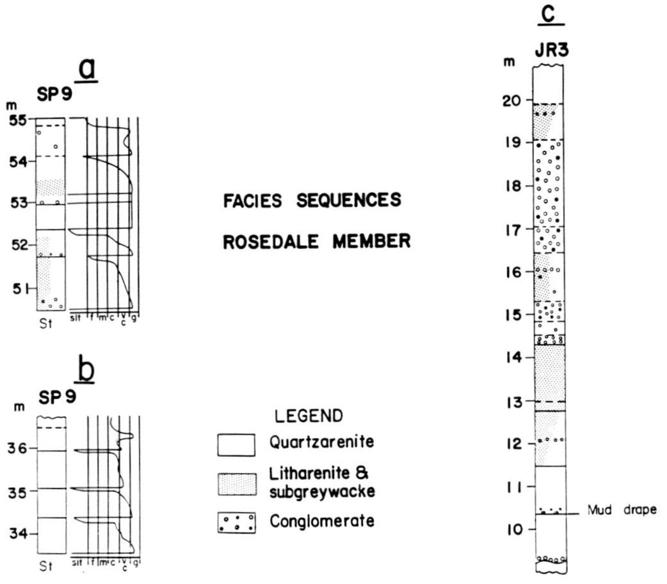

FIG. 8.—Some detailed facies sequences in SP 9. Note siltstone cappings on top of some fining-upward sedimentation units (a and b). In JR 3 note how quartzarenite at 11 m grades into litharenite and subgreywacke at 13 m underneath polymictic conglomerate (c). Positions of these columns are shown on Figure 9.

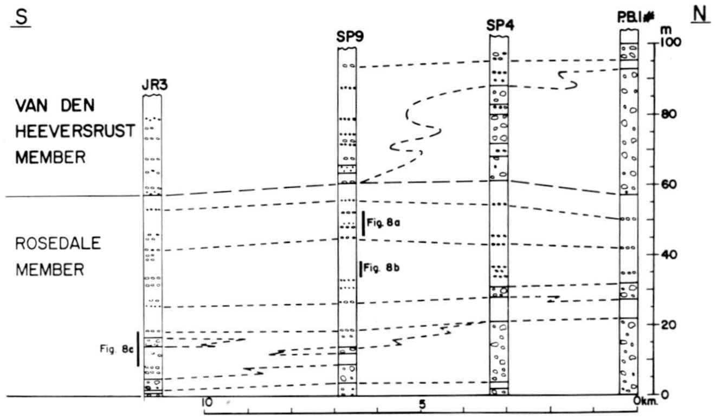

FIG. 9.—Correlation diagram between cored drill holes on President Brand and southward, showing facies changes, especially in the Van den Heeversrust Member. See Figure 1 for position of section line.

Holdings Division, where the latter thins considerably due to a marginal unconformity and is also noticeable on Loraine Mine (Fig. 3). The distribution of MPS in both the Rosedale and Van den Heeversrust conglomerates describes distinctive fans which are geographically superimposed. At any specific locality, the mean size of the largest quartz clast is smaller in the Van den Heeversrust conglomerate than in the Rosedale conglomerate (Figs. 6a, 6b). The former is only well defined in a small area of about 110 km^2 covering Holdings Division and surrounding mines President Brand and Free State Geduld. In these areas there is a pronounced decrease in conglomerate thickness down the paleoslope due to marked facies changes from conglomerate into grit and arenite. The apices of both fans as deduced from these clast size distributions lay west of St. Helena Mine.

The lower part of the Van den Heeversrust Member consists of a typical conglomeratic/subgreywacke sequence. The base is erosional in the proximal area, but in the medial to distal areas a thin transition zone 5 m thick can be recognized. The robust conglomerate changes within 5 km downfan into first a sequence of conglomerate/arenite units and then into an arenite sequence with only a few pebble lags which grade into grits beyond the limit of the conglomerate fan (Figs. 6b, 9). This unit, similar to others in the sequence, demonstrates how facies in the Eldorado Formation change dramatically over a few kilometers. Because of this phenomenon, the EC sandy unit distally becomes difficult to separate from the overlying EB unit. They have therefore been grouped together. Lithologically, there is no difference between this member and the Rosedale Member except that in the lower part of the former, yellowish-grey litharenites predominate, grading upward into quartzarenite. In the Loraine area subordinate oligomictic conglomerates, called the EC placers, occur at various horizons within this member.

An interesting feature is the change downslope in thickness of sedimentation units. The average thickness of conglomerate and overlying pebbly arenite beds decreases down the paleoslope over a distance of 15 km from 100 cm at Holdings Division to 50 cm at FSG. The significance of this change has not yet been clarified. Trough crossbedding is ubiquitous and sets very in thickness between 4 cm and 8 cm and reach 12 cm rarely. Individual beds and coset units are lenticular. Symmetrical and slightly asymmetrical ripples with a ripple length of 7 cm and a height of 1.2 cm have been observed in the arenites.

At two localities, Holdings Division No. 5 Shaft and FSG No. 2 Shaft, detailed measurements of the MPS vertically through the gravel deposit have disclosed a very interesting phenomenon. The MPS at both localities first increases gradually to a maximum of 12 cm at Holdings Division and 8 cm at FSG halfway up the conglomerate and then decreases toward the top, producing a symmetrical distribu-

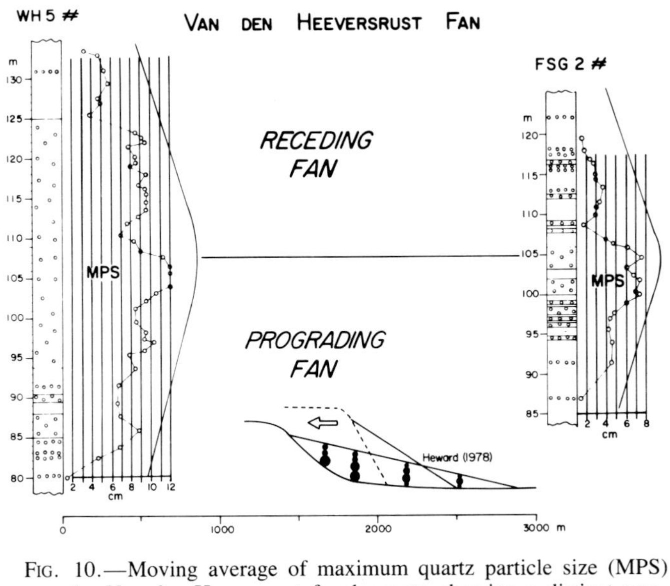

FIG. 10.—Moving average of maximum quartz particle size (MPS) across the Van den Heeversrust fanglomerate showing a distinct symmetrical distribution as emphasized by the curved lines at both localities.

tion curve (Fig. 10). The significance of this clast pattern will be discussed later.

The EA Zone

The geometry of the upper two members is reflected in the combined isopach map (Fig. 11) which depicts some important differences from those in Figure 5. First, the depoaxis in the south migrated some 4.5 km westward from President Steyn to President Brand Mine and in the north it changed its orientation from northwestward to northward. Second an anticlinal warp, oriented northeastward, developed during sedimentation of the upper two members; the geometry of the overlying Ventersdorp lava does not show any valley feature in this position, so the thinning is attributed to deposition and not to postdepositional erosion. The depoaxis continues north of this feature. Conglomerate development in these two upper members again shows a fan-shape distribution with its proximal part in the west (Fig. 12). Its elongated north-south shape is enigmatic, but it may be due to coalescence and interfingering of two fans during deposition of the EA zone, a phenomenon which is discussed in more detail below.

A marked change occurs at the base of the EA zone throughout the area, especially near the proximal western edge. Furthermore, there is also a marked lithological change within this zone between the Welkom facies and the Loraine facies. In the past this facies change has caused much confusion in local stratigraphic classification. In the proximal Welkom area it starts at the base with robust polymictic conglomerate with very few thin subgreywacke intercalations; previously, this was included in the so-called 'Boulder Beds' but it should not be confused with the true

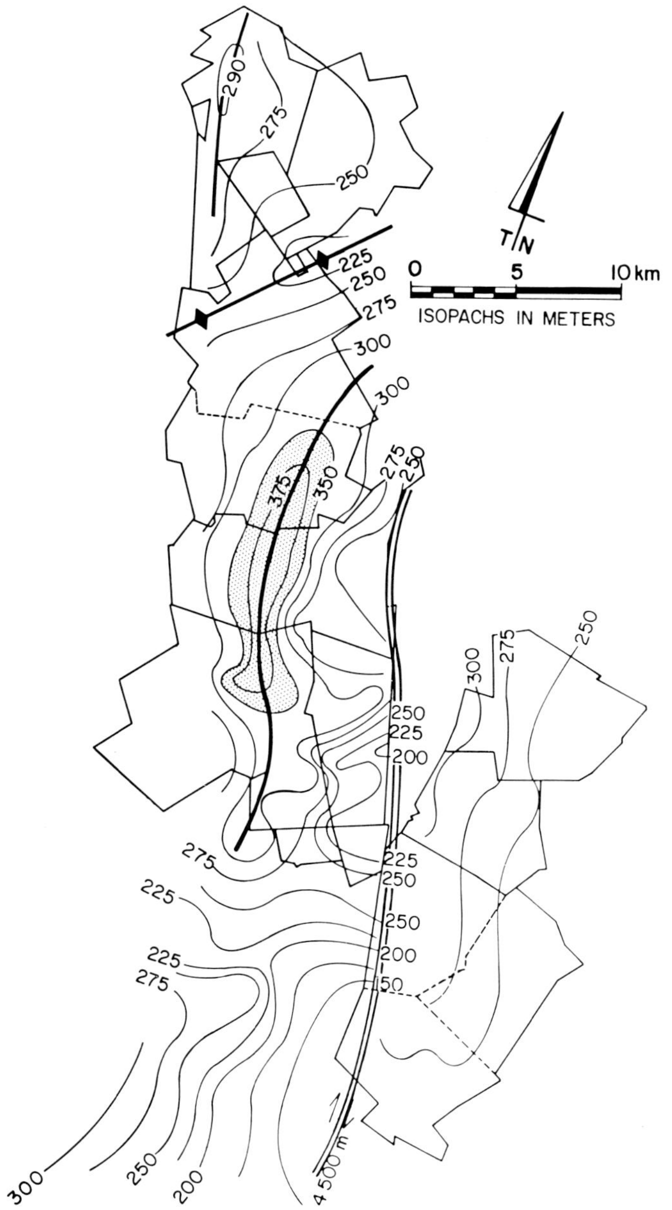

FIG. 11.—Isopach map of the combined EA zone and Uitkyk Member. Note north-south orientation of depoaxis and the discontinuation of this axis on Free State Geduld (north) as a result of the northeast-southwest anticlinal warp.

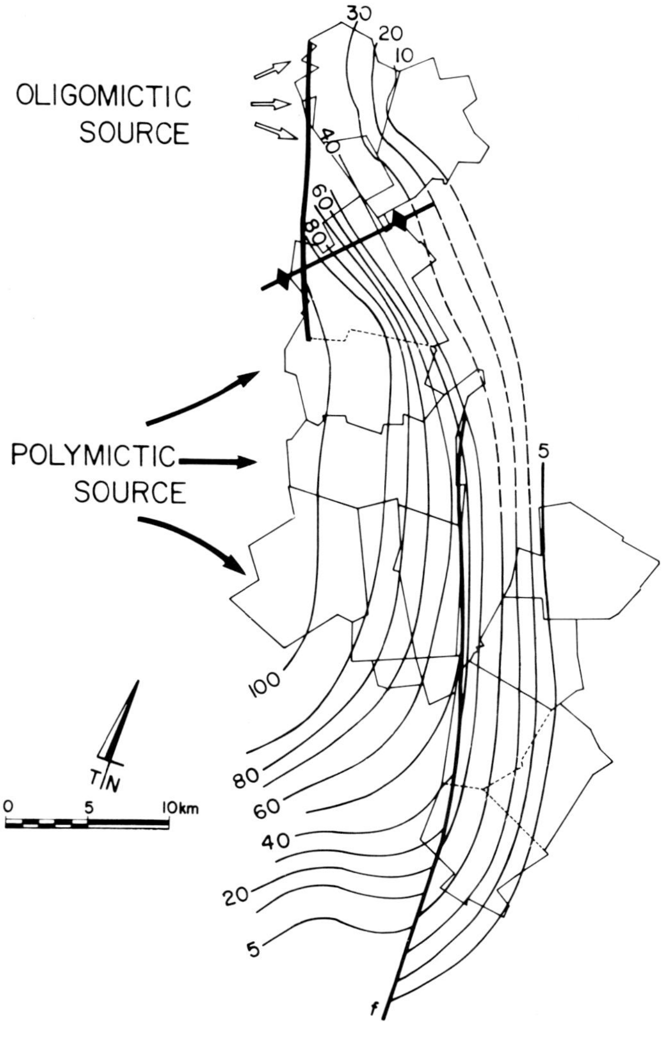

FIG. 12.—Distribution of percentage conglomerate in the combined EA zone and Uitkyk Member. The close contour spacing on Free State Geduld results from the facies changes mainly in the EA zone. The solid line on western border of Loraine Mine is the subcrop position of the EA zone.

boulder beds of the Uitkyk Member overlying this unit. From Holdings Division southeastward down the paleoslope, the individual conglomerate beds in JR 5 deteriorate into conglomerate/subgreywacke units in HAK 1 and DR 6 and grit/sandy units in BM 1 (Fig. 4). In general, the maturity of the sediments also increases, although nowhere are any oligomictic conglomerates of any significance developed in the Welkom facies.

The situation is quite different northward in the Loraine facies. In a 2.5-km zone defined by drill holes KK 5 and EV 3 the robust polymictic conglomerates decrease from 90 percent to 50 percent and change into conglomerate/subgreywacke beds (Figs. 1, 12). Simultaneously, some small pebble oligomictic conglomerates and quartzarenites make their appearance as intercalations in the immature sequence. Farther northward, significantly in the general vicinity of the anticlinal warp, the polymictic conglomerates disappear and grade into litharenite and subgreywacke beds, while intercalations of the Loraine facies become more prominent. Toward the northern part of Loraine Mine, oligomictic conglomerates and quartarenites are well developed near the western edge and predominate over, and interfinger with, the Welkom facies (Fig. 13). Here the pebble size decreases rapidly and the conglomerates grade into quartzites (arenites), becoming more mature, mineralogically and texturally, eastward and northeastward (Winter, 1964a). Winter's observation fits very well with the regional pattern described above.

These oligomictic conglomerates contain heavy mineral and gold concentrations, which is the main reason why the EA zone is economically viable in the Loraine facies, whereas its correlative in the Welkom facies is not. The detrital pyrite grains are much larger in the oligomictic conglomerates and range to as much as 6 mm, whereas the pyrite grains in the intercalated polymictic conglomerates have a maximum size of only 2 mm (Winter, 1964a). The significance of this will be discussed later.

The polymictic conglomerates of the Welkom facies have the following general pebble assemblage: quartz = 20 percent, quartzite = 35 percent, chert = 30 percent, yellow siltstone/shale = 10 percent and some porphyry fragments. In a few cases rare granite and gneiss cobbles have been observed. The porphyries were derived from the Dominion Supergroup (McKinney, 1964) underneath the Witwatersrand Supergroup. These conglomerates are polymodal, composed of small pebbles as matrix in the proximal area where the MPS is as much as 60 cm in diameter. They are dominantly clast-supported and mostly free of sand-size matrix. Although bed thicknesses are difficult to determine, it seems that individual beds are 1 m to 2 m thick and are sheetlike in appearance. The beds show possible inverse grading of clast sizes in a few places in the proximal gravels, but generally no vertical particle size trend is apparent.

The thickness of this unit varies between 100 m and 140 m in the Loraine area, but its correlative reaches 190 m in the Welkom area. In the former area the base of this unit is an angular unconformity where it transgresses the Rosedale Placer toward the west (Winter, 1964a). This relationship is also evident in drill holes RP 1 and KK 4 (Fig. 1), where as much as 120 m of the Van den Heeversrust and Rosedale Members have been eroded on this unconformity prior to deposition of the EA zone. Local angular unconformities even developed at bases of individual placers in the EA zone. For example, an angular unconformity has been recorded beneath the EA 12 placer, which changes westward within 40 m form a 10° angular unconformity to a 60° unconformity (Fig. 14).

Winter (1964a) made the following important observations in the Loraine area: the depoaxis of the EA zone is oriented in a south-southeastward direction and coincides with the present synclinal axis; economical gold placers are only developed in a zone about 320 m wide along the synclinal axis and westward as far as the respective subcrop positions of the EA placers; individual beds thicken rapidly from their respective subcrop positions toward the trough axis and conglomerates grade within a few hundred meters into arenites east of the trough axis.

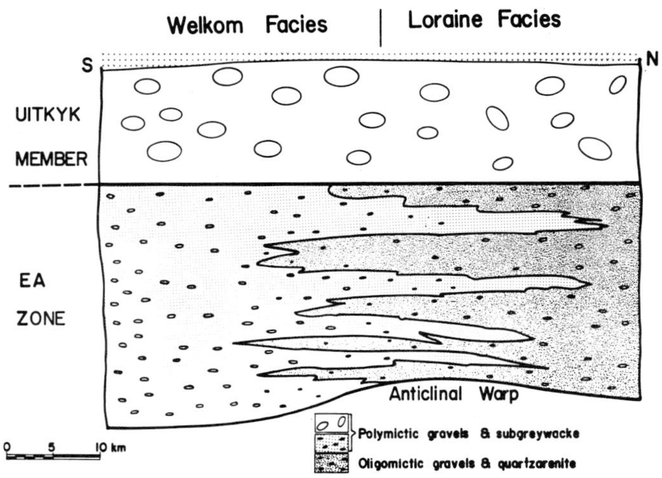

FIG. 13.—Schematic diagram showing the interfingering relationships in the EA zone of the mineralogically mature Loraine facies and the immature Welkom facies. Significantly, the maximum interfingering occurs above the anticlinal warp.

The Uitkyk Member

In the proximal Welkom area of Holdings Division, FSG and St. Helena Mines the 'typical' boulder beds of this member are hardly distinguishable from the underlying massive conglomerates of the EA zone. This is the reason why misinterpretations occurred in the past. There are, however, a few parameters that are distinctly different: first,

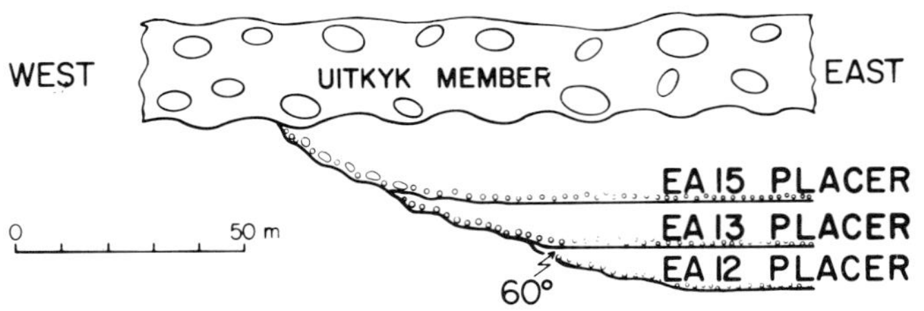

FIG. 14.—Basin edge unconformities are typically developed in the EA zone at the bases of placers, illustrating syntectonic sedimentation during their deposition.

at most localities there is an abrupt clast size increase above the base of the Uitkyk Member; second, there is a sharp decrease of quartz clasts from more than 20 percent in the EA zone to 10 percent in the Uitkyk Member; concomitantly there is an increase in blue and green chert clasts.

Southward and southeastward from the proximal area, the boulder beds of this member change facies very rapidly over 4.1 km from a robust cobble and boulder conglomerate to a litharenitic facies (Fig. 4). Toward the north and northeast, the robust boulder beds are more continuous and well developed, so that this member can be identified easily where it overlies the mainly sandy EA zone on FSG (North) and Loraine. Therefore, the conglomerates change facies more rapidly toward the south than toward the north. On the other hand, the polymictic conglomerates of the EA zone spread much farther southeastward than northeastward. In distal positions, such as on Beatrix Gold Mine, Erfdeel Gold Mine (drill hole EDPC 1) and Jeannette Gold Mine (drill hole WR 1), both members change into a sandy sequence and distinction between them becomes more difficult. These facies changes from robust conglomerate to almost 100 percent arenite take place within a distance varying between 14 km and 18 km (Fig. 12).

PALEOCURRENTS

Paleocurrent measurements were done on the lower two members, and general observations of paleocurrent vectors were made on the upper two members. Foreset aximuths and the orientation of trough axes were used as paleocurrent indicators. The mean vectors of measurements collected at 10 locations (Fig. 6a) illustrate a very consistent southeastern paleocurrent direction in the Rosedale arenites. The vectoral mean for the whole area calculated from 110 measurements is 151° TN. Unimodal patterns and high consistency ratios were obtained at all the localities. This direction is not consistent with the eastward distribution shown by attributes of the conglomerates (Fig. 6a) and an explanation must be offered for this phenomenon.

In places where arenite is intercalated with the basal conglomerate, crossbedding in the former also indicates a southeastern paleocurrent direction. Therefore, two different directions of distribution, one for the arenite and another for the conglomerate, are indicated. In the upper 6 m of the Rosedale Member, a paleocurrent transition zone exists in which an interfingering of beds derived from the northwest as well as from the southwest occurs. A definite switch in paleocurrent direction therefore takes place within the arenites of the transition zone. It is significant that the base of this zone is marked by a definite textural change as well. A sequence of medium-grained to very coarse-grained arenites is overlain at the base of the transition zone by either a conglomerate bed or a pebbly grit as much as 20 cm thick. The rest of the zone above it also consists of pebbly and gritty arenite and thin conglomerate beds.

Paleocurrents measured in the arenites above the base of the Van den Heeversrust Member in general trend north and northeast (Fig. 6b) and coincide with the distribution indicated by the clast size distribution of the basal conglomerate of this member. The vectoral mean of 46 measurements from five localities is 025° TN. Paleocurrent information is sparse in the overlying members, but a few observations indicate that some of the quartzarenites might have been distributed southeastward as well.

GOLD AND PYRITE MINERALIZATION

In general, the polymictic conglomerates of the Eldorado Formation are not economically important. The only viable gold ore carrier is the Rosedale Placer at the base of the sequence which contains low values of as much as 5 g/t Au over 100 cm. At a few localities on the western edge of the basin on Holdings Division and FSG, however, some sporadic high values as much as 30 g/t Au over 100 cm are encountered. In these cases this placer has been mined on a local scale.

The high gold values are associated with large detrital pyrite grains (Fig. 7a) which are locally concentrated in varying amounts as part of the conglomerate matrix. The amount and size of the detrital pyrite decrease rapidly eastward down the paleoslope and so does the gold value, so that within 1 km downcurrent this placer becomes completely uneconomical.

Various placer deposits within the Eldorado Formation are economically viable in the Loraine area. Almost all of them are oligomictic conglomerates that occur at various stratigraphic horizons within the sequence, although the best mineralized placers appear in the EA zone. Good descriptions of these EA placers (reefs) are to be found in Winter (1964a). They have undergone little sedimentological investigation, so that only their broad framework and crude depositional model can now be deduced.

The EA placers are quartz cobble to large pebble conglomerates with a minor amount of black chert pebbles and rare quartzite pebbles. Regarding clast assemblage, the oligomictic nature of the Loraine facies contrasts sharply with any of the polymictic conglomerates in the Welkom area. Their matrix is quartzarenite which contains a variable detrital pyrite component. Down the paleoslope to the east, the pebble size, pyrite grain size and amount drop dramatically, producing only a poorly pyrite-mineralized conglomerate 1 km from the subcrop. Concomitantly, the gold values, which are around 1000 cm g/t over 100 cm in many places in the proximal area, drop to uneconomical levels eastward when one crosses beyond the synclinal axis. The most dramatic drop in gold value occurs east of the synclinal axis, a phenomenon which has not yet been studied in any detail.

MODEL OF SEDIMENTATION

Environment of Deposition

The Eldorado conglomerates are interpreted as alluvial fanglomerates because they constitute bodies of coarse debris with radial sediment dispersal patterns. Down the paleoslope, the gravels interfinger with braided alluvial sands that display trough crossbedding and lenticularity. Most gravels are sheetflood deposits because they display crude normal clast grading, crude plane bedding, and some are capped by thin gritty sands. In rare cases, inverse grading occurs, indicating possible debris flows.

Four main pulses of debris influx coinciding with the four members can be identified. In their proximal parts conglomerate dominates the lower parts of the lower two members. Conglomerate and arenite are more complexly arranged in the upper two members and at any one locality have much higher percentages of conglomerate than the lower two members, indicating the progradational relationship of the overlying fans with the underlying ones.

The fans were distributed from west to east (Figs. 6, 12) but the regional paleoslope was toward the southeast, as clearly indicted by the paleocurrent pattern of the Rosedale sandstones (Fig. 6a), suggesting a broad braid plain with shallow low sinuosity channels superimposed upon it where these sands were redistributed. These mature sandstones were obviously derived from a silicic source in the northwest, and as the polymictic Rosedale gravel fan retreated, the braid plain was established over the same areas. Figure 15 depicts a physiographic model which combines the features described above and which is comparable with the wet alluvial fan/braid plain setting described by Rust (1979, Fig. 2), the model of the Hornli Fan (Bürgisser, 1984) and the Van Horn Fan (McGowen and Groat, 1971).

The intercalated arenites of the Van den Heeversrust fan show the same distribution as those of the associated gravels, thus suggesting fan gravel and sand prograded over the braid plain sands mainly from west to east. The fan slope was probably steeper than that of the Rosedale fan, as shown by the rapid pebble-size change and the consistent transport direction of the gravel and sand. Little paleocurrent information is available from the mature topmost arenites of this member, but the similarity of their lithology to that of the Rosedale arenites suggests resorting and possible redistribution along a braid plain oriented similarly to that of the Rosedale cycle. Thus, the lower two members are crude fining-upward sequences, each suggesting recession of the source area subsequent to rejuvenation.

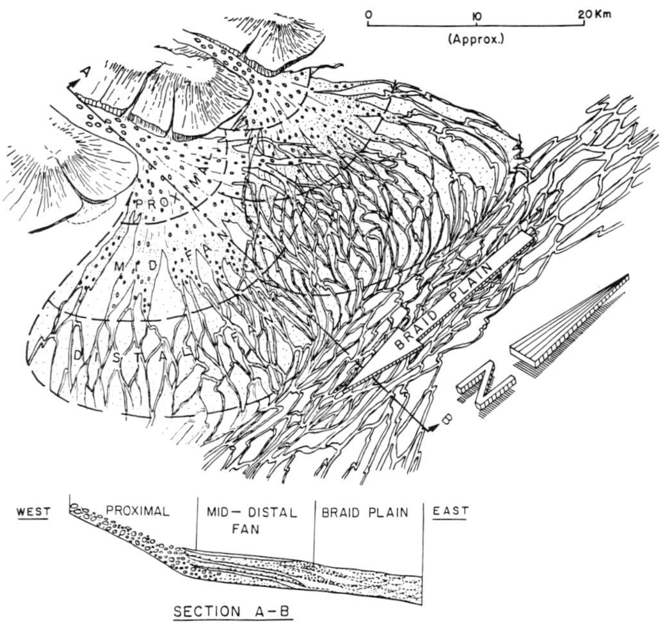

FIG. 15.—Physiographic model of sedimentation of the Eldorado sediments. Note how the alluvial fans distribute gravel and sand eastward onto an alluvial braid plain where the sediments are redistributed south and southeastward down the regional slope.

Interpretation of the EA zone is more complex. Again, there were two sources of sediment in the same areas that produced two distinctly different facies interfingering with each other. The mature Loraine facies was derived from a homogeneous silicic source west of Loraine Mine and was distributed eastward in that area. Material of the Welkom facies was shed simultaneously from a heterogeneous polymictic source west of Holdings Division and St. Helena Mines. The latter produced the polymictic gravels and mineralogically immature sands. The two distinct facies interfinger in a wide zone on FSG (North) and the southern part of Loraine Mine over an anticlinal warp in the basin. The control on sedimentation exercised by this warp is a matter for further study. This interfingering relationship is especially evident in the alternating sequence of quartzarenite and subgreywacke, as well as in the alternation of polymictic and oligomictic conglomerates of the EA zone. The differences in detrital pyrite grain sizes of the interfingering two facies at a particular locality emphasize the difference in source areas.

Tectonic Control and Sedimentation

The robust conglomerates which overlie the quartzarenites at the base of the Van den Heeversrust fan emphasize one of the transitions in the basin which is undoubtedly indicative of tectonism. This change marked a rejuvenation of the source area.

The gradual increase in clast size vertically across 27 m of this alluvial fan suggests progradation and the gradual decrease further upward is indictive of fan recession (Fig. 10). This explanation is in agreement with that by Heward (1978), where he describes the initial phases of fan-building off a steep scarp as the progradational phase (coarsening upward), and as the scarp erodes and retreats, the slope of the fan lowers, the fan recedes, a phase which is represented by fining of the clast size upward. This process was the beginning of a geomorphological cycle, a process that was probably repeated several times through the depositional history of the Eldorado Formation; the start of the main cycles is marked by the bases of the different members.

The thickening of individual Loraine placers near the depoaxis, their rapid facies change and their basin edge unconformable relationships show that syntectonic sedimentation was the regular pattern during deposition of the EA placers. The occurrence of some oligomictic gold placers in the lower two members in the Loraine area indicates that 'Loraine-type' source rocks were already active periodically during the early stages of Eldorado sedimentation. This source became more active and produced its maximum sediment input during deposition of the EA gold placers. This maximum sediment supply, in fact, is also true of the polymictic source west of Welkom, which experienced maximum uplift during this time as well; rare basement gneiss and lower Witwatersrand quartzite cobbles are proof of this.

Sediment from both sources, introduced by sheet floods and possibly some debris flows onto alluvial fans, intermixed in the medial and distal areas and along the toes of the fans were the sandy sediments were redistributed southeastward down the regional paleoslope along a broad braid plain. Most of the mature sands and some fine-grained gravels of the Eldorado Formation were deposited in this latter environment (present in drill hole EDPC 1).

The two lower members represent two cycles, each one starting off with a source area uplift. During deposition of the EA zone, several uplifts and active sediment input from both the Loraine and Welkom facies obscure and interrupt the geomorphological cycle. These uplifts are reflected in the pebble sizes and assemblages, and three different cycles between 30 m and 50 m thick can be recognized.

The onset of the fourth major uplift produced the polymictic boulder beds of the Uitkyk Member, which were distributed from west of Welkom and covered the whole area, thereby completely obliterating the oligomictic source west of Loraine Mine that had existed until that time. Furthermore, this marked syntectonic unconformity extends much farther westward across all older formations, so that the sedimentary basin eventually overlapped the fold felt in a manner very similar to the Cretaceous-Tertiary example of Utah described by Miall (1978). This major uplift occurred in three phases and was followed by a period of standstill, erosion and recession of the source area. During the initial uplift the coarse gravels were distributed more actively northeastward in contrast to the EA polymictic gravels which spread out actively toward the east and southeast. This is the reason why the sandy distal equivalents of the Uitkyk Member in the southern part of the Welkom area overlie the polymictic conglomerates of the EA zone (Fig. 4).

The rapid thinning of members and units of the Eldorado sequence westward and the increasing number of angular marginal unconformities suggest uplift in the west by folding and thrusting, in preference to the classical fault-bounded through model of Pretorious (1976) which would have produced a thick sediment pile in the west. The present model is also at variance with the epeirogenic uplift model proposed by Minter (1978) and Tankard and others (1982). The syntectonic unconformities of the Tertiary sequence in the Spanish Pyrenees described by Riba (1976) display a remarkable similarity in scale and arrangement with the unconformities of the Eldorado sequence. Miall (1978) pointed out that the position of the axis of rotation determines the type of basin edge unconformity. According to the various examples he gave, the axis of rotation during deposition of the upper Eldorado must have been within the basin and very close to its edge. Syntectonic sedimentation is also strongly suggested on a smaller scale by the marginal unconformities between the various EA placers.

In particular, the features of the lower two fans described here are different in two ways from the small fault-bounded fans described, for example, by Gloppen and Steel (1981) and Nemec and Steel (1984). Their cycles mainly show a coarsening-upward motif and the basin sequence is very thick. The much thinner Eldorado sequence, the westward wedging of individual units (and members) and the presence of mainly fining-upward stacked sequences give strength to the proposal for their tectonic setting given above. These sequences show some resemblances to the Old Red Sandstone as shown in Table 1 of Heward (1978).

The plate tectonic island arc/foreland basin model suggested by Winter (1984) may well be compatible with the model of sedimentation described above. According to Winter (1984), this model may, in very general terms, be compared with the Cretaceous island arc/foreland model of the Rocky Mountains. According to Miall (1978), the margins of active foreland molasse basins are characterized by internal unconformities, folds and faults. Internal angular unconformities, which die out laterally into bedding, indicating the occurrence of localized pulses of uplift contemporaneous with sedimentation, are very common in molasse basins. All these features are typical of the Eldorado gravels and sands, and it can best be interpreted as a foreland molasse basin setting.

ACKNOWLEDGMENTS

I gratefully acknowledge the permission and financial support granted by the management of Anglo American Corporation to publish this paper. It was presented at the Third International Fluvial Sedimentology Conference held at Fort Collins, Colorado in August 1985. I also thank G. Wheelock for access to unpublished data.

REFERENCES

BÜRGISSER, H. M., 1984, A unique mass flow marker bed in a Miocene streamflow molasse sequence, in Koster, E. H., and Steel, R. J., eds. Sedimentology of Gravels and Conglomerates: Canadian Society of Petroleum Geologists Memoir 10, p. 147–163.

BUSCH, D. A., 1971, Genetic units in delta prospecting: American Association of Petroleum Geologists Bulletin, v. 55, p. 1137–1154.

ERIKSSON, K. A., TURNER, B. R., AND VOS, R. G., 1981, Evidence of tidal processes from the lower part of the Witwatersrand Supergroup, South Africa: Sedimentary Geology, v. 29, p. 309–325.

GLOPPEN, T. G., AND STEEL, R. J., 1981, The deposits, internal structure and geometry in six alluvial fan—fan delta bodies (Devonian-Norway)—A study in the significance of bedding sequence in conglomerates, in Ethridge, F. G., and Flores, R. M., eds., Non-Marine Depositional Evironments: Models for Exploration: Society of Economic Paleontologists and Mineralogists Special Publication 31, p. 49–69.

HEWARD, A. P., 1978, Alluvial fan sequence and megasequence models, with examples from Westphalian D—Stephanian B coalfields, northern Spain, in Miall, A. D., ed., Fluvial Sedimentology: Canadian Society of Petroleum Geologists Memoir 5, p. 669–702.

KINGSLEY, C. S., 1979, The Eldorado Formation: Sedimentological aspects of the lower succession in the central part of the Welkom Goldfield: Internal AAC report, 87 p.

———, 1984, Dagbreek fan-delta: An alluvial placer to prodelta sequence in the Proterozoic Welkom Goldfield, Witwatersrand, South Africa, in Koster, E. H., and Steel, R. J., eds., Sedimentology of Gravels and Conglomerates: Canadian Society of Petroleum Geologists Memoir 10, p. 321–330.

KLEYNHANS, J. J., 1970, A sedimentological study of the conglomerates in the Elsburg Stage on the Welkom and Western Holdings Gold Mines: Unpublished M. S. thesis, University of the Orange Free State, Bloemfontein, 149 p.

MCGOWEN, J. H., AND GROAT, C. G., 1971, Van Horn Sandstone, West Texas: An alluvial fan model for mineral exploration: Bureau of Economic Geology, University of Texas, Report of Investigation 72, 57 p.

MCKINNEY, J. S., 1964, Geology of the Anglo American Group Mines in the Welkom area, Orange Free State Goldfield, in Haughton, S. H., ed., The Geology of Some Ore Deposits in Southern Africa, Vol. I: Geological Society of South Africa, p. 451–506.

MIALL, A. D., 1978, Tectonic setting and syndepositional deformation of molasse and other nonmarine-paralic sedimentary basins: Canadian Journal of Earth Sciences, v. 15, p. 1613–1632.

MINTER, W. E. L., 1978, A sedimentological synthesis of placer gold, uranium and pyrite concentrations in Proterozoic Witwatersrand sediments, in Miall, A. D., ed., Fluvial Sedimentology: Canadian Society of Petroleum Geologists Memoir 5, p. 801–829.

NEMEC, W., AND STEEL, R. J., 1984, Alluvial and coastal conglomerates: Their significant features and some comments on gravelly mass-flow deposits, in Koster, E. H., and Steel, R. J., eds., Sedimentology of Gravels and Conglomerates: Canadian Society of Petroleum Geologists Memoir 10, p. 1–31.

PRETORIUS, D. A., 1976, Gold in Proterozoic sediments of South Africa: Systems, paradigms and models, in Wolf, K. H. ed., Handbook of Strata-Bound and Stratiform Ore Deposits: Elsevier, New York, v. 7, p. 1–27.

RIBA, O., 1976, Syntectonic unconformities of the Alto Cardener, Spanish Pyrenees: A genetic interpretation: Sedimentary Geology, v. 15, p. 213–234.

RUST, B. R., 1979, Coarse alluvial deposits, in Walker, R. G., ed., Facies Models: Geoscience Canada Reprint Series 1, p. 9–21.

SIMS, J. F. M., 1969, The stratigraphy and paleocurrent history of the upper division of the Witwatersrand System on the President Steyn mine and adjacent areas in the Orange Free State Goldfield with specific reference to the origin of auriferous reefs, Vol. I and II: Unpublished Ph.D. Dissertation, University of the Witwatersrand, Johannesburg, 181 p.

TANKARD, A. J., JACKSON, M. P. A., ERIKSSON, K. A., HOBDAY, D. K. HUNTER, D. R., AND MINTER, W. E. L., 1982, Crustal Evolution of Southern Africa: Springer-Verlag, New York, 523 p.

VOS, R. G., 1975, An alluvial plain and lacustrine model for the Precambrian Witwatersrand deposits of South Africa: Journal of Sedimentary Petrology, v. 45, p. 480–493.

WINTER, H. de la R., 1964a, The geology of the northern section of the Orange Free State Goldfield, in Haughton, S. H. ed., The Geology of Some Ore Deposits in Southern Africa, Vol. I: Geological Society of South Africa, Johannesburg, p. 417–448.

———, 1964b, The geology of the Virginia section of the Orange Free State Goldfield, in Haughton, S. H., ed., The Geology of Some Ore Deposits in Southern Africa, Vol. I: Geological Society of South Africa, Johannesburg, p. 507–548.

———, 1984, Structural style of the Rocky Mountains, Utah: Kaapvaal analogues: Geobulletin, Geological Society of South Africa v. 27, p. 44.

SUBJECT INDEX

Conventions used: An *underlined* page number denotes the first page of a paper in which the word is used numerous times throughout the paper. Stratigraphic members are indexed under the formation of which they are a part. Geographic features and selected geologic structures show in parentheses the State or Country which the feature is located.

E

F

T

U

Y

Z